Remote Sensing Image Classification and Semantic Segmentation

Remote Sensing Image Classification and Semantic Segmentation

Editors

Jiaojiao Li
Qian Du
Jocelyn Chanussot
Wei Li
Bobo Xi
Rui Song
Yunsong Li

MDPI

Basel • Beijing • Wuhan • Barcelona • Belgrade • Novi Sad • Cluj • Manchester

Guest Editors
Jiaojiao Li
Xidian University
Xi'an
China

Qian Du
Mississippi State University
Starkville
USA

Jocelyn Chanussot
University Grenoble Alpes
Grenoble
France

Wei Li
Beijing Institute of Technology
Beijing
China

Bobo Xi
Xidian University
Xi'an
China

Rui Song
Xidian University
Xi'an
China

Editorial Office
MDPI AG
Grosspeteranlage 5
4052 Basel, Switzerland

This is a reprint of the Special Issue, published open access by the journal *Remote Sensing* (ISSN 2072-4292), freely accessible at: https://www.mdpi.com/journal/remotesensing/special_issues/KVCSC58HQ0.

For citation purposes, cite each article independently as indicated on the article page online and using the guide below:

Lastname, A.A.; Lastname, B.B. Article Title. *Journal Name* **Year**, *Volume Number*, Page Range.

ISBN 978-3-7258-1365-0 (Hbk)
ISBN 978-3-7258-1366-7 (PDF)
https://doi.org/10.3390/books978-3-7258-1366-7

Contents

About the Editors

Jiaojiao Li

Jiaojiao Li (S'16-M'17-SM'24) received the B.E. degree in computer science and technology, M. S. degree in software engineering, and Ph.D. degree in communication and information systems from Xidian University in 2009, 2012, and 2016, respectively. She was an exchange Ph.D. Student of Mississippi State University, supervised by Dr. Qian Du. She is currently an Associate Professor and Doctoral supervisor with the school of Telecommunication, Xidian University, China. Her research interests include hyperspectral remote sensing image analysis and processing and pattern recognition.

Qian Du

Qian Du (S'98–M'00–SM'05) received a Ph.D. degree in electrical engineering from the University of Maryland, Baltimore County, Baltimore, MD, USA, in 2000. She is currently a Bobby Shackouls Professor with the Department of Electrical and Computer Engineering, Mississippi State University, Starkville, MS, USA. Her research interests include hyperspectral remote sensing image analysis and applications, pattern classification, data compression, and neural networks.

Jocelyn Chanussot

Jocelyn Chanussot received the M.Sc. degree in electrical engineering from the Grenoble Institute of Technology (Grenoble INP), France, in 1995, and the Ph.D. degree from the Université de Savoie, Annecy, France, in 1998. From 1999 to 2023, he has been with Grenoble INP, where he was a Professor of signal and image processing. He is currently a Research Director with INRIA, Grenoble. His research interests include image analysis, hyperspectral remote sensing, data fusion and artificial intelligence. He has been a visiting scholar at the University of California, Los Angeles (UCLA), Stanford University (USA), KTH (Sweden) and NUS (Singapore). Since 2013, he is an Adjunct Professor of the University of Iceland. He holds the AXA chair in remote sensing and is an Adjunct professor at the Chinese Academy of Sciences, Aerospace Information research Institute, Beijing.

Dr. Chanussot is the founding President of IEEE Geoscience and Remote Sensing French chapter. He has received multiple outstanding paper awards. He was the Vice-President of the IEEE Geoscience and Remote Sensing Society, in charge of meetings and symposia (2017-2019). He is the founder of the IEEE GRSS Workshop on Hyperspectral Image and Signal Processing, Evolution in Remote sensing (WHISPERS). He was a member of the Machine Learning for Signal Processing Technical Committee of the IEEE Signal Processing Society (2006-2008) and the Program Chair of the IEEE International Workshop on Machine Learning for Signal Processing (2009). He is an Associate Editor for the IEEE Transactions on Geoscience and Remote Sensing, the IEEE Transactions on Image Processing and the Proceedings of the IEEE. He was the Editor-in-Chief of the IEEE Journal of Selected Topics in Applied Earth Observations and Remote Sensing (2011-2015). He is a Fellow of the IEEE, an ELLIS Fellow, a member of the Institut Universitaire de France (2012-2017) and a Highly Cited Researcher since 2018.

Wei Li

Wei Li (Senior Member, IEEE) received a B.E. degree in telecommunications engineering from Xidian University, Xi'an, China, in 2007, a M.S. degree in information science and technology from Sun Yat-sen University, Guangzhou, China, in 2009, and a Ph.D. degree in electrical and computer engineering from Mississippi State University, Starkville, MS, USA, in 2012. Subsequently, he spent one year as a Postdoctoral Researcher with the University of California, Davis, CA, USA. He is currently with the School of Information and Electronics, Beijing Institute of Technology. His research interests include hyperspectral image analysis, pattern recognition, and data compression.

Bobo Xi

Bobo Xi (Member, IEEE) received a B.E. degree in information engineering and a Ph.D. degree in information and communication engineering from Xidian University, Xi'an, China, in 2017 and 2022, respectively. He is currently a Lecturer with the State Key Laboratory of Integrated Services Networks, School of Telecommunications, Xidian University. He has published over fifteen papers in refereed journals, including the IEEE Transactions on Image Processing, the IEEE Transactions on Neural Networks and Learning Systems, and the IEEE Transactions on Geoscience and Remote Sensing. His research interests include hyperspectral image processing, machine learning, and deep learning.

Rui Song

Rui Song received his Ph.D. degree in Signal and Information Processing from Xidian University, Xián, China in 2009. He is currently a Professor and Ph.D. advisor in the State Key Laboratory of Integrate Service Network, School of Tele-communications at Xidian University. His research interests include image and video coding algorithms and VLSI architecture design, intelligent image processing, and the understanding and reconstruction of 3D scene.

Yunsong Li

Yunsong Li received a M.S. degree in telecommunication and information systems and a Ph.D. degree in signal and information processing from Xidian University, China, in 1999 and 2002, respectively. He joined the School of Telecommunications Engineering, Xidian University in 1999, where he is currently a Professor. Prof. Li is the director of the Image Coding and Processing Center at the State Key Laboratory of Integrated Service Networks. His research interests focus on image and video processing, hyperspectral image processing, and high-performance computing.

remote sensing

Article

A Deep Learning Classification Scheme for PolSAR Image Based on Polarimetric Features

Shuaiying Zhang [1], Lizhen Cui [2], Zhen Dong [1] and Wentao An [3,*]

[1] College of Electronic Science and Engineering, National University of Defense Technology (NUDT), Changsha 410073, China; dongzhen@nudt.edu.cn (Z.D.)
[2] College of Life Sciences, University of Chinese Academy of Sciences, Beijing 100049, China
[3] National Satellite Ocean Application Service, Beijing 100081, China
* Correspondence: anwentao@mail.nsoas.org.cn; Tel.: +86-1881-0991-992

Abstract: Polarimetric features extracted from polarimetric synthetic aperture radar (PolSAR) images contain abundant back-scattering information about objects. Utilizing this information for PolSAR image classification can improve accuracy and enhance object monitoring. In this paper, a deep learning classification method based on polarimetric channel power features for PolSAR is proposed. The distinctive characteristic of this method is that the polarimetric features input into the deep learning network are the power values of polarimetric channels and contain complete polarimetric information. The other two input data schemes are designed to compare the proposed method. The neural network can utilize the extracted polarimetric features to classify images, and the classification accuracy analysis is employed to compare the strengths and weaknesses of the power-based scheme. It is worth mentioning that the polarized characteristics of the data input scheme mentioned in this article have been derived through rigorous mathematical deduction, and each polarimetric feature has a clear physical meaning. By testing different data input schemes on the Gaofen-3 (GF-3) PolSAR image, the experimental results show that the method proposed in this article outperforms existing methods and can improve the accuracy of classification to a certain extent, validating the effectiveness of this method in large-scale area classification.

Keywords: polarimetric synthetic aperture radar (PolSAR); reflection symmetric decomposition (RSD); data input scheme; land classification; polarimetric scattering characteristics

Citation: Zhang, S.; Cui, L.; Dong, Z.; An, W. A Deep Learning Classification Scheme for PolSAR Image Based on Polarimetric Features. *Remote Sens.* **2024**, *16*, 1676. https://doi.org/10.3390/rs16101676

Academic Editors: Jocelyn Chanussot, Jiaojiao Li, Qian Du, Wei Li, Bobo Xi, Rui Song and Yunsong Li

Received: 2 April 2024
Revised: 2 May 2024
Accepted: 6 May 2024
Published: 9 May 2024

1. Introduction

Polarimetric synthetic aperture radar (PolSAR) is able to acquire comprehensive polarization information of land targets, and it can actively detect targets in all-weather and all-day conditions. Compared to single- and dual-polarized images, PolSAR images contain a significant amount of back-scattering information about the objects [1]. Currently, PolSAR classification methods mainly include polarization feature-based approaches, statistical distribution characteristics of PolSAR data, and deep learning classification methods [2–7].

A large number of scholars have conducted in-depth research on PolSAR classification and achieved good results. The method mainly adopts polarization decomposition to extract the polarization scattering information of the targets and further classify them based on these features. Cloude et al. have undertaken a lot of work on PolSAR classification [8,9]. C. Lardeux et al. [10] used a support vector machine (SVM) classifier to extract polarization features from PolSAR images of different frequencies and perform classification using these features. Dickinson et al. [11] classified targets in multiple scenarios using polarization decomposition. Yin et al. [12] addressed the issue of insufficient information extraction for temporal polarization spatial features in existing models by using the Vision Transformer 3D attention module to classify multi-temporal PolSAR images, effectively addressing the aforementioned issues. Similarly, Wang et al. [13] also used ViT networks to achieve effective classification of PolSAR images. Hua et al. [14] proposed using a 3D residual module to extract information from PolSAR images. These methods also combine the extracted polarization features with deep learning to achieve the classification of PolSAR images.

The classification method based on statistical features of PolSAR images mainly utilizes the difference in statistical characteristics of target objects to classify different targets in the images. Lee et al. [15] used polarization decomposition and unsupervised classification based on a complex Wishart classifier to classify PolSAR images. Silva et al. [16] used the minimum random distance and Wishart distribution to segment the targets in PolSAR images. Chen et al. [17] used the method of polarimetric similarity and maximum–minimum scattering features to improve the accuracy of classification. Wu et al. [18] used a domain-based Wishart MRF method to classify PolSAR images and produced good results compared with other methods. Dong et al. [19]. proposed the copula-based joint statistical model to extract polarization features and use it for PolSAR image classification. Statistical methods can analyze land features in the data dimension and achieve image classification, but many parameters still need to be manually determined in advance, which brings a significant workload to research.

Although the above methods have achieved good results, they are all based on pixel-level classification, ignoring the relationship between the classified pixels and their neighborhoods. Liu et al. [20] used the information from the center pixel as well as the surrounding neighborhood pixels, combined them into superpixels, and used them as the smallest classification unit to classify PolSAR images, resulting in better classification outcomes.

Many researchers have studied the polarization decomposition method, and some famous algorithms include Pauli decomposition [8], SDH decomposition [21], Freeman decomposition [22], Yamaguchi decomposition [23], and reflection symmetrical decomposition (RSD) [24]. Van Zyl decomposition [25], H/A/Alpha decomposition [9], Huynen decomposition [26], Cameron decomposition [27], and Krogager decomposition [21] are also commonly used. These polarization decomposition algorithms have been applied to PolSAR land cover classification by relevant researchers. Nie et al. [28] utilized 12 polarization features obtained from Freeman–Durden decomposition, Van Zyl decomposition, and H/A/Alpha decomposition and achieved good classification results on limited samples using an enhanced learning framework. Wang et al. [2] applied the Freeman–Durden decomposition method and used a feature fusion strategy to classify PolSAR images of the Flevoland region. Ren et al. [29] utilized polarization scattering features obtained from T-matrix, Pauli decomposition, H/A/Alpha decomposition, and Freeman decomposition. Zhang et al. [30] applied the RSD method to extract polarimetric features from Gaofen3 images and obtained good results. Quan [31] proposed two polarimetric features—scattering contribution combiner (SCC) and scattering contribution angle (SCA)—for unified scattering characterization of manmade targets. The method achieved the physical optimization of scattering modeling. He also proposed a fine polarimetric decomposition method and derived several products to finely simulate the scattering mechanisms of urban buildings, which can also fulfill its use for effective surveillance [32].

Deep learning methods extract information about land targets through a certain number of network layers and utilize deep-level features extracted from the targets to classify objects in the image. Compared to traditional classification methods or machine learning, deep learning can more fully exploit the scattering characteristics inherent in land targets. In PolSAR data analysis, deep belief networks [33], stacked autoencoders [34], generative adversarial networks [35], convolutional neural networks [36,37], and deep stacked networks have achieved tremendous success [38–40]. Deep learning is a hierarchical learning method, and features extracted through this method are more discriminative [41]. Therefore, it demonstrates excellent performance in PolSAR image classification and target detection [42–49]. It has also led scholars to use various convolutional neural networks for the classification and information extraction of PolSAR images [50–54]. Liu et al. [55] proposed the active complex value convolutional wavelet neural network, and the Markov random fields method was proposed to classify PolSAR images, extracting information from multiple perspectives and achieving high-precision image classification. Yang et al. [56].

proposed a polarization direction angle composite sequence network, which extracts phase information from nondiagonal elements through real and complex convolutional long short-term memory networks. The network performance is better than that of existing convolutional neural networks based on real or complex numbers. Chu et al. [57] proposed a two-layer multi-objective superpixel segmentation network, with one layer used to optimize network parameters and the other layer used to refine segmentation results can achieve excellent segmentation results without obtaining prior information. These studies all demonstrate that the application of deep learning in the field of PolSAR is very successful. Considering the advantages of deep learning in extracting deep features from images and automatically learning parameters, drives us to use convolutional neural networks in this paper.

The PolSAR image contains multiple polarimetric characteristics and raw information about objects. Adopting an appropriate polarimetric decomposition method could extract features that represent objects, which benefits subsequent neural networks in classifying those features. Through existing research, it has been found that the most commonly used data input scheme is the 6-parameter data input scheme [58–60]. This method uses the total power of polarization, the two main diagonal elements of the polarimetric coherence matrix ($\boldsymbol{T}$-matrix), and the correlation coefficient between the non-main diagonal elements of the matrix. Although this data input scheme has achieved good classification of objects using improved neural networks, some parameters do not have clear physical meanings at the polarimetric feature level, and from the perspective of polarimetric information content, it is not complete. This prompts us to seek a data input scheme that can have physical interpretability and a more complete utilization of polarimetric information at the polarimetric feature level.

This article presents a PolSAR deep learning classification method based on the power values of polarimetric channels. It mainly utilizes horizontal, vertical, left-handed, and right-handed polarization, as well as other equivalent power values of different polarimetric channels, as input schemes for the neural network. This data input scheme is essentially a combination of polarimetric powers. The channels are equivalent to each other and represent power values under different polarization observations, and their addition and subtraction operations have clear physical meanings. Three polarimetric data input schemes were used, and then these polarimetric features were input into the neural network model to classify objects.

The main goals of this study were, therefore, (1) to provide a method for PolSAR image classification based on polarimetric features through deep learning neural networks; (2) to examine the power of classical CNNs for the classification of back-scattering similar ground objects; (3) to investigate the generalization capacity of existing CNNs for the classification of different satellite imagery; (4) to explore polarimetric features which are helpful for wetland classification and provide comparisons with different data input schemes; (5) to compare the performance and efficiency of other two schemes. Thus, this study contributes to the CNN classification tools for complex land cover mapping using polarimetric data based on polarimetric features.

2. Method

A deep learning classification scheme for PolSAR images based on polarimetric features, which mainly includes data preprocessing, polarization decomposition, polarization feature normalization, a data input scheme, and neural network classification.

2.1. Polarization Decomposition Method Based on Polarimetric Scattering Features

Target decomposition is a primary approach in polarimetric SAR data processing, which essentially represents pixels as weighted sums of several scattering mechanisms. In 1998, scholars Anthony Freeman and Stephen L. Durden proposed the first model-based, non-coherent polarimetric decomposition algorithm [22], hereinafter referred to as the Freeman decomposition. The initial purpose of the Freeman decomposition was to facilitate viewers of multi-view SAR images in intuitively distinguishing the major scattering mechanisms of objects.

The Freeman decomposition is entirely based on the back-scattering data observed by radar, and its decomposed components have corresponding physical meanings. Therefore, it later became known as the first model-based, non-coherent polarimetric decomposition algorithm. The introduction of the Freeman decomposition was pioneering at that time. After the proposal of the Freeman decomposition, as scholars extensively utilized and further researched it, they found three main issues with the decomposition method: the overestimation of volume scattering components; the presence of negative power components in the results; and the loss of polarization information. Through research, it was discovered that these three problems are not completely independent. For example, the overestimation of volume scattering components is one of the reasons for the existence of negative power values in subsequent surface scattering and double bounce components, and the loss of polarization information is also one of the reasons for the inappropriate estimation of power values of the volume scattering component.

In 2005, Yamaguchi et al. proposed the second model-based, non-coherent polarimetric decomposition algorithm [23]. This algorithm includes four scattering components, hereinafter referred to as the Yamaguchi algorithm. The Yamaguchi decomposition introduced helical scattering as the fourth scattering component, breaking the reflection symmetry assumption of the Freeman decomposition. This expansion made the algorithm applicable to a wider range of scenarios and achieved better experimental results in urban area analysis. The improved volume scattering model proposed by Yamaguchi opened up the research direction in enhancing the performance of model-based, non-coherent polarimetric decomposition algorithms through improving the scattering model. Both of the above points were pioneering work. However, the Yamaguchi algorithm did not provide a theoretical basis for selecting helical scattering as the fourth component, and according to their paper, the selection of helical scattering was based more on the comparison and preference of multiple basic scattering objects. The main innovative aspect of the Yamaguchi decomposition focused on the scattering model itself, while no improvements were made to the decomposition algorithm itself. It still followed the processing method of the Freeman decomposition. Although the algorithm showed better experimental results, issues such as the overestimation of volume scattering, negative power components, and the loss of polarization information still persisted [24].

Compared to classical polarization decomposition methods such as Freeman decomposition and Yamaguchi decomposition, the reflection symmetric decomposition [24,61] has the advantage of obtaining polarization components with non-negative power values; the decomposed results can completely reconstruct the original polarimetric coherent matrix, and the decomposition aligns strictly with the theoretical models of volume scattering, surface scattering, and double scattering. Therefore, in this paper, we chose this method to extract the polarization features of targets from PolSAR images. The reflection symmetric decomposition (RSD) is a model-based incoherent polarization decomposition method that decomposes the polarimetric coherent matrix ($\boldsymbol{T}$) into polarization features such as the power of the surface scattering component (PV), the power of the double scattering component (P_{S}), and the power of the volume scattering component (P_{D}). The value range of these three components is $[0, +\infty)$.

2.2. *Vertical, Horizontal, Left-Handed Circular, Right-Handed Circular Polarization Methods*

Currently, radar antennas primarily use two types of polarization bases: linear polarization and circular polarization. Typical linear polarization methods include horizontal polarization (H) and vertical polarization (V), and circular polarization methods include left-handed circular polarization (L) and right-handed circular polarization (R).

When a polarimetric radar uses linear polarization bases, this method first transmits horizontally polarized electromagnetic waves and uses horizontal and vertical antennas for reception. It then transmits vertically polarized electromagnetic waves and uses horizontal and vertical antennas for reception again. In the case of a single-station radar, the back-scattering alignment convention (BSA) is usually used, and the transmitting and receiving antennas use the same coordinate system. In this coordinate system, the Z-axis points towards the target, the X-axis is horizontal to the ground, and the Y-axis, along with the X-Z plane, forms a right-handed coordinate system pointing towards the sky. This coordinate system corresponds well to the horizontal (H) and vertical (V) polarization bases. In this case, the Sinclair scattering matrix can be abbreviated as:

$$\mathbf{S} = \begin{bmatrix} S_{HH} & S_{HV} \\ S_{VH} & S_{VV} \end{bmatrix} \tag{1}$$

Upon satisfying the reciprocity theorem, the polarization coherency matrix $\boldsymbol{T}$ is derived post multi-look processing, eliminating coherent speckle noise:

$$\boldsymbol{T} = \left\langle \boldsymbol{k}\boldsymbol{k}^{\mathrm{H}} \right\rangle = \begin{bmatrix} T_{11} & T_{12} & T_{13} \\ T^{*}{}_{12} & T_{22} & T_{23} \\ T^{*}{}_{13} & T^{*}{}_{23} & T_{33} \end{bmatrix} \tag{2}$$

Among them,

$$\boldsymbol{k} = \frac{1}{\sqrt{2}} \begin{bmatrix} S_{HH} + S_{VV} \\ S_{HH} - S_{VV} \\ S_{HV} - S_{VH} \end{bmatrix} \tag{3}$$

where $\boldsymbol{k}$ represents the scattering vector of the back-scattering $\mathbf{S}$-matrix in the Pauli basis, and the superscript H denotes the Hermitian transpose. <•> represents an ensemble average. The $\boldsymbol{T}$-matrix is a positive semi-definite Hermitian matrix, which can be represented as a 9-dimensional real vector [T_{11}, T_{22}, T_{33}, Re(T_{12}), Re(T_{13}), Re(T_{23}), Im(T_{12}), Im(T_{13}), Im(T_{23})]. Tij represents the element in the i-th row and j-th column of the $\boldsymbol{T}$-matrix. Re(T_{ij}) and Im(T_{ij}) represent the real and imaginary parts of the T_{ij} element, respectively.

The Sinclair matrix can be vectorized using the Pauli basis ψ_P, which can be expressed as follows:

$$\psi_P = \left\{ \sqrt{2}\begin{bmatrix} 1 & 0 \\ 0 & 1 \end{bmatrix}, \sqrt{2}\begin{bmatrix} 1 & 0 \\ 0 & -1 \end{bmatrix}, \sqrt{2}\begin{bmatrix} 0 & 1 \\ 1 & 0 \end{bmatrix}, \sqrt{2}\begin{bmatrix} 0 & -j \\ j & 0 \end{bmatrix} \right\} \tag{4}$$

The scattering vector ψ_P under the Sinclair matrix is:

$$\boldsymbol{K}_{4P} = \frac{1}{\sqrt{2}} \left[S_{HH} + S_{VV} \quad S_{HH} - S_{VV} \quad S_{HV} + S_{VH} \quad j(S_{HV} - S_{VH}) \right]^{T} \tag{5}$$

For single-polarization radar, under the condition of satisfying the reciprocity theorem, the above equation becomes:

$$\boldsymbol{K}_{P} = \frac{1}{\sqrt{2}} \left[S_{HH} + S_{VV} \quad S_{HH} - S_{VV} \quad 2S_{HV} \right]^{T} \tag{6}$$

Therefore, in single-channel single-polarization SAR data, the polarimetric scattering characteristics of the target in the Pauli basis are represented by the polarimetric coherence matrix as follows:

$$\boldsymbol{T}_{3\times3} = \left\langle \boldsymbol{K}_{P}\boldsymbol{K}_{P}^{*T} \right\rangle = \frac{1}{2} \begin{bmatrix} \left\langle |S_{HH} + S_{VV}|^{2} \right\rangle & \langle (S_{HH} + S_{VV})(S_{HH} - S_{VV})^{*} \rangle & \langle 2(S_{HH} + S_{VV})S_{HV}^{*} \rangle \\ \langle (S_{HH} - S_{VV})(S_{HH} + S_{VV})^{*} \rangle & \left\langle |S_{HH} - S_{VV}|^{2} \right\rangle & \langle 2(S_{HH} - S_{VV})S_{HV}^{*} \rangle \\ \langle 2S_{HV}(S_{HH} + S_{VV})^{*} \rangle & \langle 2S_{HV}(S_{HH} - S_{VV})^{*} \rangle & \left\langle 4|S_{HV}|^{2} \right\rangle \end{bmatrix} \tag{7}$$

In the equation, * denotes conjugation, T represents transpose, <·> represents ensemble averaging. Thus,

$$\frac{(T_{11} + T_{22})}{2} + \mathrm{Re}(T_{12}) = |S_{HH}|^{2} = H(T_{12}) \tag{8}$$

$$\frac{(T_{11}+T_{22})}{2}-\mathrm{Re}(T_{12})=|S_{VV}|^2=V(T_{12}) \tag{9}$$

In other words, the real part of the T_{12} element information can be represented by the power values of the horizontal and vertical polarization channels. In the equation above, H(.) and V(.) are channel representation methods used in this paper. Similarly, for T_{13}, T_{21}, and T_{23}, the following four channel representation methods can be obtained.

$$\frac{(T_{11}+T_{33})}{2}+\mathrm{Re}(T_{13})=H(T_{13}) \tag{10}$$

$$\frac{(T_{11}+T_{33})}{2}-\mathrm{Re}(T_{13})=V(T_{13}) \tag{11}$$

$$\frac{(T_{22}+T_{33})}{2}+\mathrm{Re}(T_{23})=H(T_{23}) \tag{12}$$

$$\frac{(T_{22}+T_{33})}{2}-\mathrm{Re}(T_{23})=V(T_{23}) \tag{13}$$

Therefore, through equation substitution, we equivalently replace the elements in the $\boldsymbol{T}$ matrix. This can, to some extent, be represented by the horizontal and vertical polarization power components to represent the real part elements in the $\boldsymbol{T}$-matrix. Similarly, we are also looking for a polarization power method that can represent the imaginary elements in the $\boldsymbol{T}$-matrix. Under a circular polarization basis, the scattering matrix under the same method can be defined as follows:

$$\begin{bmatrix} E_{LS} \\ E_{RS} \end{bmatrix}=\begin{bmatrix} S_{LL} & S_{LR} \\ S_{RL} & S_{RR} \end{bmatrix}\begin{bmatrix} E_{LI} \\ E_{RI} \end{bmatrix} \tag{14}$$

For a single-station radar, under the condition of satisfying the reciprocity theorem (S_{LR} = S_{RL}), electromagnetic waves can be converted between a linear polarization basis and a circular polarization basis [62]. This enables the conversion of the scattering matrix between the linear polarization basis and circular polarization basis as well. The specific derivation process can be found in [63], and here only the results are given as follows:

$$\begin{bmatrix} S_{LL} \\ \sqrt{2}S_{LR} \\ S_{RR} \end{bmatrix}=\begin{bmatrix} \frac{1}{2} & \frac{j}{\sqrt{2}} & -\frac{1}{2} \\ \frac{1}{\sqrt{2}} & 0 & \frac{1}{\sqrt{2}} \\ \frac{1}{2} & -\frac{j}{\sqrt{2}} & -\frac{1}{2} \end{bmatrix}\begin{bmatrix} S_{HH} \\ \sqrt{2}S_{HV} \\ S_{VV} \end{bmatrix} \tag{15}$$

Based on Formulas (6) and (14), the corresponding transformation formula between circular polarization basis and Pauli vector is as follows:

$$\begin{bmatrix} S_{LL} \\ \sqrt{2}S_{LR} \\ S_{RR} \end{bmatrix}=\begin{bmatrix} 0 & \frac{1}{\sqrt{2}} & \frac{j}{\sqrt{2}} \\ 1 & 0 & 0 \\ 0 & \frac{1}{\sqrt{2}} & -\frac{j}{\sqrt{2}} \end{bmatrix}K_P \tag{16}$$

Then,

$$\begin{bmatrix} S_{LL} \\ S_{RR} \end{bmatrix}=\frac{1}{2}\begin{bmatrix} S_{HH}-S_{VV}+j2S_{HV} \\ S_{HH}-S_{VV}-j2S_{HV} \end{bmatrix} \tag{17}$$

By changing the two Equations (16) and (17), we can obtain the following form:

$$\begin{bmatrix} S_{HH} & S_{HV} \\ S_{VH} & S_{VV} \end{bmatrix}=S_{LR}\begin{bmatrix} 1 & 0 \\ 0 & 1 \end{bmatrix}+\frac{S_{RR}}{2}\begin{bmatrix} 1 & j \\ j & -1 \end{bmatrix}+\frac{S_{LL}}{2}\begin{bmatrix} 1 & -j \\ -j & -1 \end{bmatrix} \tag{18}$$

$$K_{\mathrm{P}} = S_{\mathrm{LR}}\begin{bmatrix}\sqrt{2}\\0\\0\end{bmatrix} + \frac{S_{\mathrm{RR}}}{\sqrt{2}}\begin{bmatrix}0\\1\\j\end{bmatrix} + \frac{S_{\mathrm{LL}}}{\sqrt{2}}\begin{bmatrix}0\\1\\-j\end{bmatrix} \tag{19}$$

From the above equation, it can be inferred that the transformation from horizontal and vertical polarization to circular polarization can also be considered as a process of decomposing the scattering matrix into certain correlation terms. This means that the Sinclair matrix could be decomposed into components such as plane waves, left-handed helices, and right-handed helices, and S_{LR}, S_{RR}, S_{LL} correspond to the phase and power levels of each constituent.

Therefore, it can be inferred the following equations:

$$\frac{(T_{11}+T_{22})}{2} + \mathrm{Im}(T_{12}) = |S_{\mathrm{LL}}|^2 = L(T_{12}) \tag{20}$$

$$\frac{(T_{11}+T_{22})}{2} - \mathrm{Im}(T_{12}) = |S_{\mathrm{RR}}|^2 = R(T_{12}) \tag{21}$$

Thus, the imaginary part of the T_{12} element information can be represented by the power values of the left-hand and right-hand polarization channels, where L(.) and R(.) are also the channel representation methods used in this article. Similarly, for T_{13}, T_{23}, four channel representation methods can be obtained as follows:

$$\frac{(T_{11}+T_{33})}{2} + Im(T_{13}) = L(T_{13}) \tag{22}$$

$$\frac{(T_{11}+T_{33})}{2} - Im(T_{13}) = R(T_{13}) \tag{23}$$

$$\frac{(T_{22}+T_{33})}{2} + Im(T_{23}) = L(T_{23}) \tag{24}$$

$$\frac{(T_{22}+T_{33})}{2} - Im(T_{23}) = R(T_{23}) \tag{25}$$

Therefore, through equation substitution, we equivalently replace the elements in the $\boldsymbol{T}$ matrix. This can, to some extent, be represented by the left-hand and right-hand polarization power components to represent the imaginary part elements in the $\boldsymbol{T}$-matrix.

From the above derivation process, it can be seen that the new classification scheme first uses the power values of the horizontal, vertical, left-hand, and right-hand polarization channels. The other channels following also essentially represent power values of a certain polarization channel; that is to say, the elements in the $\boldsymbol{T}$ matrix are equivalently represented using polarization power features, and the input elements have actual physical meanings. Moreover, the combination of the mentioned channels can fully invert all elements of the $\boldsymbol{T}$-matrix, making it comprehensive from the perspective of polarization information.

2.3. Input Feature Normalization and Design of Three Schemes

Before inputting these polarizing features into the neural network, it is necessary to normalize these physical quantities to meet the requirements of the network input. In the $\boldsymbol{T}$-matrix, the total polarized power is converted into a physical quantity in units of dB. For polarized power parameters T_{11}, T_{22}, T_{33}, P_{S}, P_{D}, P_{V}, they are all divided by *Span* to achieve normalization.

Based on the existing literature and corresponding polarized power values, this paper designs three deep learning polarization data input schemes. First, the decomposed P_{S}, P_{D}, P_{V} with reflection symmetry, and the normalized P_0 ($10\log_{10}Span$) were used as the data input Scheme 1. The elements in Scheme 1 were all based on the characteristics of polarization power and contained the main polarization information of the terrain objects. Therefore, this input scheme was used as the basic one. Then, according to references [58–60], the correlation coefficients between channels T_{12}, T_{23}, T_{23}, as well as the non-normalized P_0 (NonP_0) of the $\boldsymbol{T}$ matrix, were used as the research Scheme 2, where the correlation coefficients between channels are defined by Formulas (26)–(28).

$$coeT_{12} = |\mathrm{T}_{12}| / \sqrt{\mathrm{T}_{11} \cdot \mathrm{T}_{22}} \quad (26)$$

$$coeT_{13} = |\mathrm{T}_{13}| / \sqrt{\mathrm{T}_{11} \cdot \mathrm{T}_{33}} \quad (27)$$

$$coeT_{23} = |\mathrm{T}_{23}| / \sqrt{\mathrm{T}_{33} \cdot \mathrm{T}_{22}} \quad (28)$$

In this data input scheme, except for NonP_0, the value range of the other five parameters is between 0 and 1. Finally, a total of 16 parameters, including P_0, T_{12}, T_{23}, T_{23}, $H(T_{12})$, $H(T_{13})$, $H(T_{23})$, $L(T_{12})$, $L(T_{13})$, $L(T_{23})$, $V(T_{12})$, $V(T_{13})$, $V(T_{23})$, $R(T_{12})$, $R(T_{13})$, $R(T_{23})$, were used as data input Scheme 3, where P_0 had been normalized. The other channels were normalized by dividing them by P_0, as they represent the power values of specific polarization channels. The three data input schemes are shown in Table 1.

Table 1. List of three polarization data input schemes.

Scheme	Parameters	Polarization Features
1	4	P_0, P_S, P_D, P_V
2	6	NonP_0, T_{22}, T_{33}, coeT_{12}, coeT_{13}, coeT_{23}
3	16	P_0, T_{12}, T_{23}, T_{23}, $H(T_{12})$, $H(T_{13})$, $H(T_{23})$, $L(T_{12})$, $L(T_{13})$, $L(T_{23})$, $V(T_{12})$, $V(T_{13})$, $V(T_{23})$, $R(T_{12})$, $R(T_{13})$, $R(T_{23})$

At the same time, the data distribution of the three research schemes was also statistically analyzed. Except for the NonP_0 polarization feature of Scheme 2, the polarization characteristics of the other schemes were distributed in the range [0, 1]. In order to have a visual understanding of the polarization feature distribution of each data input scheme, we conducted a histogram analysis of the polarization feature distribution, as shown in Figures 1–3.

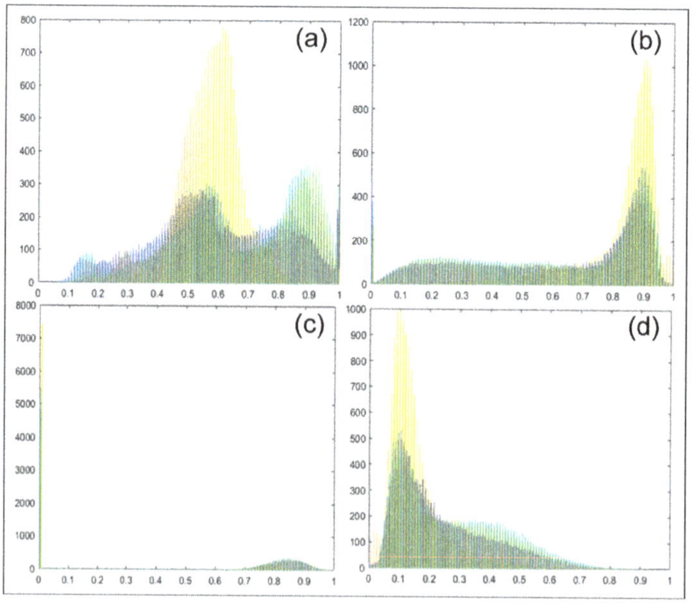

Figure 1. Histogram of polarization feature distribution for data input Scheme 1. (**a**) P_0; (**b**) P_S; (**c**) P_D; (**d**) P_V.

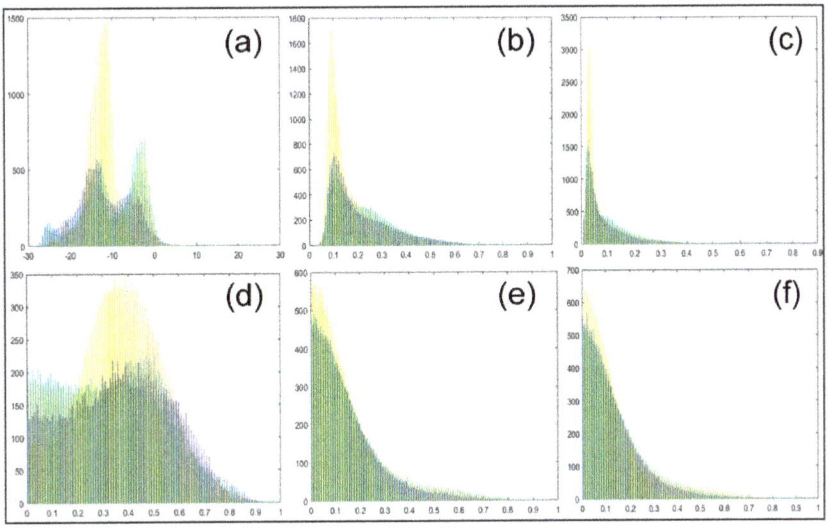

Figure 2. Histogram of polarization feature distribution for data input Scheme 2. (**a**) NonP_0; (**b**) T_{22}; (**c**) T_{33}; (**d**) coeT_{12}; (**e**) coeT_{13}; (**f**) coeT_{23}.

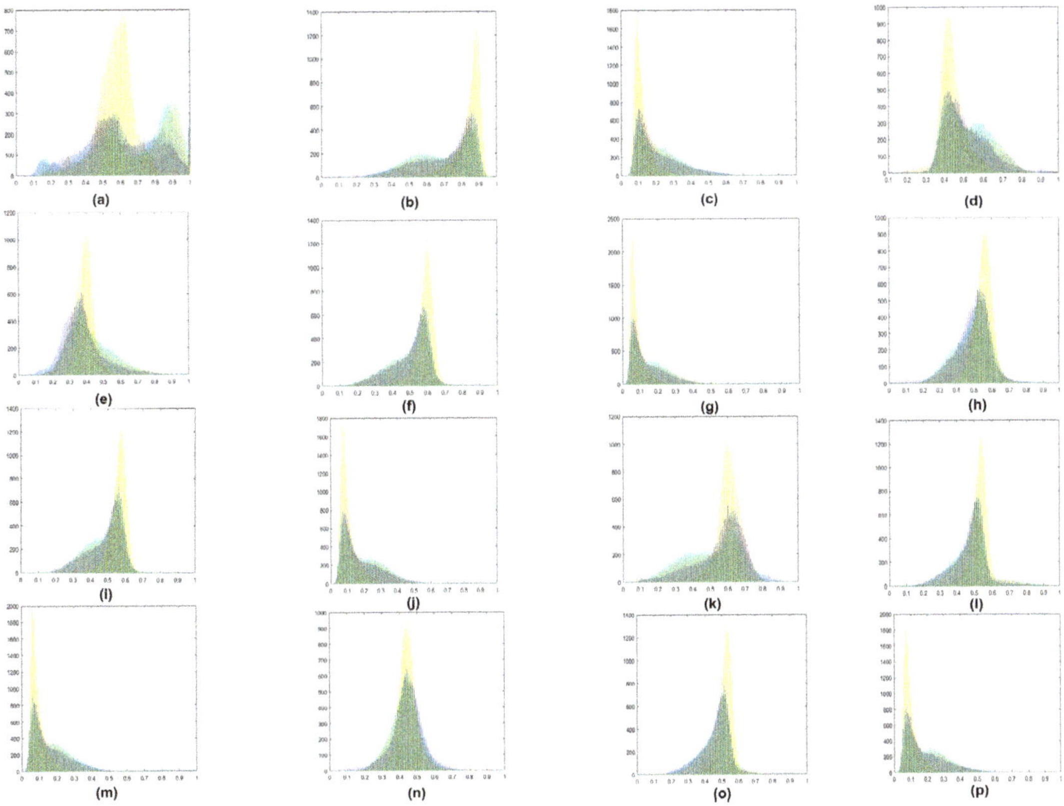

Figure 3. Histogram of polarization feature distribution for data input Scheme 3. (**a**) P_0, (**b**) T_{11}, (**c**) T_{22}, (**d**) T_{33}, (**e**) $H(T_{12})$, (**f**) $H(T_{13})$, (**g**) $H(T_{23})$, (**h**) $L(T_{12})$, (**i**) $L(T_{13})$, (**j**) $L(T_{23})$, (**k**) $V(T_{12})$, (**l**) $V(T_{13})$, (**m**) $V(T_{23})$, (**n**) $R(T_{12})$, (**o**) $R(T_{13})$, (**p**) $R(T_{23})$.

Through the experiment, we obtained the distribution histogram of P_0 and classified features of four experimental images. We discovered that P_0 was distributed at [−30 dB, 0 dB]. We obtained the distribution histogram of ground objects and discovered that it was also distributed at [−30 dB, 0 dB]. Therefore, the selected value range distribution was appropriate, and no new categories were introduced.

2.4. Experiment and Pre-Processing

After obtaining the high-resolution Gaofen-3 Level 1A QPSI data, additional data were required for radiometric calibration. The method for radiometric calibration can be found in the Gaofen-3 user manual [64]. Due to inherent speckle noise in the data, an appropriate filtering method was necessary to remove the speckle, reducing its impact on subsequent classification. Compared to traditional filtering methods, the non-local means filtering method [65] considers the influence of neighboring pixels, making it more effective. Therefore, this paper selected this method to denoise the PolSAR images. The polarization coherence matrix data and all polarization characteristic parameters of the reflection symmetry decomposition were obtained by processing the data using the polarization decomposition production algorithm mentioned in reference [24,62].

2.5. Classification Process of Polarization Scattering Characteristics Using Deep Learning

In this paper, based on the scattering mechanism, the polarization characteristics were classified into three different input schemes. Then, these three research schemes were inputted into a network model to extract the features of the objects. Finally, a Softmax classifier was used to obtain the classification results at the end of the network. The Figure 4 is a flowchart of the entire experimental process, in which different colored CNN architectures represent the extracted polarization features of different schemes. After the experiments, it was found that when the network window size was 64 × 64, various research schemes could achieve the best classification results. Therefore, this experiment selected samples of this size for experimentation. The sample dimension sizes input into the neural network were 64 × 64 × 4, 64 × 64 × 6, and 64 × 64 × 16, respectively.

The entire process of this algorithm is shown as follows (Algorithm 1).

Algorithm 1: A deep learning classification scheme for PolSAR image based on polarimetric features

Input: GF-3 PolSAR images.
Output: Predict label Y_{test} $\{y_1, y_2, \ldots, y_m\}$
1: Processing GF-3 PolSAR images.
2: Polarimetric decomposition.
3: Extract polarimetric features.
4: Feature normalization.
5: Three schemes are proposed based on the previous studies and scattering mechanisms.
6: Randomly select a certain proportion of training samples (Patch_Xtrain: {Patch_x_1, Patch_x_2, ..., Patch_x_n}, the remaining labeled samples are used as validation samples
7: Inputting Patch_x_i into CNN.
for i < N **do**
the train one time.
If good fitting, **then**
Save model, and break.
else if over-fitting or under-fitting, **then**
Adjust parameters include, i.e., learning rate, bias.
End
8: Predict Label: Y = Softmax (Patch_Xtrain)
9: Test images are input to the model and predict the patches of all pixels.
10: Do method evaluation, i.e., Statistic OA, AA, and Kappa coefficient.

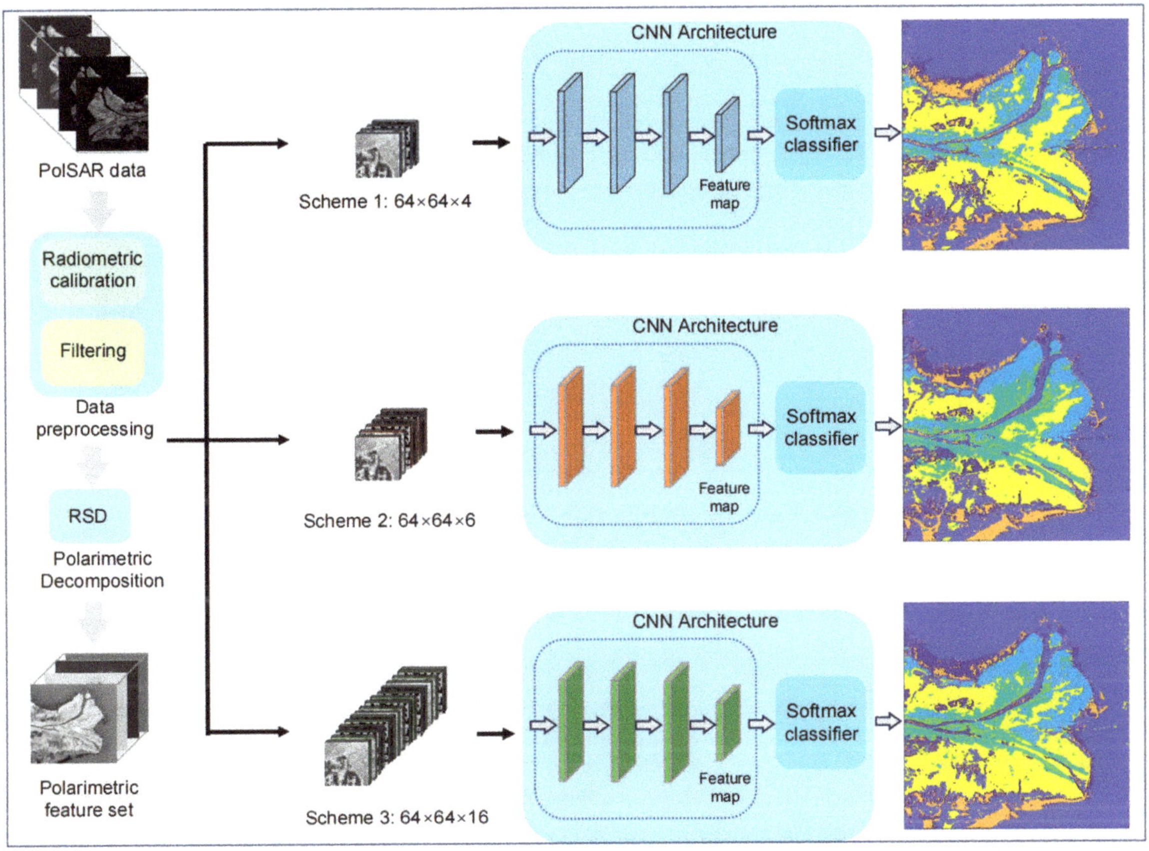

Figure 4. Data processing flowchart.

3. Experimental and Result Analysis

In this section, high-resolution PolSAR images of the Yellow River Delta area, which have undergone field surveys, were used to verify the effectiveness of the proposed approach. All experiments were conducted on an i7-10700 CPU (Intel, Santa Clara, CA, USA)and RTX 3060 Ti GPU (NVIDIA, Santa Clara, CA, USA).

3.1. Study Area and Dataset

GF-3 has a quad-polarized instrument with different modes. In this article, we used high-resolution QPSI imaging mode PolSAR images (spatial resolution is 8m) of the Yellow River Delta area (Shandong, China) for the experiment (displayed in Figure 5). There are several classical types in this area, such as nearshore water, seawater, spartina alterniflora, tamarix, reed, tidal flat, and suaeda salsa. Restricted by historical resources, we chose four images of this area. The training and validation sets were selected from three different images taken during the same quarter in this region, specifically on 14 September 2021 and 13 October 2021. The test image was taken on 12 October 2017. We used unmanned aerial vehicle (UAV) images (displayed in Figure 6), which were shot in September 2021, combined with empirical knowledge, for marking targets to guarantee the accuracy of the labeled training datasets.

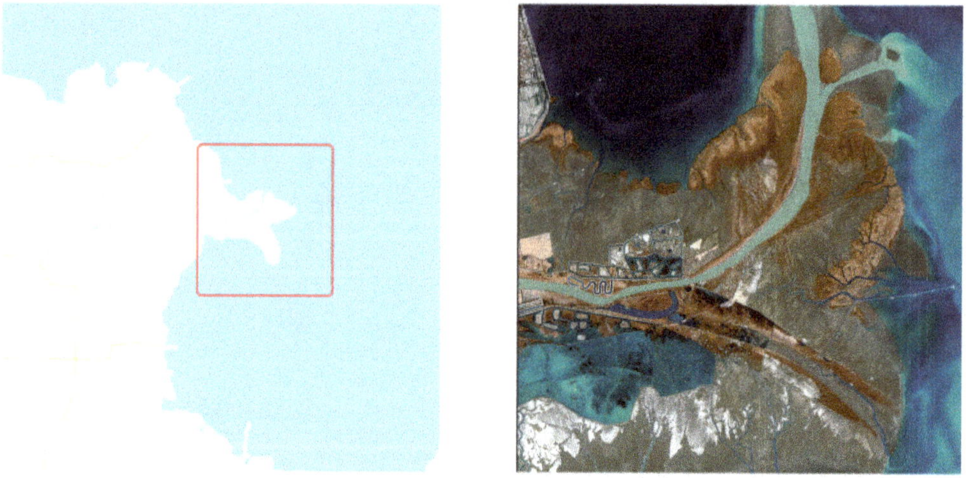

Figure 5. Study area location and its corresponding image.

Figure 6. UAV images of ground objects. (**a**) Nearshore water; (**b**) Seawater; (**c**) Spartina alterniflora; (**d**) Suaeda salsa; (**e**) Tamarix; (**f**) Reed; (**g**) Tidal flat.

In this study, we classified seven species according to the survey results: nearshore water; seawater; spartina alterniflora; tamarix; reed; tidal flat; and suaeda salsa. We labeled these targets from numbers 1 to 7, respectively. We randomly selected 800 samples of each category for training and 200 for validation. The details are shown in Table 2.

Table 2. Samples distribution.

Images	Nearshore Water	Seawater	Spartina Alterniflora	Tamarix	Reed	Tidal Flat	Suaeda Salsa
20210914_1	500	400	1000	500	500	500	500
20210914_2	500	200	0	0	0	500	0
20211013	0	400	0	500	500	0	500
Total	1000	1000	1000	1000	1000	1000	1000

3.2. Classification Results of the Yellow River Delta on AlexNet

In order to quantitatively evaluate the accuracy of the three data input schemes for classification and to avoid random results, in this paper, five independent experiments were conducted on AlexNet for each data input scheme. The overall accuracy of each classification was calculated for each experiment (the results were arranged in descending order, with the highest overall accuracy being the maximum value), as well as the average accuracy of the overall classification for the five experiments and the Kappa coefficient to evaluate the classification results. Both the accuracy of individual land cover classes and the Kappa coefficient were calculated based on the results with the highest overall accuracy.

From the classification results, obviously, it can be seen that when using Scheme 3 for polarized data input, both the highest overall accuracy and average overall accuracy were higher compared to the other two schemes, with values of 86.11% and 78.08%, respectively. We believe this is because Scheme 3 contains more polarization information than the other two schemes. The overall accuracy and Kappa coefficient of five independent experiments stayed at a relatively high level, showing the stronger robustness of Scheme 3.

It is worth noting that for the tidal flat, Scheme 2 and Scheme 3 performed poorly, with accuracies of 49.3% and 44.6%, respectively. This indicates that these two schemes did not contain polarization parameters that effectively represent the scattering characteristics of the tidal flat, resulting in low classification accuracy for this land cover. We also guess that tidal flats are unique ecosystems because the water cover changes intermittently with the tidal phase, which may also lead to low accuracy. It can also be seen from the table that the accuracy of the tamarix in Scheme 1 was lower than the other two schemes, only 40.1%, while the recognition accuracy of Scheme 2 and Scheme 3 reached 100%. This indicates that our proposed method can extract more information from PolSAR images, which is beneficial for improving the overall land use classification accuracy.

For Scheme 2, both the tidal flat and suaeda salsa had low classification accuracies of 49.3% and 50.8%, respectively. For Scheme 1, the classification accuracies for each land cover were lower than the other two schemes due to the limited number of polarization features. This indicates that neither of these two schemes effectively contained polarization information of classified features beyond a certain extent, which means that the polarization characteristics of these two schemes in terms of data input were incomplete.

In Scheme 3, the polarized features inputted into the deep learning network were the power values of the polarization channels. These channels were equivalent and contained all the polarization information using equivalent polarization power values. Apart from intertidal zones, high classification values were achieved for the other six land cover types. This indicates that using equivalent polarization power values can effectively distinguish most land cover types. However, strictly speaking, the polarization information in this scheme still cannot effectively differentiate classified land cover types, and overall classification accuracy needs further improvement. The classification accuracy and overall accuracy of land features for various schemes are shown in Table 3.

Table 3. The classification accuracy of their polarization input schemes on AlexNet.

Classification Accuracy Input Scheme	Scheme 1	Scheme 2	Scheme 3
Nearshore water	83.4	96.8	100
Seawater	98.7	96.9	99.60
Spartina alterniflora	87.0	96.8	93.3
Tamarix	40.1	100	100
Reed	50.4	94.5	68.50
Tidal flat	61.8	49.3	44.6
Suaeda salsa	98.2	50.8	96.8

Table 3. *Cont.*

Classification Accuracy Input Scheme	Scheme 1	Scheme 2	Scheme 3
Indepent experiments Overall Accuracy	74.23	83.59	86.11
	71.36	81.41	81.53
	70.41	77.83	77.04
	68	73.66	73.73
	67.84	68.87	71.99
Average Overall Accuracy	70.368	77.072	78.08
Kappa coefficient	0.6993	0.8085	0.8380

We also classified the entire test image, and it is easy to see that the classification result of Scheme 3 was better than the other two schemes. From the overall classification effect, Scheme 3 was also better than the other two schemes in terms of the number of classified objects and the classification effect between different categories.

From the classification maps, we can see that spartina alterniflora, tamarix, and reed were easily classified in Scheme 3. In the other two schemes, there were some misclassification phenomena, which indicates that compared to Scheme 3, the polarization information carried cannot distinguish these wetland vegetation types very well.

The classification results of the entire image are shown in Figure 7.

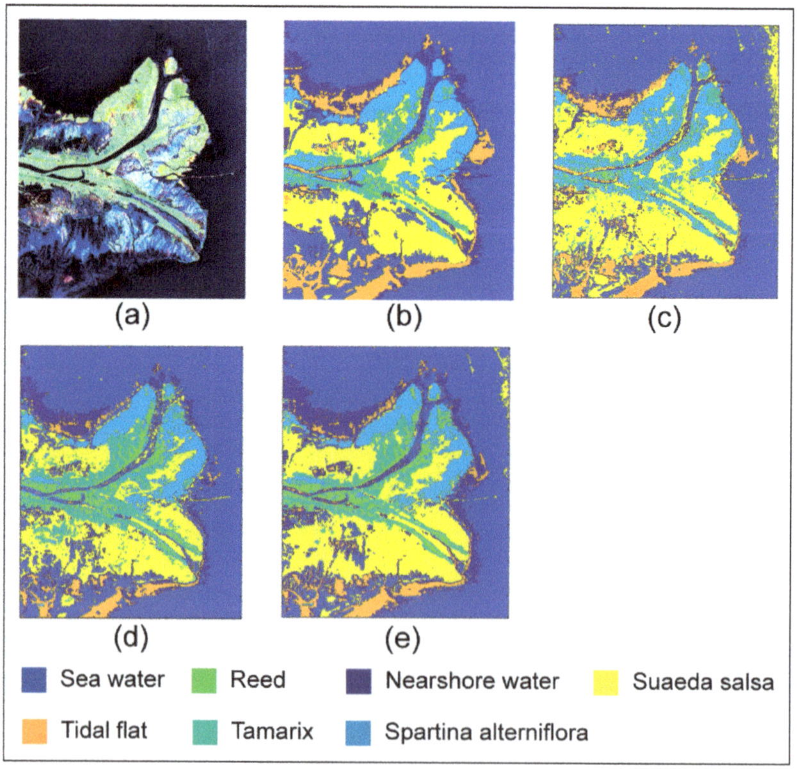

Figure 7. The classification results of the three research schemes on AlexNet. GF-3 Data (**a**) Pauli pseudo-color map. (**b**) Ground truth map. Classification results of (**c**) Scheme 1, (**d**) Scheme 2, (**e**) Scheme 3.

3.3. Classification Results of the Yellow River Delta on VGG16

To further verify the above conclusion, we also conducted comparative experiments on three schemes through VGG16. Similar to the testing results on AlexNet, when using VGG16 to test three data input schemes, there were still situations where the classification accuracy of certain land objects was low. In Scheme 1, the reed classification accuracy was 26.1%, and the tamarix accuracy was 40.2%. In Scheme 2, the tidal flat accuracy was only 28.5%, while in Scheme 3, the reed accuracy was 44.7%, and the tidal Flat accuracy was 58.3%. This indicates that none of these three schemes could fully classify the selected features beyond a certain extent, but in terms of overall classification accuracy and Kappa coefficient, Scheme 3 was still better than the other two schemes, and it was relatively complete in carrying polarization features.

Overall, Scheme 3 showed better classification performance than the other two schemes. Except for reeds and tidal flats, the classification accuracy of the other five land cover types remained consistently high. The classification accuracy of Tamarix reached 100%, and the classification accuracy of other land features was also above 94.1%. At the same time, the highest overall classification accuracy and average overall classification accuracy were also superior to the other two schemes, indicating that Scheme 3 had relatively more complete polarization information prior to inputting it into the network model.

Scheme 2 slightly underperformed Scheme 3 in the classification accuracy of suaeda salsa, with an accuracy of only 66.2%. Particularly, Scheme 2 struggled to classify tidal flats well, with a classification accuracy of only 28.5%. The polarization information contained in Scheme 2 cannot effectively characterize these two types of ground objects, resulting in relatively low classification accuracy. Instead, Scheme 1, which included surface scattering and volume scattering components, effectively characterized the scattering characteristics of tidal flat and suaeda salsa. Therefore, the effect of Scheme 1 was better than that of Scheme 2.

For Scheme 1, the overall land cover classification accuracy was lower compared to the other two schemes, mainly due to the limited number of polarimetric feature parameters, which failed to effectively represent the classified area in the input network model. The specific classification accuracies are shown in Table 4.

Table 4. The classification accuracy of their polarization input schemes on VGG16.

Classification Accuracy Input Scheme	**Scheme 1**	**Scheme 2**	**Scheme 3**
Nearshore water	89.3	95.7	95.6
Seawater	99.4	97.7	99.7
Spartina alterniflora	87.6	96.6	95.9
Tamarix	40.2	98.5	100
Reed	26.1	93.8	44.7
Tidal flat	73.2	28.5	58.3
Suaeda salsa	100	66.2	94.1
Indepent experiments overall accuracy	73.69	82.43	84.04
	72.8	82.21	83.57
	69.7	81.44	82.07
	68.66	79.44	81.54
	67.6	77.53	80.11
Average overall accuracy	70.49	80.61	82.266
Kappa coefficient	0.6930	0.7950	0.8138

Similarly, we also displayed the classification results of the entire image. The results of the three data input schemes on VGG16 are shown in Figure 8. From Figure 8, it can be observed that the classification results of Scheme 3 were still the best among the three schemes.

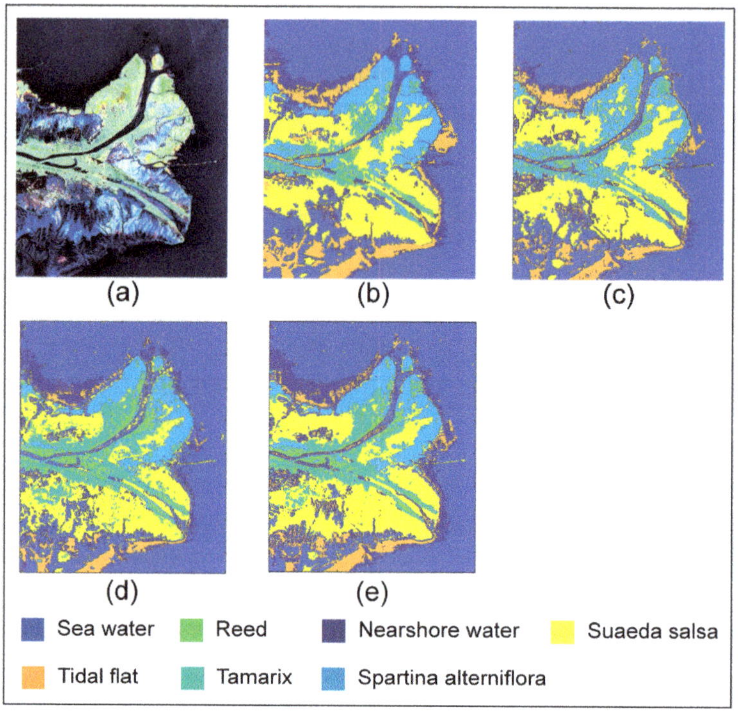

Figure 8. Classification results of three research schemes on VGG16. GF-3 Data. (**a**) Pauli pseudo-color map; (**b**) Ground truth map. Classification results of (**c**) Scheme 1, (**d**) Scheme 2, (**e**) Scheme 3.

When using VGG16 for classification, the classification results of each scheme were clearer overall than when using AlexNet, and the clustering effect of each feature was better.

4. Conclusions

In this paper, a deep learning-based classification scheme for PolSAR images using polarimetric scattering features was proposed through rigorous mathematical derivation. This scheme utilized a combination of polarimetric power features, ensuring that each channel represented power values and was equivalent to other channels. Each channel possessed practical physical meanings and clear mathematical significance. Experimental results demonstrated that compared to the 6-parameter and 4-parameter data input schemes, the proposed scheme had more comprehensive information and achieved higher classification accuracy. The proposed scheme was validated on the GF-3 dataset and showed performance improvement. However, for the classification of certain land objects, this approach lacked sufficient accuracy, and there were situations where the information was not comprehensive enough.

In future work, more comprehensive data input schemes will be explored.

Author Contributions: Methodology, W.A.; Formal analysis, S.Z.; Data curation, L.C.; Writing—original draft, S.Z.; Supervision, Z.D. All authors have read and agreed to the published version of the manuscript.

Funding: This research was funded by National Key R&D Program of China, grant number 2021YFC2803304 and 2022YFB3902404.

Data Availability Statement: Data were obtained from China Ocean Satellite Data Service System and are available at https://osdds.nsoas.org.cn/ (accessed on 30 March 2024) with the identity of protocol users.

Acknowledgments: We would like to thank the National Satellite Ocean Application Center for providing the Gaofen-3 data. Currently, these data are available to the public, and the data access address is https://osdds.nsoas.org.cn/ (accessed on 30 March 2024).

Conflicts of Interest: The authors declare no conflict of interest.

References

1. Lee, J.S.; Grunes, M.R.; Pottier, E. Quantitative comparison of classification capability: Fully polarimetric versus dual and single-polarization SAR. *IEEE Trans. Geosci. Remote Sens.* **2001**, *39*, 2343–2351.
2. Wang, Y.; Cheng, J.; Zhou, Y.; Zhang, F.; Yin, Q. A multichannel fusion convolutional neural network based on scattering mechanism for PolSAR image classification. *IEEE Geosci. Remote Sens. Lett.* **2022**, *19*, 4007805. [CrossRef]
3. Wang, X.; Cao, Z.; Cui, Z.; Liu, N.; Pi, Y. PolSAR image classification based on deep polarimetric feature and contextual information. *J. Appl. Remote Sens.* **2019**, *13*, 034529. [CrossRef]
4. Dong, H.; Zhang, L.; Lu, D.; Zou, B. Attention-based polarimetric feature selection convolutional network forPolSAR image classification. *IEEE Geosci. Remote Sens. Lett.* **2022**, *19*, 4001705. [CrossRef]
5. Lonnqvist, A.; Rauste, Y.; Molinier, M.; Hame, T. Polarimetric SAR data in land cover mapping in boreal zone. *IEEE Trans. Geosci. Remote Sens.* **2010**, *48*, 3652–3662. [CrossRef]
6. McNairn, H.; Shang, J.; Jiao, X.; Champagne, C. The contribution of ALOS PALSAR multipolarization and polarimetric data to crop classification. *IEEE Trans. Geosci. Remote Sens.* **2009**, *47*, 3981–3992. [CrossRef]
7. Qi, Z.; Yeh, A.G.-O.; Li, X.; Lin, Z. A novel algorithm for land use and land cover classification using RADARSAT-2 polarimetric SAR data. *Remote Sens. Environ.* **2012**, *118*, 21–39. [CrossRef]
8. Cloude, S.R.; Pottier, E. A review of target decomposition theorems in radar polarimetry. *IEEE Trans. Geosci. Remote Sens.* **1996**, *34*, 498–518. [CrossRef]
9. Cloude, S.R.; Pottier, E. An entropy based classification scheme for land applications of polarimetric SAR. *IEEE Trans. Geosci. Remote Sens.* **1997**, *35*, 68–78. [CrossRef]
10. Lardeux, C.; Frison, P.L.; Tison, C.; Souyris, J.C.; Stoll, B.; Fruneau, B.; Rudant, J.P. Support vector machine for multifrequency SAR polarimetric data classification. *IEEE Trans. Geosci. Remote Sens.* **2009**, *47*, 4143–4152. [CrossRef]
11. Dickinson, C.; Siqueira, P.; Clewley, D.; Lucas, R. Classification of forest composition using polarimetric decomposition in multiple landscapes. *Remote Sens. Environ.* **2013**, *131*, 206–214. [CrossRef]
12. Yin, Q.; Lin, Z.; Hu, W.; López-Martínez, C.; Ni, J.; Zhang, F. Crop Classification of Multitemporal PolSAR Based on 3-D Attention Module with ViT. *IEEE Geosci. Remote Sens. Lett.* **2023**, *20*, 4005405. [CrossRef]
13. Wang, W.; Wang, J.; Lu, B.; Liu, B.; Zhang, Y.; Wang, C. MCPT: Mixed Convolutional Parallel Transformer for Polarimetric SAR Image Classification. *Remote Sens.* **2023**, *15*, 2936. [CrossRef]
14. Hua, W.; Zhang, Y.; Zhang, C.; Jin, X. PolSAR Image Classification Based on Relation Network with SWANet. *Remote Sens.* **2023**, *15*, 2025. [CrossRef]
15. Lee, J.S.; Grunes, M.R.; Ainsworth, T.L.; Du, L.J.; Schuler, D.L.; Cloude, S.R. Unsupervised classification using polarimetric decomposition and the complex Wishart classifier. *IEEE Trans. Geosci. Remote Sens.* **1999**, *37*, 2249–2258.
16. Silva, W.B.; Freitas, C.C.; Sant'Anna, S.J.S.; Frery, A.C. Classification of segments in PolSAR imagery by minimum stochastic distances between Wishart distributions. *IEEE J. Sel. Top. Appl. Earth Observ. Remote Sens.* **2013**, *6*, 1263–1273. [CrossRef]
17. Chen, Q.; Kuang, G.Y.; Li, J.; Sui, L.C.; Li, D.G. Unsupervised land cover/land use classification using PolSAR imagery based on scattering similarity. *IEEE Trans. Geosci. Remote Sens.* **2013**, *51*, 1817–1825. [CrossRef]
18. Wu, Y.H.; Ji, K.F.; Yu, W.X.; Su, Y. Region-based classification of Polarimetric SAR imaged using Wishart MRF. *IEEE Trans. Geosci. Remote Sens. Lett.* **2008**, *5*, 668–672. [CrossRef]
19. Dong, H.; Xu, X.; Sui, H.; Xu, F.; Liu, J. Copula-Based Joint Statistical Model for Polarimetric Features and Its Application in PolSAR Image Classification. *IEEE Trans. Geosci. Remote Sens.* **2017**, *55*, 5777–5789. [CrossRef]
20. Liu, B.; Hu, H.; Wang, H.; Wang, K.; Liu, X.; Yu, W. Superpixel-based classification with an adaptive number of classes for polarimetric SAR images. *IEEE Trans. Geosci. Remote Sens.* **2013**, *51*, 907–924. [CrossRef]
21. Krogager, E. New decomposition of the radar target scattering matrix. *Electron. Lett.* **1990**, *26*, 1525–1527. [CrossRef]
22. Freeman, A.; Durden, S.L. A three-component scattering model for polarimetric SAR data. *IEEE Trans. Geosci. Remote Sens.* **1998**, *36*, 963–973. [CrossRef]
23. Yamaguchi, Y.; Moriyama, T.; Ishido, M.; Yamada, H. Four-component scattering model for polarimetric SAR image decomposition. *IEEE Trans. Geosci. Remote Sens.* **2005**, *43*, 1699–1706. [CrossRef]

24. An, W.T.; Lin, M.S. A reflection symmetry approximation of multi-look polarimetric SAR data and its application to freeman durden decomposition. *IEEE Trans. Geosci. Remote Sens.* **2019**, *57*, 3649–3660. [CrossRef]
25. van Zyl, J.J.; Arii, M.; Kim, Y. Model-based decomposition of polarimetric SAR covariance matrices constrained for nonnegative eigenvalues. *IEEE Trans. Geosci. Remote Sens.* **2011**, *49*, 3452–3459. [CrossRef]
26. Huynen, J.R. Physical reality of radar targets. *Proc. SPIE* **1993**, *1748*, 86–96.
27. Cameron, W.L.; Leung, L.K. Feature motivated polarization scattering matrix decomposition. In Proceedings of the IEEE International Conference on Radar, Arlington, VA, USA, 7–10 May 1990.
28. Nie, W.; Huang, K.; Yang, J.; Li, P. A deep reinforcement learning-based framework for PolSAR imagery classification. *IEEE Trans Geosci. Remote. Sens.* **2021**, *60*, 4403615. [CrossRef]
29. Ren, B.; Zhao, Y.; Hou, B.; Chanussot, J.; Jiao, L. A mutual information-based self-supervised learning model for PolSAR land cover classification. *IEEE Trans. Geosci. Remote. Sens.* **2021**, *59*, 9224–9237. [CrossRef]
30. Zhang, S.; An, W.; Zhang, Y.; Cui, L.; Xie, C. Wetlands Classification Using Quad-Polarimetric Synthetic Aperture Radar through Convolutional Neural Networks Based on Polarimetric Features. *Remote. Sens.* **2022**, *14*, 5133. [CrossRef]
31. Quan, S.; Qin, Y.; Xiang, D.; Wang, W.; Wang, X. Polarimetric Decomposition-Based Unified Manmade Target Scattering Characterization With Mathematical Programming Strategies. *IEEE Trans. Geosci. Remote. Sens.* **2021**, *60*, 1–18. [CrossRef]
32. Quan, S.; Zhang, T.; Wang, W.; Kuang, G.; Wang, X.; Zeng, B. Exploring Fine Polarimetric Decomposition Technique for Built-Up Area Monitoring. *IEEE Trans. Geosci. Remote. Sens.* **2023**, *61*, 1–19. [CrossRef]
33. Hinton, G.E.; Osindero, S.; Teh, Y.-W. A fast learning algorithm for deep belief nets. *Neural Comput.* **2006**, *18*, 1527–1554. [CrossRef] [PubMed]
34. Vincent, P.; Larochelle, H.; Lajoie, I.; Bengio, Y.; Manzagol, P.-A. Stacked denoising autoencoders: Learning useful representations in a deep network with a local denoising criterion. *J. Mach. Learn. Res.* **2010**, *11*, 3371–3408.
35. Goodfellow, I.; Pouget-Abadie, J.; Mirza, M.; Xu, B.; Warde-Farley, D.; Ozair, S.; Courville, A.; Bengio, Y. Generative adversarial networks. *Commun. ACM* **2014**, *63*, 139–144. [CrossRef]
36. Krizhevsky, A.; Sutskever, I.; Hinton, G.E. ImageNet Classification with Deep Convolutional Neural Networks. In Proceedings of the 25th International Conference on Neural Information Processing Systems, Lake Tahoe, NV, USA, 3–6 December 2012; pp. 1097–1105.
37. Szegedy, C.; Liu, W.; Jia, Y.; Sermanet, P.; Reed, S.; Anguelov, D.; Erhan, D.; Vanhoucke, V.; Rabinovich, A. Going deeper with convolutions. In Proceedings of the IEEE Conference on Computer Vision and Pattern Recognition (CVPR), Boston, MA, USA, 7–12 June 2015.
38. Jiao, L.; Liu, F. Wishart deep stacking network for fast POLSAR image classifification. *IEEE Trans. Image Process.* **2016**, *25*, 3273–3286. [CrossRef] [PubMed]
39. Liu, F.; Jiao, L.; Tang, X. Task-oriented GAN for PolSAR image classifification and clustering. IEEE Trans. *Neural Netw. Learn. Syst.* **2019**, *30*, 2707–2719. [CrossRef] [PubMed]
40. Guo, Y.; Wang, S.; Gao, C.; Shi, D.; Zhang, D.; Hou, B. Wishart RBM based DBN for polarimetric synthetic radar data classifification. In Proceedings of the 2015 IEEE International Geoscience and Remote Sensing Symposium (IGARSS), Milan, Italy, 26–31 July 2015.
41. Shao, Z.; Zhang, L.; Wang, L. Stacked sparse autoencoder modeling using the synergy of airborne LiDAR and satellite optical and SAR data to map forest above-ground biomass. *IEEE J. Sel. Top. Appl. Earth Obs. Remote. Sens.* **2017**, *10*, 5569–5582. [CrossRef]
42. Zhang, L.; Ma, W.; Zhang, D. Stacked sparse autoencoder in PolSAR data classifification using local spatial information. *IEEE Geosci. Remote Sens. Lett.* **2016**, *13*, 1359–1363. [CrossRef]
43. Yu, Y.; Li, J.; Guan, H.; Wang, C. Automated detection of three-dimensional cars in mobile laser scanning point clouds using DBM-Hough-forests. *IEEE Trans. Geosci. Remote. Sens.* **2016**, *54*, 4130–4142. [CrossRef]
44. Chen, Y.; Lin, Z.; Zhao, X.; Wang, G.; Gu, Y. Deep learning-based classifification of hyperspectral data. *IEEE J. Sel. Top. Appl. Earth Observ. Remote Sens.* **2014**, *7*, 2094–2107. [CrossRef]
45. Zhang, L.; Shi, Z.; Wu, J. A Hierarchical oil tank detector with deep surrounding features for high-resolution optical satellite imagery. *IEEE J. Sel. Top. Appl. Earth Obs. Remote. Sens.* **2015**, *8*, 4895–4909. [CrossRef]
46. Liang, H.; Li, Q. Hyperspectral imagery classifification using sparse representations of convolutional neural network features. *Remote Sens.* **2016**, *8*, 99. [CrossRef]
47. Yu, Y.; Li, J.; Guan, H.; Jia, F.; Wang, C. Learning hierarchical features for automated extraction of road markings from 3-D mobile LiDAR point clouds. *IEEE J. Sel. Top. Appl. Earth Obs. Remote. Sens.* **2014**, *8*, 709–726. [CrossRef]
48. Xie, H.; Wang, S.; Liu, K.; Lin, S.; Hou, B. Multilayer feature learning for polarimetric synthetic radar data classifification. In Proceedings of the 2014 IEEE Geoscience and Remote Sensing Symposium, Quebec City, QC, Canada, 13–18 July 2014; pp. 2818–2821.
49. Chen, X.; Hou, Z.; Dong, Z.; He, Z. Performance analysis of wavenumber domain algorithms for highly squinted SAR. *IEEE J. Sel. Top. Appl. Earth Obs. Remote. Sens.* **2023**, *16*, 1563–1575. [CrossRef]
50. Dong, H.; Zhang, L.; Zou, B. Exploring vision transformers for polarimetric SAR image classification. *IEEE Trans. Geosci. Remote. Sens.* **2021**, *60*, 5219715. [CrossRef]
51. Deng, P.; Xu, K.; Huang, H. When CNNs meet vision transformer: A joint framework for remote sensing scene classification. *IEEE Geosci. Remote. Sens. Lett.* **2021**, *19*, 1–5. [CrossRef]

52. Ren, S.; Zhou, F.; Bruzzone, L. Transfer-Aware Graph U-Net with Cross-Level Interactions for PolSAR Image Semantic Segmentation. *Remote. Sens.* **2024**, *16*, 1428. [CrossRef]
53. Wang, Y.; Zhang, W.; Chen, W.; Chen, C. BSDSNet: Dual-Stream Feature Extraction Network Based on Segment Anything Model for Synthetic Aperture Radar Land Cover Classification. *Remote. Sens.* **2024**, *16*, 1150. [CrossRef]
54. Shi, J.; Nie, M.; Ji, S.; Shi, C.; Liu, H.; Jin, H. Polarimetric Synthetic Aperture Radar Image Classification Based on Double-Channel Convolution Network and Edge-Preserving Markov Random Field. *Remote. Sens.* **2023**, *15*, 5458. [CrossRef]
55. Liu, L.; Li, Y. PolSAR Image Classification with Active Complex-Valued Convolutional-Wavelet Neural Network and Markov Random Fields. *Remote. Sens.* **2024**, *16*, 1094. [CrossRef]
56. Yang, R.; Xu, X.; Gui, R.; Xu, Z.; Pu, F. Composite Sequential Network With POA Attention for PolSAR Image Analysis. *IEEE Trans. Geosci. Remote. Sens.* **2021**, *60*, 5209915. [CrossRef]
57. Chu, B.; Zhang, M.; Ma, K.; Liu, L.; Wan, J.; Chen, J.; Chen, J.; Zeng, H. Multiobjective Evolutionary Superpixel Segmentation for PolSAR Image Classification. *Remote. Sens.* **2024**, *16*, 854. [CrossRef]
58. Ai, J.; Wang, F.; Mao, Y.; Luo, Q.; Yao, B.; Yan, H.; Xing, M.; Wu, Y. A fine PolSAR terrain classification algorithm using the texture feature fusion-based improved convolutional autoencoder. *IEEE Trans. Geosci. Remote. Sens.* **2021**, *60*, 5218714. [CrossRef]
59. Zhou, Y.; Wang, H.; Xu, F.; Jin, Y.-Q. Polarimetric SAR image classifification using deep convolutional neural networks. *IEEE Geosci. Remote Sens. Lett.* **2016**, *13*, 1935–1939. [CrossRef]
60. Chen, S.-W.; Tao, C.-S. PolSAR image classifification using polarimetric-feature-driven deep convolutional neural network. *IEEE Geosci. Remote Sens. Lett.* **2018**, *15*, 627–631. [CrossRef]
61. An, W.; Lin, M.; Yang, H. Modified reflection symmetry decomposition and a new polarimetric product of GF-3. *IEEE Geosci. Remote. Sens. Lett.* **2021**, *19*, 1–5. [CrossRef]
62. An, W. Polarimetric Decomposition and Scattering Characteristic Extraction of Polarimetric SAR. Ph.D. Thesis, Tusinghua University, Beijing, China, 2010.
63. Yang, J. On Theoretical Problems in Radar Polarimetry. Ph.D. Thesis, Niigata University, Niigata, Japan, 1999.
64. User Manual of Gaofen-3 Satellite Products, China Resources Satellite Application Center. 2016. Available online: https://osdds.nsoas.org.cn/ (accessed on 30 March 2024).
65. Chen, J.; Chen, Y.; An, W.; Cui, Y.; Yang, J. Nonlocal filtering for polarimetric SAR data: A pretest approach. *IEEE Trans. Geosci. Remote Sens.* **2011**, *49*, 1744–1754. [CrossRef]

remote sensing

MDPI

Article

AFMUNet: Attention Feature Fusion Network Based on a U-Shaped Structure for Cloud and Cloud Shadow Detection

Wenjie Du [1], Zhiyong Fan [1,2,*], Ying Yan [1,2], Rui Yu [1] and Jiazheng Liu [1]

1 School of Automation, Nanjing University of Information Science and Technology, Nanjing 210044, China; 202183250041@nuist.edu.cn (W.D.); ying.yan@nuist.edu.cn (Y.Y.); 20212382072@nuist.edu.cn (R.Y.); 202183250036@nuist.edu.cn (J.L.)
2 Collaborative Innovation Center on Atmospheric Environment and Equipment Technology, Nanjing University of Information Science and Technology, Nanjing 210044, China
* Correspondence: 001163@nuist.edu.cn

Abstract: Cloud detection technology is crucial in remote sensing image processing. While cloud detection is a mature research field, challenges persist in detecting clouds on reflective surfaces like ice, snow, and sand. Particularly, the detection of cloud shadows remains a significant area of concern within cloud detection technology. To address the above problems, a convolutional self-attention mechanism feature fusion network model based on a U-shaped structure is proposed. The model employs an encoder–decoder structure based on UNet. The encoder performs down-sampling to extract deep features, while the decoder uses up-sampling to reconstruct the feature map. To capture the key features of the image, Channel Spatial Attention Module (CSAM) is introduced in this work. This module incorporates an attention mechanism for adaptive field-of-view adjustments. In the up-sampling process, different channels are selected to obtain rich information. Contextual information is integrated to improve the extraction of edge details. Feature fusion at the same layer between up-sampling and down-sampling is carried out. The Feature Fusion Module (FFM) facilitates the positional distribution of the image on a pixel-by-pixel basis. A clear boundary is distinguished using an innovative loss function. Finally, the experimental results on the dataset GF1_WHU show that the segmentation results of this method are better than the existing methods. Hence, our model is of great significance for practical cloud shadow segmentation.

Keywords: cloud shadow segmentation; convolution neural network; attention mechanism; feature fusion; deep learning

Citation: Du, W.; Fan, Z.; Yan, Y.; Yu, R.; Liu, J. AFMUNet: Attention Feature Fusion Network Based on a U-Shaped Structure for Cloud and Cloud Shadow Detection. *Remote Sens.* **2024**, *16*, 1574. https://doi.org/10.3390/rs16091574

Academic Editors: Claudio Piciarelli and Benoit Vozel

Received: 24 January 2024
Revised: 19 March 2024
Accepted: 19 April 2024
Published: 28 April 2024

1. Introduction

With the decade-long development of remote sensing technology, the Gaofen series of satellites has formed a "three-high" observation system with high spatial, temporal, and spectral resolution [1], which uses sensors to acquire images by obtaining information about the Earth over long distances. In the remote sensing image, the cloud shadow area is an important identification; through the identification of the cloud shadow position in the image, we can obtain the visible light, infrared rays, and other information on the ground, used to monitor the cloud coverage, the type of cloud, and the direction of cloud movement. This provides meteorologists and weather forecasters with critical data to help them predict the weather more accurately. However, merely identifying the location of cloud cover is insufficient. The presence of cloud shadows can obstruct analysis in precision agriculture and other fields, leading to biases in the results. Therefore, applications of cloud shadow detection are increasingly widespread in meteorological forecasting, environmental monitoring, and natural disaster detection. The cloud detection technology [2] is inadequate; thus, utilizing cloud and cloud shadow detection technology to accurately detect cloud cover from remote sensing images is a crucial preprocessing step for most satellite imagery. In this paper, we propose a segmentation algorithm for separating the three components of clouds, cloud shadows, and background in remote sensing images.

Traditional cloud shadow segmentation methods can be broadly categorized into the following five types: 1. thresholding-based methods; 2. morphology-based methods; 3. statistical-based methods; 4. texture feature-based methods; and 5. machine learning-based methods. The thresholding method uses various physical methods, such as AVHRR and NIR images, to set feature thresholds such as luminance, chromaticity, etc., to detect the cloud shadows in the image. Early on in this research, people used fixed thresholds to distinguish clouds from other parts. For instance, Saunders and Kriebel [3] processed the NOAA-9 dataset over a week by determining thresholds for a range of physical parameters including cloud-top temperatures, optical depths, and liquid water content. While the fixed threshold method is straightforward and user-friendly, it lacks the adaptability needed to accommodate various meteorological conditions, lighting scenarios, geographical regions, and times of day. Additionally, it often necessitates manual threshold adjustments, which pose numerous shortcomings and limitations. Later, many researchers proposed improvements by using dynamic thresholding for cloud detection [4–7]. The dynamic thresholding method adjusts thresholds based on environmental conditions through the construction of diverse physical models, thereby enhancing the accuracy of automatic cloud analysis. However, for complex cloud and feature types, this method can be challenging to apply to the background, and it also incurs significant computational costs. Secondly, the morphological method based on set theory proposes a series of operations, such as expansion, erosion, open and close operations, and hit–hit–miss transformations for images. Danda and Xiang Liu et al. [8,9] constructed skeleton features to help analyze the morphology of the cloud and thus separate it from other regions by using a gray-level morphological edge extraction method. Moreover, Tom et al. [10] established a common method based on morphological data to create an efficient computational paradigm for the combination of simple nonlinear grayscale operations such that the cloud detection filter exhibits spatial high-pass properties, emphasizes cloud shadow regions in the data, and suppresses all other clutter. A series of methods regarding morphology are more effective for the case of blurred cloud edges and complex shapes, but they are difficult to apply directly to multispectral images. Thirdly, statistical methods use various statistical and analytical tools to establish regression equations for differences in reflectance, brightness, or temperature between picture pixels in satellite data to detect clouds. For example, Amato et al. [11] used PCA and nonparametric density estimation applied to the SEVIRI sensor dataset, and Wylie et al. [12] combined time-series analyses of more than 20 years of polar-orbiting satellite cloud data to predict future cloud trends. However, since the sample data used in regression models are historical, this type of method is not widely used and is limited to specific times and regions. Fourthly, the texture feature method identifies cloudy and non-cloudy regions by extracting the texture features of images. For example, Abuhussein et al. [13,14] conducted segmentation by analyzing the GLCM (Gray-Level Co-occurrence Matrix) to capture spatial relationships and covariance frequencies between pixels of varying gray levels in the image. This process enables the extraction of crucial information regarding the image texture. Reiter and Changhui et al. [15–17] completed segmentation by using the wavelet transform to detect texture features and edge information in the image at different spatial scales and to decompose the cloud image into details at different scales to obtain local and global features of the cloud, while Surya et al. [18] used a clustering algorithm to group texture regions similar to the cloud shadow. This method works better for texture-rich cloud shadow images. To overcome the limitations of the first four traditional methods, machine learning algorithms are proposed to realize cloud shadow segmentation by training classifiers. Support vector machines, random forests, and neural networks are typical classifiers. For instance, Li et al. [19] proposed a classifier based on support vector machines to detect clouds in images, while Ishida et al. [20] quantitatively guided the support vector machines with the help of classification effect metrics to improve the feature space used for detecting cloud shadows and to reduce the frequency of erroneous results. Fu et al. [21] combined the ensemble thresholding method and random forest for the FY-2G image set to improve the meteorological satellite cloud detection technique and Jin et al. [22] established a BP neural network backpropagation model for the MODIS dataset, which improved the learning model to a certain extent. Although these methods are indeed more effective, they necessitate manual feature engineering to select suitable labels for training and testing a large volume of data annotations. Furthermore, the quality of the model is directly influenced by the features selected.

To overcome the shortcomings of manual feature engineering, deep convolutional neural networks (CNN) gradually emerged; a variety of convolutional neural networks were proposed for remote sensing image segmentation tasks, and semantic segmentation algorithms based on deep learning began to gradually become mainstream. Long et al. [23] first proposed a fully convolutional neural network, FCN, for semantic segmentation in 2015, which can directly realize end-to-end pixel-by-pixel classification. Mohajerani et al. [24] applied the FCN network to the remote sensing image Landsat dataset cloud detection technique in 2018, which dramatically improved the efficiency of the target classification of remote sensing images; however, the results obtained were still not fine enough and not sensitive enough for the detailed parts of the image. Since then, there has been a surge in deep learning networks, with numerous CNN frameworks continuously being proposed. In 2015, Badrinarayanan et al. [25] introduced SegNet, a segmentation network based on an encoder–decoder structure, utilizing up-sampling with the unpooling operation. Subsequently, in 2019, Lu et al. [26] adapted the SegNet network model for cloud recognition in remote sensing images. Their approach improved the accuracy of cloud recognition by preserving positional indices during the pooling process, thus retaining image details through a symmetrical parallel structure. Although it demonstrated some ability in cloud–snow differentiation, its training time was found to be excessively long and inefficient. In 2016, Chen et al. [27] designed an inflated convolutional network called DeepLab, aimed at expanding the sensory field by introducing voids in the convolutional kernel. DeepLab enhances the robustness of image segmentation. However, it imposes specific requirements on the size of the segmented target. It excels in segmenting foreground targets within the general size range. Nonetheless, when faced with extreme size variations in the target, such as very small or very large targets, DeepLab exhibits poor performance and suffers from segmentation instability. In 2015, Ronneberger et al. [28] proposed the UNet image segmentation network, named because the network framework is shaped like the letter U. The contextual information is fused through feature splicing in the channel dimension during the up-sampling process to achieve a more fine-grained segmentation, which is suitable for highly detailed segmentation tasks. In 2017, Zhao et al. [29] designed a pyramidal scene parsing network structure, PSPNet, which integrates contextual information from different regions, applies convolutional kernels of different sizes, and employs a multi-scale sensory field to efficiently combine local and global cues. In 2022, Zhang et al. [30] proposed a dual pyramidal network, DPNet, inspired by PSPNet. This multi-scale feature captures features of the image from different scales, thus enhancing the network's capability in feature extraction, but it also incurs greater computational cost, making training and prediction slower.

Although existing CNNs perform better in remote sensing image segmentation tasks, there is still a general problem: due to the down-sampling nature of the convolutional operation, the network is prone to lose critical detail information during feature extraction and scale reduction, which leads to many problems, such as inaccuracy and blurred edges in segmentation results. Many studies have demonstrated that combining low-level and high-level semantic information can significantly improve model performance [31]. However, traditional feature fusion methods are usually too simple and do not pay enough attention to edge information and image features to effectively restore lost information, especially for tasks with complex backgrounds, which may lead to missed detection of fine targets and edge blurring. To address these challenges in semantic segmentation, we propose a new approach for cloud shadow segmentation—an attention mechanism feature fusion network based on the UNet framework. The encoder–decoder architecture of UNet effectively extracts and restores feature information across various scales, making it particularly suitable for smaller-scale datasets. Therefore, we adopt this U-shaped network structure as our baseline and integrate the channel attention mechanism and spatial attention mechanism module into it. This integration allows for adaptive attention to different channels of the image and feature map information, with the goal of enhancing the fine detection of cloud shadows. The addition of the new feature fusion module can effectively fuse the low-level and high-level features, restore the lost information, and segment the fine features more accurately in such a complex context as the cloud shadow segmentation task. The AFMUNet network framework is shown in Figure 1. After inputting the image, the high-level image features are initially extracted through down-sampling. Subsequently, during the up-sampling process and enhancement of feature map resolution, we progressively enhance the receptive field adaptively and employ different channel operations. In addition, the feature fusion module is utilized in each layer to integrate contextual information more accurately and fuse low-level and high-level information. Furthermore, an innovative loss function is employed during the training process, and classification results are outputted after multiple samplings. Through the combined effect of the above modules, the detection accuracy of our network was substantially improved. The main contributions of this paper's work are as follows:

- An integrated module of channel space attention mechanism, suitable for cloud shadow segmentation tasks within a U-shaped structure, is proposed. This model facilitates dynamic adjustment of feature map weights, enhancing the ability to capture crucial image features and thereby improving segmentation accuracy.
- The feature fusion operation of the original network is updated, which helps to better understand the target and background in the image, segment the image using information from different scales, and deal with cloud shadow targets of different sizes and shapes.
- An innovative weighted loss function is developed for the dataset, which improves the accuracy of model learning and optimizes the model performance to some extent.
- A network that integrates the above three features and combines them with a feature extraction network is proposed to segment high-resolution remote sensing images.

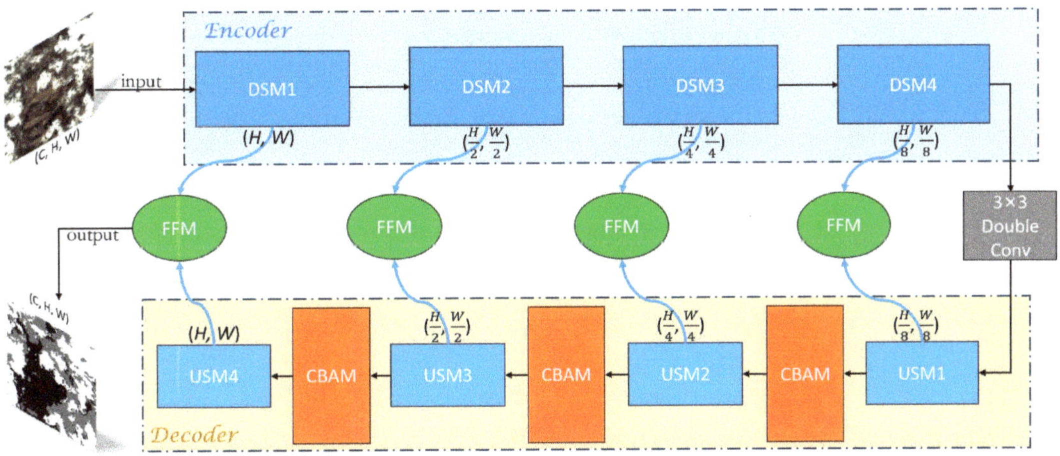

Figure 1. Attention mechanism feature fusion network framework based on U-shaped structure.

2. Methodology

Since the purpose of the cloud–shadow segmentation task is to match labels on a pixel-by-pixel basis on an image to distinguish between clouds, cloud shadows, and backgrounds, the task can be regarded as a semantic segmentation task for triple categorization. Recently, CNNs have achieved great success in the field of computer vision, especially in image segmentation tasks. As pointed out in Section 1, due to the diversity of cloud layers, irregular shapes, and variations in lighting conditions and shooting locations, cloud shadow segmentation tasks often require highly accurate models to cope with these complexities. Nevertheless, traditional machine learning algorithms may face challenges in meeting the stringent accuracy demands of cloud shadow segmentation tasks, particularly in scenarios involving snowy mountainous terrain or under low-light conditions [32]. When dealing with the cloud shadow segmentation task, we need an efficient network structure that can fully capture the detailed features of clouds while preserving the surface information. To fulfill this requirement, we choose the UNet structure as the backbone network framework, which is appropriately modified to incorporate CSAM and FFM improvement modules to further improve the performance of the model in capturing the complex structure and irregular shape of cloud shadows.

2.1. UNet—A Network Based on Encoder–Decoder Architecture (Related Work)

UNet is a classical deep-learning architecture especially suited for image segmentation tasks. It is designed as an encoder–decoder structure with special skip connections to better capture features and details at different scales in segmentation tasks. The following are the main features and working principles of UNet:

1. Encoder Part: The encoder part of UNet consists of multiple convolutional layers that gradually halve the size of the feature map while increasing the number of feature channels. This helps to extract high-level feature representations of the image and capture semantic information at different scales. The encoder part usually includes operations such as convolutional layers, pooling layers, etc.

2. Jump concatenation: UNet introduces jump concatenation to concatenate the feature maps of the encoder with the feature maps of the decoder to include more detailed information in the decoder. This helps to overcome the problem of information loss that may be introduced by pooling operations and improves the performance of the segmentation model.

3. Decoder Part: The decoder part of UNet consists of multiple convolutional and up-sampling layers that gradually recover the spatial resolution of the feature map through operations such as inverse convolution. The decoder part restores the low-resolution feature map to the size of the original input image through the up-sampling operation and, at the same time, performs feature extraction through the convolution operation.

4. Output Layer: The output layer of UNet is usually a convolutional layer whose output is a segmentation mask indicating the class or segmentation result of each pixel in the image. The number of channels in the output layer is usually equal to the number of categories in the task.

The UNet architecture has achieved excellent performance in a variety of fields, such as medical image segmentation, remote sensing image analysis, and automated driving, where it can efficiently capture semantic information and details in an image while maintaining high resolution. In our study, only the basic architecture of UNet is retained, based on which innovations and modifications are made.

2.2. CSAM (Channel Spatial Attention Module)

To better understand the key features and structures in an image and to improve the segmentation of complex scenes, we introduce the attention mechanism. The concept of attention mechanism originated in the field of natural language processing. It serves to emphasize words at different positions within an input sentence, thereby facilitating improved translation into the target language [33,34]. For instance, in machine translation, the attention mechanism helps the model focus on relevant parts of the input sentence when generating each word of the translation. This allows for more accurate and contextually appropriate translations, especially in cases where the input sentence is long or complex.

Similarly, in text summarization, the attention mechanism aids in identifying important sentences or phrases to include in the summary, resulting in more concise and informative summaries. Now, we apply it to image semantic segmentation tasks to help process image information more efficiently by focusing attention on key regions in the image while suppressing irrelevant information. This is an approach that mimics the human visual and cognitive system, which is similar to how the human cerebral cortex achieves efficient analysis by focusing on specific parts when processing image and video information in complex scenes. In general, the attention mechanism can be categorized into four dimensions—channeling, spatial, temporal, and branching [35]—which play different roles in different computer vision tasks.

As shown in Figure 2 below, we add the CSAM module to the basic structure of UNet after the end of each sample in the up-sampling phase, which skillfully combines the channel and spatial attention mechanisms. For a given feature map, the CSAM module is capable of generating feature map information in the channel and spatial dimensions [36] and multiplying them with the original input feature map to perform adaptive feature adjustment and correction. Eventually, the CSAM module outputs feature maps, adjusted by the attention mechanism, with stronger semantic information and adaptability. This module enhances our ability to focus on the channel information of the image during cloud shadow segmentation tasks, thereby improving cloud perception and segmentation accuracy.

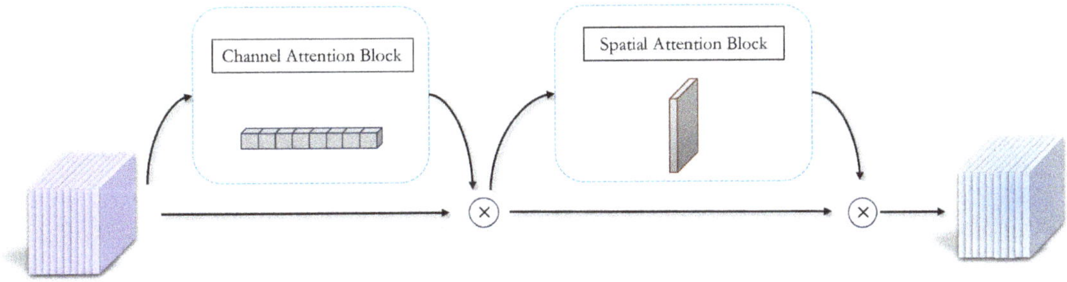

Figure 2. Channel space attention mechanism module.

2.2.1. CAB (Channel Attention Block)

CAB is an important component of the CSAM module. It focuses on weighting attention given to the channel dimensions in the feature map [37,38]. The goal of the channel attention mechanism is to enhance the attention given to different channels by dynamically adjusting the weights between channels. This is crucial to improve the model's ability to perceive different features in the image. The CAB module works as follows:

The steps of the CAB module are shown in Figure 3 below. Step 1: Firstly, the input feature map $\boldsymbol{F_{in}}$ is subjected to global average and maximum pooling operations, and the input information is compressed and downgraded to obtain two 1×1 average pooled features, $\boldsymbol{F^c_{avg}}$, and maximum pooled features, $\boldsymbol{F^c_{max}}$. Step 2: Then, they are fed into a weight-sharing two-layer neural network, MLP. Step 3: Finally, the MLP output features are subjected to an element-by-element summation operation, which is applied to the input feature map after activation by the Sigmoid function to generate the final Channel Attention Feature, M_c. The above computational process is expressed as Equation (1), shown below:

$$
\begin{aligned}
M_c(\boldsymbol{F_{in}}) &= \sigma(MLP(AvgPool(\boldsymbol{F_{in}})) + MLP(MaxPool(\boldsymbol{F_{in}}))) \\
&= \sigma(\boldsymbol{W}_1(\boldsymbol{W}_0(\boldsymbol{F^c_{avg}})) + \boldsymbol{W}_1(\boldsymbol{W}_0(\boldsymbol{F^c_{max}}))) \\
\sigma(x) &= sigmoid(x) = \frac{1}{1+e^{-x}}
\end{aligned}
\tag{1}
$$

where $\sigma(\cdot)$ is the sigmoid function and $\boldsymbol{W}_0/\boldsymbol{W}_1$ represents the weights of the hidden/output layer. The parameters of $\boldsymbol{W}_0$ and $\boldsymbol{W}_1$ are shared in MLP.

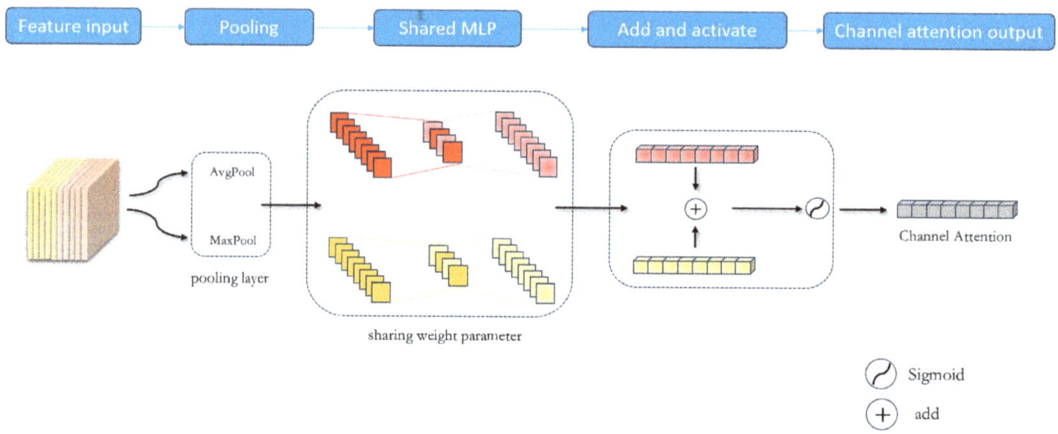

Figure 3. Channel attention block.

Attention weights on the channel dimensions, indicating the contribution of different channels to the final feature representation, were generated by CAB, and these weights were applied to the original input feature map to generate features for the input spatial attention mechanism module. Channel-level feature tuning is achieved by weighting each channel's features. This means that the model can better focus on the channel features that are important to the task at hand, improving the representation of semantic information.

2.2.2. SAB (Spatial Attention Block)

Unlike CAB, the SAB module focuses on the spatial dimension of the feature map. Its goal is to enhance the focus on different regions in the image by adjusting the weights of different spatial locations to improve the model's perception of global contextual information The SAB module works in Figure 4 as follows:

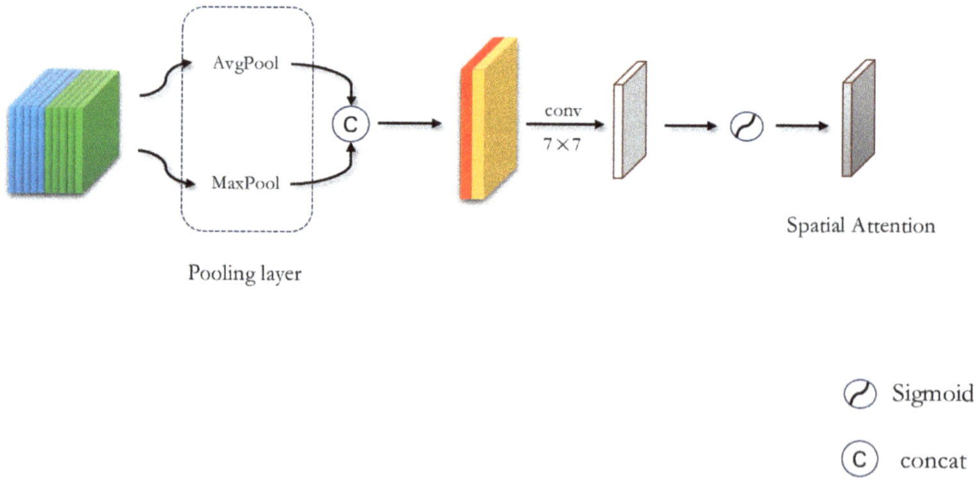

Figure 4. Spatial attention block.

Step 1: First, the feature map output from the CAB module is used as the input of this module, F_{in}, and global maximum pooling and average pooling are done on the channel dimensions; then, these two results are used in a splicing operation. Step 2: Next, a 7×7 convolution kernel is chosen to perform a convolution operation on the splicing result, and the channel dimensions are reduced to 1. Step 3: Finally, after the Sigmoid activation function maps the weights between 0 and 1 to represent the order of importance of each position, these spatial attention weights are applied to the inputs to generate the feature map of the spatial channel attention mechanism, M_s. The above computational process is expressed as Equation (2), shown below.

$$\begin{aligned} M_s(\boldsymbol{F_{in}}) &= \sigma(f^{7\times7}([AvgPool(\boldsymbol{F_{in}}); MaxPool(\boldsymbol{F_{in}})])) \\ &= \sigma(f^{7\times7}(\boldsymbol{F^s_{avg}}; \boldsymbol{F^s_{max}})) \end{aligned} \tag{2}$$

where 7×7 is the kernel of convolution. This size performs better than others.

SAB generates attention weights in the spatial dimension through a series of convolutional operations and activation functions that indicate the contribution of different locations to the final feature representation. This means that the model can better focus on key regions in the image, thus improving the perception of global contextual information. The SAB module helps us to more accurately capture the contours and structure of objects in tasks such as semantic segmentation.

2.3. FFM (Feature Fusion Module)

The introduction of the FFM module [39–41] plays a key role in the process of feature fusion of information from different feature maps obtained from deeper and shallower layers when jump connections in the original network structure are involved. The FFM module allows us to efficiently fuse features of different scales and resolutions in order to capture the complex structure and irregular shapes of cloud shadows.

The steps of the FFM module are depicted in Figure 5. Step 1: Accept two feature maps with different resolutions from the encoder and decoder sections as input. Step 2: Perform a series of operations such as splicing, convolution, and so on, to fuse them into an enhanced hybrid feature map, which strengthens the representation of the hybrid features and makes them more suitable for subsequent processing. Step 3: Perform a global averaging of the hybrid feature map pooling to reduce the spatial dimension to 1×1 to obtain global channel statistics. Step 4: Introduce two consecutive 1×1 convolution operations via Relu and Sigmoid activation functions in order to enhance the nonlinearity and show the importance of each channel. Step 5: Multiply the channel attention weights with the element-by-element hybrid feature map obtained from Step 2 to perform the mul operation to obtain a weighted feature map. Step 6: Finally, the weighted feature map obtained from Step 5 is subjected to element-by-element add-sum operation with the hybrid feature map obtained from Step 2, to produce the final fused feature map. The above computational process is expressed as Equation (3), shown below.

$$\begin{aligned} \boldsymbol{F_{conv}} &= Conv(Concat(\boldsymbol{F_1}, \boldsymbol{F_2})) \\ \alpha &= relu(f^{1\times1}(AvgPool(\boldsymbol{F_{conv}}))) \\ M_F(\boldsymbol{F_1}, \boldsymbol{F_2}) &= \boldsymbol{F_{conv}} + \boldsymbol{F_{conv}} \otimes \sigma(f^{1\times1}(\alpha)) \\ relu(x) &= max(0, x) \end{aligned} \tag{3}$$

where $\boldsymbol{F_{conv}}$ is the fusion of the input from shallow and deep layers and α represents the enhanced nonlinear result as an intermediate variable.

The FFM module is a well-designed feature fusion mechanism that effectively integrates feature maps from shallow and deep layers by means of utilizing channel complementarity, adaptively adjusting the weights of the channel features dynamically to better fuse information from different scales and semantic levels. This innovative fusion module offers an effective tool for our research and improves the performance of the capture and segmentation tasks of feature statistics.

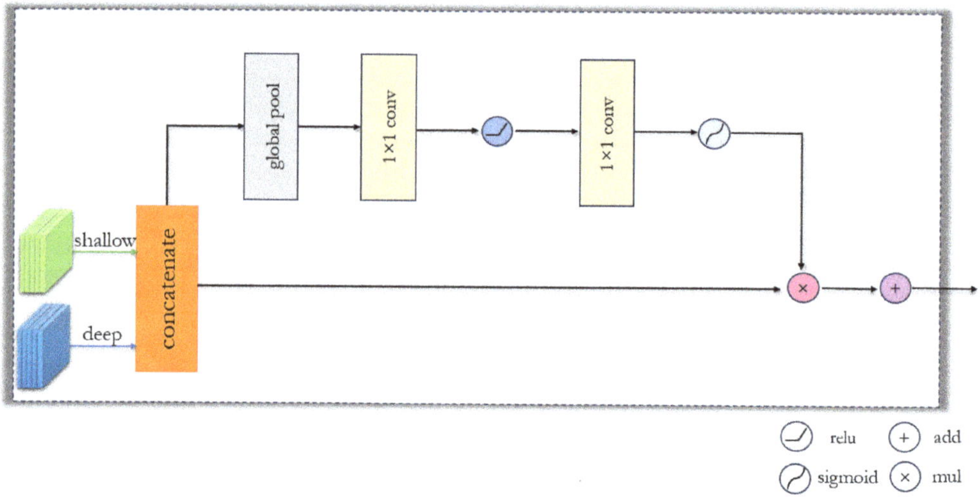

Figure 5. Feature fusion module.

2.4. Loss Function

The loss function is an important component in various segmentation network models based on deep learning [42]. It is used to measure the difference between the prediction and true values of the network and guide the model to make more accurate predictions. In the segmentation task, the reasonable selection, optimization, and innovation of the loss function can enhance the learning process of the model to achieve better segmentation results [43] as well as portability and application to other networks; thus, the study of the loss function selection is particularly important. The commonly used loss functions [44] are as follows:

1. Cross Entropy Loss Function

$$L = -\frac{1}{N}\sum_{i=1}^{N}\sum_{j=1}^{N}[y_{ij}\log(p_{ij}) + (1-y_{ij})\log(1-p_{ij})] \tag{4}$$

where N denotes the number of samples, and M denotes the number of categories. As the most commonly used loss function in image segmentation, which can be used in a large number of semantic segmentation tasks, the cross-entropy loss can help the network to correct categorization of the pixels after judging how good or bad the model is for the dataset.

2. Weighted Cross-Entropy Loss Function

$$L_w = -\frac{1}{N}\sum_{i=1}^{N}\sum_{j=1}^{N}[w_j y_{ij}\log(p_{ij}) + (1-y_{ij})\log(1-p_{ij})] \tag{5}$$

Despite being similar to the cross-entropy loss function, multiplying all positive samples by a coefficient for weighting allows the model to focus more on a smaller number of samples, thus mitigating the problem of the imbalanced number of categories.

3. Focal Loss

$$L_F = -\frac{1}{N}\sum_{i=1}^{N}\sum_{j=1}^{N}[(1-p_{ij})y_{ij}\log(p_{ij}) + {p_{ij}}^{\gamma}(1-y_{ij})\log(1-p_{ij})] \tag{6}$$

In addition to the imbalance in the number of samples from different categories, the problem of imbalance in the number of easily recognized samples and hard-to-recognize samples is often encountered, and the Focal Loss can help the network to better deal with the imbalance in the distribution of samples.

4. Dice Loss

$$L_D = 1 - \frac{2|X \cap Y|}{|X| + |Y|} \tag{7}$$

where $|X \cap Y|$ is the intersection between samples X and Y, $|X|$ represents the number of X samples, and $|Y|$ stands for the number of Y samples.

Unlike the weighted cross-entropy loss function, the Dice Loss does not require category reweighting; it calculates the loss directly from the Dice coefficients, which can help the network better handle overlaps and boundaries between categories.

5. IOU Loss

$$L_I = 1 - \frac{|X \cap Y|}{|X \cup Y|} \tag{8}$$

where $|X \cup Y|$ depicts the union between samples X and Y.

The IOU loss measures how similar the predicted segmentation results are to the true segmentation, and it helps to optimize the spatial consistency of the segmentation.

In summary, since the cloud shadows in the image are prone to overlap, and it is desired to distinguish the boundary between the two more accurately, L and L_D are selected in this experiment for proper weighting to derive an innovative loss function applicable to the task of the dataset in this paper.

$$Loss = \alpha \cdot L + \beta \cdot L_D \tag{9}$$

From Table 1, it is evident that the last row, which utilizes different weight proportions in the loss function weighted combination, achieves the best performance. This finding aligns with our initial conjecture. The Dice Loss effectively distinguishes between overlap regions and boundaries, aiding in completing the classification task more effectively. Moreover, continuous training is essential for further enhancing the model's classification accuracy.

Table 1. Effect of different combinations of weight coefficients on segmentation results.

α	β	MPA (%)	MIoU (%)
0.2	0.8	65.93	58.69
0.3	0.7	71.97	58.79
0.4	0.6	77.04	65.77
0.5	0.5	76.59	65.07
0.6	0.4	81.86	78.22
0.7	0.3	87.32	86.30
0.8	0.2	96.88	93.02

3. Experimental Analysis

3.1. Dataset

To further validate the generalization performance of the proposed model, we employed the GF1_WHU cloud shadow dataset created by Li et al. [45] as a generalization dataset. This dataset utilizes high-resolution GF-1 Wide Field of View (WFV) images with a spatial resolution of 16 m and covers four multispectral bands, spanning from visible to near-infrared spectral regions. The dataset consists of 108 GF-1 WFV 2a-level scene images, manually labeled by experts in remote sensing image interpretation at the SENDIMAGE laboratory of Wuhan University. These images encompass five main land cover types, including water, vegetation, urban areas, snow and ice, and barren land, representing different regions worldwide. During the model training process, we cropped the images to 256 × 256 pixels, removing black borders and unclear images, resulting in a total of 5428 images used for training and 1360 images for validation and testing, to evaluate the model's training results, detection accuracy, and generalization performance. To illustrate the dataset effectively, we selected images from different scenes, as shown in Figure 6.

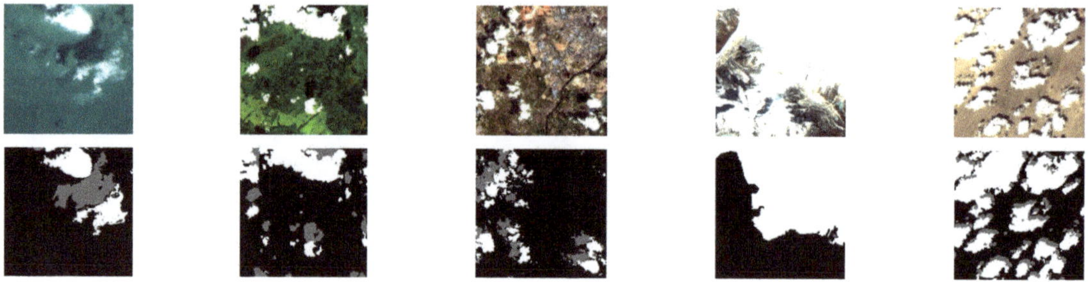

(**a**) (**b**) (**c**) (**d**) (**e**)

Figure 6. Examples from GF1_WHU Wuhan University cloud shadow dataset: (**a**) water; (**b**) vegetation; (**c**) snow; (**d**) ice; (**e**) barren.

Each original image is captured by three channels of RGB: white represents clouds, gray represents cloud shadows, and black is the background. In addition, to prevent overfitting and to enhance the robustness of the model, we also augmented the dataset by randomly flipping, clipping, rotating, scaling, and panning the images as well as adding noise interference to the images.

3.2. *Experimental Details*

In this section, using the Legion Y740 laptop sourced from Lenovo in Beijing, China, we harness PyTorch 2.0 to train and test all models on its GeForce RTX 2080Ti graphics card based on the dataset introduced in the preceding dataset section. This comprehensive evaluation aims to assess the efficiency and accuracy of our proposed network model for cloud shadow segmentation. Through a series of ablation experiments and comparison experiments, we thoroughly evaluated our model from both qualitative and quantitative perspectives [46,47]. The quantitative metrics pixel accuracy (PA), precision (PC), recall (RC), mean intersection over union (MIoU), reconciliation average (F1), and frequency weighted intersection over union (FWIoU) are calculated as follows:

$$\mathrm{PA} = \frac{\mathrm{TP}+\mathrm{TN}}{\mathrm{TP}+\mathrm{TN}+\mathrm{FP}+\mathrm{FN}} \tag{10}$$

$$\mathrm{PC} = \frac{\mathrm{TP}}{\mathrm{TP}+\mathrm{FP}} \tag{11}$$

$$\mathrm{RC} = \frac{\mathrm{TP}}{\mathrm{TP}+\mathrm{FN}} \tag{12}$$

$$\mathrm{MIoU} = \frac{\mathrm{TP}}{\mathrm{TP}+\mathrm{FP}+\mathrm{FN}} \tag{13}$$

$$\mathrm{F}_1 = \frac{2\times\mathrm{PC}\times\mathrm{RC}}{\mathrm{PC}+\mathrm{RC}} \tag{14}$$

$$\mathrm{FWIoU} = \frac{\mathrm{TP}+\mathrm{FN}}{\mathrm{TP}+\mathrm{FP}+\mathrm{TN}+\mathrm{FN}} \times \frac{\mathrm{TP}}{\mathrm{TP}+\mathrm{FP}+\mathrm{FN}} \tag{15}$$

In Equations (10)–(15) above, TP represents true positives, which correspond to the number of pixels correctly identified as positive samples. Similarly, FP denotes false positives, indicating the number of pixels incorrectly classified as positive samples. TN refers to true negatives, representing the number of pixels accurately identified as negative samples. Lastly, FN signifies false negatives, indicating the number of pixels incorrectly classified as negative samples.

In this section, we provide a comprehensive evaluation of our proposed algorithm, verifying the efficiency and sophistication of our algorithm for the task of remote sensing image change detection through ablation experiments and comparison experiments. Our experiments are conducted on the GF1_WHU dataset, with an initial learning rate of 0.001. The number of samples used in each round is 4, the number of training samples is 5428, the number of training times is 150, and the quantitative metrics used are PR, RC, MIoU and F1.

3.3. Ablation Experiment

In conducting the ablation experiments, changes in the results are observed by censoring part of the network architecture and testing the effect of different modules on the whole model. Since UNet is the basic framework of our network, UNet is used as the starting point for comparison, where we use metrics such as PA, RC, F1, and MIoU to evaluate the performance of the model. As can be seen in Table 2 below, the combination of all components achieves the optimization of the model's performance.

Table 2. Performance comparison of different combinations of modules in the model.

Method	PA (%)	RC (%)	PC (%)	F1 (%)	MIoU (%)
UNet	95.27	89.72	93.24	92.03	91.30
UNet + CSAM	96.48 (↑)	94.32	**95.02**	93.41	92.89
UNet + FFM	95.32	94.83	93.62	91.82	91.33
UNet + CSAM + FFM	96.93 (↑)	95.82	94.97	93.75	93.21
UNet + CSAM + FFM + Loss AFMUNet (Ours)	**97.12**	**96.03**	93.21	**93.90**	**93.42**

The arrow means this kind of combination improves the performance of model. The bold indicates the highest value in the column.

In order to enhance deep feature extraction, alleviate information loss resulting from constant down-sampling, and effectively capture multi-scale contextual information, as indicated by the ablation results of the deep feature sampling process, the CSAM Attention Mechanism Module proves beneficial for information recovery to capture detailed information. Additionally, the FFM module aids in better integrating contextual information, facilitating the fusion of features from different scales. Table 2 demonstrates a significant improvement in model performance following the introduction of these modules. Notably, the introduction of the Feature Fusion Module alone does not yield substantial improvements to the original model.

3.4. Comparison Experiment

In this experiment, the core of the cloud–shadow segmentation task is semantic segmentation, so our proposed network is compared with other semantic segmentation algorithms. PA, FWIoU, F1, and MIoU are selected as the evaluation metrics to comprehensively evaluate the performance of the model, as shown in Table 3.

From the comparison results of different methods in the experimental setting in Table 3, it can be seen that our proposed algorithm outperforms the current traditional segmentation methods in all five metrics and is also basically better than the latest methods. Among all the networks considered, SegNet and FCN8 exhibit the poorest performance in terms of the metrics evaluated. While the metrics of the other models show improvement over successive iterations, they still fall short of the performance achieved by the models proposed in this paper. According to Table 3, we found that the above methods can achieve high-precision segmentation of cloud shadow datasets; to further visualize the effectiveness of our methods, Figure 7 shows the visualization experiment results of cloud shadow segmentation.

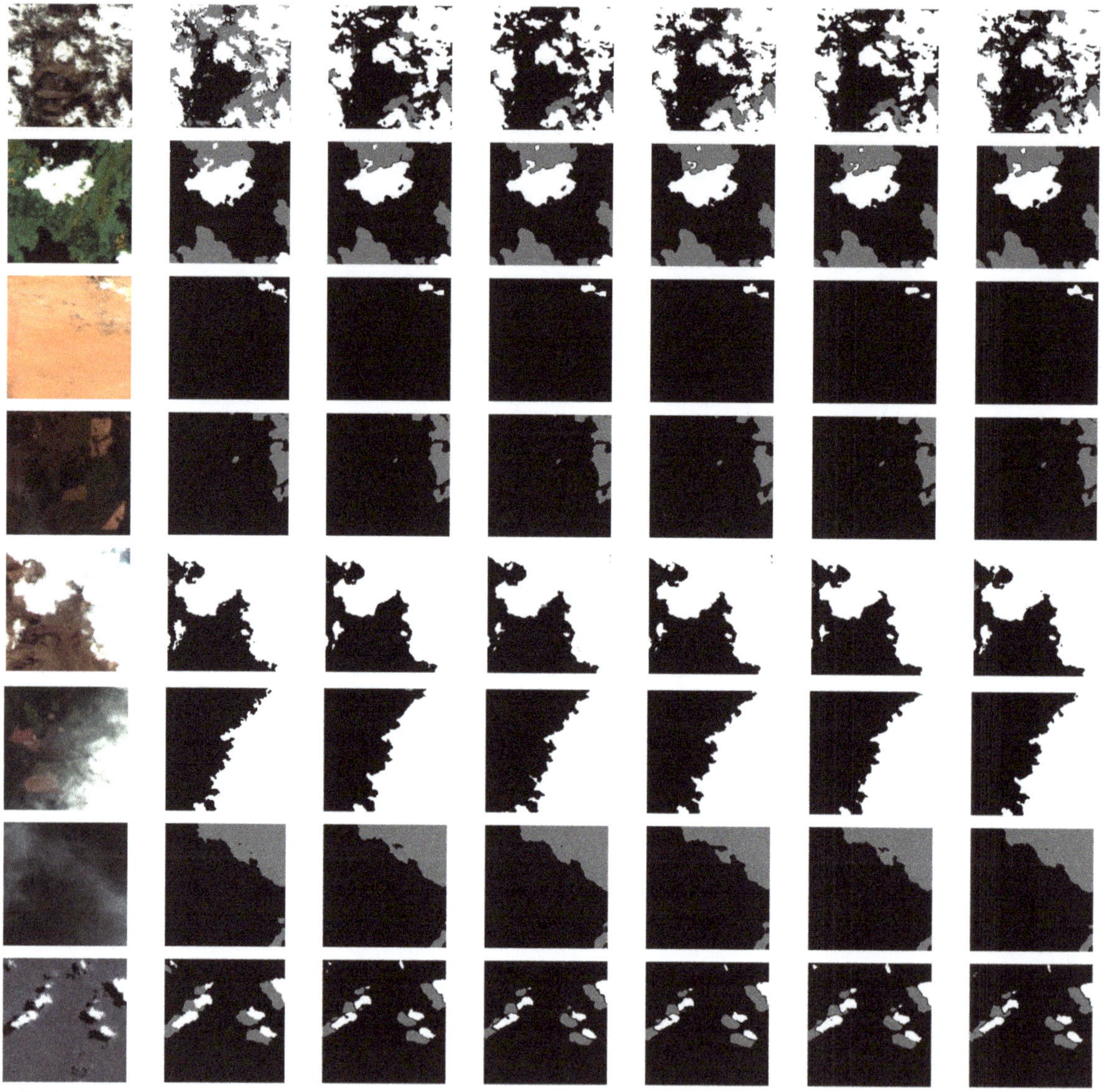

(**a**) Image (**b**) Label (**c**) UNet (**d**) PSPNet (**e**) SegNet (**f**) HRNet (**g**) Ours

Figure 7. Comparison of visualization results for different models: (**a**) the original image; (**b**) the corresponding label; (**c**) the prediction of UNet; (**d**) the prediction of PSPNet; (**e**) the prediction of SegNet; (**f**) the prediction of HRNet; (**g**) the prediction of the proposed AFMUNet.

Figure 7 shows the visualization effect of different methods for image segmentation in the cloud shadow dataset. Eight examples are selected to demonstrate the segmentation effect. It can be observed that the proposed method is more accurate for cloud segmentation, especially in the segmentation of the edge region of the cloud. However, the performance is poor for cloud shadows and thin or unclear clouds. The segmentation effect of the Segnet model is relatively rough, the edge information is incompletely obtained, and too much information is lost in the feature extraction stage. It can be found from Figure 7 that it does not perform well at the boundary of the cloud and loses a lot of shape-striped feature information. When the texture is slightly complex, PSPNet cannot completely segment the boundary of clouds and cloud shadows. HRNet, on the other hand, slightly improves the effect compared to the above two models, with more delicate processing of the edges, but still has shortcomings compared to our model. UNet is a classic segmentation network known for its superior performance in training on smaller datasets and producing smoother segmentation edges. However, it still requires improvement in processing details. Our model addresses this limitation to some extent, effectively recognizing cloud and cloud shadow boundaries while enhancing detail processing. Nonetheless, further refinement is needed to effectively handle very low light or thin cloud bodies. In summary, from a qualitative point of view, our method performs better in different environments compared with other methods, which proves the importance and effectiveness of the model proposed in this paper.

Table 3. Results on GF1_WHU dataset testing set.

Method	PA (%)	MPA (%)	MIoU (%)	F1 (%)	FWIoU (%)
SegNet	94.80	93.90	88.28	90.77	90.16
UNet	96.33	95.49	91.32	93.21	92.80
FCN8s	95.20	94.84	90.58	92.92	92.36
PSPNet	96.51	95.78	91.76	93.89	93.31
DANet [48]	94.82	94.13	89.25	91.70	91.32
DeepLab V3Plus	96.27	95.42	91.18	93.11	92.56
BiseNet V2 [49]	95.76	94.85	90.27	92.34	91.87
HRNet [50]	96.87	95.73	92.02	93.93	93.40
SP_CSANet [51]	97.33	96.01	91.34	93.12	92.63
CDUNet [52]	97.21	96.53	93.33	95.03	94.58
AFMUNet (Ours)	97.40	96.62	93.28	95.10	94.43

In order to better illustrate the model generalization and effectiveness of the model in the face of different environmental backgrounds, as shown in Figure 8 above, we chose vegetation, land, desert, barren, and snowy mountainous areas for model testing. For the images in the green vegetation environment in the first group, all are able to segment the general outline of the clouds well, but the details in the middle and background overlapping region are poor, and our model segments the edges of the clouds and the boundary well. In the second group, PSPNet, SegNet, and HRNet perform poorly for the shallow, scattered, and complex clouds, while UNet shows some improvement and recognizes the information of some thin clouds but still demonstrates a large deficiency compared to our model. By observing the third and fourth sets of images, it is not difficult to find that our model smoothly distinguishes the neighboring regions of clouds and cloud shadows and handles the edge information more naturally compared with other models. When confronted with remote sensing images containing significant noise interference, the performance of UNet, SegNet, and HRNet models is deemed insufficient. Instances of omission and misdetection, such as in the snowy mountain zones depicted in the comparative images, are observed. These models encounter challenges in accurately distinguishing between ice, snow, and clouds. Although the PSPNet segmentation effect offers some improvement, the texture features of the clouds are lost, and the boundary cannot be clearly reflected. None of the aforementioned models are suitable for the challenging task of cloud shadow segmentation across diverse and complex environments. In contrast, the algorithm proposed in this paper adeptly addresses cloud shadow segmentation in various situations and scenarios. By optimizing deeper features and leveraging the enhanced channel and feature fusion capabilities enabled by the spatial attention mechanism module, our algorithm effectively recovers high-definition remote sensing images.

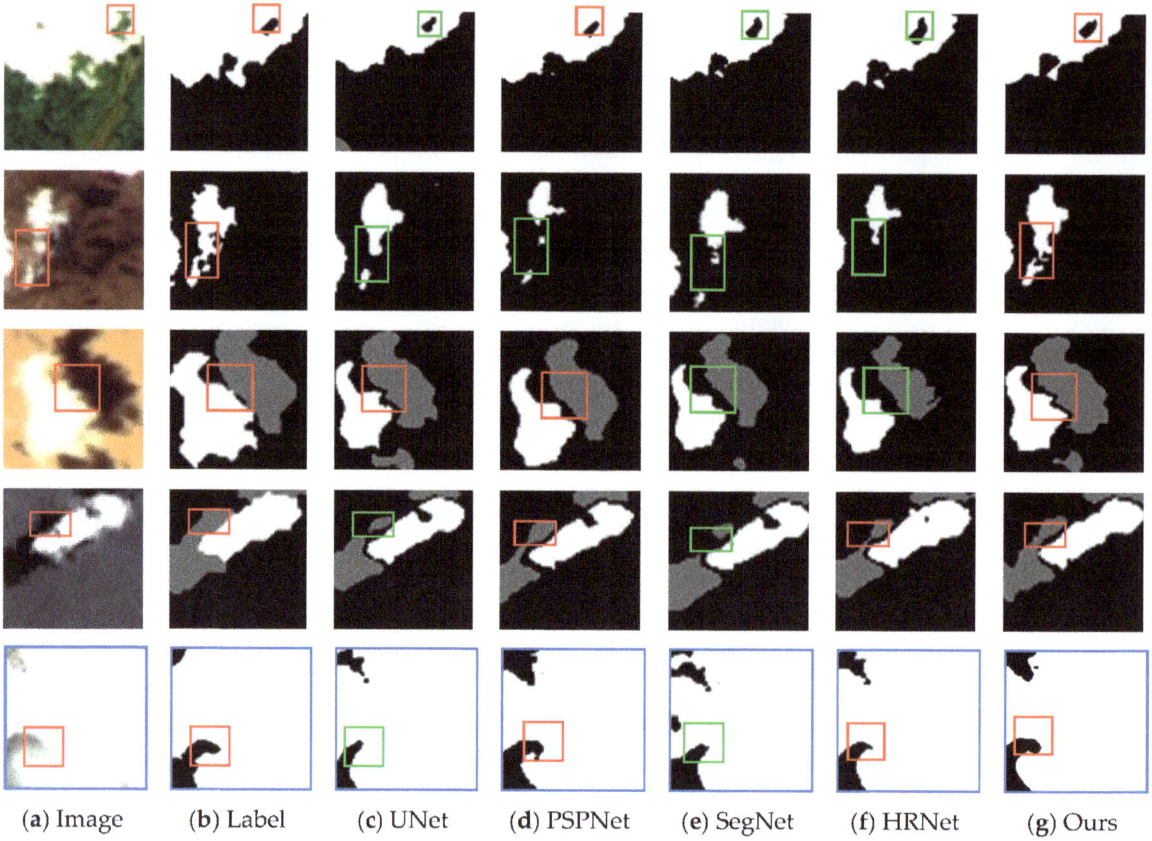

Figure 8. Fine comparison of different models in different contexts: (**a**) the original image; (**b**) the corresponding label; (**c**) the prediction of UNet; (**d**) the prediction of PSPNet; (**e**) the prediction of SegNet; (**f**) the prediction of HRNet; (**g**) the prediction of the proposed AFMUNet. (Red boxes indicate better segmentation results, while green boxes segment poorer results).

To further analyze our algorithm, we compared the segmentation results of different types of clouds, as shown in Figures 9 and 10. It can be observed that our proposed model performs well in segmenting both thin and thick clouds, effectively delineating the overall contours of the clouds and shadows and clearly distinguishing them from the background. However, upon comparing the third row on the left with the second row on the right, it is evident that AFMUNet exhibits superior segmentation performance for thick clouds compared to thin clouds. Thick clouds only lose some fine texture details, while thin clouds tend to lose fragmented point cloud and shadow information during segmentation.

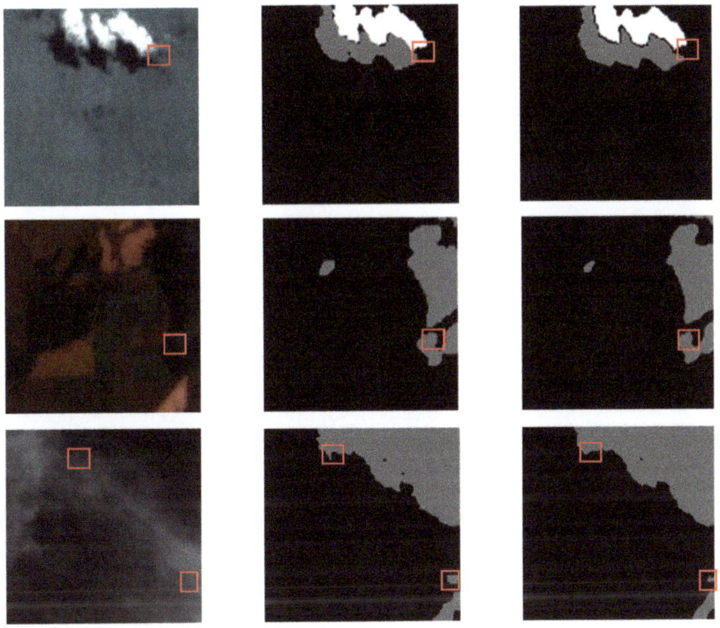

(**a**) Image (**b**) Label (**c**) Ours

Figure 9. Results of thin cloud segmentation: (**a**) the original image; (**b**) the corresponding label; (**c**) the prediction of the proposed AFMUNet. (Red boxes indicate noteworthy edge details of the result).

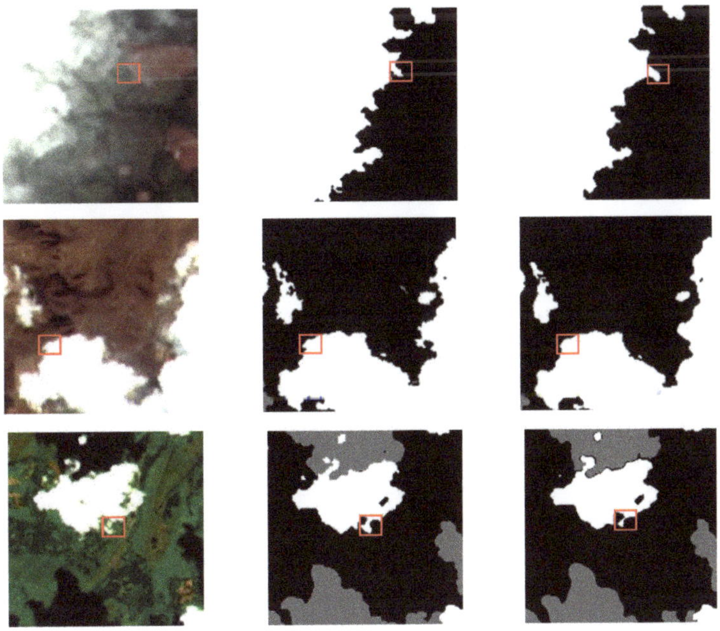

(**a**) Image (**b**) Label (**c**) Ours

Figure 10. Results of thick cloud segmentation: (**a**) the original image; (**b**) the corresponding label; (**c**) the prediction of the proposed AFMUNet. (Red boxes indicate better segmented edge details).

4. Conclusions

In remote sensing images, the accurate segmentation of cloud shadow regions is of great practical significance for practical tasks such as meteorological prediction, environmental monitoring, and natural disaster detection. In this paper, an attention mechanism feature aggregation algorithm is proposed for cloud shadow segmentation, fully leveraging the advantages of convolutional neural networks in deep learning. UNet is selected as the backbone network, an innovative loss function is employed, and two auxiliary modules, CSAM and FFM, are introduced. Our proposed model initiates constant down-sampling to extract high-level features. Adaptive improvement of sensory fields and selection of different channel operations are introduced during each up-sampling process to increase the resolution of feature maps, enabling the acquisition of rich contextual information. This facilitates the accurate fusion of low- and high-level information within each layer's feature fusion module, ultimately restoring the classification and localization of high-resolution remote sensing images. Compared with previous deep learning and segmentation methods, our approach achieves significant improvement in accuracy in cloud shadow segmentation tasks. Experiments demonstrate the remarkable noise resistance and identification capabilities of this method. It accurately locates cloud shadows and segments fine cloud crevices in complex environments, while also producing smoother edge segmentation. Particularly noteworthy is its performance in the task of identifying thick clouds. However, there are still some shortcomings in cloud shadow segmentation: (1) under the influence of light, some inconspicuous cloud seams may be incorrectly segmented into other features and thus recognized as background; (2) refinement is still needed for the segmentation of thin clouds to capture the fragmented information of cloud shadows; (3) to be better adapted to practical applications, in the future, we also need to appropriately compress and simplify the model while maintaining the accuracy and reduce the segmentation result time to improve the training speed of the network. In the future, augmented learning can be implemented by incorporating a pre-training phase into the model, aiming to enhance segmentation accuracy and reduce training time. Additionally, efforts will be made to explore its application in other domains, including river segmentation and medical tumor segmentation.

Author Contributions: Conceptualization, Z.F.; methodology, W.D. and Z.F.; validation, W.D., Y.Y. and R.Y.; writing—original draft preparation, W.D. and J.L.; writing—review and editing, Y.Y.; visualization, J.L.; supervision, R.Y. All authors have read and agreed to the published version of the manuscript.

Funding: This research received no external funding.

Data Availability Statement: The data and the code of this study are available from the corresponding author upon request (001163@nuist.edu.cn). The data are not publicly available due to restrictions (e.g., privacy, legal or ethical reasons).

Conflicts of Interest: The authors declare no conflicts of interest.

References

1. Xiong, J.H.; Wu, H.; Gao, Y.; Cai, S.; Liang, D.; Yu, W.P. Ten years of remote sensing science: NSFC program fundings, progress, and challenges. *Natl. Remote Sens. Bull.* **2023**, *27*, 821–830. [CrossRef]
2. Mahajan, S.; Fataniya, B. Cloud detection methodologies: Variants and development—A review. *Complex Intell. Syst.* **2020**, *6*, 251–261. [CrossRef]
3. Saunders, R.W.; Kriebel, K.T. An improved method for detecting clear sky and cloudy radiances from AVHRR data. *Int. J. Remote Sens.* **1988**, *9*, 123–150. [CrossRef]
4. Hutchinson, K.D.; Hardy, K.R. Threshold functions for automated cloud analyses of global meteorological satellite imagery. *Int. J. Remote Sens.* **1995**, *16*, 3665–3680. [CrossRef]
5. Xiong, Q.; Wang, Y.; Liu, D.; Ye, S.; Du, Z.; Liu, W.; Huang, J.; Su, W.; Zhu, D.; Yao, X.; et al. A cloud detection approach based on hybrid multispectral features with dynamic thresholds for GF-1 remote sensing images. *Remote Sens.* **2020**, *12*, 450. [CrossRef]
6. Derrien, M.; Farki, B.; Harang, L.; LeGléau, H.; Noyalet, A.; Pochic, D.; Sairouni, A. Automatic cloud detection applied to NOAA-11/AVHRR imagery. *Remote Sens. Environ.* **1993**, *46*, 246–267. [CrossRef]

7. Clothiaux, E.E.; Miller, M.A.; Albrecht, B.A.; Ackerman, T.P.; Verlinde, J.; Babb, D.M.; Peters, R.M.; Syrett, W.J. An evaluation of a 94-GHz radar for remote sensing of cloud properties. *J. Atmos. Ocean. Technol.* **1995**, *12*, 201–229. [CrossRef]
8. Danda, S.; Challa, A.; Sagar BS, D. A morphology-based approach for cloud detection. In Proceedings of the 2016 IEEE International Geoscience and Remote Sensing Symposium (IGARSS), Beijing, China, 10–15 July 2016; pp. 80–83. [CrossRef]
9. Liu, X.; Shen, J.P.; Huang, Y. Cloud automatic detection in high-resolution satellite images based on morphological features. In Proceedings of the Eleventh International Conference on Graphics and Image Processing (ICGIP 2019), Hangzhou, China, 12–14 October 2019; SPIE: Bellingham, WA, USA, 2020; Volume 11373, pp. 159–166. [CrossRef]
10. Tom, V.T.; Peli, T.; Leung, M.; Bondaryk, J.E. Morphology-based algorithm for point target detection in infrared backgrounds. In Proceedings of the Signal and Data Processing of Small Targets 1993, Orlando, FL, USA, 12–14 April 1993; SPIE: Bellingham, WA, USA, 1993; Volume 1954, pp. 2–11. [CrossRef]
11. Amato, U.; Antoniadis, A.; Cuomo, V.; Cutillo, L.; Franzese, M.; Murino, L.; Serio, C. Statistical cloud detection from SEVIRI multispectral images. *Remote Sens. Environ.* **2008**, *112*, 750–766. [CrossRef]
12. Wylie, D.; Jackson, D.L.; Menzel, W.P.; Bates, J.J. Trends in global cloud cover in two decades of HIRS observations. *J. Clim.* **2005**, *18*, 3021–3031. [CrossRef]
13. Abuhussein, M.; Robinson, A. Obscurant Segmentation in Long Wave Infrared Images Using GLCM Textures. *J. Imaging* **2022**, *8*, 266. [CrossRef]
14. Shao, L.; He, J.; Lu, X.; Hei, B.; Qu, J.; Liu, W. Aircraft Skin Damage Detection and Assessment from UAV Images Using GLCM and Cloud Model. *IEEE Trans. Intell. Transp. Syst.* **2023**, *25*, 3191–3200. [CrossRef]
15. Reiter, P. Cloud Detection Through Wavelet Transforms in Machine Learning and Deep Learning. *arXiv* **2020**, arXiv:2007.13678.
16. Gupta, R.; Panchal, P. Cloud detection and its discrimination using Discrete Wavelet Transform in the satellite images. In Proceedings of the 2015 International Conference on Communications and Signal Processing (ICCSP), Melmaruvathur, India, 2–4 April 2015; pp. 1213–1217. [CrossRef]
17. Changhui, Y.; Yuan, Y.; Minjing, M.; Menglu, Z. Cloud detection method based on feature extraction in remote sensing images. *Int. Arch. Photogramm. Remote Sens. Spat. Inf. Sci.* **2013**, *40*, 173–177. [CrossRef]
18. Surya, S.R.; Rahiman, M.A. Cloud detection from satellite images based on Haar wavelet and clustering. In Proceedings of the 2017 International Conference on Nextgen Electronic Technologies: Silicon to Software (ICNETS2), Chennai, India, 23–25 March 2017; pp. 163–167. [CrossRef]
19. Li, P.; Dong, L.; Xiao, H.; Xu, M. A cloud image detection method based on SVM vector machine. *Neurocomputing* **2015**, *169*, 34–42. [CrossRef]
20. Ishida, H.; Oishi, Y.; Morita, K.; Moriwaki, K.; Nakajima, T.Y. Development of a support vector machine based cloud detection method for MODIS with the adjustability to various conditions. *Remote Sens. Environ.* **2018**, *205*, 390–407. [CrossRef]
21. Fu, H.; Shen, Y.; Liu, J.; He, G.; Chen, J.; Liu, P.; Qian, J.; Li, J. Cloud detection for FY meteorology satellite based on ensemble thresholds and random forests approach. *Remote Sens.* **2018**, *11*, 44. [CrossRef]
22. Jin, Z.; Zhang, L.; Liu, S.; Yi, F. Cloud detection and cloud phase retrieval based on BP neural network. *Opt. Optoelectron. Technol.* **2016**, *14*, 74–77.
23. Long, J.; Shelhamer, E.; Darrell, T. Fully convolutional networks for semantic segmentation. In Proceedings of the IEEE Conference on Computer Vision and Pattern Recognition, San Juan, PR, USA, 17–19 June 2015; pp. 3431–3440.
24. Mohajerani, S.; Krammer, T.A.; Saeedi, P. Cloud detection algorithm for remote sensing images using fully convolutional neural networks. *arXiv* **2018**, arXiv:1810.05782.
25. Badrinarayanan, V.; Kendall, A.; Cipolla, R. Segnet: A deep convolutional encoder-decoder architecture for image segmentation. *IEEE Trans. Pattern Anal. Mach. Intell.* **2017**, *39*, 2481–2495. [CrossRef] [PubMed]
26. Lu, J.; Wang, Y.; Zhu, Y.; Ji, X.; Xing, T.; Li, W.; Zomaya, A.Y. P_SegNet and NP_SegNet: New neural network architectures for cloud recognition of remote sensing images. *IEEE Access* **2019**, *7*, 87323–87333. [CrossRef]
27. Chen, L.-C.; Papandreou, G.; Kokkinos, I.; Murphy, K.; Yuille, A.L. Deeplab: Semantic image segmentation with deep convolutional nets, atrous convolution, and fully connected crfs. *IEEE Trans. Pattern Anal. Mach. Intell.* **2017**, *40*, 834–848. [CrossRef] [PubMed]
28. Ronneberger, O.; Fischer, P.; Brox, T. U-net: Convolutional networks for biomedical image segmentation. In Proceedings of the Medical Image Computing and Computer-Assisted Intervention–MICCAI 2015: 18th International Conference, Munich, Germany, 5–9 October 2015; Proceedings, Part III 18. Springer International Publishing: Berlin/Heidelberg, Germany, 2015; pp. 234–241. [CrossRef]
29. Zhao, H.; Shi, J.; Qi, X.; Wang, X.; Jia, J. Pyramid scene parsing network. In Proceedings of the IEEE Conference on Computer Vision and Pattern Recognition, Honolulu, HI, USA, 21–26 July 2017; pp. 2881–2890.
30. Zhang, Z.; Yang, S.; Liu, S.; Cao, X.; Durrani, T.S. Ground-based remote sensing cloud detection using dual pyramid network and encoder–decoder constraint. *IEEE Trans. Geosci. Remote Sens.* **2022**, *60*, 1–10. [CrossRef]
31. Tsotsos, J.K. Analyzing vision at the complexity level. *Behav. Brain Sci.* **1990**, *13*, 423–445. [CrossRef]
32. Ding, L.; Xia, M.; Lin, H.; Hu, K. Multi-level attention interactive network for cloud and snow detection segmentation. *Remote Sens.* **2023**, *16*, 112. [CrossRef]
33. Sutskever, I.; Vinyals, O.; Le, Q.V. Sequence to sequence learning with neural networks. *Adv. Neural Inf. Process. Syst.* **2014**, *27*. [CrossRef]

34. Hu, K.; Li, Y.; Zhang, S.; Wu, J.; Gong, S.; Jiang, S.; Weng, L. FedMMD: A Federated weighting algorithm considering Non-IID and Local Model Deviation. *Expert Syst. Appl.* **2024**, *237*, 121463. [CrossRef]
35. Guo, M.-H.; Xu, T.-X.; Liu, J.-J.; Liu, Z.-N.; Jiang, P.-T.; Mu, T.-J.; Zhang, S.-H.; Martin, R.R.; Cheng, M.-M.; Hu, S.-M. Attention mechanisms in computer vision: A survey. *Comput. Vis. Media* **2022**, *8*, 331–368. [CrossRef]
36. Liu, Y.; Shao, Z.; Hoffmann, N. Global attention mechanism: Retain information to enhance channel-spatial interactions. *arXiv* **2021**, arXiv:2112.05561.
37. Hu, K.; Zhang, D.; Xia, M.; Qian, M.; Chen, B. LCDNet: Light-weighted cloud detection network for high-resolution remote sensing images. *IEEE J. Sel. Top. Appl. Earth Obs. Remote Sens.* **2022**, *15*, 4809–4823. [CrossRef]
38. Ji, H.; Xia, M.; Zhang, D.; Lin, H. Multi-supervised feature fusion attention network for clouds and shadows detection. *ISPRS Int. J. Geo-Inf.* **2023**, *12*, 247. [CrossRef]
39. Yu, C.; Wang, J.; Peng, C.; Gao, C.; Yu, G.; Sang, N. Bisenet: Bilateral segmentation network for real-time semantic segmentation. In Proceedings of the European Conference on Computer Vision (ECCV), Munich, Germany, 8–14 September 2018; pp. 325–341.
40. Hu, K.; Zhang, E.; Dai, X.; Xia, M.; Zhou, F.; Weng, L.; Lin, H. MCSGNet: A Encoder–Decoder Architecture Network for Land Cover Classification. *Remote Sens.* **2023**, *15*, 2810. [CrossRef]
41. Wang, Z.; Xia, M.; Weng, L.; Hu, K.; Lin, H. Dual encoder-decoder network for land cover segmentation of remote sensing image. *IEEE J. Sel. Top. Appl. Earth Obs. Remote Sens.* **2023**, *17*, 2372–2385. [CrossRef]
42. Wang, Q.; Ma, Y.; Zhao, K.; Tian, Y. A comprehensive survey of loss functions in machine learning. *Ann. Data Sci.* **2020**, *9*, 187–212. [CrossRef]
43. Hu, K.; Weng, C.; Shen, C.; Wang, T.; Weng, L.; Xia, M. A multi-stage underwater image aesthetic enhancement algorithm based on a generative adversarial network. *Eng. Appl. Artif. Intell.* **2023**, *123*, 106196. [CrossRef]
44. Ma, J. Segmentation loss odyssey. *arXiv* **2020**, arXiv:2005.13449. [CrossRef]
45. Li, Z.; Shen, H.; Li, H.; Xia, G.; Gamba, P.; Zhang, L. Multi-feature combined cloud and cloud shadow detection in GaoFen-1 wide field of view imagery. *Remote Sens. Environ.* **2017**, *191*, 342–358. [CrossRef]
46. Ren, H.; Xia, M.; Weng, L.; Hu, K.; Lin, H. Dual Attention-Guided Multiscale Feature Aggregation Network for Remote Sensing Image Change Detection. *IEEE J. Sel. Top. Appl. Earth Obs. Remote Sens.* **2024**, *17*, 4899–4916. [CrossRef]
47. Jiang, S.; Dong, R.; Wang, J.; Xia, M. Credit card fraud detection based on unsupervised attentional anomaly detection network. *Systems* **2023**, *11*, 305. [CrossRef]
48. Fu, J.; Liu, J.; Tian, H.; Li, Y.; Bao, Y.; Fang, Z.; Lu, H. Dual attention network for scene segmentation. In Proceedings of the IEEE/CVF Conference on Computer Vision and Pattern Recognition, Long Beach, CA, USA, 15–20 June 2019; pp. 3146–3154.
49. Yu, C.; Gao, C.; Wang, J.; Yu, G.; Shen, C.; Sang, N. Bisenet v2: Bilateral network with guided aggregation for real-time semantic segmentation. *Int. J. Comput. Vis.* **2021**, *129*, 3051–3068. [CrossRef]
50. Wang, J.; Sun, K.; Cheng, T.; Jiang, B.; Deng, C.; Zhao, Y.; Liu, D.; Mu, Y.; Tan, M.; Wang, X.; et al. Deep high-resolution representation learning for visual recognition. *IEEE Trans. Pattern Anal. Mach. Intell.* **2020**, *43*, 3349–3364. [CrossRef]
51. Qu, Y.; Xia, M.; Zhang, Y. Strip pooling channel spatial attention network for the segmentation of cloud and cloud shadow. *Comput. Geosci.* **2021**, *157*, 104940. [CrossRef]
52. Hu, K.; Zhang, D.; Xia, M. CDUNet: Cloud detection UNet for remote sensing imagery. *Remote Sens.* **2021**, *13*, 4533. [CrossRef]

remote sensing

MDPI

Article

PointMM: Point Cloud Semantic Segmentation CNN under Multi-Spatial Feature Encoding and Multi-Head Attention Pooling

Ruixing Chen [1], Jun Wu [1,*], Ying Luo [1] and Gang Xu [2]

1 School of Electronic Engineering and Automation, Guilin University of Electronic Technology, Guilin 541000, China; 19081001006@mails.guet.edu.cn (R.C.); luoying@guet.edu.cn (Y.L.)

2 Ningbo Institute of Materials Technology and Engineering, Chinese Academy of Sciences, Ningbo 315201, China; xugang@nimte.ac.cn

* Correspondence: wujun@guet.edu.cn

Abstract: For the actual collected point cloud data, there are widespread challenges such as semantic inconsistency, density variations, and sparse spatial distribution. A network called PointMM is developed in this study to enhance the accuracy of point cloud semantic segmentation in complex scenes. The main contribution of PointMM involves two aspects: (1) Multi-spatial feature encoding. We leverage a novel feature encoding module to learn multi-spatial features from the neighborhood point set obtained by k-nearest neighbors (KNN) in the feature space. This enhances the network's ability to learn the spatial structures of various samples more finely and completely. (2) Multi-head attention pooling. We leverage a multi-head attention pooling module to address the limitations of symmetric function-based pooling, such as maximum and average pooling, in terms of losing detailed feature information. This is achieved by aggregating multi-spatial and attribute features of point clouds, thereby enhancing the network's ability to transmit information more comprehensively and accurately. Experiments on publicly available point cloud datasets S3DIS and ISPRS 3D Vaihingen demonstrate that PointMM effectively learns features at different levels, while improving the semantic segmentation accuracy of various objects. Compared to 12 state-of-the-art methods reported in the literature, PointMM outperforms the runner-up by 2.3% in OA on the ISPRS 3D Vaihingen dataset, and achieves the third best performance in both OA and MioU on the S3DIS dataset. Both achieve a satisfactory balance between OA, F1, and MioU.

Keywords: point cloud semantic segmentation; CNN; multi-spatial feature encoding; multi-head attention pooling

Citation: Chen, R.; Wu, J.; Luo, Y.; Xu, G. PointMM: Point Cloud Semantic Segmentation CNN under Multi-Spatial Feature Encoding and Multi-Head Attention Pooling. *Remote Sens.* **2024**, *16*, 1246. https://doi.org/10.3390/rs16071246

Academic Editor: Andrzej Stateczny

Received: 7 February 2024
Revised: 29 March 2024
Accepted: 29 March 2024
Published: 31 March 2024

1. Introduction

Compared to 2D images, three-dimensional point clouds obtained using 3D scanners and depth sensors (such as LiDAR and RGB-D cameras) can more comprehensively and intuitively express the spatial relationships between various targets in the scene. They have been widely utilized in various industries, including 3D modeling [1], autonomous driving [2], and metaverse [3], and natural resource surveys [4]. Point cloud semantic segmentation is a crucial supporting technology for understanding and analyzing 3D scenes [5]. However, due to the spatiotemporal complexity, the irregular distribution of terrain surfaces, and the non-uniformity and disorder of point clouds themselves, achieving high-precision point cloud semantic segmentation in large-scale complex scenes remains an extremely challenging task. Designing point cloud semantic segmentation convolutional neural networks with end-to-end output capability and adaptability to various scenarios has become a current research focus [6], which can be broadly categorized into two types: indirect and direct methods. Our approach belongs to the latter.

An indirect semantic segmentation network needs to preprocess the original point cloud into a 2D/3D grid structure to leverage mature image-based CNNs for tasks such as object classification and semantic segmentation. For instance, WhuY4 [7] and NANJ2 [8] design CNNs to extract multiscale local features from projection view of point clouds. On this basis, they calculate category probabilities for each point and construct a decision tree to guide subsequent retraining. Individuals such as GERDZHEV [9] utilize convolutional kernels of varying scales to capture contextual information and aggregate feature information at different scales to obtain segmentation results. GFNet [10] employs bidirectional alignment and the propagation of complementary information to learn geometric information between different projection views. AGNet [11] introduces attention pooling on the basis of traditional graph neural network (GNN) to score feature importance. GaIA [12] autonomously learns crucial regions of point clouds based on graphical information gain and applies it to semantic segmentation tasks. However, a considerable amount of geometric structure, orientation, and other spatial relation information of target objects are lost during the point cloud projection process. Therefore, the point cloud semantic segmentation networks under multi-view projection are sensitive to changes in viewpoint and anomalies caused by occlusion. Represented by PVCNN [13], VoxSegNet [14], PVCL [15], and MPVConv [16], voxel-based 3D convolutional neural networks can effectively learn 3D spatial information and context-dependent relationships of point clouds. However, the sparsity and uneven density of point clouds can generate a large number of empty grids, resulting in low computational efficiency and high memory usage.

Direct point cloud semantic segmentation network learns features straightforwardly from 3D point clouds without the need to pre-process them into 2D/3D grids. Remarkable works have been carried out by PointNet [17] and PointNet++ [18] in solving the challenges of large-scale point cloud network computing through farthest point sampling (FPS). However, overly independent point operations in the networks hinder the capture of local spatial structures. To address this issue, PointSIFT [19], inspired by the SIFT operator, encodes the features in eight directions in the XYZ space to overcome the limitation of PointNet++ in restricting its k-nearest neighbor search to the same direction. However, this method is exceptionally sensitive to the orientation information of objects. PointWeb [20] aggregates local point cloud information through an adaptive feature adjustment module. HPRS [21] develops an adaptive spherical query module to simultaneously capture global features and finer-grained local features. MappingConvSeg [22] conducts spherical neighborhood feature learning at each downsampling layer, enhancing the network's ability to capture complex geometric structures. Zhao et al. [23] introduces dynamic convolution filters (DFConv) and an improved semantic segmentation (JISS) module into JSNet [24]. Overall, these networks aggregate neighborhood information and multiscale features through local feature encoding, resulting in improved segmentation accuracy compared to the original PointNet++. However, the feature encoding methods of such networks primarily consider position and point spacing, with limited attention to the spatial scale information of points.

Different from the PointNet++ series, direct point cloud segmentation networks based on graph convolution treat each point as a node in the graph and form directed edges with neighboring points. The challenge of obtaining such networks lies in how to construct appropriate point-to-point relationships and the advantages lie in their ability to aggregate target structural features while maintaining translation invariance in a three-dimensional space. Representative works in this category include KVGCN [25], GCN-MLP [26], RG-GCN [27], DDGCN [28], and PointCCR [29]. Some researchers attempt to learn fine-grained point cloud features by introducing self-attention mechanisms in networks. For example, Hu et al. [30] combine self-attention mechanisms with a random sampling algorithm to design the RandLA-Net network. Du et al. [31] add a dense convolutional linking layer on the basis of RandLA-Net for a more comprehensive learning of geometric shapes. LG-Net [32] achieves learning of global context information through a global correlation mining (GCM) module. Yin et al. [33], based on geometric structure and object edge integrity, design a local feature encoding network using rapid point random sampling. In order to enhance a network's ability to learn local features, Deng et al. [34] proposed PointNAC by introducing a point-pair feature encoding pattern and Copula correlation analysis module, and Wu et al. [35] developed PointConv by introducing a novel weight calculation as well. Yan et al. [36] designed an Adaptive Sampling Module and Local-Nonlocal (L-NL) Module based on attention mechanisms to mitigate noise and outliers that could disrupt the network's learning of local features. Zarzar et al. [37] designed PointRGCN for better extraction of topological structures from point clouds, employing feature encoding and aggregating context information in the form of graphs. Inspired by the breakthroughs of Transformer models in Natural Language Processing (NLP) tasks, Zhao et al. [38] applied Point Transformer and self-attention mechanisms to various point cloud classification and segmentation tasks, achieving excellent performance. Although the aforementioned networks have shown advantages in certain category-targeted segmentation tasks, they still struggle to achieve high overall segmentation accuracy (OA) and average joint intersection (MIoU) scores at the same time.

Generally speaking, compared to indirect point cloud segmentation methods, direct methods are more effective in utilizing information and are easier to capture fine-grained local features for precise segmentation. However, existing feature encoding patterns in networks only utilize relatively independent information, such as point absolute positions, point-to-point distances, and direction vectors, to express spatial structures, making it difficult to effectively extract detailed features from complex scenes. On the other hand, existing networks typically use the maximum pooling process for feature conveying. But this process may discard the local details of point cloud samples, making it difficult for the network to effectively distinguish points in different categories. In response to the above issues, this article developed a network called PointMM for the high-precision semantic segmentation of 3D point clouds. The contributions in the paper lie in two aspects, as outlined below.

Firstly, addressing the limitation of existing network feature encoding methods that only consider one-dimensional features between sampled points and their neighboring points, this paper leverages a multi-spatial feature encoding module by computing angles between point distances and normal vectors, and encoding point coordinates, distances, directional vectors, and point relationships, thereby enhancing the network's capability to learn the spatial structures of various samples more finely and completely.

Secondly, addressing the drawback of the pooling process based on symmetric functions that may discard a significant amount of detailed feature information, especially the information loss of minority class samples in 3D scene datasets under long-tailed distribution, this paper leverages a multi-head attention pooling module to score and aggregate features at different levels, thereby enhancing the network's ability to transmit information more comprehensively and accurately.

2. Our Method

2.1. Network Overview

The FPS typically employed in direct point cloud semantic segmentation networks is a "uniform" point cloud sampling method that can lead to information loss, especially for samples of the minority class. On the other hand, existing point cloud semantic segmentation networks tend to have a "unidirectional" learning process from the sampled central point to its neighboring points, which is not conducive to learning the fine local structures of point clouds. Additionally, the pooling process in existing point cloud semantic segmentation networks tends to retain the maximum values of local features, hindering the transmission of fine spatial information. This not only affects the effective learning of various sample features but also has an impact on overall segmentation accuracy to some extent. To address these issues, we use Balanced Class Sampling (BCS) to perform full sampling of minority class samples and downsampling of majority class samples in sub regions, and assign initial values to the sampled samples. When all points are sampled (given initial values) for learning, we reset all initial value information to zero and cycle this process until the set maximum batch is reached. The BCS module ensures that each class of sample points is learned by the network. Meanwhile, this article combines multi-spatial feature learning and multi-head attention pooling into PointNet++, and builds a network called PointMM, as shown in Figure 1.

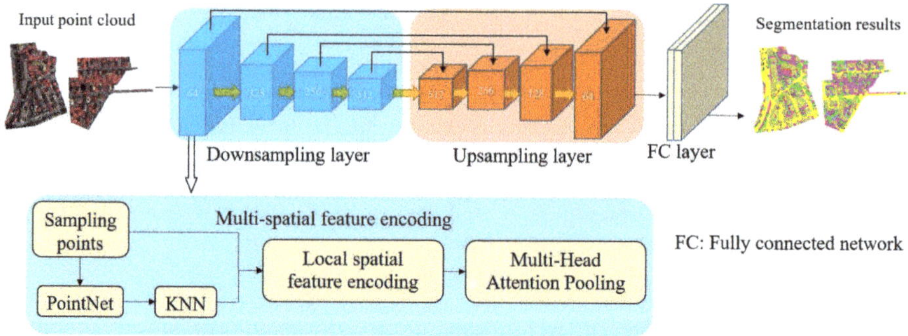

Figure 1. PointMM network structure. (The thin arrow represents the flowchart of the network framework, while the thick arrow indicates the various components of the downsampling layer).

PointMM mainly consists of four parts: the Balanced Class Sampling (BCS) module, the downsampling layer incorporating multi-spatial feature encoding and multi-head attention pooling, the up sampling layer, and the fully connected layer. Firstly, the training samples V are obtained using BCS, and each sampling center point v_i and its neighborhood points $v_{i,k}$ are extracted based on FPS and feature KNN. At this point, we obtain a point cloud of dimensions $N \times K \times D$, where N is the number of sampling center points, K is the number of neighborhood points, and D is the dimensionality of the point cloud containing coordinate and attribute information. Then, the sampling points and their neighborhood points are passed through the multi-spatial feature encoding module to obtain features η_i of dimensions $N \times K \times 13$. Subsequently, the features η_i are input into the multi-head attention pooling module, which integrates neighborhood features through pooling operations to generate a larger receptive field and more global feature vector MP(F_i). It is worth noting that we set up four downsampling layers, so the number of attention heads for each layer is 2^n ($n \in [1, 4]$). The initial input to the downsampling layer in this paper is a point cloud of dimensions $N \times K \times D$, and the number of sampled points in each subsequent layer is multiplied by 4^{-n} ($n \in [1, 4]$), where n represents the downsampling layer. Additionally, the output of the downsampling layer is feature maps of dimensions $N/4 \times 64$, $N/16 \times 128$, $N/64 \times 256$, and $N/256 \times 512$. Meanwhile, the upsampling results are cascaded with corresponding downsampling levels using 3D interpolation and skip connections to effectively fuse low-level to high-level features. Finally, a fully connected layer is utilized to establish the transformation relationship between point cloud features and label results. It should be noted that unlike PointNet++ using FPS for the indiscriminate downsampling of large-scale point cloud data, we have designed BCS to perform a complete sampling of minority class samples and downsampling of majority class samples, ensuring that the network learns each class as well as possible through the sampling points.

2.2. Multi-Spatial Feature Encoding

The feature encoding in existing point cloud semantic segmentation networks is often based on point positions and point-to-point distances. However, this relatively independent feature information is insufficient to represent the complex relationships within the neighborhood system. In addition, the use of k-nearest neighbors (KNN) in Euclidean space to extract neighborhood points tends to be limited to the same direction, preventing the comprehensive expression of spatial topological structures for a given point [39]. Inspired by RPM-Net [40] using the PPF encoding method [41] to further enhance the learning of local spatial relationships in point clouds, this article builds a local spatial feature encoding module, as shown in Figure 2, to comprehensively understand and capture spatial relationships in the local context as much as possible.

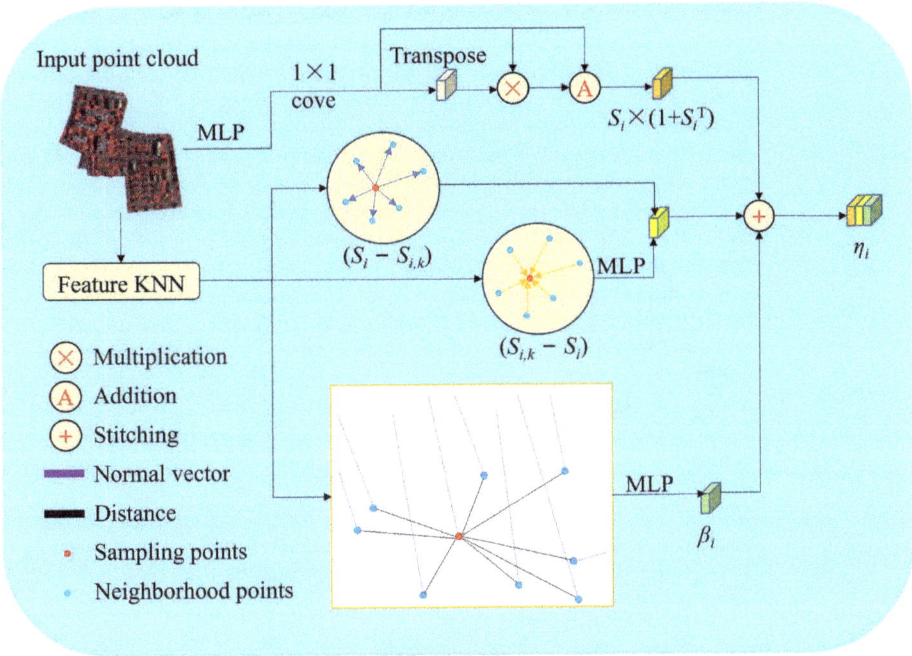

Figure 2. Multi-Spatial Feature Encoding module.

In Figure 2, the input point cloud is denoted as $V = \{n_i \mid i = 1, \ldots, N\}$, where N is the number of points, and $n_i = [v_i, r_i] \in \mathrm{R}^{3+d}$ represents the combination of coordinate information v_i and attribute information r_i. To save computational costs, point cloud V is first subjected to feature extraction using the PointNet encoding method [17]. Then, in the feature space, neighborhood point sampling is performed using k-nearest neighbors (KNN). On this basis, the neighboring points $v_{i,k}$ that are searched are more likely to belong to the same object category as the central point v_i or are on the edges between categories. Therefore, obtaining neighborhood points through feature space KNN helps the network learn the feature information of the sampled points' categories while increasing the distinctiveness of inter-class features. However, the feature encoding method of PointNet has limited capabilities in representing point cloud topological structures and spatial scales.

To further enhance the network's learning capabilities for point clouds, spatial feature encoding is applied to the sampled points and their neighborhood points using the following formula:

$$\eta_i = \alpha_i \oplus \beta_i \tag{1}$$

$$\alpha_i = S_i \times (1 + S_i^{\mathrm{T}}) + \textstyle\sum^{K}{}_{k=1,\, k\neq i} (S_i - S_{i,k}) \times [-\mathrm{MLP}(S_{i,k} - S_i)],\ S_i = \mathrm{P}(v_i),\ S_{i,k} = \mathrm{P}(v_{i,k}) \tag{2}$$

$$\beta_i = \mathrm{MLP}(v_{i,k} \oplus \sqrt{(v_i - v_{i,k})^2} \oplus (v_i - v_{i,k}) \oplus F(v_i, v_{i,k})) \tag{3}$$

$$F(v_i, v_{i,k}) = (\angle(m_i,(v_i - v_{i,k})) \oplus \angle(m_{i,k},(v_i - v_{i,k})) \oplus \angle(m_i, m_{i,k})) \tag{4}$$

In Formula (1), α_i and β_i correspond to the dual-direction feature encoding of the sampling center and neighborhood spatial structure features mentioned in Figure 2. In Formula (2), P(·) generates the sampling point features S_i and neighborhood point features $S_{i,k}$ based on the PointNet encoding method [17] and then incorporates them into α_i for enhancing the sampling point features. The calculation formula for α_i consists of three parts: 1. $S_i \times (1 + S_i^{\mathrm{T}})$ represents self-enhancement of the sampling point features. 2 $(S_i - S_{i,k})$ signifies the mutual relationship between each neighborhood point feature and the sampling point feature. They are multiplied and accumulated together to achieve the learning of enhanced sampling point features. 3. $-\mathrm{MLP}(S_{i,k} - S_i)$ first calculates the impact factors of each neighborhood point feature on the sampling point feature and projects them through a multi-layer perceptron. Formula (1) can be analyzed from a force field perspective, where each $S_{i,k}$ in the local space exerts a force on S_i. Gravity attempts to pull S_i closer to $S_{i,k}$ while repulsion pushes them apart. The strength of the force is determined by $-\mathrm{MLP}(S_{i,k} - S_i)$, and the direction is determined by $(S_i - S_{i,k})$. They adaptively learn through the difference between the two feature vectors. Therefore, α_i fully integrates the interrelationship between each neighborhood point and the sampling point, which can better describe the feature of neighborhood correlation. Formula (3) performs feature encoding based on the Euclidean distance $\sqrt{(v_i - v_{i,k})^2}$ between the sampling point and neighborhood point, the directional vector $(v_i - v_{i,k})$, the 4D point pair feature $F(v_i, v_{i,k})$), and the spatial positional information of the neighborhood points. Formula (4) calculates $\mathrm{F}(v_i, v_{i,k})$) using the 4D point pair feature encoding method from RPM-Net [42]. In this context, m_i and $m_{i,k}$ represent the normal vectors of the sampling point and neighborhood point, and the inverse trigonometric function $\angle(\cdot,\cdot)$ is used to calculate the angles between various vectors. Through Formulas (1) to (4), we not only consider the interactions between points but also describe the scale and topological structure of the sampling point's spatial environment through point distances, point normal vectors, and their angles.

2.3. Multi-Head Attention Pooling

Existing networks commonly utilize max pooling to aggregate neighborhood features for generating global feature vectors with larger receptive fields [18]. It is noteworthy that the information transmission capacity of max pooling is not only limited by the size of the pooling window but also involves a non-parametric downsampling process that results in the loss of a significant amount of information. The literature [43,44] introduces attention mechanisms to score features and aggregates them based on their importance, thereby enhancing the network model's ability to transmit local fine-grained structural information. Furthermore, the literature [45] embeds the Transformer model into point cloud semantic segmentation networks to improve the network's ability to capture dependencies between local point clouds and efficiently transmit feature information. Inspired by the above literature, this paper introduces a multi-head attention mechanism during the pooling stage to enhance the network model's capability to capture local salient structures from various samples. The overall structure is illustrated in Figure 3.

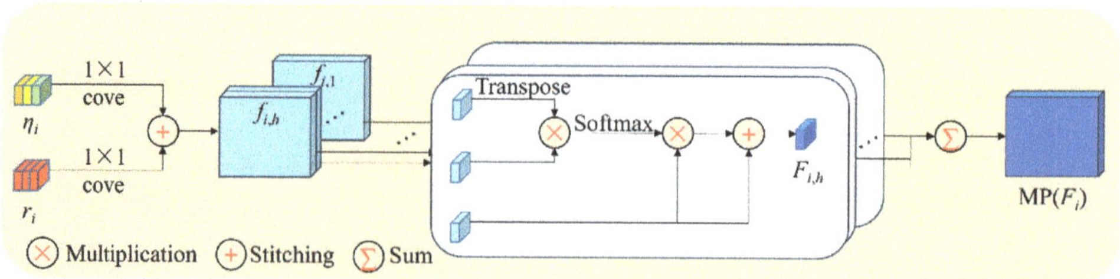

Figure 3. Multi-head attention pooling module.

We concatenate multi-spatial features with their corresponding attribute features, and after passing through multiple convolutional layers, we can obtain the following multi-head attention pooling results:

$$MP(F_i) = \sum(F_{i,1}, F_{i,2}, \ldots, F_{i,h})\cdot H_i \quad (5)$$

In Formula (5), $\sum(\cdot)$ denotes the concatenation of information $F_{i,h}$ learned from different heads of attention mechanisms, followed by fusion using the learned parameters H_i from the network. The computation process for each head of attention mechanism is derived from its self-attention scores and self-feature aggregation, expressed by the following formula:

$$F_{i,h} = [\text{SoftMax}((f_{i,h}{}^{T} \times f_{i,h})/\sqrt{C}) + 1] \times f_{i,h}, f_{i,h} = [g(\eta_i)\oplus g(r_i)]_h \quad (6)$$

In Formula (6), SoftMax refers to the normalized exponential function, C is the number of output channels, $g(\cdot)$ is a 1×1 convolution, and $[\cdot]_h$ represents the feature division according to h heads. In comparison to the max-pooling downsampling output pattern that retains predominant features, the pooling method in this paper not only utilizes attention mechanisms to emphasize fine-grained features of the point cloud's spatial structure but also reduces the loss of various sample features during information transmission through the aggregation of features based on multi-head attention scores.

3. Results

3.1. Experimental Environment and Evaluation

The proposed network is deployed on a deep learning workstation with NVIDIA GPU TiTAN XP 12G, Ubuntu 18.04 operating system and PyTorch1.10.0. The key parameters for the network were set as follows: batch size = 16, momentum = 0.9, decay steps = 300,000, decay rate = 0.5, optimizer: Adam, learning rate = 0.001, max epoch = 100, point number = 4096, the number of KNN = 32, and the radius of KNN = 0.1×2^n ($n \in [0, 3]$). The performance evaluation of the network in this study was conducted using three metrics: balanced F score (F_1 score), mean of class-wise intersection over union (MIoU), and overall point-wise accuracy (OA). The specific formulas for calculating these metrics are as follows:

$$F_1 = 2p_{ii}/\sum^{k}{}_{j=0} (p_{ij} + p_{ji}), \text{MIoU} = (1/k)\sum^{k}{}_{i=0} p_{ii}/(\sum^{k}{}_{j=0} p_{ij} + \sum^{k}{}_{j=0} p_{ji} - p_{ii}), \text{OA} = p_{ii}/p \quad (7)$$

In the above equations, 'k' represents the number of classes in the dataset. 'p_{ii}' stands for the number of point clouds correctly predicted for class 'i'; 'p_{ij}' represents the number of point clouds belonging to class 'j' but predicted as class 'i', while 'p_{ij}' represents the number of point clouds belonging to class 'i' but predicted as class 'j'. The F_1 and MIoU metrics produce values within the range of 0 to 1, with values closer to 1 indicating better segmentation results for class 'i'. On the other hand, OA is an overall segmentation evaluation metric for the model. It calculates the ratio of correctly labeled point clouds to the total number of point clouds in the model, where 'p' represents the total number of points in the point cloud model. This section may be divided by subheadings. It should provide a concise and precise description of the experimental results, their interpretation, as well as the experimental conclusions that can be drawn.

3.2. Semantic Segmentation of S3DIS Dataset

In this section, we conducted experiments to validate the effectiveness of PointMM using the publicly available 3D point cloud semantic segmentation indoor dataset, S3DIS. The S3DIS dataset comprises six areas from three different buildings, totaling 271 individual rooms. In each scene, every point corresponds to a fixed label, and these labels belong to 13 different categories such as ceiling, floor, wall, door, and others. The distribution of point clouds for each category within areas 1 to 5 is presented in Table 1.

In Table 1, the categories "ceiling", "floor", and "wall" constitute the majority class samples, while "clutter" represents the intermediate class samples (just slightly more than the sample mean but less than the majority class samples). The remaining categories belong to the minority class samples. Within the minority class samples, there are five categories "window", "column", "beam", "board", and "sofa" with an extremely low number of point clouds. Therefore, the segmentation task based on the S3DIS dataset not only faces challenges related to large data volume and high scene complexity but also involves an extremely imbalanced long-tailed distribution issue.

Table 1. Aera1~5 dataset introduction (%).

Class	Number	Proportion	Class	Number	Proportion
ceiling	5,721,636	21.6	table	715,205	2.7
floor	5,138,877	19.4	chair	953,606	3.6
wall	6,887,155	26.0	sofa	105,956	0.4
beam	317,869	1.2	Bookcase	1,456,898	5.5
column	397,336	1.5	board	264,890	1.0
window	529,781	2.0	clutter	2,595,927	9.8
door	1,403,920	5.3	All	26,489,056	100

3.2.1. Ablation Experiment

We aim to validate the effectiveness of the modules proposed in this paper. Point spatial coordinates along with their RGB information are used as input features to the network. For training samples, regions 1 to 5 of the dataset are utilized. Specifically, experiments were conducted based on PointNet++ with the addition of multi-head attention pooling (+MHP), multi-spatial feature encoding (+MSF), and a comprehensive evaluation of all modules combined, as shown in Table 2. Additionally, Table 3 presents the segmentation results of these modules in region 6. Meanwhile, the training time of each module during a single epoch is shown in Table 4.

Table 2. Each module introduction.

Name	Module
PointNet++	Baseline
+MHP	Multi-head attention pooling
+MSF	Multi-spatial feature encoding
ALL	PointMM

Table 3. Segmentation results of each module on the S3DIS dataset (Area-6) (%).

Module	MIoU	OA	Ceiling	Floor	Wall	Beam	Column	Window	Door	Table	Chair	Sofa	Bookcase	Board	Clutter
Baseline	70.2	87.7	93.0	97.3	74.8	68.7	43.2	77.8	78.9	72.4	76.8	41.9	58.7	66.2	63.2
MHP	73.3	90.7	91.4	97.9	76.9	68.0	46.5	72.6	79.2	75.3	83.6	63.2	64.7	65.3	67.8
MSF	78.0	92.7	93.3	97.2	80.6	76.4	59.5	73.5	83.8	74.5	83.5	76.8	68.5	77.0	69.7
ALL	80.4	94.0	94.6	97.8	82.7	76.2	52.9	77.5	83.6	77.8	86.6	83.7	79.1	76.9	75.8

Table 4. The training of each module per epoch (seconds).

Module	Training Duration for One Epoch
Baseline	233.3703
+MHP	1104.0414
+MSF	681.2939
ALL	1604.5551

Table 3 indicates that, compared to the baseline MIoU (approximately 70.2%), when the model only considers the MHP, the segmentation accuracy of most categories improves, except for ceiling, beam, board, and window. The reason lies in the fact that the baseline, using max-pooling modules for downsampling and feature transmission, results in the loss of a considerable amount of detailed information during the network training process. As a result, the network tends to sacrifice the segmentation accuracy of minority classes to ensure the overall segmentation accuracy with a majority class bias. The MHP module captures feature information at different levels through a multi-head attention mechanism and aggregates the features extracted by the network through weighted pooling, ensuring the complete preservation of feature information for various samples. On the other hand, the combination of the MSF module with the baseline leads to improvements of 7.9% and 5% in MIoU and OA, respectively. This demonstrates that the MSF module's ability to search for similar neighborhoods, learn the salient structural features of sampled point neighborhoods, and transmit crucial information is superior to the baseline. When both modules are loaded onto the baseline, except for the beam, column, window, door, and board, which did not achieve the best results, the segmentation accuracy for all other categories is optimal. The overall segmentation accuracy and MIoU also achieve the best results at 94% and 80.4%, respectively. Among the five categories that did not achieve the best results, only the accuracy of the column fluctuates the most, with the differences for the other four categories being only 0.1–0.3% from the optimal accuracy. This is because the column is spatially close to the wall, and their structures and spectral features are highly similar. On the other hand, the column surface is usually relatively smooth and structurally simple, with corresponding point cloud coordinates being relatively regular and a strong spectral feature consistency. The multi-head attention mechanism for modeling the geometric multi-spatial features of the target space does not achieve significant improvement in the accuracy of point clouds with regular arrangement (simple structure). This ultimately leads to confusion between the two in the neighborhood point search and feature learning stages. It should be noted that the wall belongs to the majority of targets, so its accuracy is not easily disturbed by the column, while the column belongs to the minority class targets, so its accuracy fluctuates more significantly. Usually, it is challenging for a semantic segmentation CNN to achieve optimal OA and MIoU simultaneously, as it tends to sacrifice minority class targets to achieve the overall optimal segmentation accuracy (OA). On the other hand, focusing on the learning features of minority class targets may lead to overfitting and limit overall segmentation accuracy. The PointMM in this article achieved an acceptable balance on the IoU of various class samples, while improving overall accuracy by 6.3%.

It is worth noting that the MSF module fully learns the local fine-grained structural features of the diluted point cloud from two aspects: the inter-point relationship α_i and the neighborhood spatial topology structure β_i. Meanwhile, the MHP module scores and aggregates features based on different heads of attention, allowing the network to consolidate the segmentation accuracy of the majority class targets while also considering learning minority class targets. On the other hand, according to Table 4, the training time for each epoch in the baseline is the shortest, only 233 s. Due to the more complex feature encoding in the MSF module, its duration is almost three times longer than the baseline. At the same time, as the number of downsampling layers increases, the computational complexity of the MHP module increases exponentially, resulting in a duration of 1104 s. When both modules are stacked on the baseline, PointMM shows the maximum duration (1604 s).

To demonstrate the effects of the ablative experiments more intuitively on each module in this paper, segmentation results from three different scenes in region 6 are selected for display, as shown in Figure 4. The three columns of segmentation results in Figure 4, from left to right, correspond to lounge, hallway, and office. The gray boxes in each image indicate areas of segmentation errors for comparison. Each row in Figure 4, from top to bottom, represents the segmentation results of the baseline, baseline with the MHP module, baseline with the MSF module, PointMM, and ground truth. Observing the images on the left side of Figure 4, it can be observed that due to the significant similarities in geometric structure, spatial location, and spectral information between wall and column, door, clutter, and window, the baseline misclassifies wall as door, window, and column. MHP, through multi-head attention pooling, fully preserves the features of various samples, correctly segmenting the wall at the corner of the room, but still missegments some parts of the wall as door and window. This is because MHP can only ensure the effective transmission of various sample information by pooling, but cannot extract significant features of local geometric structures. MSF, based on the original data preprocessing, effectively captures fine-grained structural features of points in space, greatly reducing the phenomenon of missegmenting the wall as other targets. However, MSF still missegments a small portion of the wall as clutter and door. PointMM, which combines the advantages of MHP and MSF, essentially achieves the correct segmentation of the wall, with only a small portion of the point cloud missegmented as a door at the corner of the two wall surfaces. On the other hand, in the left gray box of the baseline, there is also mutual missegmentation between sofa, table, and clutter. With the integration of each module, the segmentation accuracy in this local area gradually improves. For the various types of targets in the right gray box with sparse distribution or extremely low data volume, the baseline can only correctly segment some chairs, while the rest of the categories are segmented incorrectly. MHP, based on the baseline, achieves the correct segmentation of tables and clutter. MSF, based on the baseline, achieves the correct segmentation of chairs as much as possible. PointMM, based on MHP and MSF, completes the correct segmentation of all targets, with only a small amount of missegmentation in the edge area.

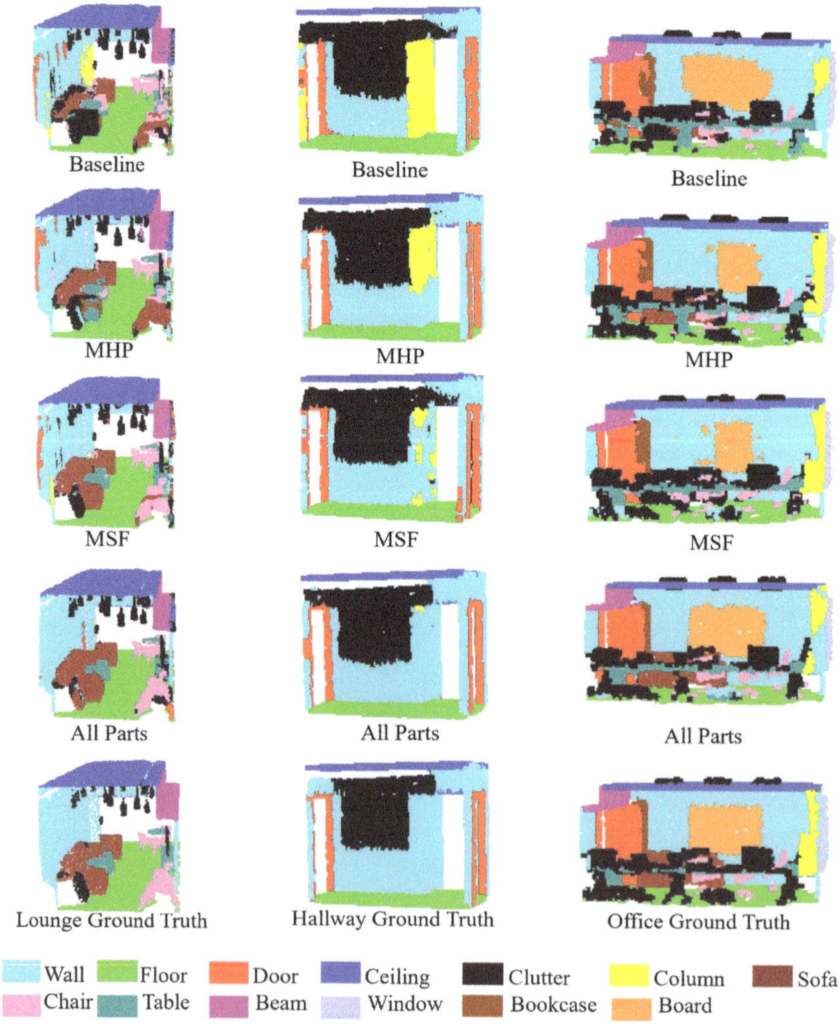

Figure 4. Segmentation results of each module.

In the hallway scene depicted in Figure 4, the wall, door, and clutter exhibit highly similar spectral information, contour structures, and spatial positions. The baseline missegments a significant amount of the wall in the left gray box as column, clutter, and door, while completely missegmenting the wall in the right gray box as column. MHP, employing the original feature encoding approach, learns various sample information, significantly reducing the missegmentation of the wall as column in the left gray box, while correctly segmenting half of the wall in the right gray box. MSF comprehensively learns the geometric relationships between sampled points and their neighboring points, leading to a substantial reduction in the missegmentation of the wall as column in the right gray box. PointMM, on the other hand, is capable of accurately identifying the aforementioned targets, achieving segmentation results highly consistent with the annotated data. PointMM only exhibits a small amount of missegmentation in the region where door, wall, and clutter intersect (left gray box), as well as an extremely small amount of missegmentation as column in the corner formed by the two wall surfaces.

In the third column of Figure 4, an office scene containing 13 categories is depicted. Due to the close connection between the open door and the bookcase, both of which are wooden structures with approximate spectral information, the baseline exhibits missegmentation at the border junction of these two objects. Similarly, the baseline missegments the wall as a board and missegments the column as a wall. MHP and MSF both show varying degrees of missegmentation between the door and bookcase in the left gray box, with both also missegmenting some boards as walls. PointMM achieves segmentation results close to the ground truth in the region where the door meets the bookcase and in the board area, except for missegmenting some columns as walls in the right gray box. The experimental results in Figure 4, combined with the segmentation accuracy from Table 3, reveal that the combination of multi-head attention pooling and the adaptive spatial feature encoding module significantly enhances the model's ability to describe features of various sample types. Additionally, PointMM proves effective in handling targets with complex local geometric or spectral features. On the other hand, the S3DIS dataset contains instances of objects of the same class sparsely and discretely distributed in the scene. In this context, the introduced neighborhood point search module based on feature KNN demonstrates clear advantages in capturing the ability of the same class point clouds. By integrating various amounts of sample information through feature KNN and thoroughly learning their neighborhood salient structural features, the network model's semantic segmentation capability is effectively improved under conditions of sparse point cloud density and complex local structures.

3.2.2. Six-Fold Cross-Validation

This section of the experiment aims to demonstrate the learning capability and generalization of the method proposed in this paper on the entire dataset. The proposed method is subjected to a standard six-fold cross-validation experiment on the S3DIS data set, and it is compared with 12 currently popular and classical deep learning methods for point cloud semantic segmentation. The evaluation metrics for each method, including overall accuracy (OA) and mean intersection over union (MIoU), are presented in Table 5.

From Table 5, it can be observed that the proposed method achieves the highest MIoU for ceiling, floor, window, table, chair, and clutter, with values of 95.4%, 97.5%, 66.5%, 73.0%, 84%, and 69.5%, respectively. These values are higher than the second highest by 0.9%, 0.2%, 0.3%, 2.2%, 7.6%, and 9.2%. The MIoU for door and bookcase ranks second, with values of 73.9% and 68.1%, lower than the first by 2.7% and 6.8%, respectively. Wall ranks third in MIoU, while beam's MIoU ranks fifth, and column, sofa, and board all rank sixth, placing them at a moderate level among the listed literature network models.

GSIP [46] proposed a method based on PointNet that performs downsampling on a per-room basis, significantly reducing computational costs. However, this network loses a considerable amount of detailed information, resulting in an OA and MIoU of only 79.8% and 48.5%, respectively. HPRS [21] has a feature encoding pattern that is too singular, limiting its applicability to large-scale complex indoor scenes, resulting in an OA and MIoU of only 84.7% and 61.3%. MCS [22] introduced MappingConv based on the spherical neighborhood feature learning pattern, showing a noticeable improvement over HPRS in accuracy. However, this method only optimizes the feature encoding of the downsampling layer and does not consider the promoting effect of the self-attention mechanisms in deep learning, resulting in an OA and MIoU of only 86.8% and 66.8%.

Table 5. Semantic segmentation accuracy on S3DIS dataset.

Method	GSIP	HPRS	MCS	KVGCN	RGGCN	LG-Net	JSNet++	KPConv	RandLA-Net	BSH-Net	PointNAC	PointTr	Ours
OA	79.8	84.7	86.8	87.4	88.1	88.3	88.7	-	88.0	90.5	90.9	90.2	90.4
Miou	48.5	61.3	66.8	60.9	63.7	70.8	62.4	70.6	70.0	66.1	67.4	73.5	70.7
Ceiling	91.8	92.7	92.4	94.5	94.0	93.7	94.1	93.6	93.1	-	-	-	95.4
Floor	89.8	94.5	95.8	94.1	96.2	96.4	97.3	92.4	96.1	-	-	-	97.5
Wall	73.0	76.3	79.5	79.5	79.1	81.3	78.0	83.1	80.6	-	-	-	81.1
Beam	26.3	30.1	55.8	53.4	60.4	65.2	41.3	63.9	62.4	-	-	-	59.5
Column	24.0	25.5	43.6	36.3	44.3	51.8	32.2	54.3	48.0	-	-	-	38.8
Window	44.6	63.1	59.6	56.8	60.1	66.2	52.0	66.1	64.4	-	-	-	66.5
Door	55.8	61.8	63.4	63.2	65.9	69.7	70.0	76.6	69.4	-	-	-	73.9
Table	55.5	65.6	67.3	64.3	70.8	69.1	69.9	57.8	69.4	-	-	-	73.0
Chair	51.1	69.3	70.2	67.5	64.9	75.1	72.7	64.0	76.4	-	-	-	84.0
Sofa	10.2	47.0	63.1	54.3	30.8	63.9	37.9	69.3	60.0	-	-	-	53.3
Bookcase	43.8	56.1	59.3	23.6	51.9	63.5	54.1	74.9	64.2	-	-	-	68.1
Board	21.8	60.1	61.8	43.1	52.6	66.0	51.3	61.3	65.9	-	-	-	58.6
Clutter	43.2	55.1	56.2	53.2	56.4	58.4	60.2	60.3	60.1	-	-	-	69.5

KVGCN [25] aggregated local–global context features to achieve a higher OA (87.4%) than GCN. However, it overlooks the impact of minority class features on MIoU (60.9%). The OA (88.0%) of RandLA-Net [30] was only at a moderate level, even if a random sampling strategy was used to increase the chances of capturing minority class samples. Although KPConv [47] achieves the best segmentation accuracy for the two categories of sofa and bookcase with extremely low point cloud counts, it overly focuses on minority class sample features, leading the model into an overfitting state, causing a substantial decline in segmentation accuracy for ceiling and floor. While LG-Net [32] achieved good results in regions with high similarity for features such as column, beam, and wall, like KPConv, it overly focuses on certain features and leads to a loss in overall segmentation accuracy. Instead, RGGCN [27], BSH-Net, PointNAC, and JSNet++ [23] overly emphasize the features of majority class targets and lose competitiveness in MIoU. Point Transformer achieved the best MIoU (73.5%) and ranking fourth in OA (90.2%). Overall, the introduction of MSF in this paper addresses the dilution of majority class samples, thereby improving the feature extraction and learning efficiency of the network model for all samples. MHP assigns attention scores to features extracted by MSF at different levels (heads) and clusters various features based on attention scores. These two components enable PointNAC to achieve impressive performance, ranking third both in OA (90.4%) and MIoU (70.7%).

3.2.3. The Experiments of Sampling Points and Neighborhood Points

To further validate the feature learning capabilities of the proposed network at different sampling densities, this section conducts experiments with different numbers of sampled points, specifically 2048, 4096, 8192, 16,384, and 32,768 points. Additionally, we compare our PointNet++, RandLA-Net, and the proposed method, and the MIoU scores for each model are shown in Figure 5.

From Figure 5, it can be observed that the performance of RandLA-Net is entirely dependent on the density of the sampled points. When the point density is not higher than 4096, RandLA-Net's segmentation performance is significantly inferior to PointNet++ and the approach proposed in this paper, with a maximum MIoU of only 65.1%. In contrast, when the number of sampled points is 2048, the proposed method exhibits a remarkable improvement, surpassing PointNet++ by 10.2%. Even when the number of sampled points is increased to 32,768, the proposed method still achieves an improvement of 8.7%. This indicates that the network model in this paper can effectively learn features from sparse point clouds through the MSF module, while MHP emphasizes the importance of the main features in the feature pooling stage through attention scores. On the other hand, as the number of sampled points in RandLA-Net gradually increases, its network MIoU also grows, eventually reaching 78.5%. However, comparing RandLA-Net with the network proposed in this paper, it is evident that the MIoU difference for RandLA-Net within the sampled point range is 18.1%, while the difference for this paper's network is only 0.4%. This indicates that the network in this paper has a stronger feature learning capability on point clouds with uneven density distribution compared to RandLA-Net.

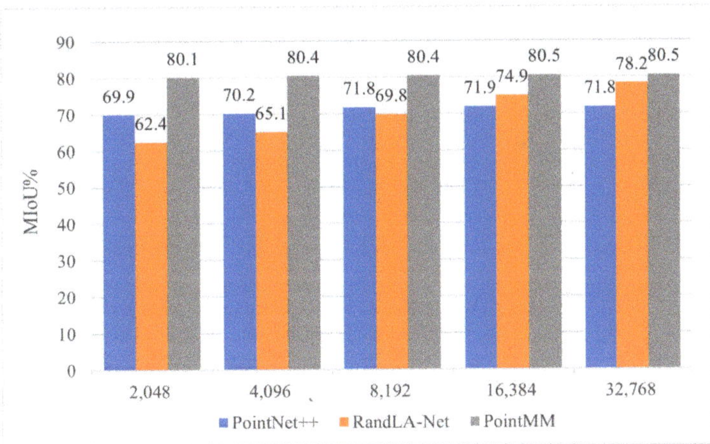

Figure 5. The MIoU of different sampling densities based on area 6.

To further investigate the influence of different numbers of neighboring points on the network's feature learning capability, this section conducts experiments using varying numbers of neighboring points, including 8, 16, 32, and 64. Additionally, we compare the OA of Euclidean k-nearest neighbors (KNN) and feature KNN, as shown in Figure 6. From Figure 6, it can be observed that the maximum difference in overall accuracy (OA) for feature KNN is 1.6%, while for spatial KNN it is 1.9%. Moreover, in terms of the segmentation accuracy with 32 neighboring points, feature KNN outperforms spatial KNN by 1.1%. This clearly demonstrates that the neighborhood points extracted by feature KNN are closer in category to their sampling center points, thereby enhancing the network model's ability to distinguish between points belonging to different classes.

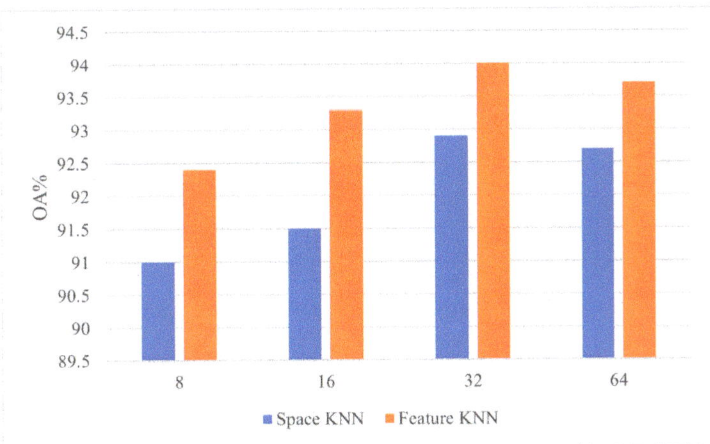

Figure 6. The OA of different neighborhood points based on area 6.

3.3. Semantic Segmentation of Vaihingen Dataset

The International Society for Photogrammetry and Remote Sensing (ISPRS) Vaihingen 3D Semantic Labeling Challenge dataset consists of five training areas and two testing areas. The dataset comprises a total of 1,181,017 points. The original 3D point cloud data is composed of nine categories of objects, including power line, car, facade, and hedge Each point within the dataset contains both 3D coordinates and RGB information. The distribution of points among these object categories, along with their respective proportions is presented in Table 6.

Table 6. Details of Vaihingen 3D dataset.

Model	Power Line	Car	Facade	Hedge	Impervious Surface	Low Vegetation	Roof	Shrub	Tree
Training-N	546	4614	27,250	12,070	193,723	180,850	152,045	47,605	135,173
Training-P	0.072%	0.612%	3.615%	1.601%	25.697%	23.989%	20.168%	6.315%	17.931%
Testing-N	600	3708	11,224	7422	101,986	98,690	109,048	24,818	54,226
Testing-P	0.146%	0.900%	2.726%	1.803%	24.770%	23.970%	26.486%	6.027%	13.170%

From this table, it is evident that, the Vaihingen 3D dataset similar to the S3DIS dataset, it also exhibits a highly imbalanced long-tail distribution. Specifically, objects such as trees, building roofs, low vegetation, and road surfaces represent the majority class samples, while power lines, cars, and hedges are extremely rare minority class samples with very few points. Since the Vaihingen 3D dataset is a large-scale outdoor scene dataset, the minority class samples are highly likely to be lost during sub-area partitioning and FPS sampling. To address this issue, in the training data sampling phase, our network first performs full sampling for minority class point clouds, then downsamples the majority class point clouds, and finally employs the BCS module to assign values to point clouds of various categories. Additionally, in this section, we compare our method with 11 recently published outdoor point cloud semantic segmentation methods, using the F1 score and OA as standard metrics for all categories, as shown in Table 7.

Table 7. Segmentation effects of different methods (%).

Model	Power Line	Car	Facade	Hedge	Impervious Surface	Low Vegetation	Roof	Shrub	Tree	OA	Average F_1
HDA	64.2	68.9	36.5	19.2	99.2	85.1	88.2	37.7	69.2	81.2	63.1
DPE	68.1	75.2	44.2	19.5	99.3	86.5	91.1	39.4	72.6	83.2	66.2
NANJ2	62.0	66.7	42.6	40.7	91.2	88.8	93.6	55.9	82.6	85.2	69.3
BSH-NET	46.5	77.8	57.9	37.9	92.9	82.3	94.8	48.6	86.3	85.4	69.5
PointNAC	52.9	76.7	57.5	41.1	93.6	83.2	94.9	50.5	85.2	85.9	70.6
Randla-Net	68.8	76.6	61.9	43.8	91.3	82.1	91.1	45.2	77.4	82.1	70.9
D-FCN	70.4	78.1	60.5	37.0	91.4	80.2	93.0	46.0	79.4	82.2	70.7
Dance-Net	68.4	77.2	60.2	38.6	92.8	81.6	93.9	47.2	81.4	83.9	71.2
GACNN	76.0	77.7	58.9	37.8	93.0	81.8	93.1	46.7	78.9	83.2	71.5
GANet	75.4	77.8	61.5	44.2	91.6	82.0	94.4	49.6	82.6	84.5	73.2
GraNet	67.7	80.9	62.0	51.1	91.7	82.7	94.5	49.9	82.0	84.5	73.6
PointMM	60.6	77.3	62.3	37.0	93.5	84.0	96.1	57.8	86.4	87.7	72.7

From Table 7, it is evident that, compared to other network models on the Vaihingen 3D dataset, PointMM achieves the best OA, ranks third in average F1 score, with a difference of only 0.9% from the top average F1 score. The proposed method excels in the segmentation accuracy of the façade, roof, shrub, and tree categories, with only a lower segmentation accuracy for power line and hedge. One reason for this is the extremely sparse point cloud count and low geometric feature saliency of these two classes. For instance, the power line consists of sporadic non-continuous line segments distributed on the roof, resembling outliers similar to the roof. As a result, PointMM is likely to confuse power line with the roof during the feature KNN step. However, employing an encoding method with the capability of extracting local fine-grained features in the feature KNN stage, as GACNN [44] does, would not only increase computational costs but also focus too much on extremely scarce minority class targets, restricting the overall OA (83.2%).

Nevertheless, methods such as DPE [48] and HAD [49] sacrifice the segmentation accuracy of other land cover types to enhance the segmentation accuracy of majority classes, particularly impervious surfaces. Their performance in OA and average F1 score needs improvement. NANJ2 [8] projects 3D point clouds onto 2D images and utilizes a mature CNN network to learn target features. This method effectively improves the segmentation accuracy of hedge, low vegetation, and shrub. However, the process of multi-view projection results in the loss of a significant amount of local spatial structure information, making it challenging to further improve the average F1 score (69.3%) and OA (85.2%). D-FCN [50], similar to the former, focuses on learning minority class targets, resulting in an improvement in the average F1 score (70.7%) but a loss in OA (82.2%).

While the random sampling of Randla-Net improves the network's ability to capture features of minority class samples, it hampers the model's comprehensive learning of majority class sample features, especially in the scenarios involving spatial overlap and high feature similarity among impervious surface, shrub, tree, and low vegetation. As a result, the overall segmentation accuracy is compromised, reaching only 82.1%. The learning ability of BSH-Net [34] for features of minority class samples is weak, resulting in an unsatisfactory average F1 score (69.5%). PointNAC builds upon the BSH-Net framework by introducing a 4D point pair feature encoding scheme, thereby enhancing the segmentation accuracy of the network. DANCE-Net [51] acknowledges the importance of elevation-remote features but has weak segmentation capabilities for hedge and shrub with overlapping low-level features. Therefore, this method fails to achieve further breakthroughs in OA (83.9%) and average F1 score (71.2%). GANet [52] and GraNet [42] introduce attention mechanisms on top of GCN to enhance the network's ability to learn local fine-grained structural features, obtaining the second and first average F1 scores, respectively.

Overall, for large-scale outdoor scenes with point cloud data, the proposed method not only effectively learns spatial scale information and intra-class semantic information for various samples through adaptive spatial feature encoding but also achieves a satisfying balance between OA and average F1 score by efficiently transmitting multi-level feature information through multi-head attention pooling. On the other hand, in Figure 7, we present the visualization results of PointNet++, NANJ2, BSH-Net, and the proposed method.

In Figure 7, the first row of images shows the segmentation results of ground truth and the four methods in the testing area. The second to sixth rows display visualizations of local areas, with segmentation errors marked by red circles. Observing the images in the first column of Figure 7, it can be seen that, except for PointMM, the other methods all to some extent misclassify car as roof. Additionally, except for PointMM, the other methods misclassify facade points as tree and roof, while PointMM only misclassifies a small portion of facade points as roof and tree. This strongly indicates that our method outperforms the other three methods in terms of the selection of sampling center points and their neighborhood points, as well as feature learning capabilities.

Comparing the images in the second column of Figure 7 with the data in Table 7, it can be observed that only NANJ2 and PointMM correctly segment the power line within the left red box. The right red box contains roof, low vegetation, and tree. In this context, our method's segmentation results closely resemble the ground truth dataset. However, BSH-Net misclassifies the roof as a car, NANJ2 misclassifies low vegetation as impervious surface, and PointNet++ exhibits all of the above-mentioned misclassification cases. This demonstrates the effectiveness of our method in learning the spatial scale, positional information, and neighborhood relationships of the point clouds.

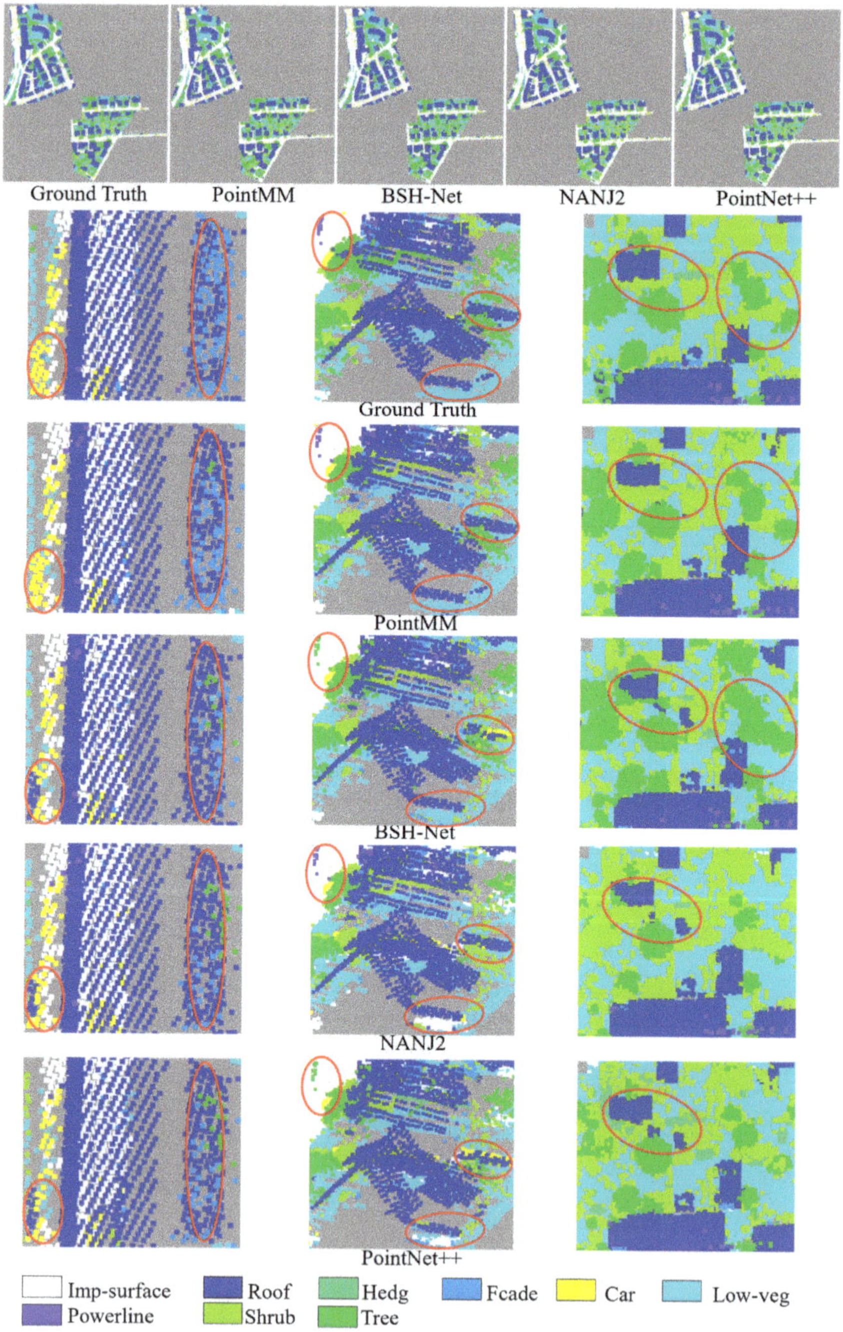

Figure 7. Segmentation results of different methods. (The red circle represents the incorrectly segmented area).

Further examination of the images in the third column of Figure 7 reveals that this area is mainly composed of three categories of low-level features: low vegetation, tree, and shrub. These features are similar and spatially close to each other. In the left red box, only PointMM incorrectly classifies a few parts of tree as shrub, while the other methods misclassify some shrub as roof. On the other hand, in the right red box, all four methods misclassify some shrub as roof. However, PointMM correctly segments tree for the most part, while BSH-NET completely misclassifies shrub as tree, and NANJ2 and PointNet++ misclassify half of the tree as shrub. Overall, our method achieves good segmentation performance on the Vaihingen 3D semantic segmentation dataset and maintains consistency with the ground truth in areas with overlapping and stacked features of various land cover types.

4. Conclusions

Although the PointNet++ series of networks consider information about sampled points and their neighborhoods, as well as local–global context information, they often lack attention to the topological structure information of the categories to which the sampled points belong. The proposed PointMM overcomes these limitations by extensively leveraging the topology information of the category to which the sampled points belong through feature KNN. It searches for neighborhood points belonging to the same category as the sampled point, focusing on more detailed spatial relationships, scales, and coordinate information. Additionally, the use of multi-head attention pooling ensures the maximal preservation of features for various sample points. This method effectively enhances the network's ability to learn fine-grained features of various sample categories from complex scenes. Compared to the literature mentioned in this paper, although PointMM achieved the best OA, the second-best MIoU, and the third-best average F1 score on both the indoor S3DIS dataset and the outdoor Vaihingen 3D dataset, it requires high computation and longer training time. Theoretically, adding the multi-head attention mechanism to the multi-spatial feature encoding module will help extract more accurate features from intra-class neighborhood points, which has not been discussed in this article. Future work will delve into this topic and test the proposed network on a larger scale and in more scenarios.

Author Contributions: Conceptualization, J.W.; methodology, J.W.; software, R.C. and Y.L.; validation, J.W.; formal analysis, Y.L.; investigation, G.X.; resources, R.C.; data curation, R.C.; writing—original draft preparation, R.C. and J.W.; writing—review and editing, R.C. and J.W.; visualization, Y.L.; supervision, J.W.; project administration, R.C. and G.X.; funding acquisition, J.W. and G.X. All authors have read and agreed to the published version of the manuscript.

Funding: The Natural Science Foundation of China under Grant: 42361071 (Funder: Jun Wu); Ningbo Science and Technology Innovation Project under Grant: 2023Z016 (Funder: Gang Xu); Innovation Project of Guangxi Graduate Education under Grant: YCBZ2023136 (Funder: Ying Luo); National Key Research and Development Program of China under Grant: 2023YFB4607000 (Funder: Gang Xu).

Data Availability Statement: The Stanford Large-Scale 3D Indoor Spaces (S3DIS) data set can be found at: http://buildingparser.stanford.edu/dataset.html (accessed on 7 February 2024) The ISPRS Vaihingen data set can be found at: https://www.isprs.org/education/benchmarks/UrbanSemLab/Default.aspx (accessed on 7 February 2024).

Conflicts of Interest: The authors declare no conflicts of interest.

References

1. Zhang, J.; Xie, H.; Zhang, L.; Lu, Z. Information Extraction and Three-Dimensional Contour Reconstruction of Vehicle Target Based on Multiple Different Pitch-Angle Observation Circular Synthetic Aperture Radar Data. *Remote Sens.* **2024**, *16*, 401. [CrossRef]
2. Jiang, Z.; Zhang, Y.; Wang, Z.; Yu, Y.; Zhang, Z.; Zhang, M.; Zhang, L.; Cheng, B. Inter-Domain Invariant Cross-Domain Object Detection Using Style and Content Disentanglement for In-Vehicle Images. *Remote Sens.* **2024**, *16*, 304. [CrossRef]
3. Caciora, T.; Jubran, A.; Ilies, D.C.; Hodor, N.; Blaga, L.; Ilies, A.; Grama, V.; Sebesan, B.; Safarov, B.; Ilies, G.; et al. Digitization of the Built Cultural Heritage: An Integrated Methodology for Preservation and Accessibilization of an Art Nouveau Museum. *Remote Sens.* **2023**, *15*, 5763. [CrossRef]
4. Muumbe, T.P.; Singh, J.; Baade, J.; Raumonen, P.; Coetsee, C.; Thau, C.; Schmullius, C. Individual Tree-Scale Aboveground Biomass Estimation of Woody Vegetation in a Semi-Arid Savanna Using 3D Data. *Remote Sens.* **2024**, *16*, 399. [CrossRef]
5. Yu, H.; Yang, Z.; Tan, L.; Wang, Y.; Sun, W.; Sun, M.; Tang, Y. Methods and datasets on semantic segmentation: A review. *Neurocomputing* **2018**, *304*, 82–103. [CrossRef]

6. Guo, Y.; Wang, H.; Hu, Q.; Liu, H.; Liu, L.; Bennamoun, M. Deep learning for 3d point clouds: A survey. *IEEE Trans. Pattern Anal. Mach. Intell.* **2020**, *43*, 4338–4364. [CrossRef] [PubMed]
7. Yang, Z.; Tan, B.; Pei, H.; Jiang, W. Segmentation and multi-scale convolutional neural network-based classification of airborne laser scanner data. *Sensors* **2018**, *18*, 3347. [CrossRef]
8. Zhao, R.; Pang, M.; Wang, J. Classifying airborne LiDAR point clouds via deep features learned by a multi-scale convolutional neural network. *Int. J. Geogr. Inf. Sci.* **2018**, *32*, 960–979. [CrossRef]
9. Gerdzhev, M.; Razani, R.; Taghavi, E.; Bingbing, L. Tornado-net: Multiview total variation semantic segmentation with diamond inception module. In Proceedings of the 2021 IEEE International Conference on Robotics and Automation (ICRA), Xi'an, China, 30 May–5 June 2021; IEEE: Piscataway, NJ, USA, 2021; pp. 9543–9549.
10. Qiu, H.; Yu, B.; Tao, D. GFNet: Geometric Flow Network for 3D Point Cloud Semantic Segmentation. *arXiv* **2022**, arXiv:2207.02605
11. Jing, W.; Zhang, W.; Li, L.; Di, D.; Chen, G.; Wang, J. AGNet: An attention-based graph network for point cloud classification and segmentation. *Remote Sens.* **2022**, *14*, 1036. [CrossRef]
12. Lee, M.S.; Yang, S.W.; Han, S.W. Gaia: Graphical information gain based attention network for weakly supervised point cloud semantic segmentation. In Proceedings of the IEEE/CVF Winter Conference on Applications of Computer Vision, Waikoloa, HI, USA, 2–7 January 2023; pp. 582–591.
13. Liu, Z.; Tang, H.; Lin, Y.; Han, S. Point-voxel cnn for efficient 3d deep learning. *Adv. Neural Inf. Process. Syst.* **2019**, *32*. [CrossRef]
14. Wang, Z.; Lu, F. *VoxSegNet: Volumetric CNNs for Semantic Part Segmentation of 3D Shapes*; Institute of Electrical and Electronics Engineers (IEEE): Piscataway, NJ, USA, 2020. [CrossRef]
15. Liu, M.; Zhou, Q.; Zhao, H.; Li, J.; Du, Y.; Keutzer, K.; Du, L.; Zhang, S. Prototype-Voxel Contrastive Learning for LiDAR Point Cloud Panoptic Segmentation. In Proceedings of the 2022 International Conference on Robotics and Automation (ICRA), Philadelphia, PA, USA, 23–27 May 2022; IEEE: Piscataway, NJ, USA, 2022; pp. 9243–9250.
16. Zhou, W.; Zhang, X.; Hao, X.; Wang, D.; He, Y. Multi point-voxel convolution (MPVConv) for deep learning on point clouds *Comput. Graph.* **2023**, *112*, 72–80. [CrossRef]
17. Qi, C.R.; Su, H.; Mo, K.; Guibas, L.J. PointNet: Deep Learning on Point Sets for 3D Classification and Segmentation. In Proceedings of the 2017 IEEE Conference on Computer Vision and Pattern Recognition (CVPR), Honolulu, HI, USA, 21–26 July 2017; IEEE: Piscataway, NJ, USA, 2017.
18. Qi, C.R.; Yi, L.; Su, H.; Guibas, L.J. PointNet++: Deep Hierarchical Feature Learning on Point Sets in a Metric Space. *Adv. Neural Inf. Process. Syst.* **2017**. [CrossRef]
19. Jiang, M.; Wu, Y.; Zhao, T.; Zhao, Z.; Lu, C. Pointsift: A sift-like network module for 3d point cloud semantic segmentation. *arXiv* **2018**, arXiv:1807.00652.
20. Zhao, H.; Jiang, L.; Fu, C.W.; Jia, J. Pointweb: Enhancing local neighborhood features for point cloud processing. In Proceedings of the IEEE/CVF Conference on Computer Vision and Pattern Recognition, Long Beach, CA, USA, 16–17 June 2019; pp. 5565–5573.
21. Su, Z.; Zhou, G.; Luo, F.; Li, S.; Ma, K.K. Semantic Segmentation of 3D Point Clouds Based on High Precision Range Search Network. *Remote Sens.* **2022**, *14*, 5649. [CrossRef]
22. Yan, K.; Hu, Q.; Wang, H.; Huang, X.; Li, L.; Ji, S. Continuous mapping convolution for large-scale point clouds semantic segmentation. *IEEE Geosci. Remote Sens. Lett.* **2021**, *19*, 1–5. [CrossRef]
23. Zhao, L.; Tao, W. Jsnet++: Dynamic filters and pointwise correlation for 3d point cloud instance and semantic segmentation. *IEEE Trans. Circuits Syst. Video Technol.* **2022**, *33*, 1854–1867. [CrossRef]
24. Zhao, L.; Tao, W. JSNet: Joint instance and semantic segmentation of 3D point clouds. In Proceedings of the AAAI Conference on Artificial Intelligence, New York, NY, USA, 7–12 February 2020; Volume 34, pp. 12951–12958.
25. Luo, N.; Yu, H.; Huo, Z.; Liu, J.; Wang, Q.; Xu, Y.; Gao, Y. KVGCN: A KNN searching and VLAD combined graph convolutional network for point cloud segmentation. *Remote Sens.* **2021**, *13*, 1003. [CrossRef]
26. Wang, Y.; Zhang, Z.; Zhong, R.; Sun, L.; Leng, S.; Wang, Q. Densely connected graph convolutional network for joint semantic and instance segmentation of indoor point clouds. *ISPRS J. Photogramm. Remote Sens.* **2021**, *182*, 67–77. [CrossRef]
27. Zeng, Z.; Xu, Y.; Xie, Z.; Wan, J.; Wu, W.; Dai, W. RG-GCN: A random graph based on graph convolution network for point cloud semantic segmentation. *Remote Sens.* **2022**, *14*, 4055. [CrossRef]
28. Chen, L.; Zhang, Q. DDGCN: Graph convolution network based on direction and distance for point cloud learning. *Vis. Comput.* **2023**, *39*, 863–873. [CrossRef]
29. Zhang, F.; Xia, X. Cascaded Contextual Reasoning for Large-Scale Point Cloud Semantic Segmentation. *IEEE Access* **2023**, *11*, 20755–20768. [CrossRef]
30. Hu, Q.; Yang, B.; Xie, L.; Rosa, S.; Guo, Y.; Wang, Z.; Trigoni, N.; Markham, A. Randla-net: Efficient semantic segmentation of large-scale point clouds. In Proceedings of the IEEE/CVF Conference on Computer Vision and Pattern Recognition, Seattle, WA, USA, 14–19 June 2020; pp. 11108–11117.
31. Du, J.; Cai, G.; Wang, Z.; Huang, S.; Su, J.; Junior, J.M.; Smit, J.; Li, J. ResDLPS-Net: Joint residual-dense optimization for large-scale point cloud semantic segmentation. *ISPRS J. Photogramm. Remote Sens.* **2021**, *182*, 37–51. [CrossRef]
32. Zhao, Y.; Ma, X.; Hu, B.; Zhang, Q.; Ye, M.; Zhou, G. A large-scale point cloud semantic segmentation network via local dual features and global correlations. *Comput. Graph.* **2023**, *111*, 133–144. [CrossRef]
33. Yin, F.; Huang, Z.; Chen, T.; Luo, G.; Yu, G.; Fu, B. Dcnet: Large-scale point cloud semantic segmentation with discriminative and efficient feature aggregation. *IEEE Trans. Circuits Syst. Video Technol.* **2023**, *33*, 4083–4095. [CrossRef]

34. Deng, C.; Peng, Z.; Chen, Z.; Chen, R. Point Cloud Deep Learning Network Based on Balanced Sampling and Hybrid Pooling. *Sensors* **2023**, *23*, 981. [CrossRef] [PubMed]
35. Wu, W.; Qi, Z.; Fuxin, L. Pointconv: Deep convolutional networks on 3d point clouds. In Proceedings of the IEEE/CVF Conference on Computer Vision and Pattern Recognition, Long Beach, CA, USA, 15–20 June 2019; pp. 9621–9630.
36. Yan, X.; Zheng, C.; Li, Z.; Wang, S.; Cui, S. Pointasnl: Robust point clouds processing using nonlocal neural networks with adaptive sampling. In Proceedings of the IEEE/CVF Conference on Computer Vision and Pattern Recognition, Seattle, WA, USA, 13–19 June 2020; pp. 5589–5598.
37. Zarzar, J.; Giancola, S.; Ghanem, B. PointRGCN: Graph convolution networks for 3D vehicles detection refinement. *arXiv* **2019**, arXiv:1911.12236.
38. Zhao, H.; Jiang, L.; Jia, J.; Torr, P.H.; Koltun, V. Point transformer. In Proceedings of the IEEE/CVF International Conference on Computer Vision, Montreal, BC, Canada, 11–17 October 2021; pp. 16259–16268.
39. Bai, X.; Luo, Z.; Zhou, L.; Chen, H.; Li, L.; Hu, Z.; Fu, H.; Tai, C.L. Pointdsc: Robust point cloud registration using deep spatial consistency. In Proceedings of the IEEE/CVF Conference on Computer Vision and Pattern Recognition, Nashville, TN, USA, 20–25 June 2021; pp. 15859–15869.
40. Yew, Z.J.; Lee, G.H. Rpm-net: Robust point matching using learned features. In Proceedings of the IEEE/CVF Conference on Computer Vision and Pattern Recognition, Seattle, WA, USA, 13–19 June 2020; pp. 11824–11833.
41. Deng, H.; Birdal, T.; Ilic, S. Ppfnet: Global context aware local features for robust 3d point matching. In Proceedings of the IEEE Conference on Computer Vision and Pattern Recognition, Salt Lake City, UT, USA, 18–23 June 2018; pp. 195–205.
42. Huang, R.; Xu, Y.; Stilla, U. GraNet: Global relation-aware attentional network for semantic segmentation of ALS point clouds. *ISPRS J. Photogramm. Remote Sens.* **2021**, *177*, 1–20. [CrossRef]
43. Wen, C.; Li, X.; Yao, X.; Peng, L.; Chi, T. Airborne LiDAR point cloud classification with global-local graph attention convolution neural network. *ISPRS J. Photogramm. Remote Sens.* **2021**, *173*, 181–194. [CrossRef]
44. Gao, Y.; Liu, X.; Li, J.; Fang, Z.; Jiang, X.; Huq, K.M. LFT-Net: Local feature transformer network for point clouds analysis. *IEEE Trans. Intell. Transp. Syst.* **2022**, *24*, 2158–2168. [CrossRef]
45. Zhang, M.; Kadam, P.; Liu, S.; Kuo, C.C. GSIP: Green semantic segmentation of large-scale indoor point clouds. *Pattern Recognit. Lett.* **2022**, *164*, 9–15. [CrossRef]
46. Thomas, H.; Qi, C.R.; Deschaud, J.E.; Marcotegui, B.; Goulette, F.; Guibas, L.J. Kpconv: Flexible and deformable convolution for point clouds. In Proceedings of the IEEE/CVF International Conference on Computer Vision, Seoul, Republic of Korea, 27 October–2 November 2019; pp. 6411–6420.
47. Huang, R.; Xu, Y.; Hong, D.; Yao, W.; Ghamisi, P.; Stilla, U. Deep point embedding for urban classification using ALS point clouds: A new perspective from local to global. *ISPRS J. Photogramm. Remote Sens.* **2020**, *163*, 62–81. [CrossRef]
48. Ye, Z.; Xu, Y.; Huang, R.; Tong, X.; Li, X.; Liu, X.; Luan, K.; Hoegner, L.; Stilla, U. LASDU: A Large-Scale Aerial LiDAR Dataset for Semantic Labeling in Dense Urban Areas. *Int. J. Geo-Inf.* **2020**, *9*, 450. [CrossRef]
49. Wen, C.; Yang, L.; Li, X.; Peng, L.; Chi, T. Directionally constrained fully convolutional neural network for airborne LiDAR point cloud classification. *ISPRS J. Photogramm. Remote Sens.* **2020**, *162*, 50–62. [CrossRef]
50. Li, X.; Wang, L.; Wang, M.; Wen, C.; Fang, Y. DANCE-NET: Density-aware convolution networks with context encoding for airborne LiDAR point cloud classification. *ISPRS J. Photogramm. Remote Sens.* **2020**, *166*, 128–139. [CrossRef]
51. Li, W.; Wang, F.D.; Xia, G.S. A geometry-attentional network for ALS point cloud classification. *ISPRS J. Photogramm. Remote Sens.* **2020**, *164*, 26–40.
52. Deng, C.; Chen, R.; Tang, W.; Chu, H.; Xu, G.; Cui, Y.; Peng, Z. PointNAC: Copula-Based Point Cloud Semantic Segmentation Network. *Symmetry* **2023**, *15*, 2021. [CrossRef]

remote sensing

MDPI

Article

ASPP⁺-LANet: A Multi-Scale Context Extraction Network for Semantic Segmentation of High-Resolution Remote Sensing Images

Lei Hu *, Xun Zhou, Jiachen Ruan and Supeng Li

School of Computer and Information Engineering, Jiangxi Normal University, Nanchang 330022, China; 202141600066@jxnu.edu.cn (S.L.)
* Correspondence: hulei@jxnu.edu.cn

Abstract: Semantic segmentation of remote sensing (RS) images is a pivotal branch in the realm of RS image processing, which plays a significant role in urban planning, building extraction, vegetation extraction, etc. With the continuous advancement of remote sensing technology, the spatial resolution of remote sensing images is progressively improving. This escalation in resolution gives rise to challenges like imbalanced class distributions among ground objects in RS images, the significant variations of ground object scales, as well as the presence of redundant information and noise interference. In this paper, we propose a multi-scale context extraction network, ASPP$^+$-LANet, based on the LANet for semantic segmentation of high-resolution RS images. Firstly, we design an ASPP$^+$ module, expanding upon the ASPP module by incorporating an additional feature extraction channel, redesigning the dilation rates, and introducing the Coordinate Attention (CA) mechanism so that it can effectively improve the segmentation performance of ground object targets at different scales Secondly, we introduce the Funnel ReLU (FReLU) activation function for enhancing the segmentation effect of slender ground object targets and refining the segmentation edges. The experimental results show that our network model demonstrates superior segmentation performance on both Potsdam and Vaihingen datasets, outperforming other state-of-the-art (SOTA) methods.

Keywords: high-resolution remote sensing images; semantic segmentation; ASPP module; local attention network model; activation function

Citation: Hu, L.; Zhou, X.; Ruan, J.; Li, S. ASPP$^+$-LANet: A Multi-Scale Context Extraction Network for Semantic Segmentation of High-Resolution Remote Sensing Images. *Remote Sens.* **2024**, *16*, 1036. https://doi.org/10.3390/rs16061036

Academic Editor: Melanie Vanderhoof

Received: 18 October 2023
Revised: 9 March 2024
Accepted: 12 March 2024
Published: 14 March 2024

1. Introduction

Remote sensing (RS) images can be used to observe natural and artificial phenomena on the Earth's surface. In the field of RS, semantic segmentation of high-resolution RS images entails a pixel-level classification task where the objective is to assign a semantic label to each pixel in the image [1]. These semantic labels mean different ground objects.

Recently, RS images have achieved spatial resolution at the centimeter scale, empowering the discernment of minute details and targets within high-resolution RS imagery. The challenge of semantic segmentation in RS images persists due to issues such as redundant information, noise interference, misclassification of tiny targets, and insufficient smoothness in the edges of ground objects. To solve this problem, this paper proposes a multi-scale context network ASPP$^+$-LANet based on LANet, which improves the segmentation performance of ground object targets at different scales and refines the edges of ground object targets.

The rapid progress of deep neural networks, especially Convolutional Neural Networks (CNNs), has greatly advanced semantic segmentation in RS images. In 2015, Long et al. [2] first proposed the concept of Fully Convolutional Networks (FCNs), an encoder-decoder structure network, which used an anti-convolutional layer instead of the fully connected layer in traditional CNNs. In the same year, the Unet network was introduced by Ronneberger et al. [3], featuring a U-shaped encoder–decoder architecture with inter-layer skip connections. In 2017, the SegNet network was proposed by Badrinarayanan et al. [4], which implemented an encoder–decoder structure. The key innovation was situated in the decoder, where instead of using deconvolution for upsampling, pooling indices were utilized to conduct non-linear upsampling during the respective encoder's max-pooling steps. The mentioned networks all employed an encoder–decoder structure with robust feature extraction capabilities. Nevertheless, without further refinement, the direct connection between shallow texture information and deep semantic information causes underutilization of feature information, leading to insufficient discrimination between shallow information and deep information. To address these issues, a multi-scale feature extraction module was introduced into the convolutional network by researchers. In 2017, the Pyramid Scene Parsing Network (PSPNet) was introduced by Zhao et al. [5], which proposed the Pyramid Pooling Module (PPM) to aggregate diverse regional contexts. In addition, Chen et al. [6–9] successively proposed DeepLab series networks for extracting multi-scale contextual features. Among them, based on DeepLab v3 [8], a decoder structure was added to DeepLab v3+ [9], which integrated the low-level features of the encoder output with the high-level features of the Atrous Spatial Pyramid Pooling (ASPP) output. Furthermore, attention mechanisms have been extensively employed in semantic segmentation networks. In 2020, a Local Attention Network (LANet) was proposed by Ding et al. [10], introducing a patch-level-based attention mechanism for extracting contextual information. Two approaches were suggested for enhancing the feature representation: the chunked attention module enhances the embedding of contextual information, while the attention embedding module enriches the semantic information of the underlying features by embedding the local focus of the high-level features. The differences in physical information content and spatial distribution are effectively addressed, the disparities between high-level and low-level features are bridged, and significant success in the field of remote sensing image segmentation is achieved. Due to these excellent features, we chose it as our benchmark network. In 2021, Li et al. [11] proposed a Multi-Attention Network (MANet), which designed a novel linear-complexity kernel attention mechanism to alleviate the computational demands of attention. In 2023, a novel three-branch network architecture, PIDNet, was proposed by Xu et al. [12]. PIDNet comprises three branches designed to parse detailed, contextual, and boundary information. Additionally, boundary attention is employed to facilitate the fusion of detailed and contextual branches.

In recent years, the Vision Transformer (ViT) [13] has demonstrated remarkable performance in the field of RS image segmentation due to its powerful self-attention-based global context modeling capability [14–18]. Among them, in 2022, Wang et al. [18] proposed the UnetFormer network for real-time urban scene segmentation in RS images. An efficient global–local attention mechanism known as the Global–Local Transformer Block (GLTB) was implemented by the network to integrate both global and local information within the decoder. A lightweight transformer-based decoder was developed using GLTB and Feature refinement head, which aimed to enhance the network's capability to extract multi-scale contextual features and effectively improve the network's segmentation performance in semantic segmentation of RS images. In 2022, Zhang et al. [19] proposed a hybrid deep neural network, Swin-CNN, combining a transformer and a CNN. The model follows an encoder–decoder structure. A novel universal backbone dual transformer is employed in the encoder module to extract features, thus aiming to enhance long-range spatial dependency modeling. The decoder module leverages some effective blocks and successful strategies from a CNN-based remote sensing image segmentation model. In the middle of the framework, spatial pyramid pooling blocks based on depthwise separable convolutions are applied to obtain multi-scale context.

As previously noted, the incorporation of multi-scale and attention modules into the semantic segmentation network of RS images has been shown to effectively enhance the network's segmentation performance. Accordingly, we designed a new $ASPP^{+}$ module by augmenting an additional feature extraction channel to the ASPP module, redesigning the dilation rates, and introducing the CA mechanism [20], thereby effectively enhancing the network's segmentation capability. The utilization of parallel dilated convolutions has been found to enhance the receptive field and capture target features of varying scales. Additionally, the incorporation of the attention module allows the model to prioritize meaningful features and acquire contextual information more effectively. Furthermore, we introduced the FReLU activation function [21] to enhance the network's generalization capability, filter out noise and low-frequency information, and retain more higher-frequency information so as to effectively improve the segmentation performance of slender ground object targets and refine the segmentation edges.

In conclusion, the main contributions of this paper include the following three aspects as follows:

(1) We propose a multi-scale context extraction network for semantic segmentation of high-resolution RS images, ASPP^{+}-LANet, by improving the LANet structure, which effectively tackles the issue of unclear segmentation in various-sized ground objects, slender ground objects, and ground object edges. By adding a new multi-scale module, the segmentation accuracy of ground objects at different scales has been improved. By introducing the activation function, the segmentation accuracy of slender ground objects and ground object edges has been improved.

(2) We designed a novel ASPP^{+} module to effectively enhance the segmentation accuracy of ground objects at different sizes. This module adds an additional feature extraction channel to ASPP. In addition, we redesigned its dilation rates and introduced the CA mechanism. The attention mechanism can focus on more meaningful areas, improving the overall segmentation progress.

(3) We introduced the FReLU activation function. By integrating it with the LANet network, the performance of ASPP^{+}-LANet has been improved. The activation function can filter out noise and low-frequency information and retain more higher-frequency information so as to effectively enhance the segmentation accuracy of slender ground objects and ground object edges.

2. Materials and Methods

2.1. Materials

In this paper, we design a series of comparative experiments using Potsdam and Vaihingen from the ISPRS dataset [22] in order to evaluate our proposed method.

2.1.1. Potsdam Datasets

The Potsdam dataset consists of 38 images, each with a size of 6000 × 6000 pixels and a spatial resolution of approximately 5 cm [23]. In the Potsdam region, there are six land cover classes, as shown in Figure 1: impervious surfaces, buildings, low vegetation, trees, cars, and clutter/background. The clutter/background class primarily includes water bodies and objects defined as outside the designated classes, which are typically irrelevant semantic objects in urban scenes. To ensure sufficient experimental data, the dataset was preprocessed prior to the experiments, involving image cropping and data augmentation. The images were uniformly cropped into 512 × 512 pixels and subjected to horizontal and vertical flipping for data augmentation. After filtering out images with problematic labels, the dataset was divided into training, validation, and testing sets in a 6:2:2 ratio.

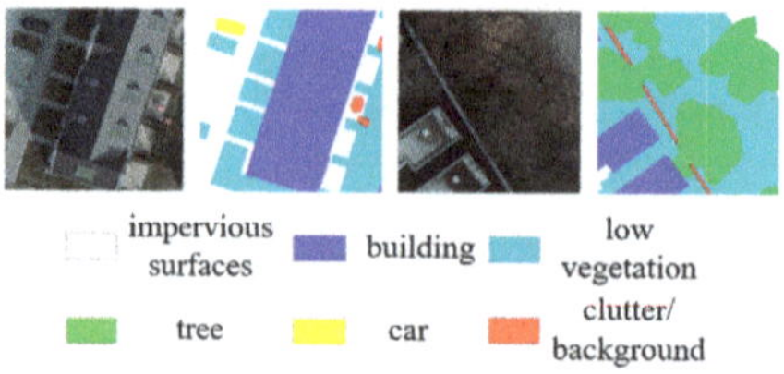

Figure 1. Partial example plot of the Potsdam dataset.

2.1.2. Vaihingen Datasets

The Vaihingen dataset consists of a total of 33 images, each of which has a varying size, with an average dimension of 2496 × 2064 pixels and a spatial resolution of approximately 9 cm [23]. The label categories and color representations are the same as those in the Potsdam dataset, as shown in Figure 2. Prior to the experiments, the images were cropped into 512 × 512 pixels and augmented by horizontal and vertical flips. After filtering out images with problematic labels, the dataset was split into a training set with 6020 image blocks, a validation set with 2006 image blocks, and a test set with 2052 image blocks.

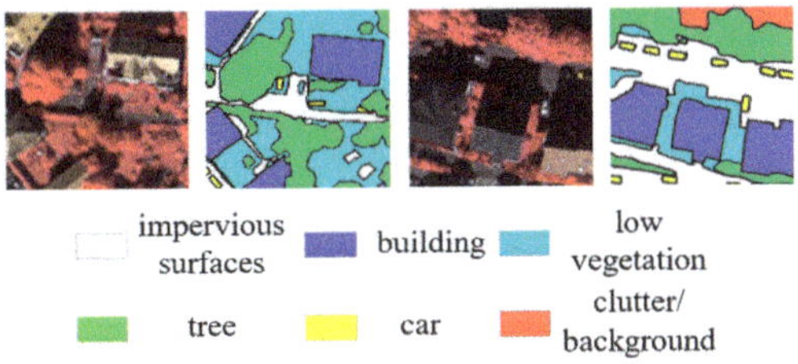

Figure 2. Partial example plot of the Vaihingen dataset.

2.2. Methods

In this section, we provide a comprehensive overview of the proposed network model, ASPP^{+}-LANet. Firstly, we present a concise summary of the network structure, highlighting the general motivation and structure. Subsequently, we explore the intricacies of two pivotal modules: the ASPP+ module and the FReLU activation function. Through the examination of these components, we aim to present a thorough understanding of the ASPP^{+}-LANet network.

2.2.1. Overall Network Structure

We propose a multi-scale context extraction network for semantic segmentation of high-resolution RS images, ASPP^{+}-LANet, as illustrated in Figure 1. Like LANet [10], our network is built upon the FCN framework [2] and employs the pre-trained ResNet50 [24] as the backbone network. It consists of two parallel branches for high-level and low-level feature extraction, incorporating multiple feature extraction and enhancement modules within these branches.

There are two motives in this paper: (1) improving the segmentation performance of ground object targets at different scales and (2) enhancing the segmentation effect of slender ground object targets and refining the segmentation edges. To achieve these goals, we added two independent modules to the LANet network: (1) the ASPP^{+} module, which facilitates the fusion of multi-scale features; (2) the FReLU activation function [21], which enhances the network's generalization ability, and filters out noise as well as low-frequency information.

Specifically, we integrated the FReLU activation function into the activation layer at the residual module of the backbone network ResNet50 and added an ASPP^{+} module on the high-level feature extraction branch, as indicated by the green dashed box in Figure 3. In the branch of high-level feature extraction, the high-level features generated by ResNet50 extract multi-scale contextual information through the ASPP^{+} module and then enhance their feature representation through the Patch Attention Module (PAM) [10]. In the low-level feature extraction branch, the low-level features generated by convolution are first feature-enhanced by the PAM, and then the semantic information of the low-level features is enriched by embedding the local focus of the high-level features through the Attention Embedding Module (AEM) [10], which enables the low-level features to enhance the high-level semantic without losing spatial information. Ultimately, the features produced by the upper and lower parallel branches are merged to derive our conclusive segmentation output.

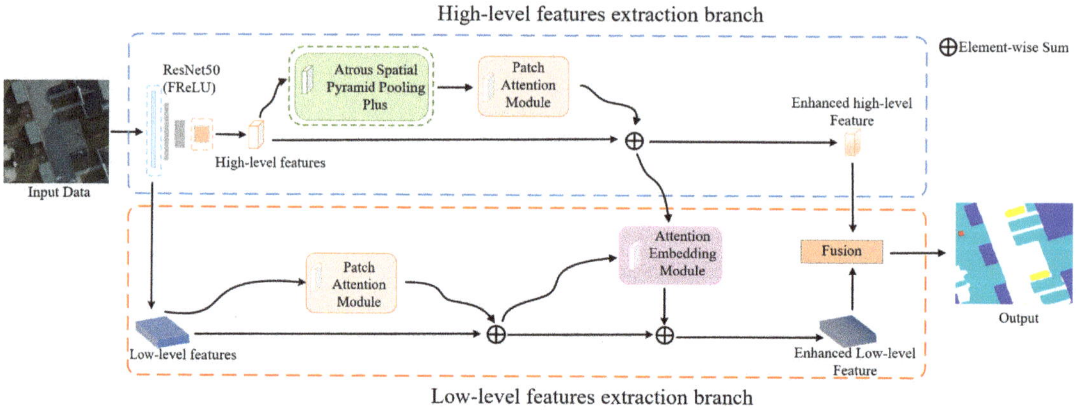

Figure 3. Overall structure of the ASPP+-LANet network.

2.2.2. ASPP+ Module

In RS images, challenges such as imbalanced ground object classes and significant variations in ground object scales exist. In such scenarios, it is difficult to extract target features only by a single scale. To address this issue, the paper proposes an improved multi-scale context extraction module, ASPP+, with the structure shown in Figure 4. It mainly consists of two components: the first component is the parallel dilated convolution multi-scale feature extraction module, employing five parallel dilated convolution branches to capture feature information of different scales; the second component is the global feature and context extraction module, taking charge of acquiring global feature and contextual information. Ultimately, the output features from both components are concatenated to form a multi-scale feature map.

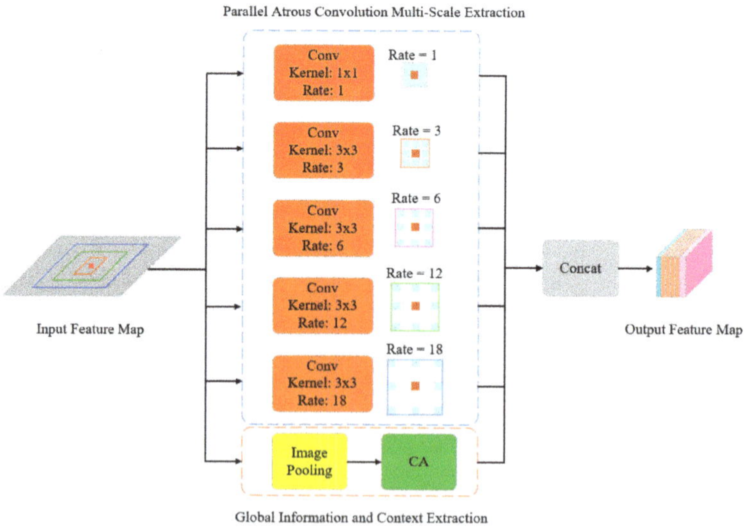

Figure 4. ASPP+ module structure.

In Figure 4, the orange box represents dilated convolutions with different dilation rates, where except for the first convolution kernel with a size of 1×1, the remaining four convolution kernels are all 3×3. Additionally, they have a stride of 1 and no padding. The Image Pooling module performs global average pooling. CA [20] refers to the Coordinate Attention module. "Concat" represents the operation of concatenating features. By concatenating the output features from these two parts, the model achieves the functionality of multi-scale context feature extraction. The dilated convolutions with different dilation rates improve the receptive field and capture target features at different scales. The attention module allows the model to focus more on meaningful features and acquire contextual information.

(1) Parallel Dilated Convolution Multi-Scale Feature Extraction

Parallel convolution can alter the receptive field of the convolutional kernel, acquiring the feature information of different scales. However, multiple parallel convolutions can increase the number of parameters and computational complexity of the network. Inspired by the ASPP module [9], using extended convolution instead of standard convolution can obtain feature information at different scales while reducing the number of parameters and computational complexity.

While the replacement of dilated convolutions has played a significant role, the setting of the dilation rate remains a challenge. The consecutive use of the same dilation rate in atrous convolutions will result in discontinuity of the convolution kernel, leading to a "grid effect", as shown in Figure 5a. On the other hand, a reasonable dilation rate, as depicted in Figure 5b, not only avoids the loss of relevant information but also captures the target context of different scales [25]. According to the literature [25], the dilation rate should follow the following principles:

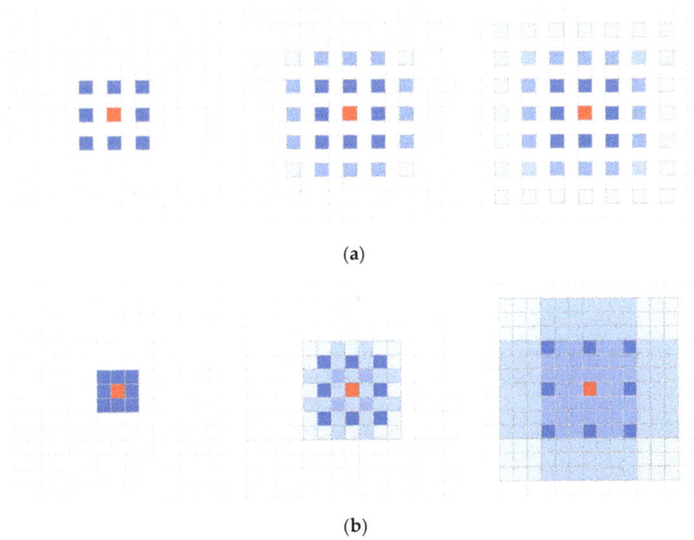

Figure 5. Schematic diagram of the "grid effect" [25]. (**a**) The "grid effect" in atrous convolutions. (**b**) Reasonable combination of dilation rates in atrous convolutions.

(a) The combination of dilation rates should not have a common factor greater than 1, as it would still lead to the occurrence of the "grid effect".

(b) Assuming that dilation rates corresponding to N convolutional kernel sizes $k \times k$ of atrous convolutions are $[r_1, \ldots, r_i, \ldots, r_n]$, it is required that Equation (1) satisfies $M_2 \leq k$.

$$M_i = max[M_{i+1} - 2r_i, M_{i+1} - 2(M_{i+1} - r_i), r_i] \quad (1)$$

where r_i represents the dilation rate of the i-th atrous convolution and M_i represents the maximum dilation rate for the i-th layer of atrous convolution, with a default value of $M_n = r_n$.

Therefore, this paper follows the aforementioned design principles and obtains a set of most appropriate dilation rates (1, 2, 4, 8, 12) through several comparative experiments (detailed experimental procedures described in Section 3.3.4), which significantly enhances the segmentation performance of ground object targets at different sizes.

Additionally, to enhance the model's generalization ability, we incorporate batch normalization and ReLU activation functions [26] after each convolutional layer. Finally, we connect the five parallel dilated convolution branches to form the parallel dilated convolution multi-scale feature extraction module, as depicted by the blue dashed box in Figure 4. The expression is represented as:

$$\left\langle C^1_{1\times1}(X)\cdot C^d_{3\times3}(X)\right\rangle, d = 2,4,8,12 \tag{2}$$

where $\langle\cdot\rangle$ represents feature concatenation, which refers to each feature being spliced along the channel dimension. $C^1_{1\times1}$ denotes a 1 × 1 convolution with a dilation rate of 1. $C^d_{3\times3}$ represents a 3 × 3 convolution with a dilation rate of d. X denotes the input feature.

(2) Global features and contextual information extraction

Global feature extraction refers to the generalization and integration of features from the entire feature map to obtain global contextual information. The global feature and context extraction module, as illustrated by the orange dashed box in Figure 4, begins by performing global average pooling on the input feature. It then utilizes a CA module to emphasize meaningful features, thereby capturing global contextual information. The expression can be represented as:

$$CA(GAP(X)) \tag{3}$$

where $CA(\cdot)$ represents Coordinate Attention. $GAP(\cdot)$ denotes Global Average Pooling. X represents the input features.

In conclusion, based on the aforementioned information, we can obtain an improved multi-scale context extraction module, referred to as the ASPP$^+$ module. Its overall representation is illustrated by Equation (4).

$$\left\langle C^1_{1\times1}(X)\cdot C^d_{3\times3}(X)\cdot CA(GAP(X))\right\rangle,\ d = 3,6,12,18 \tag{4}$$

where $\langle\cdot\rangle$ represents feature concatenation, which refers to the concatenation of each feature along the channel dimension. $C^1_{1\times1}$ denotes a 1 × 1 convolution with a dilation rate of 1. $C^d_{3\times3}$ represents a 3 × 3 convolution with a dilation rate of d. X denotes the input feature. $CA(\cdot)$ represents Coordinate Attention. $GAP(\cdot)$ denotes Global Average Pooling.

2.2.3. FReLU

In RS images, there always exists interference from noise and low-frequency information, which makes it challenging for existing image semantic segmentation networks to achieve satisfactory results for slender and limbic ground object targets. Activation functions, on the other hand, play a crucial role in enhancing network generalization, filtering out noise and low-frequency information, and preserving high-frequency information, which can help resolve this issue. Therefore, this paper conducted comparative experiments with different activation functions on the LANet network (detailed in Section 3.3.5), and the results indicate that incorporating FReLU into the LANet network yields the best performance.

This paper focuses on improving the bottleneck residual module within the ResNet50 backbone network, as illustrated in Figure 6. The activation functions in each convolutional layer of the bottleneck module are replaced with FReLU. Similar to ReLU [27] and PReLU [28], FReLU utilizes the max() function as a simple non-linear function. Whereas ReLU is defined as $y = \max(x, 0)$ and PReLU as $y = \max(x, px)$, FReLU adds a negligible spatial condition overhead and extends the conditional part to a two-dimensional condition that depends on the spatial context of each pixel, as illustrated in Figure 7. It can be represented as $y = \max(x, T(x))$, where T(·) denotes the two-dimensional spatial representation. The function definition of FReLU is as follows:

$$f(x_{c,i,j}) = max(x_{c,i,j}, T(x_{c,i,j})) \quad (5)$$

$$T(x_{c,i,j}) = x^{w}_{c,i,j} \cdot p^{w}_{c} \quad (6)$$

where (i, j) represents the pixel position in two-dimensional space; c denotes the c-th channel; $T(x_{c,i,j})$ represents the two-dimensional condition; $x^{w}_{c,i,j}$ denotes the parameterized pool window centered on the input pixel of the nonlinear activation function on the c-th channel at position (i, j) in two-dimensional space; and p^{w}_{c} represents coefficients that are shared by this window in the same channel.

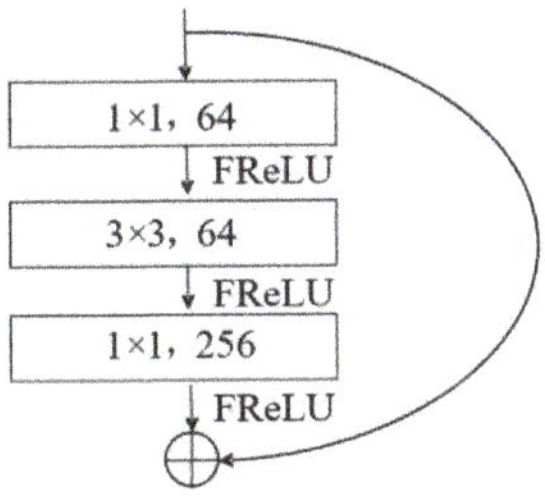

Figure 6. BottleNeck structure of ResNet50.

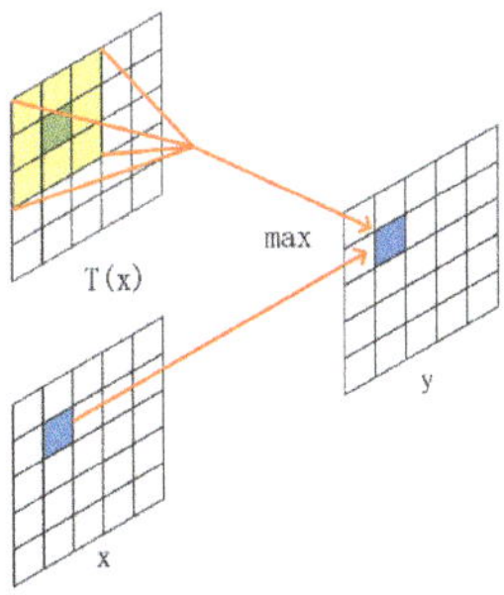

Figure 7. Schematic diagram of FReLU.

High-resolution RS images often exhibit complex backgrounds, leading to challenges in achieving accurate semantic segmentation, especially for slender and limbic ground object targets. FReLU, by incorporating spatial context information as a non-linear function condition, possesses superior contextual capturing capabilities. It effectively filters out noise and low-frequency information while preserving high-frequency details. The results show that FReLU can significantly improve the segmentation effect of slender ground objects and refine the segmentation edges.

3. Experiments and Results

In this section, we conducted a series of comparative experiments and ablation studies to validate the effectiveness of our proposed method. Initially, we delineated three evaluation metrics utilized for quantitative analysis. Following that, we furnished comprehensive details regarding the network's parameter configurations and experimental setups. Subsequently, we performed comparative experiments with other SOTA methods to assess and compare the performance of our proposed network. Additionally, we conducted ablation studies to evaluate the performance of our network under various configuration settings.

We analyzed the experimental results in terms of segmentation accuracy, visual effects, and ablation studies. Ultimately, to bolster the credibility of our experiments, we conducted an investigation into the optimal dilation rate for the ASPP+ module. Furthermore, we undertook experiments to evaluate the performance differences of various activation functions on the baseline network, LANet.

3.1. Evaluation Criteria

To quantitatively evaluate the efficacy of our proposed method, this paper utilizes three evaluation metrics for comprehensive comparison and analysis: Pixel Accuracy (PA), F1 Score (F1), and Mean Intersection over Union (MIoU). The formulas for these metrics are as follows:

PA refers to the proportion of correctly predicted pixels of a certain category to the total number of pixels.

$$PA = \frac{TP + TN}{TP + TN + FP + FN} \tag{7}$$

F1 takes into account both the precision and recall of a classification model and enables it to be seen as the harmonic mean of precision and recall.

$$Precision = \frac{TP}{TP + FP} \tag{8}$$

$$Recall = \frac{TP}{TP + FN} \tag{9}$$

$$F1 = \frac{2 * \text{Precision} * \text{Recall}}{Precision + Recall} \tag{10}$$

MIoU refers to the average Intersection over Union (IoU) of each class in the dataset.

$$IoU = \frac{TP}{TP + FP + FN} \tag{11}$$

$$MIoU = \frac{1}{n}\sum_{i=1}^{n} IoU \tag{12}$$

where TP stands for True Positive, indicating the number of pixels in the predicted results that belong to a certain class and are indeed of that class; FP stands for False Positive, signifying the number of pixels in the predicted results that belong to other classes but are mistakenly classified as that class; TN stands for True Negative, depicting the number of pixels in the predicted results that belong to other classes and are indeed of other classes; FN stands for False Negative, referring to the number of pixels in the predicted results that belong to a certain class but are mistakenly classified as other classes. n represents the number of classes. i represents the i-th class.

3.2. Implementation Details

In this article, our network and other comparative networks are implemented in the PyTorch deep learning framework, and experiments are conducted on a 64-bit Windows 10 system server. The server is equipped with an Intel Core i9-12900k CPU (3.20 GHz), 128 GB of memory, and an NVIDIA GeForce RTX 4090 graphics card.

During the training process, referring to some model [18] and synthesizing our hardware and our experimental results, the experimental parameters were set as follows: the batch size was set to 6, the learning rate was set to 0.025, the total epochs was set to 400, the momentum was set to 0.9, adaptive moment estimation optimizer (Adam) [29] was used to optimize our model, and stochastic gradient descent (SGD) was employed for the optimization training. Additionally, a "poly" learning rate decay strategy was utilized to dynamically adjust the learning rate using the following expression:

$$l = l_{ini}\left(1 - \frac{e}{e_{max}}\right)^{0.9} \quad (13)$$

where l represents the current learning rate, l_{ini} stands for the initial learning rate, e denotes the current training epoch, and e_{max} refers to the maximum number of training epochs.

3.3. Experiment Results

3.3.1. Segmentation Precision Analysis

To validate the efficacy of our proposed method, we conducted comparisons with several classical network models, including UNet [3], SegNet [4], DeepLab V3+ [9], LANet [10], MANet [11], UnetFormer [18], and Swin-CNN [19] on the Potsdam and Vaihingen datasets. The evaluation metrics for each method are presented in Tables 1 and 2. The tables clearly demonstrate that LANet's experimental results outperform classical semantic segmentation networks such as UNet, SegNet, and so on. Nonetheless, the utilization of a single-scale feature extraction approach in LANet results in diminished segmentation performance when confronted with ground object targets of varied sizes. Consequently, we implemented enhancements to LANet by integrating the ASPP^{+} module and the FReLU activation function. This integration effectively enhances the segmentation performance for ground object targets at different scales, as well as slender ground objects and ground objects' edges.

Table 1. Segmentation accuracy of different methods on the Potsdam dataset.

Method	Parameters(M)	PA/%	F1/%	MIoU/%	Kappa
UNet	17.27	92.66	78.08	71.35	0.9492
SegNet	29.45	92.61	77.61	70.84	0.9491
DeepLab V3+	21.94	90.00	72.24	64.17	0.8913
LANet	23.81	93.29	78.77	72.29	0.9496
MANet	35.86	92.06	76.89	69.86	0.9256
UNetFormer	11.28	91.23	75.01	67.51	0.9138
Swin-CNN	66	94.56	81.68	76.62	0.9521
ASPP^{+}-LANet	27.46	95.53	82.57	77.81	0.9552

Table 2. Segmentation accuracy of different methods on the Vaihingen dataset.

Method	Parameters(M)	PA/%	F1/%	MIoU/%	Kappa
UNet	17.27	98.03	81.83	79.53	0.9637
SegNet	29.45	96.82	80.21	76.77	0.9433
DeepLab V3+	21.94	92.77	73.31	67.33	0.8721
LANet	23.81	97.55	80.82	77.77	0.9465
MANet	35.86	98.08	81.81	79.55	0.9677
UNetFormer	11.28	96.73	80.08	76.52	0.9429
Swin-CNN	66	97.98	81.66	78.86	0.9625
ASPP^{+}-LANet	27.46	98.24	81.99	79.83	0.9689

As shown in Table 1, our proposed method, ASPP^{+}-LANet, achieves the following performance metrics on the Potsdam dataset: PA reaches 95.53%, F1 reaches 82.57%, and MIoU reaches 77.81%, which is improved by 2.24%, 3.80%, and 5.52%, respectively, compared to the baseline LANet network. Furthermore, our method demonstrates superior performance compared to existing semantic segmentation networks. This notable performance can be attributed to two key factors. Primarily, our proposed ASPP^{+} module enhances the network's ability to extract multi-scale features by setting appropriate dilation rates, thereby effectively improving the segmentation accuracy for ground object targets of different sizes. Moreover, the introduction of the FReLU activation function filters out noise and low-frequency information while preserving high-frequency information, thereby improving segmentation performance for slender and limbic ground object targets.

As shown in Table 2, our proposed method, ASPP$^+$-LANet, achieves the following performance metrics on the Vaihingen dataset: PA reaches 98.24%, F1 reaches 81.99%, and MIoU reaches 79.83%, which is improved by 0.69%, 1.17%, and 2.06%, respectively, compared to the baseline LANet network. Furthermore, our method demonstrates superior performance compared to existing semantic segmentation networks.

3.3.2. Renderings Analysis

To better highlight the feasibility of the proposed method in this paper, we selected six representative test targets for analysis on the Potsdam and Vaihingen datasets. Additionally, we conducted a subjective visual comparison analysis among the classical semantic segmentation methods, as illustrated in the figure below.

By comparing the visualization results on the Potsdam dataset, as shown in Figures 8 and 9, it can be observed that our proposed method achieves superior segmentation accuracy on ground object targets of different scales compared to other comparative methods. Additionally, Figures 10 and 11 demonstrate that our proposed method achieves superior segmentation accuracy on slender ground objects and ground object edges compared to other comparative methods. Moreover, Figures 12 and 13 reveal that our proposed method outperforms other comparative methods in terms of missing detections and false detections. The above experimental results validate the efficacy of our proposed ASPP$^+$-LANet model. After integrating the ASPP$^+$ module and the FReLU activation function, there was indeed a noticeable improvement in the segmentation performance of ground object targets at varying scales in the Potsdam dataset. Moreover, it also enhances the segmentation effect for slender ground object targets, refining the segmentation edges. These results demonstrate the effectiveness of our approach.

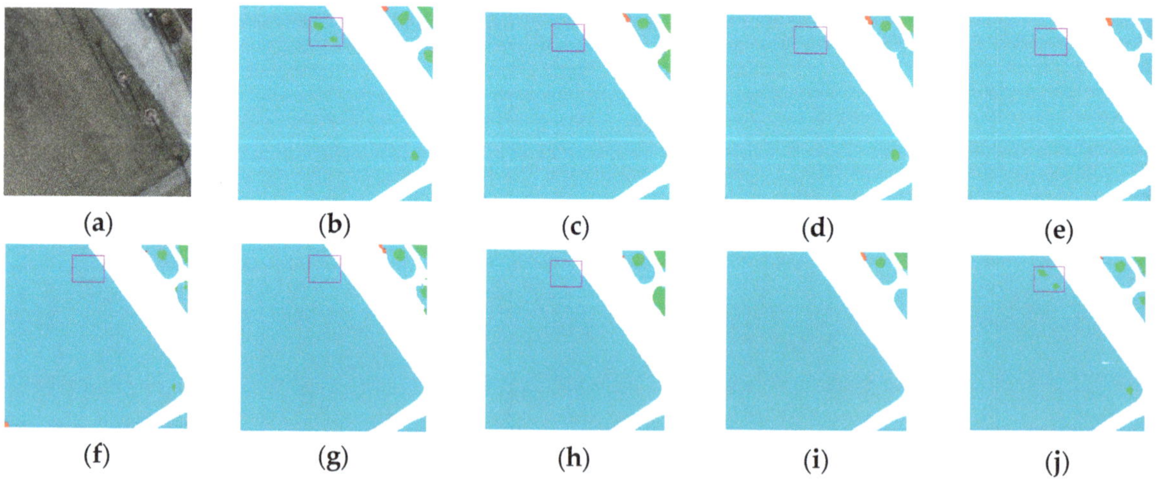

Figure 8. Visual comparison of semantic segmentation for small object features on the Potsdam dataset. (**a**) Image, (**b**) Ground truth, (**c**) UNet, (**d**) SegNet, (**e**) DeepLabv3+, (**f**) LANet, (**g**) MANet, (**h**) UnetFormer, (**i**) Swin-CNN, (**j**) ASPP$^+$-LANet. The colors represent the same types of ground object as shown in Figure 1, and the same applies to other similar images.

By analyzing the visualization results on the Vaihingen dataset, as shown in Figures 14 and 15, it can be observed that our proposed method achieves superior segmentation accuracy on ground object targets of different scales compared to other comparative methods. Additionally, Figures 16 and 17 demonstrate that our proposed method achieves better segmentation accuracy on slender ground objects and ground object edges compared to other comparative methods. Moreover, Figures 18 and 19 reveal that our proposed method outperforms other comparative methods in terms of missing detections and false detections. The above experimental results validate the effectiveness of our proposed ASPP+-LANet model. After integrating the ASPP+ module and the FReLU activation function, there was indeed a noticeable improvement in the segmentation performance of ground object targets at varying scales in the Vaihingen dataset. Moreover, it also enhances the segmentation effect for slender ground object targets, refining the segmentation edges. These results further demonstrate the effectiveness of our approach.

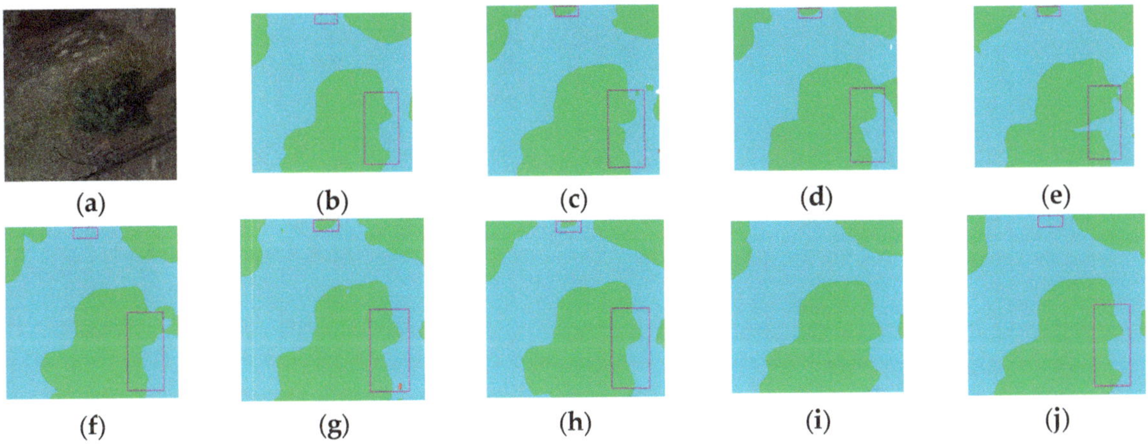

Figure 9. Visual comparison of semantic segmentation for large object features on the Potsdam dataset. (**a**) Image, (**b**) Ground truth, (**c**) UNet, (**d**) SegNet, (**e**) DeepLabv3+, (**f**) LANet, (**g**) MANet, (**h**) UnetFormer, (**i**) Swin-CNN, (**j**) ASPP+-LANet.

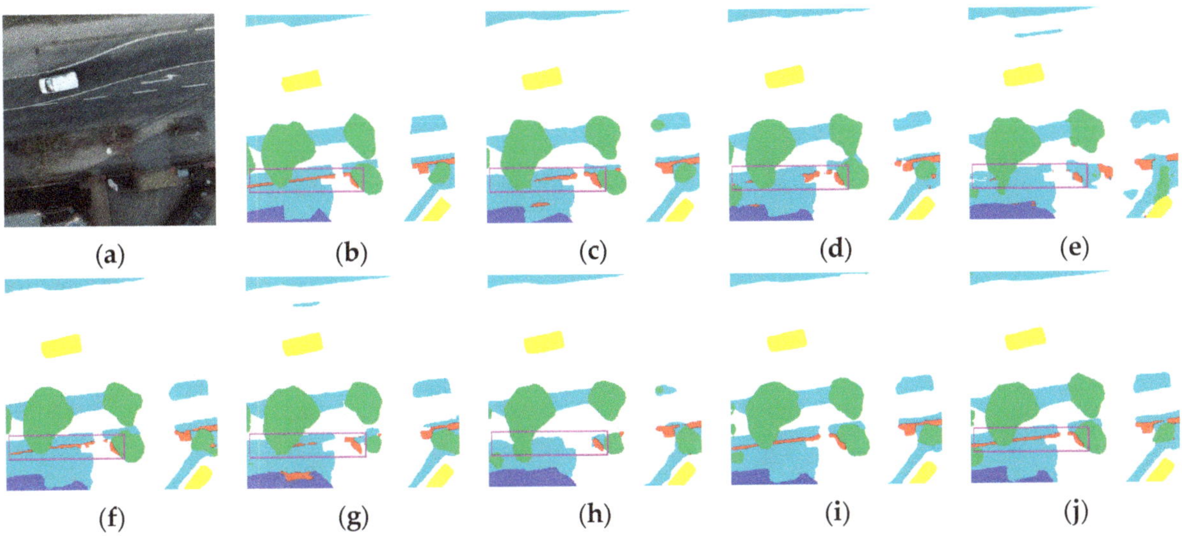

Figure 10. Visual comparison of semantic segmentation for slender ground objects features on the Potsdam dataset. (**a**) Image, (**b**) Ground truth, (**c**) UNet, (**d**) SegNet, (**e**) DeepLabv3+, (**f**) LANet, (**g**) MANet, (**h**) UnetFormer, (**i**) Swin-CNN, (**j**) ASPP+-LANet.

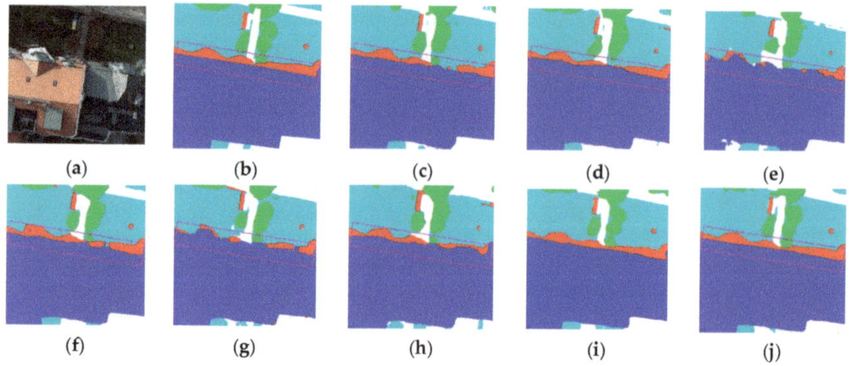

Figure 11. Visual comparison of semantic segmentation for limbic ground objects features on the Potsdam dataset. (**a**) Image, (**b**) Ground truth, (**c**) UNet, (**d**) SegNet, (**e**) DeepLabv3+, (**f**) LANet, (**g**) MANet, (**h**) UnetFormer, (**i**) Swin-CNN, (**j**) ASPP^{+}-LANet.

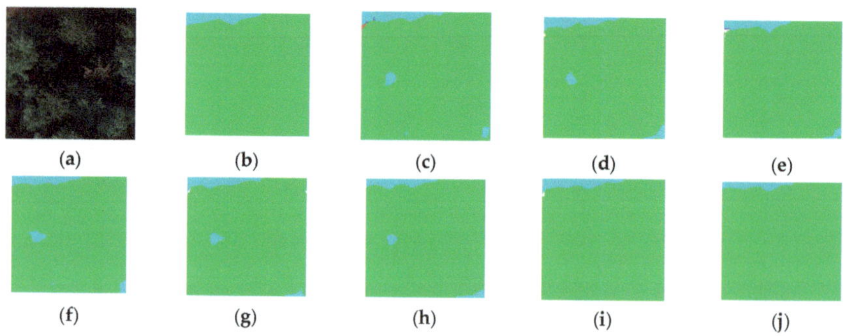

Figure 12. Visual comparison of semantic segmentation for the missing detection of object features in the Potsdam dataset. (**a**) Image, (**b**) Ground truth, (**c**) UNet, (**d**) SegNet, (**e**) DeepLabv3+, (**f**) LANet, (**g**) MANet, (**h**) UnetFormer, (**i**) Swin-CNN, (**j**) ASPP^{+}-LANet.

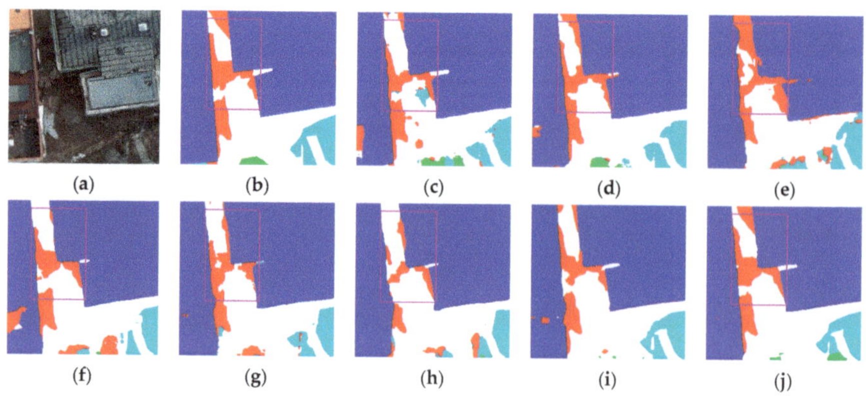

Figure 13. Visual comparison of semantic segmentation for the false detection of object features in the Potsdam dataset. (**a**) Image, (**b**) Ground truth, (**c**) UNet, (**d**) SegNet, (**e**) DeepLabv3+, (**f**) LANet, (**g**) MANet, (**h**) UnetFormer, (**i**) Swin-CNN, (**j**) ASPP^{+}-LANet.

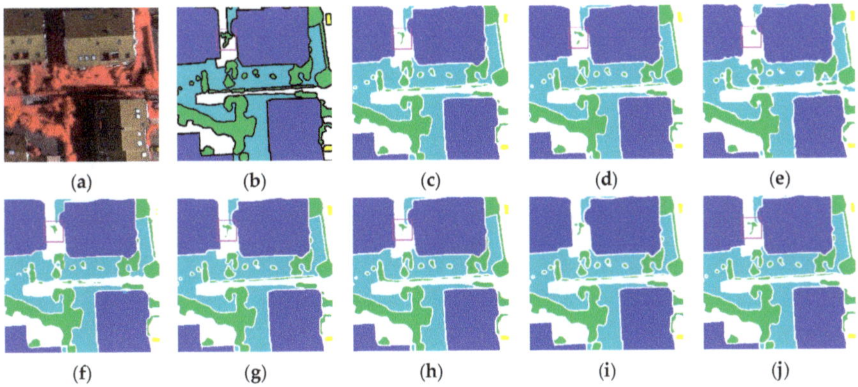

Figure 14. Visual comparison of semantic segmentation for small object features on the Vaihingen dataset. (**a**) Image, (**b**) Ground truth, (**c**) UNet, (**d**) SegNet, (**e**) DeepLabv3+, (**f**) LANet, (**g**) MANet, (**h**) UnetFormer, (**i**) Swin-CNN, (**j**) ASPP^{+}-LANet.

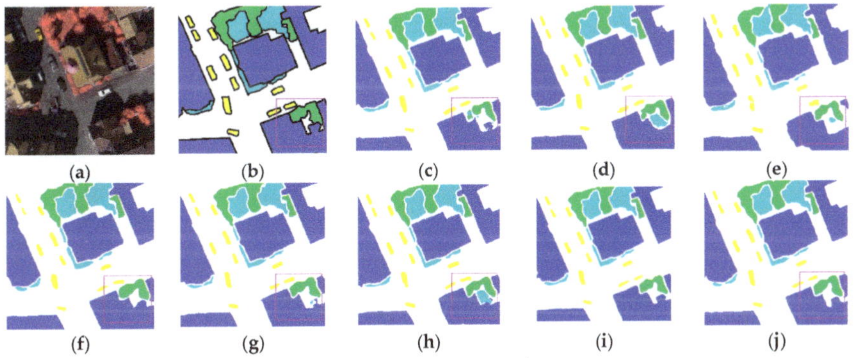

Figure 15. Visual comparison of semantic segmentation for large object features on the Vaihingen dataset. (**a**) Image, (**b**) Ground truth, (**c**) UNet, (**d**) SegNet, (**e**) DeepLabv3+, (**f**) LANet, (**g**) MANet, (**h**) UnetFormer, (**i**) Swin-CNN, (**j**) ASPP^{+}-LANet.

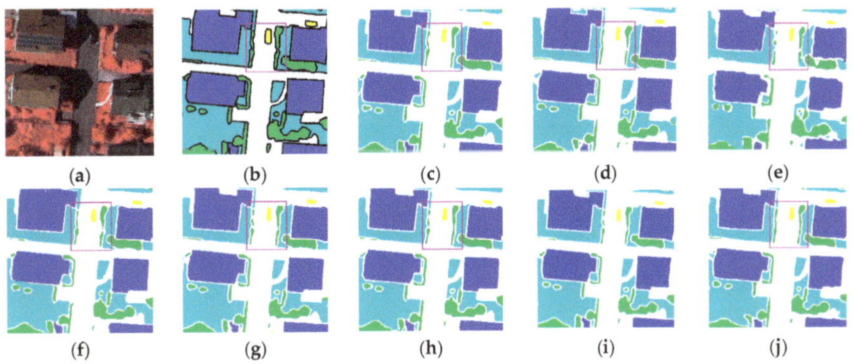

Figure 16. Visual comparison of semantic segmentation for slender ground objects features on the Vaihingen dataset. (**a**) Image, (**b**) Ground truth, (**c**) UNet, (**d**) SegNet, (**e**) DeepLabv3+, (**f**) LANet, (**g**) MANet, (**h**) UNetFormer, (**i**) Swin-CNN, (**j**) ASPP^{+}-LANet.

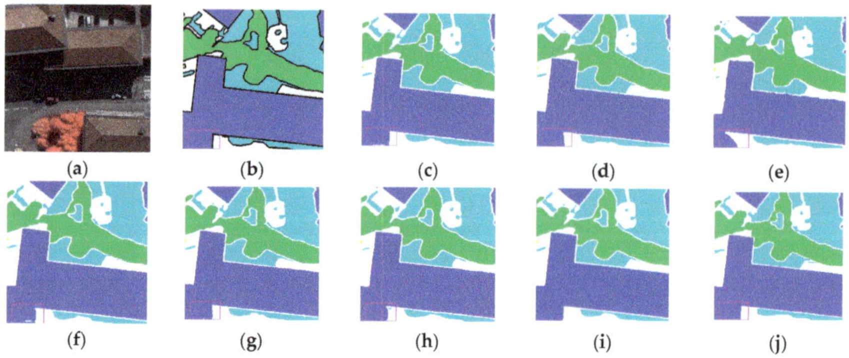

Figure 17. Visual comparison of semantic segmentation for limbic ground objects features on the Vaihingen dataset. (**a**) Image, (**b**) Ground truth, (**c**) UNet, (**d**) SegNet, (**e**) DeepLabv3+, (**f**) LANet, (**g**) MANet, (**h**) UnetFormer, (**i**) Swin-CNN, (**j**) ASPP$^+$-LANet.

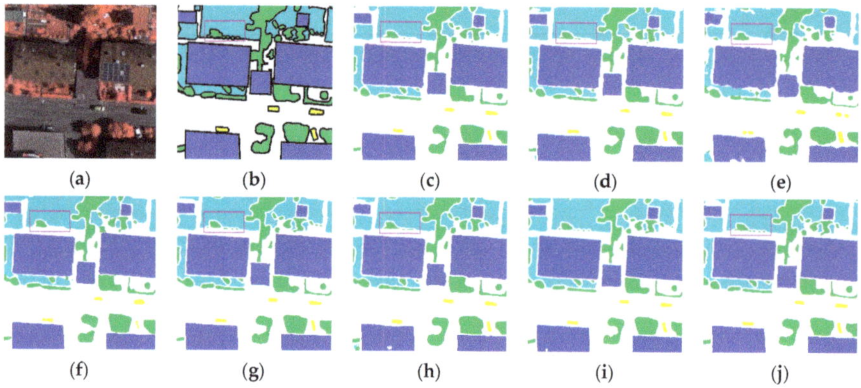

Figure 18. Visual comparison of semantic segmentation for the missing detection of object features in the Vaihingen dataset. (**a**) Image, (**b**) Ground truth, (**c**) UNet, (**d**) SegNet, (**e**) DeepLabv3+, (**f**) LANet, (**g**) MANet, (**h**) UnetFormer, (**i**) Swin-CNN, (**j**) ASPP$^+$-LANet.

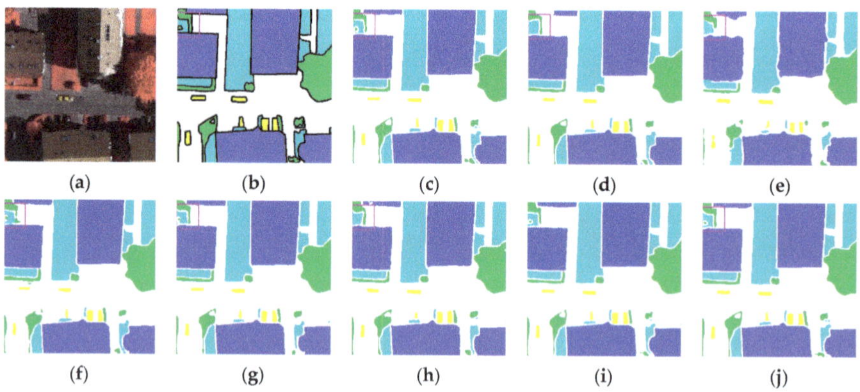

Figure 19. Visual comparison of semantic segmentation for the false detection of object features in the Vaihingen dataset. (**a**) Image, (**b**) Ground truth, (**c**) UNet, (**d**) SegNet, (**e**) DeepLabv3+, (**f**) LANet, (**g**) MANet, (**h**) UnetFormer, (**i**) Swin-CNN, (**j**) ASPP$^+$-LANet.

3.3.3. Ablation Experiments Analysis

To effectively capture detailed features from high-resolution RS images and overcome the technical challenges in accurately segmenting ground object targets at various scales, we propose the ASPP$^+$ module. Building upon the ASPP module, the ASPP$^+$ module adds a feature extraction channel, redefines the dilation rates, and introduces CA mechanisms, thereby effectively improving the segmentation performance of ground object targets at different scales. Moreover, in order to enhance the segmentation performance of slender ground object targets and refine the segmentation edges, we replaced the activation function on the backbone network (ResNet50) with FReLU. This alteration assists in filtering out noise and low-frequency information while preserving more high-frequency information, thereby further improving the segmentation accuracy of RS images. We conducted corresponding ablation experiments to individually verify the effectiveness of the ASPP module, ASPP$^+$ module, and FReLU activation function. The results of the ablation experiments on the Potsdam and Vaihingen datasets are presented in Tables 3 and 4.

Table 3. Results of ablation experiments on the Potsdam dataset.

Method	PA/%	F1/%	MIoU/%
LANet	93.29	78.77	72.29
LANet + ASPP	93.71	79.46	73.29
LANet + ASPP$^+$	93.86	79.80	73.75
LANet + FReLU	95.22	82.05	77.06
ASPP$^+$-LANet	95.53	82.57	77.81

Table 4. Results of ablation experiments on the Vaihingen dataset.

Method	PA/%	F1/%	MIoU/%
LANet	97.55	80.82	77.77
LANet + ASPP	97.77	81.31	78.65
LANet + ASPP$^+$	97.80	81.42	78.79
LANet + FReLU	97.76	81.30	78.59
ASPP$^+$-LANet	98.24	81.99	79.83

According to Table 3, it can be observed that the inclusion of the ASPP module leads to improvements in all performance metrics compared to the baseline network, LANet. Furthermore, by further refining the ASPP module, we were able to achieve even more significant enhancements in the performance metrics compared to the initial inclusion of the ASPP module. By incorporating the FReLU activation function, significant improvements can be observed in all performance metrics compared to the baseline network, LANet. Finally, by integrating the ASPP$^+$ module and the FReLU activation function into the LANet network, we further improved the overall performance metrics. The metrics such as PA, F1, and MioU reached 95.53%, 82.57%, and 77.81% respectively. Compared to the baseline network, LANet, there were increases of 2.24%, 3.80%, and 5.52% in PA, F1, and MioU, respectively.

According to Table 4, it is evident that the addition of the ASPP module leads to improvements in all metrics compared to the baseline LANet. Subsequent modifications made to the ASPP module result in slight enhancements in the metrics compared to the initial implementation. Furthermore, the inclusion of the FReLU activation function leads to improvements in all metrics compared to the LANet baseline. However, it is worth noting that the improvement achieved by incorporating FReLU is not as significant as that observed in the Potsdam dataset. This discrepancy could be attributed to the presence of a higher number of RS images related to narrow streets in the Potsdam dataset, a characteristic absent in the Vaihingen dataset. Finally, by integrating the ASPP$^+$ module and FReLU activation function into the LANet network, we further enhanced the overall performance metrics. The metrics, including PA, F1, and MioU, reached 98.24%, 81.99%, and 79.83%, respectively. Compared to the baseline LANet, there were improvements of 0.69%, 1.17%, and 2.06% in PA, F1, and MioU metrics, respectively.

In addition, to further visually represent the impact of each module in the ablation experiments on the results of semantic segmentation, we present the visualization of the core component ablation experiments of our method on the Potsdam and Vaihingen datasets, as shown in Figures 20 and 21. Among them, the first and second rows are used to verify the efficacy of large object detection and small object detection, respectively. From the figures, it can be observed that incorporating the ASPP$^+$ module into LANet improves the detection performance for ground object targets of different scales, surpassing both LANet alone and the results of incorporating the FReLU activation function in LANet The third and fourth rows are used to evaluate the detection performance of slender and limbic ground object targets. The figures demonstrate that integrating the FReLU activation function into LANet enhances the detection of slender and limbic ground object targets, outperforming both LANet alone and the results of incorporating the ASPP$^+$ module in LANet. Thus, we can conclude that the efficacy of our integration of the ASPP$^+$ module and FReLU activation function in LANet has been validated.

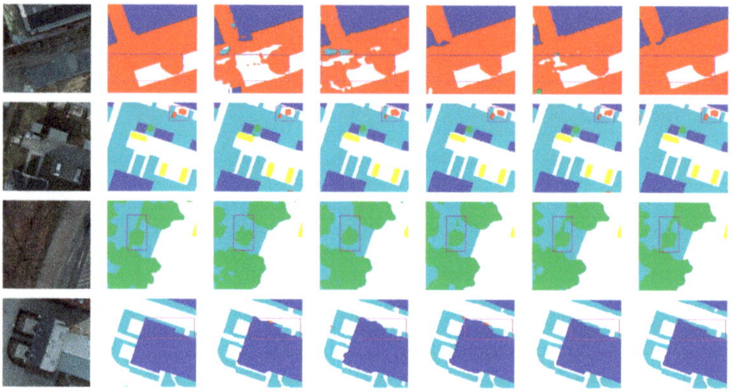

(a) (b) (c) (d) (e) (f) (g)

Figure 20. Visual comparisons of the ablation experiments conducted on the Potsdam dataset: (**a**) Image, (**b**) Grond Truth, (**c**) LANet, (**d**) LANet + ASPP, (**e**) LANet + ASPP$^+$, (**f**) LANet + FReLU, (**g**) ASPP$^+$-LANet.

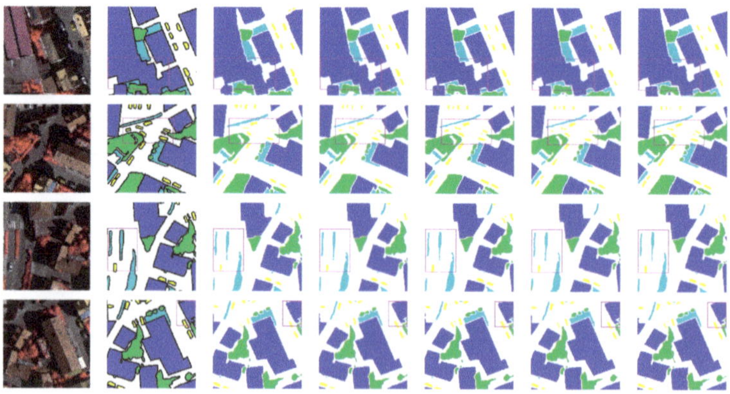

(a) (b) (c) (d) (e) (f) (g)

Figure 21. Visual comparisons of the ablation experiments conducted on the Vaihingen dataset: (**a**) Image, (**b**) Grond Truth, (**c**) LANet, (**d**) LANet + ASPP, (**e**) LANet + ASPP$^+$, (**f**) LANet + FReLU, (**g**) ASPP$^+$-LANet.

3.3.4. Dilation Rates Analysis of ASPP⁺ Module

The ASPP⁺ module is a fusion of the enhanced ASPP [9] module and the CA [20] module. This fusion facilitates the efficient extraction of multi-scale semantic features in RS images. Due to the addition of an extra feature extraction channel in ASPP, as the backbone network performs feature extraction, the resolution of the feature maps gradually decreases. The combination of (1, 6, 12, 18) is not optimal for effectively extracting multi-resolution feature maps. Insufficient utilization of smaller dilation rates hinders the segmentation capability of small targets, resulting in weaker segmentation ability of the network for ground object targets at different scales. Therefore, it is necessary to readjust the dilation rates of the atrous convolution. Considering our two distinct datasets, to avoid redundant experiments, we exclusively conducted the experimentation on the Potsdam dataset for readjusting the dilation rates. In order to effectively extract multi-scale contextual features and enhance the segmentation performance for ground object targets of varying scales, this paper follows the guidelines outlined in Section 2.2 to determine rational dilation rates. To this end, we devised five groups of experiments with different dilation rates for comparison within the ASPP⁺-LANet network, which comprise of (1, 2, 4, 6, 8), (1, 2, 4, 8, 12), (1, 3, 6, 12, 18), (1, 3, 8, 16, 18), and (1, 3, 8, 18, 24). The experimental results are presented in Table 5. According to the evaluation metrics obtained from different combinations of dilation rates, the experiment achieved optimal results when the dilation rates were (1, 2, 4, 8, 12). This is because such dilation rate settings are well-suited for feature extraction of ground object targets at different scales in the Potsdam dataset. When the dilation rate is too large or too small, it adversely affects the effectiveness of feature extraction.

Table 5. Comparative experiments with different dilation rates of ASPP⁺ on the ASPP⁺-LANet.

Dilation Rate	PA/%	F1/%	MIoU/%
(1, 2, 4, 6, 8)	95.50	82.51	77.74
(1, 2, 4, 8, 12)	95.53	82.57	77.81
(1, 3, 6, 12, 18)	95.47	82.39	77.60
(1, 3, 8, 16, 18)	95.46	82.51	77.74
(1, 3, 8, 18, 24)	95.51	82.52	77.73

3.3.5. Comparative Analysis of Activation Functions

In order to validate the effectiveness of the FReLU activation function, this paper conducted experimental comparisons of different activation functions on the benchmark network, LANet. Considering the availability of two datasets, to avoid redundant experiments, we exclusively performed activation function comparisons on the Potsdam dataset. The results are summarized in the following table.

As depicted in Table 6, among the numerous activation functions examined, the incorporation of the FReLU activation function into the baseline LANet network yielded the most favorable segmentation results on the Potsdam dataset. The evaluation metrics, including PA, F1, and MIoU, exhibited remarkable values of 95.22%, 82.05%, and 77.06%, respectively. These findings highlight the superiority of the FReLU activation function in enhancing the segmentation performance, specifically for RS tasks.

Table 6. Experimental Comparisons of Different Activation Functions on the LANet Network.

Activation Function	PA/%	F1/%	MIoU/%
LANet + LeakyReLU [26]	93.34	78.73	72.31
LANet + PReLU [28]	94.37	80.65	74.94
LANet + ELU [30]	90.23	72.58	64.83
LANet + Mish [31]	89.99	73.50	65.74
LANet + DY-ReLU [32]	94.10	80.26	74.40
LANet + FReLU	95.22	82.05	77.06

4. Discussion

According to the ablation experiments, the improved model effectively improves the accuracy of building extraction, as indicated in Tables 3 and 4. Moreover, from the first and second plots of Figures 20 and 21, we can see that our proposed network performs outstandingly well in segmenting large ground object targets as well as small ground object targets. In addition, the segmentation effect of the effect map with the ASPP module alone is much better than that of the effect map with FReLU alone, which indicates that the ASPP module can indeed effectively improve the segmentation effect of ground object targets at different scales. This is due to the fact that ASPP is a multi-scale module, which can effectively enhance the network's ability to extract multi-scale contexts. From the third and fourth plots of Figures 20 and 21, we can see that our proposed network performs outstandingly well in segmenting slender ground object targets and ground object edges. However, the segmentation effect of the effect map with the ASPP module alone is much lower than that of the effect map with FReLU alone, which indicates that the FReLU module can indeed effectively improve the segmentation effect of the slender ground object targets and ground object edges. This is because FReLU is able to filter noise and low-frequency information and retain more high-frequency information, while slender ground object targets, as well as ground object edges, mostly belong to high-frequency information.

Regarding the ASPP$^+$ module, we conducted detailed experiments on its dilation rate settings, as shown in Section 3.3.4. We found that the setting of the dilation rate is not the larger or smaller as being better for different sizes of feature targets; larger segmentation targets can be segmented by convolutional kernels with larger dilation rates; on the contrary, smaller targets can be segmented by convolutional kernels with smaller dilation rates. Therefore, the dilation rate should be set reasonably and appropriately in order to make the segmentation targets of different sizes achieve effective feature extraction.

Regarding the selection of the activation function, we also conducted detailed experiments on it, as shown in Section 3.3.5. The activation function can enhance the generalization ability of the network, filter noise and low-frequency information, and retain more high-frequency information, which can effectively improve the performance of the network. However, different activation functions do not improve the performance of the network in the same way; therefore, in this paper, the activation functions proposed in recent years are compared and tested, and the most suitable activation function for the network in this paper is derived.

In order to improve the robustness of the model, in this paper, we use the data enhancement method to perform operations such as random flipping on the Potsdam and Vaihingen datasets. We also discuss the impact of the data enhancement method on the semantic segmentation results and, based on the analysis in Figures 22 and 23, it can be seen that the use of the data enhancement method improves the combined performance metrics over the non-use of the data enhancement method on both semantic segmentation datasets, provided that all other conditions remain consistent. This further indicates that data enhancement is one of the factors that improve the semantic segmentation results of the method proposed in this paper.

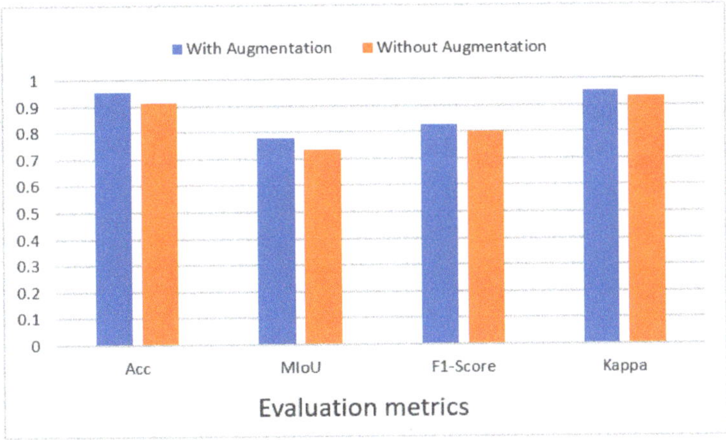

Figure 22. Effect of data augmentation on semantic segmentation results for the Potsdam dataset.

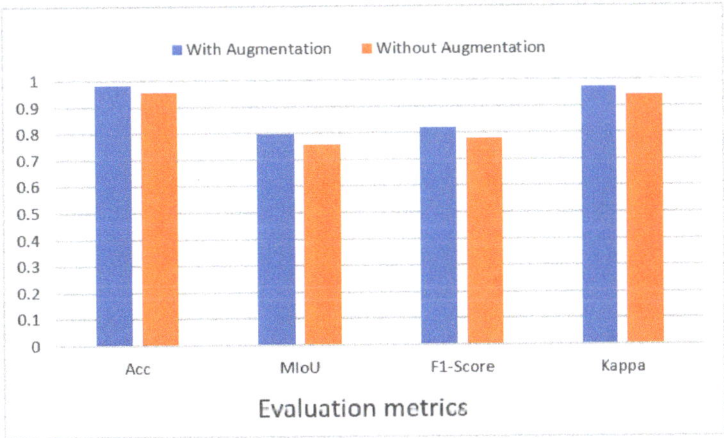

Figure 23. Effect of data augmentation on semantic segmentation results for the Vaihingen dataset.

5. Conclusions

In this paper, we propose a multi-scale context extraction network for semantic segmentation of high-resolution RS images, ASPP^{+}-LANet, aiming to fully capture the rich characteristics of ground object features. Firstly, we design a new ASPP^{+} module, expanding upon the ASPP module by incorporating an additional feature extraction channel and redesigning the dilation rate, which effectively improves the segmentation effect of ground object features at different scales by controlling the size of the dilation rate. Furthermore, the CA mechanism has been introduced to extract meaningful features and acquire contextual information. The FReLU activation function has been incorporated to enhance the segmentation effect of slender ground object targets and refine the segmentation edges. Therefore, on the Potsdam and Vaihingen datasets, ASPP^{+}-LANet achieves superior segmentation performance for ground object targets at different scales, as well as slender and limbic ground object targets.

Nevertheless, certain limitations of our current approach must be acknowledged, especially concerning the influence of shadows on the segmentation accuracy of buildings, vegetation, and other objects, as well as the segmentation boundaries of non-smooth objects. Changes in the color of objects like buildings and vegetation can be induced by shadows. To address this issue, a more precise color division is required to distinguish between shadows and actual objects, aiming to enhance accuracy levels. Furthermore, in the detection o non-smooth object edges, there is a need to enhance the network's capability to identify small target objects. This is crucial as object edges with jagged features can be perceived as tiny targets.

In the future, we will explore better methods to achieve higher accuracy and ef ficiency in RS image segmentation tasks. Firstly, we will be more specific in dividing the colors to distinguish the shadows from the actual objects; secondly, we will use the lightweight module to better optimize the network model and improve the network mode segmentation efficiency and segmentation accuracy to solve the non-smooth ground objects edges problem.

Author Contributions: Conceptualization, L.H. and X.Z.; funding acquisition, L.H.; investigation X.Z. and L.H.; methodology, X.Z. and L.H.; project administration, L.H. and S.L.; software, X.Z. and J.R.; writing–original draft, X.Z., L.H., S.L. and J.R.; writing–review and editing, L.H., X.Z. and J.R supervision, L.H.; All authors have read and agreed to the published version of the manuscript.

Funding: This research was funded by the National Natural Science Foundation of China, grant number 61662033.

Data Availability Statement: This data can be found here: https://www.isprs.org/education/benchmarks/UrbanSemLab/2d-sem-label-potsdam.aspx and https://www.isprs.org/education/benchmarks/UrbanSemLab/2d-sem-label-vaihingen.aspx, all accessed on 15 October 2022.

Acknowledgments: The authors would like to thank the editor and the anonymous reviewers who provided insightful comments on improving this article.

Conflicts of Interest: The authors declare no conflicts of interest.

References

1. Xu, S.; Pan, X.; Li, E.; Wu, B.; Bu, S.; Dong, W.; Xiang, S.; Zhang, X. Automatic Building Rooftop Extraction from Aerial Images via Hierarchical RGB-D Priors. *IEEE Trans. Geosci. Remote Sens.* **2018**, *56*, 7369–7387. [CrossRef]
2. Long, J.; Shelhamer, E.; Darrell, T. Fully Convolutional Networks for Semantic Segmentation. In Proceedings of the IEEE Conference on Computer Vision and Pattern Recognition, Boston, MA, USA, 7–12 June 2015; pp. 3431–3440.
3. Ronneberger, O.; Fischer, P.; Brox, T. U-net: Convolutional Networks for Biomedical Image Segmentation. In Proceedings of the Medical Image Computing and Computer-Assisted Intervention–MICCAI 2015: 18th International Conference, Proceedings, Part III 18, Munich, Germany, 5–9 October 2015; Springer: Cham, Switzerland, 2015; pp. 234–241.
4. Badrinarayanan, V.; Kendall, A.; Cipolla, R. Segnet: A Deep Convolutional Encoder-Decoder Architecture for Image Segmentation. *IEEE Trans. Pattern Anal. Mach. Intell.* **2017**, *39*, 2481–2495. [CrossRef] [PubMed]
5. Zhao, H.; Shi, J.; Qi, X.; Wang, X.; Jia, J. Pyramid Scene Parsing Network. In Proceedings of the IEEE Conference on Computer Vision and Pattern Recognition, Honolulu, HI, USA, 21–26 July 2017; pp. 2881–2890.
6. Chen, L.C.; Papandreou, G.; Kokkinos, I.; Murphy, K.; Yuille, A.L. Semantic image segmentation with deep convolutional nets and fully connected crfs. *arXiv* **2014**, arXiv:1412.7062.
7. Chen, L.C.; Papandreou, G.; Kokkinos, I.; Murphy, K.; Yuille, A.L. DeepLab: Semantic Image Segmentation with Deep Convolutional Nets, Atrous Convolution, and Fully Connected CRFs. *IEEE Trans. Pattern Anal. Mach. Intell.* **2017**, *40*, 834–848. [CrossRef] [PubMed]
8. Chen, L.C.; Papandreou, G.; Schroff, F.; Adam, H. Rethinking Atrous Convolution for Semantic Image Segmentation. *arXiv* **2017**, arXiv:1706.05587.
9. Chen, L.C.; Zhu, Y.; Papandreou, G.; Schroff, F.; Adam, H. Encoder-Decoder with Atrous Separable Convolution for Semantic Image Segmentation. In Proceedings of the Computer Vision—ECCV 2018, Munich, Germany, 8–14 September 2018; Ferrari, V., Hebert, M., Sminchisescu, C., Weiss, Y., Eds.; Springer: Cham, Switzerland, 2018; pp. 833–851.
10. Ding, L.; Tang, H.; Bruzzone, L. LANet: Local Attention Embedding to Improve the Semantic Segmentation of Remote Sensing Images. *IEEE Trans. Geosci. Remote Sens.* **2020**, *59*, 426–435. [CrossRef]
11. Li, R.; Zheng, S.; Zhang, C.; Duan, C.; Su, J.; Wang, L.; Atkinson, P.M. Multiattention Network for Semantic Segmentation of Fine-Resolution Remote Sensing Images. *IEEE Trans. Geosci. Remote Sens.* **2021**, *60*, 5607713. [CrossRef]
12. Xu, J.; Xiong, Z.; Bhattacharyya, S.P. PIDNet: A Real-Time Semantic Segmentation Network Inspired by PID Controllers. In Proceedings of the IEEE/CVF Conference on Computer Vision and Pattern Recognition, Vancouver, BC, Canada, 17–24 June 2023; pp. 19529–19539.
13. Xu, M.; Wang, W.; Wang, K.; Dong, S.; Sun, P.; Sun, J.; Luo, G. Vision Transformers (ViT) Pretraining on 3D ABUS Image and Dual-CapsViT: Enhancing ViT Decoding via Dual-Channel Dynamic Routing. In Proceeding of the IEEE International Conference on Bioinformatics and Biomedicine (BIBM), Istanbul, Turkiye, 5–8 December 2023; pp. 1596–1603.

4. Zheng, S.; Lu, J.; Zhao, H.; Zhu, X.; Luo, Z.; Wang, Y.; Fu, Y.; Feng, J.; Xiang, T.; Torr, P.H.; et al. Rethinking Semantic Segmentation from a Sequence-to-Sequence Perspective with Transformers. In Proceedings of the IEEE/CVF Conference on Computer Vision and Pattern Recognition, Nashville, TN, USA, 20–25 June 2021; pp. 6881–6890.
5. Strudel, R.; Garcia, R.; Laptev, I.; Schmid, C. Segmenter: Transformer for Semantic Segmentation. In Proceedings of the IEEE/CVF International Conference on Computer Vision, Montreal, QC, Canada, 10–17 October 2021; pp. 7242–7252.
6. Xie, E.; Wang, W.; Yu, Z.; Anandkumar, A.; Alvarez, J.M.; Luo, P. SegFormer: Simple and Efficient Design for Semantic Segmentation with Transformers. *Adv. Neural Inf. Process. Syst.* **2021**, *34*, 12077–12090.
7. He, X.; Zhou, Y.; Zhao, J.; Zhang, D.; Yao, R.; Xue, Y. Swin Transformer Embedding UNet for Remote Sensing Image Semantic Segmentation. *IEEE Trans. Geosci. Remote Sens.* **2022**, *60*, 4408715. [CrossRef]
8. Wang, L.; Li, R.; Zhang, C.; Fang, S.; Duan, C.; Meng, X.; Atkinson, P.M. UNetFormer: A UNet-Like Transformer for Efficient Semantic Segmentation of Remote Sensing Urban Scene Imagery. *ISPRS J. Photogramm. Remote Sens.* **2022**, *190*, 196–214. [CrossRef]
9. Zhang, C.; Jiang, W.; Zhang, Y.; Wang, W.; Zhao, Q.; Wang, C. Transformer and CNN Hybrid Deep Neural Network for Semantic Segmentation of Very-High-Resolution Remote Sensing Imagery. *IEEE Trans. Geosci. Remote Sens.* **2022**, *60*, 4408820. [CrossRef]
20. Hou, Q.; Zhou, D.; Feng, J. Coordinate Attention for Efficient Mobile Network Design. In Proceedings of the IEEE Conference on Computer Vision and Pattern Recognition (CVPR), Nashville, TN, USA, 20–25 June 2021; pp. 13708–13717.
21. Ma, N.; Zhang, X.; Sun, J. Funnel Activation for Visual Recognition. In Proceedings of the European Conference on Computer Vision (ECCV), Glasgow, UK, 23–28 August 2020; pp. 351–368.
22. Rottensteiner, F.; Sohn, G.; Gerke, M.; Wegner, J.D.; Breitkopf, U.; Jung, J. Results of the ISPRS Benchmark on Urban Object Detection and 3D Building Reconstruction. *ISPRS J. Photogramm. Remote Sens.* **2014**, *93*, 256–271. [CrossRef]
23. Lyu, Y.; Vosselman, G.; Xia, G.S.; Yilmaz, A.; Yang, M.Y. UAVid: A Semantic Segmentation Dataset for UAV Imagery. *ISPRS J. Photogramm. Remote Sens.* **2020**, *165*, 108–119. [CrossRef]
24. He, K.; Zhang, X.; Ren, S.; Sun, J. Deep Residual Learning for Image Recognition. In Proceedings of the IEEE Conference on Computer Vision and Pattern Recognition, Las Vegas, NV, USA, 27–30 June 2016; pp. 770–778.
25. Wang, P.; Chen, P.; Yuan, Y.; Liu, D.; Huang, Z.; Hou, X.; Cottrell, G. Understanding Convolution for Semantic Segmentation. In Proceedings of the IEEE Winter Conference on Applications of Computer Vision (WACV), Lake Tahoe, NV, USA, 12–15 March 2018; pp. 1451–1460.
26. Maas, A.L.; Hannun, A.Y.; Ng, A.Y. Rectifier Nonlinearities Improve Neural Network Acoustic Models. *Proc. ICML* **2013**, *30*, 3.
27. Glorot, X.; Bordes, A.; Bengio, Y. Deep Sparse Rectifier Neural Networks. *J. Mach. Learn Res.* **2011**, *15*, 315–323.
28. He, K.; Zhang, X.; Ren, S.; Sun, J. Delving Deep into Rectifiers: Surpassing Human-Level Performance on Imagenet Classification. In Proceedings of the IEEE International Conference on Computer Vision, Santiago, Chile, 7–13 December 2015; pp. 1026–1034.
29. Kingma, D.P.; Ba, J. Adam: A Method for Stochastic Optimization. *arXiv* **2014**, arXiv:1412.6980.
30. Clevert, D.; Unterthiner, T.; Hochreiter, S. Fast and Accurate Deep Network Learning by Exponential Linear Units (elus). *arXiv* **2015**, arXiv:1511.07289.
31. Misra, D. Mish: A Self Regularized Non-monotonic Activation Function. *arXiv* **2019**, arXiv:1908.08681.
32. Chen, Y.; Dai, X.; Liu, M.; Chen, D.; Yuan, L.; Liu, Z. Dynamic Relu. In Proceedings of the European Conference on Computer Vision, Online, 23–28 August 2020; pp. 351–367.

remote sensing

Article

Multi-Scale Feature Fusion Network with Symmetric Attention for Land Cover Classification Using SAR and Optical Images

Dongdong Xu [1,*], Zheng Li [1,2], Hao Feng [1,2], Fanlu Wu [1] and Yongcheng Wang [1]

[1] Changchun Institute of Optics, Fine Mechanics and Physics, Chinese Academy of Sciences, Changchun 130033, China; lizheng20@mails.ucas.ac.cn (Z.L.); fenghao21@mails.ucas.ac.cn (H.F.); flwu@ciomp.ac.cn (F.W.); wangyc@ciomp.ac.cn (Y.W.)
[2] University of Chinese Academy of Sciences, Beijing 100049, China
* Correspondence: xudongdong@ciomp.ac.cn

Abstract: The complementary characteristics of SAR and optical images are beneficial in improving the accuracy of land cover classification. Deep learning-based models have achieved some notable results. However, how to effectively extract and fuse the unique features of multi-modal images for pixel-level classification remains challenging. In this article, a two-branch supervised semantic segmentation framework without any pretrained backbone is proposed. Specifically, a novel symmetric attention module is designed with improved strip pooling. The multiple long receptive fields can better perceive irregular objects and obtain more anisotropic contextual information. Meanwhile, to solve the semantic absence and inconsistency of different modalities, we construct a multi-scale fusion module, which is composed of atrous spatial pyramid pooling, varisized convolutions and skip connections. A joint loss function is introduced to constrain the backpropagation and reduce the impact of class imbalance. Validation experiments were implemented on the DFC2020 and WHU-OPT-SAR datasets. The proposed model achieved the best quantitative values on the metrics of OA, Kappa and mIoU, and its class accuracy was also excellent. It is worth mentioning that the number of parameters and the computational complexity of the method are relatively low. The adaptability of the model was verified on RGB–thermal segmentation task.

Keywords: land cover classification; SAR and optical images; attention mechanism; multi-scale feature fusion; semantic segmentation

Citation: Xu, D.; Li, Z.; Feng, H.; Wu, F.; Wang, Y. Multi-Scale Feature Fusion Network with Symmetric Attention for Land Cover Classification Using SAR and Optical Images. *Remote Sens.* **2024**, *16*, 957. https://doi.org/10.3390/rs16060957

Academic Editor: Andrea Garzelli

Received: 10 January 2024
Revised: 23 February 2024
Accepted: 6 March 2024
Published: 8 March 2024

1. Introduction

Semantic segmentation refers to pixel-level annotations and different types of objects can be distinguished in segmented maps. It is a more refined task than classification and detection, and has been widely developed in assisted driving, geological detection and medical image analysis, among other scenarios. In particular, in Earth observation (EO) missions, land use and land cover (LULC) classification has become a key link in remote sensing (RS) data interpretation. Such classification results are already used for crop monitoring, urban development planning, disaster response and other tasks [1,2]. However, most of the common segmentation methods are based on unimodal RS images [3], which are insufficient for complex scene representation [4]. Spectral confusion or noise interference often affects the accuracy of classification. With the continuous development of sensors and imaging techniques, it becomes slightly easier to acquire multi-modal remote RS images of the same region simultaneously [5]. More comprehensive information about land cover can be acquired, further meeting the needs of advanced vision tasks. Optical images such as multi-spectral and hyperspectral images are still the primary data used for remote sensing classification [6]. The spatial resolution of these images is high, and more details of ground objects can be preserved. However, the imaging process is often disturbed by weather factors, especially frequent occlusion by clouds and fog. This is the drawback of catoptric imaging. In contrast, radar devices such as synthetic aperture radar (SAR) generate images by continuously transmitting microwaves and using scattered echoes. They are not easily disturbed and can almost work in all-day and all-weather conditions [7,8]. Therefore, SAR images can provide structural and electromagnetic scattering information but suffer from severe speckle noise, resulting in lower resolution [9]. It is easy to see that optical and SAR images are obviously complementary. Some objects that are difficult to recognize in a unimodal image might be clearly identified in another modal image. Therefore, the joint use of multi-modal image data is beneficial in improving the accuracy of land cover classification [8,10].

The joint application of optical and SAR images for semantic segmentation has been of great interest. The multi-modal classification methods can be roughly grouped into two categories, which are conventional machine learning and deep learning methods [2,11]. In the first category, support vector machines (SVMs) [12], conditional random fields (CRFs) [13], random forest (RF) [14], K-nearest neighbors (KNNs) [15] and other nonparametric approaches have been applied to classification tasks [16]. The above methods have achieved some classification accuracy. However, due to their weak feature extraction ability and insufficient high-dimensional information representation, they cannot obtain better classification results. Recently, deep neural networks with powerful feature extractors have shown great advantages in multi-modal classification tasks [17–19]. These methods can be subdivided according to the fusion level of the inherent information. Pixel-level fusion exists in early networks. Original pixels are fed to the multi-layer perceptron to aggregate the predictions [7]. The contextual information and the correlation between pixels are ignored. Decision-level fusion is applied in the late stage and depends on the results of several methods. The drawback is that the multi-dimensional features from different modalities are not considered [2]. At present, intermediate feature-level fusion, which focuses on the extraction and transformation of semantic features, is a research hotspot. The most dominant methods to effectively obtain and fuse the multi-modal information from optical and SAR images are the two-branch end-to-end segmentation models without weight-sharing [2,3,5,7,9,20]. These supervised methods have received the most attention and are constantly improving. The attention-based MCAM [2] module, the SACSM [3] module, the SaC [7] module and the SEPP [9] module have been used for salient feature extraction. Employed fusion strategies include the gate methods of GHFM [5] and CRGs [9], the cross-fusion method reported in [1], etc. Other optimizations with respect to pre-processing and loss functions are also considered. The ultimate goal is to realize the representation and complementary utilization of high-dimensional semantic features of different modalities. The supervised methods can achieve high accuracy, but they rely on registered images with semantic labels when training and testing. In reality, we may not obtain usable optical images immediately or there may be only a single unimodal image available at a time. This could lead to a rapid degradation of multi-modal classification performance. To this end, semantic knowledge distillation has been introduced for knowledge transfer and aggregation [21,22]. To address the scarcity of labeled data, some researchers have adopted self-supervised learning to realize joint segmentation with SAR and multi-spectral images [23–25]. In addition, with the further innovation of deep learning frameworks, graph convolution networks [26] and transformers [27] are gradually emerging in LULC classification tasks.

As described above, although the supervised two-branch models have some constraints, they are of important research significance for the joint classification of multi-modal SAR and optical images. After analysis and comparison, we think that there are still several challenges to be overcome. First, multi-modal semantic features are not effectively extracted. Attention modules [2,3,7,9] with square convolution kernels are defective for the representation of irregular objects. Anisotropic contextual information should be further integrated. Secondly, the fusion strategies for multi-scale features need to be improved. Existing methods [1,5,9] usually focus on high-level features, while low-level features and other complementary information are ignored. The semantic inconsistency of different modalities cannot be mitigated. Thirdly, multi-modal registered datasets with SAR and optical images for multi-class segmentation are extremely scarce [2,24,25]. The generalization and adaptability of the models have to be considered. Finally, the network structures have become gradually complicated to obtain higher classification accuracy. It is worth thinking about how to ensure the performance of the models with as few parameters and computation as possible.

The work of this paper is aimed at the situation and existing difficulties, and the main contributions are summarized below.

1. We propose a multi-modal segmentation model for the classification of optical and SAR images. It is an end-to-end network (SAMFNet) based on a multi-layer symmetric attention module and multi-scale feature fusion module. There are no other pretrained backbones in the framework.
2. A novel symmetric attention module is constructed with strip pooling. Multiple long receptive fields help to obtain more complementary and contextual information from the two branches. Atrous spatial pyramid pooling, varisized convolutions and skip connections are tactfully combined to fuse the multi-scale and multi-level semantic features.
3. The proposed model achieves the best numerical and visual results on two available datasets. The applicability of the model is proven on another RGB–thermal segmentation task. The designed network is relatively lightweight, and the computational costs and parameters are low, considering its classification accuracy.

2. Materials and Methods

2.1. Data Preparations

The SEN12MS is a curated dataset composed of dual-polarimetric SAR, multi-spectral images and MODIS (Moderate-Resolution Imaging Spectroradiometer)-derived land cover maps [28]. The first two are from Sentinel-1 and Sentinel-2. Multi-modal data were collected from regions of interest around the world with four seed values. There are 180,662 image patches in total with a size of 256 × 256. The ground sampling distance (GSD) of the original data can reach 10 m, but the land cover maps with labeled classes are at a lower resolution of 500 m. They are relatively crude for specific classification or detection tasks. In the 2020 IEEE GRSS Data Fusion Contest [29], the source images are the same as SEN12MS, and some high-resolution (10 m) labels were semi-manually generated for validation based on the original MODIS maps. As a result, 6114 image patches with high-resolution labels were obtained to construct the DFC2020 dataset. Figure 1 shows visual examples of the DFC2020 data.

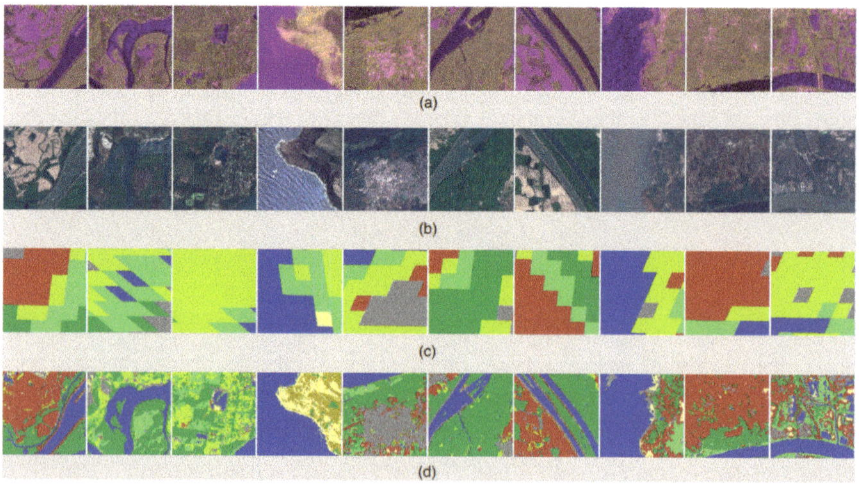

Figure 1. Source images and labels of DFC2020 [29]. (**a**) SAR images of Sentinel-1. (**b**) Optical images of Sentinel-2. (**c**) Low-resolution semantic labels. (**d**) High-resolution semantic labels.

As to WHU-OPT-SAR [2], there are 100 pairs of SAR and optical images from GF-3 and GF-1 with a size of 5556 × 3704. The imaging areas are all in Hubei Province, and the resolutions of source images and labels were unified to 5m. Figure 2 shows two examples of WHU-OPT-SAR data.

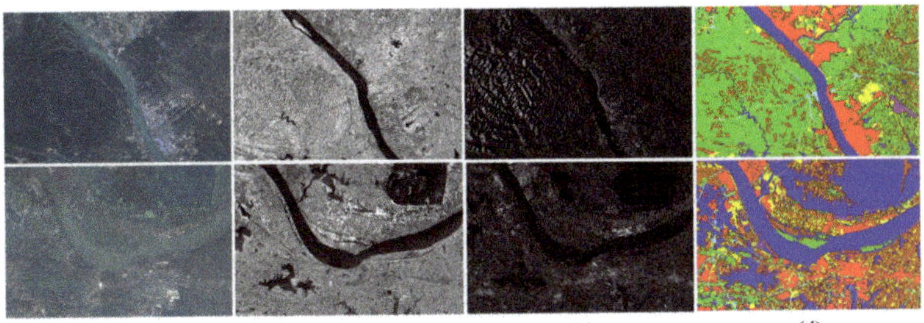

Figure 2. Source images and labels of WHU-OPT-SAR [2]. (**a**) RGB images of GF-1. (**b**) NIR images of GF-1. (**c**) SAR images of GF-3. (**d**) Semantic labels.

In order to train the model on these two datasets under a unified framework, the pre-processing of the data should be carried out in advance. Firstly, the number of images in WHU-OPT-SAR is small, but the image size is quite large. Therefore, augmentation was performed by cropping with a proper stride. Then, tens of thousands of patches with a size of 256 × 256 were obtained. Secondly, the number of channels of the multi-modal inputs needs to be consistent. Single-channel gray SAR images and four-channel (RGB and NIR) optical images were adopted. Since the multi-spectral images of DFC2020 have 13 bands, B4, B3, B2 and B8 were chosen to combine the corresponding RGBN inputs. Furthermore, flipping and scaling at multiple scales were also executed when importing the training data to improve the generalization capability of the model. Other details are presented in the subsequent experiments.

2.2. Attention Mechanisms for SAR and Optical Image Classification

In complex scenes, semantic segmentation networks usually face challenges such as information redundancy and poor relevance. In order to overcome these problems, over the years, researchers have introduced several attention mechanisms to the semantic segmentation models. They are able to extract more salient features adaptively, so the performances of diverse networks are effectively promoted [30–33]. For the joint classification of SAR and optical images, methods with attention mechanisms can be mainly classified as two types. Some methods are built with channel attention [3], spatial attention [34] or their combined block modules [7], such as CBAM [35]. By introducing these mechanisms, the models are able to learn the importance of channels and regions of the source images automatically. They pay more attention to the target itself and ignore the contents that are detrimental to the classification task. The multi-scale information is also focused [36,37]. Other methods are based on self-attention [2,9,16]. The essence of the self-attention mechanism is to perform a mapping from a query to a sequence of key-value pairs. The correlations between the matrices are adequately calculated. Its advantage lies in the ability to establish global dependencies and capture internal correlations of features.

In fact, increasing the receptive field is the main purpose of the attention-based semantic segmentation methods. Self-attention can establish long-range dependencies, but such methods require large memory for complex matrix calculations. Dilated convolution [38] and global/pyramid pooling [39,40] have also been used to improve the receptive field, but they are both confined to the square convolution kernels [41]. To this end, strip pooling (SP) with two-cross long kernels has been proposed for salient feature extraction, which can be seen as an effective attention mechanism [41–43]. Encoders with strip pooling are able to probe the input feature maps through long windows so that objects with irregular shapes are easy to process and more anisotropic contextual information in complex scenes can be obtained. In view of the above advantages, this paper introduces the use of pooling into multi-modal semantic segmentation to solve the problems of detail errors and class confusion in land cover classification.

2.3. Proposed SAMFNet

2.3.1. Framework of the SAMFNet

The proposed SAMFNet (Figure 3) is a concise end-to-end model, and the two inputs are single-channel SAR images and four-channel optical images. The symmetric attention extraction module (SAEM) is embedded at the medial axis of the network. It can extract and supplement distinguishing features to each convolutional group of the two branches for subsequent calculations. With the deepening of the network layers, features obtained by the attention mechanism are gradually refined. More contextual information can be gathered for feature encoding. Atrous spatial pyramid pooling (ASPP) [38] and convolutions with varisized kernel sizes are combined as the multi-scale semantic extraction module (MSEM) to obtain more high-level features. To complement more low-level features, skip connections are added to realize information transmission from shallow layers to deep layers. Then, the multi-level semantic features are concatenated and upsampled for decoding. The encoding and decoding processes of the proposed structure are relatively simple, but the performance of the network is pretty good. Meanwhile, the SAEM and MSEM can be replaced flexibly if other better feature extraction modules are explored.

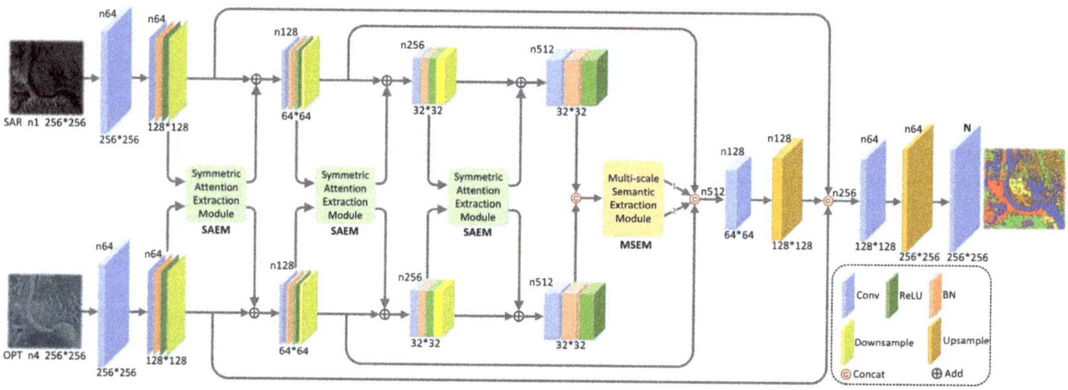

Figure 3. The framework of the proposed SAMFNet. n represents the number of channels.

2.3.2. Symmetric Attention Extraction Module

The SAEM designed in this paper is inspired by the multi-modal transfer module (MMTM) [44]. The reason it is called symmetric attention is that the inputs of this module are the multi-modal features from the two branches. Then, the obtained salient features are fed back to the original branch. Figure 4 shows the detailed form of the module. The transformation process is annotated with symbols.

The leftmost part of the figure shows preliminary feature convolution with 1×1 kernels, halving the number of the concatenated channels. F_{in1} and F_{in2} represent the inputs from the two branches. F_B is the basic feature map, which is actually a four-dimensional tensor. A single-channel and two-dimensional map with a size of $H \times W$ is taken as an example for explanation of the pooling process.

$$F_B = Conv(F_{in1}, F_{in2}) \tag{1}$$

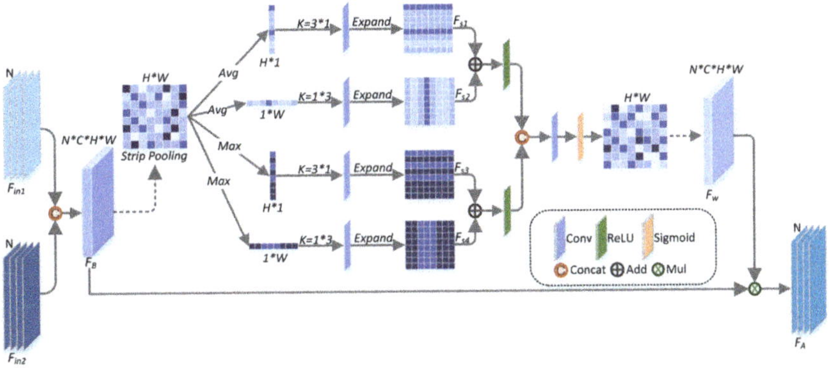

Figure 4. The structure of the symmetric attention extraction module.

It can be seen that four strip pooling branches (two average pooling and two max pooling) were designed for feature representation. The map is compressed into a single row or column after pooling. More global information is obtained through these long receptive fields. After each pooling, strip convolution with a specific kernel is performed for further feature transformation. Then, the single row or column features are expanded to a size of $H \times W$. F_{s1}, F_{s2}, F_{s3} and F_{s4} are acquired as the outputs of strip pooling.

$$F_s = Exp(Conv(Pool(F_B))) \quad (2)$$

The F_{s1} and F_{s2} are fused, then activated to obtain the final feature maps after average strip pooling. Similarly, the F_{s3} and F_{s4} are combined to obtain the feature maps after max strip pooling. Then, they are concatenated, and the feature weights (F_w) are acquired after convolution and nonlinear computation. At last, element-wise multiplication operations between F_B and F_w are implemented to calculate the ultimate highlighted feature (F_A) maps after the attention extraction module.

$$F_w = Sigmoid(Conv(ReLU(F_{s1} + F_{s2}), ReLU(F_{s3} + F_{s4}))) \quad (3)$$

$$F_A = F_B \otimes F_w \quad (4)$$

The unique advantage of strip pooling is that the long-range dependencies can be established easily. Average and max pooling with row and column transforms are introduced together, so land objects with different shapes and scales can be portrayed more accurately. Simultaneously, the particular strip forms can also remove unnecessary connections between feature maps, which greatly reduces the computational complexity compared to other attention-based algorithms.

2.3.3. Multi-Scale Semantic Extraction Module

After multiple groups of convolution transformation and salient feature extraction, high-level features of SAR and optical branches can be built. In fact, multi-scale information is essential for accurate semantic segmentation and other computer vision tasks. Many researchers have focused on this topic. Figure 5 shows the specific structure of the designed MSEM module.

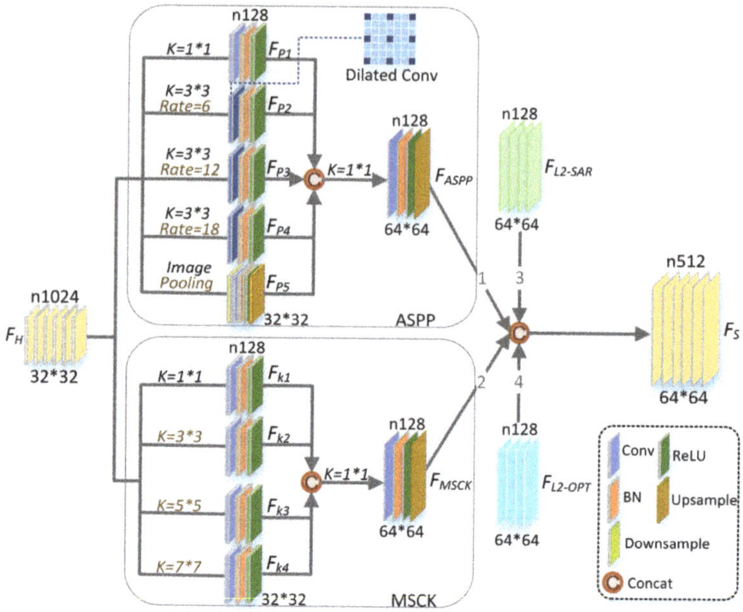

Figure 5. The structure of the multi-scale semantic extraction module.

In the figure above, F_H represents the concatenated result of the high-level features from the two branches. ASPP [38] and its variants have been widely adopted in segmentation tasks. They employ several dilated convolutions with different sampling rates to obtain multi-scale information. Such changes allow the network to utilize larger receptive fields without using regular pooling. They can also reduce information loss. The receptive fields are adjusted when setting different dilation rates, and corresponding multi-scale features are easily acquired. The rate group $[6, 12, 18]$ was used in this MSEM module. F_{P1} represents the features calculated by the normal convolution group (Conv + BN + ReLU), while F_{P2}, F_{P3} and F_{P4} are obtained by the dilated convolution group. F_{P5} is acquires by the pooling group and upsampling. Then, the feature maps of the five branches are concatenated together. The following convolution group is used to reduce the number of channels to the single branch. It should be noted that we performed an interpolation behind the convolution to maintain consistency with the size of low-level features from shallow layers.

The bottom half of the figure is called the multi-scale convolution kernel (MSCK) module. In the four branches, four different convolution kernels are used to extract specific features. F_{k1}, F_{k2}, F_{k3} and F_{k4} are constructed with increasing receptive fields. This helps to capture richer multi-scale information and can solve the problem of features of diverse ground objects not being distinctly distinguished in complex scenes. The convolution and interpolation operations are also executed to obtain the F_{MSCK}, which has the same tensor form as ASPP. At the end of the module, F_{ASPP} and F_{MSCK} represent the high-level features. F_{L2-SAR} and F_{L2-OPT} from the second convolutional group of multi-modal branches represent the low-level features. Then, they are combined in sequence to build the final multi-scale semantic feature maps (F_S). So far, we have completed the feature encoding process of SAR and optical images.

2.3.4. Decoding Process

Compared with feature encoding, the decoding process of semantic segmentation is relatively simple. The common methods use multiple groups of convolutions and interpolations to restore the features to the size of original inputs.

In Figure 6, F_{UP1} is acquired directly through the first pair of convolution and interpolation transformation. In the proposed method, in order to better realize feature mapping between the multi-modal inputs and the output, the low-level features (F_{L1-SAR} and F_{L1-OPT}) after the first convolutional group of dual branches are added through skip connections. Therefore, more foundational information can be utilized, which is beneficial to improve the classification accuracy. F_{UP2} represents the results of the second interpolation. F_{Map} is achieved after the final convolution operation. N represents the total number of classes of ground objects. Finally, F_{Map} is used to facilitate the calculation of the confusion matrix.

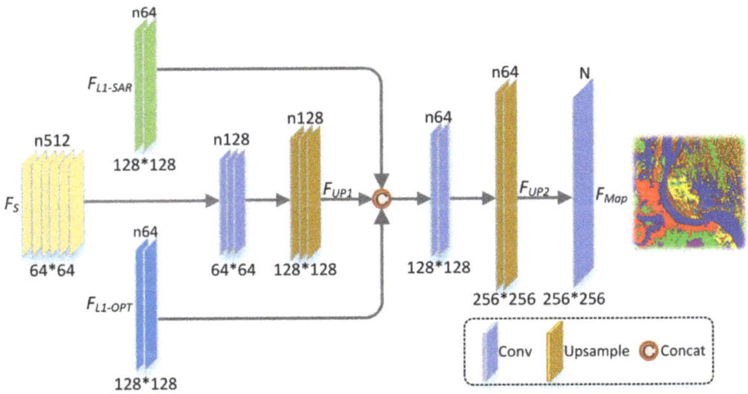

Figure 6. The structure of the decoder.

2.3.5. Joint Loss Function

In semantic segmentation tasks, pixel-level classification needs to be achieved, so the corresponding pixel-level loss functions are used to constrain network training. However, a single loss function usually cannot describe the segmentation result objectively and comprehensively. The introduction of multiple functions to form a joint loss function [7,15,21] has been proven to be conducive to obtaining better numerical and graphic results. In this article, fully considering the multi-modal characteristics of the input images and the complexity of the classification scenes, the following three loss functions are combined.

$$L = L_{CE} + L_{Focal} + L_{SIoU} \tag{5}$$

where L_{CE} represents the cross-entropy loss, which has been widely adopted in image segmentation and other classification tasks. L_{CE} is defined as follows:

$$L_{CE} = -\frac{1}{N}\sum_{n=1}^{N} y_n log(p_n) \tag{6}$$

where y_n and p_n denote the label and the predicted output of image n, respectively. N is the total number of inputs for computation. The definitions of the symbols in the following formulas are the same. The loss focuses on pixel-level information and inevitably ignores the spatial consistency between prediction regions. This may lead to scattered and discontinuous segmentation regions when the amount of different samples is unbalanced. Some researchers proposed using the focal loss (L_{Focal}) [45] to address this class imbalance problem. L_{Focal} is defined as follows:

$$L_{Focal} = -\frac{1}{N}\sum_{n=1}^{N}\alpha(1 - p_n e^{-y_n})^{\gamma} y_n log(p_n) \quad (7)$$

where α denotes the conditioning weight of the positive samples. The exponent γ is the decay factor of L_{CE}. These two hyperparameters are adjusted to balance the positive and negative samples. Moreover, the IoU loss [46] is also taken into account. IoU is the intersection over union. It can be seen as an acknowledged metric for evaluating the effectiveness of object detection and segmentation algorithms. The IoU loss is able to better constrain the similarity between the segmentation results and the true segmentation labels. The soft IoU loss (L_{SIoU}) additionally adds softmax to the predicted output for smoothing. L_{SIoU} is defined as follows:

$$L_{SIoU} = -\frac{1}{C}\sum_{c=1}^{C}\frac{\sum_{n=1}^{N} y_n p_n}{\sum_{n=1}^{N}(y_n + p_n - y_n p_n)} \quad (8)$$

where C is the total number of classes. It has been proven that the joint loss function can balance the various optimization objectives and achieve better segmentation results.

3. Results

The experiments are presented in detail in the following subsections. More specific analysis of the images and numerical results are provided.

3.1. Experimental Settings

The experiments reported in the article were all performed on a server with two RTX3090, and the GPU memory was 24 GB. Pytorch (1.13) on an Ubuntu (18.04) system was used to build the network framework. The Adam optimizer was adopted for parameter updating, and the weight decay was 0.0001. The step size and the gamma of StepLR are 30 and 0.1. The basic learning rate was set to 0.001. To keep the image sizes consistent, all the inputs of the two datasets were cropped to 256 × 256. The batch size for training was set to 32. Limited by the capabilities of the server, the image size and batch size were set to 128 × 128 and 4 when TAFFN was implemented. A total of 6114 pairs of images of the DFC2020 dataset and 29,400 pairs of the WHU-OPT-SAR dataset were obtained. The ratio of the training set to the testing set was 4:1.

3.2. Evaluation Metrics

The segmentation results should be evaluated more accurately, so overall accuracy (OA), Kappa coefficient (Kappa) and mean intersection over union (mIoU) were used for numerical measurements. OA focuses on how well all samples are classified. Kappa is used for consistency checking which can also measure the classification accuracy. mIoU is the average of the ratio of the intersection and union of the true and predicted pairs for each class. It can be seen as a standard metric for semantic segmentation. The formulas of the above three metrics are defined one by one as follows:

$$OA = \frac{\sum_{i=1}^{K} p_{ii}}{\sum_{i=1}^{K}\sum_{j=1}^{K} p_{ij}} \times 100\% \quad (9)$$

$$p_e = \frac{\sum_{i=1}^{K}(\sum_{j=1}^{K} p_{ji} \times \sum_{j=1}^{K} p_{ij})}{(\sum_{i=1}^{K}\sum_{j=1}^{K} p_{ij})^2} \tag{10}$$

$$Kappa = \frac{(OA - p_e)}{(1 - p_e)} \times 100\% \tag{11}$$

$$mIoU = \frac{1}{K}\sum_{i=1}^{K} \frac{p_{ii}}{\sum_{j=1}^{K} p_{ij} + \sum_{j=1}^{K} p_{ji} - p_{ii}} \times 100\% \tag{12}$$

where K is the total number of classes. It is equal to the width of the confusion matrix. p_{ij} is the amount of pixels in class i predicted as class j. Higher values of the three metrics indicate better results.

3.3. Experimental Analysis

In order to illustrate the feasibility of the proposed method, DeeplabV3+ [47], DenseASPP [48], DANet [30], CCNet [31], TAFFN [34] and MCANet [2] were selected for comparison. Among them, the attention mechanism is not introduced in DeeplabV3+ and DenseASPP. They mainly perform semantic feature transformation through dense connections and ASPP. Different attention methods are separately introduced in the remaining four methods for salient feature extraction. The codes used for validation are mainly provided by the authors or codebase. For the sake of fairness, the data loader, training process and loss function of these methods are consistent with the proposed method.

3.3.1. Experiments on DFC2020 Dataset

Comparative experiments were implemented on DFC2020, and the numerical results, including the accuracy of each class, are listed in Table 1. As shown in the last row, the proposed model achieved the best numerical results on OA, Kappa and mIoU among all the compared methods. These three metrics increased by 3.4%, 4.2% and 7.3%, respectively. The reasons for the better results lie in the following aspects. In the stage of feature encoding, the series wound symmetric attention module on the central axis is a contributing factor. The horizontal and vertical receptive fields are conducive to acquiring and transmitting multi-size contextual information. The gradually refined features are then fed back to each branch for secondary learning. Another important point is the integration of multi-scale features. This improves the semantic understanding of images. Both the enhanced high-level features and the shallow low-level features from the two branches are used to fuse the final output of the encoder. All these lay the foundation for decoding accurate segmentation maps. In a word, the effective extraction and interaction of complementary features are key to improving the numerical metrics. CCNet obtained suboptimal OA and Kappa values. The recurrent criss-cross attention focuses on full-image contextual information, but the mIoU is low. Partly because the number of parameters is large, it is difficult to train a stable network on a smaller dataset. MCANet performed well on all three metrics. The multi-modal cross-attention module and feature fusion module were introduced together to retain more features. DeepLabV3+ and DANet achieved similar results. They benefit from the use of ASPP and dual attention, respectively. The values of DenseASPP were slightly lower. The special structure with dense connections needs more data to optimize the weights. As to TAFFN, due to its high computational complexity, the existing server cannot support training when the images are large. Therefore, both experimental calculations and classification maps are based on the central part of the images (128 × 128). The image size of other methods is 256 × 256. Therefore, the results may be improved if the server has enough capacity. The comparison of parameters and computation is explained in the Section 4.

Meanwhile, the proposed method can also achieve maximum classification accuracy for each land cover target. This is particularly prominent in the classification of shrubland, grassland and barren. These three objects have many interaction areas with other objects (shrubland and forest, grassland and forest, and barren and cropland). They can be segmented more accurately because the multi-modal complementary information and high-level semantic features are all taken into account by the proposed method. From the numerical analysis, we can see that the proposed model has certain advantages.

The source images and classification maps of diverse models are demonstrated in Figure 7. Different land covers are distinguished by colors. The four selected groups of typical images contain all the categories. In the first group, in addition to the large area of forest and grassland in the label image, the meandering water body and its surrounding scattered wetland and cropland are the focuses of segmentation. DeepLabV3+ and DANet can mainly segment the obvious target area, and the degree of fine classification is not enough. In the image of DenseASPP, there are confusions between forest and grassland, wetland and water. The proposed method, MCANet and CCNet achieve better results. In the latter three groups of images, the interactions between different ground objects are more serious. In the second group, the proposed method, CCNet and DenseASPP work well. As to MCANet, large areas of wetland are misclassified as cropland. The superiority of the proposed model is more obvious in the third group. The forest and shrubland are almost mixed into one class in DeepLabV3+, DenseASPP, DANet and MCANet. At the same time, the distribution of barren is not well reflected. In the last group, all kinds of ground objects can be well reflected by MCANet and the proposed method. The accuracies of forest and barren need to be improved in other methods. The four segmentation images of TAFFN can generally reflect different classes of ground objects, but there are some obvious misclassified areas.

Table 1. Experimental results on DFC2020 dataset (Optimal values are in bold).

Models	Class Accuracy								OA	Kappa	mIoU
	Forest	Shrubland	Grassland	Wetland	Cropland	Built-Up	Barren	Water			
DeepLabV3+	0.8802	0.5803	0.6421	0.5638	0.8410	0.8320	0.4996	0.9890	0.8297	0.7920	0.6106
DenseASPP	0.8115	0.4729	0.7517	0.5612	0.7774	0.8911	0.4681	0.9866	0.8109	0.7706	0.5831
DANet	0.8856	0.5513	0.6617	0.5593	0.8257	0.8347	0.4367	0.9881	0.8267	0.7882	0.6022
CCNet	0.9159	0.5784	0.7330	0.5985	0.8290	0.8407	0.4398	0.9723	0.8419	0.8070	0.5028
TAFFN	0.9137	0.1022	0.4898	0.3616	0.7219	0.8258	0.0230	0.9904	0.7491	0.6897	0.4439
MCANet	0.8880	0.5655	0.7372	0.5541	0.8149	0.8699	0.3978	0.9927	0.8374	0.8018	0.6127
Proposed	**0.9291**	**0.6324**	**0.8247**	**0.6257**	**0.8423**	**0.8922**	**0.5890**	**0.9956**	**0.8763**	**0.8492**	**0.6853**

We can see that the obtained numerical results are basically consistent with the results of the classification maps. Generally, the proposed method and CCNet achieved the best performances. The segmented maps are closer to semantic labels. On the contrary, the forest and cropland accuracy of DenseASPP, the wetland and barren accuracy of MCANet and the grassland accuracy of DeepLabV3+ and DANet are relatively low. All these drawbacks are well reflected in the final classification maps.

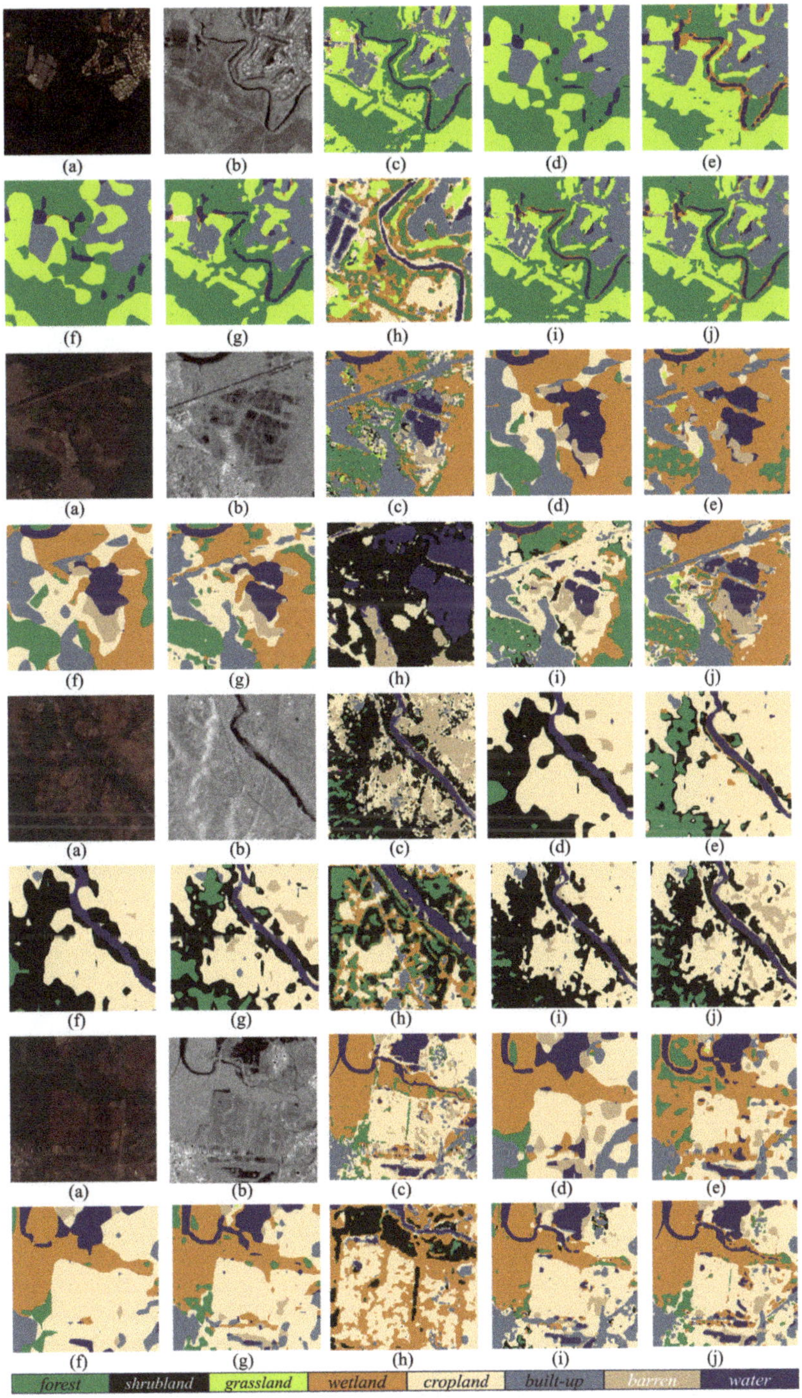

Figure 7. Land cover classification maps of different models on DFC2020. (**a**) Optical images. (**b**) SAR images. (**c**) Semantic labels. (**d**) DeepLabV3+. (**e**) DenseASPP. (**f**) DANet. (**g**) CCNet. (**h**) TAFFN (128 × 128). (**i**) MCANet. (**j**) Proposed SAMFNet.

3.3.2. Experiments on WHU-OPT-SAR Dataset

The WHU-OPT-SAR dataset is larger than DFC2020 and is more imbalanced. The relevant numerical results are listed in Table 2. The proposed model also obtained the best results on OA, Kappa and mIoU and achieved excellent performance in class accuracy. Except the OA of DenseASPP and TAFFN, the overall numerical results of these methods are somewhat reduced compared with the first dataset. The increase in data helps to optimize weights and can better leverage the advantages of dense connections and ASPP. Although the number of parameters of TAFFN is small, the computational complexity is very high. More data means more accurate inference. Therefore, the performance of these two methods was improved. CCNet still obtained suboptimal values of OA and Kappa, and the mIoU was also acceptable. DenseASPP obtained a suboptimal mIoU value and exceeded the results of MCANet. The values of DeepLabV3+ and DANet were slightly lower. The results of TAFFN can be improved with the proper setting of parameters and a large amount of computation.

For class accuracy, only the value of villages of the proposed method was slightly lower (−0.0015) than MCANet. The classification effects on water, road and other land use types were more excellent. This is attributed mainly to the efficient extraction and fusion of multi-modal salient features and multi-scale contextual information. These objects are linear, curving or scattered. The improvement of their accuracy is conducive to fine segmentation.

Table 2. Experimental results on WHU-OPT-SAR dataset (Optimal values are in bold).

Models	Class Accuracy							OA	Kappa	mIoU
	Farmland	City	Village	Water	Forest	Road	Others			
DeepLabV3+	0.8277	0.7636	0.6289	0.7777	0.8988	0.5420	0.3132	0.8180	0.7427	0.4786
DenseASPP	0.8290	0.7632	0.6560	0.8088	0.8998	0.5433	0.3307	0.8251	0.7532	0.4912
DANet	0.8255	0.7632	0.6246	0.7856	0.8952	0.4970	0.2860	0.8158	0.7397	0.4726
CCNet	0.8370	0.7527	0.6380	0.8175	0.8990	0.5613	0.2913	0.8268	0.7551	0.4908
TAFFN	0.7951	0.7298	0.4018	0.7146	0.8681	0.1424	0.0005	0.7622	0.6598	0.3602
MCANet	0.8269	0.7367	**0.6713**	0.8039	0.8990	0.5423	0.3003	0.8225	0.7497	0.4837
Proposed	**0.8379**	**0.7645**	0.6698	**0.8257**	**0.9025**	**0.5865**	**0.3646**	**0.8334**	**0.7652**	**0.5049**

Figure 8 shows the semantic segmentation results of different methods on WHU-OPT-SAR. We also used four groups of images of complex scenes for comparison. In the first group, the intersecting roads, scattered water and villages are hard to classify. The results of DenseASPP, CCNet, MCANet and the proposed method are relatively good. DeepLabV3+ and DANet are not accurate enough to describe the scattered regions and boundaries between different classes. In the second group, since the characteristics of cities and villages are similar, the results of DenseASPP and MCANet are somehow affected, so parts of city regions are regarded as villages. The proposed method and CCNet achieved preferable segmentation. In the third group, the distribution of objects in the source image is more complex. In the forest, there are pairwise interactions among water, farmland and villages, and there are multiple intersecting roads. The roads and water are not well-represented by DeepLabV3+ and DANet. Other methods work well in this scenario. In the last group, the roads, cities and villages are densely distributed, and water crisscrosses farmland. The ground objects are reflected more accurately by MCANet and the proposed method. For TAFFN, objects like narrow roads, similar villages and cities and irregular water cannot be reflected well in multiple groups of images. We can see that the numerical results and the classification maps of this dataset are also matched.

After the experimental analysis on the two datasets, it can be seen that the proposed model behaves well both on numerical results and classification maps. Meandering water, scattered villages and intersecting roads are all well segmented. More importantly, there are rarely large areas of misclassification or missing objects. The method has strong adaptability in complex scenes with various ground objects. The proposed framework, feature extraction strategies and joint loss function do play a big role in classification. The ablation experiments are explained in detail in the next section.

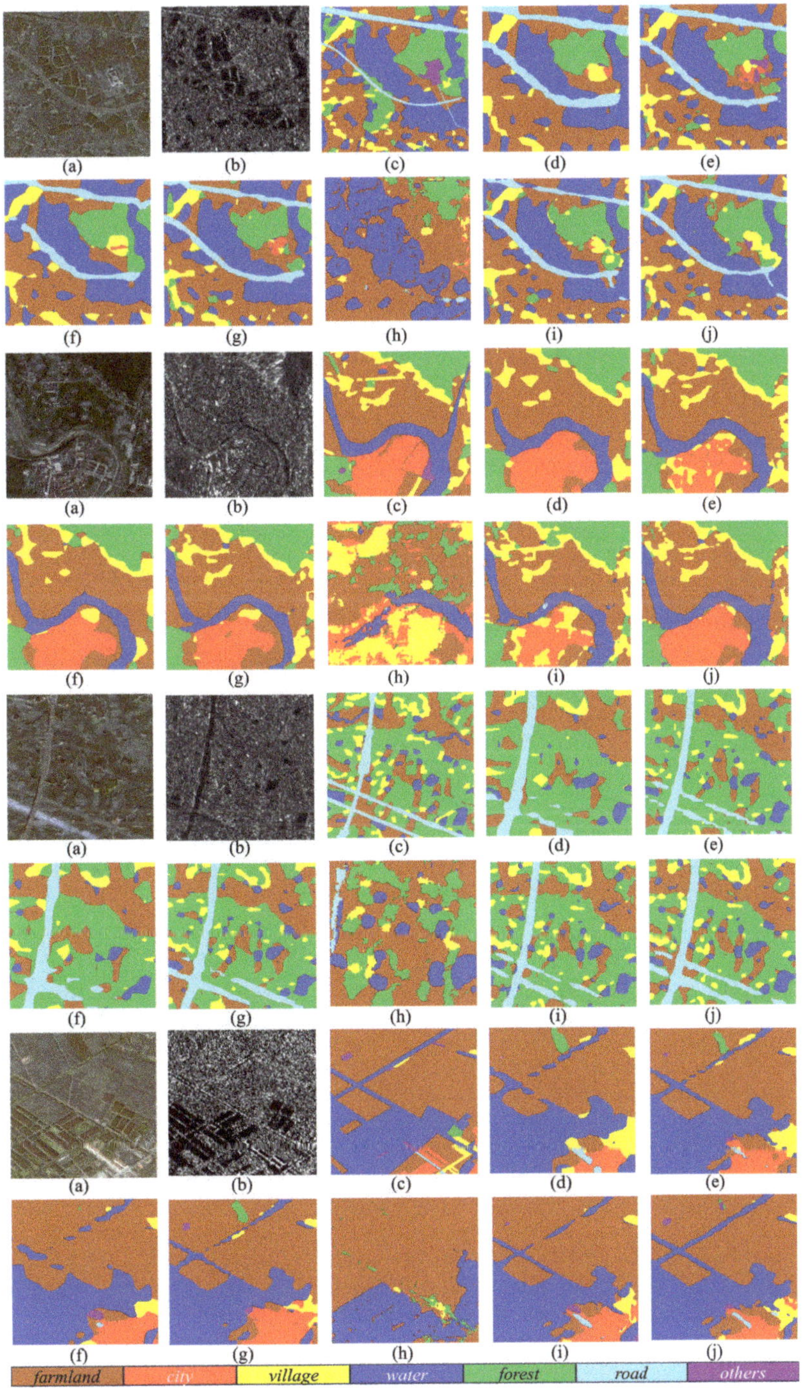

Figure 8. Land cover classification maps of different models on WHU-OPT-SAR. (**a**) Optical images. (**b**) SAR images. (**c**) Semantic labels. (**d**) DeepLabV3+. (**e**) DenseASPP. (**f**) DANet. (**g**) CCNet. (**h**) TAFFN (128 × 128). (**i**) MCANet. (**j**) Proposed SAMFNet.

4. Discussion

4.1. Computational Complexity

To further illustrate the efficiency of different methods, the number of parameters and the amount of computation are counted in Table 3. When training the network, growing the parameters increases the space complexity of the model. Devices with large video memory are needed. The amount of computation determines the execution time. It depends on the computing power of the GPU, so there are requirements for the flops of the hardware chip. We assume that for all models, the inputs are the coupled SAR and optical images with a size of 256 × 256. The values obtained under the two datasets are very close, since only the number of output channels of the last layer is different. For parameters, the proposed method is only inferior to TAFFN. There are no other pretrained backbones like ResNet or VGGNet added to the proposed model. It should be noted that although TAFFN has few parameters, its computational complexity is relatively high. Once the batch size increases, it is difficult for general GPUs to meet the requirements of flops. CCNet and MCANet performed well in the previous experimental analysis, but the number of parameters was indeed large. As to computation, although the flops of DeepLabV3+ and DANet are smaller, it is difficult for them to achieve desirable outcomes. The result of the proposed model is suboptimal. The computational complexity of CCNet and MCANet is relative high. In summary, the parameters and computational costs of the proposed method are low, and we can also obtain the best segmentation results. This will greatly improve the practicability of the model.

Table 3. Comparison of parameters and computation.

Models	Params	FLOPs	Input Tensor
DeepLabV3+	39.05 M	13.25G	SAR: [1,1,256,256] OPT: [1,4,256,256]
DenseASPP	35.39 M	39.40G	
DANet	47.45 M	14.41G	
CCNet	70.95 M	79.99G	
TAFFN	0.31 M	33.13G	
MCANet	85.93 M	102.39G	
Proposed	19.60M	28.78G	

4.2. Analysis of Different Attention Mechanisms

In this subsection, CBAM [35], SP [41] and the designed SAEM are compared under the unified framework proposed in Figure 3. Only part of the attention mechanism is different. In CBAM, the attention module consists of channel attention and spatial attention. Conventional max pooling and average pooling are used for feature transformation. In the original SP, only strip average pooling is introduced to generate the long receptive fields. We added two more branches with strip max pooling to SP. Then, multi-dimensional feature superposition, fusion and conversion were implemented to obtain more contextual information. From the numerical results in Table 4, we can see that SP performed slightly better than CBAM. The proposed model with SAEM achieved the best results on the three metrics, which were basically 1% higher.

Figure 9 further illustrates the effect of different attention mechanisms through three groups of classification maps. In the first group, although all three models obtained good classification results, the proposed method can describe the distribution of grassland more accurately. In the second group, the water is the focus to be classified, and it is clearly delineated by the proposed method. In the last group, the scattered barren are clearly classified by the proposed method, while the other two models only focus on large area. It can be seen that the designed SAEM is conducive to salient feature extraction and improves the fineness of segmentation.

Table 4. Comparison of attention mechanisms.

Models	OA	Kappa	mIoU
Att-CBAM	0.8658	0.8363	0.6640
Att-SP	0.8680	0.8390	0.6681
Proposed-SAEM	0.8763	0.8492	0.6853

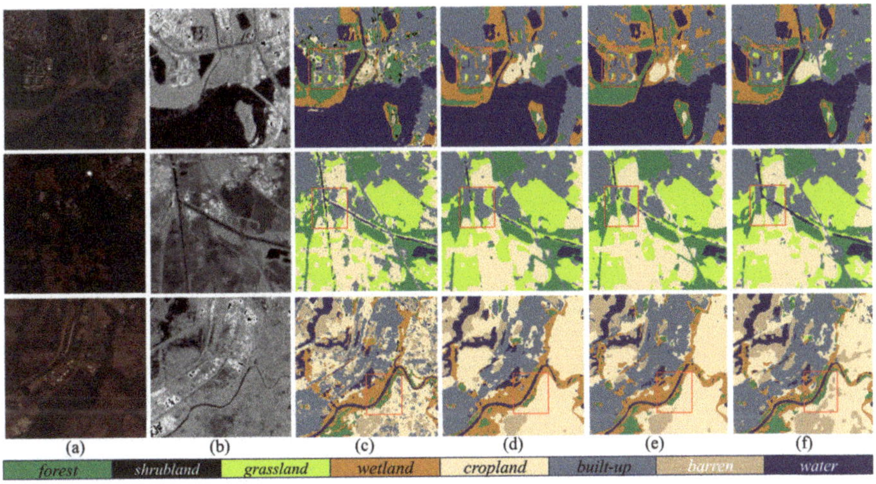

Figure 9. Land cover classification maps on DFC2020. (**a**) Optical images. (**b**) SAR images. (**c**) Semantic labels. (**d**) Att-CBAM. (**e**) Att-SP. (**f**) Proposed-SAEM.

4.3. Analysis of Multi-Scale Feature Extraction

Multi-scale semantic information is very important for image segmentation. The designed MSEM combines the ASPP [38] and convolutions with different kernels (MSCK) to obtain high-level features. We want to verify the effect of the two parts. Three cases of contrast experiments were carried out. Modules with ASPP or MSCK and with both were compared. It should be noted that the number of feature maps outputted by the original MSEM module is 256 channels. Therefore, when implementing the above two experiments, the output of each part was also adjusted to 256 channels. In Table 5, the validation values of the first two experiments are close, but the model with joint modules is superior. This set of experiments proves that the combination of multi-scale information is helpful to improve the classification accuracy.

Table 5. Comparison of multi-scale modules.

Cases	Multi-Scale Module		OA	Kappa	mIoU
	ASPP	MSCK			
1	✓	-	0.8734	0.8456	0.6784
2	-	✓	0.8724	0.8441	0.6739
3	✓	✓	0.8763	0.8492	0.6853

4.4. Analysis of the Joint Loss Function

The joint loss function used in this paper contains three loss terms. L_{CE} is used as the basic loss to constrain and guide the training process. Therefore, it was retained in all experiments. L_{Focal} is a commonly used loss function to solve class imbalance. L_{SIoU} can further smoothly update the gradients, which, in turn, reduces oscillation during training. Then, there are four combinations for comparison (L_{CE}, $L_{CE} + L_{Focal}$, $L_{CE} + L_{SIoU}$ and $L_{CE} + L_{Focal} + L_{SIoU}$). Table 6 shows the results of models with different compositions of losses. We can see that the model with only L_{CE} can help to obtain good accuracy. The joint application of L_{Focal} or L_{SIoU} can effectively improve the segmentation effect, but only one of them is insufficient. The joint loss function with the three terms achieved the best results. In follow-up research, we are going to learn other losses or adjust the proportion of each loss to achieve better constraints.

Table 6. Comparison of loss functions.

Cases	Loss Function			OA	Kappa	mIoU
	L_{CE}	L_{Focal}	L_{SIoU}			
1	✓	-	-	0.8691	0.8404	0.6730
2	✓	✓	-	0.8738	0.8460	0.6781
3	✓	-	✓	0.8736	0.8455	0.6758
4	✓	✓	✓	0.8763	0.8492	0.6853

4.5. Application in Other Multi-Modal Segmentation Tasks

In order to show the applicability of the proposed method, supplementary experiments were carried out on the MFNet dataset [49]. This RGB–thermal dataset consists of 1569 pairs of RGB and thermal images and has been applied in autonomous driving systems. It should be noted that the number of pixels of each class is unbalanced, and there is a large amount of unlabeled pixels. Since the number of classes and the channels of the multi-modal images are inconsistent with the previous datasets, the inputs and the size of confusion matrix need to be modified accordingly when executing the proposed model. Figure 10 shows the segmentation maps of four urban scenes. The first two were taken during daytime, and the others were taken at nighttime. The results of MFNet were generated by the demo code provided by the author. It can be seen that the proposed method can achieve better classification results. Both color cones and bumps can be clearly segmented. Meanwhile, numerical metrics were also compared. The class accuracy and mIoU of the proposed method reached 0.8144 and 0.6912, respectively, which are both higher than the optimal values of the original MFNet. The proposed model does have a certain application potential in other multi-modal segmentation tasks.

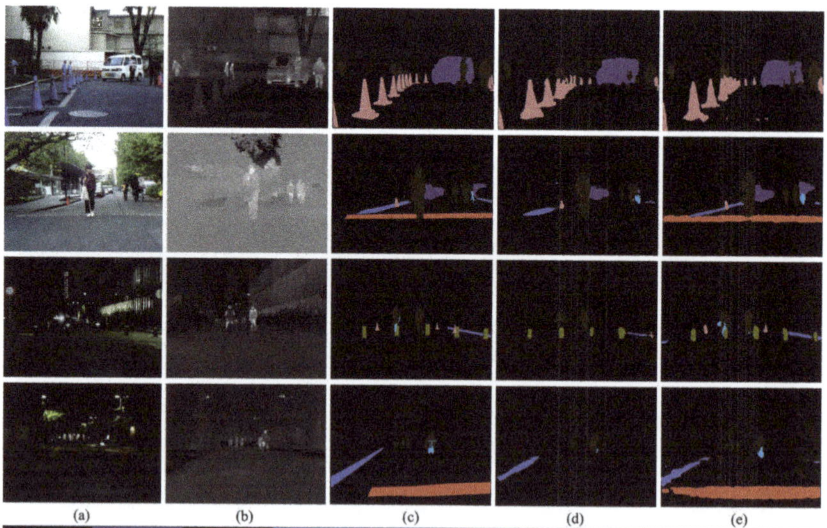

Figure 10. Urban scene segmentation maps on the RGB–thermal dataset. (**a**) RGB images. (**b**) Thermal images. (**c**) Semantic labels. (**d**) MFNet. (**e**) Proposed SAMFNet.

5. Conclusions

In this paper, we proposed a two-branch semantic segmentation network for land cover classification with multi-modal optical and SAR images. The numerical results and segmentation maps demonstrated various advantages of the proposed method. First, the novel symmetric attention mechanism with multiple long receptive fields can extract more contextual information. Objects with different shapes in the original images are perceived well. Secondly, multi-scale semantic fusion is implemented to enrich complementary information. High-level features extracted by dilated and varisized convolutions and low-level features from shallow layers are all considered and integrated together. Thirdly, a symmetrical structure and multiple plug-and-play modules were adopted to build the model. It has strong flexibility and adaptability. This was verified on an RGB–thermal dataset. Furthermore, the computational complexity of the proposed model is relatively low, and high classification accuracy was achieved. All these advantages prove the effectiveness of the method. However, the current research still depends heavily on the labeled dataset. In the future, we will deeply explore the implementation of semi-supervised and weakly supervised methods and study a lightweight network so that the model can be applied to more practical scenarios.

Author Contributions: Conceptualization, D.X.; methodology, D.X. and Z.L.; software, D.X. and Z.L.; validation, D.X., Z.L. and H.F.; formal analysis, D.X. and Z.L.; investigation, D.X., Z.L. and H.F.; resources, D.X.; data curation, D.X. and F.W.; writing—original draft preparation, D.X.; writing—review and editing, F.W., Z.L. and H.F.; visualization, D.X.; supervision, Y.W.; project administration, D.X.; funding acquisition, F.W. All authors have read and agreed to the published version of the manuscript.

Funding: This research was funded by the National Key R&D Program of China (No. 2022YFB3902300).

Data Availability Statement: The relevant methods, data and results of this study can be exchanged and shared in depth after communicating with the corresponding author.

Acknowledgments: The authors thank the 2020 IEEE GRSS Data Fusion Contest and Wuhan University for providing the aligned multi-modal data.

Conflicts of Interest: The authors declare no conflicts of interest.

References

1. Hong, D.; Gao, L.; Yokoya, N.; Yao, J.; Chanussot, J.; Du, Q.; Zhang, B. More Diverse Means Better: Multimodal Deep Learning Meets Remote-Sensing Imagery Classification. *IEEE Trans. Geosci. Remote Sens.* **2021**, *59*, 4340–4354. [CrossRef]
2. Li, X.; Zhang, G.; Cui, H.; Hou, S.; Wang, S.; Li, X.; Chen, Y.; Li, Z.; Zhang, L. MCANet: A joint semantic segmentation framework of optical and SAR images for land use classification. *Int. J. Appl. Earth Obs.* **2022**, *106*, 102638. [CrossRef]
3. Li, X.; Lei, L.; Sun, Y.; Li, M.; Kuang, G. Multimodal Bilinear Fusion Network With Second-Order Attention-Based Channel Selection for Land Cover Classification. *IEEE J. Sel. Top. Appl. Earth Observ. Remote Sens.* **2020**, *13*, 1011–1026. [CrossRef]
4. Tang, L.; Yuan, J.; Ma, J. Image fusion in the loop of high-level vision tasks: A semantic-aware real-time infrared and visible image fusion network. *Inform. Fusion* **2022**, *82*, 28–42. [CrossRef]
5. Li, X.; Lei, L.; Sun, Y.; Li, M.; Kuang, G. Collaborative Attention-Based Heterogeneous Gated Fusion Network for Land Cover Classification. *IEEE Trans. Geosci. Remote Sens.* **2021**, *59*, 3829–3845. [CrossRef]
6. Meng, H.; Li, C.; Liu, Y.; Gong, Y.; He, W.; Zou, M. Corn Land Extraction Based on Integrating Optical and SAR Remote Sensing Images. *Land* **2023**, *12*, 398. [CrossRef]
7. Li, W.; Sun, K.; Li, W.; Wei, J.; Miao, S.; Gao, S.; Zhou, Q. Aligning semantic distribution in fusing optical and SAR images for land use classification. *ISPRS J. Photogramm. Remote Sens.* **2023**, *199*, 272–288. [CrossRef]
8. Li, X.; Lei, L.; Zhang, C.; Kuang, G. Dense Adaptive Grouping Distillation Network for Multimodal Land Cover Classification With Privileged Modality. *IEEE Trans. Geosci. Remote Sens.* **2022**, *60*, 1–14. [CrossRef]
9. Kang, W.; Xiang, Y.; Wang, F.; You, H. CFNet: A Cross Fusion Network for Joint Land Cover Classification Using Optical and SAR Images. *IEEE J. Sel. Top. Appl. Earth Observ. Remote Sens.* **2022**, *15*, 1562–1574. [CrossRef]
10. Zhang, H.; Wan, L.; Wang, T.; Lin, Y.; Lin, H.; Zheng, Z. Impervious Surface Estimation From Optical and Polarimetric SAR Data Using Small-Patched Deep Convolutional Networks: A Comparative Study. *IEEE J. Sel. Top. Appl. Earth Observ. Remote Sens.* **2019**, *12*, 2374–2387. [CrossRef]
11. Xu, X.; Li, W.; Ran, Q.; Du, Q.; Gao, L.; Zhang, B. Multisource Remote Sensing Data Classification Based on Convolutional Neural Network. *IEEE Trans. Geosci. Remote Sens.* **2018**, *56*, 937–949. [CrossRef]

12. Melgani, F.; Bruzzone, L. Classification of hyperspectral remote sensing images with support vector machines. *IEEE Trans. Geosci. Remote Sens.* **2004**, *42*, 1778–1790. [CrossRef]
13. Talukdar, S.; Singha, P.; Mahato, S.; Shahfahad; Pal, S.; Liou, Y.A.; Rahman, A. Land-Use Land-Cover Classification by Machine Learning Classifiers for Satellite Observations—A Review. *Remote Sens.* **2020**, *12*, 1135. [CrossRef]
14. Ham, J.; Chen, Y.; Crawford, M.; Ghosh, J. Investigation of the random forest framework for classification of hyperspectral data. *IEEE Trans. Geosci. Remote Sens.* **2005**, *43*, 492–501. [CrossRef]
15. Blanzieri, E.; Melgani, F. Nearest Neighbor Classification of Remote Sensing Images With the Maximal Margin Principle. *IEEE Trans. Geosci. Remote Sens.* **2008**, *46*, 1804–1811. [CrossRef]
16. Li, K.; Wang, D.; Wang, X.; Liu, G.; Wu, Z.; Wang, Q. Mixing Self-Attention and Convolution: A Unified Framework for Multisource Remote Sensing Data Classification. *IEEE Trans. Geosci. Remote Sens.* **2023**, *61*, 5523216. [CrossRef]
17. Ienco, D.; Interdonato, R.; Gaetano, R.; Ho Tong Minh, D. Combining Sentinel-1 and Sentinel-2 Satellite Image Time Series for land cover mapping via a multi-source deep learning architecture. *ISPRS J. Photogramm. Remote Sens.* **2019**, *158*, 11–22. [CrossRef]
18. Liu, S.; Qi, Z.; Li, X.; Yeh, A.G.O. Integration of Convolutional Neural Networks and Object-Based Post-Classification Refinement for Land Use and Land Cover Mapping with Optical and SAR Data. *Remote Sens.* **2019**, *11*, 690. [CrossRef]
19. Feng, Q.; Yang, J.; Zhu, D.; Liu, J.; Guo, H.; Bayartungalag, B.; Li, B. Integrating Multitemporal Sentinel-1/2 Data for Coastal Land Cover Classification Using a Multibranch Convolutional Neural Network: A Case of the Yellow River Delta. *Remote Sens.* **2019**, *11*, 1006. [CrossRef]
20. Li, X.; Lei, L.; Kuang, G. Locality-Constrained Bilinear Network for Land Cover Classification Using Heterogeneous Images. *IEEE Geosci. Remote Sens. Lett.* **2022**, *19*, 2501305. [CrossRef]
21. Gao, M.; Xu, J.; Yu, J.; Dong, Q. Distilled Heterogeneous Feature Alignment Network for SAR Image Semantic Segmentation. *IEEE Geosci. Remote Sens. Lett.* **2023**, *20*, 4004705. [CrossRef]
22. Kang, J.; Wang, Z.; Zhu, R.; Xia, J.; Sun, X.; Fernandez-Beltran, R.; Plaza, A. DisOptNet: Distilling Semantic Knowledge From Optical Images for Weather-Independent Building Segmentation. *IEEE Trans. Geosci. Remote Sens.* **2022**, *60*, 4706315. [CrossRef]
23. Chen, Y.; Bruzzone, L. Self-Supervised SAR-Optical Data Fusion of Sentinel-1/-2 Images. *IEEE Trans. Geosci. Remote Sens.* **2022**, *60*, 5406011. [CrossRef]
24. Jain, P.; Schoen-Phelan, B.; Ross, R. Self-Supervised Learning for Invariant Representations From Multi-Spectral and SAR Images. *IEEE J. Sel. Top. Appl. Earth Observ. Remote Sens.* **2022**, *15*, 7797–7808. [CrossRef]
25. Liu, C.; Sun, H.; Xu, Y.; Kuang, G. Multi-Source Remote Sensing Pretraining Based on Contrastive Self-Supervised Learning. *Remote Sens.* **2022**, *14*, 4632. [CrossRef]
26. Gao, L.; Hong, D.; Yao, J.; Zhang, B.; Gamba, P.; Chanussot, J. Spectral Superresolution of Multispectral Imagery With Joint Sparse and Low-Rank Learning. *IEEE Trans. Geosci. Remote Sens.* **2021**, *59*, 2269–2280. [CrossRef]
27. Zhao, X.; Tao, R.; Li, W.; Li, H.C.; Du, Q.; Liao, W.; Philips, W. Joint Classification of Hyperspectral and LiDAR Data Using Hierarchical Random Walk and Deep CNN Architecture. *IEEE Trans. Geosci. Remote Sens.* **2020**, *58*, 7355–7370. [CrossRef]
28. Schmitt, M.; Hughes, L.H.; Qiu, C.; Zhu, X.X. SEN12MS—A Curated Dataset of Georeferenced Multi-Spectral Sentinel-1/2 Imagery for Deep Learning and Data Fusion. *arXiv* **2019**, arXiv:1906.07789.
29. Yokoya, N.; Ghamisi, P.; Haensch, R.; Schmitt, M. 2020 IEEE GRSS Data Fusion Contest: Global Land Cover Mapping With Weak Supervision [Technical Committees]. *IEEE Geosci. Remote Sens. Mag.* **2020**, *8*, 154–157. [CrossRef]
30. Fu, J.; Liu, J.; Tian, H.; Li, Y.; Bao, Y.; Fang, Z.; Lu, H. Dual Attention Network for Scene Segmentation. In Proceedings of the 2019 IEEE/CVF Conference on Computer Vision and Pattern Recognition (CVPR), Long Beach, CA, USA, 15–20 June 2019; pp. 3141–3149. [CrossRef]
31. Huang, Z.; Wang, X.; Wei, Y.; Huang, L.; Shi, H.; Liu, W.; Huang, T.S. CCNet: Criss-Cross Attention for Semantic Segmentation. *IEEE Trans. Pattern Anal. Mach. Intell.* **2023**, *45*, 6896–6908. [CrossRef]
32. Zhong, Z.; Lin, Z.Q.; Bidart, R.; Hu, X.; Daya, I.B.; Li, Z.; Zheng, W.S.; Li, J.; Wong, A. Squeeze-and-Attention Networks for Semantic Segmentation. In Proceedings of the 2020 IEEE/CVF Conference on Computer Vision and Pattern Recognition (CVPR), Seattle, WA, USA, 13–19 June 2020; pp. 13062–13071. [CrossRef]
33. Yang, X.; Li, S.; Chen, Z.; Chanussot, J.; Jia, X.; Zhang, B.; Li, B.; Chen, P. An attention-fused network for semantic segmentation of very-high-resolution remote sensing imagery. *ISPRS J. Photogramm. Remote Sens.* **2021**, *177*, 238–262. [CrossRef]
34. Xu, Z.; Zhu, J.; Geng, J.; Deng, X.; Jiang, W. Triplet Attention Feature Fusion Network for SAR and Optical Image Land Cover Classification. In Proceedings of the 2021 IEEE International Geoscience and Remote Sensing Symposium IGARSS, Brussels, Belgium, 11–16 July 2021; pp. 4256–4259. [CrossRef]
35. Woo, S.; Park, J.; Lee, J.Y.; Kweon, I.S. CBAM: Convolutional Block Attention Module. In *Computer Vision—ECCV 2018, Proceedings of the 15th European Conference, Munich, Germany, 8–14 September 2018*; Springer International Publishing: Cham, Switzerland, 2018; pp. 3–19.
36. Tao, A.; Sapra, K.; Catanzaro, B. Hierarchical Multi-Scale Attention for Semantic Segmentation. *arXiv* **2020**, arXiv:2005.10821.
37. Yuan, M.; Ren, D.; Feng, Q.; Wang, Z.; Dong, Y.; Lu, F.; Wu, X. MCAFNet: A Multiscale Channel Attention Fusion Network for Semantic Segmentation of Remote Sensing Images. *Remote Sens.* **2023**, *15*, 361. [CrossRef]
38. Chen, L.C.; Papandreou, G.; Schroff, F.; Adam, H. Rethinking Atrous Convolution for Semantic Image Segmentation. *arXiv* **2017**, arXiv:1706.05587.

9. Lazebnik, S.; Schmid, C.; Ponce, J. Beyond Bags of Features: Spatial Pyramid Matching for Recognizing Natural Scene Categories. In Proceedings of the 2006 IEEE Computer Society Conference on Computer Vision and Pattern Recognition (CVPR'06), New York, NY, USA, 17–22 June 2006; Volume 2, pp. 2169–2178. [CrossRef]
0. Zhao, H.; Shi, J.; Qi, X.; Wang, X.; Jia, J. Pyramid Scene Parsing Network. In Proceedings of the 2017 IEEE Conference on Computer Vision and Pattern Recognition (CVPR), Honolulu, HI, USA, 21–26 July 2017; pp. 6230–6239. [CrossRef]
1. Hou, Q.; Zhang, L.; Cheng, M.M.; Feng, J. Strip Pooling: Rethinking Spatial Pooling for Scene Parsing. In Proceedings of the 2020 IEEE/CVF Conference on Computer Vision and Pattern Recognition (CVPR), Seattle, WA, USA, 13–19 June 2020; pp. 4002–4011. [CrossRef]
2. Song, Q.; Mei, K.; Huang, R. AttaNet: Attention-Augmented Network for Fast and Accurate Scene Parsing. *arXiv* **2021**, arXiv:2103.05930.
3. Chen, K.; Dai, X.; Xia, M.; Weng, L.; Hu, K.; Lin, H. MSFANet: Multi-Scale Strip Feature Attention Network for Cloud and Cloud Shadow Segmentation. *Remote Sens.* **2023**, *15*, 4853. [CrossRef]
4. Vaezi Joze, H.R.; Shaban, A.; Iuzzolino, M.L.; Koishida, K. MMTM: Multimodal Transfer Module for CNN Fusion. In Proceedings of the 2020 IEEE/CVF Conference on Computer Vision and Pattern Recognition (CVPR), Seattle, WA, USA, 13–19 June 2020; pp. 13286–13296. [CrossRef]
5. Lin, T.Y.; Goyal, P.; Girshick, R.; He, K.; Dollár, P. Focal Loss for Dense Object Detection. In Proceedings of the 2017 IEEE International Conference on Computer Vision (ICCV), Venice, Italy, 22–29 October 2017; pp. 2999–3007. [CrossRef]
6. Yu, J.; Jiang, Y.; Wang, Z.; Cao, Z.; Huang, T. UnitBox: An Advanced Object Detection Network. In Proceedings of the 24th ACM International Conference on Multimedia, Amsterdam, The Netherlands, 15–19 October 2016. [CrossRef]
7. Chen, L.C.; Zhu, Y.; Papandreou, G.; Schroff, F.; Adam, H. Encoder-Decoder with Atrous Separable Convolution for Semantic Image Segmentation. In *Computer Vision—ECCV 2018, Proceedings of the 15th European Conference, Munich, Germany, 8–14 September 2018*; Springer International Publishing: Cham, Switzerland, 2018; pp. 833–851.
48. Yang, M.; Yu, K.; Zhang, C.; Li, Z.; Yang, K. DenseASPP for Semantic Segmentation in Street Scenes. In Proceedings of the 2018 IEEE/CVF Conference on Computer Vision and Pattern Recognition, Salt Lake City, UT, USA, 18–23 June 2018; pp. 3684–3692. [CrossRef]
49. Ha, Q.; Watanabe, K.; Karasawa, T.; Ushiku, Y.; Harada, T. MFNet: Towards real-time semantic segmentation for autonomous vehicles with multi-spectral scenes. In Proceedings of the 2017 IEEE/RSJ International Conference on Intelligent Robots and Systems (IROS), Vancouver, BC, Canada, 24–28 September 2017. [CrossRef]

remote sensing

Article

An Overlay Accelerator of DeepLab CNN for Spacecraft Image Segmentation on FPGA

Zibo Guo [1], Kai Liu [1,*], Wei Liu [2], Xiaoyao Sun [1], Chongyang Ding [1] and Shangrong Li [1]

[1] School of Computer Science and Technology, Xidian University, Xi'an 710071, China; zbguo@stu.xidian.edu.cn (Z.G.)
[2] Smart Earth Key Laboratory, Beijing 100094, China; liuwei45_2000@163.com
* Correspondence: kailiu@mail.xidian.edu.cn

Abstract: Due to the absence of communication and coordination with external spacecraft, non cooperative spacecraft present challenges for the servicing spacecraft in acquiring information about their pose and location. The accurate segmentation of non-cooperative spacecraft components in images is a crucial step in autonomously sensing the pose of non-cooperative spacecraft. This paper presents a novel overlay accelerator of DeepLab Convolutional Neural Networks (CNNs) for spacecraft image segmentation on a FPGA. First, several software–hardware co-design aspects are investigated: (1) A CNNs-domain COD instruction set (Control, Operation, Data Transfer) is presented based on a Load–Store architecture to enable the implementation of accelerator overlays. (2) An RTL-based prototype accelerator is developed for the COD instruction set. The accelerator incorporates dedicated units for instruction decoding and dispatch, scheduling, memory management, and operation execution. (3) A compiler is designed that leverages tiling and operation fusion techniques to optimize the execution of CNNs, generating binary instructions for the optimized operations Our accelerator is implemented on a Xilinx Virtex-7 XC7VX690T FPGA at 200 MHz. Experiments demonstrate that with INT16 quantization our accelerator achieves an accuracy (mIoU) of 77.84%, experiencing only a 0.2% degradation compared to that of the original fully precision model, in accelerating the segmentation model of DeepLabv3+ ResNet18 on the spacecraft component images (SCIs) dataset. The accelerator boasts a performance of 184.19 GOPS/s and a computational efficiency (Runtime Throughput/Theoretical Roof Throughput) of 88.72%. Compared to previous work, our accelerator improves performance by 1.5× and computational efficiency by 43.93%, all while consuming similar hardware resources. Additionally, in terms of instruction encoding, our instructions reduce the size by 1.5× to 49× when compiling the same model compared to previous work.

Keywords: image semantic segmentation; instruction set architecture (ISA); field programmable gate array (FPGA); spacecraft component images

Citation: Guo, Z.; Liu, K.; Liu, W.; Sun, X.; Ding, C.; Li, S. An Overlay Accelerator of DeepLab CNN for Spacecraft Image Segmentation on FPGA. *Remote Sens.* **2024**, *16*, 894. https://doi.org/10.3390/rs16050894

Academic Editor: Gemine Vivone

Received: 25 January 2024
Revised: 28 February 2024
Accepted: 28 February 2024
Published: 2 March 2024

1. Introduction

Recently, the exploration of deep space has gained extensive support from various countries and enterprises [1]. Vision-based Artificial Intelligence (AI) applications are crucial for current and upcoming space missions, such as automation navigation systems for collision avoidance [2], asteroid classifications [3], and debris removal [4]. One notable application of these technologies is the accurate recognition of spacecraft feature components in images [5]. In scenarios where the target spacecraft lacks sensors or communication capabilities, such as during debris removal operations [6], it is desirable to implement an object recognition payload that can segment **spacecraft component images (SCIs)** obtained from visual sensors to locate the target object of interest.

As a fundamental problem in computer vision, semantic segmentation aims to assign semantic labels (class labels) to every pixel in an image. Early segmentation algorithms relied on handcrafted feature matching [7,8], but these methods have been shown to exhibit poor generalization and stability. In recent decades, deep learning methods based on convolutional neural networks (CNNs) have become the mainstream approach for almost all vision tasks, including semantic segmentation [9]. Compared to previous methods, CNNs exhibit higher reliability in the presence of noisy interference or previously unseen scenarios [6]. Therefore, CNNs are now being applied to recognize space targets from spacecraft images, which are more susceptible to interference than natural images from common datasets such as COCO [10,11]. Several studies have demonstrated the promising performance of CNNs-based approaches for spacecraft component image semantic segmentation [10].

However, the CNN deployment on resource-constraint embedded hardware systems onboard also poses significant challenges due to their compute-intensive and memory-intensive characteristics. Typically, CNN-based approaches can be delineated into two distinct phases: the training phase and the inference phase. During the training phase, a CNN model learns to discern the relationships between input data and their corresponding labels. Through iterative processes, the CNN refines its parameters, progressively improving its ability to capture task-relevant features. Upon completion of the training phase, the CNN model is prepared for the inference phase, during which it generates predictions for new unseen data. As the parameters remain fixed once the training is complete, the training phase can be performed offline at a data center on the ground. The key challenge lies in efficiently implementing inference for CNNs using onboard hardware, a crucial aspect in deploying CNN-based semantic segmentation approaches onboard.

Field Programmable Gate Arrays (FPGAs) with high parallelism and reconfigurability are widely employed in exploration missions [12]. For instance, onboard science data processing systems like Spacecube, based on the Xilinx Virtex family of FPGAs, have been utilized to implement data processing requirements for robotic servicing [12]. In this paper, we design an accelerator on an FPGA to aid processor acceleration CNNs computation for the SCIs segmentation task in a space scene.

Several studies have investigated the deployment of CNNs for semantic image segmentation onto FPGAs. Shen et al. proposed a model called LNS-Net [13] based on U-Net [14] for lung nodule segmentation and accelerated this CNN model on four Xilinx VCU118 FPGAs using a proposed mapping scheme that took advantage of the massive parallelism. Bai et al. designed RoadNet-RT [15], a lightweight CNN segmentation model for road scenarios, and implemented an accelerator for this model on a Xilinx ZCU102 FPGA to perform inference with an 8-bit quantized model.

In addition to the networks-specific custom accelerators, some studies have explored overlay accelerators. Liu et al. designed an efficient custom deconvolution (DeCONV) architecture and designed a U-Net CNN accelerator to support the acceleration of semantic segmentation tasks on FPGAs [16]. They later optimized this architecture and proposed a unified processing engine to address the problem of convolution (CONV) and DeConv modules not being able to share computational resources. The optimized architecture shows remarkable performance on remote sensing image segmentation tasks [17]. Wu et al. proposed a reconfigurable FPGA hardware accelerator for various CNN-based vision tasks including semantic segmentation [18]. They implemented diverse operator modules including CONV, depthwise convolution (DwCONV), and others, and proposed efficient data flow scheduling and processing schemes under the constraint of limited computing resources. The evaluation results showed that the accelerator can efficiently accelerate the semantic segmentation model ENet [19], which is common for embedded devices.

Most of the previous works have either designed U-Net-specific accelerators on FPGAs or evaluated U-Net on FPGA-based CNNs domain-specific accelerators. While U-Net's **Encoder–Decoder** architecture addresses the issue of missing low-level features, its encoder network lacks a component that captures multi-scale features, leading to a loss of contextual information. To overcome this limitation, the Pyramid Scene Parsing Network (PSPNet) [20] was proposed, which leverages different downsample rates of pooling followed by CONV operations to extract abundant multi-scale semantic features. Furthermore, The DeepLabV3 [21] introduced an **Atrous Spatial Pyramid Pooling (ASPP)** module which reduced the feature response loss caused by down and up samples in PSPNet converting the Pooling-CONV-Upsample operation to an Atrous CONV. The computational principle of Atrous CONV is shown in Figure 1, and it can be seen that adjusting the rate can achieve convolution with a larger receptive field without increasing the convolution kernel parameters and computational effort. The convolution with different receptive fields facilitates the capture of features at various scales.

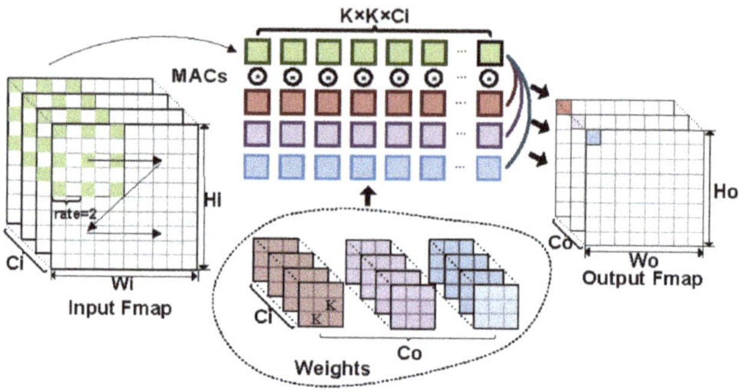

Figure 1. The computational principle of atrous convolution. (* denotes a set of multiply-accumulate (MAC) operations. Dark red, purple, and blue represent 3 different convolutional kernel parameters and the output feature maps of the corresponding channels, respectively. Green represents the input feature maps of the involved operations).

DeepLabV3+ [22] extended DeepLabV3 by adding a decoder to refine the segmentation result, allowing it to take into account multi-scale contextual information and low-level sharper boundaries information through the ASPP module and Encoder–Decoder structure. Table 1 compares the accuracy and complexity of the aforementioned CNNs on our SCIs dataset. It can be seen that DeepLabv3+ has better accuracy at lower complexity instead.

Table 1. The structures used in different CNN segmentation algorithms (backbone is VGG16) and the complexity and accuracy of the SCIs set of each algorithm. (SCIs dataset consists of 8833 spacecraft simulated images, including 5 feature component types [23]).

Model	U-Net [14]	PSPNet [20]	DeeplabV3 [21]	DeeplabV3+ [22]
Structure	E-D	ASPP	ASPP	ASPP and E-D
Parameter (M)	24.89	139.82	19.44	19.56
Complexity (GOPS)	112.76	40.82	42.64	48.42
Accuracy (mIoU)	65.15	53.66	61.47	**81.62**

For the acceleration of the Deeplabv3+ model, Morì et al. devised a hardware-aware pruning method based on genetic algorithms to reduce model operations and parameters [24]. Furthermore, they implemented an overlay CNN accelerator on an Intel Arria 10 GX1150 FPGA platform, evaluating its acceleration performance with the DeepLabv3+ ResNet18 model. Im et al. designed a DT-CNN ASIC accelerator [25] supporting variant convolution based on 65 nm CMOS technology. This accelerator efficiently accelerates dilated and transposed convolution by skipping redundant zero computations. The acceleration performance of ENet, Deeplabv3+, and FCN [9] models was also evaluated. However, these efforts are still lacking in terms of acceleration efficiency and model adaptation.

This paper aims to map a DeepLabv3+ CNN onto a flight-like hardware FPGA for the purpose of a semantic SCIs segmentation task. There are two main challenges involved in this process: (1) Accelerators that are specifically designed for certain CNN models require FPGA reconfiguration when switching to other models, a process which is not practical for onboard scenarios. (2) The extensive intermediate results generated by the complicated skip-connection of the Encoder–Decoder structure must be cached in the limited on-chip SRAM or require additional external memory access, posing a significant challenge for a resource-constrained onboard FPGA.

To address these challenges, this paper presents a comprehensive flow for mapping CNNs onto FPGAs as is illustrated in Figure 2. To decouple the hardware architecture from the specific CNN model structure, we designed a customized instruction set architecture called COD (Control, Operation, and Data transfer). During the offline stage, we quantized and tiled the model parameters, and converted and compiled the computation graph to generate COD instruction sequences. (processes: ❶ and ❷) At this stage, we employed a quantization method that effectively halved the model size (32 bits to 16 bits) while incurring an accuracy loss of less than 0.5%. Our proposed COD instruction set and compiler have a 1.5× to 49× size reduction compared to previous work, and a 26% reduction in DRAM accesses compared to the primitive design. During the online stage, we design the hardware accelerator architecture corresponding to the COD and implement it on the Xilinx Virtex-7 VX690T FPGA to achieve the task of segmentation of SCI images. (process: ❸) The performance and computational efficiency of our accelerator was 1.5× and 43.93% higher than previous work, respectively, with a 5.1× increase in energy efficiency compared to an NVIDIA RTX 2080Ti GPU.

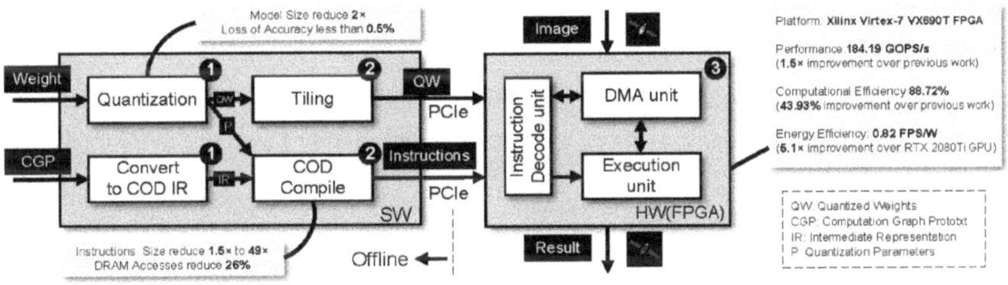

Figure 2. The overview of the mapping flow.

The main contributions of this work are as follows:

1. To facilitate network replacement and decouple the accelerator micro-architecture from a specific network, we propose a COD instruction set based on load–store. This enables re-compiled instruction sequences to overlay the accelerator without the need for hardware re-burn.
2. We propose an accelerator micro-architecture based on a COD instruction set, which contains an instruction decoder and dispatch unit, data scheduler unit, and unified Execution Unit (EU). The first two guarantee the coarse-grained parallel data transfer based on dependency of instructions. The unified EU for CONV and Atrous CONV ensures the fine-grained parallel data operation leveraging spatial and temporal data reuse.
3. We develop a compiler for COD instruction generation to convert the computational graph of an input CNN model into a sequence of COD instructions and produce corresponding binary signals. The compiler was designed to incorporate tiling and operation fusion techniques, aimed at optimizing the execution of the CNN.
4. We implemented our accelerator on the Xilinx VC709 development board with an XC7VX690T FPGA chip, which is commonly used on spacecraft. Our accelerator runs at 200 MHz and achieves a performance of 184.19 GOPS/s and a detection accuracy (mIoU) of 77.84% for the SCI dataset when accelerating the Deeplabv3+ ResNet18 CNN model.

The remaining parts of this paper are organized as follows: Section 2 introduces the preliminaries about CNNs and DeepLabv3+. Section 3 describes the COD instruction set. The accelerator micro-architecture is proposed in Section 4. Section 5 presents optimization strategies for instruction sequence compilation. Section 6 presents our experimental results in the SCIs segmentation task. Finally, Section 7 concludes this paper.

2. Deeplab CNN Preliminaries

The flow of SCIs segmentation using DeepLabv3+ is illustrated in Figure 3. It employs a classical CNN backbone and ASPP as the encoder module to capture multi-scale high-level features, and a simple decoder to merge detailed low-level features. In the encoder, operations that involve 'Rate' refer to atrous convolution operations, where 'Rate' determines the dilation rate. Our overlay accelerator supports all the basic operations involved in the CNN-process depicted in Figure 3.

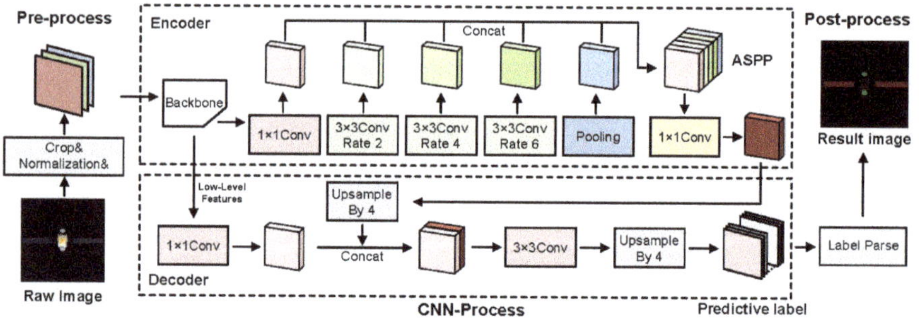

Figure 3. Overview of the DeepLabV3+ semantic SCIs segmentation. (Green and red areas are antenna and panel components, respectively, in the result image).

Below, we provide a brief explanation and mathematical notation for these operations. In the following notations, X and Y represent the input and output tensors, respectively, having shapes of (C_i, W_i, H_i) and (C_o, W_o, H_o), where w stands for width, h for height, and c for the number of channels in the feature maps.

Convolution: It takes as inputs a set of nonlinear functions of spatially nearby regions of outputs from the prior layer, which are multiplied by weights and added with bias. (The input to first layer is a tensor of image pixels.) It is equationally described in Equation (1) [26].

$$Y = \mathrm{Conv}(X)_{W,b} = X \otimes W + b \tag{1}$$

The tensors $W(wk, hk, ci, co)$ and $b(co)$ represent the weight and bias parameters for the convolution operation, respectively, acquired through training. Here, wk denotes kernel width, hk denotes kernel height, ci represents the number of input feature map channels, and co indicates the number of output feature map channels.

Atrous (Dilated) Convolution: Its operation mode functions in the same manner as standard convolution, but with the addition of a dilation rate that adjusts the receptive field (the size of the region of the input feature map that produces each output element) without increasing the number of convolution parameters.

Max Pooling: This operation is a commonly used convex function for downsampling. Its mathematical representation is given by Equation (2) [26].

$$Y = \mathrm{Maxpool}(X) \rightarrow y_{i,j,k} = \max_{(p,q,k)\in\Re ijk} (x_{p,q,k}) \tag{2}$$

$y_{i,j,k}$ represents the values at the (i, j, k) position within the Y, while $x_{p,q,k}$ denotes the values at the (p, q, k) position within the X. $\Re ijk$ signifies the sliding window region in which y aligns with the input tensor X where the pooling operation is executed.

Element-Wise Addition: It is the operation of summing two identically shaped tensors by position and is commonly used for residual structures and feature fusion. Its mathematical representation is given by Equation (3) [26].

$$Y = X^1 + X^2 \rightarrow y_{w,h,c} = x^1{}_{w,h,c} + x^2{}_{w,h,c} \tag{3}$$

Upsampling (Nearest Interpolation): It is the operation to expand the feature resolution. Its mathematical representation is given by Equation (4) [26].

$$Y = \mathrm{Upsample}_s(X) \rightarrow y_{i,j,k} = x_{[i/s],[j/s],k} \quad (4)$$

The variable s represents the upsampling factor. Additionally, $x_{[i/s],[i/s],k}$ specifies the value located at the nearest position in the input tensor X corresponding to the position (i, j, k) of the output tensor Y.

ReLU/LeakyReLU: It is an activation function placed after a convolution.

Concatenation: It is a tensor concatenation operation. It is equationally described in Equation (5) [26].

$$Y = \mathrm{Concate}(X^1, X^2, \ldots X^n) \quad (5)$$

$X^1, X^2, \ldots X^n, Y$ are tensors of the same shape in w and h dimensions and Y comes from the concatenation of $X^1, X^2, \ldots X^n$ along the c-dimension.

Batch Normalization: Batch Normalization (BN) is commonly used following a convolution layer to improve model training [27]. The operations of BN can be expressed using Equation (6).

$$Y = \mathrm{BN}(X) \rightarrow y = \gamma \frac{x - \mu}{\sqrt{\sigma^2 + \varepsilon}} + \beta \quad (6)$$

Here, γ is the scaling factor and β is the shift factor, both of which are learnable parameters used to adjust the normalized scale and mean, respectively. μ and σ^2 represent the mean and variance of the input X calculated during training, with ε being a small constant for numerical stability.

In the sequel, we will show the data path of the aforementioned basic operations for their spatial or temporal parallel compute. Meanwhile, their instruction coding and the parallelism schedule between operations will also be described in detail.

3. COD Instruction Set Architecture

Our accelerator does not rely on fixed data scheduling based on a specific network (SN) [28]. Instead, it drives the data stream by reading and executing instructions, effectively decoupling the hardware micro-architecture from the SN by **Instruction Set Architecture (ISA)**. As shown in Figure 4, when the network is replaced, our overlay accelerator only requires re-compiling the computing graph to the new instruction sequence. However, for an SN accelerator, a new hardware micro-architecture (RTL or HLS code) based on the new network must be designed and the FPGA re-burned, which is an inefficient task in a space environment. Hence, we propose a novel ISA called COD in this section, which integrates three types of instructions for **control, operation**, and **data transfer**, covering all the CNN basic operations discussed in Section 2.

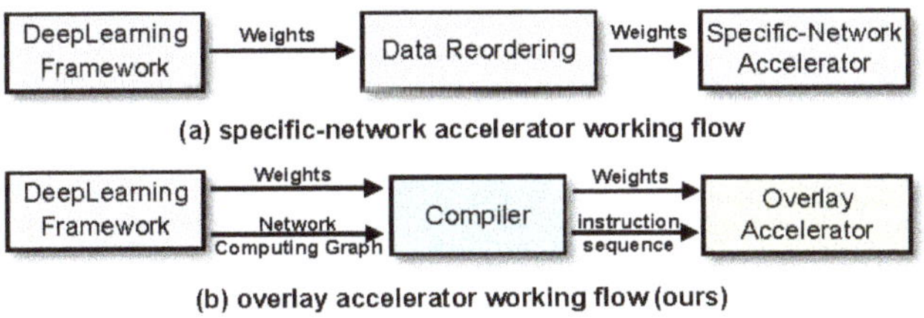

Figure 4. Workflow for SN accelerator versus overlay accelerator.

3.1. Control Flow

The Instruction Set (IS) refers to the vocabulary of commands that is understood by a specific hardware architecture. A control logic structure is employed in the hardware to facilitate an explicit Control Flow (CF), with the IS being decoded as a crucial signal in the CF that controls the sequential execution of tasks. Therefore, prior to discussing the IS design, it is imperative to clarify the CF of our accelerator.

Our accelerator follows a load–store architecture, wherein the CF schedules the data from memory to the Execution Unit (EU) and subsequently manages the storage of results from the EU back to the memory. It is evident that the efficacy of data storage and load represents a significant bottleneck in the overall performance of this architecture [29]. However, there is a large gap between the memory-intensive characteristics of CNNs and the insufficient on-chip memory resources of FPGAs. Full avoidance of external memory(DRAM) access is unfeasible. Figure 5 shows the memory footprint of intermediate results and convolution kernels in each layer of the DeeplabV3+ CNN model. It can be seen that the memory space requirement of some layers even exceeds 5 MB, while for FPGAs commonly used on satellites, most of their on-chip memory resources (SRAM) are below 7 MB, such as the Xilinx XC7VX690T 6.6 MB and XC7K325T 3.2 MB.

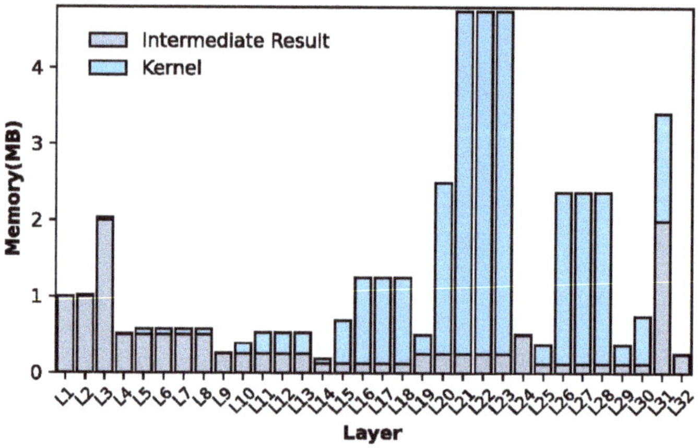

Figure 5. The memory footprint of DeeplabV3+ ResNet18 CNN model with INT16 quantization (Input shape: 256 × 256 × 3).

For minimizing DRAM access, we designed a dynamic memory hierarchy (DMH), as shown in Figure 6. If the intermediate results of a layer can be stored in the on-chip buffer, then the storing of DRAM on this layer and the reading of DRAM on the next layer can be skipped. Of course, the selection of a branch path depends on the signal decoded from the instruction. We can substantially reduce the consumption of external communications via optimizing instruction compilation in certain on-chip buffer space constraints. For example, if we have a 1 MB on-chip buffer, for the network shown in Figure 5, there will be 30 layers that do not require storing feature maps by external memory.

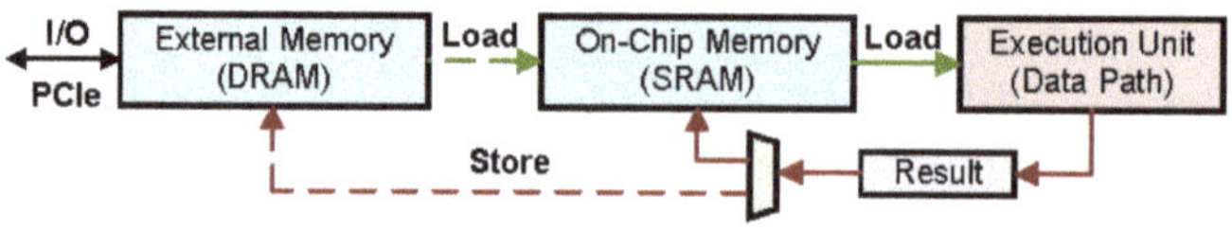

Figure 6. The control flow of our load–store architecture.

3.2. COD Instruction Set

IS is a collection of control information in CF. Instruction length and granularity are the two main factors that impact the performance of ISs. Prior specialized ISs developed for CNN domains can be broadly classified into two categories based on their execution granularity, as illustrated in Table 2.

Fine-grained ISs such as Cambricon [30] and OPU [31] feature instructions with a fixed length and separate instruction parsing and control units in their hardware architecture. Such ISs typically require a group of instructions to execute an entire load–compute–store flow with higher execution parallelism per instruction. However, fine-grained ISs can lead to complex CFs with numerous branch paths, necessitating careful consideration of instruction dependencies by both the relevant compiler and hardware control logic to ensure the correct execution of instruction sequences. As a result, fine-grained ISs require more FPGA logic resources for command control, which is not friendly to resource-constrained flight-FPGAs. Therefore, we opt for a concise coarse-grained IS, similar to SLC [32] and Xilinx DPU [33,34], to identify the CF.

Table 2. Comparison of some previous CNN-domain instruction sets.

	Cambricon [30]	SLC [32]	DPU [33,34]	OPU [31]	COD (Ours)
Year	ISCA16	TRTS18	TCAD19	TVLSI20	**2024**
Hardware	ASIC	FPGA	FPGA	FPGA	**FPGA**
Instruction length	64 bit	128 bit	128 bit/192 bit	32 bit	**256 bit**
Instruction granularity	Fine	Coarse	Coarse	Coarse	**Coarse**

We analyze all data transmissions in CF and design a Data Transfer Instruction (DTI) to identify the data transfer path. In the case of the access branch in DMH, we design a Control Instruction (CTI) to schedule the data flow. Furthermore, we design an Operation Instruction (OPI) to specify the parameters of the EU runtime. Together, CTI, OPI, and DTI form a 256-bit COD instruction. The number of bits and information details occupied by each instruction type are illustrated in Figure 7a. We introduce each instruction type as follows:

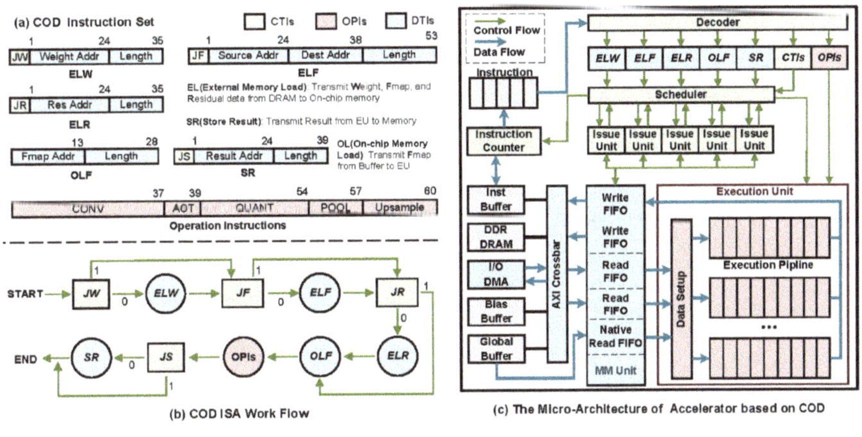

Figure 7. Overview of the COD ISA and prototype accelerator.

DTI: The DTI consists of four loading instructions (ELW, ELF, ELR, OLF) and one storing instruction (SR). ELW, ELF, and ELR handle the loading of weight, Feature Map (Fmap), and Residual data from DRAM to the on-chip buffer, respectively. The Residual contains data from Fmap that needs to skip some layers during delivery. These data are not involved in the convolution operation and are moved to the on-chip Addition FIFO for the element-wise addition operation. The OLF instruction is used to load Fmap from the on-chip buffer to the EU. The SR instruction is used to store the result data derived from the EU to Memory (DRAM or on-chip buffer).

CTI: CTI is the branch control command mentioned in Section 3.1. To control three DTI instructions that may access DRAM, we have designed three selector instructions: Jumping ELW (JW), Jumping ELF (JF), and Jumping ELR (JR). Additionally, we have designed the Jumping Store (JS) instruction to handle the situation where the result may store FIFO in EU. Furthermore, a 1-bit interrupt instruction has been designed to remind the host of the timing of reading the result.

OPI: The relevant operations in CONV, ACT (ReLU), QUANT (Quantization), POOL and Upsample instructions are identified by their parameters. The QUANT instruction contains parameters related to bias and partial sum in addition to the quantization parameters.

3.3. COD Work Flow

We integrated CTI, OPI, and DTI into a single 256-bit COD very long instruction word (VLIW) and designed its decoder and parallel EU in the accelerator. In a typical VLIW superscalar processor, the compiler explicitly specifies the control dependencies between instructions. However, CNN inference with forwarding propagation in layers has a clear layer order. Therefore, we design a fixed depth pipeline at the accelerator micro-architecture level to ensure the sequential execution of all instruction types to reduce the complexity of the compiler. The execution flow of instructions, as shown in Figure 7b, indicates that CTIs act as decision nodes that determine the path for each execution branch. In the loading data stage, ELW, ELF, and ELR do not have dependencies on each other, and they are executed concurrently, sharing DRAM bandwidth in our accelerator. In the computation and data storing stage, OPIs and SR are also executed by a parallel pipeline. The parallel architecture of the accelerator is described in Section 4.

4. Prototype Accelerator

In this section, we present our prototype accelerator for COD, which comprises a series of instruction decode and dispatch units, a memory management unit (MM Unit), and an EU. The micro-architecture is illustrated in Figure 7c.

The workflow of the accelerator is as follows: During the preliminary stage, the instruction sequence generated by the compiler, the quantized weight, and the image are sent from the host to an on-chip buffer or DRAM using I/O DMA with AXI4 bus protocol. The accelerator subsequently operates through six major instruction pipeline stages, namely, fetching, decoding, issuing, memory accessing, execution, and writing back. The CF and Data Flow (DF) of these stages are depicted in Figure 7c. The instruction counter (IC) fetches instructions sequentially from the buffer and passes them to the decoder until an interrupt signal is received. The decoder disassembles COD VLIW into DTIs, CTIs, and OPIs using a bit-wise approach. OPIs are issued directly to the EU, while DTIs are transmitted to the MM Unit via the scheduler and the issue unit. The issue unit synchronizes the transfer status to the scheduler while issuing DTI to the MM Unit. Memory accessing, execution, and writing back form a coarse-grained parallel pipeline that is controlled by the scheduler. Additionally, we have designed a spatial parallel fine-grained execution pipeline to accelerate the OPIs in EU.

4.1. Control Logic

In the instruction pipeline of serial execution, as depicted in Figure 8a, two execution bottlenecks, caused by communication and computation, have to be endured. However, in the domain of CNNs, computation does not rely on global data, as the output of each computation is only related to the data corresponding to the sliding windows. Consequently, we designed a Coarse-Grain Temporal Pipeline (CTP) at the instruction level to enable the simultaneous execution of DTIs and OPIs in a single clock cycle.

To guarantee proper instruction execution, we categorize the dependencies of DTIs and OPIs into three levels: independent, partially dependent (p-dependent), and globally dependent (g-dependent). Table 3 illustrates how instruction X depends on instruction Y. When an instruction is independent of another instruction, the execution of the former does not need to take into account the execution process of the latter. When an instruction is p-dependent on another instruction, it has to wait for the latter to be executed for a certain amount of time before it can be executed (signal is generated and distributed by the scheduler). When an instruction is g-dependent on another instruction, it must wait for the latter to be executed before it can be executed. Subsequently, based on the COD instruction workflow and the dependencies, we design an instruction execution CTP, as shown in Figure 8b. The ELoad stage contains the ELW, ELR, and ELF instructions; the OLoad stage contains the OLF instruction; the Compute stage contains the CONV, ACT, QUANT, POOL, and UPSAMPLE instructions, and the SR stage contains the SR instruction.

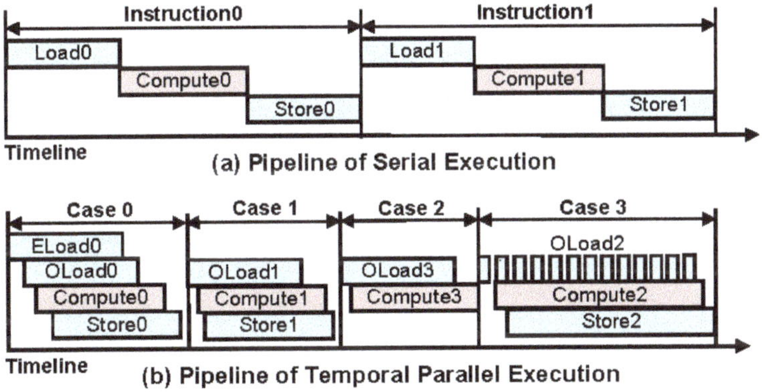

Figure 8. The temporal parallel instruction pipeline.

Table 3. Dependency table between DTIs and OPIs.

X \ Y	ELW(R)	ELF	OLF	OPIs	SR
ELW(R)	/	independent	independent	independent	independent
ELF	g-dependent	/	independent	independent	independent
OLF	g-dependent	p-dependent	/	independent	independent
OPIs	g-dependent	p-dependent	p-dependent	/	independent
SR	g-dependent	p-dependent	p-dependent	p-dependent	/

In our implementation strategy, weights and residuals are preloaded into the on-chip buffer, so all other instructions are g-dependent on the ELW and ELR instructions. These two instructions, on the other hand, have no dependency on each other and are executed simultaneously through multiple ports of the MM Unit. After ELW(R) is executed, feature maps start to be loaded while OPIs and SRs are executed one after another. Figure 8b shows the timing diagram of the instruction execution CTP for four typical cases. Case 0 is the case when JW, JR, and JF are 0. After the ELF instruction has loaded a certain amount of data, the subsequent stages are executed in parallel one after another. The Eload stage is jumped in case 1, and the SR stage is jumped in case 2. Different from the communication-bound in the previous three cases, the execution of computation-bound occurs in some instruction species with high data reuse, as shown in case 3.

To ensure the correct execution of CTP, we designed a multi-port shared DRAM bandwidth MM Unit and a scheduler, as illustrated in Figure 9. Four on-chip buffers and external memory DRAM are interconnected via AXI crossbar and are uniformly addressed between each memory. Each on-chip buffer is implemented with dual-port block RAM, writing data through AXI port and reading data through native port. Multiple AXI ports provide support for accessing data from different banks of DRAM, ensuring the concurrent execution of DTIs. Moreover, our MM Unit not only receives DTIs from the Issue unit, but also synchronizes the instruction execution process to the scheduler through the Issue unit. The scheduler will proceed to read the subsequent COD instruction only after all DTIs have been executed.

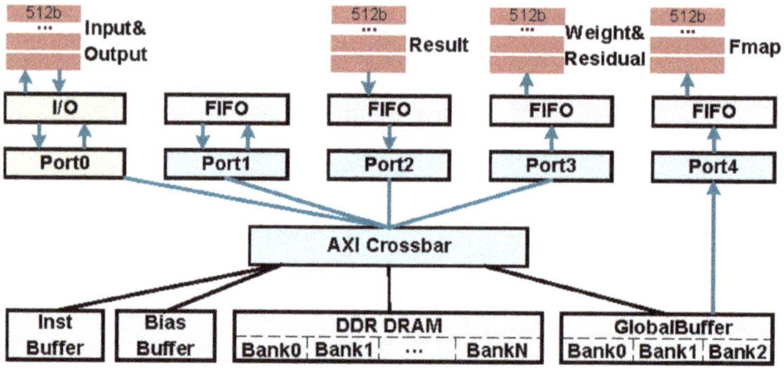

Figure 9. Memory Management Unit (MM Unit).

4.2. Execution Logic

In addition to the instruction-level parallelism enabled by CTP, there are further opportunities for parallelism in numerical operations pertaining to OPIs. In this section, we propose an EU capable of performing the parallel computation of OPIs, utilizing both spatial and temporal parallelism methodologies.

Spatially Parallel Structure (SPS): The CONV is computed as described in Section 2. We exploit the C_o-dimensional irrelevance of the CONV result $Y(F_{out})$ to design a SPS that enables parallel computation of 1-POC (Parallel Output Channel) channels. The choice of parallelism POC determines the hardware architecture design, which we determine in this paper based on burst transmission width and data quantization width. Our accelerator connects to the DRAM via AXI4 channels, where each channel typically supports 64 bytes per cycle through burst transmission mode in state-of-the-art FPGA platforms [35]. In addition, our data format is 16 bit. Thus, to match the access speed of the AXI4 bus (64 Bytes/cycle), we must implement 32 (64 Bytes/16 bits) computations per clock cycle, which we choose as our POC.

Figure 10 illustrates the SPS of the EU, where we use 16 spatially parallel FTPs (0–15 lines) to process each of the 32 output channels of F_{out}. To exploit this feature more effectively, we operate the DSP48 at twice the clock frequency of the system. Meanwhile, we design two sets of LUTRAM for each FTP to cache weight, which matches the DSP48. In this way, each FTP can perform two output channels at the system clock frequency, effectively saving DSP48 resources. The ELW instruction drives the weight fetch unit to load two weights into LUTRAMs in each FTP along the C_o dimension before the OPIs start executing. With the execution of OPIs in CTP, F_{in} is broadcast to 16 FTPs, and the 32 channels of F_{out} are computed in parallel.

Fine-grain Temporal Pipeline (FTP): Opportunities for parallelism arise for each input channel that the FTP is responsible for, as the multiplication operations within each kernel sliding window are uncorrelated. In Figure 10, we employ 32 cascaded DSP48s to form a 1D systolic array, creating a computational pipeline for parallel computation of 1-PIC (Parallel Input Channel) channels. Subsequently, two quant units and two pool units, collectively forming a FTP, follow this array. Once the FTP is established, it can handle the computation of two output channels within each system clock cycle (100 MHz).

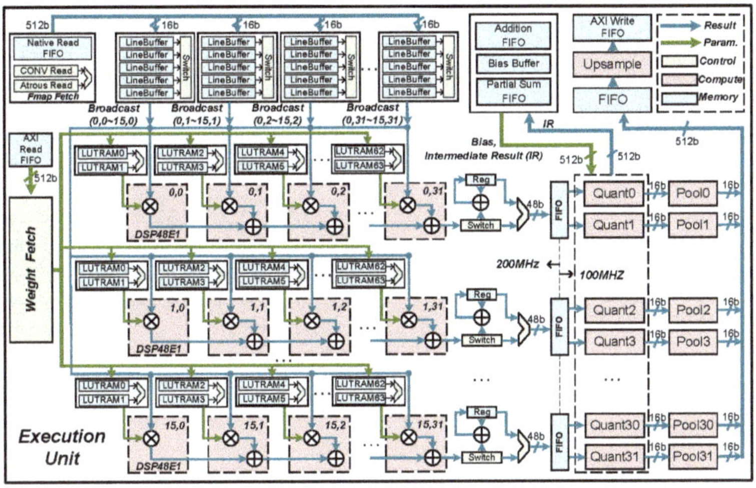

Figure 10. The overview of the execution unit.

Algorithm 1 presents the computational flow of the pipeline with $K = 1$ and $S = 1$ for $F_{in}(32, 4, 4)$, illustrating the operations at each clock cycle for each level of DSP. It is observed that the pipeline is established and one F_{out} can be output for each clock cycle after 31 cycles. The implementation of 1024 MACs (Multiply Accumulate) operations utilizes 63 clock cycles, resulting in a 16-fold efficiency improvement over a naive serial design.

Algorithm 1 CONV Operation Pipeline

Input: F_{in}(32,4,4), W(2,32,1,1)
//Due to K = 1, the indexes of the 3rd and 4th dimensions of W are omitted in the following description
Output: F_{out}(2,4,4)
Clock Cycle 00: **DSP L0:** W**[0][0]** × F_{in}**[0][0,0] = P0,0;**
Clock Cycle 01: DSP L0:W[1][0] × F_{in}[0][0,0] = P1,0;
DSP L1: P0,0 + W**[0][1]** × F_{in}**[1][0,0] = P1,1;**
Clock Cycle 02: DSP L0: W[0][0] × F_{in}[0][0][1] = P2,0;
DSP L1: P1,0 + W[1][1] × F_{in}[1][0][0] = P2,1;
DSP L2: P1,1 + W**[0][2]** × F_{in}**[2][0,0] = P2,2;**
......
// Pipeline setup
Clock Cycle 31: DSP L0:W[1][0] × F_{in}[0][3,3] = P31,0;
......
DSP L31: P30,30 + W**[0][31]** × F_{in}**[31][0,0] =** F_{out}**[0][0,0];**
Clock Cycle 32: DSP L1:P31,0 + W[1][0] × F_{in}[1][0,0] = P32,1;
......
DSP L31: P31,30 + W[1][31] × F_{in}[31][0,0] = F_{out}[1][0,0];
......
Clock Cycle 62: DSP L31 P61,30 + W[1][31] × F_{in}[31][3,3] = F_{out}[1][3,3];

As shown in Figure 10, to ensure the accuracy of FTPs data fetching, weight and Fmap caches are designed separately. Two sets of weight caches composed of LUTRAM are allocated for each FTP, and the two sets of cache alternate in inputting weight for DSP during operation. To reuse the Fmap, 16 FTPs share 32 Fmap caches, where each cache stores one channel of F_{in}, and five line buffers alternate write reads, broadcasting the correct Fmap to all FTPs. For atrous CONV, unnecessary rows in the Fmap fetch unit and unnecessary columns in the line buffers are skipped by the read logic, enabling the atrous CONV to share the same FTP as the CONV. Moreover, a temporary cache logic is incorporated after the systolic array, which is used to accumulate the result of multiple clock cycles to support the instruction of kernel size greater than 1. The intermediate result of the array is accumulated and stored in a reg type variable, and the result is output when the count reaches the size of the kernel (W). For instance, when K = 3, the output of the array is summed with the data from Reg and the result is re-stored in Reg until the ninth output completes the sum.

Following the convolution unit, we designed the Quant, Pool, and Upsample units to execute other OPI instructions. The Quant unit is shown in Figure 11a. First, it quantizes the input data from 48 bits to 16 bits by performing a bit shift operation. The exact shift parameter, denoted as Fl, is determined by parsing the Quant instruction. Additionally this instruction defines the operation mode of the Add Partial Sum (Psum) module. There are three modes: (1) When the input data represents the final result, it is directly fed into the next module. (2) When the data is an intermediate result (IR) and corresponds to the first tile, it is stored in the Psum FIFO. (3) Subsequent tiles read the data of Psum FIFO and accumulate it. (More details about tiling will be discussed in Section 5.2). The final result of the convolution is then directed to the Add bias and ReLU modules for the corresponding logical operations. Following this, there is an Element-Wise Addition module. It functions similarly to the Add Psum module, with the key difference being that the Addition FIFO can also be initially loaded with data via the ELR instruction. This feature is useful when dealing with situations where the amount of residual data exceeds the FIFO capacity. Finally, the result of the upsample is sent to the MM Unit to execute the RS instruction.

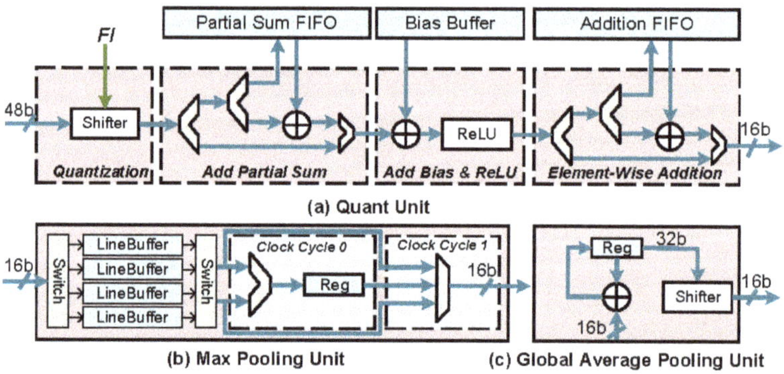

Figure 11. The overview of the Quant and Pool units.

5. COD Compiler

We develop a specialized compiler based on the COD encoding rule to translate high-level language CNN computation graphs into a COD instruction sequence composed of binary digits that the accelerator can understand and execute. Additionally, we perform optimizations, including BN folding and fixed-point quantization, on the input CNN before compiling it. Figure 12 depicts the entire process of deploying a CNN received from a DL framework into our accelerator. After optimization, the fixed-point weights, computation graph prototxt (CGP), and quantization information files are sent to the compiler. In the tiling phase, the CONV layers of the CGP are divided into multiple sub-blocks to fit the FTP mentioned in Section 4, and the weights are rearranged according to the tiling rules. In the fusion phase, the operations of other layers are merged into each sub-block. In the assembly phase, the COD instruction information is converted into binary digits. All COD instructions are arranged to form the instruction sequence corresponding to the input CNN.

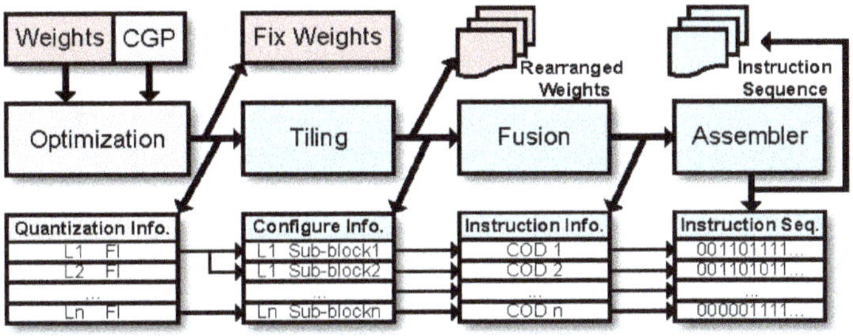

Figure 12. The workflow of the compiler.

5.1. Optimizations

BN Folding: The coefficients $\gamma, \sigma, \varepsilon, \beta$, and μ in the BN operation described in Equation (6) are explicitly determined during the inference stage. When we substitute Equation (1) into Equation (6), it results in Equation (7), representing the convolution merge BN operation. This equation can be simplified to Equation (8). It is evident that the computational pattern in Equation (8) is the same as that used in convolution. Therefore, BN folding can be achieved by modifying the weight and bias of the CONV layer to incorporate the BN coefficients, resulting in new weight $\hat{W}$ and new bias $\hat{b}$ as shown in Equations (9) and (10). This technique eliminates the need for computing BN, thereby reducing the inference time.

$$y = \gamma \frac{(Wx + b) - \mu}{\sqrt{\sigma^2 + \varepsilon}} + \beta \tag{7}$$

$$y = \frac{\gamma W}{\sqrt{\sigma^2 + \varepsilon}} x + \frac{\gamma}{\sqrt{\sigma^2 + \varepsilon}}(b - \mu) + \beta \tag{8}$$

$$\hat{W} = \frac{\gamma W}{\sqrt{\sigma^2 + \varepsilon}} \tag{9}$$

$$\hat{b} = \frac{\gamma}{\sqrt{\sigma^2 + \varepsilon}}(b - \mu) + \beta \tag{10}$$

Data Quantization: Our post-training quantization scheme is based on the fusion of methods proposed in [36,37]. It involves a linear mapping of integers x to floats $\hat{x}$ using Equation (11).

$$X_f \approx \hat{X}_i = 2^{-f_l} \cdot X_i \tag{11}$$

where $-f_l$ and $\hat{X}_i$ represent the fraction length parameter and the floating point value from the de-quantization of X_i, respectively. Substituting the original CONV Equation (1) each term with (11), we can obtain the full integers CONV Equation (12).

$$\hat{o}_i = \frac{2^{-f_{l}x} \cdot 2^{-f_{l}w}}{2^{-f_{l}o}} \sum x_i \cdot w_i + \frac{2^{-f_{l}b}}{2^{-f_{l}o}} b_i \tag{12}$$

The fraction length parameter f_l is pre-computed offline on the calibration set using the method proposed in [37], as shown in Equation (13).

$$\arg\max \sum \cos(\hat{o}_i, o_f) \tag{13}$$

The resulting array of quantization information, consisting of f_l for each layer, is fed to the compiler, and these parameters are compiled into Quant instructions. At runtime, only a simple shift operation is required in the Quant unit.

5.2. Tiling

Tiling Rule: The tiling rule presented in Equation (14) and Figure 13a slices the CONV operation into sub-blocks along the C_i and C_o dimensions to fit the parallelism capability of the accelerator. The parameter Sn represents the total number of sub-blocks, which is determined by the amount of parallelism in the C_i and C_o dimensions, i.e., PIC and POC respectively.

$$Sn = \lceil C_i/PIC \rceil \cdot \lceil C_o/POC \rceil \quad (14)$$

To ensure that the size of data scheduled by an instruction does not exceed the on-chip buffer capacity, the tiling rule can be extended to consider the H dimension as well. The parameter T_h determines the height of each sub-block, and it should satisfy the constraint in Equation (15), where C(GlobalBuffer) represent the size of the on-chip buffer. This constraint guarantees that the feature map of each sub-block can fit into the on-chip buffer.

$$T_h \times W \times PIC < \mathrm{C(GlobalBuffer)} \quad (15)$$

However, it is unnecessary to perform K dimensional tiling of weights since the on-chip buffer of weights typically has sufficient capacity to cache the weight data tiled in the C_i and C_o dimensions. Therefore, the tiling rule presented in Figure 13a only slices the CONV operation along the C_i and C_o dimensions.

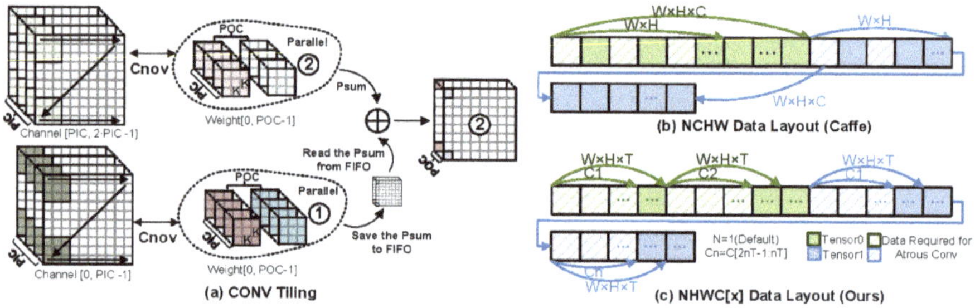

Figure 13. CONV Tiling and Data layout.

Data Layout: To optimize the utilization of the 64 bytes of data accessed from the AXI4 channel per clock cycle via burst mode, a specific data layout must be designed, which differs from the generic DL framework. As illustrated in Figure 13b, a classic DL framework like Caffe arranges data in a three-dimensional tensor based on the channel (C), height (H), and width (W). However, for Atrous CONV, this arrangement leads to numerous non-contiguous data accesses, thereby wasting the bandwidth of the AXI4 bus. To avoid this issue, we propose a NHWC[x] scheme based on NHWC, as depicted in Figure 13c. In this scheme, the tensor is sliced along the C dimension based on the maximum amount of data accessed in one burst (T). The sliced block is then arranged in order, with the HWC order used within each block. Since the design of tiling unifies T and the POC and PIC, the 64 bytes of data accessed in one burst precisely contain the data needed for all FTPs.

5.3. Fusion and Assembler

To minimize unnecessary data movement, we integrate the Quant ReLU, Pool, and Upsample operations into the sub-block CONV operation and execute them in parallel in the FTP of our accelerator. The parameters of these fused-operations are combined to form the OPI information for each sub-block. Using this OPI information, we generate DTI and CTI, with the main objective being to find the optimal data scheduling path that minimizes the latency of the load–store process. The load-related DTIs depend on SR instructions in the previous layer of the instruction sequence. To reduce the external memory load (ELoad) as much as possible, the SR instruction address is directed towards the on-chip cache address, as illustrated in Figure 8b case 1, 2, 3.

The assembler is responsible for converting the COD instruction information generated by each fused-operation into binary digits, based on the encoding format described in Section 3.2. When switching between different CNNs, our accelerator can simply overlay a new COD instruction sequence into the instruction buffer, without the need to re-burn the FPGA.

6. Experiments

The workflow of our accelerator is illustrated in Figure 2. In the offline phase, we employ PyTorch for model training and quantification. Subsequently, the compiler generates instruction sequences and rearranged weights based on FIs and CGPs. During the runtime phase, the Host PC transmits instructions, weight files, and preprocessed images to the external DRAM of the FPGA via the PCIe bus. The accelerator initiates the CNN inference process, and upon completion, the Host PC retrieves the inference results from the DRAM. It should be noted that this work focused on accelerating the CNN process, and other operations such as image preprocessing and result display were implemented on the CPU. Further reports and details of the evaluation are provided below.

In this section, we conduct experiments based on the aforementioned process. Initially, we train and quantize the segmentation model using PyTorch 1.11.0 and the CUDA 11.3 toolkit on an NVIDIA RTX 3090 GPU. Next, we developed the proposed compiler in C++ to transform the CGP into a sequence of COD instructions. Lastly, we implement the prototype accelerator on a Xilinx VC709 development board with a XC7VX690T FPGA. All the accelerator hardware modules are developed using Verilog HDL. The accelerator is synthesized and implemented with Vivado 2018.3.

6.1. SCIs Segmentation

Dataset: In this subsection, we evaluate the performance of our segmentation models on two datasets.

Satellite Dataset [5]: This dataset consists of 3117 images collected from the internet, all having a consistent resolution of 1280 × 720. It is divided into training (2516 images) and test subsets (600 images). The dataset includes three main feature component types: Body, Solar Panel, and Antenna.

SCIs Dataset [23]: This newly created dataset contains 8833 simulated spacecraft images, with 7061 images designated for training and the remaining 1772 for testing. The dataset spans 26 different image resolutions, ranging from 90 × 82 to 1015 × 1015. It encompasses 16 diverse spacecraft types and five crucial feature component types: Panel, Antenna, Thruster, Optical load, and Mechanical arm. This dataset closely aligns with the actual segmentation needs of space scenes, setting it apart from the Satellite Dataset.

Preprocessing and Hyperparameters: For all images, we apply uniform resizing to 256 × 256 both during training and inference. Additionally, for the training set, we employ standard data augmentation techniques, including random scaling (0.5, 2.0), random horizontal fliping, and normalization.

The training hyperparameters are as follows: the learning rate schedule "poly" policy [38] and initial learning rate 0.005, weight decay of 1×10^{-4}, number of iterations 20,000, batch size of 32, and cross-entropy loss type. Hyperparameters without mentioned task-related training were adopted from the CNN's base model.

Benchmark: We configure six benchmark CNN models for the SCIs segmentation task, based on the Deeplabv3 series of algorithms. These models consist of two head networks: Deeplabv3+ [22] and DeepLabv3 [21], paired with three backbone networks: VGG16 [39], ResNet18 [40], and SqueezeNet1.1 [41]. The head network with ASPP module has dilation rates of 1, 2, 4, 6. Table 4 displays the model sizes and complexities. The GOPS (Giga-operations) column in the table represents the number of operations (multiplication or addition operations) included in each model.

Table 4. The model size and complexity of the DeeplabV3 series model on the satellite dataset.

Model	Backbone	Model Size (MB)		Complexity
		FP32	INT16	(GOPS)
DeepLabv3	VGG16	77.96	38.88	42.64
DeepLabv3	ResNet18	63.72	31.86	11.06
DeepLabv3	SqueezeNet1.1	21.80	10.90	2.84
DeepLabv3+	VGG16	78.24	39.12	48.42
DeepLabv3+	ResNet18	64.16	32.08	17.18
DeepLabv3+	SqueezeNet1.1	22.28	11.14	9.28

Segmentation Result: We employed both mIoU (mean Intersection over Union) and PA (Pixel Accuracy) [42] metrics to assess the segmentation accuracy of the six models across the two datasets, as demonstrated in Table 5. Figure 14 shows a visualization of the segmentation result obtained using the Deeplabv3+ ResNet18 model. To reduce the computational complexity and memory footprint of these models, we adopt an INT16 quantization scheme, as discussed in Section 5.1. We observe that the quantized models achieve almost the same accuracy as the original float (FP32) models, with accuracy degradation ranging between −0.14 and +0.09 for the mIOU on the Satellite dataset and between −0.5 and +0.54 on the SCI dataset. The degradation in quantification accuracy typically arises from two sources: clipping error and rounding error, which are mutually exclusive. Retaining a larger quantitation range, such as the maximum and minimum values, reduces clipping error to zero but significantly increases rounding error, especially when quantifying activations. Activations, having more outliers than weights, are particularly susceptible to this effect. The EasyQuant quantitation framework [37] used in this paper iteratively retains the quantitation parameters with the highest cosine similarity between the inverse quantized data and the original data during the quantitation process. This implies that the clipping range of quantization may not strictly follow the maximum and minimum of the data, leading to some outliers not being considered within the quantization range. Consequently, outliers in the quantized activation for each layer may have a comparatively lesser impact on forward propagation. In fact, these outliers may not always have a positive effect on the final accuracy, since in cases where the outliers are noise, the quantized model may bring unexpected accuracy gains, as is the case for some models in Table 5. However, these marginal gains are also influenced by the convergence degree of the model. When the model is trained with more rounds of higher accuracy, the noise in the forward propagation is reduced, and consequently, this accuracy gain may be diminished as well.

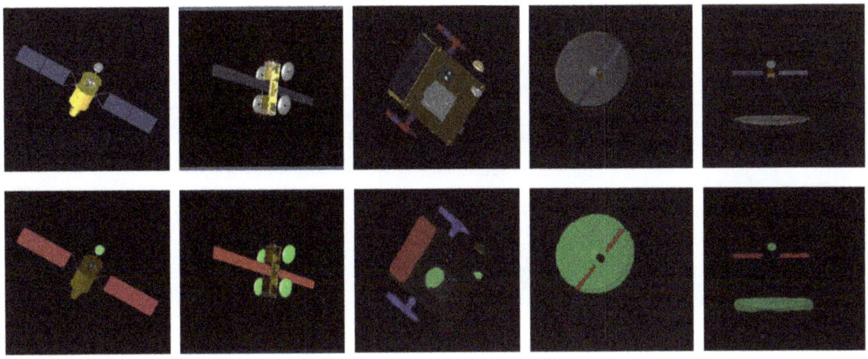

Figure 14. Result on the SCI image based on our model (DeepLabv3+ ResNet18): input image (**top**) and segmentation result (**bottom**). Green, blue, and red areas are antenna, mechanical arm, and panel components, respectively.

Table 5. The accuracy of the DeeplabV3 series model.

Model	**Accuracy**			
	mIoU		**PA**	
	FP32	**INT16**	**FP32**	**INT16**
Satellite Dataset				
DeepLabv3 VGG16	67.32%	67.31%	95.12%	95.12%
DeepLabv3 ResNet18	60.57%	60.66%	93.30%	93.33%
DeepLabv3 SqueezeNet1.1	54.93%	54.98%	91.31%	91.34%
DeepLabv3+ VGG16	67.46%	67.32%	95.50%	95.50%
DeepLabv3+ ResNet18	62.63%	62.71%	93.99%	94.01%
DeepLabv3+ SqueezeNet1.1	56.06%	56.05%	92.47%	92.49%
SCIs Dataset				
DeepLabv3 VGG16	69.72%	69.42%	99.00%	99.00%
DeepLabv3 ResNet18	64.06%	63.56%	98.86%	98.84%
DeepLabv3 SqueezeNet1.1	61.09%	61.63%	98.70%	98.70%
DeepLabv3+ VGG16	81.62%	81.65%	99.56%	99.55%
DeepLabv3+ ResNet18	78.04%	77.84%	99.45%	99.43%
DeepLabv3+ SqueezeNet1.1	74.14%	74.36%	99.35%	99.35%

6.2. Accelerator Performance Analysis

In this subsection, we provide information about the implementation details of the accelerator and then analyze its performance. Considering the model complexity, we focus on Deeplabv3+ ResNet18 and SqueezeNet1.1 for model acceleration in this subsection.

Implementation Details: Table 6 displays the parameters and resource utilization of our prototype accelerator. The global buffer is 1 MB implemented by BRAM resource for caching intermediate feature maps. The weight buffer is distributed adjacent to each DSP, and we configure two 64 B LUTRAM caches for each DSP, which allows our DSP to operate at two times the system clock frequency. This design allows the EU using 512 DSP resource to achieve the computational efficiency of 1024 multiplier and adder equivalents.

Table 6. Parameters and resource utilization of our accelerator.

Parameters of Our Accelerator					
Buffer	Global Buffer		512 KB		
	Weight Buffer		64 KB (1024 × 64 B)		
	Bias Buffer		16 KB		
	Instruction Buffer		79 KB (32 B × 2500)		
Operation	Operations in EU		512 (32 × 16) multipliers and adders		
Bus	AXI bus width		512 bits		
Data	Width		16 bits (fixed point)		
Resource Utilization					
Resource	LUT	FF	LUTRAM	DSP	BRAM
Used	198,262	185,839	42,097	519	724
Total	433,200	866,400	174,200	3600	1470
Utilization	45.77%	21.45%	24.17%	14.42%	49.25%

Reducing External Memory Access: Enhancing energy efficiency and throughput can be achieved by reducing off-chip data movement and enhancing EU utilization [24]. The DMH introduced in Section 3.1 effectively utilizes the on-chip buffer and minimizes DRAM accesses. To illustrate, we consider the DeepLabv3+ ResNet18 model as an example, which we compiled into 2424 COD instructions. A comparison of DRAM accesses between our COD CF and the primitive CF case is presented in Figure 15. In the primitive CF, DRAM accesses involve inputs, output feature maps, and weights. (Thanks to our instruction buffer, we can cache all instructions on-chip.) The DMH structure of the COD control flow avoids DRAM accesses for intermediate feature maps by directly caching them in the on-chip Global Buffer. For the DeepLabv3+ ResNet18 model, we achieve an impressive 26% reduction in DRAM accesses overall. Notably, in the most efficient RES1 layer, we achieve a remarkable 95% reduction in DRAM accesses. These savings in access time contribute to the high performance of our accelerator.

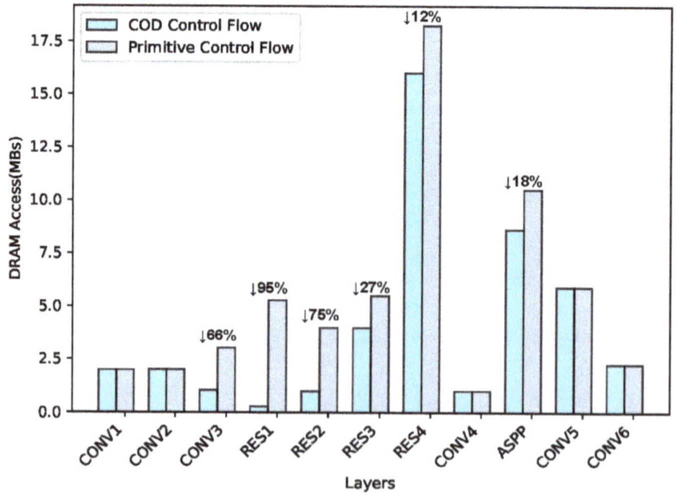

Figure 15. The comparison of external memory access between primitive control flow and our COD control flow on the DeepLabv3+ ResNet18 model.

Performance Analysis: To evaluate the performance of our accelerator, we employed a roofline model [29], as depicted in Equation (16), where the TTR represents the Theoretical Roof Throughput. This model considers both memory and compute bottlenecks, providing a valuable representation of the hardware performance.

$$P = \begin{cases} \beta \cdot I, & I < \mathrm{I_{max}} \\ \mathrm{TTR}, & I \geq \mathrm{I_{max}} \end{cases} \tag{16}$$

Within the equation, P represents performance, measured in throughput (GOPS/s, Giga-operations per second). Additionally, β corresponds to DRAM access bandwidth (GB/s, Giga-bytes per second), I denotes operation density (OPS/Byte, operations per byte), and I_{max} signifies the point of intersection between computational and bandwidth bottlenecks, calculable using Equation (17).

$$I_{max} = \frac{\mathrm{TTR}}{\beta} \tag{17}$$

Furthermore, Theoretical Roof Throughput (TTR) of hardware is calculated according to Equation (18), where $\mathrm{MAC_{num}}$ represents the number of MAC units (DSP48E1) in hardware and f is the working clock frequency of MAC units. To convert the unit of operations from MACs (multiply-accumulate operations) to OPS (multiplication or addition operations), it is necessary to multiply by a factor of 2.

$$\mathrm{TTR} = \mathrm{MAC_{num}} \times 2 \times f \tag{18}$$

The TTR of our accelerator is calculated at 207.6GOPS/s ($519 \times 200 \times 2$), while actual testing revealed a bandwidth (β) of approximately 6.7 GB/s. To assess the accelerator's runtime performance, we added a global clock cycle counter and a Xilinx ILA (Integrated Logic Analyzer) IP into the design. When the accelerator is running, the ILA can be triggered to view the counter number based on the instruction address and state machine ID, and the delay of each stage can be calculated based on the running clock frequency and the clock cycle number. The actual performance of the accelerator can then be calculated from the operations and delays. Utilizing roof throughput data and runtime performance data, we constructed the roofline model for our accelerator, as illustrated in Figure 16.

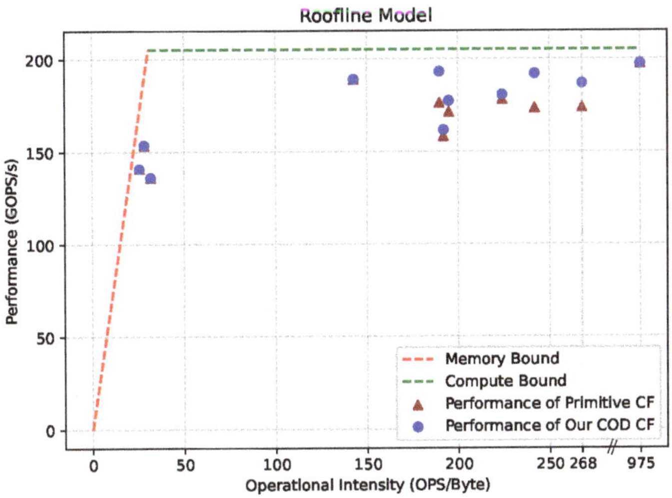

Figure 16. The roofline model of our accelerator.

In the figure, the dotted line illustrates the hardware acceleration limit of our accelerator. The bandwidth bottleneck is highlighted in red, and the computational bottleneck is depicted in green. Scattered dots represent the acceleration performance of each layer in the DeepLabv3+ ResNet18 model. Closeness of the dots to the bounding line indicates higher hardware utilization. The primitive CF case represents a scenario where all layer data is fetched from DRAM. Our COD CF reduces unnecessary DRAM accesses, bringing our performance closer to the boundary.

In total, we achieved model acceleration with a latency of 93.27 ms and a performance of 184.19 GOPS/s, representing 88.72% of the TTR. This indicates that 88.72% of the clock cycles are effectively utilized for computation.

6.3. Comparison with Related Works

In this subsection, we compare the efficiency of our COD instructions and accelerator with prior research in terms of instruction set coding and computational efficiency, respectively.

Instruction Coding Efficiency Comparison: Despite our COD ISA having a 256-bit word length for a single instruction list, our scheme maintains excellent coding efficiency due to the high parallelism strategy of our hardware accelerator. Table 7 provides an instruction size comparison between our COD instructions and previous works for the same CNN models.

Table 7. Comparison of total instruction size for different accelerators.

Model	Instruction Size (KB)				Reduction Rate
	SLC [32]	IUU [43]	LIS [44]	Ours	
VGG-11	1620	270	—	**33**	**49/8.2/—**
VGG-16	2650	450	106	**54**	**49/8.3/1.9**
VGG-19	—	600	108	**73**	**—/8.2/1.5**

The hardware parallelism for IUU [43] and SLC [32] is limited to 64 (PIC, POC = 8). This parameter is directly correlated with the number of instructions because the CONV operation is sliced according to this parameter, with each tiling requiring one instruction to drive it. In contrast, our COD accelerator features a parallelism of 1024 (PIC, POC = 32), enabling us to encode the same model with fewer instructions. As a result, our COD reduces the instruction size by a factor of 8× compared to IUU [43] and 49× compared to SLC [32], respectively. LIS [43] is a lightweight instruction set that supports dilated convolution and mixed-precision operands. However, its execution depends on a RISC-V processor, requiring the inclusion of a 96 KB program within the instructions. In contrast, our instruction parsing unit and instruction encoding are co-designed, making our instructions independent of RISC-V or other processors for execution. As a result, our COD reduces the instruction size by a factor of 1.9× and 1.5× compared to LIS [43].

While instructions constitute a relatively small amount of data compared to weights and feature maps, it is crucial to consider the constraints of bandwidth and storage resources in space applications.

Computational Efficiency Comparison: Table 8 presents a performance comparison of our accelerator with previous CNN-based image segmentation accelerators. The "—" in the table indicates that the accelerator did not report that parameter or performance. Computational Efficiency reflects how efficiently the accelerator utilizes computational resources and is calculated as Performance divided by TTR. Note that in the comparison we uniformly use the number of DSPs used to denote the MAC_{num} in the TTR. The model abbreviations in the table represent DLV3P-X (DeepLabv3+ Xception [45]), DLV3P-B (DeepLabv3+ ResNet18), and DLV3P-C (DeepLabv3+ SqueezeNet1.1).

Table 8. Comparison with previous image segmentation accelerators.

	Liu et al. [16] in TRETS 2018	Wu et al. [18] in TCASI 2022	Bai et al. [15] in TCASI 2020	Im et al. [25] in TCASI 2020	Morì et al. [24] in DAC 2022		Ours	
Accelerator Type	Overlay	Overlay	SN	Overlay	Overlay		**Overlay**	
Model	U-Net	ENet	RoadNet-RT	DL3P-X	DL3P-B		**DL3P-B**	**DL3P-C**
Platform	Xilinx XC7Z045	Intel Arria 10	Xilinx ZCU102	65 nm CMOS	Intel Arria 10		**Xilinx XC7VX690T**	
Frequency (MHz)	200	200	250	200	189.81	148.44	**200**	
Precision	16-bit	8-bit	8-bit	8-bit	16-bit		**16-bit**	
DSPs used	900	607	1560	—	690	1362	**519**	
Performance (GOPS/s)	107.00	200.31 *	331.00	65.23 **	117.31	183.3	**184.19**	**159.48**
Computational Efficiency (GOPS/s/TTR)	29.72%	82.5%	42.43%	—	44.78%	45.33%	**88.72%**	**76.82%**

* The data calculated based on the computational efficiency and used DSP in [18]. ** The data calculated based on the latency and model architecture in [25].

Morì et al. introduced a hardware-aware pruning method using a genetic algorithm [24], effectively reducing the complexity of the benchmark model DL3P-B. However, when accelerating the original model, our accelerator outperforms theirs with similar resource consumption. In the acceleration of the DL3P-B model, our computational efficiency is 43.93% better than that of their accelerator. In addition to [43], Im et al. designed the DT-CNN accelerator [25], which also supports the ASPP structure of DeepLabv3+. We obtained a performance of approximately 65.23 GOPS/s for DT-CNN when accelerating the DL3P-X model based on the delay and network structure parameters they provided. Compared to this, our accelerator achieves higher performance.

In addition to the DeepLabv3+ model, we also compared other similar segmentation task models. Bai et al. introduced a lightweight road segmentation model, RoadNet-RT [18], and implemented an SN-type model accelerator on a ZCU102 FPGA with an acceleration performance of 331GOPS/s. However, it consumes more computational resources, resulting in lower computational efficiency. In comparison, our computational efficiency is 46.29% higher than [18]. Wu et al. proposed an efficient accelerator [18] supporting multiple convolution types. For the semantic segmentation task, they accelerated the ENet model, achieving a performance of 200.31 GOPS/s and a computational efficiency of 82.5%. Our accelerator outperforms theirs with a 6.22% higher computational efficiency compared to [18]. Liu et al. [16] designed a custom architecture for DeCONV in the U-Net model and implemented the image segmentation task at 107 GOPS/s. We outperform them with a performance that is 77.91 GOPS/s higher and a computational efficiency that is 59% higher.

Comparison with Other Overlay Accelerators: In addition to addressing semantic segmentation tasks, more previous accelerators are catered to more fundamental assignments, including classification. Consequently, to gauge the efficiency of our accelerator in comparison to previous overlay accelerators, we assess both the processing efficiency and resource consumption of the classical VGG-16 model, as summarized in Table 9.

Compared to fpgaConvNet [46], our work uses less computational resources and achieves higher performance. Compared to Angel-eye [47], we use similar LUT resources and achieve similar performance, but our DSP usage is significantly reduced and the overall computational resource efficiency is improved by 8.51%. While we may not possess a performance advantage compared to Caffeine [48] and FlexCNN [49], our work uses far fewer resources. In fact, we demonstrate a resource efficiency improvement of 15.16% and 19.80% compared to Caffeine [48] and FlexCNN [49], respectively. Furthermore, given that Xilinx's Vitis AI tool employs 8-bit quantization, the Xilinx B4096 DPU [34,50] exhibits reduced LUT resource consumption. However, its computational resource efficiency is comparatively lower at 57.59%, potentially attributed to multi-core DDR sharing. In contrast, our work boasts a more substantial efficiency improvement at 30.82%. The DPU's inference performance is sourced from the official Xilinx document [34], while its resource consumption data is extracted from the official document [50].

Comparison with GPU (Graphics Processing Unit): In addition to FPGAs, GPUs are a prevalent hardware platform for CNN acceleration. In Table 10, we present a comparison of the acceleration performance between our accelerator and a GPU. It is evident that the GPU, equipped with more computational resources and higher frequencies, demonstrates faster processing speeds, but it also brings higher power consumption. Considering energy efficiency as a crucial metric for onboard computing platforms, our dedicated accelerator showcases a noteworthy 5.1× improvement in energy efficiency when performing SCD segmentation tasks compared to a general-purpose GPU.

Table 9. Performance and computational efficiency comparison with previous overlay accelerators (Model: VGG 16, Image Size: 224 × 224).

	fpgaConvNet [46]	Caffeine [48]	Angel-Eye [47]	Xilinx B4096 DPU [34,50] *	FlexCNN [49]	COD(Ours)
Platform	Zynq Z045	XC7VX690T	Zynq Z045	ZCU102	Alveo U250	**XC7VX690T**
Precision	16-bit	16-bit	16-bit	8-bit	16-bit	**16-bit**
Frequency (MHz)	125	150	150	281	241	**200**
Batch Size	1	1	1	3	1	**1**
DSPs used	900	2833	780	1926	4667	**519**
LUTs used	218,600	350,892	182,616	111,798	682,732	**198,262**
Performance (GOPS/s)	155.81	488.00	187.80	623.10	1543.40	**183.54**
Computational Efficiency (GOPS/s/TTR)	69.25%	73.25%	80.26%	57.59%	68.61%	**88.41%**

* Xilinx DPU's VGG16 model contains fully connected layers, whereas the other work in the table contains only convolutional layers. It is worth noting that the convolutional layer accounts for 99.6% of all computation in VGG16.

Table 10. Energy efficiency comparison with GPU (Model: DL3P-C, Image Size: 256 × 256).

Platform	RTX 2080 Ti GPU	XC7VX690T FPGA
Framework	Pytorch-GPU	-
Frequency (MHz)	1635	**200**
External Memory	11 GB GDDR6	**4 GB DDR3**
Speed (Frames/s)	39.6	**17.2**
Power (W)	250	**21** *
Energy Efficiency (Frames/s/W)	0.16	**0.82**

* The power consumption is measured from the board using a power meter during FPGA inference.

7. Conclusions and Future Work

This paper introduces an innovative workflow for deploying DeepLabv3+ CNN onto FPGAs, comprising a tailored COD instruction set, an RTL-based overlay CNNs accelerator, and a specialized compiler. Our accelerator was implemented on a Xilinx Virtex

XC7VX690T FPGA at 200 MHz. In our experiments, the accelerator achieved an accuracy of 77.84% with INT16 quantization, exhibiting only a 0.2% degradation compared to the fully precision model on the SCIs dataset. Notably, the accelerator delivered a performance of 184.19 GOPS/s with a computational efficiency of 88.72%. In contrast to prior work, our accelerator exhibited a 1.5× performance improvement and a remarkable 43.93% boost in computational efficiency. Moreover, our COD instruction set demonstrated a substantial reduction in size, ranging from 1.5× to 49× when compiling the same model compared to previous methodologies.

The experiments presented in this paper are conducted on the ground. The PC serves as the analog source for sending and receiving data, while the FPGA development board functions as the implementation platform for the accelerator, performing CNN inference computations. For deployment in the actual space environment, it also is essential to consider engineering experiments, including mechanical tests, high- and low-temperature tests, radiation resistance tests, etc., to verify the reliability of the accelerator.

Random bit-bias feature faults (RBFFs) [51] caused by single and multiple event upsets is an issue to be considered during the migration of our design to an actual hardware platform in a space environment. From an architectural design perspective, the impact of the radiation environment on the accelerator can be mitigated through the implementation of logical redundancy. In subsequent work, we will add parity bits to the COD instruction and use the triple modular redundancy (TMR) approach to increase the fault tolerance of instruction set execution in hardware. Moreover, different CNN models have different tolerances for RBFF, and due to our overlay design we can explore highly fault-tolerant CNN models for deployment without redesigning the hardware.

Author Contributions: Conceptualization, Z.G.; Methodology, Z.G. and X.S.; Software, Z.G. and W.L.; Validation, Z.G., X.S. and W.L.; Formal analysis, K.L. and C.D.; Investigation, Z.G., K.L., S.L. and X.S.; Resources, W.L. and K.L.; Data curation, Z.G., X.S., W.L. and S.L.; Writing—original draft, Z.G.; Writing—review editing, K.L. and W.L.; Visualization, Z.G. and X.S.; Supervision, K.L., C.D. and W.L.; Project administration, W.L., C.D. and K.L.; Funding acquisition, W.L. All authors have read and agreed to the published version of the manuscript.

Funding: This work was supported by the National Natural Science Foundation of China under Grant 62171342 and the State Key Laboratory of Geo-Information Engineering under Grant SKLGIE2023-M-3-1.

Data Availability Statement: Data are contained within this article.

Conflicts of Interest: The authors declare no conflicts of interest.

References

1. Yin, S.; Xiao, B.; Ding, S.X.; Zhou, D. A Review on Recent Development of Spacecraft Attitude Fault Tolerant Control System. *IEEE Trans. Ind. Electron.* **2016**, *63*, 3311–3320. [CrossRef]
2. Uriot, T.; Izzo, D.; Simões, L.F.; Abay, R.; Einecke, N.; Rebhan, S.; Martinez-Heras, J.; Letizia, F.; Siminski, J.; Merz, K. Spacecraft Collision Avoidance Challenge: Design and results of a machine learning competition. *arXiv* **2020**, arXiv:2008.03069.
3. Carruba, V.; Aljbaae, S.; Domingos, R.C.; Lucchini, A.; Furlaneto, P. Machine learning classification of new asteroid families members. *Mon. Not. R. Astron. Soc.* **2020**, *496*, 540–549. [CrossRef]
4. Forshaw, J.L.; Aglietti, G.S.; Navarathinam, N.; Kadhem, H.; Salmon, T.; Pisseloup, A.; Joffre, E.; Chabot, T.; Retat, I.; Axthelm, R.; et al. RemoveDEBRIS: An in-orbit active debris removal demonstration mission. *Acta Astronaut.* **2016**, *127*, 448–463. [CrossRef]
5. Dung, H.A.; Chen, B.; Chin, T.J. A Spacecraft Dataset for Detection, Segmentation and Parts Recognition. In Proceedings of the Conference on Computer Vision and Pattern Recognition Workshops (CVPRW), Nashville, TN, USA, 19–25 June 2021; pp. 2012–2019.
6. Black, K.; Shankar, S.; Fonseka, D.; Deutsch, J.; Dhir, A.; Akella, M.R. Real-Time, Flight-Ready, Non-Cooperative Spacecraft Pose Estimation Using Monocular Imagery. *arXiv* **2021**, arXiv:2101.09553.
7. Shotton, J.; Winn, J.; Rother, C.; Criminisi, A. Textonboost for image understanding: Multi-class object recognition and segmentation by jointly modeling texture, layout, and context. *Int. J. Comput. Vis.* **2009**, *81*, 2–23. [CrossRef]
8. Ladický, L.; Russell, C.; Kohli, P.; Torr, P.H. Associative hierarchical crfs for object class image segmentation. In Proceedings of the International Conference on Computer Vision(ICCV), Kyoto, Japan, 27 September–4 October 2009; pp. 739–746.
9. Long, J.; Shelhamer, E.; Darrell, T. Fully convolutional networks for semantic segmentation. In Proceedings of the Conference on Computer Vision and Pattern Recognition (CVPR), Boston, MA, USA, 7–12 June 2015; pp. 3431–3440.

10. Liu, Y.; Zhu, M.; Wang, J.; Guo, X.; Yang, Y.; Wang, J. Multi-Scale Deep Neural Network Based on Dilated Convolution fo Spacecraft Image Segmentation. *Sensors* **2022**, 22, 4222. [CrossRef] [PubMed]
11. Lin, T.Y.; Maire, M.; Belongie, S.; Hays, J.; Perona, P.; Ramanan, D.; Dollár, P.; Zitnick, C.L. Microsoft coco: Common object in context. In *Proceedings of the Computer Vision–ECCV 2014: 13th European Conference, Zurich, Switzerland, 6–12 September 2014* Proceedings, Part V 13; Springer: Berlin/Heidelberg, Germany, 2014; pp. 740–755..
12. Petrick, D.; Geist, A.; Albaijes, D.; Davis, M.; Sparacino, P.; Crum, G.; Ripley, R.; Boblitt, J.; Flatley, T. SpaceCube v2.0 space fligh hybrid reconfigurable data processing system. In Proceedings of the IEEE the Aerospace Conference, Big Sky, MT, USA, 1–8 March 2014; pp. 1–20.
13. Shen, J.; Wang, D.; Huang, Y.; Wen, M.; Zhang, C. Scale-out Acceleration for 3D CNN-based Lung Nodule Segmentation on a Multi-FPGA System. In Proceedings of the Design Automation Conference (DAC), Las Vegas, NV, USA, 2–6 June 2019; pp. 1–6
14. Ronneberger, O.; Fischer, P.; Brox, T. U-net: Convolutional networks for biomedical image segmentation. In Proceedings of the International Conference on Medical Image Computing and Computer-Assisted Intervention (MICCAI), Munich, Germany, 5–9 October 2015; pp. 234–241.
15. Bai, L.; Lyu, Y.; Huang, X. Roadnet-rt: High throughput cnn architecture and soc design for real-time road segmentation. *IEEE Trans. Circuits Syst. I Regul. Pap.* **2020**, 68, 704–714. [CrossRef]
16. Liu, S.; Fan, H.; Niu, X.; Ng, H.C.; Chu, Y.; Luk, W. Optimizing CNN-Based Segmentation with Deeply Customized Convolutional and Deconvolutional Architectures on FPGA. *ACM Trans. Reconfig. Technol. Syst.* **2018**, 11, 1–22. [CrossRef]
17. Liu, S.; Luk, W. Towards an Efficient Accelerator for DNN-Based Remote Sensing Image Segmentation on FPGAs. In Proceedings of the International Conference on Field Programmable Logic and Applications (FPL), Barcelona, Spain, 8–12 September 2019 pp. 187–193.
18. Wu, X.; Ma, Y.; Wang, M.; Wang, Z. A Flexible and Efficient FPGA Accelerator for Various Large-Scale and Lightweight CNNs *IEEE Trans. Circuits Syst. I Regul. Pap.* **2022**, 69, 1185–1198. [CrossRef]
19. Adam, P.; Abhishek, C.; Sangpil, K.; Eugenio, C. ENet: A Deep Neural Network Architecture for Real-Time Semantic Segmentation *arXiv* **2016**, arXiv:1606.02147.
20. Zhao, H.; Shi, J.; Qi, X.; Wang, X.; Jia, J. Pyramid scene parsing network. In Proceedings of the Conference on Computer Vision and Pattern Recognition (CVPR), Honolulu, HI, USA, 21–26 July 2017; pp. 2881–2890.
21. Chen, L.C.; Papandreou, G.; Schroff, F.; Adam, H. Rethinking atrous convolution for semantic image segmentation. *arXiv* **2017** arXiv:1706.05587.
22. Chen, L.C.; Zhu, Y.; Papandreou, G.; Schroff, F.; Adam, H. Encoder-decoder with atrous separable convolution for semantic image segmentation. In Proceedings of the European Conference on Computer Vision (ECCV), Munich, Germany, 8–14 September 2018 pp. 801–818.
23. SCIs Segmentation Dataset. Available online: https://github.com/ZiBoGuo/SCIs-Dataset (accessed on 5 October 2023).
24. Morì, P.; Vemparala, M.R.; Fasfous, N.; Mitra, S.; Sarkar, S.; Frickenstein, A.; Frickenstein, L.; Helms, D.; Nagaraja, N.S.; Stechele, W.; et al. Accelerating and pruning CNNs for semantic segmentation on FPGA. In Proceedings of the 59th ACM/IEEE Design Automation Conference, San Francisco, CA, USA, 10–14 July 2022; pp. 145–150.
25. Im, D.; Han, D.; Choi, S.; Kang, S.; Yoo, H.J. DT-CNN: An energy-efficient dilated and transposed convolutional neural network processor for region of interest based image segmentation. *IEEE Trans. Circuits Syst. I Regul. Pap.* **2020**, 67, 3471–3483. [CrossRef]
26. Alzubaidi, L.; Zhang, J.; Humaidi, A.J.; Al-Dujaili, A.; Duan, Y.; Al-Shamma, O.; Santamaría, J.; Fadhel, M.A.; Al-Amidie, M.; Farhan, L. Review of deep learning: Concepts, CNN architectures, challenges, applications, future directions. *J. Big Data* **2021**, 8, 1–74. [CrossRef]
27. Ioffe, S.; Szegedy, C. Batch normalization: Accelerating deep network training by reducing internal covariate shift. In Proceedings of the International Conference on Machine Learning, Lille, France, 6–11 July 2015; pp. 448–456.
28. Nguyen, D.T.; Nguyen, T.N.; Kim, H.; Lee, H.J. A High-Throughput and Power-Efficient FPGA Implementation of YOLO CNN for Object Detection. *IEEE Trans. Very Large Scale Integr. (VLSI) Syst.* **2019**, 27, 1861–1873. [CrossRef]
29. Williams, S.; Waterman, A.; Patterson, D. Roofline: An insightful visual performance model for multicore architectures. *Commun. ACM* **2009**, 52, 65–76. [CrossRef]
30. Liu, S.; Du, Z.; Tao, J.; Han, D.; Luo, T.; Xie, Y.; Chen, Y.; Chen, T. Cambricon: An Instruction Set Architecture for Neural Networks. In Proceedings of the Annual International Symposium on Computer Architecture (ISCA), Seoul, Republic of Korea, 18–22 June 2016; pp. 393–405.
31. Yu, Y.; Wu, C.; Zhao, T.; Wang, K.; He, L. OPU: An FPGA-Based Overlay Processor for Convolutional Neural Networks. *IEEE Trans. Very Large Scale Integr. (VLSI) Syst.* **2020**, 28, 35–47. [CrossRef]
32. Yu, J.; Ge, G.; Hu, Y.; Ning, X.; Qiu, J.; Guo, K.; Wang, Y.; Yang, H. Instruction driven cross-layer cnn accelerator for fast detection on fpga. *ACM Trans. Reconfig. Technol. Syst. (TRETS)* **2018**, 11, 1–23. [CrossRef]
33. Xing, Y.; Liang, S.; Sui, L.; Jia, X.; Qiu, J.; Liu, X.; Wang, Y.; Shan, Y.; Wang, Y. Dnnvm: End-to-end compiler leveraging heterogeneous optimizations on fpga-based cnn accelerators. *IEEE Trans.-Comput.-Aided Des. Integr. Circuits Syst.* **2019**, 39, 2668–2681. [CrossRef]
34. Vitis AI Library User Guide (UG1354). Available online: https://docs.xilinx.com/r/1.4.1-English/ug1354-xilinx-ai-sdk/ZCU102-Evaluation-Kit (accessed on 2 January 2024).

35. Cong, J.; Wei, P.; Yu, C.H.; Zhang, P. Automated accelerator generation and optimization with composable, parallel and pipeline architecture. In Proceedings of the ACM/ESDA/IEEE Design Automation Conference (DAC), IEEE, San Francisco, CA, USA, 24–28 June 2018; pp. 1–6.
36. Qiu, J.; Wang, J.; Yao, S.; Guo, K.; Li, B.; Zhou, E.; Yu, J.; Tang, T.; Xu, N.; Song, S.; et al. Going Deeper with Embedded FPGA Platform for Convolutional Neural Network. In Proceedings of the ACM/SIGDA International Symposium on Field-Programmable Gate Arrays (FPGA), Monterey, CA, USA, 21–24 February 2016; pp. 26–35.
37. Wu, D.; Tang, Q.; Zhao, Y.; Zhang, M.; Fu, Y.; Zhang, D. EasyQuant: Post-training Quantization via Scale Optimization. *arXiv* **2020**, arXiv:2006.16669.
38. Liu, W.; Rabinovich, A.; Berg, A.C. Parsenet: Looking wider to see better. *arXiv* **2015**, arXiv:1506.04579.
39. Simonyan, K.; Zisserman, A. Very deep convolutional networks for large-scale image recognition. *arXiv* **2014**, arXiv:1409.1556.
40. He, K.; Zhang, X.; Ren, S.; Sun, J. Deep residual learning for image recognition. In Proceedings of the IEEE Conference on Computer Vision and Pattern Recognition (CVPR), Las Vegas, NV, USA, 27–30 June 2016; pp. 770–778.
41. Iandola, F.N.; Han, S.; Moskewicz, M.W.; Ashraf, K.; Dally, W.J.; Keutzer, K. SqueezeNet: AlexNet-level accuracy with 50x fewer parameters and< 0.5 MB model size. *arXiv* **2016**, arXiv:1602.07360.
42. Ulku, I.; Akagündüz, E. A survey on deep learning-based architectures for semantic segmentation on 2d images. *Appl. Artif. Intell.* **2022**, *36*, 2032924. [CrossRef]
43. Hu, Y.; Liang, S.; Yu, J.; Wang, Y.; Yang, H. On-chip instruction generation for cross-layer CNN accelerator on FPGA. In Proceedings of the 2019 IEEE Computer Society Annual Symposium on VLSI (ISVLSI), Miami, FL, USA, 15–17 July 2019; pp. 7–12.
44. Friedrich, S.; Sampath, S.B.; Wittig, R.; Vemparala, M.R.; Fasfous, N.; Matúš, E.; Stechele, W.; Fettweis, G. Lightweight instruction set for flexible dilated convolutions and mixed-precision operands. In Proceedings of the 2023 24th International Symposium on Quality Electronic Design (ISQED), San Francisco, CA, USA, 5–7 April 2023; pp. 1–8.
45. Chollet, F. Xception: Deep learning with depthwise separable convolutions. In Proceedings of the IEEE Conference on Computer Vision and Pattern Recognition (CVPR), Honolulu, HI, USA, 21–26 July 2017; pp. 1251–1258.
46. Venieris, S.I.; Bouganis, C.S. fpgaConvNet: Mapping regular and irregular convolutional neural networks on FPGAs. *IEEE Trans. Neural Netw. Learn. Syst.* **2018**, *30*, 326–342. [CrossRef]
47. Guo, K.; Sui, L.; Qiu, J.; Yu, J.; Wang, J.; Yao, S.; Han, S.; Wang, Y.; Yang, H. Angel-Eye: A Complete Design Flow for Mapping CNN Onto Embedded FPGA. *IEEE Trans. Comput.-Aided Des. Integr. Circuits Syst.* **2018**, *37*, 35–47. [CrossRef]
48. Zhang, C.; Sun, G.; Fang, Z.; Zhou, P.; Pan, P.; Cong, J. Caffeine: Toward Uniformed Representation and Acceleration for Deep Convolutional Neural Networks. *IEEE Trans. Comput.-Aided Des. Integr. Circuits Syst.* **2019**, *38*, 2072–2085. [CrossRef]
49. Basalama, S.; Sohrabizadeh, A.; Wang, J.; Guo, L.; Cong, J. FlexCNN: An End-to-End Framework for Composing CNN Accelerators on FPGA. *ACM Trans. Reconfig. Technol. Syst.* **2023**, *16*, 1–32. [CrossRef]
50. Zynq DPU Product Guide (PG338). Available online: https://docs.xilinx.com/r/3.2-English/pg338-dpu/Advanced-Tab (accessed on 2 January 2024).
51. Ning, X.; Ge, G.; Li, W.; Zhu, Z.; Zheng, Y.; Chen, X.; Gao, Z.; Wang, Y.; Yang, H. FTT-NAS: Discovering fault-tolerant convolutional neural architecture. *ACM Trans. Des. Autom. Electron. Syst. TODAES* **2021**, *26*, 1–24. [CrossRef]

remote sensing

MDPI

Article

A Joint Convolutional Cross ViT Network for Hyperspectral and Light Detection and Ranging Fusion Classification

Haitao Xu [1,2], Tie Zheng [1,†], Yuzhe Liu [3,†], Zhiyuan Zhang [3], Changbin Xue [1,*] and Jiaojiao Li [3]

1 National Space Science Center, Chinese Academy of Sciences, Beijing 100190, China; xuhaitao@nssc.ac.cn (H.X.); zhengtie@nssc.ac.cn (T.Z.)
2 University of Chinese Academy of Sciences, Beijing 100049, China
3 The State Key Laboratory of Integrated Service Networks, School of Telecommunications Engineering, Xidian University, Xi'an 710200, China; 21011210556@stu.xidian.edu.cn (Y.L.); zzy@stu.xidian.edu.cn (Z.Z.); jjli@xidian.edu.cn (J.L.)
* Correspondence: xuechangbin@nssc.ac.cn
† These authors contributed equally to this work.

Abstract: The fusion of hyperspectral imagery (HSI) and light detection and ranging (LiDAR) data for classification has received widespread attention and has led to significant progress in research and remote sensing applications. However, existing common CNN architectures suffer from the significant drawback of not being able to model remote sensing images globally, while transformer architectures are not able to capture local features effectively. To address these bottlenecks, this paper proposes a classification framework for multisource remote sensing image fusion. First, a spatial and spectral feature projection network is constructed based on parallel feature extraction by combining HSI and LiDAR data, which is conducive to extracting joint spatial, spectral, and elevation features from different source data. Furthermore, in order to construct local–global nonlinear feature mapping more flexibly, a network architecture coupling together multiscale convolution and a multiscale vision transformer is proposed. Moreover, a plug-and-play nonlocal feature token aggregation module is designed to adaptively adjust the domain offsets between different features, while a class token is employed to reduce the complexity of high-dimensional feature fusion. On three open-source remote sensing datasets, the performance of the proposed multisource fusion classification framework improves about 1% to 3% over other state-of-the-art algorithms.

Keywords: hyperspectral; LiDAR; fusion classification; transformer; feature fusion

Citation: Xu, H.; Zheng, T.; Liu, Y.; Zhang, Z.; Xue, C.; Li, J. A Joint Convolutional Cross ViT Network for Hyperspectral and Light Detection and Ranging Fusion Classification. *Remote Sens.* **2024**, *16*, 489. https://doi.org/10.3390/rs16030489

Academic Editor: Kevin Tansey

Received: 10 December 2023
Revised: 22 January 2024
Accepted: 24 January 2024
Published: 26 January 2024

1. Introduction

Hyperspectral sensors are capable of capturing images in dozens or hundreds of narrow bands, thereby combining spectral and spatial information effectively. With their unique spectral spatial combination structure, they are suitable for a wide range of applications, such as agriculture, aerospace, mineral exploration, etc. [1–3]. Hyperspectral image classification technology aims to assign a class label to each pixel, which can effectively improve the interpretation perception of hyperspectral images. With advancements in sensor capability, more types of optical data can be acquired, such as LiDAR elevation images, synthetic aperture radar (SAR), panchromatic images, and infrared images [4–6], to name a few. Meanwhile, to improve the perception of hyperspectral images, combining different source data for joint classification is a straightforward and effective method [7,8]. Hyperspectral images reflect the material spectral information of objects, but different objects of the same material cannot be accurately distinguished from spectral information. Typically, a concrete pavement and a concrete roof in a captured image share the same spectral profile but have significant differences in spatial elevation features. In this study, the elevation information from LiDAR is aggregated with hyperspectral images to aid in classification and to reduce the aforementioned phenomenon of homospectral dissimilarity by utilizing the accurate height information from LiDAR [9].

In early research on hyperspectral image fusion classification, various machine-learning-based approaches proved to be successful, including support vector machines (SVMs) based on kernel function theory [10], logistic regression (LR) [11], and random forest algorithms (RF) [12]. Despite the excellent classification performance of these machine learning methods, they rely heavily on hand-designed features and fall short in the ability to extract deep features from hyperspectral images.

Since the advent of deep learning (DL) in the last decade, deep-learning-based classification techniques for hyperspectral image fusion have evolved rapidly [13]. Deep-learning-based methods improve the understanding of remote sensing images by learning the internal patterns of the data samples and mining the deep feature representation of the data [14]. Likewise, deep learning networks have demonstrated powerful advantages over traditional methods in many visual tasks. Representative deep learning frameworks include recurrent neural networks (RNNs) [15], convolutional neural networks (CNNs) [16], long short-term memory (LSTM) networks [17], etc. In particular, CNNs are commonly employed in hyperspectral image processing tasks owing to the kernel acceptance field. Furthermore, the field of hyperspectral image fusion classification has also experienced the rapid development of deep learning technology based on convolutional neural networks. Li et al. proposed a dual-branch network [18], which uses different branches to extract features from hyperspectral images and LiDAR images and enhances the ability to extract features from different sources. On this basis, the hierarchical random walk network (HRWN) [19] utilizes the random walk algorithm to fuse the dual-branch features, which improves the fusion effect and efficiency. In addition, Hong et al. designed the Couple CNN network [20], which employs a spatial–spectral two-branch parameter sharing strategy to reduce the semantic difference between the spatial–spectral features extracted from different sources and to reduce the difficulty in fusing HSI and LiDAR image features. The hashing-based deep metric learning (HDML) proposed by Song et al. employs an attention approach with metric learning loss and also achieved excellent classification performance [21].

However, deep classification networks suffer from network degradation, especially when dealing with high-dimensional hyperspectral data [22]. In the classification task, too deep a network structure leads to feature dispersion and incomplete feature extraction, thus reducing the classification accuracy. To address this problem, several studies have employed attention mechanisms to restrict features and reuse features from different layers to prevent feature degradation. Typically, the FusAtNet [23] network extracts features from hyperspectral and LiDAR data using multilayer attention modules, then merges the extracted features, resulting in excellent classification performance. And Li et al. proposed the Sal^{2}RN network and designed a feature-forward multiplexing module to fully integrate features from different levels and overcome the problem of deep feature degradation [9]. Additionally, the convolutional network still suffers from a defect that prevents it from effectively representing global features, and the fixed-size convolutional kernel limits its ability to model global features. To counter this challenge, Yang et al. creatively proposed the cascaded dilated convolutional network (CDCN) in their work [24], which utilizes the stacked dilated convolution method to extend the receptive field of the convolution kernel and to realize the interaction of features at different scales. And the CDCN enhances the performance of the network when it comes to classification.

Recently, transformer architectures have become the backbone of many vision tasks, and vision transformers have demonstrated a powerful performance in a variety of remote sensing tasks [25]. Compared to CNN-based networks, the vision transformer architecture can deal with the long-range dependency problem among data and better model the contextual information of the data [26]. The transformer achieves global image modeling through data slice embedding and self-attention mechanisms [27]. As a revolutionary paradigm for hyperspectral image classification, SpectralFormer introduces the transformer architecture network for the first time and adopts additional class tokens for feature representation [28].

For the purpose of enhancing the feature aggregation ability of transformer networks, many methods combine convolution with the characteristics of transformers in an effort to further improve the accuracy of hyperspectral fusion classification. For instance, DHViT [29] incorporates convolution and a vision transformer into its LiDAR and hyperspectral feature extraction branches, which significantly enhances the robustness of the network. However, for the hyperspectral patch input paradigm [30], the above ViT-based network can only simulate the correlation between the current patch sizes and still lacks much feature interaction between different scales to effectively perceive the spatial diversity in the complex geographic environment, which greatly affects the final performance of fusion classification. Furthermore, the vanilla feature fusion method mostly performs feature concatenation, ignoring the differences between different source features [31,32]. Specifically, the spectral, spatial, and elevation features are spliced in the channel dimension, and there are semantic differences among different features, which cannot effectively improve the fusion performance [33]. For the purpose of reducing the feature drift between different modalities, a more flexible fusion method should be developed to improve the efficiency of utilizing multisource features.

To address the above challenges, this paper proposes a fusion hyperspectral and LiDAR classification architecture based on convolution and a transformer. The proposed multibranch interaction structure captures features from three perspectives: spectral, spatial, and elevation. This improves the effectiveness of the feature extraction network. Specifically, our research focuses on analyzing both hyperspectral and LiDAR images simultaneously. The transformer network framework combining multiscale convolution with multiscale cross-attention is proposed for joint feature extraction. Finally, a multiscale token fusion strategy is used to aggregate the extracted features. Overall, the main contributions of this paper are summarized as follows:

(1) We propose a multisource remote sensing image classification framework that integrates multiscale feature extraction with cross-attention learning representation based on spectral–spatial feature tokens. This approach greatly improves the joint classification performance, outperforming state-of-the-art (SOTA) methods with advanced analytical capabilities.
(2) To consider the differences in spatial scale information of different classes, we propose a Multi-Conv-Former Block (MCFB), a backbone feature extractor that combines convolutional networks with multiscale transformer feature extraction. This strategy skillfully captures complex edge details in HSI and LiDAR images and identifies the spatial dependencies of multiscale transformer features, which facilitates the mining of more representative perceptual features from different scales.
(3) We design a Cross-Token Fusion Module (CTFM) to maximize the fusion of HSI and LiDAR feature tokens through a nonlocal cross-learning representation. This strategy elevates shallow feature extraction to deep feature fusion, enhances the synergy among multisource remote sensing image data, and realizes more cohesive information integration.

The remainder of this article is organized as follows. Section 2 introduces the related work in the research field, Section 3 introduces the network structure proposed in this paper in detail, Section 4 demonstrates the experimental setup and analysis, Section 5 discusses the results, and Section 6 concludes this paper.

2. Related Work

Within remote sensing image fusion classification, researchers have explored numerous approaches to improve the accuracy and efficiency of multisource data integration. These developments, from traditional to advanced algorithms, mark considerable progress in addressing the complexities of multisource data fusion classification. Zhang et al. [34] proposed the Adaptive Locality-Weighted Multi-source Joint Sparse Representation model for multiple remote sensing data fusion classification. The method employs an adaptive locality weight, calculated for each data source, to constrain sparse coefficients and address the instability in sparse decomposition, thereby enhancing the fusion of information from various sources. Although the sparse representation yields better fusion performance, the need for sparse optimization solving during fusion leads to its low efficiency, which may limit the application of sparse representation fusion methods. Considering the differences in data structure between HSI and LiDAR and the presence of non-negligible noise in remotely sensed images, the two data sources are more suitably fused at the feature level or decision level for delicate scene classification tasks. Rasti B et al. [35] proposed an orthogonal total variation component fusion method. This method employs extinction profiles to extract spatial and elevation information from HSI and LiDAR features. However, simple concatenation or stacking of high-dimensional features may lead to the Hughes phenomenon during the feature-level fusion [36]. In order to solve this problem, most studies utilize principal component analysis (PCA) to reduce the HSI data dimensionality [37]. Liao et al. [38] employed a SVM to classify spectral features, spatial features, elevation features, and fusion features separately and then, based on the results of the four classifications, to complete the decision-level fusion through the weighted vote. Although traditional methods such as the above can achieve effective fusion of features, they rely on efforts to design suitable extractors, which are otherwise prone to local differences due to mismatches between images from multiple sources.

Deep learning can extract high-level semantic features from data end to end, achieving more accurate classification results [39]. Xu et al. [18] proposed a novel two-tunnel CNN framework for extracting spectral–spatial features from HSI. A CNN with a cascade block was designed for feature extraction from other remote sensing data. The spatial and spectral information of the HSI data was extracted using two-tunnel CNN branching, whereas the spatial information of the other source data was extracted using cascaded network blocks. Although the dual-branching network can extract information separately, it overlooks the complementarity between multiple source images, which may lead to poor classification performance after feature fusion.

Recent innovations in transformer architectures have opened new avenues in remote sensing image processing. The ViT [25] introduces a groundbreaking approach to image recognition by adapting attention mechanisms, treating images as sequences of patches. It applies the transformer encoder directly to these sequences, preceding traditional convolutional layers. Based on these architectures, DHViT [29] and FusAtNET [23] have introduced remote sensing data processing changes by incorporating the transformer architecture. DHViT's innovation lies in its architecture that utilizes the powerful modeling capability of long-range dependencies and strong generalization ability across different domains of the transformer network, based exclusively on the self-attention mechanism. In comparison, FusAtNET employs a dual-attention-based spectral–spatial multimodal fusion network, which effectively utilizes a "self-attention" mechanism in HSI and a "cross-attention" mechanism using LiDAR modality. This approach allows for extracting and fusing spectral and spatial features, improving fusion classification. Additionally, the HRWN [19] introduced a two-branch CNN structure to extract spectral and spatial features. After that, the predictive distributions and pixel affinities of the two-branch CNNs act as global prior and local similarity, respectively, in the subsequent hierarchical random walk layers. This model improves boundary localization and reduces spatial fragmentation in classification maps to improve classification performance. However, despite their advancements, these transformer-based methods face challenges such as potential overfitting from augmented feature dimensionality and lack of research on the interactive perception of different modal remote sensing data information, which may cause performance degradation.

3. Methodology

In this section, the proposed fusion classification network is reviewed in detail, and the innovations are presented separately.

3.1. Overall Network Framework

The overall network framework of the proposed method is shown in Figure 1. In contrast to traditional methods, this paper innovates a multibranch interactive feature extraction structure to avoid the disadvantages of the separate extraction of each branch of the multibranch network and adopts an interactive feature extraction method in the extraction of LiDAR elevation information and hyperspectral spatial information. And an additional spectral feature extraction branch is added to carry out the spectral information modeling of hyperspectral data. To be specific, due to the high channel dimension of hyperspectral images, it is necessary to reduce the dimensions of the data. In this paper principal component analysis is utilized to reduce the dimensions of the original data. For the hyperspectral image $\mathbf{X} \in \mathbb{R}^{H \times W \times D}$, where D is the number of dimensions of the original data, there are $\mathbf{X} = [\mathbf{X}_1, \mathbf{X}_2, \mathbf{X}_3, ..., \mathbf{X}_D]$. Where $\mathbf{X}_j (1 \leq j \leq D)$ represents the data value at each channel, the zero-centered data $\tilde{\mathbf{X}}$ are first obtained by de-meaning.

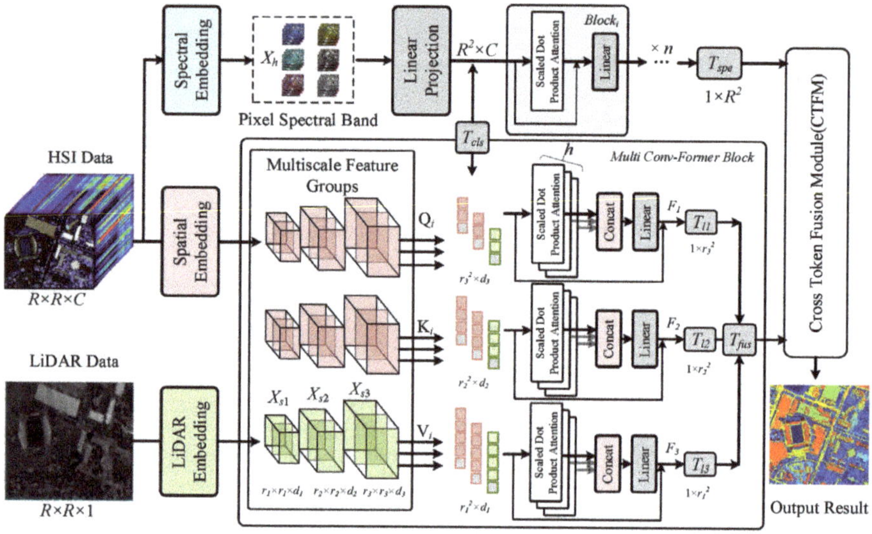

Figure 1. The overall network framework of the proposed algorithm, in which the multiple data flow processes are spectral feature extraction, spatial feature extraction, and LiDAR elevation feature extraction. In the figure, "T" represents the class token, and "concat" is the feature concatenation operation.

To decompose the covariance matrix using singular value decomposition (SVD) [40], we need to construct and solve the following symmetric matrix:

$$\mathcal{M} = (\mathbf{V}\Sigma^T\mathbf{U}^T)(\mathbf{U}\Sigma\mathbf{V}^T) = \mathbf{V}\Sigma^T(\mathbf{U}^T\mathbf{U})\Sigma\mathbf{V}^T = \mathbf{V}\Sigma^T\Sigma\mathbf{V}^T, \tag{1}$$

$$\mathcal{M}' = (\mathbf{U}\Sigma\mathbf{V}^T)(V\Sigma^T\mathbf{U}^T) = \mathbf{U}\Sigma(\mathbf{V}^T\mathbf{V})\Sigma^T\mathbf{U}^T = \mathbf{U}\Sigma\Sigma^T\mathbf{U}^T. \tag{2}$$

The matrix $\mathbf{V}$ is the matrix of eigenvalues corresponding to the original data $\mathbf{X}$; take the first C eigenvalues to form the matrix $\mathbf{P}$; then, the data after dimension reduction are $\mathbf{X}_h = \mathbf{PX}$.

For the hyperspectral image input $\mathbf{X}_h$ as well as the LiDAR elevation input $\mathbf{X}_l$, the patch partition strategy is first to divide them into $\mathbf{X}_i^h \in \mathbb{R}^{r \times r \times C}$ and $\mathbf{X}_i^l \in \mathbb{R}^{r \times r \times 1}$, where r is a hyperparameter representing the size of the input patch and C is the number of channels for hyperspectral image dimensionality reduction. For the spatial part, we use the Multi-Conv-Former Block (MCFB) for feature extraction, and in this block, we process both hyperspectral spatial information and LiDAR elevation information:

$$\mathbf{F}_{spa} = \Gamma\{\mathbf{X}_i^h, \mathbf{X}_i^l\}, \tag{3}$$

where $\mathbf{F}_{spa}$ represents the final spatial feature output, and Γ represents the MCFB feature extraction module processing. The structure of this module will be explained in detail in the next section.

For spectral dimension feature extraction, we adopt the ViT network with an additional class token as the feature extractor, unlike the traditional ViT network; the pixel values within different patches are divided in the embedding part, according to the data values of different channel dimensions. The specific process is as follows.

First, for the hyperspectral data $\mathbf{X}_h$, we divide them into a number of patches along the channel dimension, denoted as $\mathbf{X}_i^{spe}$, and then, we have

$$\mathbf{X}_h = \{\mathbf{X}_1^{spe}, \mathbf{X}_2^{spe}, \ldots, \mathbf{X}_i^{spe}\}, 1 \leq i \leq C. \tag{4}$$

After each set of patches is embedded by feature mapping, an additional set of class tokens of the same scale is added as the input data for subsequent feature extraction:

$$\mathbf{S} = \{\xi(\mathbf{X}_i^{spe})||\mathbf{T}_i\}. \tag{5}$$

In the formula, ξ represents the feature-mapping operation, which aims to map the channel dimension data and convert the spatial features, and $\mathbf{T}_i$ represents the additional class token, which is a vector of random initial values and is constantly updated with the learning of the network to represent the category information of the group of features. The subsequent linear transformations used for self-attention feature extraction are denoted as $\mathbf{W}_q$, $\mathbf{W}_k$, and $\mathbf{W}_v$:

$$\mathbf{Q} = \mathbf{S} \cdot \mathbf{W}_q, \mathbf{K} = \mathbf{S} \cdot \mathbf{W}_k, \mathbf{V} = \mathbf{S} \cdot \mathbf{W}_v. \tag{6}$$

To summarize, the self-attention layer can be represented as follows:

$$\mathbf{F}_{spe} = Attention(\mathbf{Q}, \mathbf{K}, \mathbf{V}) = softmax(\frac{\mathbf{QK}^T}{\sqrt{d}}). \tag{7}$$

The extracted features in this part are denoted as $\mathbf{F}_{spe}$. Then, $\mathbf{F}_{spa}$ and $\mathbf{F}_{spe}$ penetrate the proposed Cross-Token Fusion Module for feature fusion to generate a more robust feature output.

$$Output = \sigma(\Phi(\mathbf{F}_{spa}, \mathbf{F}_{spe})), \tag{8}$$

where Φ represents the proposed CTFM method, and σ represents the classification head output.

3.2. Multi-Conv-Former Feature Extraction

The CNN architecture lacks global modeling capability, and the transformer architecture lacks local spatial feature extraction capability. In this section, the proposed Multi-Conv-Former feature extraction module will be introduced in detail. This module includes a hierarchical multiscale convolution as a shallow feature extraction network and a multi-scale cross-attention feature extraction module for multiscale features. The combination of the two structures improves the feature sensing capability and the robustness of the extracted features. Specifically, the overall process is as follows.

For hyperspectral image input $\mathbf{X}_h$ and LiDAR elevation input $\mathbf{X}_l$, two-dimensional convolution is first used for multiscale feature extraction. In this work, three levels of multiscale feature output are used to achieve spatial size reduction and channel-scale high-dimensional mapping. The initially selected patch input size is 11×11. In the first stage, two consecutive convolutional layers are used with kernel sizes of 7×7 and 3×3 and a padding size of 1. At the same time, Batch Norm is applied for normalization. In both the second and third stages, two consecutive convolutional layers are of size 3×3, with padding size 1. The final feature sizes of the three scales obtained are $\mathbf{X}_{s1} \in \mathbb{R}^{1\times1\times256}$, $\mathbf{X}_{s2} \in \mathbb{R}^{3\times3\times128}$, and $\mathbf{X}_{s3} \in \mathbb{R}^{7\times7\times64}$. It is worth noting that a global averaging pooling layer is employed after each layer for sizing. Finally, depth-separable convolution is utilized to map the extracted hierarchical multiscale features and transform them into data input patterns for the transformer architecture.

$$\mathbf{Q}_i, \mathbf{K}_i, \mathbf{V}_i = PointWise(DepthWise(\mathbf{X}_{si})), i = 1, 2, 3. \tag{9}$$

Similar to the spectral branching operation, class tokens are added to the feature embedding for each scale. After that, the multiattention mechanism is used to extract features at different scales.

$$head_n = Attention(\mathbf{Q}_i, \mathbf{K}_i, \mathbf{V}_i), 1 \leq n \leq h, \tag{10}$$

$$\mathbf{F}_i = MultiHead(\mathbf{Q}_i, \mathbf{K}_i, \mathbf{V}_i) = Concat(head_1, head_2, \ldots, head_h), \tag{11}$$

where $\mathbf{F}_i$ denotes the feature output of Conv-Former, whose dimensions are consistent with the input dimensions.

After extracting the features at different scales by multiple attention, in order to reduce the complexity of subsequent fusion, we choose the previously added learnable class tokens for feature representation. The randomly generated $\mathbf{T}_{cls}$ at the time of input embedding is continuously updated with network training and has the ability to represent features. Therefore, we utilize this $\mathbf{T}_{cls}$ alone for subsequent processing. Finally, class tokens of different scales are concatenated along the channel dimension to generate the final classification token $\mathbf{T}_{fus}$.

$$\mathbf{T}_{fus} = \{\mathbf{T}_{l1} || \mathbf{T}_{l2} || \mathbf{T}_{l3}\}, \tag{12}$$

where the symbol $||$ represents the concatenation operation along the channel dimension. The subsequent $\mathbf{T}_{fus}$ is passed through the data stream as an input to the feature fusion module.

3.3. Cross-Token Fusion Module

In this subsection, we introduce the token fusion method. Ordinary fusion strategies are fused in the channel dimension, ignoring the distinction between features from diverse sources and modalities. Based on the different classes of markers extracted in the feature extraction part, we design the nonlocal token fusion module, which models the relationship between diverse sources, reduces the intra-class variance, and avoids the phenomenon of excessive differences in the features of various modalities.

The specific flow of the proposed Cross-Token Fusion Module is shown in Figure 2. Specifically, for the $\mathbf{T}_{spe}$ and $\mathbf{T}_{fus}$ extracted previously, linear transformations are used to obtain linear mappings Query($\mathbf{Q}$), Key($\mathbf{Q}$), and Value($\mathbf{V}$). For different features, we denote the spectral feature as $\mathbf{Q}_{spe}$, $\mathbf{K}_{spe}$, and $\mathbf{V}_{spe}$ and the spatial fusion feature as $\mathbf{Q}_{fus}$, $\mathbf{K}_{fus}$, and $\mathbf{V}_{fus}$. Unlike the traditional self-attention mechanism, the values of the two types of features are exchanged in order to realize the attentional interaction between different features. After that, a convolution with a kernel size of 1×1 is adopted for linear transformation. This operation is denoted as the Conv Flow. The Conv Flow is used for the two obtained groups of $\mathbf{Q}$, $\mathbf{K}$, and $\mathbf{V}$ values. Matrix multiplication is then performed on $\mathbf{K}$ and $\mathbf{Q}$ to obtain the self-attention matrix ξ. This process can be described as follows:

$$\xi_{spe} = \mathbf{K}_{spe} \cdot \mathbf{Q}_{spe}, \tag{13}$$

$$\xi_{fus} = \mathbf{K}_{fus} \cdot \mathbf{Q}_{fus}. \tag{14}$$

Next, multiply the mixed attention matrix with the extracted $\mathbf{V}$ features to obtain the attention-enhanced mixed features.

$$O_{spe} = \mathbf{V}_{fus} \cdot Soft(\xi_{spe}) + \mathbf{T}_{spe}, \tag{15}$$

$$O_{fus} = \mathbf{V}_{spe} \cdot Soft(\xi_{fus}) + \mathbf{T}_{fus}, \tag{16}$$

where O_{spe} and O_{fus} denote the spatial and spectral feature outputs of the spatial feature modulation enhancement, respectively. The final feature outputs are concatenated along the channel dimension:

$$Output = \{O_{spe} \| O_{fus}\}, \tag{17}$$

where $\|$ is a concatenation operation that joins the features from the Cross-Token Fusion in the channel dimension to obtain the final output features, which are then processed by the classification header of the fully connected layer for the final output.

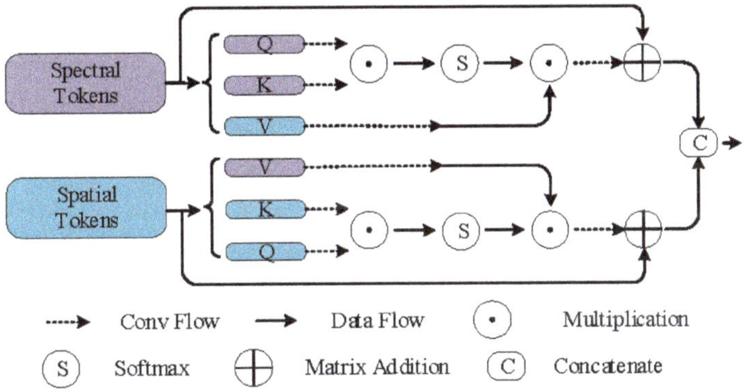

Figure 2. The structure diagram of the Cross-Token Fusion Module.

4. Experiments and Analysis

Three publicly available multisource remote sensing datasets were employed to evaluate the performance of the proposed network experimentally. First, a description of the selected datasets employed in the experiment is provided. An elaboration on the specific experimental settings follows this. Then, the ablation experiments performed on the roles and functionalities of different modules within the proposed framework are described. Finally, the experimental outcomes underscore the superior performance of the proposed framework relative to existing techniques.

4.1. Data Descriptions

In order to evaluate the effectiveness of the proposed network framework, three datasets containing HSI and LiDAR data were selected for the experiments: Houston2013, Trento, and MUUFL. Table 1 details the names of land-cover categories, the number of training samples, and the number of test samples for these datasets.

(1) Houston2013 Dataset:

The Houston2013 dataset, sourced from the 2013 IEEE GRSS Data Fusion Contest, encompasses the University of Houston campus and its adjoining regions [41]. The Compact Airborne Spectrographic Imager collected the HSI, and the NSF-funded Center for Airborne Laser Mapping captured the LiDAR. The dataset's dimensions stand at 349 × 1905 pixels, boasting a spatial resolution of 2.5 m. The HSI data feature 144 spectral bands spanning a wavelength range of 0.38 to 1.05 μm. For the same region, the LiDAR data for the identical region comprise a single band. This scene contains fifteen different classes of interest. To enhance clarity and comprehensive understanding, Figure 3 shows supplemental visual depictions, including a pseudo-color composite of the HSI data, a grayscale rendition of the LiDAR data, and an associated ground-truth map.

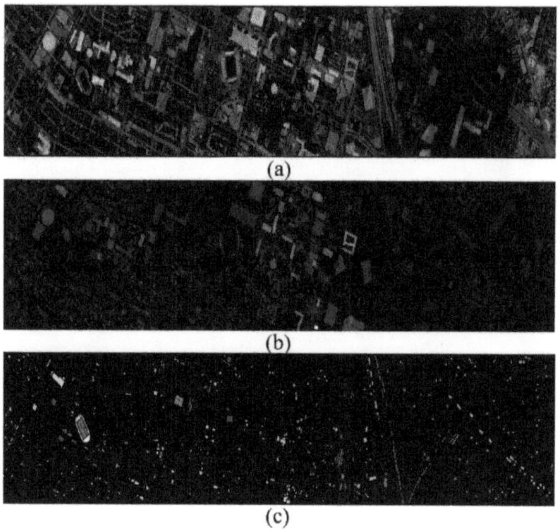

Figure 3. Houston dataset. (a) Pseudo-color composite image based on bands 59, 26, and 18 for HSIs. (b) Grayscale image for LiDAR-based DSM. (c) Ground-truth map.

Table 1. Training and test sample numbers for Houston2013, Trento, and MUUFL.

Houston2013 Dataset				Trento Dataset			MUUFL Dataset		
No.	Class Name	Training	Test	Class Name	Training	Test	Class Name	Training	Test
1	Healthy grass	198	1053	Apple trees	129	3905	Trees	100	23,146
2	Stressed grass	190	1064	Buildings	125	2778	Mostly grass	100	4170
3	Synthetic grass	192	505	Ground	105	374	Ground surface	100	6782
4	Trees	188	1056	Woods	154	8969	Dirt	100	1726
5	Soil	186	1056	Vineyard	184	10,317	Road	100	6587
6	Water	182	143	Roads	122	3052	Water	100	366
7	Residential	196	1072				Building shadow	100	2133
8	Commercial	191	1053				Building	100	6140
9	Road	193	1059				Sidewalk	100	1285
10	Highway	191	1036				Yellow curb	100	83
11	Railway	181	1054				Cloth panels	100	169
12	Parking lot1	192	1041						
13	Parking lot2	184	285						
14	Tennis court	181	247						
15	Running track	187	473						
	Total	2832	12,197	Total	819	29,395	Total	1100	52,587

(2) Trento Dataset:

The Trento dataset, captured over a rural landscape in southern Trento, Italy, was sourced using the AISA Eagle hyperspectral imaging system [35,42]. This system is equipped with the AISA Eagle sensor, which captures 63 spectral bands across a wavelength spectrum of 0.42 to 0.99 μm. Complementing the HSI, LiDAR data were gathered using the Optech Airborne Laser Terrain Mapper (ALTM) 3100EA sensor, represented in a single raster format. This dataset spans 600 × 166 pixels, maintaining a spatial resolution of 1 m and containing six different classes of interest. For visualization and analytical purposes, Figure 4 shows a pseudo-color composite of the HSI data, a grayscale representation of the LiDAR data, and an associated ground-truth map, respectively.

(3) MUUFL Dataset:

The MUUFL Gulfport dataset was captured over the Gulf Park campus of the University of Southern Mississippi in November 2010 by the reflective optics system imaging spectrometer sensor [43]. The HSI was collected by the ITRES Compact Airborne Spectrographic Imager (CASI-1500) sensor, and the ALTM sensor captured LiDAR data. Initially, the HSI imagery incorporated 72 bands, but the first and last 4 bands were excluded due to noise considerations, leading to 64 bands. The LiDAR component comprises 2 elevation rasters with a 1.06 μm wavelength. Both modalities are coregistered, rendering a dataset dimension of 325 × 220 pixels, with a spatial resolution of 0.54 m × 1 m. There are eleven different classes of interest in this scene. Figure 5 shows the HSI data, LiDAR imagery, and the corresponding ground-truth map, respectively.

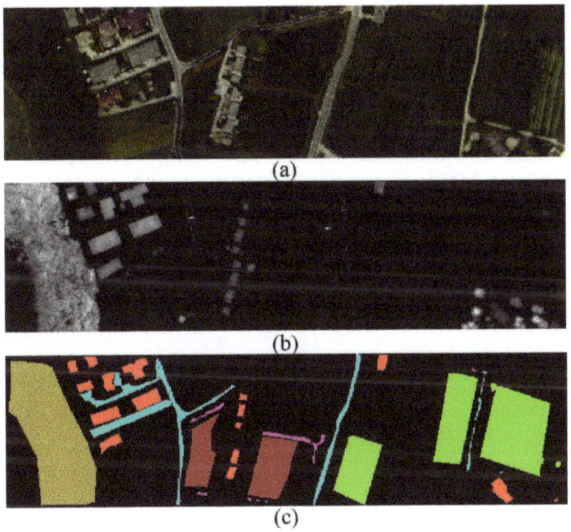

Figure 4. Trento dataset. (**a**) Pseudo-color composite image based on bands 20, 16, and 4 for HSIs. (**b**) Grayscale image for LiDAR-based DSM. (**c**) Ground-truth map.

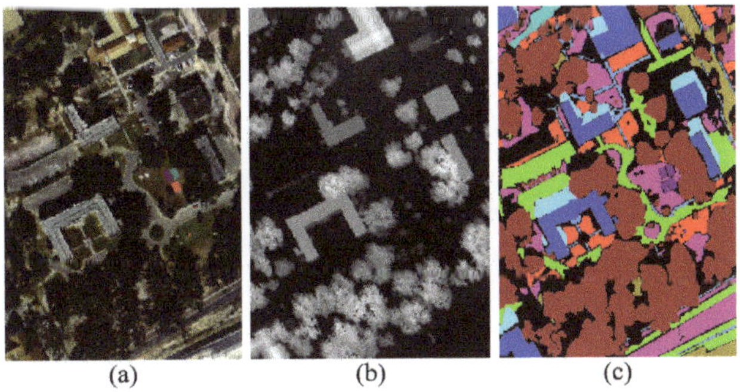

Figure 5. MUUFL dataset. (**a**) Pseudo-color composite image based on bands 30, 20, and 10 for HSIs. (**b**) Grayscale image for LiDAR-based DSM. (**c**) Ground-truth map.

4.2. Experimental Settings

Four widely used quantitative metrics were computed to measure the classification performance of the proposed methodology compared to other existing models. These metrics include the overall accuracy (OA), average accuracy (AA), Kappa coefficient (Kappa), and per-class accuracy. A superior score for these indicators signifies enhanced classification accuracy. To eliminate the bias caused by random initialization factors of framework parameters in learning-based models, each experiment was repeated ten times to obtain the average value of each quantitative metric.

Experimentation was conducted on a desktop PC with an Intel Core i9-12900 processor 2.40 GHz CPU, 64 GB RAM, and an NVIDIA GeForce RTX 3080 GPU. All experiment operations were facilitated using the PyTorch framework version 2.0.

4.3. Parameter Analysis

The classification performance and the training process are closely related to several hyperparameters, which were analyzed, including the patch size, reduced spectral dimension, attention heads, multiscale spatial feature extraction, and learning rate. In the following experiments, the settings and tuning of hyperparameters depended on the training dataset. Specifically, after setting the hyperparameters, the model was trained using the training dataset, and then the performance of the network on the test dataset was evaluated.

(1) Patch Size:

The patch size refers to the size of a small square area for HSI or LiDAR data input denoted as r. Other hyperparameter values were fixed when evaluating the effect of r. Then, r was selected from a candidate set {9, 11, 13, 15, 17} to evaluate its effect. Since the Multi-Conv-Former Block module combines maximum pooling with convolutional layers to accomplish multiscale feature extraction, the network cannot achieve multiscale effects if the patch size is less than 9. Based on our empirical study, the features extracted by various values of r yield different classification performances. Figure 6a shows the Kappa coefficient of the proposed network framework at different patch sizes. As can be seen, when r is set to 11, the optimal Kappa is achieved in the three datasets.

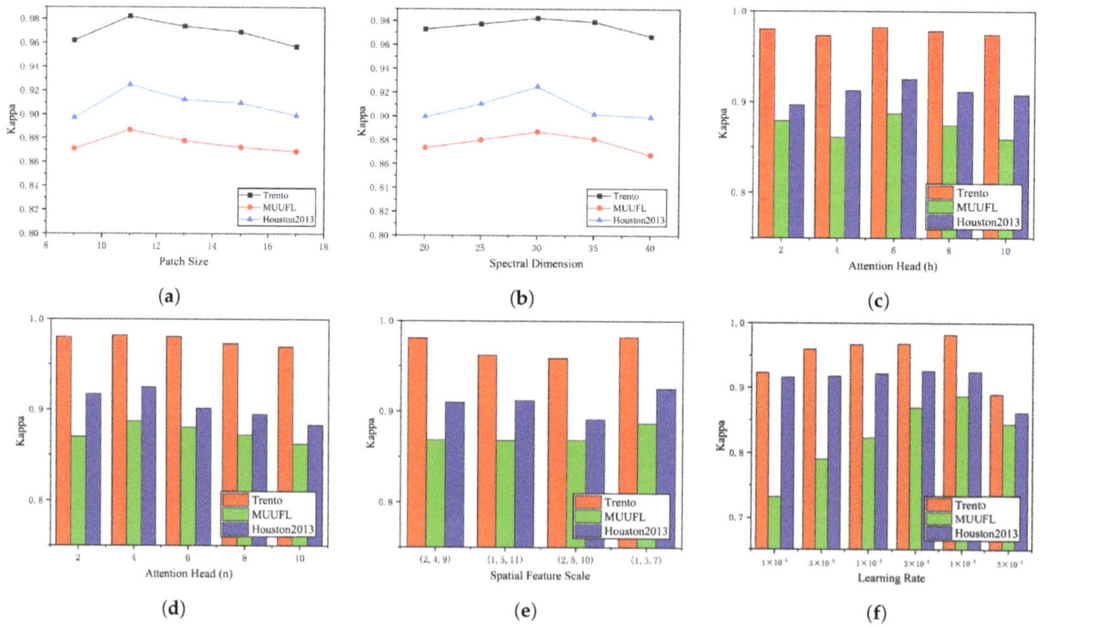

Figure 6. Influence of different parameters on the Kappa coefficient. (**a**) Patch size. (**b**) Reduced spectral dimension. (**c**) Spectral feature extraction module attention heads. (**d**) Multiscale cross-attention spatial feature extraction module attention heads. (**e**) Multiscale spatial feature extraction. (**f**) Learning rate.

(2) Reduced Spectral Dimension:

Reduced Spectral Dimension means using the SVD method to reduce the spectral dimension and extracting only the first c principal components. c was selected from a candidate set {20, 25, 30, 35, 40} to evaluate its effect. Figure 6b shows the Kappa coefficient of the proposed network framework at different reduced spectral dimensions. This trend shows that as c increases, the Kappa value initially increases and then decreases. When the spectral dimension equals 30, the proposed network can achieve the best classification results.

(3) Attention Heads:

Both the spectral feature extraction module and the multiscale cross-attention spatial feature extraction module utilize the multihead attention mechanism, and the attention heads are represented by h and n, respectively. Multihead attention is employed to learn the correspondences between different representational subspaces, where each head corresponds to an independent subspace of feature representation. Therefore, the number of attention heads can affect the capacity of the transformer to represent features and, thus, the classification performance. Figure 6c,d shows the changes in Kappa with h and n on the three datasets, and the candidate set of attention heads is {2, 4, 6, 8, 10, 12}. The experimental results show that the reasonable h and n are 6 and 4, respectively.

(4) Multiscale Spatial Feature Extraction:

The multiscale spatial feature extraction technique is employed in the backbone network to capture the complex unstructured edge details of different target classes. Three levels of downsampling of spatial dimensions are performed on HSI and LiDAR images. The multistage downsampling ratios are $(s1 \times s1)$, $(s2 \times s2)$, and $(s3 \times s3)$. Since maximum pooling and convolutional layers are used by multiscale feature extraction, $s1, s2$, and $s3$ are selected from the candidate set $\{(1,3,7),(2,4,9),(1,5,11),(2,5,10)\}$ to evaluate the effect of different spatial scales. Figure 6e shows the Kappa coefficient of the proposed network framework at different scales of spatial feature. It is obvious that the Kappa value reaches the optimum when the multispatial feature sizes are $s1 = 1$, $s2 = 3$, and $s3 = 7$.

(5) Learning Rate:

The learning rate L is a critical hyperparameter that controls the speed at which the objective function converges to the local optimum. In the experiments, the learning rate was methodically searched for in a candidate set: $\{1 \times 10^{-5}, 3 \times 10^{-5}, 1 \times 10^{-4}, 3 \times 10^{-4}, 1 \times 10^{-3}, 3 \times 10^{-3}\}$. The experimental results obtained by setting different values of L are shown in Figure 6f. It can be observed that the optimal learning rate is 1×10^{-3}.

4.4. Ablation Analysis

(1) Ablation Analysis of Different Modal Data Inputs

Two experimental frameworks were established to analyze the impact of different source data inputs on the model classification performance. The first experiment only used HSI data as an input, while the second was limited to LiDAR data input. The experimental results are shown in Table 2. HSI data can be used to distinguish targets of different materials, while LiDAR data provide rich spatial domain elevation information, enhancing the characterization of scenes in HSI. The comparison of OA, Kappa, and AA on the three datasets shows that the backbone network proposed in this paper based on multisource data fusion has a better classification performance. These experimental results confirm that customized fusion networks can effectively utilize information from multisource data to improve classification performance.

Table 2. Ablation analysis of different modal data inputs.

Cases	Houston2013			Trento			MUUFL		
	OA	Kappa	AA	OA	Kappa	AA	OA	Kappa	AA
HSI	89.51%	0.8866	90.94%	95.37%	0.9386	95.22%	89.42%	0.8630	91.18%
LiDAR	58.04%	0.5480	60.24%	89.25%	0.8564	79.68%	54.31%	0.4414	59.78%
HSI + LiDAR	**93.10%**	**0.9251**	**93.65%**	**98.67%**	**0.9822**	**98.28%**	**91.41%**	**0.8869**	**90.96%**

(2) Ablation Analysis of Multiscale cross-attention Spatial Feature

The proposed spatial feature extractor module, Multi-Conv-Former Block, injects texture features from HSI and LiDAR at three scales (i.e., 1 × 1, 3 × 3, and 7 × 7 spatial downsampling resolutions). To demonstrate the advantages of the backbone network at multiple spatial scales, we conducted an ablation study, and the results are shown in Table 3. Note that the first three rows of the table are equivalent to using usual feature extraction methods when using single-scale spatial features. From Table 3, it can be seen that, when multiscale spatial feature extraction is utilized, the classification performance is improved as when injecting the 7 × 7 spatial scale feature into the backbone for the Houston2013 dataset. Furthermore, as seen from the last row of Table 3, the classification performance is best when we implement three different spatial scales for the backbone. Specifically, the utilization of multiscale feature extraction has resulted in a noteworthy improvement in the overall classification accuracy of the backbone network. The improvement ranges from a minimum of 0.22% to a maximum of 3.20% across the three datasets compared to the feature extraction backbone network that solely relied on a single scale. This finding highlights the potential of multiscale feature extraction in enhancing the backbone network's classification accuracy.

Table 3. Ablation analysis of multiscale spatial feature scale.

Case			Houston2013			Trento			MUUFL		
1 × 1	3 × 3	7 × 7	OA	Kappa	AA	OA	Kappa	AA	OA	Kappa	AA
✓	-	-	90.41%	0.8959	90.74%	97.98%	0.9731	96.98%	89.26%	0.8596	89.93%
-	✓	-	92.54%	0.9137	92.75%	97.53%	0.9669	93.97%	89.52%	0.8633	90.80%
-	-	✓	92.88%	0.9140	92.46%	95.47%	0.9401	94.75%	89.85%	0.8676	90.59%
✓	✓	-	91.53%	0.9080	92.64%	97.82%	0.9709	95.08%	90.21%	0.8707	88.10%
✓	-	✓	93.08%	0.9237	93.19%	98.17%	0.9814	98.00%	89.89%	0.8680	90.23%
-	✓	✓	92.48%	0.9183	93.22%	98.62%	0.9816	93.83%	89.72%	0.8691	90.63%
✓	✓	✓	**93.10%**	**0.9251**	**93.65%**	**98.67%**	**0.9822**	**98.28%**	**91.41%**	**0.8869**	**90.96%**

(3) Ablation Analysis of Feature Fusion

To fully utilize and fuse the spectral and spatial information, a Cross-Token Fusion Module combines cross-attention and is designed to learn spectral and multiscale spatial features. This section evaluates the impact of the Cross-Token Fusion Module within our proposed classification network. The baseline module for this analysis is established by omitting the Cross-Token Fusion Module and instead employing a simple cascaded approach. The baseline employs a cascade-based feature flattened and concatenated network. Table 4 lists the classification performance experimental results of using two different fusion modules. The proposed model exhibits a significant improvement in comparison to the baseline network, particularly on the Houston2013 dataset. The performance of the model is reflected in the observed OA gain of 3.76%, K gain of 0.0401, and AA gain of 3.42%. The proposed model can combine shallow features with deep features, effectively integrate the spectral and multiscale spatial feature information of HSI and LiDAR, enhance the collaboration between multisource remote sensing impact data, and significantly improve the classification results.

Table 4. Ablation analysis of feature fusion.

Cases	Houston2013			Trento			MUUFL		
	OA	Kappa	AA	OA	Kappa	AA	OA	Kappa	AA
Baseline	89.34%	0.8843	90.23%	98.35%	0.9780	97.04%	90.27%	0.8710	89.31%
Proposed	**93.10%**	**0.9251**	**93.65%**	**98.67%**	**0.9822**	**98.28%**	**91.41%**	**0.8869**	**90.96%**

4.5. Classification Results and Analysis

Comparative experiments were conducted to evaluate the effectiveness of the proposed model. For this purpose, several representative classification methods were selected, including classical methods such as CNN-PPF [44] and 3DCNN [45]. The two-branch CNN network [18], known for its ability to process both spectral and spatial information simultaneously, was also included. Additionally, ViT [25] and SpectralFormer [28] were integrated to highlight the superior performance of the proposed network. These models are based on advanced transformer architecture. Finally, advanced fusion and classification networks such as Couple CNN [20] and HRWN [19] were incorporated to evaluate multi-source fusion models extensively, ensuring a comprehensive assessment against current state-of-the-art methodologies.

(1) Quantitative Results and Analysis

The OA, Kappa, AA, and per-class accuracy of the proposed method and each comparative method are reported in Tables 5–7 for the Houston2013, Trento, and MUUFL datasets, respectively. The optimal results are highlighted in bold in each table, while the second best results are underlined. The values of the evaluation indicators clearly show that the proposed framework outperforms comparison methods, often reporting results with higher accuracy.

Table 5. Classification performance obtained using different methods for the Houston2013 dataset.

No.	CNN-PPF	3D CNN	Two-Branch	Couple CNN	HRWN	ViT	Spectral Former	Proposed
1	83.00%	**98.30%**	83.10%	82.43%	85.31%	82.72%	81.86%	82.34%
2	84.12%	98.68%	84.87%	84.87%	83.79%	80.45%	**100.00%**	93.70%
3	**100%**	99.53%	**100%**	99.80%	99.05%	99.60%	95.25%	99.60%
4	88.54%	94.30%	92.14%	92.06%	92.30%	92.42%	96.12%	**98.58%**
5	**100%**	98.82%	97.73%	**100%**	**100%**	97.73%	99.53%	99.81%
6	97.20%	89.45%	68.53%	97.20%	97.28%	95.80%	94.41%	**100%**
7	83.40%	79.89%	87.33%	**92.91%**	89.33%	74.44%	83.12%	76.40%
8	46.25%	82.41%	70.75%	**96.01%**	93.74%	42.55%	76.73%	94.11%
9	84.04%	79.36%	84.51%	84.99%	88.66%	65.25%	79.32%	**93.77%**
10	56.37%	84.96%	62.64%	67.47%	86.17%	50.77%	78.86%	**90.73%**
11	80.08%	72.32%	76.47%	**98.57%**	92.75%	71.44%	88.71%	97.34%
12	87.42%	80.55%	91.26%	96.15%	96.47%	56.00%	87.32%	**99.71%**
13	82.81%	89.73%	8.12%	84.91%	**91.93%**	64.21%	72.63%	78.60%
14	**100%**	99.74%	**100%**	**100%**	**100%**	**100%**	**100%**	**100%**
15	98.94%	99.34%	98.93%	99.58%	**100%**	98.52%	**100%**	**100%**
AA(%)	84.81%	89.82%	82.70%	91.79%	90.47%	78.13%	88.91%	**93.65%**
OA(%)	81.69%	88.54%	80.42%	90.58%	89.67%	74.36%	88.01%	**93.10%**
Kappa	0.803	0.8761	0.8124	0.8978	0.8828	72.43	0.8699	**0.9251**

Concretely, Table 5 shows that for the Houston dataset, the OA, Kappa, and AA values of the proposed framework were 93.10%, 0.9251, and 93.65%, respectively, which are competitive in the HSI and LiDAR joint classification task. Furthermore, the proposed framework outperformed other state-of-the-art methods such as Couple CNN, HRWN, and SpectralFormer. Specifically, the proposed framework achieved a classification average accuracy that was 1.86% higher than Couple CNN. Additionally, it outperformed HRWN and SpectralFormer by 3.18% and 4.74%, respectively. The proposed network integrates multiscale convolution with cross-attention, effectively addressing the limitations of global modeling and local feature extraction. As a result, the network can simultaneously extract spatial features from diverse sources and capture the delicate edge intricacies of the object under scrutiny. For instance, in Table 5, the Houston2013 datasets No. 9 and No. 10 represent "road" and "highway", respectively. The proposed model achieves per-class classification accuracy of 93.77% and 90.73% for these two datasets, which is significantly higher than other methods.

Table 6. Classification performance obtained using different methods for the MUUFL dataset.

No.	CNN-PPF	3D CNN	Two-Branch	Couple CNN	HRWN	ViT	Spectral Former	Proposed
1	88.34%	82.27%	86.88%	94.66%	**95.20%**	87.62%	88.83%	93.95%
2	81.49%	81.04%	77.19%	**85.08%**	84.72%	81.29%	66.62%	82.09%
3	77.25%	67.58%	83.57%	77.04%	72.93%	59.99%	71.73%	**88.60%**
4	93.57%	85.64%	95.71%	97.45%	**98.20%**	82.73%	88.47%	96.76%
5	88.90%	83.00%	**94.55%**	86.18%	85.35%	80.12%	84.21%	90.18%
6	99.18%	91.34%	61.20%	**100%**	**100%**	84.97%	92.62%	99.73%
7	90.06%	86.12%	83.54%	**95.59%**	94.33%	79.79%	86.45%	87.29%
8	81.12%	71.94%	94.79%	**96.16%**	92.82%	82.88%	83.37%	95.44%
9	72.14%	71.39%	63.97%	**74.86%**	64.36%	71.67%	74.24%	74.32%
10	80.72%	91.73%	54.22%	96.39%	85.54%	**97.59%**	89.16%	93.98%
11	97.63%	95.89%	94.08%	**99.41%**	97.63%	95.86%	96.45%	98.22%
AA(%)	86.40%	82.54%	80.87%	90.44%	88.28%	82.24%	83.83%	**90.96%**
OA(%)	85.53%	79.32%	86.95%	91.17%	89.32%	81.23%	83.24%	**91.41%**
Kappa	0.8122	0.7364	0.8301	0.8745	0.8589	0.7564	0.7818	**0.8869**

Table 7. Classification performance obtained using different methods for the Trento dataset.

No.	CNN-PPF	3D CNN	Two-Branch	Couple CNN	HRWN	ViT	Spectral Former	Proposed
1	97.13%	99.22%	91.45%	99.13%	89.29%	87.35%	96.08%	**99.64%**
2	92.12%	90.50%	97.83%	95.43%	91.22%	81.21%	95.86%	**99.28%**
3	98.93%	97.90%	92.48%	**99.73%**	83.72%	96.79%	95.99%	98.93%
4	99.10%	97.05%	98.31%	99.51%	98.08%	97.42%	97.99%	**100.00%**
5	96.71%	94.09%	99.86%	98.84%	**100%**	74.66%	95.25%	98.57%
6	68.32%	79.48%	83.08%	93.25%	87.27%	69.95%	57.76%	**93.28%**
AA(%)	94.14%	93.04%	96.19%	98.19%	95.55%	84.57%	92.37%	**98.28%**
OA(%)	92.05%	93.86%	93.84%	97.24%	91.60%	83.70%	89.82%	**98.67%**
Kappa	0.9216	0.9183	0.9419	0.9758	0.9403	0.7844	0.8982	**0.9822**

The proposed framework has demonstrated promising results for the MUUFL dataset, achieving an OA of 91.41%, Kappa of 0.8869, and AA of 90.96%, as presented in Table 6. These results indicate a slight advantage over the Couple CNN method. However, the classification results of the advanced HRWN method are unsatisfactory, with an overall accuracy that is more than 2% lower than that of the proposed method. This lower performance can be attributed to the spatial features, which may cause overfitting or even misclassification of the image under limited training sample conditions. However, the adjacent intervals of different land cover classes within MUUFL images are relatively small, and the distribution of the same land cover class needs to be more scattered, which may lead to highly mixed pixels in the boundary areas, thus complicating classification. This problem caused each method to obtain a low level of accuracy when classifying the No. 9 class, "sidewalks", in the MUUFL dataset.

As for the Trento dataset, Table 7 shows that the proposed method not only produces the highest OA (98.67%), Kappa (0.9822), and AA (98.28%), but also most of the classes surpass other methodologies in terms of classification accuracy (e.g., "Apple Trees",

"Buildings", "Woods", "Roads"). The above results directly indicate that multiscale feature extraction using a cross-learning representation based on spectral–spatial feature labeling can significantly improve the classification performance.

(2) Visual Evaluation and Analysis

The classification maps obtained by various comparison methods and the proposed method using the MUUFL, Houston2013, and Trento datasets are presented in Figures 7, 8 and 9, respectively. The proposed method exhibits more distinct boundaries compared to other methods, indicating its superior classification performance. This observation is consistent with the overall accuracy results of the quantitative analysis.

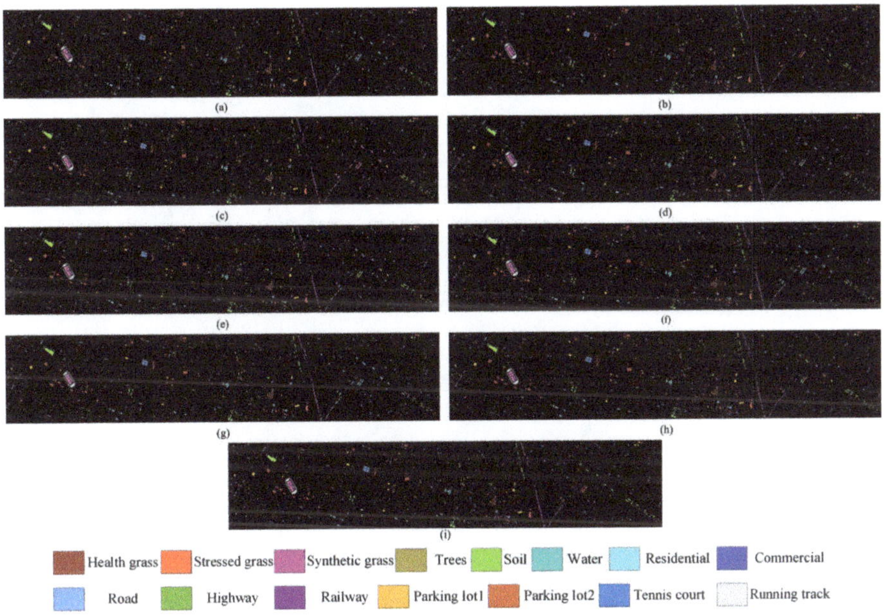

Figure 7. Classification maps using different methods on the Houston2013 dataset. (**a**) CNN-PPF (81.69%). (**b**) 3D CNN (88.54%). (**c**) Two-Branch (80.42%). (**d**) Couple CNN (90.58%). (**e**) HRWN (89.67%). (**f**) ViT (74.36%). (**g**) SpectralFormer (88.01%). (**h**) Proposed (93.10%). (**i**) Ground-truth map.

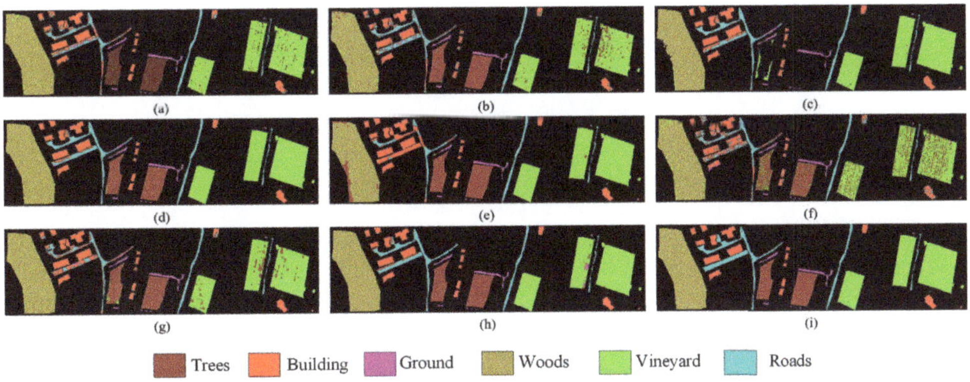

Figure 8. Classification maps using different methods on the Trento dataset. (**a**) CNN-PPF (92.05%). (**b**) 3D CNN (93.86%). (**c**) Two-Branch (93.84%). (**d**) Couple CNN (97.24%). (**e**) HRWN (91.60%). (**f**) ViT (83.70%). (**g**) SpectralFormer (89.82%). (**h**) Proposed (98.67%). (**i**) Ground-truth map.

Specifically, the proposed method is more accurate in classifying irregularly distribute small scene features because it employs the Multi-Conv-Former Block to extract multiscal spatial features. For instance, in Figure 8, the strip distribution terrain in the Trent dataset No. 6 is shown in blue, which represents "Roads". The classification boundar of the proposed model is significantly better than the remaining models. On the righ side of Figure 7, the long strip-shaped terrain in the Houston2013 dataset No. 11 i represented in purple, representing "Railway". The classification completeness of th proposed model is significantly better than the remaining models. Certain classification within the remaining datasets also manifested analogous visual outcomes. Howeve the proposed model requires further improvement in accurately classifying extensiv continuous features. For instance, in the Trento dataset, a small section of the No. "vineyard" that is depicted in green is wrongly classified as "apple trees" or "ground". To address this issue, the design of the shallow CNN needs to be carefully considered.

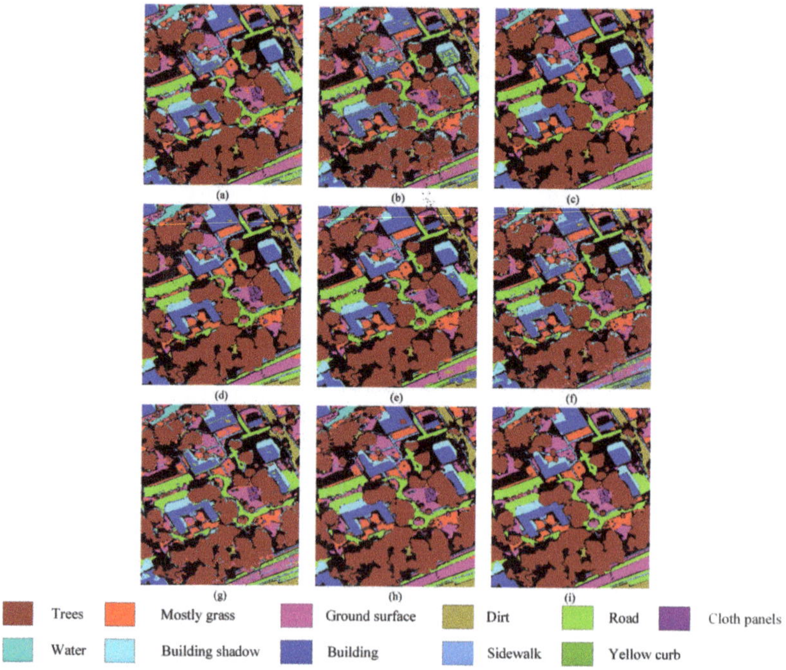

Figure 9. Classification maps using different methods on the MUUFL dataset. (**a**) CNN-PPF (85.53%) (**b**) 3D CNN (79.32%). (**c**) Two-Branch (86.95%). (**d**) Couple CNN (91.17%). (**e**) HRWN (89.32%) (**f**) ViT (81.23%). (**g**) SpectralFormer (83.24%). (**h**) Proposed (91.41%). (**i**) Ground-truth map.

5. Discussion

While remote sensing hyperspectral data capture abundant spectral information, it can be challenging to differentiate between ground objects with similar spectral characteristics. However, LiDAR data can offer additional context to overcome this challenge. This paper explores the structural relationships between various data types and proposes a feature-level fusion technique that blends HSI and LiDAR data. This innovative approach enables us to extract and fuse features efficiently, significantly improving the classification accuracy.

Our research proposed a novel joint convolutional cross-ViT framework for HSI and LiDAR data fusion classification. The proposed framework was tested for classification accuracy on three publicly available datasets, as reported in Tables 5–7.

(1) Our study compared the proposed framework with several state-of-the-art methods, including Coupled CNN, HRWN, and SpectralFormer. According to Tables 5–7,

the proposed model shows superior classification accuracy compared to the other models. The Houston2013 dataset has the most classes of interest, and each class is spatially dispersed. However, the proposed framework effectively captures complex edge details from three perspectives (spectral, spatial, and elevation) by adopting the multibranch interaction structure of MCFB, achieving good classification accuracy. For the MUUFL dataset, the spatial complexity of class distribution may lead to misclassification. As a result, the proposed model only slightly outperformed the other methods on this dataset. The Trento dataset features easily distinguishable contours for each class; thus, our framework and others show notable classification accuracy. However, our framework uses CTFM to maximize the fusion of HSI and LiDAR feature tokens through a nonlocal cross-learning representation. This strategy significantly enhances the synergy among multisource remote sensing image data, elevating shallow feature extraction to deep feature fusion and enhancing the efficacy of feature extraction. As a result, our framework outperforms others in terms of classification accuracy.

(2) The difference in the classification accuracy of the proposed model on the Houston2013, Trento, and MUUFL datasets can be attributed to the unusual characteristics of each dataset. The urban and semi-urban environments in the Houston2013 and MUUFL datasets pose more complex classification challenges to the classification model than the rural Trento dataset. The Trento dataset exhibits higher performance metrics, primarily due to its data characteristics and land cover distribution. As illustrated in Figure 8, each class in the Trento dataset exhibits a more blocky and concentrated distribution pattern. In contrast, the Houston2013 dataset, shown in Figure 7, contains 15 different classes that are spatially dispersed, and the MUUFL dataset, depicted in Figure 9, contains 11 classes that are more messy and intertwined, making the classification task more difficult. Moreover, these datasets have specific differences in spatial resolution and spectral quality. With its multiscale feature extraction, the proposed algorithm effectively utilizes spatial and spectral features of varying scales, showing adaptability to different datasets. This approach allows the algorithm to maintain high classification accuracy across various environments, especially in datasets with complex urban structures.

(3) Although the proposed framework performs well in HSI and LiDAR data fusion classification, its computational complexity still needs to be improved. The data processing approach, which combines the MCFB and the CTFM, effectively improves classification accuracy but requires substantial computational resources. This challenge points to our future work focusing on optimizing the network architecture to enhance the model's usability in processing remote sensing images.

6. Conclusions

In this paper, a multisource fusion classification paradigm for hyperspectral and LiDAR images is proposed, which achieves excellent classification accuracy. In order to solve the chronic defect of CNN architecture that lacks global modeling capability, this work designed the excellent Multi-Conv-Former Block to combine the advantages of convolutional and transformer architectures and, at the same time, introduces a multiscale structure so that the network perceives the global–local joint information at different scales, which improves the classification accuracy. In addition, in order to further improve the feature fusion effect of multisource information, this work designed a Cross-Token Fusion Module feature fusion architecture, which uses the nonlocal structure to fuse the features of different modalities, and the lightweight category token used for fusion reduces the complexity of the high-dimensional features, improves the fusion efficiency, and at the same time provides more robust features for the final classification. Overall, the fusion classification network proposed in this paper achieves excellent classification performance on three publicly available hyperspectral datasets, proving the effectiveness and innovation of this method.

Author Contributions: H.X. and J.L. conceived and designed the study; H.X. and Z.Z. performed the experiments; Y.L. and Z.Z. shared part of the experiment data; H.X. and T.Z. analyzed the data; H.X., T.Z. and Y.L. wrote the paper; C.X. and J.L. reviewed and edited the manuscript. All authors have read and agreed to the published version of the manuscript.

Funding: This work was partially supported by the Key Research Program of the Chinese Academy of Sciences, grant No. KGFZD-145-2023-15, and the National Nature Science Foundation of China under grant 62371359.

Data Availability Statement: The data presented in this study are available on request from the corresponding author.

Conflicts of Interest: The authors declare that we have no financial or personal relationships with people or organizations that could inappropriately influence our work, and this paper was not published before. The founding sponsors had no role in the design of the study; in the collection, analyses, or interpretation of data; in the writing of the manuscript; or in the decision to publish the results.

References

1. Fauvel, M.; Tarabalka, Y.; Benediktsson, J.A.; Chanussot, J.; Tilton, J.C. Advances in spectral–spatial classification of hyperspectral images. *Proc. IEEE* **2012**, *101*, 652–675. [CrossRef]
2. Ghamisi, P.; Yokoya, N.; Li, J.; Liao, W.; Liu, S.; Plaza, J.; Rasti, B.; Plaza, A. Advances in hyperspectral image and signal processing: A comprehensive overview of the state of the art. *IEEE Geosci. Remote Sens. Mag.* **2017**, *5*, 37–78. [CrossRef]
3. Ghamisi, P.; Plaza, J.; Chen, Y.; Li, J.; Plaza, A.J. Advanced spectral classifiers for hyperspectral images: A review. *IEEE Geosci Remote Sens. Mag.* **2017**, *5*, 8–32. [CrossRef]
4. Li, W.; Gao, Y.; Zhang, M.; Tao, R.; Du, Q. Asymmetric feature fusion network for hyperspectral and SAR image classification *IEEE Trans. Neural Netw. Learn. Syst.* **2022**, *34*, 8057–8070. [CrossRef] [PubMed]
5. Chen, Z.; Pu, H.; Wang, B.; Jiang, G.M. Fusion of hyperspectral and multispectral images: A novel framework based on generalization of pan-sharpening methods. *IEEE Geosci. Remote Sens. Lett.* **2014**, *11*, 1418–1422. [CrossRef]
6. Arad, B.; Ben-Shahar, O. Sparse recovery of hyperspectral signal from natural RGB images. In Proceedings of the Computer Vision–ECCV 2016: 14th European Conference, Amsterdam, The Netherlands, 11–14 October 2016; Proceedings, Part VII 14; Springer: Berlin/Heidelberg, Germany, 2016; pp. 19–34.
7. Ngiam, J.; Khosla, A.; Kim, M.; Nam, J.; Lee, H.; Ng, A.Y. Multimodal deep learning. In Proceedings of the 28th International Conference on Machine Learning (ICML-11), Bellevue, WA, USA, 28 June–July 2011; pp. 689–696.
8. Sun, W.; Ren, K.; Meng, X.; Yang, G.; Peng, J.; Li, J. Unsupervised 3D tensor subspace decomposition network for spatial-temporal-spectral fusion of hyperspectral and multispectral images. *IEEE Trans. Geosci. Remote Sens.* **2023**, *61*, 5528917. [CrossRef]
9. Li, J.; Liu, Y.; Song, R.; Li, Y.; Han, K.; Du, Q. Sal^{2}RN: A Spatial–Spectral Salient Reinforcement Network for Hyperspectral and LiDAR Data Fusion Classification. *IEEE Trans. Geosci. Remote Sens.* **2022**, *61*, 5500114. [CrossRef]
10. Melgani, F.; Bruzzone, L. Classification of hyperspectral remote sensing images with support vector machines. *IEEE Trans. Geosci. Remote Sens.* **2004**, *42*, 1778–1790. [CrossRef]
11. Li, J.; Bioucas-Dias, J.M.; Plaza, A. Spectral–spatial hyperspectral image segmentation using subspace multinomial logistic regression and Markov random fields. *IEEE Trans. Geosci. Remote Sens.* **2011**, *50*, 809–823. [CrossRef]
12. Samat, A.; Persello, C.; Liu, S.; Li, E.; Miao, Z.; Abuduwaili, J. Classification of VHR multispectral images using extratrees and maximally stable extremal region-guided morphological profile. *IEEE J. Sel. Top. Appl. Earth Obs. Remote Sens.* **2018**, *11*, 3179–3195. [CrossRef]
13. Shi, Y.; Han, L.; Huang, W.; Chang, S.; Dong, Y.; Dancey, D.; Han, L. A Biologically Interpretable Two-Stage Deep Neural Network (BIT-DNN) for Vegetation Recognition From Hyperspectral Imagery. *IEEE Trans. Geosci. Remote Sens.* **2021**, *60*, 4401320. [CrossRef]
14. Li, S.; Song, W.; Fang, L.; Chen, Y.; Ghamisi, P.; Benediktsson, J.A. Deep learning for hyperspectral image classification: An overview. *IEEE Trans. Geosci. Remote Sens.* **2019**, *57*, 6690–6709. [CrossRef]
15. Mou, L.; Ghamisi, P.; Zhu, X.X. Deep recurrent neural networks for hyperspectral image classification. *IEEE Trans. Geosci. Remote Sens.* **2017**, *55*, 3639–3655. [CrossRef]
16. Sun, Y.; Xue, B.; Zhang, M.; Yen, G.G.; Lv, J. Automatically designing CNN architectures using the genetic algorithm for image classification. *IEEE Trans. Cybern.* **2020**, *50*, 3840–3854. [CrossRef] [PubMed]
17. Shi, X.; Chen, Z.; Wang, H.; Yeung, D.Y.; Wong, W.K.; Woo, W.c. Convolutional LSTM network: A machine learning approach for precipitation nowcasting. *Adv. Neural Inf. Process. Syst.* **2015**, *28*, 802–810.
18. Xu, X.; Li, W.; Ran, Q.; Du, Q.; Gao, L.; Zhang, B. Multisource remote sensing data classification based on convolutional neural network. *IEEE Trans. Geosci. Remote Sens.* **2017**, *56*, 937–949. [CrossRef]
19. Zhao, X.; Tao, R.; Li, W.; Li, H.C.; Du, Q.; Liao, W.; Philips, W. Joint classification of hyperspectral and LiDAR data using hierarchical random walk and deep CNN architecture. *IEEE Trans. Geosci. Remote Sens.* **2020**, *58*, 7355–7370. [CrossRef]
20. Hang, R.; Li, Z.; Ghamisi, P.; Hong, D.; Xia, G.; Liu, Q. Classification of hyperspectral and LiDAR data using coupled CNNs. *IEEE Trans. Geosci. Remote Sens.* **2020**, *58*, 4939–4950. [CrossRef]

21. Song, W.; Dai, Y.; Gao, Z.; Fang, L.; Zhang, Y. Hashing-based deep metric learning for the classification of hyperspectral and LiDAR data. *IEEE Trans. Geosci. Remote Sens.* **2023**, *61*, 5704513. [CrossRef]
22. He, K.; Zhang, X.; Ren, S.; Sun, J. Deep residual learning for image recognition. In Proceedings of the IEEE Conference on Computer Vision and Pattern Recognition, Las Vegas, NV, USA, 26 June–1 July 2016; pp. 770–778.
23. Mohla, S.; Pande, S.; Banerjee, B.; Chaudhuri, S. Fusatnet: Dual attention based spectrospatial multimodal fusion network for hyperspectral and lidar classification. In Proceedings of the IEEE/CVF Conference on Computer Vision and Pattern Recognition Workshops, Seattle, WA, USA, 14–19 June 2020; pp. 92–93.
24. Yang, J.; Wu, C.; Du, B.; Zhang, L. Enhanced multiscale feature fusion network for HSI classification. *IEEE Trans. Geosci. Remote Sens.* **2021**, *59*, 10328–10347. [CrossRef]
25. Dosovitskiy, A.; Beyer, L.; Kolesnikov, A.; Weissenborn, D.; Zhai, X.; Unterthiner, T.; Dehghani, M.; Minderer, M.; Heigold, G.; Gelly, S.; et al. An image is worth 16x16 words: Transformers for image recognition at scale. *arXiv* **2020**, arXiv:2010.11929.
26. Peng, Z.; Huang, W.; Gu, S.; Xie, L.; Wang, Y.; Jiao, J.; Ye, Q. Conformer: Local features coupling global representations for visual recognition. In Proceedings of the IEEE/CVF International Conference on Computer Vision, Montreal, QC, Canada, 10–17 October 2021; pp. 367–376.
27. Mei, S.; Song, C.; Ma, M.; Xu, F. Hyperspectral image classification using group-aware hierarchical transformer. *IEEE Trans. Geosci. Remote Sens.* **2022**, *60*, 5539014. [CrossRef]
28. Hong, D.; Han, Z.; Yao, J.; Gao, L.; Zhang, B.; Plaza, A.; Chanussot, J. SpectralFormer: Rethinking hyperspectral image classification with transformers. *IEEE Trans. Geosci. Remote Sens.* **2021**, *60*, 5518615. [CrossRef]
29. Xue, Z.; Tan, X.; Yu, X.; Liu, B.; Yu, A.; Zhang, P. Deep hierarchical vision transformer for hyperspectral and LiDAR data classification. *IEEE Trans. Image Process.* **2022**, *31*, 3095–3110. [CrossRef]
30. Chen, H.; Wang, T.; Chen, T.; Deng, W. Hyperspectral image classification based on fusing S3-PCA, 2D-SSA and random patch network. *Remote Sens.* **2023**, *15*, 3402. [CrossRef]
31. Mu, C.; Liu, Y.; Liu, Y. Hyperspectral image spectral–spatial classification method based on deep adaptive feature fusion. *Remote Sens.* **2021**, *13*, 746. [CrossRef]
32. Yang, L.; Yang, Y.; Yang, J.; Zhao, N.; Wu, L.; Wang, L.; Wang, T. FusionNet: A convolution–transformer fusion network for hyperspectral image classification. *Remote Sens.* **2022**, *14*, 4066. [CrossRef]
33. Zhang, M.; Ghamisi, P.; Li, W. Classification of hyperspectral and LiDAR data using extinction profiles with feature fusion. *Remote Sens. Lett.* **2017**, *8*, 957–966. [CrossRef]
34. Zhang, Y.; Prasad, S. Multisource geospatial data fusion via local joint sparse representation. *IEEE Trans. Geosci. Remote Sens.* **2016**, *54*, 3265–3276. [CrossRef]
35. Rasti, B.; Ghamisi, P.; Gloaguen, R. Hyperspectral and LiDAR fusion using extinction profiles and total variation component analysis. *IEEE Trans. Geosci. Remote Sens.* **2017**, *55*, 3997–4007. [CrossRef]
36. Khodadadzadeh, M.; Li, J.; Prasad, S.; Plaza, A. Fusion of hyperspectral and LiDAR remote sensing data using multiple feature learning. *IEEE J. Sel. Top. Appl. Earth Obs. Remote Sens.* **2015**, *8*, 2971–2983. [CrossRef]
37. Zare, A.; Ozdemir, A.; Iwen, M.A.; Aviyente, S. Extension of PCA to higher order data structures: An introduction to tensors, tensor decompositions, and tensor PCA. *Proc. IEEE* **2018**, *106*, 1341–1358. [CrossRef]
38. Liao, W.; Bellens, R.; Pižurica, A.; Gautama, S.; Philips, W. Combining feature fusion and decision fusion for classification of hyperspectral and LiDAR data. In Proceedings of the 2014 IEEE Geoscience and Remote Sensing Symposium, Quebec City, QC, Canada, 13–18 July 2014; pp. 1241–1244.
39. Song, D.; Gao, J.; Wang, B.; Wang, M. A Multi-Scale Pseudo-Siamese Network with an Attention Mechanism for Classification of Hyperspectral and LiDAR Data. *Remote Sens.* **2023**, *15*, 1283. [CrossRef]
40. Gerbrands, J.J. On the relationships between SVD, KLT and PCA. *Pattern Recognit.* **1981**, *14*, 375–381. [CrossRef]
41. Debes, C.; Merentitis, A.; Heremans, R.; Hahn, J.; Frangiadakis, N.; van Kasteren, T.; Liao, W.; Bellens, R.; Pižurica, A.; Gautama, S.; et al. Hyperspectral and LiDAR data fusion: Outcome of the 2013 GRSS data fusion contest. *IEEE J. Sel. Top. Appl. Earth Obs. Remote Sens.* **2014**, *7*, 2405–2418. [CrossRef]
42. Dalponte, M.; Bruzzone, L.; Gianelle, D. Fusion of hyperspectral and LIDAR remote sensing data for the estimation of tree stem diameters. In Proceedings of the 2009 IEEE International Geoscience and Remote Sensing Symposium, Cape Town, South Africa, 12–17 July 2009; Volume 2, p. II–1008.
43. Du, X.; Zare, A. *Technical Report: Scene Label Ground Truth Map for MUUFL Gulfport Data Set*; University of Florida: Gainesville, FL, USA, 2017.
44. Li, W.; Wu, G.; Zhang, F.; Du, Q. Hyperspectral image classification using deep pixel-pair features. *IEEE Trans. Geosci. Remote Sens.* **2016**, *55*, 844–853. [CrossRef]
45. Roy, S.K.; Krishna, G.; Dubey, S.R.; Chaudhuri, B.B. HybridSN: Exploring 3-D–2-D CNN feature hierarchy for hyperspectral image classification. *IEEE Geosci. Remote Sens. Lett.* **2019**, *17*, 277–281. [CrossRef]

remote sensing

MDPI

Article

Self-Distillation-Based Polarimetric Image Classification with Noisy and Sparse Labels

Ningwei Wang [1], Haixia Bi [1,*], Fan Li [1], Chen Xu [2,3] and Jinghuai Gao [1]

[1] School of Information and Communications Engineering, Xi'an Jiaotong University, Xi'an 710049, China; ningweiwang@stu.xjtu.edu.cn (N.W.); lifan@mail.xjtu.edu.cn (F.L.); jhgao@mail.xjtu.edu.cn (J.G.)
[2] School of Mathematics and Statistics, Xi'an Jiaotong University, Xi'an 710049, China; xnolimited@hotmail.com
[3] Department of Mathematics and Fundamental Research, Peng Cheng Laboratory, Shenzhen 518055, China
* Correspondence: haixia.bi@xjtu.edu.cn

Abstract: Polarimetric synthetic aperture radar (PolSAR) image classification, a field crucial in remote sensing, faces significant challenges due to the intricate expertise required for accurate annotation, leading to susceptibility to labeling inaccuracies. Compounding this challenge are the constraints posed by limited labeled samples and the perennial issue of class imbalance inherent in PolSAR image classification. Our research objectives are to address these challenges by developing a novel label correction mechanism, implementing self-distillation-based contrastive learning, and introducing a sample rebalancing loss function. To address the quandary of noisy labels, we proffer a novel label correction mechanism that capitalizes on inherent sample similarities to rectify erroneously labeled instances. In parallel, to mitigate the limitation of sparsely labeled data, this study delves into self-distillation-based contrastive learning, harnessing sample affinities for nuanced feature extraction. Moreover, we introduce a sample rebalancing loss function that adjusts class weights and augments data for small classes. Through extensive experiments on four benchmark PolSAR images, our approach demonstrates its effectiveness in addressing label inaccuracies, limited samples and class imbalance. Through extensive experiments on four benchmark PolSAR images, our research substantiates the robustness of our proposed methodology, particularly in rectifying label discrepancies in contexts marked by sample paucity and imbalance. The empirical findings illuminate the superior efficacy of our approach, positioning it at the forefront of state-of-the-art PolSAR classification techniques.

Keywords: label correction; self-distillation contrastive learning; sample rebalancing; polarimetric synthetic aperture radar (PolSAR) image classification

Citation: Wang, N.; Bi, H.; Li, F.; Xu, C.; Gao, J. Self-Distillation-Based Polarimetric Image Classification with Noisy and Sparse Labels. *Remote Sens.* **2023**, *15*, 5751. https://doi.org/10.3390/rs15245751

Academic Editor: Giuseppe Scarpa

Received: 13 October 2023
Revised: 12 December 2023
Accepted: 12 December 2023
Published: 15 December 2023

1. Introduction

Polarimetric synthetic aperture radar (PolSAR) is an advanced and important remote sensing technique owing to its distinctive ability to transmit and receive electromagnetic waves across various polarization modes [1]. This unique capability enables PolSAR to provide richer information on the scattering properties of Earth's surface. Consequently, PolSAR image classification, which is oriented towards categorizing image pixels into corresponding terrain classes, becomes instrumental for a spectrum of applications ranging from sea monitoring and agriculture to geological mapping and strategic governmental decisions [2]. PolSAR image classification has evolved, leading to diverse methodologies categorized into three main types: (1) physical-scattering-mechanism-based methods [2–4], (2) statistics-based methods [5,6], and (3) machine-learning-based methods [7–10]. Deep learning, with its superior feature representation, has significantly advanced PolSAR image classification [11].

However, PolSAR classification faces challenges, particularly noisy and sparse labels. These distortions misguide the model to assimilate noise patterns instead of the authentic features. Limited annotations further challenge model accuracy and generalization. This paper seeks to unravel the following conundrum: How to improve the accuracy and robustness of the DNN-based PolSAR image classification method in a weak label scenario, i.e., with noisy and sparse labels?

Addressing noisy labels has engendered the inception of two predominant methodologies [12–14]. The first strategy focuses on the identification and purgation of these erroneous labels prior to model training [15,16]. This rectification can be accomplished through manual scrutiny, clustering, or the deployment of outlier detection algorithms. The alternative approach pivots towards the direct training of noise-robust models on corrupted datasets [17,18]. This necessitates the modification of the conventional loss function, accounting for the noisy labels. Ensemble learning, epitomized by methodologies like bootstrapping [19], self-training [20], and co-teaching [21], emerges as a robust tool. Such strategies harness the predictive prowess of an array of models, thereby refining overarching performance.

The field of image processing has traditionally seen a surge of research focusing on mitigating the challenges posed by noisy labels. In the specific domain of PolSAR image classification, the investigation into noisy labels remains comparatively nascent. Ni et al. [22] pioneered an insightful difference distribution diagram, articulating the intrinsic probability of a training sample being untainted. This probabilistic assessment paved the way for distinguishing clean labels from their noisy counterparts. Further innovation was heralded by Hou et al. [23] through their generative classification framework, adeptly tackling both the predicaments of unfaithful limited labels and the perturbations introduced by outliers in PolSAR pixels. Nevertheless, contemporary algorithms harbor intrinsic limitations. In contexts enriched with labels, eliminating detected noisy labels might not inflict significant harm. Yet, in scenarios marked by label paucity, such removal intensifies the small-sample dilemma, leading to potential algorithmic performance deterioration. Furthermore, an evident lacuna remains, as these methodologies overlook the potential leverage that can be garnered from the inherent similarity between training samples, which is quintessential for labeling.

To address these challenges, our research introduces a relabeling mechanism. This endeavor is grounded in the pivotal assertion that the discriminative model features extracted from neighboring samples with the same label play a vital role in driving the relabeling mechanism's efficacy.

Parallelly, the PolSAR image classification domain grapples with the issue of label scarcity. With the progress of deep learning, many PolSAR image classification methods [24–28] have been proposed to alleviate this problem. Semisupervised learning [29–31] ambitiously seeks to optimize classifier generalization, leveraging both labeled and unlabeled data. Active learning [11,32], in its quest, adopts a selective approach to acquire salient samples for labeling, aiming for maximized learning efficiency. Transfer learning [33,34], drawing from affluent source domains, endeavors to uplift the performance in target domains characterized by data scarcity. Reinforcement learning [35,36], albeit less prevalent in PolSAR terrains, adopts a unique perspective, emphasizing sequential decision making and reward maximization.

Venturing into a distinct trajectory, self-supervised learning [20,24,25] exploits the data's inherent properties to formulate alternative guidance signals, often involving pretext tasks for model training. This paradigm notably circumvents the label reliance in semisupervised learning, human intervention in active learning, domain-specific insights in transfer learning, and environmental interactions in reinforcement learning. However, self-supervised learning's capability to harness the intrinsic label information positions it advantageously, enabling nuanced feature extraction. Such prowess is manifested through its "pseudolabel" generation, correlating closely with true labels, and thus fostering meaningful data interpretations without extensive manual annotations [24].

Contrastive learning, as an important branch of self-supervised learning, while achieving commendable success in natural image classifications, remains scarcely explored within the domain of PolSAR images. TCSPANet, as delineated by [37], integrates a dual-stage methodology: Initially, TCNet, rooted in contrastive learning, facilitates unsupervised representation learning. Subsequently, a subpatch attention encoder (SPAE), structured upon the transformer paradigm, models the context within patch samples. In a distinct approach, Zhang et al. [26] introduced the PolSAR-specific contrastive learning network (PCLNet). This network employs an unsupervised pretraining phase, anchored on instance discrimination [38], to harness valuable representations from unlabeled PolSAR data. Further, the self-supervised PolSAR representation learning (SSPRL) method [25] draws inspiration from the accomplishments of BYOL [19]. It is pertinent to note the following differences: TCSPANet operates through a bifurcated framework encompassing TCNet and SPAE, PCLNet capitalizes on an instance-discrimination-based pretraining phase, and SSPRL deploys a twin network structure alongside positive pairs, aiming for optimal efficiency across varied domains.

DINO [39] distinguishes itself by leveraging an exponential moving average (EMA) and central updates to fortify knowledge distillation. Unlike SSPRL, DINO uses EMA to seamlessly integrate the parameters of the online network with its target counterpart, an innovation that curtails parameter oscillation, thereby augmenting model stability. Within the DINO architecture, the teacher model's output serves to refine a center vector, which subsequently modulates the teacher model's results. This innovative step considerably bolsters the training efficacy of the student model. Recognizing its potential, we meld it into our framework, aiming to address the persistent issue of limited PolSAR-labeled data availability.

A pivotal concern in real-world datasets is the unequal distribution of object types, culminating in sample imbalance challenges. This imbalance frequently translates to suboptimal performance for minority classes. To address this, our research introduces a novel Self-Distillation-Based Correction Strategy (SDBCS), which integrates a label correction strategy, a sample rebalancing loss function, and data augmentation targeted for minority classes, enhancing overall classification accuracy. Our research proffers three pivotal contributions:

(1) We propose a new method using a feature distance matrix to correct label inaccuracies. This matrix, derived from contrastive learning principles, helps identify and rectify mislabeled samples by analyzing pixel similarities.
(2) We explore self-distillation learning to overcome the scarcity of labeled data in PolSAR. This approach utilizes inherent sample similarities for discriminative representation and achieves effective results, even with limited labels.
(3) Our strategy includes a rebalancing loss function and a data augmentation method for minority classes, significantly improving classification accuracy for minority classes.

2. Literature Review

2.1. Noisy Label Correction

The challenge of noisy labels in deep learning has become particularly critical in recent times. Models trained on noisy datasets can become susceptible to suboptimal representations, causing degraded performance in subsequent tasks. Addressing the noisy label issue, the research community has primarily focused on two solutions: (1) methods that train noise-resilient models directly on corrupted datasets and (2) methods that detect and rectify noisy labels before model training.

The former strategy involves modeling noise patterns directly, employing techniques such as robust loss functions [40,41], and noise corrections via noise transition matrices [15]. For instance, Ma et al. [18] developed a loss function that augments the resilience of DNNs against noisy labels. However, these methods often falter in the face of intricate noise patterns. Conversely, the latter strategy, gaining traction in recent years, particularly emphasizes sample selection. While some early approaches focused on curtailing the influence of noisy samples by training on selected clean subsets [42,43], more contemporary methods exploit semisupervised learning techniques [44]. Nonetheless, these techniques frequently rely on assumptions about noise patterns, which can be detrimental if real-world noise deviates from these assumptions.

The intricacy of labeling PolSAR data, given the specialized expertise it demands, cannot be underestimated. This involves conferring precise class labels to specific pixels or regions within a PolSAR image, thereby setting the stage for frequent mislabeling. Such mislabeling, i.e., noisy labels, will inevitably undermine model performance. Notably, the differential distribution diagram delineated by [22] offered insights into clean sample probabilities, assisting in discerning between clean and noisy labels. Hou et al. [23] tackled the quandary of unreliable limited labels using a blended generative classification framework, wherein both labeled and unlabeled pixels were harnessed to derive high-level features.

2.2. *Label Scarcity Problem with Contrastive Learning*

PolSAR image classification, powered by supervised CNNs, has shown notable success. Yet, amassing large labeled datasets is both costly and time-intensive. Furthermore, limited training data can lead to model overfitting and reduced generalization. Given these issues, recent efforts, including label scarcity learning [45,46], aim to extract meaningful knowledge from minimal labeled samples. Specifically, methods under label scarcity learning, such as those cited, either harness learned optimization [47] or execute a feed-forward pass [48–50] without weight modifications. However, the methods employing a feed-forward pass often necessitate intricate inference protocols, reliance on RNN architectures, or task-specific fine-tuning [51,52].

Remarkable advancements in unsupervised representation learning have been realized via the advent of contrastive learning methodologies. By juxtaposing positive and negative samples in a self-supervised fashion, these strategies seek to derive salient data representations. For instance, the InstDisc [38] technique was the first to innovate a discrimination task, leveraging a memory bank to accumulate negative samples, thereby creating an expansive and consistent dictionary. Meanwhile, methods like CPC v1 [53], CMC [54], and MoCo v1 [55] have offered a multitude of contrasting and clustering tasks. Grill et al. [19] introduced BYOL, which employs one view's extracted feature to predict the feature of another view from the same instance, utilizing a momentum-based moving average for updating both encoder and representation. Yet, for all their success, contrastive learning techniques still grapple with achieving pinnacle accuracy on certain downstream assignments, particularly when benchmarked against supervised methods. Building upon prior successes, DINO emerged as a proposed solution to address these challenges, showcasing enhanced quality in learned representations. Notably, Caron et al. [39], drawing inspiration from BYOL, introduced several innovative techniques to elevate the performance metrics of self-supervised learning strategies.

Despite the evident potential of contrastive learning in generic image classification, its application remains conspicuously underrepresented in PolSAR imagery. Noteworthy explorations by Cui et al. [37] and Zhang et al. [25,26] have begun harnessing the merits of methods such as SimCLR, InstDisc, and BYOL for self-supervised PolSAR representation learning. These trailblazers proposed an avant-garde, self-supervised PolSAR representation learning paradigm, underscoring the potential synergy between contrastive learning and PolSAR imagery, especially in scenarios punctuated by label paucity.

To summarize, despite the widespread use of deep learning in PolSAR image classification, its effectiveness heavily relies on extensive annotations. This study aims to bridge the noticeable gap in applying contrastive learning within the PolSAR context. Our work differentiates itself by introducing a label correction strategy that utilizes inherent similarities among training samples to correct erroneous labels, which effectively solves the dilemma of noisy labels. Furthermore, we integrate self-distillation-based contrastive learning and a sample rebalancing loss function into an integrated framework, remarkably improving the classification performance on the PolSAR dataset, which presents label scarcity and class imbalance challenges.

3. Methodology

3.1. Overview of Our Method

In the subsequent sections, we delineate our methodology, beginning with the establishment of pertinent notations, followed by an exposition of the proposed framework. Given a PolSAR image, the PolSAR feature data are represented as $X \in R^{H \times W \times D}$, where H and W are the height and width of the PolSAR image, respectively, and D signifies the dimension of the chosen raw feature vector. The objective of our approach is to allocate a class label to each pixel in the image.

Figure 1 encapsulates the architecture of our proposed model, integrating modules for self-distillation-based feature extraction, label correction, and classification. Our approach commences with a finite set of randomly chosen pixels possessing noisy labels. In the initial phase, a convolutional neural network (CNN) is trained employing self-distillation-based deep representation learning. Following this, a global distance matrix is constructed, facilitating the identification of pixels bearing the highest resemblance for each sample. The labeling process then ensues, wherein labels are attributed based on the prevalence of a particular label within each cohort of similar pixels. Conclusively, to address class imbalances, a sample rebalancing loss function is introduced, which duly modulates the weights designated to varying classes, thereby refining classification accuracy.

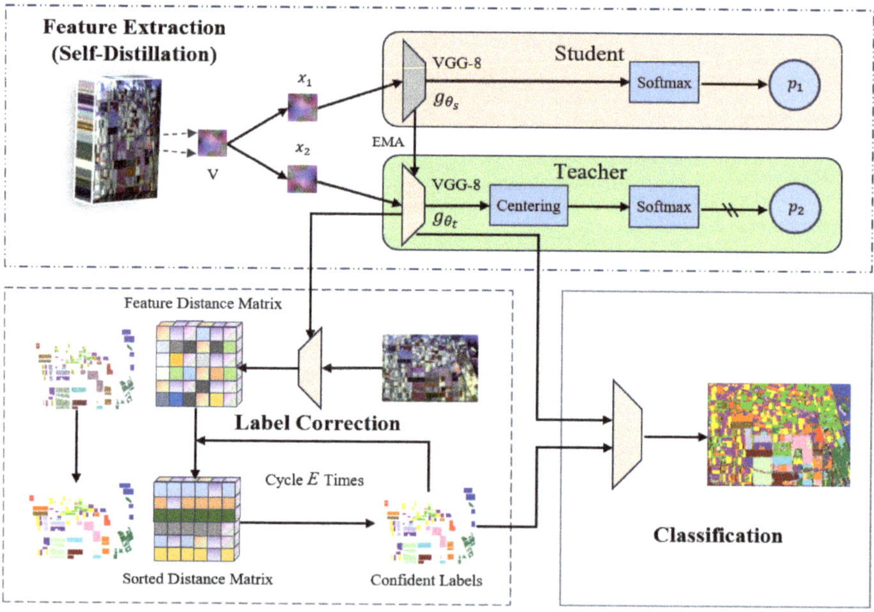

Figure 1. The proposed methodological pipeline encompasses three distinct modules: self-distillation-based feature extraction, label correction, and classification.

3.2. Raw Feature Extraction

We initiate by procuring the unprocessed polarimetric attributes, serving as the foundational input for our methodology. The resultant 6D feature set, symbolized as RF-i for i in the range 1 to 6, is derived from the complex coherency polarimetric matrix T, constructed using the Pauli basis of the PolSAR scattering matrix [56]. These attributes encapsulate critical information about the scattering mechanisms and are crucial for effective PolSAR image analysis.

As illustrated in Table 1, within this 6D feature set, RF-1 represents the total polarimetric power, known as SPAN (SPAN = $T_{11} + T_{22} + T_{33}$), expressed in decibel units. This feature provides a baseline measure of the total reflected energy, fundamental in understanding the overall scattering characteristics of the observed scene. RF-2 and RF-3 symbolize the normalized power ratios of T_{22} and T_{33}, respectively. In the coherency matrix T, T_{22}, and T_{33} represent the power received in different polarization channels, such as horizontal–horizontal (HH) or vertical–vertical (VV), depending on the orientation of the PolSAR system. These elements are essential for analyzing the scattering behavior of different surface types in PolSAR imagery. By normalizing these power values against the SPAN, we obtain a relative measurement that is more robust to variations in absolute signal strength. RF-4 to RF-6 denote the relative correlation coefficients linked to the cross-polarization components T_{12}, T_{13}, and T_{23}. These coefficients measure the degree of correlation between different polarimetric channels, providing insights into the geometrical and dielectric properties of the scattering targets. They are particularly useful in distinguishing various surface types and man-made structures, which often exhibit unique polarimetric signatures.

Table 1. Raw polarimetric features employed in the proposed method.

Designation	Description
$RF\text{-}1 = 10\log_{10}(\text{SPAN})$	Polarimetric total power in decibel
$RF\text{-}2 = \frac{T_{22}}{\text{SPAN}}$	Normalized ratio of power T_{22}
$RF\text{-}3 = \frac{T_{33}}{\text{SPAN}}$	Normalized ratio of power T_{33}
$RF\text{-}4 = \frac{\lvert T_{12}\rvert}{\sqrt{T_{11}\cdot T_{22}}}$	Relative correlation coefficient of T_{12}
$RF\text{-}5 = \frac{\lvert T_{13}\rvert}{\sqrt{T_{11}\cdot T_{33}}}$	Relative correlation coefficient of T_{13}
$RF\text{-}6 = \frac{\lvert T_{23}\rvert}{\sqrt{T_{22}\cdot T_{33}}}$	Relative correlation coefficient of T_{23}

The necessity of this raw feature extraction process stems from its capability to convert complex and multidimensional PolSAR data into a format that is interpretable and applicable to machine learning algorithms. Features like T_{22} and T_{33} help in understanding the scattering behavior of different surfaces, crucial for accurate image classification. The feature extraction process thus translates PolSAR data into a form that machine learning algorithms can more effectively process and analyze. The selection of these particular features is informed by their established effectiveness in extracting meaningful information from PolSAR data, as highlighted in the existing literature [57]. These features assist in distinguishing different surface types and physical properties in the observed area, enhancing the classification accuracy. By employing these specific features, our approach not only capitalizes on the intrinsic properties of PolSAR data but also significantly enhances the potential for precise and robust classification outcomes. The scaling of RF-2 through RF-6 to the interval [0, 1] ensures uniformity in feature magnitude, which aids the learning algorithm in effectively processing and interpreting the data. This methodical approach to feature extraction lays a solid foundation for the subsequent machine learning processes, enabling our model to more accurately interpret and classify the intricate patterns inherent in PolSAR imagery.

3.3. *Self-Supervised Learning with Knowledge Distillation*

As we navigate through the challenges instigated by noisy labels and a scant quantity of labeled samples, we explore avant-garde techniques to bolster the discriminative prowess of our model. The presence of label noise and limited labeled samples present a dichotomy; while we require robustly discriminative features for label correction and subsequent classification, using these labels directly for learning might culminate in procuring misleading discriminative features. Enter contrastive learning, which offers a resolute solution by gleaning more illuminative supervised signals from raw unlabeled PolSAR data in an unsupervised fashion. To amplify the discriminative capacity of the model, we enlist knowledge distillation methodologies. At its core, knowledge distillation conceives a streamlined student model and hones it through the mentorship of a superior-performing teacher model. The quintessence of this paradigm lies in transmitting knowledge from the teacher to the student, optimizing performance.

Our approach heralds a more kinetic interaction between teacher and student models. This synergy is materialized by gauging the disparity between the outcomes of the student and teacher models. This ushers in our feature extraction technique based on self-distillation contrastive learning. In the subsequent sections, we delve deep into aspects encompassing pretraining tasks, loss functions, and the architecture of the encoder and self-distillation module.

3.3.1. Pretext Task and Loss Function

In traditional supervised learning, models are honed to discern the intricate relationships between input data and their associated output labels, necessitating the availability of class information. Diverging from this paradigm, we propose an approach grounded in instance discrimination tasks. Within this framework, a neural network is self-supervised, training itself on two distinct data augmentation views. This methodology capacitates the network to concurrently project two variant views of an identical sample to a congruent representation space while projecting views from distinct samples to separate representation spaces. The inherent advantage is that the samples intrinsically act as their own supervisors, obviating the need for manual labeling. This strategy paves the way for harnessing vast repositories of unlabeled PolSAR images. Furthermore, by pretraining this network, we establish a deep feature network that is transferable. The network exhibits strong discriminative feature extraction capabilities, facilitating accurate label correction. Additionally, it adeptly addresses the small-sample challenges often encountered in classification tasks.

Figure 1 illustrates our proposed self-distillation contrastive learning model tailored for PolSAR data. This model is architecturally segmented into two networks: a student network, g_{θ_s}, and a teacher network, g_{θ_t}, visually discernible through orange and green modules, respectively. Both these networks, characterized by their respective parameters θ_s and θ_t, are intrinsically structured into three foundational components: an encoder, a projection head, and a predictor. Upon the sequential processing through these components, each network computes a probability distribution over Q dimensions, respectively denoted as P_s and P_t. Within the framework of our self-distillation contrastive learning approach, the designed loss function plays a pivotal role. It serves to nudge the neural networks into aligning similar instances in close proximity within the feature representation space, while simultaneously pushing apart dissimilar instances. This strategic configuration aids in fostering the extraction of robust discriminative features. A key element in this mechanism is the temperature parameter, denoted as $\tau_s > 0$, which dictates the acuteness of the distribution contour of P_s as

$$P_s(x)^{(i)} = \frac{\exp\left(\frac{g_{\theta_s}(x)^{(i)}}{\tau_s}\right)}{\sum_{k=1}^{Q} \exp\left(\frac{g_{\theta_n}(x)^{(k)}}{\tau_s}\right)} \quad (1)$$

In a parallel fashion, the temperature parameter τ_t governs the sharpness of P_t. To harmonize these distributions, we adopt a strategy of minimizing the cross-entropy loss concerning the parameters θ_s of the student network, all the while maintaining the teacher network g_{θ_t} in a static state. The objective function can be formally expressed as

$$\min_{\theta_s} H(P_t(x), P_s(x)) \quad (2)$$

where the relationship $H(a,b) = -a \log b$ holds true. We generate a set of views, V, from the PolSAR images, where views x_1 and x_2 are two randomly augmented views. Our primary pursuit is encapsulated in the minimization of the loss, articulated as

$$\min_{\theta_s} \sum_{x \in \{x_1, x_2\}} \sum_{x' \in V x' \neq x} H(P, P(x')) \tag{3}$$

To refine the parameters θ_s, we employ the stochastic gradient descent method, targeting the minimization of Equation (3).

3.3.2. Architecture of Encoder and Self-Distillation Module

In light of the aforementioned principles, we architected a network for self-distillation contrastive learning. The encoder in our model incorporates the VGGNet-8 structure, serving as a convolutional feature extractor designed for processing input images. It is composed of three convolutional blocks, each containing two layers that use 3 × 3 convolutional kernels, followed by a ReLU activation function and 2 × 2 max-pooling, effectively capturing and processing image features. In parallel, the projection head transforms the input feature vectors into a lower-dimensional space through dense layers, enabling the learning of more compact yet abstract data representations while preserving crucial feature information. Additionally, the predictor utilizes a fully connected layer to map these feature vectors into Q dimensions. This dimensionality reduction is achieved using a softmax activation function, which calculates the probability distribution across various classes, ensuring an effective and efficient classification process.

During the training regime, neither network updates its parameters based on labeled data. An input image, denoted as x, undergoes random augmentations to yield two distinct variants, x_1 and x_2. Subsequently, these variants are independently channeled into both the student and teacher networks. It is imperative to note that while these networks architecturally mirror each other, they possess unique parameters, thus fostering independent learning and nuanced data comprehension. To achieve consistent representations, the output of the teacher network is centralized by computing its mean over the entire batch, subsequently normalizing these features across individual samples. Both networks yield an M-dimensional feature vector, which undergoes further normalization via a temperature-regulated softmax operation across its dimensions. The congruence between the feature vectors from the student and teacher networks is ascertained using a cross-entropy loss. This loss function measures the discrepancy between the predicted probability distributions of the two networks. By striving to minimize this loss, we compel the networks to generate analogous representations for equivalent input samples, thus enhancing the knowledge transfer from the teacher to the student. It is paramount during training to restrict the flow of gradients solely to the student network. To achieve this, we deploy a stop-gradient operator on the teacher network, ensuring its immunity from external updates and guaranteeing that only the student network receives iterative refinements.

Our methodology presents a notable divergence from traditional knowledge distillation practices, especially in its approach to temperature scaling. Conventionally, the teacher temperature parameter is held invariant throughout the training, serving to temper the fluctuations in its output probabilities. In contrast, our approach harnesses a temperature scheduling mechanism that methodically diminishes the temperature of the teacher model as training advances. The initiation phase employs a heightened temperature to ensure a robust training foundation, which is progressively tapered to bolster the distillation impact. By refashioning the teacher model's knowledge, manifested as soft targets or feature representations, we aim to effectively shepherd the student's learning trajectory. Furthermore, we introduce a mechanism for updating the center vector based on the outputs of the teacher model. This innovation not only enhances knowledge distillation but also marks a distinction from traditional methodologies.

Unlike the conventional approach of initializing the teacher network by directly copying the student network's weights, our strategy crafts the teacher network based on antecedent iterations of the student network. This process is refined using the nuances of the exponential moving average, as demonstrated by the following rule: $\theta_t \leftarrow \lambda\theta_t + (1-\lambda)\theta_s$.

As training ensues, we adopt a λ value that commences at 0.996 and ascends, tracing a cosine trajectory until it culminates at unity. Consequently, in the nascent stages, the teacher network's parameters gravitate swiftly toward their student counterparts. Yet, as the training journey evolves, this adaptation pace decelerates, culminating in a poised equilibrium. This meticulously crafted strategy strikes a harmonious balance between maintaining the stability of the teacher network and optimizing its directive potency on the student network's representations. To encapsulate, our proposed self-distillation contrastive learning method undergoes cyclical refinements, capitalizing on variances between views to adeptly mediate the knowledge transference between the student and teacher constructs. The outcome of this innovative methodology is the adept extraction of discerningly potent features, leading to a marked enhancement in model proficiency.

4. Enhancing Classification Accuracy

In this section, we address two pivotal aspects of classification accuracy: label correction and addressing class imbalance. The label correction module corrects mislabeled instances, while our class imbalance strategy ensures a fair representation of all classes. This dual approach is crucial for the precise categorization of PolSAR data, where both label quality and balanced class representation significantly impact the classifier's performance.

As illustrated in Figure 2, our proposed label correction strategy capitalizes on the inherent affinities among training samples to amend erroneously assigned labels. Within this strategy, the backbone network of a contrastive learning framework is employed to distill features and create a comprehensive distance matrix encompassing all training samples. For each pixel, we then identify its top-K nearest samples, based on the predefined distance metric. The label that exhibits the highest frequency among these nearest samples is then designated to the pixel under consideration. This approach adeptly harnesses representational affinities to ameliorate the classification of incorrectly labeled instances.

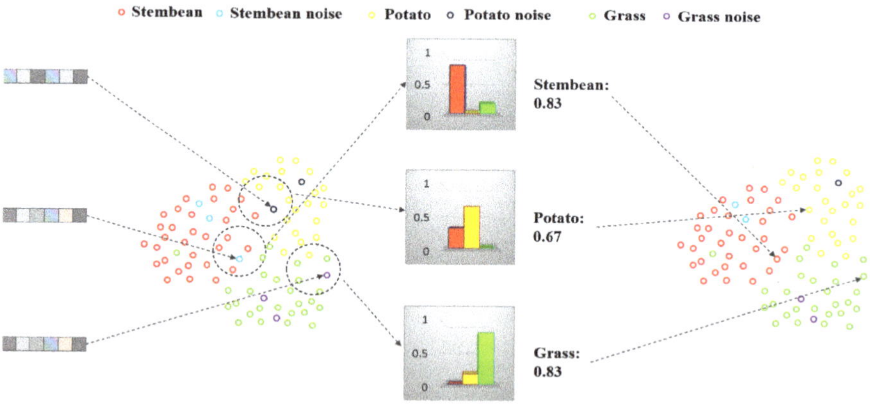

Figure 2. The label correction procedure can be delineated as depicted in this figure. Initially, a global distance matrix is constructed to discern pixels demonstrating the paramount similarity to individual samples. Subsequent to this step, the label for each pixel is determined based on the predominant label within its associated cluster of similar pixels.

Consider a scenario wherein a sample, erroneously labeled under a *non-Stembean* category, requires rectification, given that its ground truth designation is *Stembean*. Assuming a top-K threshold of 6, the six proximal samples in the feature space relative to this label are selected. The distribution among these reveals 1 *Grass* label, 5 *Stembean* labels, and no other categories. As a result, the *Stembean* category emerges with a probability of 0.83, surpassing the stipulated threshold for label correction. Consequently, the label is rectified to *Stembean*

To better understand the pseudocode of our label correction algorithm presented in Algorithm 1, it is essential to define some key variables used within it. The total pixel count is denoted as N, calculated as $H \times W$, where H and W represent the height and width, respectively. We denoted the set of labeled pixel pairs as $L = \{(x_1,t_1),(x_2,t_2),\ldots,(x_n,t_n)\}$, where $\mathcal{X}$ and $\mathcal{Y}$ represent the data and label parts of L, respectively, and n is significantly smaller than N. Within this context, M denotes the total number of classes. The label of each sample x_i corresponding to the one-hot label vector y_i is expressed as $l_i = \arg_j[y_i(j) = 1] \in \{1,\ldots,M\}$. The objective of our approach is to allocate a class label y_i to each pixel i, where $i \in \{1,2,\ldots,n\}$. Algorithm 1 delineates the pseudocode of our advanced label correction algorithm, which intakes both original features and augmented image labels. The primary objective of this module is to redress noisy labels. Its foundational architecture, denoted as f, is sculpted through self-distillation rooted in contrastive learning. For brevity, we define f_i as the representational feature of the sample x_i. p_{f_i} is the projection head derived from p_f, which exemplifies the encoder's prowess in capturing intricate, high-dimensional features. Additionally, p_f is employed to construct a K-Nearest Neighbors Classifier (KNNC) k_q, with $k_{q_i} \triangleq k_q\left(p_{f_i}\right)$ representing its predictive vector.

Algorithm 1: Label Correction Algorithm

Input:
$(\mathcal{X},\mathcal{Y})$
n represents the size of the training set
Sample relabelling threshold θ_s
Max epochs E
p_f represents the feature extractor
$\mathcal{Y}_E$ is a list of elements denoted by $\mathcal{Y}_e$

Output:
The clean label of $\mathcal{Y}$

Data augmentation on small classes:
for $i = 1$ **to** n **then**
  Extract feature
end for
for $i = 1$ **to** n **then**
  for $j = 1$ **to** n **then**
    Calculate similarity between each representation: Equation (4)
end for
for $e = 1$ **to** E **then**
  for $i = 1$ **to** n **then**
    Measure of consistency c_i: Equation (6)
    if $c_i < \theta_s$ **then**
      l_i^r is likely to be wrong
    else
      $y_i^r \leftarrow l_i^r$
    end if
    $\mathcal{Y}_{Ei} = y_i^r$
end for
for $j = 1$ **to** E **then**
  $\mathcal{Y}_j = Max_j \sum_{i=1}^{n} \mathcal{Y}_{Ej}$
end for

The affinity between the representations p_{f_i} and p_{f_j} of samples x_i and x_j is articulated as s_{ij}, where both i and j iterate from 1 to n. The cosine similarity is computed as

$$\mathbf{s}_{ij} = \frac{p_{f_i}^T p_{f_j}}{\left\|p_{f_i}\right\|_2 \left\|p_{f_j}\right\|_2} \quad (4)$$

This remains our measure of choice. The index set for the S-nearest neighbors of sample x_i in $\mathcal{X}$, predicated on this similarity, is denoted as N_i. For every sample x_i, the normalized label distribution is computed as

$$k'_{q_i} = \frac{1}{S} \sum_{n \in N_i} y_n^r \quad (5)$$

A subsequent balanced version, $k_{q_i} \in R^M$, adjusts for the label distribution $\pi = \sum_{i=1}^{N} yi^r$ inherent to the dataset, with $kq_i = \pi^{-1} k'_{q_i}$, where π^{-1} comprises the inverse of π's entries, compensating for potential sample selection biases arising from class imbalances.

For each specific sample, we ascertain instances manifesting maximal similarity using their respective distance metrics, and based on these proximate samples, we proceed to refine the associated labels. For every pixel, the foremost top-K nearest samples delineated by the designated distance metric are identified. We introduce a consistency metric, represented as c_i, which gauges the congruence between sample label $l_i^r = \arg\max jy_i^r(j)$ and the prediction sourced from KNNC:

$$c_i = \frac{k_{q_i}(l_i^r)}{\max jkq_i(j)} \quad (6)$$

This metric is derived by dividing the value of the distribution k_{q_i} corresponding to the label l_i^r by its predominant peak $\max jkq_i(j)$. A pronounced c_i value for a given sample x_i insinuates a consensus among its neighboring samples in favor of its prevailing label l_i^r, suggesting its likely accuracy. Applying a threshold θ_s to c_i, a pristine subset $(\mathcal{X}_c, \mathcal{Y}_c^r)$ is derived. By default, we utilize $\theta s = 0.65$, implying that a sample x_i is deemed pristine when the consensus, as reflected in kq_i, among its neighbors corroborates its extant label y_i^r.

In light of limited labeling, we propose a data augmentation strategy that capitalizes on the original features and labeled image pairs. Specifically, our approach adopts an offline data augmentation technique tailored for underrepresented or minor-category samples, ensuring that transformations are conducted on the training data prior to their introduction into the label correction module. Historically, popular data augmentation methodologies have included translation, image flipping, rotation, and cropping, as corroborated by Hernandez et al. [58] and Wong et al. [59]. In alignment with these practices, we implement five cardinal data augmentation operations, represented as AUG-i (where $i \in 1, \ldots, 4$): AUG-1 denotes horizontal flipping, AUG-2 implies a 90° clockwise rotation, AUG-3 indicates a 180° clockwise rotation, and AUG-4 pertains to a 270° clockwise rotation. Subsequently, each training image patch pair, $(x_i, y_i) \in (\mathcal{X}, \mathcal{Y})$, where $i \in 1, 2, \ldots, n$, is extended into a series of eight image patch pairs. These include (x_i, y_i), (x_i^{R90}, y_i), (x_i^{R180}, y_i), (x_i^{R270}, y_i), (x_i^F, y_i), (x_i^{FR90}, y_i), (x_i^{FR180}, y_i), and (x_i^{FR270}, y_i). The subsequent seven pairs in this sequence correspond to transformations driven by the operations AUG-1 through AUG-4.

With the refined labels in place, the primary objective of the classification module is the categorization of PolSAR data. A projection head is utilized within this module, projecting the representations gleaned from the network onto a dimensionality defined by the class number. This is mathematically represented as

$$d_i(\vec{x}) = \frac{\exp\left(\vec{W}_i^T \vec{x} + \vec{b}_i\right)}{\sum_{j=1}^{M} \exp\left(\vec{W}_j^T \vec{x} + \vec{b}_j\right)} \quad (7)$$

Here, $\vec{x}$ is the output of the projection head, with $\vec{W}_\bullet^T$ and $\vec{b}_\bullet$ representing the associated weight and bias, respectively. Furthermore, to effectively confront sample imbalance, we introduce a rebalancing loss, denoted as $\mathcal{L}_{CACE}$, encapsulated in Equation (5). The foundational loss function employed is the categorical cross-entropy [60], $\mathcal{L}_{CCE}$. The derivation of $\mathcal{L}_{CACE}$ necessitates averaging two error magnitudes, both of which are scaled by the categorical weight W.

$$\mathcal{L}_{CACE} = -W \times [\mathcal{L}_{CCE}(\vec{y}, y)], \tag{8}$$

Here, W is formulated as $[\frac{1}{N_1}, \frac{1}{N_2}, \ldots, \frac{1}{N_M}]^T$. In this equation, $\vec{y}$ symbolizes the predicted label, y stands for the ground truth label, and N_k represents the count of labels in the kth class.

5. Experimental Results

In this section, we provide a rigorous evaluation of our proposed method on four PolSAR datasets, both from quantitative and qualitative perspectives. We initially detail the experimental datasets and our chosen parameter settings in Section 5.1. In Section 5.2, an ablation study is presented to highlight the significance of the four pivotal components of our method: self-distillation backbone, noise label correction approach, sample rebalancing loss function, and augmented dataset.

For clarity, we elucidate that the classification average accuracy (AA) for a class is the proportion of accurately classified pixels for that class to the total pixels of the class, whereas the overall accuracy (OA) refers to the proportion of all correctly classified pixels in the entire image to the overall pixels in the image. Data with the highest accuracy are highlighted in bold for emphasis.

5.1. Experimental Data and Parameter Setting

Figure 3 offers visual insights, and Table 2 presents a summary of the PolSAR images employed in our experiments. The first dataset is composed of L-band four-look PolSAR data, captured by the NASA/JPL AIRSAR system over the Flevoland region, the Netherlands, in August 1989, whose PauliRGB image is portrayed in Figure 3(a1). Spanning an area of 750 × 1024 pixels, it offers a resolution of 6.6 m in the slant range and 12.1 m in the azimuth direction. The dataset delineates 15 distinct land cover classes, as illustrated in Figure 3(a2), with color codings that represent the legend of the ground truth map. The number of pixels for each class is listed as below: *Water* (12,671), *Barley* (7156), *Peas* (9111), *Stembean* (6103), *Beet* (10,050), *Forest* (14,822), *Bare soil* (3078), *Grass* (6269), *Rapeseed* (12,690), *Lucerne* (9477), *Wheat 1* (17,283), *Wheat 2* (10,591), *Wheat 3* (21,300), *Building* (476), and *Potato* (15,292). For model training, a random subset comprising 1% of the labeled samples is utilized. We then proceed to extract image patches of dimensions 12 × 12 × 6, where 12 × 12 signifies the window size and 6 represents the channel count.

Table 2. Summary of PolSAR datasets.

Dataset	Size	Spatial Resolution (m)	Bands	Classes
Flevoland (Dataset 1)	750 × 1024	6.6 × 12.1	L-band	15
Oberpfaffenhofen	1300 × 1200	1.5 × 1.8	L-band	3
Flevoland (Dataset 2)	1020 × 1024	6 × 12	L-band	14
San Francisco	1800 × 1380	3 to 100	C-band	5

Our second dataset, as illustrated in Figure 3(b1), comprises an E-SAR L-band image, which covers a 1300 × 1200 pixel area in the Oberpfaffenhofen region, Germany. This dataset includes several distinct land categories, with the number of pixels for each as follows: *Build-up* areas (333,955), *Wood Land* (265,516), and *Open Area* (760,769). Its diversity renders it apt for gauging the robustness of our method in varied landscapes. The ground truth class labels and their associated color legends for this area are delineated in Figure 3(b2), serving as a benchmark for our model's predictions and enabling classification accuracy quantification.

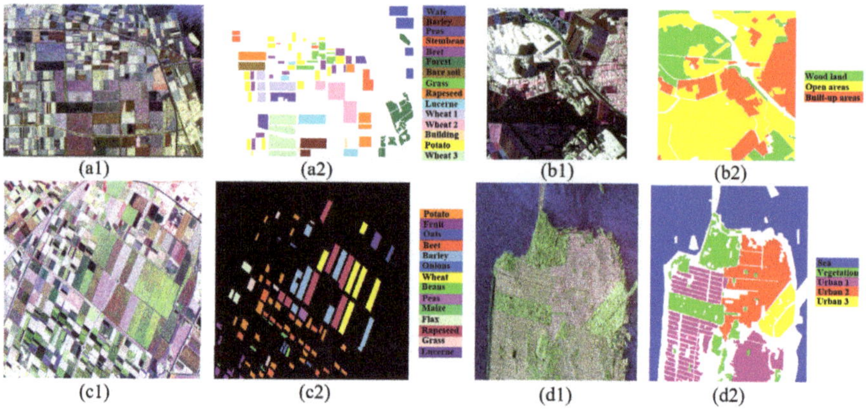

Figure 3. In this study, a series of experimental images were employed to rigorously assess the efficacy of our proposed method. The selected images encompass the following: Flevoland area dataset 1: (**a1**) A PauliRGB depiction of the region. (**a2**) The associated ground truth class labels, supplemented by their corresponding color codes. Oberpfaffenhofen Area Data Set: (**b1**) The PauliRGB representation of the aforementioned area. (**b2**) Ground truth class labels, paired with their relevant color codes. Flevoland area dataset 2: (**c1**) Another distinct PauliRGB portrayal from the Flevoland region. (**c2**) Its affiliated ground truth class labels, along with the matching color codes. San Francisco Area Data Set: (**d1**) The PauliRGB visualization of this iconic urban landscape. (**d2**) The ground truth class labels, harmonized with their specified color codes.

Figure 3(c1) showcases the third dataset: an L-band AIRSAR image captured over the Flevoland region in 1991. This dataset, spanning dimensions of 1020 × 1024 pixels, is indispensable for discerning the radar responses of different land cover types and augmenting our grasp of PolSAR data interpretation. Figure 3(c2) manifests the corresponding ground truth labels and color codings. This dataset, encapsulating 14 classes, is referred to as Flevoland area dataset 2 in Section 5. This dataset includes a diverse range of land types, with the number of pixels for each being *Potato* (21,539), *Fruit* (4062), *Oats* (1394), *Beet* (10,795), *Barley* (24,543), *Onions* (2130), *Wheat* (26,277), *Beans* (1082), *Peas* (2160), *Maize* (1290), *Flax* (4301), *Rapeseed* (28,235), *Grass* (4204), and *Bare Soil* (2952) pixels.

The fourth dataset entails a 25-look Radarsat-2 image of the San Francisco region from 2008, with a size of 1800 × 1380 pixels. This dataset features five classes, with the number of pixels for each being *Sea* (841,489), *Vegetation* (236,715), *Urban 1* (80,616), *Urban 2* (348,056), and *Urban 3* (282,975). Figure 3(d1) renders the PauliRGB image, while Figure 3(d2) displays the ground truth class labels. Notably, in Figure 3(d2), void regions are apparent, symbolizing unlabeled classes or interclass boundaries. These void zones are excluded from experimental consideration and analysis.

The optimization algorithm was parameterized with a learning rate (τ) set at 0.001, complemented by a momentum parameter of 0.9. During training, we utilized a batch size of 128. For all experiments, we initialized with a noisy label rate of 20%. All experiments were orchestrated within the TensorFlow framework, leveraging a Dell Z690 workstation equipped with a GeForce RTX 3090 GPU and a memory capacity of 64 GB.

5.2. Ablation Study

The proposed method, predicated on the robust self-distillation mechanism for correcting noisy labels, was rigorously tested on various prominent PolSAR images, as delineated in earlier sections. This study bifurcates into three critical experimental segments, each elucidating distinctive facets of the model's capabilities. Initially, the research accentuates the advantages of harnessing self-distillation for feature extraction, particularly when maneuvering high-dimensional vector distance computations in PolSAR imagery. For this purpose, two contrasting experimentations were devised: one incorporating contrastive learning and the other omitting it. To testify the effectiveness of each component of our proposed SDBCS, we conducted four groups of experiments as follows: We start with VGGNet-8 as our baseline, which trains directly on noisy-labeled samples. We then examine the influence of our label correction module with the VGGNet-8+CS model. Advancing further, SDVGGNet-8+CS enriches the previous model by adding self-distillation-based contrastive learning, aiming for enhanced feature extraction. The penultimate step in our experimental series, SDVGGNet-8+CS+Aug, integrates data augmentation into the SDVGGNet-8+CS framework to further improve the model's resilience to noisy data and enhance generalization. The culmination of our experimental series, the SDBCS framework, incorporates data augmentation and balanced loss into SDVGGNet-8+CS, specifically designed to overcome class imbalance and enhance the model's classification efficacy.

We leverage the Oberpfaffenhofen dataset to verify the efficacy of our method. Table 3 elucidates the foundational methodology, wherein a VGGNet-8 neural network was trained directly on the dataset, inclusive of the noise-labeled samples, sans any modification. This primary approach served as a litmus test for gauging model performance. Resultant accuracies across various classes were as follows: *Build-up* at 65.69%, *Wood Land* at 68.55%, and *Open Area* at 84.87%. Consequently, the OA was pegged at 76.98%, with an AA of 73.04%. The Precision, which indicates the accuracy of positive predictions, was recorded at 73.09%. The F1-Score, which balances precision and recall, was 73.05%, indicating a moderate balance in the model's ability to correctly identify classes and its robustness in terms of recall. The Kappa statistic, measuring agreement beyond chance, stood at 60.96%, suggesting a fair level of agreement. The Mean IoU, crucial for assessing the model's performance in segmenting classes, was 58.34%.

Table 3. OA values (%) of Oberpfaffenhofen area data for our proposed method.

Method	Build-up	Wood Land	Open Area	OA	AA
VGGNet-8	65.69	68.55	84.87	76.98	73.04
VGGNet-8+CS	63.89	76.28	87.02	79.25	75.73
SDVGGNet-8+CS	72.94	91.13	88.22	85.04	84.10
SDVGGNet-8+CS+Aug	81.07	92.19	87.48	86.82	86.91
SDBCS	79.08	89.12	92.38	88.48	86.86
Method	**Precision**	**F1-Score**	**Kappa**	**Mean IoU**	
VGGNet-8	73.09	73.05	60.96	58.34	
VGGNet-8+CS	75.15	75.37	64.85	61.30	
SDVGGNet-8+CS	81.57	82.58	75.08	70.91	
SDVGGNet-8+CS+Aug	84.02	85.25	78.23	74.84	
SDBCS	85.87	86.35	80.58	76.49	

5.2.1. Noisy Label Correction

To elevate the established baseline, we incorporated a correction mechanism into the VGGNet-8 model. For confident label determination, an intricate global distance matrix encompassing all pixels was constructed. The objective was to discern the most congruent pixels for each sample and subsequently adopt the predominant label within its pixel cohort. Each training sample was aligned to the label of the nearest k training data points. After computing the feature distance, the samples were sorted based on proximity. This strategy was devised to counteract the detriments of noisy labels and bolster classification accuracy. Despite inevitable trade-offs, the method showcased an upswing in performance. The achieved accuracies were *Build-up* at 63.89%, *Wood Land* at 76.28%, and *Open Area* at

87.02%. OA increased to 79.25%, with an AA of 75.73%. Additionally, precision improved to 75.15%, F1-Score to 75.37%, Kappa to 64.85%, and Mean IoU to 61.30%.

5.2.2. Self-Distillation Feature Extraction

The model's performance was further augmented by embedding a self-distillation technique, thereby enabling the model to introspectively refine from its own predictions. This adaptation yielded notable enhancements, with the accuracies for *Build-up*, *Wood Land* and *Open Area* classes registering at 72.94%, 91.13%, and 88.22%, respectively. The OA marked an impressive 85.04%, culminating in an AA of 84.10%. Precision increased to 81.57%, F1-Score to 82.58%, Kappa to 75.08%, and Mean IoU to 70.91%.

In the realm of PolSAR data analysis, initial steps involve feature extraction from PolSAR data using VGGNet-8 and self-distillation methodologies. A supplementary set termed VGGNet-8, was introduced for a comparative evaluation, which essentially trains without noise labels. These methodologies illuminate intricate relationships within the data rendering high-dimensional features that encapsulate pivotal backscattering properties of PolSAR data. The subsequent phase emphasizes dimensionality mitigation.

Efficient dimensionality reduction is pivotal for interpreting high-dimensional data One salient technique in this domain is t-distributed Stochastic Neighbor Embedding (t-SNE) [61], a sophisticated nonlinear algorithm grounded in neighborhood graphs, tailored to preserve the data's intrinsic local structure. This is achieved by t-SNE's transformation of interpoint distances into congruent probability distributions spanning various dimensions Leveraging t-SNE, we embarked on visualizing both raw and quantized feature spaces Given its design as an unsupervised algorithm tailored for dimensionality reduction and 3D data projection, t-SNE demonstrates exceptional prowess in rendering visualizations of intricate, high-dimensional datasets, thereby enhancing the interpretation of PolSAR data The utility of t-SNE is further accentuated when amalgamated with visual aids like scatter plots and pseudocolor images, facilitating a lucid conveyance of intricate data relationships and patterns.

Figure 4 presents a detailed visual exposition of the spatial and polarimetric attributes across three preselected regions from the Oberpfaffenhofen dataset. The visualization unmistakably illustrates a clear delineation of three terrain typologies within the three-dimensional space charted by t-SNE. In particular, Figure 4a,d underscores the aptitude of the feature extraction network, shining light on its innate ability to capture and epitomize the quintessential characteristics of terrain surfaces. A deeper foray into Figure 4b,c provides a comparative purview against Figure 4a. Significantly, Figure 4c, harmonized with the self-distillation paradigm, exhibits heightened alignment with the ground truth, especially in the positionings pertaining to the three distinct categories.

In parallel with our assessment of Figure 4d, a detailed comparative analysis is presented in Figure 4e,f, bringing forth salient observations. Notably, Figure 4f, emblematic of the self-distillation-based method, highlights a pronounced aggregation in the central positions associated with various categories. This stands in stark contrast to the more scattered distribution observed within the VGGNet-8 influenced outcomes, as delineated in Figure 4e. Collectively, these observations underscore the superior discriminative capacity of the self-distillation approach, adeptly capturing inherent class distinctiveness and intricate intercategory dynamics. This fortifies the assertion of its pivotal role in elevating feature representation in the analyzed PolSAR dataset.

5.2.3. Data Augmentation and Balanced Loss

We further refined the SDVGGNet-8+CS model by integrating data augmentation, resulting in the SDVGGNet-8+CS+Aug configuration. This intermediate step was crucial in assessing the incremental benefits brought by data augmentation to the self-distillation process. The SDVGGNet-8+CS+Aug model demonstrated a significant improvement in dealing with noisy data and generalization capabilities, as evidenced by the following accuracies: *Build-up* at 81.07%, *Wood Land* at 92.19%, and *Open Area* at 87.48%. The OA and AA were recorded at 86.82% and 86.91%, respectively. Additionally, the model saw improvements in Precision (84.02%), F1-Score (85.25%), Kappa (78.23%), and Mean IoU (74.84%). These advancements highlight the method's impact in not only improving accuracy but also precision, consistency, and segmentation effectiveness.

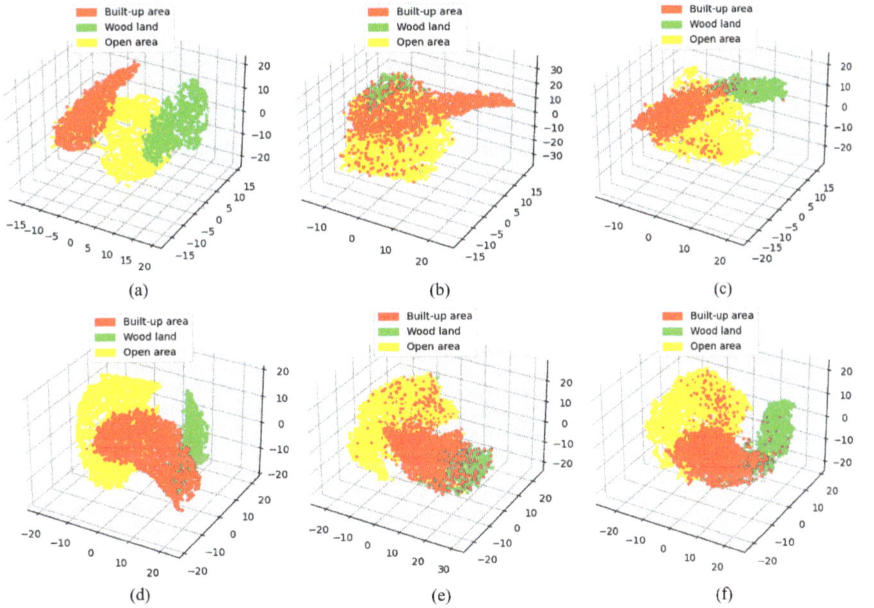

Figure 4. The presented figures showcase t-SNE plots derived from the Oberpfaffenhofen images. For enhanced visualization fidelity, the dataset is judiciously bifurcated into two subsets according to their sample proportions. Specifically, subsets (**a–c**) constitute 0.5% of the overarching samples, whereas subsets (**d–f**) account for 1%. In terms of methodological delineation, subsets (**a**,**d**) resonate with the features from the VGGNet-8 backbone trained devoid of noise labels, and subsets (**b**,**e**) are aligned with the VGGNet-8 training approach. Conclusively, subsets (**c**,**f**) are emblematic of the feature extraction facilitated through the self-distillation-based paradigm. Such a structured presentation aids in an in-depth comparison and assessment of the respective methodologies across varied sample sizes.

As illustrated in Table 4, we explored different loss functions in order to find a robust option. The studied loss functions include $\mathcal{L}_{CCE}$ [60], Label Smoothing Categorical Cross-Entropy Loss [62] ($\mathcal{L}_{SCCE}$), Focal Loss [63] ($\mathcal{L}_{focal}$), and our proposed $\mathcal{L}_{CACE}$.

It is evident that both $\mathcal{L}_{SCCE}$ and $\mathcal{L}_{focal}$ demonstrate promising results under certain parameter settings. However, it is crucial to note that slight changes in their parameters can lead to significant drops in classification performance. For instance, when the ϵ parameter in $\mathcal{L}_{SCCE}$ changes from 0.3 to 0.2, there is a notable decrease in OA by 2.11%. Similarly, in Focal Loss, a change in the γ parameter from 1.8 to 2.0 results in a reduction in OA by 2.17%. This sensitivity to parameter adjustments indicates that both $\mathcal{L}_{SCCE}$ and Focal Loss may not be robust across different categories or datasets, as their effectiveness heavily relies on fine-tuning specific parameters.

In contrast, our $\mathcal{L}_{CACE}$, meticulously designed to overcome the limitations of existing methods, demonstrated remarkable results. Significantly, $\mathcal{L}_{CACE}$ stands out due to its parameter-free design, eliminating the need for meticulous parameter tuning that plagues other loss functions. This unique feature enhances its robustness, making it exceptionally suitable for a wide range of PolSAR datasets. It achieved impressive classification accuracies and showcased enhanced Precision (85.87%), F1-Score (86.35%), Kappa (80.58%), and Mean

IoU (76.49%). The absence of parameters in $\mathcal{L}_{CACE}$ not only simplifies its application but also ensures consistent performance across various scenarios in PolSAR datasets.

In conclusion, this investigative endeavor presents a holistic exploration of innovative methodologies tailored for optimizing neural-network-centric classifiers within the remote sensing land cover classification domain. The empirical findings highlight the paramount importance of bespoke strategies, especially when confronting challenges like label noise and constrained data availability. The integration of self-distillation, data augmentation and balanced loss within the SDBCS framework emerges as a testament to this. Such revelations not only augment our contemporary understanding of effective strategies within this discipline but also establish an empirical benchmark, poised to guide and inspire subsequent research trajectories in analogous domains.

Table 4. Performance comparison of different loss functions with SDVGGNet-8+CS+Aug architecture on Oberpfaffenhofen area data, utilizing 0.05% of ground truth labels as the training set.

Loss Function	Build-up	Wood Land	Open Area	OA	AA
$\mathcal{L}_{CCE}$	81.07	92.19	87.48	86.82	86.91
$\mathcal{L}_{SCCE}$ $\epsilon(0.2)$	81.42	**92.94**	87.58	87.11	87.31
$\mathcal{L}_{SCCE}$ $\epsilon(0.3)$	80.97	91.15	92.16	**89.22**	**88.09**
$\mathcal{L}_{SCCE}$ $\epsilon(0.5)$	**81.82**	92.23	87.75	87.17	87.27
$\mathcal{L}_{focal}$ $\gamma(2.0)$ $\alpha(0.37)$	81.10	90.82	88.02	86.87	86.65
$\mathcal{L}_{focal}$ $\gamma(2.0)$ $\alpha(0.50)$	81.10	91.26	87.97	86.93	86.78
$\mathcal{L}_{focal}$ $\gamma(1.8)$ $\alpha(0.50)$	75.87	90.93	**94.27**	89.10	87.02
$\mathcal{L}_{CACE}$	79.08	89.12	92.38	88.48	86.86
Loss Function	**Precision**	**F1-Score**	**Kappa**	**Mean IoU**	
$\mathcal{L}_{CCE}$	84.02	85.25	78.23	74.84	
$\mathcal{L}_{SCCE}$ $\epsilon = 0.2$	84.17	85.50	78.73	75.17	
$\mathcal{L}_{SCCE}$ $\epsilon = 0.3$	86.39	**87.18**	**81.92**	**77.73**	
$\mathcal{L}_{SCCE}$ $\epsilon = 0.5$	84.24	85.52	78.82	75.21	
$\mathcal{L}_{focal}$ $\gamma(2.0)$ $\alpha(0.37)$	84.13	85.21	78.25	74.78	
$\mathcal{L}_{focal}$ $\gamma(2.0)$ $\alpha(0.50)$	84.09	85.23	78.39	74.84	
$\mathcal{L}_{focal}$ $\gamma(1.8)$ $\alpha(0.50)$	**86.53**	86.69	81.51	77.06	
$\mathcal{L}_{CACE}$	85.87	86.35	80.58	76.49	

6. Discussion

In this section, we provide a rigorous evaluation of our proposed method on four PolSAR datasets, both from quantitative and qualitative perspectives. Section 6.1 delves into a sensitivity analysis, assessing the robustness of the proposed SDBCS framework on the Oberpfaffenhofen dataset. Section 6.2 furnishes a comparison between our proposed method and four contemporary state-of-the-art competitors employing deep learning techniques, namely Sel-CL [64], SSR [65], PASGS [22], and Auto-PASGS [22].

6.1. Sensitivity Analysis

To elucidate the robustness of the proposed method across varying proportions of training data, this section meticulously evaluates the SDBCS framework on the Oberpfaffenhofen dataset. This dataset served as a canvas for a rigorous appraisal of the land cover classification efficacy of SDBCS, with the Average Correction Rate (ACR) as the evaluation metric. Table 5 presents the corrected label rate for three principal land cover classes—*Build-up* area, *Wood Land*, and *Open Area*—across 0.05%, 0.1%, and 0.2% data proportions.

In juxtaposition with Sel-CL and SSR, the supremacy of SDBCS was consistently evident. It is noteworthy that, particularly in the *Build-up* area class, SDBCS was adept at maintaining commendable classification accuracy, even with limited data, underscoring its potent capacity for generalization relative to other methods. A salient aspect of the study was the discernible prowess of SDBCS in classifying the *Wood Land* segment, even when confronted with constrained data volumes. For the *Open Area* category, SDBCS's consistency in distinguishing between diverse land cover types was evident, signifying its resilience and robustness in comparison with alternative methodologies.

Table 5. Comparative analysis of the performance of SDBCS on varying proportions of the Oberpfaffenhofen dataset (Corrected label rates %).

	0.05%			
	Build-up	**WoodLand**	**OpenArea**	**ACR**
Initial	81.82	79.84	79.13	80.01
Sel-CL	83.96	81.45	84.01	83.53
SSR	83.95	82.26	83.74	83.53
SDBCS	85.56	93.55	88.35	88.53
	0.1%			
	Build-up	**WoodLand**	**OpenArea**	**ACR**
Initial	80.39	78.24	80.43	80.01
Sel-CL	85.36	82.05	83.83	83.90
SSR	85.64	82.06	81.11	82.50
SDBCS	83.15	96.18	84.78	86.54
	0.2%			
	Build-up	**WoodLand**	**OpenArea**	**ACR**
Initial	81.40	78.23	79.96	80.00
Sel-CL	82.98	90.37	80.96	83.27
SSR	83.55	88.63	80.89	83.05
SDBCS	83.69	97.69	84.29	86.69

SDBCS's consistently superior performance, relative to Sel-CL and SSR, across categories and proportions, accentuates the method's robustness and efficiency. Its capacity to sustain high accuracy, especially evident in the *Wood Land* category, underscores its potential for precise classification even in resource-constrained scenarios.

6.2. Results and Comparisons

Figure 5 provides a visual representation of the efficacy of each method across the Flevoland area dataset 1, Oberpfaffenhofen dataset, Flevoland area dataset 2, and San Francisco dataset. Following this visual exploration, an intricate analysis aligned with the associated tables is provided. Table 6 furnishes an exhaustive evaluation of the experimental outcomes from the Flevoland area dataset 1. Our SDBCS method is benchmarked against the prevalent state-of-the-art techniques: Sel-CL, SSR, PASGS, and Auto-PASGS. The core of this evaluation revolves around classification accuracy across diverse land cover categories, elucidating the subtle yet pivotal advantages proffered by SDBCS.

Table 6. Classification performances (%) of Flevoland area dataset 1 for the proposed method.

Method	Stembeans	Peas	Forest	Lucerne	Wheat	Beet
Sel-CL	85.94	85.35	89.88	93.76	90.08	83.23
SSR	87.87	87.25	89.01	93.66	93.31	80.74
PASGS	93.77	94.66	99.26	93.01	93.82	86.85
Auto-PASGS	96.03	92.32	97.39	95.03	96.38	89.27
SDBCS	93.48	85.62	97.23	91.53	96.88	79.91
Method	**Potatoes**	**Bare Soil**	**Grass**	**Rapeseed**	**Barley**	**Wheat 2**
Sel-CL	81.48	82.09	71.01	60.42	94.49	69.82
SSR	84.33	84.73	66.56	57.22	92.27	64.69
PASGS	92.52	81.03	70.87	70.55	95.08	81.72
Auto-PASGS	95.57	34.19	80.85	75.31	95.40	78.01
SDBCS	94.29	99.94	88.53	84.00	99.68	92.03

Table 6. *Cont.*

Method	Wheat 3	Water	Building	OA	AA	Precision
Sel-CL	95.01	87.59	72.68	84.81	82.86	79.94
SSR	94.23	76.51	71.63	83.37	81.61	79.04
PASGS	95.65	90.56	40.76	89.74	85.34	84.21
Auto-PASGS	83.92	95.58	59.24	88.88	84.30	89.93
SDBCS	97.52	83.14	84.24	91.87	91.21	88.89
Method	**F1-Score**	**Kappa**	**Mean IoU**			
Sel-CL	80.43	83.36	69.46			
SSR	79.34	81.67	67.57			
PASGS	84.52	88.83	75.78			
Auto-PASGS	85.61	87.86	76.50			
SDBCS	89.47	91.13	81.64			

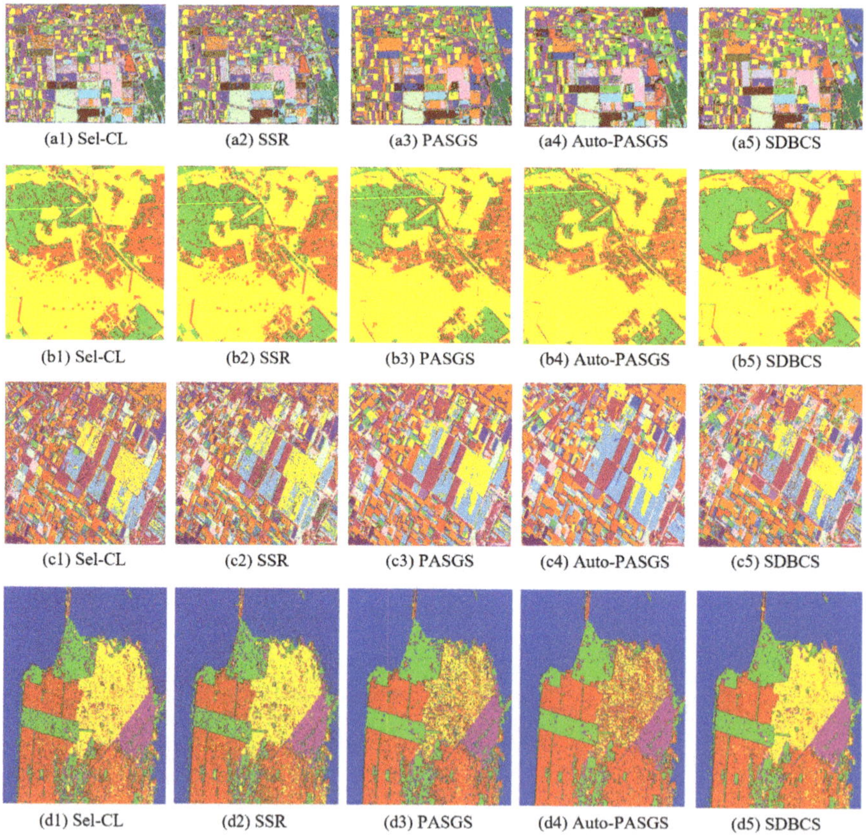

Figure 5. The figures presented offer a comprehensive visualization of the following: (**a1–a5**) class label predictions for the Flevoland area dataset 1, as forecasted by Sel-CL [64], SSR [65], PASGS [22], Auto-PASGS [22], and SDBCS. Subsequently, (**b1–b5**) showcases outcomes from the Oberpfaffenhofen dataset, (**c1–c5**) presents findings associated with the Flevoland area dataset 2, and, lastly, the San Francisco area dataset is elucidated in (**d1–d5**). Such systematic representation facilitates an insightful comparison and evaluation across the diverse methodologies and datasets.

Dissecting individual land cover classes reveals the consistent preeminence of SDBCS. As a case in point, within the *Stembeans* category, SDBCS registers a commendable accuracy of 93.48%, surpassing Sel-CL and SSR, which have accuracies of 85.94% and

87.87%, respectively. Further, SDBCS achieves accuracies of 99.68% for *Barley* and 84.00% for *Rapeseed*, outperforming its competitors. This performance accentuates the capability of SDBCS to address intricate and multifaceted land cover types. Aggregating results across all classes, SDBCS achieves a commendable OA of 91.87%, overshadowing Sel-CL (84.81%), SSR (83.37%), PASGS (89.74%), and Auto-PASGS (88.88%). SDBCS not only excels in overall accuracy but also demonstrates superior performance in other metrics. It attains the highest Precision (88.89%), F1-Score (89.47%), Kappa (91.13%), and Mean IoU (81.64%).

These empirical findings highlight SDBCS's paramount stance in land cover classification, particularly amidst noise-induced challenges. Its unwavering performance across a range of land cover categories substantiates its potential to enhance the accuracy of land cover classification in remote sensing.

Table 7 illustrates the performance metrics of various agricultural land cover classification methodologies applied to the 1% Flevoland area dataset 2. SDBCS emerges as the superior method, outstripping competitors across several categories. It achieves stellar accuracy rates, exemplified by *Potatoes* (98.33%) and *Beet* (94.95%), underscoring its finesse in discerning pivotal agricultural variants. Its proficiency further extends to nuanced categories like *Oats* (92.47%) and *Barley* (81.80%). When compared with methods like Sel-CL, SSR, PASGS, and Auto-PASGS, SDBCS's superiority in accuracy remains evident. This exemplary performance, even in formidable land cover classes like *Bare Soil* (94.17%) and *Rapeseed* (96.70%), reinforces SDBCS's promise in remote sensing agricultural land cover classification. Additionally, SDBCS demonstrates robust performance in Precision (83.33%), F1-Score (85.88%), Kappa (90.47%), and Mean IoU (76.64%).

Table 8 sheds light on the performance assessment of multiple methodologies, including Sel-CL, SSR, PASGS, Auto-PASGS, and our proposed SDBCS, applied to the 0.05% San Francisco area dataset. This dataset focuses on diverse land cover classifications, including *Sea*, *Vegetation*, and three *urban* categories. The results in the table accentuate SDBCS's commendable adaptability, especially under sample-limited circumstances. While methods like Sel-CL and SSR display varying accuracies, Auto-PASGS manifests an intriguing trend, exhibiting a high accuracy for one category but faltering in others. SDBCS leads with the highest Precision (87.24%), F1-Score (88.09%), Kappa (88.50%), and Mean IoU (79.22%) SDBCS, however, consistently exhibits robustness across distinct land cover types, further cementing its efficacy in challenging classification scenarios.

Table 7. Classification performances (%) of the Flevoland area dataset 2 for the proposed method.

Method	Potatoes	Fruit	Oats	Beet	Barley	Onions
Sel-CL	87.93	90.10	73.60	91.56	82.98	42.39
SSR	85.10	89.96	64.56	92.02	84.97	59.81
PASGS	97.33	89.19	85.08	93.09	93.04	29.53
Auto-PASGS	99.57	98.01	88.95	91.18	96.33	40.47
SDBCS	98.33	89.36	92.47	94.95	81.80	75.77
Method	**Wheat**	**Beans**	**Peas**	**Maize**	**Flax**	**Rapeseed**
Sel-CL	86.57	70.89	91.20	72.25	94.12	89.87
SSR	86.45	71.90	81.99	66.59	88.17	85.19
PASGS	92.14	78.28	97.36	63.88	91.86	94.05
Auto-PASGS	80.48	22.09	100	90.47	93.75	96.15
SDBCS	91.89	72.83	99.95	83.80	92.84	96.70
Method	**Grass**	**Bare Soil**	**OA**	**AA**	**Precision**	**F1-Score**
Sel-CL	85.01	95.05	86.69	82.39	71.05	75.17
SSR	79.76	91.73	85.19	80.59	69.65	73.53
PASGS	75.95	91.67	91.64	83.75	83.84	82.50
Auto-PASGS	74.43	94.68	91.01	83.33	85.75	82.58
SDBCS	88.25	94.17	91.87	89.51	83.33	85.88

Table 7. *Cont.*

Method	Kappa	Mean IoU
Sel-CL	84.49	62.85
SSR	82.76	60.75
PASGS	90.16	73.32
Auto-PASGS	89.42	74.02
SDBCS	90.47	76.64

Table 8. Classification performances (%) of San Francisco area data for the proposed method.

Method	Sea	Vegetation	Urban 2	Urban 3	Urban 1	OA	AA
Sel-CL	91.13	78.43	75.30	82.28	81.38	84.53	81.70
SSR	91.22	75.89	75.49	81.83	81.86	84.23	81.26
PASGS	99.89	88.14	82.57	66.61	84.5	89.01	84.34
Auto-PASGS	99.70	86.98	90.54	53.40	88.42	88.41	83.81
SDBCS	99.75	86.53	80.96	86.57	93.84	92.00	89.53
Method	**Precision**	**F1-Score**	**Kappa**	**Mean IoU**			
Sel-CL	76.63	78.60	78.19	65.74			
SSR	76.14	78.14	77.76	65.14			
PASGS	84.18	83.87	84.07	73.13			
Auto-PASGS	83.66	82.29	83.25	71.35			
SDBCS	87.24	88.09	88.50	79.22			

Table 9 provides an analytical performance overview of diverse methodologies on the 0.05% Oberpfaffenhofen area dataset. Our model, SDBCS, consistently excels, outpacing counterparts in pivotal categories. Illustratively, in the *Build-up* category, SDBCS achieves an accuracy of 79.08%, superseding Sel-CL's 69.32%. In the *Wood Land* category, SDBCS's accuracy peaks at 89.12%, transcending SSR's 78.47%. Notably, in the OA metric, SDBCS's performance at 88.48% distinctly eclipses Sel-CL's 84.55% and SSR's 82.63%. SDBCS's prowess becomes manifest in the AA metric, with an accuracy of 86.86%, surpassing both SSR's 77.99% and PASGS's 78.70%. Furthermore, the superiority of SDBCS is underlined by its leading Precision of 85.88%, F1-Score of 86.35%, Kappa of 80.58%, and Mean IoU of 76.49%.

In summation, these empirical outcomes robustly underscore the innate capability of SDBCS to adeptly navigate challenges engendered by noisy labels, restricted sample sizes, and a gamut of land cover classifications. The intrinsic proficiency of SDBCS in rectifying label inaccuracies and capitalizing on limited annotations underscores its pivotal role in the evolutionary trajectory of research within remote sensing, with a particular emphasis on land cover classification endeavors.

6.3. Limitations and Enhancements

In Table 3, when the VGGNet-8 model was enhanced with our label correction module, there was a drop in prediction accuracy for the built-up land type. This decline can be linked to the unique properties of built-up areas in the Oberpfaffenhofen dataset, which are characterized by complex spatial structures and varied spectral signatures. The label correction process involves aligning each training sample with the label of its nearest k training data points based on feature distance. However, due to the spectral resemblance of some built-up areas to other land types, mislabeling may occur. This challenge is intrinsic to handling complex urban environments in remote sensing imagery.

While our method effectively addresses noisy labels, its performance might be influenced by the quality of the feature representations. If the feature extraction process fails to adequately distinguish between different classes, the label correction might not be as effective.

Table 9. Classification performances (%) of Oberpfaffenhofen area data for the proposed method.

Method	Build-up	Wood Land	Open Area	OA	AA	Precision
Sel-CL	69.32	79.14	93.13	84.55	80.53	81.46
SSR	62.66	78.47	92.84	82.63	77.99	79.42
PASGS	61.20	79.55	95.36	83.89	78.70	80.42
Auto-PASGS	59.24	75.99	95.87	82.99	77.03	79.73
SDBCS	79.08	89.12	92.38	88.48	86.86	85.88
Method	**F1-Score**	**Kappa**	**Mean IoU**			
Sel-CL	80.96	73.51	68.83			
SSR	78.57	70.00	65.73			
PASGS	79.29	72.07	66.88			
Auto-PASGS	78.10	70.27	65.40			
SDBCS	86.35	80.58	76.49			

To improve the performance in boundary areas and in general, we could consider integrating additional context-aware mechanisms. For instance, incorporating attention mechanisms could enable the model to focus on more relevant features, thereby improving the accuracy of the label correction, especially in complex regions. Another potential enhancement is to use multiscale feature representations. This approach could help in capturing both fine-grained details and broader contextual information, thereby improving the model's ability to handle diverse and challenging scenarios, including boundary regions.

7. Conclusions

Confronting the complexities of PolSAR image classification, our study introduces a novel label correction approach, designed for managing noisy labels, and leverages unsupervised contrastive learning to enhance polarimetric representation ability and further classification accuracy in label scarcity scenarios. The innovative label correction technique we developed employs similarities among training samples with a feature distance matrix derived from contrastive learning, which identifies and rectifies mislabeled samples, thereby addressing the noisy label issue. In addition, by adopting self-supervised representation learning, we significantly enhance the model's robustness and accuracy, especially in the context of limited labels in PolSAR image classification. Our method also includes strategic rebalancing and data augmentation techniques to tackle the class imbalance problem, improving the classification accuracy of minority classes. Extensive evaluations on four benchmark datasets have proven the effectiveness and superiority of the proposed method. To sum up, our approach effectively improves the accuracy and robustness of DNN-based PolSAR image classification methods in noisy and sparse label scenarios, addressing the initial challenges we set out to overcome.

Author Contributions: Conceptualization, N.W. and H.B.; Methodology, N.W. and H.B.; Software, N.W.; Validation, N.W., H.B. and F.L.; Formal analysis, H.B.; Investigation, N.W., H.B. and C.X.; Resources, H.B., F.L., C.X. and J.G.; Data curation, H.B., C.X. and J.G.; Writing—original draft, N.W.; Writing—review & editing, H.B.; Visualization, H.B., F.L. and C.X.; Supervision, H.B., F.L., C.X. and J.G.; Project administration, H.B. and J.G.; Funding acquisition, H.B. and J.G. All authors have read and agreed to the published version of the manuscript.

Funding: This work was supported by the National Key R&D Program of China under Grant 2022YFA1003800, NSFC under Grant 42201394, Major Key Project of Peng Cheng Laboratory under Grant PCL2023AS1-2, and Qinchuangyuan High-level Innovation and Entrepreneurial Talent Program under Grant QCYRCXM-2022-30.

Data Availability Statement: No new data were created or analyzed in this study. Data sharing is not applicable to this article.

Conflicts of Interest: The authors declare no conflict of interest.

References

1. Chen, S.W.; Tao, C.S. PolSAR image classification using polarimetric-feature-driven deep convolutional neural network. *IEEE Geosci. Remote Sens. Lett.* **2018**, *15*, 627–631. [CrossRef]
2. Lee, J.S.; Grunes, M.R.; Ainsworth, T.L.; Du, L.J.; Schuler, D.L.; Cloude, S.R. Unsupervised classification using polarimetric decomposition and the complex Wishart classifier. *IEEE Trans. Geosci. Remote Sens.* **1999**, *37*, 2249–2258.
3. Van Zyl, J.J. Unsupervised classification of scattering behavior using radar polarimetry data. *IEEE Trans. Geosci. Remote Sens.* **1989**, *27*, 36–45. [CrossRef]
4. Bi, H.; Sun, J.; Xu, Z. A graph-based semisupervised deep learning model for PolSAR image classification. *IEEE Trans. Geosci. Remote Sens.* **2018**, *57*, 2116–2132. [CrossRef]
5. Yu, P.; Qin, A.K.; Clausi, D.A. Unsupervised polarimetric SAR image segmentation and classification using region growing with edge penalty. *IEEE Trans. Geosci. Remote Sens.* **2011**, *50*, 1302–1317. [CrossRef]
6. Chen, Q.; Kuang, G.; Li, J.; Sui, L.; Li, D. Unsupervised land cover/land use classification using PolSAR imagery based on scattering similarity. *IEEE Trans. Geosci. Remote Sens.* **2012**, *51*, 1817–1825. [CrossRef]
7. Tu, S.T.; Chen, J.Y.; Yang, W.; Sun, H. Laplacian eigenmaps-based polarimetric dimensionality reduction for SAR image classification. *IEEE Trans. Geosci. Remote Sens.* **2011**, *50*, 170–179. [CrossRef]
8. Ersahin, K.; Scheuchl, B.; Cumming, I. Incorporating texture information into polarimetric radar classification using neural networks. In Proceedings of the IGARSS 2004, 2004 IEEE International Geoscience and Remote Sensing Symposium, Anchorage, AK, USA, 20–24 September 2004; Volume 1.
9. Wu, Y.; Ji, K.; Yu, W.; Su, Y. Region-based classification of polarimetric SAR images using Wishart MRF. *IEEE Geosci. Remote Sens. Lett.* **2008**, *5*, 668–672. [CrossRef]
10. Bi, H.; Yao, J.; Wei, Z.; Hong, D.; Chanussot, J. PolSAR Image Classification Based on Robust Low-Rank Feature Extraction and Markov Random Field. *IEEE Geosci. Remote Sens. Lett.* **2022**, *19*, 4005205. [CrossRef]
11. Bi, H.; Xu, F.; Wei, Z.; Xue, Y.; Xu, Z. An active deep learning approach for minimally supervised PolSAR image classification. *IEEE Trans. Geosci. Remote Sens.* **2019**, *57*, 9378–9395. [CrossRef]
12. Xu, Y.; Li, Z.; Li, W.; Du, Q.; Liu, C.; Fang, Z.; Zhai, L. Dual-channel residual network for hyperspectral image classification with noisy labels. *IEEE Trans. Geosci. Remote Sens.* **2021**, *60*, 5502511. [CrossRef]
13. Xiao, T.; Xia, T.; Yang, Y.; Huang, C.; Wang, X. Learning from massive noisy labeled data for image classification. In Proceedings of the IEEE Conference on Computer Vision and Pattern Recognition, Boston, MA, USA, 7–12 June 2015; pp. 2691–2699.
14. Lee, K.H.; He, X.; Zhang, L.; Yang, L. Cleannet: Transfer learning for scalable image classifier training with label noise. In Proceedings of the IEEE Conference on Computer Vision and Pattern Recognition, Salt Lake City, UT, USA, 18–23 June 2018; pp. 5447–5456.
15. Goldberger, J.; Ben-Reuven, E. Training deep neural-networks using a noise adaptation layer. In Proceedings of the International Conference on Learning Representations, San Juan, Puerto Rico, 2–4 May 2016.
16. Han, J.; Luo, P.; Wang, X. Deep self-learning from noisy labels. In Proceedings of the IEEE/CVF International Conference on Computer Vision, Seoul, Republic of Korea, 27 October–2 November 2019; pp. 5138–5147.
17. Kim, Y.; Yim, J.; Yun, J.; Kim, J. Nlnl: Negative learning for noisy labels. In Proceedings of the IEEE/CVF International Conference on Computer Vision, Seoul, Republic of Korea, 27 October–2 November 2019; pp. 101–110.
18. Ma, X.; Huang, H.; Wang, Y.; Romano, S.; Erfani, S.; Bailey, J. Normalized loss functions for deep learning with noisy labels. In Proceedings of the International Conference on Machine Learning, PMLR, Virtual, 13–18 July 2020; pp. 6543–6553.
19. Grill, J.B.; Strub, F.; Altché, F.; Tallec, C.; Richemond, P.; Buchatskaya, E.; Doersch, C.; Avila Pires, B.; Guo, Z.; Gheshlaghi Azar, M.; et al. Bootstrap your own latent-a new approach to self-supervised learning. In Proceedings of the Advances in Neural Information Processing Systems, Virtual, 6–12 December 2020; Volume 33, pp. 21271–21284.
20. Li, Y.; Xing, R.; Jiao, L.; Chen, Y.; Chai, Y.; Marturi, N.; Shang, R. Semi-supervised PolSAR image classification based on self-training and superpixels. *Remote Sens.* **2019**, *11*, 1933. [CrossRef]
21. Han, B.; Yao, Q.; Yu, X.; Niu, G.; Xu, M.; Hu, W.; Tsang, I.; Sugiyama, M. Co-teaching: Robust training of deep neural networks with extremely noisy labels. In Proceedings of the Advances in Neural Information Processing Systems, Montréal, QC, Canada, 3–8 December 2018; Volume 31.
22. Ni, J.; Xiang, D.; Lin, Z.; López-Martínez, C.; Hu, W.; Zhang, F. DNN-based PolSAR image classification on noisy labels. *IEEE J. Sel. Top. Appl. Earth Obs. Remote Sens.* **2022**, *15*, 3697–3713. [CrossRef]
23. Hou, B.; Wu, Q.; Wen, Z.; Jiao, L. Robust semisupervised classification for PolSAR image with noisy labels. *IEEE Trans. Geosci. Remote Sens.* **2017**, *55*, 6440–6455. [CrossRef]
24. Qiu, W.; Pan, Z.; Yang, J. Few-Shot PolSAR Ship Detection Based on Polarimetric Features Selection and Improved Contrastive Self-Supervised Learning. *Remote Sens.* **2023**, *15*, 1874. [CrossRef]
25. Zhang, W.; Pan, Z.; Hu, Y. Exploring PolSAR images representation via self-supervised learning and its application on few-shot classification. *IEEE Geosci. Remote Sens. Lett.* **2022**, *19*, 4512605. [CrossRef]
26. Zhang, L.; Zhang, S.; Zou, B.; Dong, H. Unsupervised deep representation learning and few-shot classification of PolSAR images. *IEEE Trans. Geosci. Remote Sens.* **2020**, *60*, 5100316. [CrossRef]
27. Zhang, P.; Liu, C.; Chang, X.; Li, Y.; Li, M. Metric-based Meta-Learning Model for Few-Shot PolSAR Image Terrain Classification. In Proceedings of the 2021 CIE International Conference on Radar (Radar), Haikou, China, 15–19 December 2021; pp. 2529–2533.

28. Bi, H.; Xu, F.; Wei, Z.; Han, Y.; Cui, Y.; Xue, Y.; Xu, Z. Unsupervised PolSAR image factorization with deep convolutional networks. In Proceedings of the IGARSS 2019–2019 IEEE International Geoscience and Remote Sensing Symposium, Yokohama, Japan, 28 July–2 August 2019; pp. 1061–1064.
29. Hu, J.; Hong, D.; Zhu, X.X. MIMA: MAPPER-induced manifold alignment for semi-supervised fusion of optical image and polarimetric SAR data. *IEEE Trans. Geosci. Remote Sens.* **2019**, *57*, 9025–9040. [CrossRef]
30. Xin, X.; Li, M.; Wu, Y.; Zheng, M.; Zhang, P.; Xu, D.; Wang, J. Semi-Supervised Classification of Dual-Frequency PolSAR Image Using Joint Feature Learning and Cross Label-Information Network. *IEEE Trans. Geosci. Remote Sens.* **2022**, *60*, 5235716. [CrossRef]
31. Wei, B.; Yu, J.; Wang, C.; Wu, H.; Li, J. PolSAR image classification using a semi-supervised classifier based on hypergraph learning. *Remote Sens. Lett.* **2014**, *5*, 386–395. [CrossRef]
32. Liu, W.; Yang, J.; Li, P.; Han, Y.; Zhao, J.; Shi, H. A novel object-based supervised classification method with active learning and random forest for PolSAR imagery. *Remote Sens.* **2018**, *10*, 1092. [CrossRef]
33. Qin, X.; Yang, J.; Zhao, L.; Li, P.; Sun, K. A Novel Deep Forest-Based Active Transfer Learning Method for PolSAR Images. *Remote Sens.* **2020**, *12*, 2755. [CrossRef]
34. Doz, C.; Ren, C.; Ovarlez, J.P.; Couillet, R. Large Dimensional Analysis of LS-SVM Transfer Learning: Application to Polsar Classification. In Proceedings of the ICASSP 2023—2023 IEEE International Conference on Acoustics, Speech and Signal Processing (ICASSP), Rhodes Island, Greece, 4–10 June 2023; pp. 1–5.
35. Nie, W.; Huang, K.; Yang, J.; Li, P. A deep reinforcement learning-based framework for PolSAR imagery classification. *IEEE Trans. Geosci. Remote Sens.* **2021**, *60*, 4403615. [CrossRef]
36. Huang, K.; Nie, W.; Luo, N. Fully polarized SAR imagery classification based on deep reinforcement learning method using multiple polarimetric features. *IEEE J. Sel. Top. Appl. Earth Obs. Remote Sens.* **2019**, *12*, 3719–3730. [CrossRef]
37. Cui, Y.; Liu, F.; Liu, X.; Li, L.; Qian, X. TCSPANET: Two-staged contrastive learning and sub-patch attention based network for polsar image classification. *Remote Sens.* **2022**, *14*, 2451. [CrossRef]
38. Wu, Z.; Xiong, Y.; Yu, S.X.; Lin, D. Unsupervised feature learning via non-parametric instance discrimination. In Proceedings of the IEEE Conference on Computer Vision and Pattern Recognition, Salt Lake City, UT, USA, 18–23 June 2018; pp. 3733–3742.
39. Caron, M.; Touvron, H.; Misra, I.; Jégou, H.; Mairal, J.; Bojanowski, P.; Joulin, A. Emerging properties in self-supervised vision transformers. In Proceedings of the IEEE/CVF International Conference on Computer Vision, Montreal, BC, Canada, 11–17 October 2021; pp. 9650–9660.
40. Ghosh, A.; Kumar, H.; Sastry, P.S. Robust loss functions under label noise for deep neural networks. In Proceedings of the AAAI Conference on Artificial Intelligence, San Francisco, CA, USA, 4–9 February 2017; Volume 31.
41. Wang, Y.; Ma, X.; Chen, Z.; Luo, Y.; Yi, J.; Bailey, J. Symmetric cross entropy for robust learning with noisy labels. In Proceedings of the IEEE/CVF International Conference on Computer Vision, Seoul, Republic of Korea, 27 October–2 November 2019; pp. 322–330.
42. Jiang, L.; Zhou, Z.; Leung, T.; Li, L.J.; Fei-Fei, L. Mentornet: Learning data-driven curriculum for very deep neural networks on corrupted labels. In Proceedings of the International Conference on Machine Learning, PMLR, Stockholm, Sweden, 10–15 July 2018; pp. 2304–2313.
43. Yu, X.; Han, B.; Yao, J.; Niu, G.; Tsang, I.; Sugiyama, M. How does disagreement help generalization against label corruption? In Proceedings of the International Conference on Machine Learning, PMLR, Long Beach, CA, USA, 9–15 June 2019; pp. 7164–7173.
44. Berthelot, D.; Carlini, N.; Goodfellow, I.; Papernot, N.; Oliver, A.; Raffel, C.A. Mixmatch: A holistic approach to semi-supervised learning. In Proceedings of the Advances in Neural Information Processing Systems, Vancouver, BC, Canada, 8–14 December 2019; Volume 32.
45. Edwards, H.; Storkey, A. Towards a Neural Statistician. In Proceedings of the International Conference on Learning Representations, Toulon, France, 24–26 April 2017.
46. Kaiser, Ł.; Nachum, O.; Roy, A.; Bengio, S. Learning to remember rare events. *arXiv* **2017**, arXiv:1703.03129.
47. Ravi, S.; Larochelle, H. Optimization as a model for few-shot learning. In Proceedings of the International Conference on Learning Representations, San Juan, Puerto Rico, 2–4 May 2016.
48. Bertinetto, L.; Henriques, J.F.; Valmadre, J.; Torr, P.; Vedaldi, A. Learning feed-forward one-shot learners. In Proceedings of the Advances in Neural Information Processing Systems, Barcelona, Spain, 5–10 December 2016; Volume 29.
49. Santoro, A.; Bartunov, S.; Botvinick, M.; Wierstra, D.; Lillicrap, T. Meta-learning with memory-augmented neural networks. In Proceedings of the International Conference on Machine Learning, PMLR, New York, NY, USA, 20–22 June 2016; pp. 1842–1850.
50. Snell, J.; Swersky, K.; Zemel, R. Prototypical networks for few-shot learning. In Proceedings of the Advances in Neural Information Processing Systems, Long Beach, CA, USA, 4–9 December 2017; Volume 30.
51. Finn, C.; Abbeel, P.; Levine, S. Model-agnostic meta-learning for fast adaptation of deep networks. In Proceedings of the International Conference on Machine Learning, PMLR, Sydney, Australia, 6–11 August 2017; pp. 1126–1135.
52. Lake, B.; Salakhutdinov, R.; Gross, J.; Tenenbaum, J. One shot learning of simple visual concepts. In Proceedings of the Annual Meeting of the Cognitive Science Society, Boston, MA, USA, 20–23 July 2011; Volume 33.
53. Oord, A.v.d.; Li, Y.; Vinyals, O. Representation learning with contrastive predictive coding. *arXiv* **2018**, arXiv:1807.03748.
54. Tian, Y.; Krishnan, D.; Isola, P. Contrastive multiview coding. In *Computer Vision–ECCV 2020, Proceedings of the 16th European Conference, Glasgow, UK, 23–28 August 2020*; Proceedings, Part XI 16; Springer: Berlin/Heidelberg, Germany, 2020; pp. 776–794.

55. He, K.; Fan, H.; Wu, Y.; Xie, S.; Girshick, R. Momentum contrast for unsupervised visual representation learning. In Proceeding of the IEEE/CVF Conference on Computer Vision and Pattern Recognition, Seattle, WA, USA, 13–19 June 2020; pp. 9729–9738
56. Cloude, S.R.; Pottier, E. A review of target decomposition theorems in radar polarimetry. *IEEE Trans. Geosci. Remote Sens.* **199** *34*, 498–518. [CrossRef]
57. Zhou, Y.; Wang, H.; Xu, F.; Jin, Y.Q. Polarimetric SAR image classification using deep convolutional neural networks. *IEEE Geosc Remote Sens. Lett.* **2016**, *13*, 1935–1939. [CrossRef]
58. Hernández-García, A.; König, P. Do deep nets really need weight decay and dropout? *arXiv* **2018**, arXiv:1802.07042.
59. Wong, S.C.; Gatt, A.; Stamatescu, V.; McDonnell, M.D. Understanding data augmentation for classification: When to warp? I Proceedings of the 2016 International Conference on Digital Image Computing: Techniques and Applications (DICTA), Gol Coast, Australia, 30 November–2 December 2016; pp. 1–6.
60. De Boer, P.T.; Kroese, D.P.; Mannor, S.; Rubinstein, R.Y. A tutorial on the cross-entropy method. *Ann. Oper. Res.* **2005**, *134*, 19–6 [CrossRef]
61. Van der Maaten, L.; Hinton, G. Visualizing data using t-SNE. *J. Mach. Learn. Res.* **2008**, *9*, 2579–2605.
62. Müller, R.; Kornblith, S.; Hinton, G.E. When does label smoothing help? In Proceedings of the Advances in Neural Informatio Processing Systems, Vancouver, BC, Canada, 8–14 December 2019; Volume 32.
63. Lin, T.Y.; Goyal, P.; Girshick, R.; He, K.; Dollár, P. Focal loss for dense object detection. In Proceedings of the IEEE Internationa Conference on Computer Vision, Venice, Italy, 22–29 October 2017; pp. 2980–2988.
64. Li, S.; Xia, X.; Ge, S.; Liu, T. Selective-supervised contrastive learning with noisy labels. In Proceedings of the IEEE/CV Conference on Computer Vision and Pattern Recognition, New Orleans, LA, USA, 18–24 June 2022; pp. 316–325.
65. Wang, Y.; Sun, X.; Fu, Y. Scalable penalized regression for noise detection in learning with noisy labels. In Proceedings of th IEEE/CVF Conference on Computer Vision and Pattern Recognition, New Orleans, LA, USA, 18–24 June 2022; pp. 346–355.

remote sensing

MDPI

Article

SDAT-Former++: A Foggy Scene Semantic Segmentation Method with Stronger Domain Adaption Teacher for Remote Sensing Images

Ziquan Wang, Yongsheng Zhang, Zhenchao Zhang *, Zhipeng Jiang, Ying Yu, Li Li and Lei Zhang

School of Geospatial Information, PLA Strategic Support Force Information Engineering University, Zhengzhou 450001, China; aresdrw@163.com (Z.W.); yszhang2001@vip.163.com (Y.Z.); jiangzp0803@163.com (Z.J.); yuying5559104@163.com (Y.Y.); lili315114@163.com (L.L.); zhang295498@126.com (L.Z.)

* Correspondence: zhzhc_1@163.com; Tel.: +86-150-9330-3012

Abstract: Semantic segmentation based on optical images can provide comprehensive scene information for intelligent vehicle systems, thus aiding in scene perception and decision making. However, under adverse weather conditions (such as fog), the performance of methods can be compromised due to incomplete observations. Considering the success of domain adaptation in recent years, we believe it is reasonable to transfer knowledge from clear and existing annotated datasets to images with fog. Technically, we follow the main workflow of the previous SDAT-Former method, which incorporates fog and style-factor knowledge into the teacher segmentor to generate better pseudo-labels for guiding the student segmentor, but we identify and address some issues, achieving significant improvements. Firstly, we introduce a consistency loss for learning from multiple source data to better converge the performance of each component. Secondly, we apply positional encoding to the features of fog-invariant adversarial learning, strengthening the model's ability to handle the details of foggy entities. Furthermore, to address the complexity and noise in the original version, we integrate a simple but effective masked learning technique into a unified, end-to-end training process. Finally, we regularize the knowledge transfer in the original method through re-weighting. We tested our SDAT-Former++ on mainstream benchmarks for semantic segmentation in foggy scenes, demonstrating improvements of 3.3%, 4.8%, and 1.1% (as measured by the mIoU) on the ACDC, Foggy Zurich, and Foggy Driving datasets, respectively, compared to the original version.

Keywords: semantic segmentation in foggy scenes; unsupervised domain adaptation; UDA; self-training

Citation: Wang, Z.; Zhang, Y.; Zhang, Z.; Jiang, Z.; Yu, Y.; Li, L.; Zhang, L. SDAT-Former++: A Foggy Scene Semantic Segmentation Method with Stronger Domain Adaption Teacher for Remote Sensing Images. *Remote Sens.* **2023**, *15*, 5704. https://doi.org/10.3390/rs15245704

Academic Editors: Qian Du, Gemine Vivone, Jiaojiao Li, Wei Li, Jocelyn Chanussot, Rui Song, Yunsong Li and Bobo Xi

Received: 2 November 2023
Revised: 8 December 2023
Accepted: 10 December 2023
Published: 12 December 2023

1. Introduction

Among the various perception methods, vision-based methods have attracted interest due to their comprehensive, intuitive, and cost-effective advantages [1,2]. In particular, robust semantic segmentation [3–10] based on visual images is important for autonomous driving, as it can save on the huge costs of installing auxiliary sensors (like LiDAR), thereby effectively aiding intelligent vehicles.

However, the segmentation models trained on clear-scene datasets often generalize poorly under adverse weather conditions (such as foggy scenes [11]) due to the degradation of visibility [12]. Meanwhile, the cost of directly producing annotations for foggy images is much higher than for clear ones, which makes it difficult to address the problem of semantic segmentation in foggy scenes (SSFS) using a traditional fully supervised training strategy. At present, the most common way is to transform it into a domain adaptation (DA) problem [13], which uses finely annotated datasets containing clear scenes (such as Cityscapes [14]) as the source domain and foggy scenes as the target domain (with no labels) to transfer the segmentation knowledge by training a DA model. Domain adaptation methods are often based on Generative Adversarial Networks (GANs) [15] and self-training [16]. GAN-based DA methods regard domain differences as noise that needs to be aligned across the input [15,17,18], feature [19], and output spaces [20,21]. Self-training methods [22–25] use the current model to generate pseudo-labels on the target domain and perform self-guidance. But directly using DA methods makes it challenging to handle large dual-domain differences (such as style differences between cities and visual degradation caused by haze), resulting in poor-quality pseudo-labels. These methods tend to easily generate a large area of classification error at the boundary between fog and objects [11]. Some methods [26–28] introduce intermediate domains to reduce the domain gap by collecting or generating a set of images with different degrees of haze or from different time periods using curriculum learning strategies. But they require a large amount of data and are prone to accumulating errors. Recently, introducing a single clear domain as an intermediate domain [29] has gained attention, as this approach only requires collecting clear images from the target city to serve as the intermediate domain. Cycle training or spatial alignment can then be used on this domain to guide the segmentation of target domain images. However, the intermediate domain and target domain information are still treated independently and not fully utilized. In contrast, our method integrates information from various domains through cyclical training, thus achieving the organic integration of information.

Despite the importance of both style gap and fog gap, most methods still focus on only one of them, resulting in little improvement when facing real foggy scenes. This may be due to the different training paradigms. When dealing with the fog gap, adversarial training strategies or explicit fog modeling approaches are often used, whereas excellent, newly developed methods mainly adopt self-training strategies [22,23,25,30] when dealing with the style gap. Simply combining the two strategies can cause interference between sub-modules due to chaotic backward gradients. Recently, the authors of SDAT-Former [1] proposed a strong teacher for foggy road scene semantic segmentation, which differs from previous domain adaptation methods, as it considers both style and fog knowledge, successfully transferring style-invariant knowledge and fog-invariant knowledge to the teacher segmentor [25,31]. This enables the teacher segmentor to have a broader perspective and generate superior pseudo-labels in the target domain, thereby guiding the training of the student segmentor (the main segmentor to be published). Specifically, this method divides the entire training process into several mini-epochs, each consisting of four iterations that perform fog-invariant adversarial learning, intermediate domain style feature learning, information integration, and target domain mask domain adaptation, respectively. This effectively solves the mutual interference between gradients and successfully handles the problem of significant style and fog differences, surpassing the previous year's state-of-the-art solutions on mainstream foggy scene semantic segmentation benchmarks.

However, SDAT-Former [1] still has many drawbacks. Firstly, the extraction of style features in the intermediate domain is cumbersome and cannot be integrated into an end-to-end training process. SDAT-Former first trains an LSGAN [17] to apply the source domain style to the intermediate domain images, then uses DAFormer [25] to predict the labels of the transformed images. These training steps are performed offline and consume significant computational resources and time. Additionally, when the intermediate domain changes, the corresponding models need to be retrained to generate new data. The style features learned by the GAN-based models may not be comprehensive due to down-sampling operations for calculating discrimination probabilities [17] and artifacts [1]. In this case, the label-based learning approach is prone to introducing noise, which can damage the model. Secondly, in the fog-invariant feature learning step, the original feature dimension is too low, but the actual variations in fog may be subtle, leading to the extracted features not being representative enough. Furthermore, the three components of SDAT-Former contribute equally to the parameters of the teacher segmentor, but in reality, they should be assigned weights or dynamically adjusted. Finally, the performance of each component eventually converges to a stable condition, but the SDAT-Former method does not take this factor into account or adopt appropriate consistency constraints to accelerate convergence.

Based on the above, we propose the improved "SDAT-Former++" which is shown in Figure 1. This new version retains the cyclical training strategy from SDAT-Former [1] but incorporates substantial optimizations. To address the complexity of intermediate domain learning, we introduce a simple but effective strategy using masked autoencoder learning [32,33], which can align the context by predicting masked images. This approach enables the model to better distinguish similar categories such as roads and sidewalks. By directly recovering the masked intermediate domain images, we use a basic backbone to learn the style features of the intermediate domain in a complete and artifact-free manner. Moreover, the knowledge is directly saved in the model's parameters, thus facilitating an end-to-end training process without the need for extra offline operations. Additionally, the model can start training directly when the intermediate domain changes, achieving a complete separation between the model and the data. To tackle the problem of low feature dimensions and inadequate representations in fog-invariant learning, we introduce positional encoding [34,35] to separate more high-dimensional details, making the fog discriminator more sensitive and compelling the fog-invariant feature extractor to be robust. To address the issue of evenly distributed knowledge transfer, we introduce weight perturbations based on a random distribution for regularization.

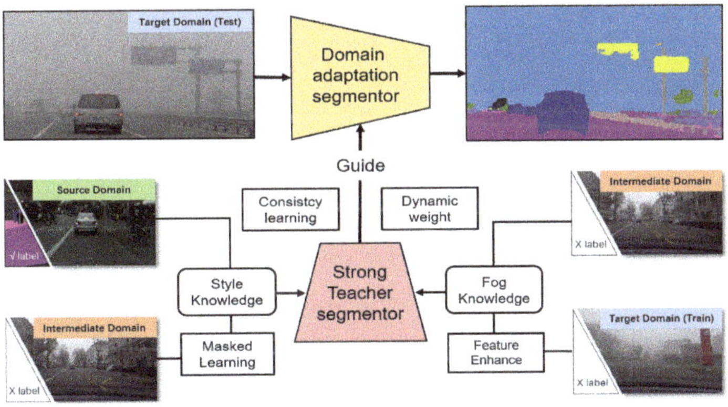

Figure 1. The main idea of the proposed method. Unlike the original SDAT-Former, we optimize the learning of style information and add feature enhancement for fog-invariant feature learning, greatly reducing the computing consumption and integrating the processing pipeline. We also add consistency learning and dynamic weighting when processing the knowledge transfer.

Compared to the original SDAT-Former publication, this paper provides more comprehensive experimental results and technical details. In addition to the existing ACDC [36] and Foggy Zurich [27] datasets, a more challenging dataset, Foggy Driving Dense [37], is also included. We also conduct extensive ablation experiments and provide favorable entropy analysis evidence.

The contributions of this work can be summarized as follows:

- To the best of our knowledge, this work is the first to propose an end-to-end cyclical training domain adaptation semantic segmentation method that considers both style-invariant and fog-invariant features.
- Our method proves the importance of masked learning and feature enhancement in foggy road scene segmentation and demonstrates their mechanisms through visualizations.
- Our method significantly outperforms SDAT-Former on mainstream benchmark datasets for foggy road scene segmentation and exhibits strong generalization in rainy and snowy scenes. Compared to the original method, SDAT-Former++ pays more attention to the more important categories in road scenes and is more suitable for applications in intelligent vehicles. We test our SDAT-Former++ method on mainstream benchmarks for semantic segmentation in foggy scenes and demonstrate improve-

ments of 3.3%, 4.8%, and 1.1% (as measured by the mIoU) on the ACDC, Foggy Zurich and Foggy Driving datasets, respectively, compared to the original method.

2. Method

2.1. Overview

Suppose there are N_s labeled images $\{(x_s^i, y_s^i)\}_{i=1}^{N_s}$ from the clear source domain s where y_s^i is the pixel-level segmentation label for x_s^i, and N_t unlabeled images $\left\{x_t^k\right\}_{k=1}^{N_t}$ from the target foggy domain t. Our goal is to transfer segmentation knowledge from the clear source domain s to the foggy target domain t using our proposed SDAT-Former++ method. Motivated by the success of DAFormer [25], we use a similar framework including a "student" segmentor and a "teacher" segmentor to train in a self-training manner. However, since the images in domain s and domain t were taken in different cities and under different weather conditions, they exhibit a large domain gap caused by two factors, i.e., the style factor and the fog factor, which poses a challenge to this method. Therefore, we introduce an intermediate domain m with N_m unlabeled images $\left\{\left(x_m^j\right)\right\}_{j=1}^{N_m}$. This domain shares similar fog influence (no fog) to the source domain and similar style variation to the target domain (imaged in the same city). We also call these images the "reference images" I^{ref} of the foggy images I^{fog}. Thus, our main goal is to cumulatively transfer four kinds of knowledge to the "teacher" segmentor to generate more robust pseudo-labels of t, thereby empowering the "student" segmentor to complete the segmentation tasks: (a) segmenting the knowledge from s, (b) segmenting the style knowledge from m, (c) segmenting the knowledge from t, and (d) identifying and removing fog. Among these, (c) and (d) focus on overcoming the "fog gap" between s and t, whereas (b) focuses on the "style gap". Figure 2 depicts the framework of our proposed method.

Concretely, we reorganize the training workflow cyclically, where every four iterations constitute a "mini-epoch". The segmentation knowledge from s can be learned from labels $\{(y_s^i)\}_{i=1}^{N_s}$ in a supervised way (Figure 2A), and we train it throughout the process. In the first iteration of a mini-epoch, a fog-pass filter [38] is trained for discriminating fog factors from the clear source domain m and foggy target domain t (Figure 2B.1). Here, we use positional encoding (PE) [34,35] to enhance the features and capture more high-frequency information. In the second iteration, the segmentor backbone is trained to generate features that fool the fog-pass filter (Figure 2B.2). These two iterations aim to train a robust extractor for fog-invariant features in an adversarial manner. For the third iteration, we abandon the complex operation mode in the original version of the method [1] and use a feature extractor with a decoder to recover the masked images $\{(\tilde{x}_m^j)\}_{j=1}^{N_m}$ from the intermediate domain and extract style features. In the last iteration, the parameters stored in the teacher segmentor can be updated by the "student" segmentor (containing knowledge from s), the "teaching-assistant" backbone (containing style knowledge from m), and the fog-invariant backbone (containing fog-invariant knowledge) in an exponential moving average (EMA) [31] way with dynamic weight (DW) (Figure 2D. Then, the self-training process is performed on the target foggy domain t. Thus, the "teacher" can be "strong" enough to handle the domain gap and guide the student (main) segmentor.

2.2. Sub-Modules

The main workflow includes 6 sub-modules: (a) "student" segmentor f_θ (can be published as the final segmentor), (b) "teaching assistant" backbone b_ψ^{si} (learns the style knowledge), (c) decoder d_r for reconstruction, (d) "teacher" segmentor h_φ, (learns knowledge from the target domain), (e) fog-invariant backbone b_ω^{fi}, and (f) fog-pass filter $\mathcal{F}$ (learns to recognize fog factors).

All the segmentors contain a backbone and decoder head. The backbone follows the design of Mix Transformers (MiT) [39] to produce multi-level feature maps, whereas the decoder head follows ASPP [40] to predict segmentation maps. The fog-invariant backbone

b_{ω}^{fi} shares the same architecture as MiT for subsequent knowledge transfer. The fog-pass filter $\mathcal{F}$ follows the design in FIFO [38]. The detailed architectures are described later.

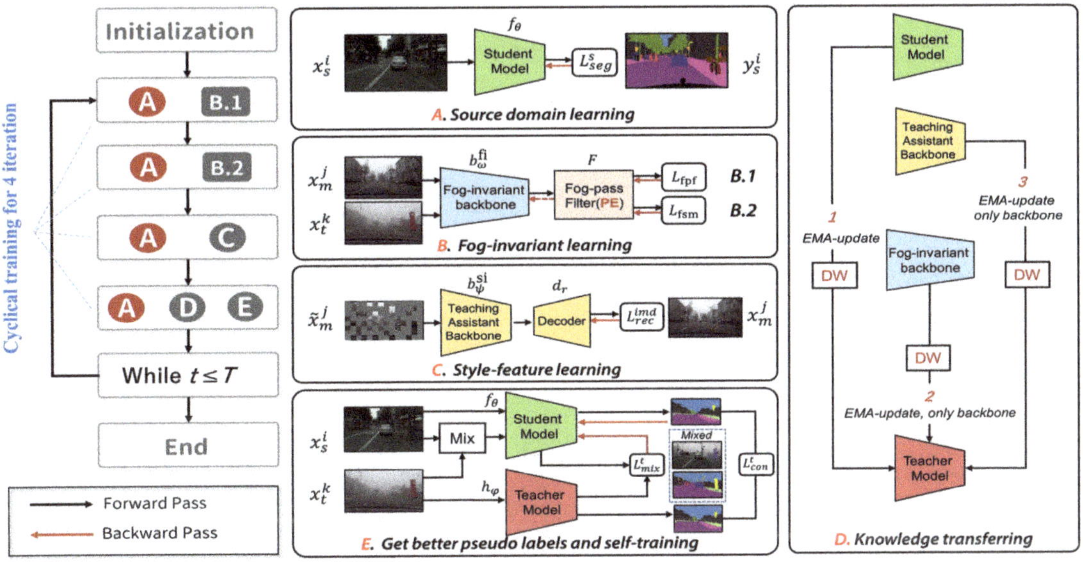

Figure 2. The overall workflow of our method. (**Left**) Training flow within a mini-epoch that can be repeated as the base training unit. (**Right**) The sub-process (**A–E**) includes learning segmentation and style knowledge from the source and intermediate domains (**A**,**C**), attempting to train the backbone producing fog-invariant features adversarially (**B.1**,**B.2**), transferring all knowledge to the teacher (**D**), and compelling it to generate better pseudo-labels for supervision (**E**).

2.3. *Supervised Training on Source Domain*

Denote H and W as the height and width of the input image size and C as the number of object categories. First, we can use f_θ to learn the segmentation knowledge from the labeled source domain $\{(x_s^i, y_s^i)\}_{i=1}^{N_s}$ using a categorical cross-entropy loss function:

$$\mathcal{L}_s^i = -\sum_{p=1}^{H\times W}\sum_{c=1}^{C} y_s^{(i,p,c)} \log f_\theta(x_s^i)^{(p,c)} \tag{1}$$

2.4. *Masked Learning on the Intermediate Domain*

In the original version of SDAT-Former [1], an LSGAN [17] is used to transfer styles between the source domain and the intermediate domain. Then, the source styles are applied to the later images to narrow the domain gap. Next, a DAFormer [25] model is used to predict the transformed images $\{(\tilde{x}_m^j)\}_{j=1}^{N_m}$ and generate pseudo-labels, which have the same spatial layout as the original images $\{(x_m^j)\}_{j=1}^{N_m}$. This method adds two offline training steps and results in a significant loss in the resolution and details of the predicted values, even leading to artifacts. Training based on such pseudo-labels inevitably introduces noise. Moreover, when changing the intermediate domain, we have to reconfigure two pre-trained networks, influencing the deployment.

Since learning based on intermediate domain data aims to capture style features, pseudo-labels may not be necessary. In this section, we introduce a more concise method to model masked images. Specifically, we employ a uniform distribution to randomly sample a mask:

$$M_{\substack{mb+1:(m+1)b \\ nb+1:(n+1)b}} = [v \geq r] \quad \text{with} \quad v \sim U(0,1) \tag{2}$$

where $[*]$ is the Iverson bracket, b is the patch size, r is the mask ratio, and $m \in [0..W/b-1]$ and $n \in [0..H/b-1]$ are the patch indices. Thus, we obtain the masked intermediate image $\tilde{x}^j_m$ through element-wise multiplication of the mask and image:

$$\tilde{x}^j_m = M \odot x^j_m \tag{3}$$

Then, we try to use encoder b^{si}_{ψ} and decoder d_r to recover the original image:

$$x^{j,rec}_m = b^{\mathrm{si}}_{\psi}(d_r(\tilde{x}^j_m)) \tag{4}$$

We force the model to adopt the L1 loss function to recover the original image information. As a result, the feature extraction network obtains more realistic and context-aware style features, which are difficult to achieve through label-based approaches and do not lead to any resolution loss or noise:

$$\mathcal{L}^{rec}_m = |x^{j,rec}_m - x^j_m| \tag{5}$$

The knowledge from the intermediate domain can be stored in the parameters of b^{si}_{ψ} which can be passed to the "teacher" segmentor rather than being directly transferred to the final segmentor. This part is described in Section 2.7.

2.5. Fog-Invariant Feature Learning

Here, we focus on overcoming the fog gap between the intermediate domain and the target domain. Since Section 2.4 described the learning of cross-style knowledge, now, we only need to process the fog factor. That is, the final segmentor should output the fog-invariant features from the pair of foggy and non-foggy images. To achieve this, we design a fog-invariant feature extractor b^{fi}_{ω} and a fog-pass filter $\mathcal{F}$ based on the architecture of FIFO [38].

2.5.1. Training the Fog-Pass Filter

Given a pair of images (I^a, I^b) from the mini-batch, b^{fi}_{ω} can output L layer features of each image. We follow FIFO [38] to calculate these features' Gram matrix to capture a holistic fog representation denoted as $\{(\mathbf{u}^{a,l}, \mathbf{u}^{\mathrm{b,l}})\}^L_{l=1}$. Denote $\mathcal{F}^l$ as the fog-pass filter attached to the lth layer feature. The fog factors of these two images can be computed by $\mathbf{f}^{\mathrm{a,l}} = \mathcal{F}^l(\mathbf{u}^{\mathrm{a,l}})$ and $\mathbf{f}^{\mathrm{b,l}} = \mathcal{F}^l(\mathbf{u}^{\mathrm{b,l}})$, respectively.

To enhance the representation of the fog factors, we follow previous works [34,35,41] and adopt a sinusoidal positional encoding scheme to capture the high-frequency details:

$$\psi(\mathbf{f}) = (\sin(\omega_1\mathbf{f}), \cos(\omega_1\mathbf{f}), ..., \sin(\omega_n\mathbf{f}), \cos(\omega_n\mathbf{f})) \tag{6}$$

where the frequencies $\omega_1, \omega_2, ...\omega_n$ are learnable during training and n is the positional encoding dimension. The role of the fog-pass filter is to inform the fog-invariant backbone b^{fi}_{ω} about how I^a and I^b are different in terms of fog conditions through $\psi(\mathbf{f}^{\mathrm{a,l}})$ and $\psi(\mathbf{f}^{\mathrm{b,l}})$. For this purpose, the fog-pass filter learns a space of fog factors, where those of the same fog domain are grouped closely together and those of different domains are far apart. The loss function for $\mathcal{F}^l$ is designed as follows:

$$\begin{aligned}\mathcal{L}_{\mathcal{F}^l} = \sum_{(a,b)} (1 - \Pi(a,b))\left[m - d(\psi(\mathbf{f}^{\mathrm{a,l}}), \psi(\mathbf{f}^{\mathrm{b,l}}))\right]^2 \\ +\Pi(a,b)\left[d(\psi(\mathbf{f}^{\mathrm{a,l}}), \psi(\mathbf{f}^{\mathrm{b,l}})) - m\right]^2\end{aligned} \tag{7}$$

where $d()$ is the cosine distance, m is the margin, and $\Pi(a,b)$ denotes the indicator function that returns 1 if I^a and I^b are of the same fog domain and 0 otherwise.

2.5.2. Fog Factor Matching Loss

In contrast to the function of the fog-pass filter, which attempts to separate the fog factors of different fog domains, the fog-invariant backbone b^{fi}_{ω} learns to close the distance between the fog factors. To this end, the second loss matches the two fog factors given by frozen fog-pass filters:

$$\mathcal{L}^{l}_{fsm}(\psi(\mathbf{f}^{a,l}),\psi(\mathbf{f}^{b,l}))=\frac{1}{4d_l^2n_l^2}\sum_{i=1}^{d_l}\left(\psi(\mathbf{f}_i^{a,l})-\psi(\mathbf{f}_i^{b,l})\right)^2 \tag{8}$$

where d_l and n_l denote the dimensions of their fog factors and the spatial size of the lth feature map, respectively. The knowledge from fog-invariant training can be also stored in the parameters in b^{fi}_{ω} and can be passed to the "teacher" segmentor, as described in Section 2.7.

2.6. Self-Training on the Target Domain and Consistency Learning

We use a teacher segmentor h_φ to directly address the two gaps (style + fog) between the source domain and the target domain to obtain better domain adaptation performance. Specifically, h_φ can first produce pseudo-labels for the foggy target domain data

$$\widetilde{y}_t^{(k,p,c)}=\left[c=\arg\max_{c'}h_\varphi(x_t^k)^{(p,c')}\right] \tag{9}$$

Additionally, a quality (confidence) estimation is produced for the pseudo-labels. Here, we use the ratio of pixels exceeding a threshold τ of the maximum softmax probability

$$q_t^k=\frac{\sum_{p=1}^{H\times W}\left[\max_{c'}h_\varphi(x_t^k)^{(p,c')}\geq\tau\right]}{H\times W} \tag{10}$$

The pseudo-labels and their quality estimates are used to additionally train the segmentor h_φ on the target domain

$$\mathcal{L}_t^k=-\sum_{p=1}^{H\times W}\sum_{c=1}^{C}q_t^k\widetilde{y}_t^{(k,p,c)}\log h_\varphi(x_t^k)^{(p,c)} \tag{11}$$

The self-training process can be more efficient if the segmentor is trained on augmented data [42]. In this work, we follow DACS [23] and employ color jitter, Gaussian blur, and ClassMix [43] for data augmentation to learn more domain features. To accelerate the training, we introduce a consistency learning strategy between teacher h_φ and student f_θ. Specifically, for one specific sample x, we use the Kullback–Leibler divergence as a consistency loss, forcing convergence between the teacher and student

$$\mathcal{L}_{con}(x)=\sum_i \text{KLdiv}(f_\theta(x),h_\varphi(x)) \tag{12}$$

2.7. Cyclical Training with Knowledge Transferring

The above steps facilitate domain adaptation learning from different levels, but they need to be organically combined. If we include so many backward processes in a single iteration, the gradient propagation could be easily confused and the sub-modules could face potential interface issues Thus, we use cyclical training and build a "strong teacher" to merge the above-mentioned knowledge. We divide every four iterations into a "mini-epoch". Considering that fog-invariant feature learning works adversarially, we allocate the $1st$ and $2nd$ iterations to train the fog-pass filter $\mathcal{F}$ and fog-invariant backbone b^{fi}_{ω} successively. The $3rd$ iteration is allocated to intermediate domain learning using the teaching-assistant segmentor g_ψ. For the $4th$ iteration, since the intermediate feature extractor b^{si}_{ψ} does not need to complete segmentation, we remove the pre-updating used in [1] to prevent interface issues. All the knowledge can be transferred to the teacher segmentor through an optimized three-step exponentially moving average (EMA[31]) update (Figure 2D):

$$\begin{aligned} h_{\varphi}^{t+1} &= \alpha_1 h_{\varphi}^{t} + (1-\alpha_1) b_{\omega}^{\mathrm{fi}|t} \\ h_{\varphi}^{t+2} &= \alpha_2 h_{\varphi}^{t+1} + (1-\alpha_2) f_{\theta}^{t} \\ h_{\varphi}^{t+3} &= \alpha_3 h_{\varphi}^{t+2} + (1-\alpha_3) b_{\psi}^{\mathrm{si}|t} \end{aligned} \quad (13)$$

where $\alpha_i = \alpha + \delta_i \quad \delta_i \sim N(0, V)$, i.e., the parameters are perturbed by a normal distribution and thus the knowledge can be regularized. Then, we conduct self-training on the target domain, as described in Section 2.6. In our proposed method, we use EMA [31] to update the model parameters because it can transmit domain knowledge while protecting the segmentor from the noise in the pseudo-labels [44]. Thus, the teacher segmentor can be powerful enough to guide the student segmentor in the domain adaptation training. In the ablation study, we discuss EMA updating in detail.

3. Results

3.1. The Network Parameters

Our implementation was based on the mmsegmentation framework [45] and PyTorch [46]. The MiT-b5 backbone (used in f_θ, h_φ, g_ψ, and b_ω^{fi}) produced a feature pyramid with channels of 64, 128, 320, and 512. The ASPP decoder used $n_{ch} = 256$ and dilation rates of 1, 6, 12, and 18. All encoders were pre-trained on the ImageNet-1k [47] dataset. The fog-pass filters $\mathcal{F}$ were composed of a fully connected layer and LeakyReLU layer to convert the Gram matrix of the feature maps into fog vectors.

3.2. Implementation Details

The main workflow was trained by AdamW [48], the learning rate was 6×10^{-5} with a weight-decay of 0.01, and linear learning rate warm-up followed the "poly" strategy after 1.5k iterations. All the input images and labels were cropped to 512×512, and the maximum number of training iterations was 40k. Following DACS [23], we used the same data augmentation parameters and set $\alpha = 0.99$, $\tau = 0.968$, and the perturbation variance $V = 0.1$. We set the weight of the source domain supervised learning loss (Equation (1)) to 1 and the weight of the intermediate domain style feature learning loss (Equation (5)) to 0.5. Following FIFO [38], we set the loss weights for both the fog-pass-filter loss (Equation (7)) and the fog factor matching loss (Equation (8)) to 0.001, with $m = 0.1$. We set the weight of the consistency learning loss (Equation (12)) to 0.1 to avoid learning errors from the teacher network. The weight of the loss function in the target domain had already been determined based on confidence and did not need to be set manually. The dimension n for positional encoding was set to 512. All the experiments were conducted on a single Tesla-v100 GPU with a memory of 32 GB and equipped with CUDA 10.2.

3.3. Datasets

Cityscapes [14] is a real-world dataset composed of street scenes captured in 50 different cities. The data split includes 2975 training images and 500 validation images with pixel-level labels. The Cityscapes dataset is the source domain and shares the same class set with all the datasets mentioned in this paper.

ACDC [36] contains four categories of adverse conditions (fog, snow, rain, and nighttime) with pixel-level annotations. Each category contains 1000 images and is split into a train set, validation set, and test set at a ratio of about 4:1:5. The annotations of the test set were withheld for online testing. We mainly used the foggy images. Moreover, the ACDC dataset also provides clear reference images of each foggy image, which can be used as the intermediate domain.

Foggy Zurich [11] contains 3808 real foggy road views from the city of Zurich and its suburbs. It is split into two categories of fog density—light and medium—consisting of 1552 and 1498 images, respectively. It has a test set, Foggy Zurich-test, which includes 40 images with labels that are compatible with those of Cityscapes.

Foggy Driving [11] contains 101 real-world foggy images collected from the Internet with different sizes and fog densities, including a challenging subset of 21 images with "dense fog" (referred to as Foggy Driving Dense) [37]. The dataset can only be used for evaluation.

The comparison results are shown in Table 1 and Table 2.

Table 1. Performance comparison I. Experiments were conducted on the ACDC [36] and Foggy Zurich-test (FZ) [27] dataset, measuring the mean intersection over union (mIoU) (%) across all 19 classes following the Cityscapes [14] benchmark.

Experiment	Method	Backbone	ACDC	FZ	Experiment	Method	Backbone	ACDC	FZ
Backbone	-	DeepLabv2 [49]	33.5	25.9	DA-based	LSGAN [17]	DeepLabv2	29.3	24.4
	-	RefineNet [50]	46.4	34.6		Multi-task [51]	DeepLabv2	35.4	28.2
	-	MPCNet [4]	45.9	39.4		AdaptSegNet [20]	DeepLabv2	31.8	26.1
	-	SegFormer [39]	47.3	37.7		ADVENT [21]	DeepLabv2	32.9	24.5
Dehazing	DCPDN [52]	DeepLabv2	33.4	28.7		CLAN [22]	DeepLabv2	38.9	28.3
	MSCNN [53]	RefineNet	38.5	34.4		BDL [30]	DeepLabv2	37.7	30.2
	DCP [54]	RefineNet	34.7	31.2		FDA [55]	DeepLabv2	39.5	22.2
	Non-local [56]	RefineNet	31.9	27.6		DISE [19]	DeepLabv2	42.3	40.7
	SGLC [57]	RefineNet	39.2	34.5		ProDA [24]	DeepLabv2	38.4	37.8
Synthetic	SFSU [11]	RefineNet	45.6	35.7		DACS [23]	DeepLabv2	41.3	28.7
	CMAda [27]	RefineNet	51.1	46.8		DAFormer [25]	SegFormer	48.9	44.4
	FIFO [38]	RefineNet	54.1	48.4		CuDA-Net [26]	DeepLabv2	55.6	<u>49.1</u>
SDAT	SDAT-Former [1]	SegFormer	<u>56.0</u>	49.0	Ours	SDAT-Former++	SegFormer	**59.3**	**53.8**

3.4. Performance Comparison

We compared our method to several categories of methods, including:

- *Backbones*: RefineNet [50], Deeplabv2 [49], MPCNet [4], and SegFormer [39].
- *Dehazing methods*: MSCNN [53], DCP [54], SGLC [57], DCPDN [52], and non-local [56].
- *DA-based methods*: LSGAN [17], AdaptSegNet [20], Multi-Task [51], ADVENT [21], CLAN [22], BDL [30], FDA [55], DISE [19], ProDA [24], DACS [23], DAFormer [25], and CuDA-Net [26].
- *Synthetic methods*: SFSU [11], CMAda [27], and FIFO [38]

The configuration of each type of method was as follows. We trained the *backbone* methods on the Cityscapes dataset with labels and tested them on the ACDC and Foggy Zurich datasets to evaluate their performance across the domains. We set the source domain for the *DA-based* methods as clear Cityscapes, representing the s domain in our method. We used the fog images from ACDC and Foggy Zurich (with medium-level fog) as the target domain data. For the intermediate domain m, we combined the ACDC fog reference set (1000 images) with a manual selection of 600 clear images from the Foggy Zurich dataset (light-level fog). For the *synthetic* methods, the paradigm was to fine-tune the segmentation model pre-trained on clear weather images from Cityscapes. This fine-tuning used synthetic foggy images, such as those from Foggy Cityscapes DBF [11], along with labels corresponding to their clear weather images. We first used the *dehazing* methods to dehaze the foggy images and then used the corresponding backbone segmentor for predictions.

We compared our method to other outstanding works on the relatively easy ACDC-test [36] and Foggy Zurich-test [27] datasets. Table 1 shows the results, and the results from the ACDC dataset can be found on the https://acdc.vision.ee.ethz.ch/benchmarks# semanticSegmentation (accessed on accessed on 11 February 2023). ACDC-fog benchmark website (with our method named "SDAT-Former++"). Our method significantly outperformed the baseline algorithm DAFormer [25], yielding 10.4% and 9.4% higher mIoU values on the two datasets, respectively. Our method also outperformed the recently proposed MPCNet (in RS 2023 [4]) and SGLC (in CVPR23) [57], thus demonstrating the necessity of developing semantic segmentation methods for foggy scenarios. Compared to the original SDAT-Former [1], our method achieved improvements of 3.3% and 3.4%. This indicates that our method is robust without any special operations or removal of fog. Since the ground truths from the ACDC-test dataset were withheld, we used the Foggy Zurich-test [27] and Foggy Driving Dense datasets for qualitative comparison. The upper three rows in Figure 3 show the results on the challenging Foggy Driving Dense dataset [11], and the bottom three rows correspond to Foggy Zurich images output by DAFormer [25] (our baseline), CuDA-Net [26], MPCNet [4], SDAT-Former [1], and SDAT-Former++. Due to DAFormer's inability to handle style differences in intermediate domains, it failed to handle the sky in foggy conditions. CuDA-Net removed these artifacts but made mistakes in identifying objects occluded by fog (as shown by the yellow box). MPCNet tended to classify fog as buildings or fences. In contrast, our method was highly accurate in segmenting details and handling fog.

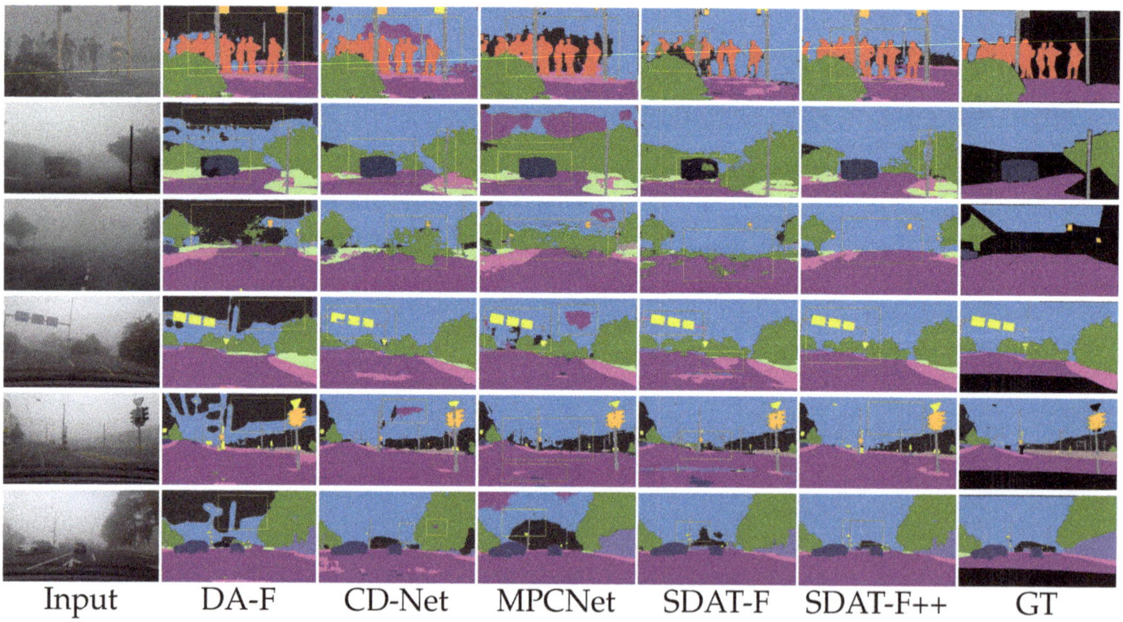

Figure 3. Qualitative comparison with other methods. Since the ground truths from the ACDC-test dataset were withheld and the fog in the images from the Foggy Driving dataset was light, we randomly selected images from the challenging Foggy Driving Dense dataset (top three lines) and Foggy Zurich-test dataset (bottom three lines) with dense fog to compare the performance of our method with that of other methods.

Then, we tested our method on the Foggy Driving (FD) [11] and the more challenging Foggy Driving Dense (FDD) [37] datasets. Many methods lost competitiveness or were completely ineffective on these datasets, so only a subset of methods was chosen for comparison. In Table 2, it can be seen that our method achieved improvements of 8.1% and 12.6% in terms of the mIoU over the baseline algorithm DAFormer [25] on FD and FDD, respectively. Our method also outperformed CuDA-Net (with improvements of 1.9%

and 3.0%) and FIFO (with improvements of 4.7% and 2.4%). In Figure 3, it can be seen that our method better preserved the segmentation of small objects in the images, for example, the "pole" in the second row, the traffic lights in the third row, and the road signs in the fourth row. This indicates that our method can effectively distinguish small objects while removing the effects of fog, which is crucial for the stability of segmentation.

Table 2. Performance comparison II. Experiments were conducted on the Foggy Driving [11] and Foggy Driving Dense [37] datasets, measuring the mean intersection over union (mIoU) (%) across all classes.

Experiment	Method	Backbone	FD	FDD
Backbone	-	DeepLabv2 [49]	26.3	17.6
	-	RefineNet [50]	34.6	35.8
	-	SegFormer [39]	36.2	37.4
Synthetic	CMAda3 [27]	RefineNet	49.8	43.0
	FIFO [38]	RefineNet	50.7	48.9
DA-based	AdaptSegNet [20]	DeepLabv2	29.7	15.8
	ADVENT [21]	DeepLabv2	46.9	41.7
	FDA [55]	DeepLabv2	21.8	29.8
	DAFormer [25]	SegFormer	47.3	39.6
	CuDA-Net [26]	DeepLabv2	53.5	48.2
Ours	SDAT-Former[1]	SegFormer	54.3	50.8
	SDAT-Former++	SegFormer	**55.4**	**51.2**

4. Discussion

4.1. Effectiveness of Fog-Invariant Feature Learning

In Table 3, it can be seen that the non-modified DAFormer, which is also the baseline of the original SDAT-Former, only yielded an mIoU of 48.92% on ACDC. Since we used adversarial training to acquire fog-invariant features, cyclical training was necessary to avoid gradient interference. This shows that the segmentor achieved an mIoU gain of +4.92% after the addition of this component, which was the most significant contribution to the performance improvement.

Table 3. Ablation study. We conducted an ablation study on the ACDC-test dataset, measuring the mIoU (%) across all classes.

Experiment					mIoU	Gain
Initialization	DAFormer				48.92	+0.00
	Cyclical(w/o DW [1])	imd(ls+da) [2]	fog_inv [3](w/o PE [4])		mIoU	Gain
SDAT-F [1]					10.23	−38.69
	✓				49.88	+0.96
		✓			50.52	+1.60
	✓	✓			51.61	+2.69
	✓		✓		53.84	+4.92
	✓	✓	✓		55.98	+7.06
	Cyclical(w/ DW)	imd(masked)	con_learn [5]	fog_inv(w/ PE)	mIoU	Gain
SDAT-F++	✓				50.34	+1.42
		✓			52.63	+3.71
			✓		51.33	+2.41
	✓	✓			56.19	+7.27
	✓		✓	✓	58.42	+9.50
	✓	✓	✓	✓	59.28	+10.36

[1] Indicates dynamic weight allocation. [2] Indicates use of LSGAN [17] and DAFormer [25] to obtain pseudo-labels. [3] Note that cyclical training is necessary for fog-invariant learning; we did not experiment with fog-invariant learning alone. [4] Indicates positional encoding. [5] Consistency learning.

As depicted in the qualitative results in Figure 4, without fog-invariant learning, the segmentor exhibited prediction drift in foggy conditions, such as misidentifying the sky as vegetation and road, which is consistent with the reports in FIFO [38].

For SDAT-Former++, a 9.50% improvement in the mIoU was achieved after performing fog-invariant feature learning, and the incorporation of positional encoding resulted in a further performance improvement (4.58% higher), indicating that positional encoding effectively enhanced the depiction of fog-related details in images. Figure 5 demonstrates this in two aspects: (1) capturing motion blur and (2) improving the identification of obscured objects within the fog. As shown in the first row, the original SDAT-Former exhibited incomplete segmentation of nearby objects, whereas SDAT-Former++ effectively overcame motion blur, thereby contributing to safer vehicle behavior. In the second row, SDAT-Former failed to detect a tree hidden in the dense fog, whereas the new version with positional encoding accurately captured this obscured element.

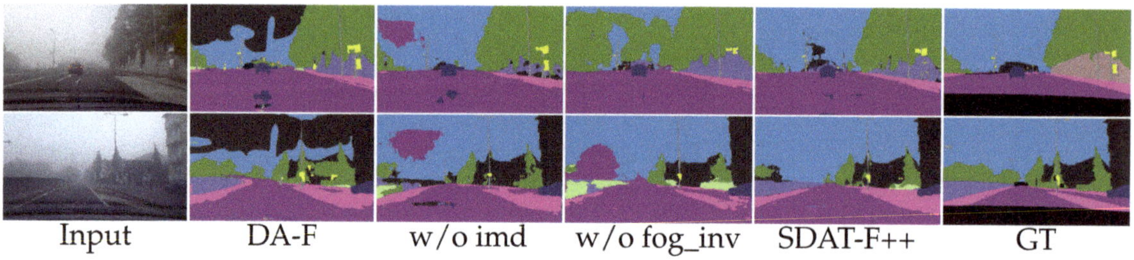

Figure 4. Qualitative results of ablation study. These experiments were conducted on the Foggy Zurich-test dataset. Both points (i.e., intermediate domain style learning (Column 3) and fog-invariant feature learning (Column 4)) yielded significant improvements compared to the baseline.

Figure 5. Qualitative results for the incorporation of positional encoding in fog-invariant learning. From left to right: input image, performance without/with position encoding. Compared to the original method [1], our method can better overcome incomplete segmentation caused by motion blur and effectively identify objects obscured by dense fog.

4.2. *Effectiveness of Style-Invariant Features Learning*

In Figure 4, without the help of the intermediate domain, the segmentor misjudged the sky and some ground categories, even with the fog-invariant module. Interestingly, the original DAFormer identified the sky as buildings, but after adding the intermediate domain information, this prediction became vegetation and road. This also illustrates the influence of style information implicitly.

The knowledge from the intermediate domain was mainly used to help the segmentor address the style gap. For SDAT-Former, the segmentor achieved an mIoU gain of +1.60%

by learning on the intermediate domain. For SDAT-Former++, this gain was 3.71%. As mentioned before, pseudo-label learning based on style transformations introduces noise. Figure 6 shows some bad pseudo-labels with artifacts and incomplete segmentation of entities. This can inevitably affect training. After SDAT-Former++ adopted mask learning, these problems were avoided.

Figure 6. Qualitative comparison of using masked learning in the intermediate domain. From left to right: Input image, bad prediction by SDAT-Former [1], refined prediction by our method. The original version uses style transfer, which can inevitably lead to artifacts in predictions, whereas SDAT-Former++ does not.

4.3. Effectiveness of Cyclical Training

The main purpose of cyclical training was to integrate different training paradigms. It did not significantly improve the performance of the segmentor, but its absence could have been fatal. In Table 3, it can be seen that our segmentor obtained an mIoU gain of +0.96% using cyclical training because no changes happened in the sub-modules. After using dynamic weight allocation, the performance improved by +1.42%. However, without cyclical training, our model only achieved an mIoU of 10.2, which means that training failed. In addition, cyclical training was also necessary for fog-invariant feature learning. This method effectively prevents gradient confusion in the temporal dimension and is a promising training strategy for the future.

4.4. What Does SDAT-Former++ Learn?

To further investigate the roles of masked learning and fog-invariant learning, we visualized the feature maps of the style-invariant backbone b_{ψ}^{si} and the fog-invariant backbone b_{ω}^{fi}. We averaged the second dimension of the multi-channel tensor, where brighter pixels indicate higher values. In Figure 7, from left to right are the intermediate domain image, the output of b_{ψ}^{si} without masked learning (SDAT-Former [1]), its target domain image, and the output of b_{ω}^{fi} without and with positional encoding. Qualitatively, the model b_{ψ}^{si} focused more on contextual information and extracted more complete features after using masked learning, which was mostly domain-independent (such as edges and contours). On the other hand, the fog-invariant backbone performed a distinct "binary classification" on objects and fog, with the classification becoming more refined after the use of feature enhancement through positional encoding. Both of these knowledge transfer processes were handed over to the teacher network h_{φ}, demonstrating the robust recognition ability of SDAT-Former++.

Figure 7. Qualitative feature maps of b_{ψ}^{si} and b_{ω}^{fi}. From left to right: intermediate domain image, output of b_{ψ}^{si} without masked learning (SDAT-Former [1]) and the case with it, target domain image, output of b_{ω}^{fi} without and with positional encoding.

4.5. Sensitivity Analysis/Adaptability to Fog

We did not design additional modules specifically for fog processing, but our method demonstrated excellent anti-fog interference performance, which was analyzed using entropy. The brighter the pixels in the entropy map, the higher the uncertainty, indicating that the model was more likely to make incorrect judgments. Conversely, the model output more certain segmentation results. However, the model also generated high-certainty but incorrect segmentation. Therefore, only the segmentation models that resulted in low entropy predictions and conformed to the distribution of the real-world scenario were truly notable. We performed predictive entropy analysis on the images from the Foggy Driving Dense dataset [37], as shown in Figure 8. The baseline model DAFormer [25] made highly uncertain predictions on fog-obscured pixels, potentially leading to unsafe situations. SDAT-Former alleviated this but still retained uncertainty. In contrast, our model generated lower uncertainty in dense fog conditions while still producing accurate road and sky segmentation results, demonstrating the exceptional reliability of our method.

Figure 8. Entropy analysis. From left to right: input images (dense fog), entropy map output by DAFormer [25], entropy map output by SDAT-Former [1], and entropy map output by our method. Our method resulted in lower prediction entropy for the pixels occupied by fog, indicating higher confidence in its predictions.

4.6. Number of Images from the Intermediate Domain

We explored the effect of intermediate domain images with varying quantities from different datasets, which is shown in Table 4. Firstly, using an exclusive intermediate domain led to optimal results on the current dataset but did not achieve the same performance on another dataset. For example, using intermediate domain images from the ACDC dataset resulted in a segmentor mIoU of 47.42% on the Foggy Zurich dataset. This was due to the style variations between the datasets. Secondly, in the same dataset, the number of images from the intermediate domain had little influence on the final performance. In other words, the corresponding relationship between the clear domain and the foggy domain does not need to be very strict, indicating the segmentor has adaptability in both fog-invariant feature learning and intermediate domain segmentation learning.

Table 4. Discussion about the usage of intermediate domain images. We chose different numbers of clear images from the different datasets, denoted as $\mathcal{M}$. The results are measured by the mIoU (%).

Discussion of Numbers					mIoU	
	400 [1]	600 [2]	1000 [3]	1600 [4]	ACDC	FZ
Number of images from intermediate domain	✓				56.19	47.42
		✓			54.17	51.61
			✓		59.28	53.82
				✓	58.34	53.97

[1] Clear reference images from the training set of the ACDC fog dataset. [2] Manually selected images from the "light fog" category in the Foggy Zurich dataset. [3] Combination of 400 images from the ACDC dataset and 600 images from the FZ dataset. [4] Remaining 600 reference images from the ACDC fog validation/test set.

4.7. Generalization to Rainy and Snowy Scenes

We found that SDAT-Former++ could make better predictions for clear images (Figure 9). We used the trained SDAT-Former++ to re-predict the intermediate domain images and obtained surprisingly high-quality pseudo-labels. This indicates that the target domain is also an "extension domain" to the intermediate domain, forcing the model to complete more difficult tasks, potentially improving performance in the current task. Furthermore, we tested our method on the rain and snow validation sets of the ACDC dataset (Table 5 and Figure 9), showing improvements compared to DAFormer, indicating the potential of our method in addressing the understanding of different adverse scenes.

Table 5. Generalization to other adverse scenes. We conducted zero-shot testing on the snowy and rainy validation sets of the ACDC dataset.

Generalization on ACDC Validation Subsets		Rain	Snow
Method	SegFormer(no UDA) [39]	40.62	42.03
	DAFormer(baseline) [25]	48.27	49.19
	SDAT-Former [1]	53.99	58.04
	SDAT-Former++	56.83	60.14

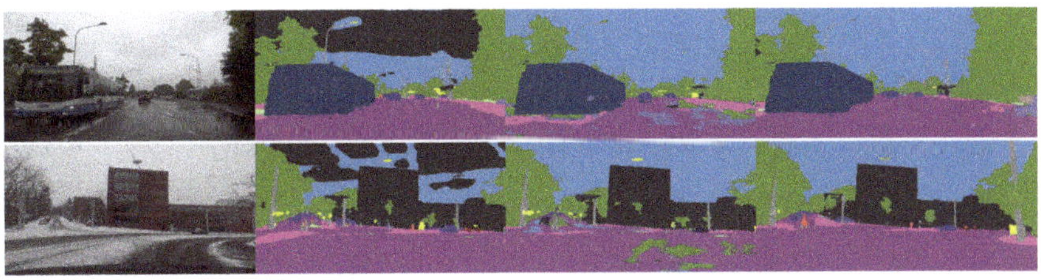

Figure 9. Qualitative results of generalization on rainy and snowy images. From left to right: input images, predictions of DAFormer [25], predictions of SDAT-Former [1], and predictions of our method. These experiments were conducted on the ACDC rain and snow subsets. We directly used the checkpoint acquired by this paper to test without any extra training. The newly proposed SDAT-Former++ greatly improved segmentation compared to DAFormer and the original SDAT-Former.

4.8. Order of EMA Updating

EMA updating is a temporal ensemble algorithm, signifying that $(a(x+b) \neq ax+b)$; thus, different sequences of EMA updating may affect the final parameters of the segmentor.

In Table 6, we present the results of an ablation study on the order of EMA updating. The results show that altering the sequence of EMA updating concerning the teacher segmentor had little effect on performance, which can be attributed to cyclical training.

Table 6. The order of EMA updating. We designed three different sequences for parameter updating.

	Order of EMA Updating			mIoU	Gain
	Fi [2]→T [3]	S [1]→T	TA [4]→T	ACDC	FZ
Configuration	1	2	3	58.14	52.78
	2	1	3	59.24	53.80
	1	3	2	59.17	53.68
	1	2	3	59.28	53.82

[1] "S" represents the student segmentor f_θ. [2] "Fi" represents the fog-invariant backbone b^{fi}_{ω}. [3] "T" represents the teacher segmentor h_φ. [4] "TA" represents the teaching-assistant backbone b^{si}_{ψ}.

4.9. Memory Consumption Comparison

Our method does not require all modules to work simultaneously. We adopt cyclical training where every four iterations constitute one mini-epoch, and only two–three modules need to be executed in each iteration. Specifically, in the first and second iterations, only the student segmentor f_θ and the fog-related modules (b^{fi}_{ω} and $\mathcal{F}$) are involved. The third iteration needs f_θ, b^{si}_{ψ}, and d_r, whereas the fourth iteration needs f_θ and h_φ. The transferring of EMA parameters does not increase memory consumption. Due to the introduction of new loss functions, our method consumes more memory compared to previous methods, but it does not exceed the limit of a Tesla V100 (32 GB). During the testing phase, our method only deploys f_θ; thus, the consumption is consistent with the original SegFormer [39]. In this context, our method is more like online knowledge distillation, aiming to train a better student network. We provide a comparison of the memory consumption between our method and DAFormer [25], SegFormer [39], and SDAT-Former [1] during the training and testing phases in Table 7.

Table 7. Memory consumption comparison. We recorded the memory consumption during training and testing when batch_size =1, with an input size of 512 × 512 for both the source domain and target domain images, measured in GB.

	Memory Consumption Comparison (GB)				
	Train				Test
Mini-epoch	Iter 4n	Iter 4n + 1	Iter 4n + 2	Iter 4n + 3	
SegFormer [39]	5.7				5.7
DAFormer [25]	11.3				
SDAT-Former [1]	5.9	7.7	8.3	11.9	
SDAT-Former++	6.4	8.5	9.4	13.3	

5. Conclusions

We propose a stronger domain-adaptive teacher-guided semantic segmentation method called SDAT-Former++. It improves both style-invariant and fog-invariant feature learning. Specifically, we replace the strategy of generating pseudo-labels using supervised learning with a simple yet effective masked learning strategy. This integrates all training processes into an end-to-end framework, greatly simplifying the training process and improving performance. Furthermore, we enhance the fog-invariant feature learning module by introducing positional encoding, guiding the model to learn more refined fog-related features and scene contours. In the information integration part, we use consistency learning to accelerate model convergence and narrow the gap between the student and teacher segmentors.

Experimental results demonstrate that SDAT-Former++ surpasses the baseline methods on mainstream foggy road scene datasets. It achieves improvements of 3.3%, 4.8%, 1.1%, and 0.4% on the ACDC Fog, Foggy Zurich, Foggy Driving, and Foggy Driving Dense datasets, respectively. Through analysis of the model outputs, we find that both intermediate domain learning and fog-invariant feature learning in SDAT-Former++ have positive effects, alleviating the issue of prediction artifacts in the baseline methods. When facing dense fog, the proposed method exhibits lower uncertainty and demonstrates good safety performance. Visualizing the model's feature maps also reveals that intermediate domain data primarily focuses on learning domain-style independent features (such as contours and edges), whereas fog-invariant feature learning differentiates between fog and entities in the images. Masked learning enables the model to better capture contextual information rather than specific details, and positional encoding generates better contour information, assisting the main segmentation model in producing better edges. Our method also shows generalization ability to other adverse scenes such as rainy and snowy scenes.

In future studies, we plan to further research the fog factor and attempt to more accurately avoid its influence. We also plan to research the unified segmentor, which is suitable for all adverse conditions.

Author Contributions: Conceptualization, Z.W. and Z.Z; methodology, Z.W., Z.Z.m and Z.J.; software, Z.W. and Z.J.; validation, Z.W., Y.Z., and Y.Y.; formal analysis, Y.Z., L.L., and L.Z.; investigation, Z.W., Z.Z., Y.Y., and L.L.; data curation, Z.Z., Z.J., Y.Y., L.L., and L.Z.; writing—original draft preparation, Z.W., Z.J., and L.Z; writing—review and editing, Z.W., Z.Z, Z.J., L.L., and L.Z.; visualization, Y.Z. and Z.Z.; supervision, Y.Z. and Z.Z.; project administration, Y.Z.; funding acquisition, Y.Z. and Y.Y. All authors have read and agreed to the published version of the manuscript.

Funding: This work was supported by the National Natural Science Foundation of China under Grant 42071340 and the Program of Song Shan Laboratory (included in the management of Major Science and Technology of Henan Province) under Grant 2211000211000-01.

Data Availability Statement: Data are contained within the article .

Conflicts of Interest: The authors declare no conflict of interest.

References

1. Wang, Z.; Zhang, Y.; Yu, Y.; Jiang, Z. SDAT-Former: Foggy Scene Semantic Segmentation Via A Strong Domain Adaptation Teacher. In Proceedings of the 2023 IEEE International Conference on Image Processing (ICIP), Kuala Lumpur, Malaysia, 8–11 October 2023; pp. 1760–1764. [CrossRef]
2. Ranft, B.; Stiller, C. The Role of Machine Vision for Intelligent Vehicles. *IEEE Trans. Intell. Veh.* **2016**, *1*, 8–19. [CrossRef]
3. Dai, Y.; Li, C.; Su, X.; Liu, H.; Li, J. Multi-Scale Depthwise Separable Convolution for Semantic Segmentation in Street–Road Scenes. *Remote Sens.* **2023**, *15*, 2649. [CrossRef]
4. Liu, Q.; Dong, Y.; Jiang, Z.; Pei, Y.; Zheng, B.; Zheng, L.; Fu, Z. Multi-Pooling Context Network for Image Semantic Segmentation. *Remote Sens.* **2023**, *15*, 2800. [CrossRef]
5. Šarić, J.; Oršić, M.; Šegvić, S. Panoptic SwiftNet: Pyramidal Fusion for Real-Time Panoptic Segmentation. *Remote Sens.* **2023**, *15*, 1968. [CrossRef]
6. Lv, K.; Zhang, Y.; Yu, Y.; Zhang, Z.; Li, L. Visual Localization and Target Perception Based on Panoptic Segmentation. *Remote Sens.* **2022**, *14*, 3983. [CrossRef]
7. Li, X.; Xu, F.; Lyu, X.; Gao, H.; Tong, Y.; Cai, S.; Li, S.; Liu, D. Dual attention deep fusion semantic segmentation networks of large-scale satellite remote-sensing images. *Int. J. Remote Sens.* **2021**, *42*, 3583–3610. [CrossRef]
8. Li, X.; Xu, F.; Xia, R.; Li, T.; Chen, Z.; Wang, X.; Xu, Z.; Lyu, X. Encoding contextual information by interlacing transformer and convolution for remote sensing imagery semantic segmentation. *Remote Sens.* **2022**, *14*, 4065. [CrossRef]
9. Li, X.; Xu, F.; Liu, F.; Xia, R.; Tong, Y.; Li, L.; Xu, Z.; Lyu, X. Hybridizing Euclidean and Hyperbolic Similarities for Attentively Refining Representations in Semantic Segmentation of Remote Sensing Images. *IEEE Geosci. Remote Sens. Lett.* **2022**, *19*, 1–5. [CrossRef]
10. Li, X.; Xu, F.; Liu, F.; Lyu, X.; Tong, Y.; Xu, Z.; Zhou, J. A Synergistical Attention Model for Semantic Segmentation of Remote Sensing Images. *IEEE Trans. Geosci. Remote Sens.* **2023**, *61*, 1–16. [CrossRef]
11. Sakaridis, C.; Dai, D.; Van Gool, L. Semantic foggy scene understanding with synthetic data. *Int. J. Comput. Vis.* **2018**, *126*, 973–992. [CrossRef]
12. Narasimhan, S.G.; Nayar, S.K. Contrast restoration of weather degraded images. *IEEE Trans. Pattern Anal. Mach. Intell.* **2003**, *25*, 713–724. [CrossRef]
13. Michieli, U.; Biasetton, M.; Agresti, G.; Zanuttigh, P. Adversarial Learning and Self-Teaching Techniques for Domain Adaptation in Semantic Segmentation. *IEEE Trans. Intell. Veh.* **2020**, *5*, 508–518. [CrossRef]
14. Cordts, M.; Omran, M.; Ramos, S.; Rehfeld, T.; Enzweiler, M.; Benenson, R.; Franke, U.; Roth, S.; Schiele, B. The cityscapes dataset for semantic urban scene understanding. In Proceedings of the IEEE conference on Computer Vision and Pattern Recognition, Las Vegas, NV, USA, 27–30 June 2016; pp. 3213–3223.

15. Goodfellow, I.; Pouget-Abadie, J.; Mirza, M.; Xu, B.; Warde-Farley, D.; Ozair, S.; Courville, A.; Bengio, Y. Generative adversarial networks. *Commun. ACM* **2020**, *63*, 139–144. [CrossRef]
16. Lee, D.H. Pseudo-label: The simple and efficient semi-supervised learning method for deep neural networks. In Proceedings of the Workshop on Challenges in Representation Learning, ICML , 2013; Volume 3, p. 896.
17. Mao, X.; Li, Q.; Xie, H.; Lau, R.Y.K.; Wang, Z.; Smolley, S.P. Least Squares Generative Adversarial Networks. *arXiv* **2016**, arXiv:1611.04076 .
18. Hoffman, J.; Tzeng, E.; Park, T.; Zhu, J.Y.; Isola, P.; Saenko, K.; Efros, A.A.; Darrell, T. CyCADA: Cycle-Consistent Adversarial Domain Adaptation. Computer Vision and Pattern Recognition *arXiv* **2017**, arXiv:1711.03213v3.
19. Chang, W.L.; Wang, H.P.; Peng, W.H.; Chiu, W.C. All about structure: Adapting structural information across domains for boosting semantic segmentation. In Proceedings of the IEEE/CVF Conference on Computer Vision and Pattern Recognition Long Beach, CA, USA, 15–20 June 2019; pp. 1900–1909.
20. Tsai, Y.H.; Hung, W.C.; Schulter, S.; Sohn, K.; Yang, M.H.; Chandraker, M. Learning to adapt structured output space for semantic segmentation. In Proceedings of the IEEE Conference on Computer Vision and Pattern Recognition, Salt Lake City, UT, USA 18–23 June 2018; pp. 7472–7481.
21. Vu, T.H.; Jain, H.; Bucher, M.; Cord, M.; Pérez, P. Advent: Adversarial entropy minimization for domain adaptation in semantic segmentation. In Proceedings of the IEEE/CVF Conference on Computer Vision and Pattern Recognition, Long Beach, CA, USA 15–20 June 2019; pp. 2517–2526.
22. Zou, Y.; Yu, Z.; Liu, X.; Kumar, B.V.; Wang, J. Confidence Regularized Self-Training. In Proceedings of the IEEE International Conference on Computer Vision (ICCV), Seoul, Republic of Korea, 27 October–2 November 2019.
23. Tranheden, W.; Olsson, V.; Pinto, J.; Svensson, L. Dacs: Domain adaptation via cross-domain mixed sampling. In Proceedings of the Proceedings of the IEEE/CVF Winter Conference on Applications of Computer Vision, Virtual, 5–9 January 2021; pp. 1379–1389.
24. Zhang, P.; Zhang, B.; Zhang, T.; Chen, D.; Wang, Y.; Wen, F. Prototypical pseudo label denoising and target structure learning for domain adaptive semantic segmentation. In Proceedings of the IEEE/CVF Conference on Computer vision and Pattern Recognition, Nashville, TN, USA, 20–25 June 2021; pp. 12414–12424.
25. Hoyer, L.; Dai, D.; Van Gool, L. Daformer: Improving network architectures and training strategies for domain-adaptive semantic segmentation. In Proceedings of the IEEE/CVF Conference on Computer Vision and Pattern Recognition, New Orleans, LA, USA, 18–24 June 2022; pp. 9924–9935.
26. Ma, X.; Wang, Z.; Zhan, Y.; Zheng, Y.; Wang, Z.; Dai, D.; Lin, C.W. Both style and fog matter: Cumulative domain adaptation for semantic foggy scene understanding. In Proceedings of the IEEE/CVF Conference on Computer Vision and Pattern Recognition, New Orleans, LA, USA, 18–24 June 2022; pp. 18922–18931.
27. Dai, D.; Sakaridis, C.; Hecker, S.; Van Gool, L. Curriculum model adaptation with synthetic and real data for semantic foggy scene understanding. *Int. J. Comput. Vis.* **2020**, *128*, 1182–1204. [CrossRef]
28. Dai, D.; Gool, L.V. Dark Model Adaptation: Semantic Image Segmentation from Daytime to Nighttime. Computer Vision and Pattern Recognition. *arXiv* **2018**, arXiv:1810.02575 .
29. Bruggemann, D.; Sakaridis, C.; Truong, P.; Gool, L.V. Refign: Align and Refine for Adaptation of Semantic Segmentation to Adverse Conditions. In Proceedings of the IEEE/CVF Winter Conference on Applications of Computer Vision, Waikoloa, HI, USA, 2–7 January 2023; pp. 3174–3184.
30. Li, Y.; Yuan, L.; Vasconcelos, N. Bidirectional learning for domain adaptation of semantic segmentation. In Proceedings of the IEEE/CVF Conference on Computer Vision and Pattern Recognition, Long Beach, CA, USA, 16–17 June 2019; pp. 6936–6945.
31. Laine, S.; Aila, T. Temporal ensembling for semi-supervised learning. *arXiv* **2016**, arXiv:1610.02242.
32. He, K.; Chen, X.; Xie, S.; Li, Y.; Dollár, P.; Girshick, R. Masked autoencoders are scalable vision learners. In Proceedings of the IEEE/CVF Conference on Computer Vision and Pattern Recognition, New Orleans, LA, USA, 18–24 June 2022; pp. 16000–16009.
33. Hoyer, L.; Dai, D.; Wang, H.; Van Gool, L. MIC: Masked image consistency for context-enhanced domain adaptation. In Proceedings of the IEEE/CVF Conference on Computer Vision and Pattern Recognition, Vancouver, BC, Canada, 18–22 June 2023; pp. 11721–11732.
34. Dosovitskiy, A.; Beyer, L.; Kolesnikov, A.; Weissenborn, D.; Zhai, X.; Unterthiner, T.; Dehghani, M.; Minderer, M.; Heigold, G.; Gelly, S.; et al. An image is worth 16x16 words: Transformers for image recognition at scale. *arXiv* **2020**, arXiv:2010.11929.
35. Wang, Z.; Wu, S.; Xie, W.; Chen, M.; Prisacariu, V.A. NeRF–: Neural radiance fields without known camera parameters. *arXiv* **2021**, arXiv:2102.07064.
36. Christos, S.; Dengxin, D.; Luc, V.G. ACDC: The adverse conditions dataset with correspondences for semantic driving scene understanding. In Proceedings of the IEEE/CVF International Conference on Computer Vision, Montreal, BC, Canada, 11–17 October 2021; pp. 10765–10775.
37. Sakaridis, C.; Dai, D.; Hecker, S.; Van Gool, L. Model adaptation with synthetic and real data for semantic dense foggy scene understanding. In Proceedings of the European Conference on Computer Vision (ECCV), Munich, Germany, 8–14 September 2018; pp. 687–704.
38. Lee, S.; Son, T.; Kwak, S. Fifo: Learning fog-invariant features for foggy scene segmentation. In Proceedings of the IEEE/CVF Conference on Computer Vision and Pattern Recognition, New Orleans, LA, USA, 18–24 June 2022; pp. 18911–18921.

39. Xie, E.; Wang, W.; Yu, Z.; Anandkumar, A.; Alvarez, J.M.; Luo, P. SegFormer: Simple and efficient design for semantic segmentation with transformers. *Adv. Neural Inf. Process. Syst.* **2021**, *34*, 12077–12090.
40. Chen, L.C.; Zhu, Y.; Papandreou, G.; Schroff, F.; Adam, H. Encoder-decoder with atrous separable convolution for semantic image segmentation. In Proceedings of the European Conference on Computer Vision (ECCV), Munich, Germany, 8–14 September 2018; pp. 801–818.
41. Gong, R.; Wang, Q.; Danelljan, M.; Dai, D.; Van Gool, L. Continuous Pseudo-Label Rectified Domain Adaptive Semantic Segmentation With Implicit Neural Representations. In Proceedings of the IEEE/CVF Conference on Computer Vision and Pattern Recognition, Vancouver, BC, Canada, 18–22 June 2023; pp. 7225–7235.
42. French, G.; Laine, S.; Aila, T.; Mackiewicz, M.; Finlayson, G. Semi-supervised semantic segmentation needs strong, varied perturbations. *arXiv* **2019**, arXiv:1906.01916.
43. Olsson, V.; Tranheden, W.; Pinto, J.; Svensson, L. Classmix: Segmentation-based data augmentation for semi-supervised learning. In Proceedings of the IEEE/CVF Winter Conference on Applications of Computer Vision, Virtual, 5–9 January 2021; pp. 1369–1378.
44. Jin, Y.; Wang, J.; Lin, D. Semi-supervised semantic segmentation via gentle teaching assistant. *Adv. Neural Inf. Process. Syst.* **2022**, *35*, 2803–2816.
45. Contributors, M. MMSegmentation: Openmmlab semantic segmentation toolbox and benchmark, 2020. Available online: https://gitee.com/open-mmlab/mmsegmentation (accessed on 9 December 2023).
46. Paszke, A.; Gross, S.; Massa, F.; Lerer, A.; Bradbury, J.; Chanan, G.; Killeen, T.; Lin, Z.; Gimelshein, N.; Antiga, L.; et al. Pytorch: An imperative style, high-performance deep learning library. *Adv. Neural Inf. Process. Syst.* **2019**, *32* .
47. Deng, J.; Dong, W.; Socher, R.; Li, L.J.; Li, K.; Fei-Fei, L. Imagenet: A large-scale hierarchical image database. In Proceedings of the 2009 IEEE Conference on Computer Vision and Pattern Recognition, Miami, FL, USA, 20–25 June 2009; IEEE: Manhattan, NY, USA , 2009; pp. 248–255.
48. Loshchilov, I.; Hutter, F. Decoupled weight decay regularization. *arXiv* **2017**, arXiv:1711.05101.
49. Chen, L.C.; Papandreou, G.; Kokkinos, I.; Murphy, K.; Yuille, A.L. Deeplab: Semantic image segmentation with deep convolutional nets, atrous convolution, and fully connected crfs. *IEEE Trans. Pattern Anal. Mach. Intell.* **2017**, *40*, 834–848. [CrossRef] [PubMed]
50. Lin, G.; Milan, A.; Shen, C.; Reid, I. Refinenet: Multi-path refinement networks for high-resolution semantic segmentation. In Proceedings of the IEEE Conference on Computer Vision and Pattern Recognition, Honolulu, HI, USA, 21–26 July 2017; pp. 1925–1934.
51. Kerim, A.; Chamone, F.; Ramos, W.; Marcolino, L.S.; Nascimento, E.R.; Jiang, R. Semantic Segmentation under Adverse Conditions: A Weather and Nighttime-aware Synthetic Data-based Approach. *arXiv* **2022**, arXiv:2210.05626.
52. Zhang, H.; Patel, V.M. Densely Connected Pyramid Dehazing Network. *arXiv* **2018**, arXiv:1803.08396 .
53. Ren, W.; Liu, S.; Zhang, H.; Pan, J.; Cao, X.; Yang, M.H. Single image dehazing via multi-scale convolutional neural networks. In *Proceedings of the Computer Vision—ECCV 2016: 14th European Conference, Amsterdam, The Netherlands, 11–14 October 2016*, Proceedings, Part II 14; Springer: Berlin/Heidelberg, Germany , 2016; pp. 154–169.
54. He, K.; Sun, J.; Tang, X. Single image haze removal using dark channel prior. *IEEE Trans. Pattern Anal. Mach. Intell.* **2010**, *33*, 2341–2353.
55. Yang, Y.; Soatto, S. Fda: Fourier domain adaptation for semantic segmentation. In Proceedings of the IEEE/CVF Conference on Computer Vision and Pattern Recognition, Seattle, WA, USA, 13–19 June 2020; pp. 4085–4095.
56. Berman, D.; Avidan, S. Non-local image dehazing. In Proceedings of the IEEE Conference on Computer Vision and Pattern Recognition, Las Vegas, NV, USA, 27–30 June 2016; pp. 1674–1682.
57. Benjdira, B.; Ali, A.M.; Koubaa, A. Streamlined Global and Local Features Combinator (SGLC) for High Resolution Image Dehazing. In Proceedings of the IEEE/CVF Conference on Computer Vision and Pattern Recognition, Vancouver, BC, Canada, 18–22 June 2023; pp. 1854–1863.

remote sensing

MDPI

Article

Learn to Few-Shot Segment Remote Sensing Images from Irrelevant Data

Qingwei Sun [1,2], Jiangang Chao [2,3,*], Wanhong Lin [2,3], Zhenying Xu [2,3], Wei Chen [2,3] and Ning He [2,3]

1 Department of Aerospace Science and Technology, Space Engineering University, Beijing 101416, China; sunqw@alumni.nudt.edu.cn
2 China Astronaut Research and Training Center, Beijing 100094, China
3 National Key Laboratory of Human Factors Engineering, China Astronaut Research and Training Center, Beijing 100094, China
* Correspondence: xjtucjg@139.com

Abstract: Few-shot semantic segmentation (FSS) is committed to segmenting new classes with only a few labels. Generally, FSS assumes that base classes and novel classes belong to the same domain which limits FSS's application in a wide range of areas. In particular, since annotation is time consuming, it is not cost-effective to process remote sensing images using FSS. To address this issue we designed a feature transformation network (FTNet) for learning to few-shot segment remote sensing images from irrelevant data (FSS-RSI). The main idea is to train networks on irrelevant already labeled data but inference on remote sensing images. In other words, the training and testing data neither belong to the same domain nor category. The FTNet contains two main modules: a feature transformation module (FTM) and a hierarchical transformer module (HTM). Among them the FTM transforms features into a domain-agnostic high-level anchor, and the HTM hierarchically enhances matching between support and query features. Moreover, to promote the development of FSS-RSI, we established a new benchmark, which other researchers may use. Our experiments demonstrate that our model outperforms the cutting-edge few-shot semantic segmentation method by 25.39% and 21.31% in the one-shot and five-shot settings, respectively.

Keywords: meta-learning; cross-domain segmentation; few-shot semantic segmentation; transformer

Citation: Sun, Q.; Chao, J.; Lin, W.; Xu, Z.; Chen, W.; He, N. Learn to Few-Shot Segment Remote Sensing Images from Irrelevant Data. *Remote Sens.* **2023**, *15*, 4937. https://doi.org/10.3390/rs15204937

Academic Editor: Shuying Li

Received: 21 August 2023
Revised: 27 September 2023
Accepted: 11 October 2023
Published: 12 October 2023

1. Introduction

Deep learning-based semantic segmentation is widely used in remote sensing [1,2]. Generally, semantic segmentation provides pixel-level classification for downstream applications, which is a fundamental computer vision task. Many models have been built by adopting fully convolutional networks and have achieved satisfactory results [3–5]. On this basis, novel modules such as encoder–decoder [6,7], dilated convolution [8], and atrous spatial pyramid pooling [9] have been proven to be effective. Indeed, pre-trained backbones, such as the ResNet [10] and VGG [11], have been utilized in various semantic segmentation models [8–10] for feature extraction, which has gradually become a stereotype. By contrast, VIT [12], SETR [13], and SegFormer [14] divide images into a patch sequence. In these works, transformers were used to extract image features [15,16], and their results surpassed traditional methods to some extent.

However, a large dataset is needed during training, thus limiting semantic segmentation's application in a broader field. FSS has been proposed [17,18] to solve this limitation. Unlike supervised learning-based methods, FSS requires only a few annotations to segment new classes.

There are no overlap categories between training and inference for FSS [19], which is the main difference between few-shot semantic segmentation and semantic segmentation. Most FSS methods follow meta-learning [20], where episodes are formed by image and label pairs [17,21,22] to mimic few-shot scenes. Currently, FSS is mainly divided into two groups:

relation-based methods and metric-based methods. Among them, relation-based methods [18,19,22–24] share the same backbone and freeze their parameters during training. The main idea is to design a practical decoder to compare the query and support data. In contrast, metric-based methods [21,25] tend to develop effective encoders to separate foreground and background classes. Furthermore, some works [26,27] bring transformers into FSS tasks with excellent results. As for remote sensing, it is time-consuming to obtain numerous annotated data. Therefore, some works [28–30] have aimed to reduce the need for annotations or use semi-supervised methods [31] to handle unknown categories.

Generally, FSS is conducted in a cross-validation manner with four splits [32]. Although there is no class overlap between the training and testing sets, they belong to the same domain. For example, there are 20 categories in PASCAL-5^i [17], but each class's pixel distribution is similar, called the in-domain dataset. Additionally, although FSS is named "few-shot", a large, labeled dataset is still needed during training, which is inconvenient for remote sensing. We aim to train such a network on a large but irrelevant dataset and to predict masks on remote sensing images.

This work extends few-shot semantic segmentation to a new task called FSS-RSI. As we know, FSS's training and testing sets contain different categories within the same domain. By contrast, FSS-RSI's data differ not only in classes but also in image acquisition sensor and pixel distributions, which belong to irrelevant/cross-domain data.

To achieve the goal of FSS-RSI, the FTNet was designed. The meta-learning method [20] was adopted to train our network. Specifically, the FTNet transforms the support and query features into a domain-agnostic space with the learnable FTM. In this way, the gap between the support data and the query data is narrowed. In addition, the HTM is used to parse the correlations between the support and query features, which fully promotes the fitting capability of the support and query features.

To validate our network and provide convenience for other researchers, we established a new benchmark. The images used came from four different datasets, DeepGlobe [33], Potsdam [34], Vaihingen [35], and AISD [36], which were captured by satellites or drones. All four datasets are typical in remote sensing and contain commonly used categories in engineering. We combined these datasets into an FSS-format dataset and used them as a benchmark for FSS-RSI.

PASCAL-5^i and our benchmark were used for our experiments. The FTNet achieves comparable accuracy to the cutting-edge method on the in-domain dataset. As for FSS-RSI, the FTNet performs at an absolute advantage. The mIoUs in the one-shot and five-shot settings were 25.39% and 21.31%, respectively, higher than the state-of-the-art (SOTA) method.

In summary, our main contributions lie in the following aspects:

- We extend the FSS to FSS-RSI, which aims to utilize irrelevant domain data to guide the segmentation of remote sensing images.
- A new benchmark is proposed. This benchmark may promote the development of FSS-RSI and serve as a tool for researchers.
- We propose an effective network with the FTM and the HTM. Our method significantly outperforms the cutting-edge few-shot semantic segmentation method in the FSS-RSI task.

2. Method

2.1. Problem Setting

Table 1 shows the differences between semantic segmentation (SS), FSS, and FSS-RSI. We define the training and testing data as domains $\boldsymbol{D}_{\text{train}}$ and $\boldsymbol{D}_{\text{test}}$ and their semantic categories as $\boldsymbol{C}_{\text{train}}$ and $\boldsymbol{C}_{\text{test}}$, respectively.

Table 1. Differences between SS, FSS, and FSS-RSI.

Task	Data Source	Categories	Example	
			Training Pair	Testing Pair
SS	$D_{\text{train}} = D_{\text{test}}$	$C_{\text{train}} \cap C_{\text{test}} = 1$	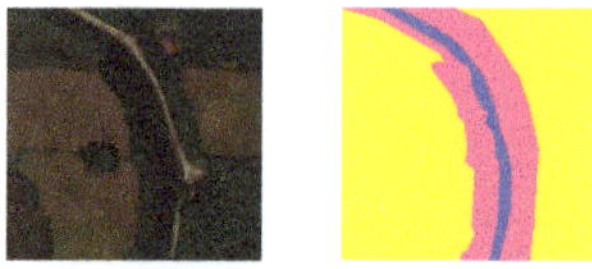	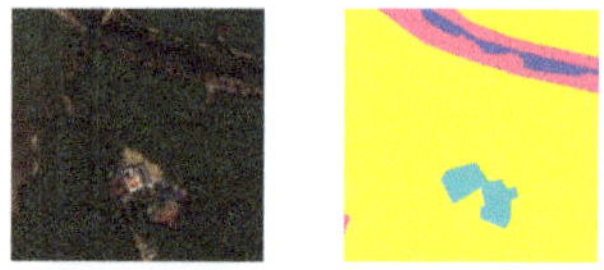
FSS	$D_{\text{train}} = D_{\text{test}}$	$C_{\text{train}} \cap C_{\text{test}} = \varnothing$	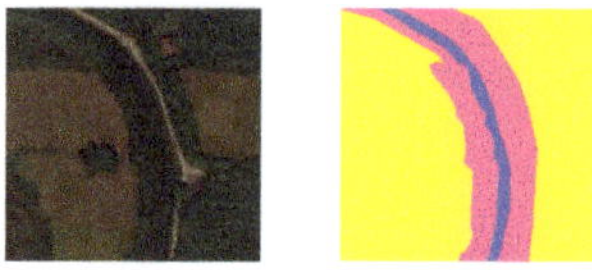	
FSS-RSI	$D_{\text{train}} \neq D_{\text{test}}$	$C_{\text{train}} \cap C_{\text{test}} = \varnothing$	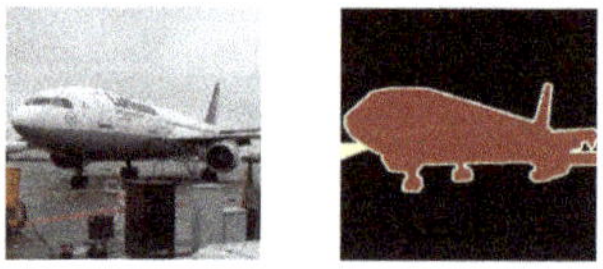	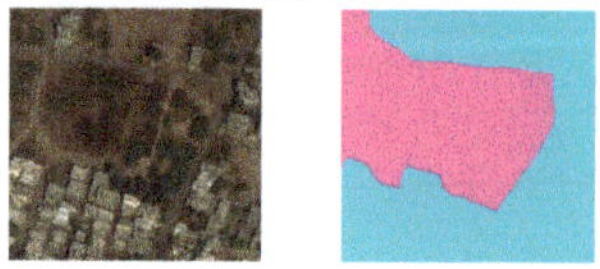

For SS, D_{train} and D_{test} belonged to the same domain, specifically remote sensing, in our task. The training and testing categories were the same. That is, SS only handles classes that have appeared in training. For FSS, both D_{train} and D_{test} were derived from remote sensing, but their categories did not overlap. That is, FSS can process classes with no appearance during training. FSS-RSI was the most challenging task, with D_{train} and D_{test} originating from different domains. The two domains have different classes and pixel distributions, which we call irrelevant data.

In the FSS-RSI task, episodes [18] were used to mimic few-shot scenes. Each episode consisted of a query set $Q = \{(I^q, M^q)\}$ and a support set $S = \{(I_i^s, M_i^s)\}_{i=1}^K$. In our study, $(\cdot,\cdot)$ represents image pairs consisting of RGB images and corresponding masks. $I^s, I^q \in \mathbb{R}^{H\times W\times 3}$ denotes the RGB images. $M^s, M^q \in \mathbb{R}^{H\times W}$ represents their masks. K means K pairs of images and masks were used, which we call the K-shot. S_{train} and $I^q \in Q_{\text{train}}$ are the inputs during training. The proposed network predicts a binary mask to compute loss with $M^q \in Q_{\text{train}}$. In the testing phase, the network predicted a new mask with S_{test} and $I^q \in Q_{\text{test}}$ as the inputs. It should be noted that $S_{\text{train}}, Q_{\text{train}} \subset D_{\text{train}}$, and $S_{\text{test}}, Q_{\text{test}} \subset D_{\text{test}}$, respectively.

2.2. Model

The FTNet is designed to deal with FSS-RSI tasks. As shown in Figure 1, the network is built in a meta-learning manner [20]. Specifically, we used ResNet50 [10], which was pre-trained by ImageNet [37], as the backbone and froze its parameters during training. The query and support branches share the same backbone to extract multi-layered features. Furthermore, a prior mask [18] from high-level feature maps was introduced to strengthen the connection between the query and support data. It should be noted that support masks are important for FSS. Therefore, the FTNet adopted a support mask several times to enhance its guidance for the query images. In particular, the FTM is designed to transform the middle query feature, support feature, and prototype into a domain-independent, high-level feature space called the feature anchor. The FTNet achieves better performance when processing FSS-RSI tasks with the FTM. In addition, we input the fused query feature, support feature, and mask into the HTM, which enhanced information fusion within and between features. Figure 1 shows the model architecture of a one-shot structure, which can be easily expanded to a five-shot structure.

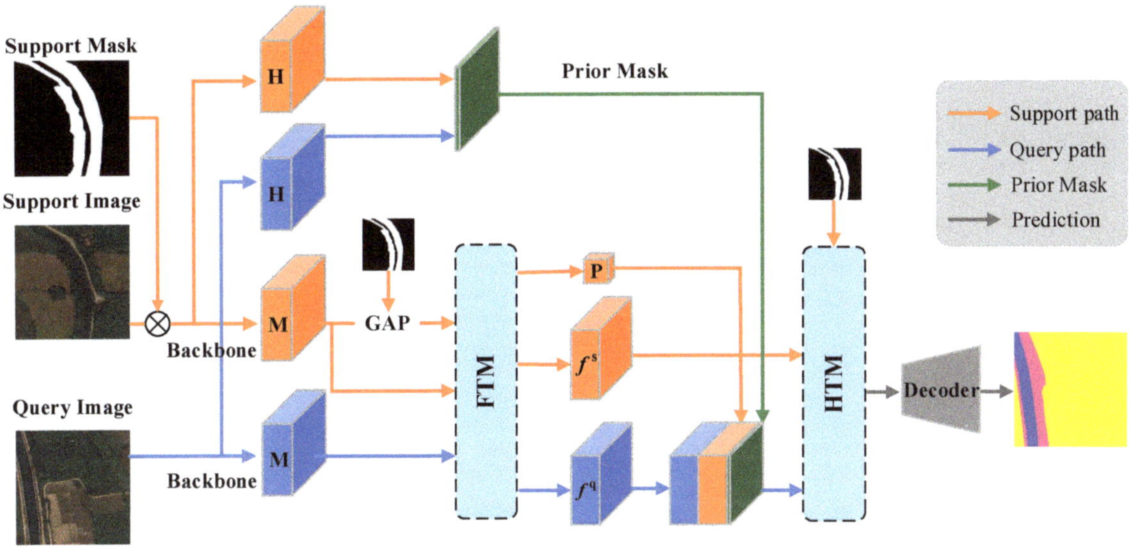

Figure 1. The architecture of the FTNet. This network was built in a meta-learning manner with a prior mask [18]. The FTM and HTM are designed for better performance. H is the high-level feature, M is the middle-level feature, P is the prototype, f^s is the support feature, f^q is the query feature, $\otimes$ is the element-wise multiplication, and GAP denotes the global average pooling.

2.2.1. Feature Extraction

During the training phase, we froze the backbone's parameters, which was the strategy employed by other methods [18,19,26]. There are five stages included in ResNet50. The FTNet mainly adopts feature maps for stage 3, stage 4, and stage 5, which are denoted as f_{s3}, f_{s4}, and f_{s5}. In order to enhance the performance of high-level feature maps, PPM [38] was used to refactor stage 5. Thus, we obtained f_{s6} as the following:

$$f_{s6} = \underset{i=1,2,3,4}{\mathcal{F}_{\text{cat}}} \left(\mathcal{U}^i \left(\mathcal{F}^i_{\text{conv}} \left(\mathcal{F}^i_{\text{pool}}(f_{s5}) \right) \right) \right) \tag{1}$$

where $\mathcal{F}_{\text{pool}}$ means the average pooling and $\mathcal{F}_{\text{conv}}$ denotes the convolution, followed by the BatchNorm [39] and ReLU functions. $\mathcal{U}$ is the upsampling and $\mathcal{F}_{\text{cat}}$ represents the concatenation. Pyramids with the sizes 1×1, 2×2, 3×3, and 6×6 were used, and i is the level of pyramid. f_{s4}, f_{s5}, and f_{s6} were resized to be the same as f_{s3}.

According to [22,40–42], middle-level features contain more semantic information, such as the outline and color. Therefore, the FTNet concatenates f_{s3} and f_{s4} to obtain a middle-level feature map:

$$f^s_m = \mathcal{F}_{\text{conv}}(\mathcal{F}_{\text{cat}}(f^s_{s3}, f^s_{s4})) \in \mathbb{R}^{c \times h \times w} \tag{2}$$

$$f^q_m = \mathcal{F}_{\text{conv}}\left(\mathcal{F}_{\text{cat}}\left(f^q_{s3}, f^q_{s4}\right)\right) \in \mathbb{R}^{c \times h \times w} \tag{3}$$

where f^s_m and f^q_m are the middle-level features of the support and query data. $\mathcal{F}_{\text{conv}}$ and $\mathcal{F}_{\text{cat}}$ are the same as in Function (1) with different parameters. Furthermore, we calculated the prototype using the support mask and f^s_m as the following:

$$p^s = \mathcal{F}_{\text{gap}}(f^s_m \odot \mathcal{R}(M^s)) \in \mathbb{R}^c \tag{4}$$

where $\odot$ means the Hadamard product, and $\mathcal{R}$ represents the operation to reshape the initial query mask from $\mathbb{R}^{H\times W}$ to $\mathbb{R}^{c\times h\times w}$, with the same size as f_m^s. $\mathcal{F}_{\text{gap}}$ is the global average pooling to reshape the feature map from $\mathbb{R}^{c\times h\times w}$ to $\mathbb{R}^{c\times 1}$.

In addition, the prior mask generated by the high-level feature boosts performance in a training-class-insensitive way [18]. We used f_{s4} and f_{s5} to generate prior masks and merge them as the following:

$$M^p = \mathcal{F}_{\text{cat}}(\mathcal{P}(f_{s4}), \mathcal{P}(f_{s5})) \tag{5}$$

where $\mathcal{P}$ denotes the generation of a prior mask M^p.

2.2.2. Feature Transformation Module

Motivation. Features extracted by convolutions have an excellent characterization within category and domain. As for FSS-RSI, the parameters learned during training tend to segment the categories that appear during training. Therefore, the FTNet transforms features into a space independent of classes and domains. This strategy reduces the influence of the source domain and training data. Inspired by a task-adaptive feature transformer (TAF) [43], we propose a simple learnable transformation matrix that transforms f_m^s, f_m^q and p^s to a domain-agnostic space.

For the feature matrix $\boldsymbol{F}$, the goal was to find a matrix $\boldsymbol{T}$ that transforms $\boldsymbol{F}$ to a domain-independent feature matrix $\boldsymbol{W}$, called the feature anchor, as the following:

$$\boldsymbol{TF} = \boldsymbol{W} \tag{6}$$

In general, $\boldsymbol{F}$ is a non-square matrix with no inverse. One solution is to calculate the pseudo-inverse [43] of $\boldsymbol{F}$ is $\boldsymbol{F}^+ = \{\boldsymbol{F}^T\boldsymbol{F}\}^{-1}\boldsymbol{F}^T$. Thus, the transformation matrix was obtained as the following:

$$\boldsymbol{T} = \boldsymbol{WF}^+ \tag{7}$$

The parameters of $\boldsymbol{W}$ were initialized randomly and changed with the gradient's backpropagation; therefore, the matrix $\boldsymbol{T}$ was constantly optimized.

Specifically, for the prototype p^s, we obtained $p^s_{\text{new}} = \boldsymbol{T}p^s$. As for f_m^s, we needed to transform it as the following:

$$f_m^{s'} = \mathcal{R}(f_m^s) \tag{8}$$

where $\mathcal{R}(\cdot)$ represents the reshape operation: $\mathbb{R}^{c\times h\times w} \to \mathbb{R}^{c\times(h\times w)}$. We used the transformation matrix to multiply Formula (8) to obtain $f^{s'}_{m,\text{new}} = \boldsymbol{T}f_m^{s'}$. Furthermore, $f^{s'}_{m,\text{ new}}$ was transformed to the original shape as f_m^s, that is,

$$f^s_{m,\text{ new}} = \mathcal{R}^{-1}\left(f^{s'}_{m,\text{ new}}\right) \tag{9}$$

where the inverse reshape is included. The same operation was performed for f_m^q. Finally, p^s_{new}, $f^s_{m,\text{ new}}$, and $f^q_{m,\text{ new}}$ were obtained. They are domain-agnostic features. In other words, the gap between p^s_{new} and $f^q_{m,\text{ new}}$ and the gap between $f^s_{m,\text{ new}}$ and $f^q_{m,\text{ new}}$ were significantly reduced. We provide a more detailed explanation in Appendix A.

We further merged p^s_{new}, $f^q_{m,\text{ new}}$, and M^p to obtain f^q_{merge} as the following:

$$f^q_{\text{merge}} = \mathcal{F}_{\text{cat}}\left(f^q_{m,\text{ new}}, M^p, \mathcal{J}(p^s_{\text{new}})\right) \tag{10}$$

where $\mathcal{J}(\cdot)$ is the repeat operation: $\mathbb{R}^{c\times 1} \to \mathbb{R}^{c\times h\times w}$. It should be noted that only the parameters of $\boldsymbol{W}$ were learnable. The transformation matrix was calculated directly. This method does not add too many parameters.

2.2.3. Hierarchical Transformer Module

Motivation. A transformer is used in many works to extract features [12–14]. It establishes relationships within and between features to mine the connections between image blocks. As illustrated in [27], prototype-based FSS models are committed to providing class-wise clues rather than pixel-wise clues. We adopted self- and cross-attention paradigms to mine deep matching correlations.

To strengthen the support data's performance, we again used support masks. We define $Q = W^q f^q$, $K = W^k(f^s \odot M^s)$, and $V = W^v(f^s \odot M^s)$. We followed the usual transformer calculation, which is formulated as the following:

$$Trans(Q, K, V) = \text{softmax}\left(\frac{QK^T}{\sqrt{d}}\right)V \tag{11}$$

where W^q, W^k, and W^v are the learnable parameters, and d is the hidden dimension. Equation (11) is a general form of cross-attention. As shown in Figure 2, when the features f^s in K and V are replaced with f^V, Equation (11) represents the self-attention manner, which represents the relationship among query features. The main task of the HTM is to calculate an informative query feature. Thus, we only performed self-attention within the query path in a standard multi-head manner [12]. In addition, a cross-attention layer follows self-attention. Similar to [27], Q was obtained from the query features, and K and V were obtained from the support features. Inspired by the ResNet [10] and SegFormer [14], we design a hierarchical architecture with four scale blocks. Each block contains self- and cross-attention, followed by downsampling. At the end of the HTM process, we concatenated the four blocks after scaling them to the same resolution. In a nutshell, our model extracts abundant information within query features and obtains pixel-wise matching correlations using a cross-attention layer. We demonstrate the role of the HTM in mining this matching relationship, as detailed in Appendix B.

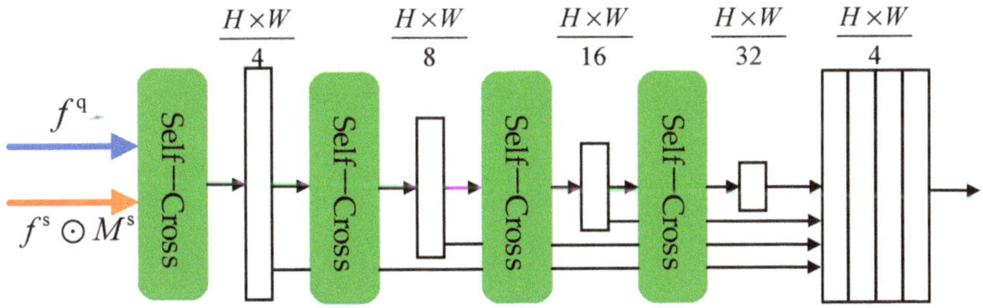

Figure 2. The architecture of the HTM used in our network.

Finally, the FTNet adopts a simple decoder to generate predictions, mainly including stacked convolutional layers and upsampling. Because FSS is a binary classification task, binary cross-entropy loss (BCE) was used to optimize our model, which is formulated as the following:

$$L = \frac{1}{n}\sum_{i=1}^{n} \text{BCE}\left(\mathcal{P}_i^q, M^q\right) \tag{12}$$

where n is the number of episodes in each batch and $\mathcal{P}$ is the prediction of the query image.

2.2.4. Extension to K-Shot

Extending our model to K-shot ($K > 1$) format was straightforward. The K-shot setting means that there are K support pairs in one episode. Specifically, the support pair is $\mathbf{S} = \left\{\left(I_i^s, M_i^s\right)\right\}_{i=1}^{K}$ and the query pair is still $\mathbf{Q} = \{(I^q, M^q)\}$. In order not to change the model's settings, we concatenated the K groups in the channel dimension directly as the following:

$$S = \left\{ \left(\mathcal{F}_{\text{cat}}(I_i^s)_{i=1}^K, \mathcal{F}_{\text{cat}}(M_i^s)_{i=1}^K \right) \right\} \tag{13}$$

Therefore, the FTNet obtained an input with the same structure as a one-shot structure with simple convolutions.

2.3. Benchmark

The FSS-RSI benchmark was derived from four datasets of remote sensing, including DeepGlobe [33], Potsdam [34], Vaihingen [35], and AISD. These datasets differ from the commonly used FSS datasets regarding their pixel distributions and categories.

DeepGlobe [33] consists of natural landscape images taken by satellites. This dataset is annotated into seven categories: unknown, urban, aquatic, agricultural, forested, barren, and rangeland areas. The ground sampling distance was 50 cm. However, only 803 training images were labeled with a size of 2448 × 2448. Fortunately, this dataset was adopted as a tool for FSS-RSI, so DeepGlobe's training set could meet the need. Specifically, we divided each image into 36 equal blocks and resized them to 400 × 400. The category "unknown" was set as the background. Images that contained only a single category were filtered out. Finally, we obtained 9175 pairs containing six categories. We named this dataset FSS-RSI-DeepGlobe; some samples are shown in Figure 3a.

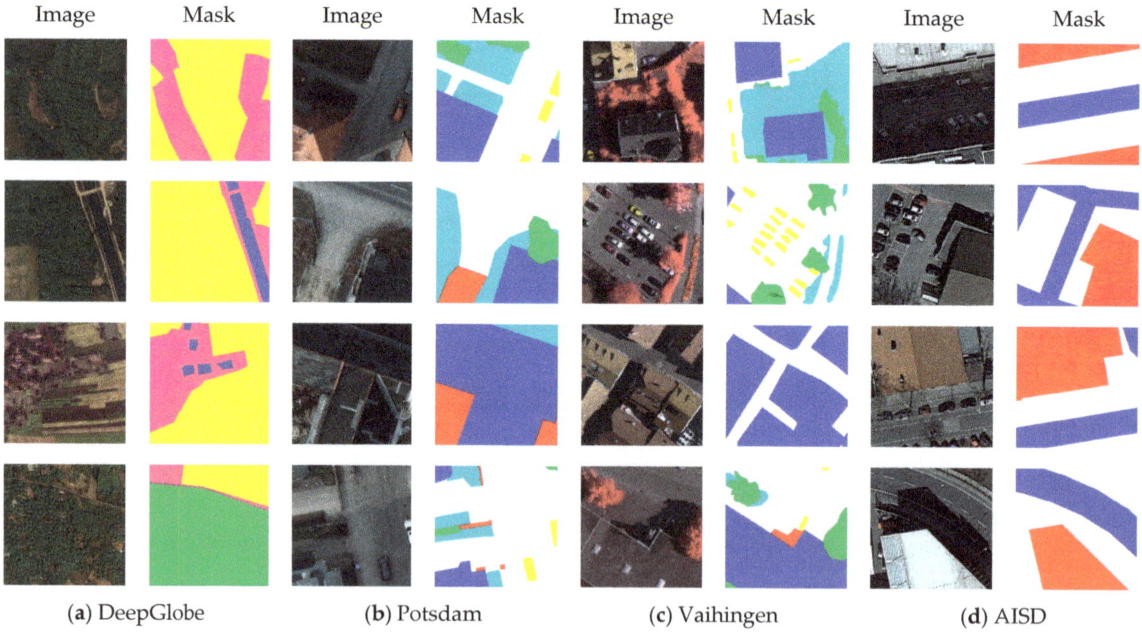

Figure 3. Some images and their corresponding masks of our benchmark: (**a**) data from DeepGlobe; (**b**) data from Potsdam; (**c**) data from Vaihingen; and (**d**) data from AISD.

Potsdam [34] was captured over Potsdam in Germany by aerial cameras. This dataset is annotated into six categories: clutter, tree, low vegetation, building, car, and impervious surface. The ground sampling distance of the images was 5 cm. We removed the category "car" because of the overlap with the source domain used in our work. The buildings in Potsdam are scattered and the category distribution is more balanced. Potsdam contains 38 image patches. The size of all images is 6000 × 6000. Each image was divided into 225 equal pieces. Similar to the DeepGlobe dataset, we removed pairs with a single category.

Finally, we obtained 1896 pairs containing five categories. We named this dataset FSS-RSI-Potsdam; some samples are shown in Figure 3b.

Vaihingen [35] was captured over Vaihingen in Germany by aerial cameras and includes five categories (after removing "car"), like Potsdam. The ground sampling distance of the images was 9 cm. Unlike Potsdam, the class distribution is more compact, with dense settlement structures, narrow streets, and large buildings. Vaihingen contains 33 patches of different sizes. We resized the images to 2800 × 2000 pixels and divided each image into 35 equal pieces. We removed images with a single category. As already known, episodes needed to be built in each category. However, the filtered Vaihingen dataset contains only 6 "clutter" samples, which was insufficient to build a rich episode. Thus, images containing the category "clutter" were discarded. Finally, we obtained 308 pairs containing four categories. We named this dataset FSS-RSI-Vaihingen; some images are shown in Figure 3c.

AISD [36] is an aerial image segmentation dataset obtained using the OpenStreetMap [44–46] and Google Maps [47]. AISD covers parts of different cities, of which Berlin was selected for our experiment. There are only two categories in AISD: road and building. However, their appearance is very similar within and between the two categories. Thus, we believe AISD is a challenging task for FSS. AISD contains 200 patches of the same size, at 2611 × 2453. We resized the images with 2800 × 2400 pixels and divided each image into 42 equal pieces. We removed images with a single category similar to the other three datasets. Finally, we obtained 5640 pairs containing two categories. We named this dataset FSS-RSI-AISD; some samples are shown in Figure 3d. Table 2 provides a detailed description of PASCAL-5^i and our benchmark.

Table 2. Details of our benchmark. The FID was calculated between each dataset and PASCAL-5^i.

Dataset	Numbers	Classes	FID
PASCAL-5^i	17,125	person, bird, dog, cat, cow, chair, dining table, potted plant, sheep, horse, airplane, bicycle, boat, car, bottle, sofa, tv/monitor bus, motorbike, and train	–
FSS-RSI-DeepGlobe	9175	agricultural, forested, barren, urban, rangeland, and aquatic areas	186.55
FSS-RSI-Potsdam	1896	clutter, tree, low vegetation, building, and impervious surface	151.86
FSS-RSI-Vaihingen	308	tree, low vegetation, building, and impervious surface	328.08
FSS-RSI-AISD	5640	building and road	194.90

As shown in Figure 4, we further calculated the pixel distribution of each category. The pixel distribution was relatively balanced, except for "urban areas" in FSS-RSI-DeepGlobe. Because of the richness of this dataset, we kept the class "urban areas".

As shown in Table 2, the Fréchet inception distance (FID) [48] was reported to measure the different data distributions between our benchmark and PASCAL-5^i. The FID is the Fréchet inception distance between the Gaussians obtained from the distributions of two datasets:

$$d^2((\mu_1, c_1), (\mu_2, c_2)) = \|\mu_1 - \mu_2\|_2^2 + Tr\left(c_1 + c_2 - 2(c_1 c_2)^{\frac{1}{2}}\right) \quad (14)$$

where (μ_1, c_1) and (μ_2, c_2) are means and covariances of the two distributions, and Tr is the matrix trace. The larger the FID, the greater the difference between the datasets, and vice versa.

As shown in Table 3, the same method was used to calculate the FID within PASCAL-5^i. We followed the strategy of FSS, that is, using a standard cross-training manner. Specifically, the FID was calculated between each fold and the other three folds. Compared to the data within PASCAL-5^i, the distribution gaps between our benchmark and PASCAL-5^i were vast, where the FID was more than twice that of the in-domain data. In particular, the FID of FSS-RSI-Vaihingen was 328.08. Therefore, it is further proven that our benchmark and PASCAL-5^i belong to different domains.

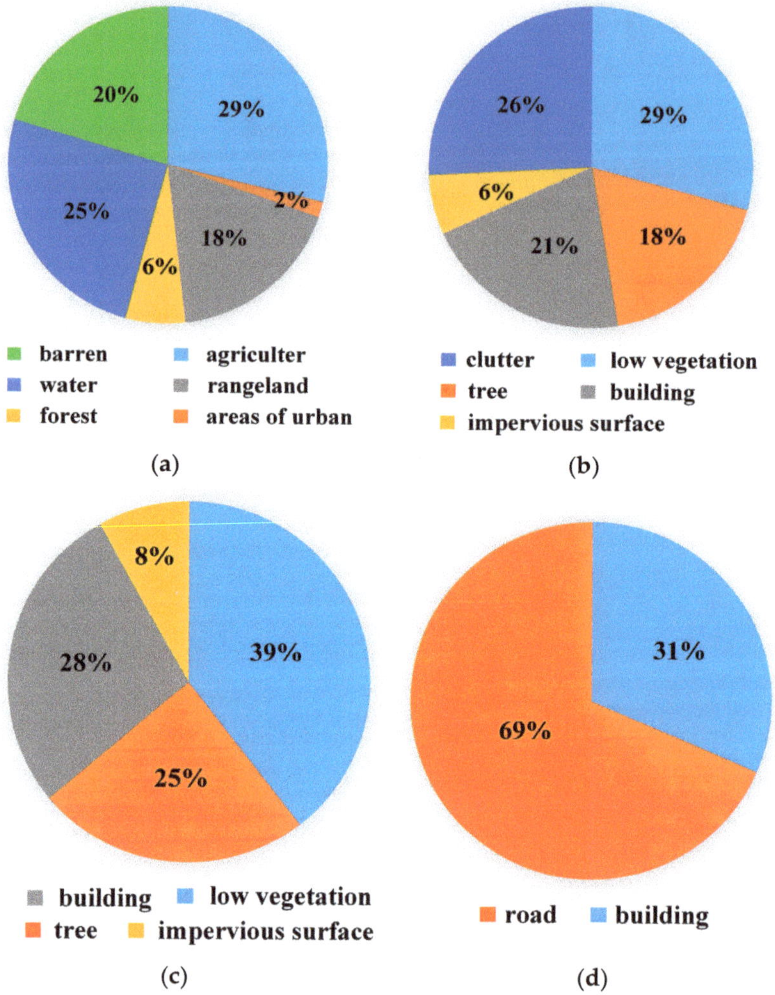

Figure 4. Pixel distributions of our benchmark: (**a**) FSS-RSI-DeepGlobe; (**b**) FSS-RSI-Potsdam; (**c**) FSS-RSI-Vaihingen; and (**d**) FSS-RSI-AISD.

Table 3. FIDs of different PASCAL-5^i splits.

Split-1	Split-2	FID
Fold0	Fold1 + Fold2 + Fold3	79.47
Fold1	Fold0 + Fold2 + Fold3	47.47
Fold2	Fold0 + Fold1 + Fold3	41.45
Fold3	Fold0 + Fold1 + Fold2	61.48

2.4. Experiments

2.4.1. Datasets and Metric

Datasets. PASCAL-5^i is the irrelevant domain training set created from PASCAL VOC 2012 [49] with SDS [50] augmentation. The benchmark we proposed is the testing set in the remote sensing domain.

Metric. The mean intersection over union (mIoU) [19,26] was adopted in our experiment, as a standard metric in semantic segmentation. The IoU is defined as the following:

$$\mathrm{IoU} = \frac{\mathrm{TP}}{\mathrm{FN} + \mathrm{FP} + \mathrm{TP}} \tag{15}$$

where FN, FP, and TP represent the number of false negatives, false positives, and true positives of the predictions, respectively. Furthermore, the mIoU is the average IoU of all categories.

2.4.2. Training and Testing Strategy

We used a generic meta-learning manner for training and testing. That is, each batch contained an episode. Unlike FSS, the entire PASCAL-5^i dataset with all four splits was used as the training data. Indeed, we use a supervised learning strategy and fixed the backbone's pre-training parameters during training. The Adam optimizer was adopted with a learning rate of 10^{-4}, and the weight decay was 0.01. Furthermore, the size of the images was reshaped to 400 × 400 pixels, which was followed by random scaling, rotation, and cropping. A mini-batch of 16 was utilized in the experiment. We trained each model on four 2080 Ti GPUs with 50 epochs.

The test was performed on a single GPU. It should be noted that we tested the benchmark with the model trained on the PASCAL-5^i without transfer. The mIoU was calculated for each dataset based on the average of 5 runs with different random seeds. A total of 600 tasks were contained in each run.

3. Result

3.1. Models for Comparison

To prove the performance of the FTNet, we selected several representative FSS models, including RPMMs [23], the PFENet [18], HSNet [24], BAM [19], and HDMNet [26]. Among them, RPMMs and the PFENet are classic prototype-based architectures, especially the PFENet, which is the most similar model to the FTNet. The HSNet uses 4D convolution to push meta-learning-based FSS to new heights. BAM and the HDMNet are cutting-edge in-domain FSS methods based on meta-learning and base-learning. For the RPMMs, PFENet, and HSNet, their released codes were used with the same settings. For BAM and the HDMNet, their meta paths were adopted, as there were no base classes in our benchmark. The testing method was exactly the same as ours.

3.2. Main Results

The results are shown in Table 4. The mIoU of the FTNet significantly exceeded that of the existing FSS model, including the SOTA model. Specifically, on the FSS-RSI-DeepGlobe dataset, the FTNet outperformed the suboptimal HSNet by 30.18% and 25.98% in the one-shot and five-shot settings, respectively. On the FSS-RSI-Potsdam dataset, the FTNet outperformed the suboptimal method by 37.57% and 8.90% in the one-shot and five-shot settings, respectively. On the FSS-RSI-AISD dataset, the FTNet outperformed the suboptimal methods by 17.48% and 13.61% in the one-shot and five-shot, respectively. In addition, our one-shot result was 13.53% higher than the HDMNet on the FSS-RSI-Vaihingen dataset. But the FTNet obtained a value that was 2.00% lower than BAM in the five-shot setting. For the mean results of all datasets, the mIoU significantly exceeded the suboptimal model, which was 25.39% and 21.31% higher than the HSNet in the one-shot and five-shot settings, respectively.

As can be seen, the PFENet achieved an mIoU that was 10% lower on the FSS-RSI-DeepGlobe and FSS-RSI-Potsdam datasets, especially for the value of only 2.42% in the FSS-RSI-DeepGlobe's five-shot setting. This proves that the PFENet had no FSS-RSI performance on these two datasets. However, as illustrated in Appendix C, the PFENet's performance within the domain was good. The same is true for RPMMs on the FSS-RSI-Potsdam dataset. As already known, the HDMNet is a cutting-edge FSS model, even with only its meta branch. The results can be seen in Appendix C. However, for the FSS-RSI task, the HDMNet did not perform well. The obtained mIoUs were 17.70 and 23.08 in the one-shot and five-shot settings, which were 41.16% and 17.56% lower than the FTNet, respectively. Performing slightly worse than our method, the HSNet performed suboptimally on FSS-RSI-DeepGlobe's one-shot and five-shot settings and on FSS-RSI-Potsdam's one-shot setting. We found that except for the PFENet, the five-shot results of all models were better than the one-shot results. This phenomenon indicates that when there are more support data, FSS-RSI performs better, which is similar to FSS. This experiment showed that the FTNet achieved the best result in FSS-RSI with absolute advantages over the other cutting-edge FSS methods.

Table 4. The mIoUs (%) of different methods experimented on our benchmark. The best results are denoted in bold. Suboptimal results are underlined.

Method	FSS-RSI-DeepGlobe		FSS-RSI-Potsdam		FSS-RSI-Vaihingen		FSS-RSI-AISD		Average	
	1-Shot	5-Shot	1-Shot	5-Shot	1-Shot	5-Shot	1-Shot	5-Shot	1-Shot	5-Shot
PFENet	3.97	2.42	6.26	4.84	12.58	12.29	25.03	25.18	11.96	11.18
RPMMs	10.66	13.17	6.76	7.56	16.24	16.45	25.12	23.86	14.70	15.26
HSNet	31.78	35.38	17.94	19.31	21.78	24.16	24.47	26.41	23.99	26.32
BAM	12.23	26.65	12.53	17.37	17.19	**26.98**	23.35	27.41	16.33	24.60
HDMNet	16.68	17.57	11.27	23.49	21.80	25.53	21.04	25.74	17.70	23.08
FTNet	**41.37**	**44.57**	**24.68**	**25.58**	**24.75**	26.44	**29.51**	**31.14**	**30.08**	**31.93**

Some qualitative results of our methods are shown in Figure 5. Classes with regular shapes, such as buildings in the FSS-RSI-Potsdam and FSS-RSI-Vaihingen datasets, obtained better results. However, FSS-RSI is challenging for irregular categories, such as trees and low vegetation. In particular, FSS-RSI did not work well for categories with similar appearances, such as the barren and rangeland areas in the FSS-RSI-DeepGlobe dataset and all categories in the FSS-RSI-AISD dataset. These challenging cases also need to be solved using semantic segmentation. Indeed, compared to commonly handled categories, such as cars, people, and animals, it is more difficult to segment remote sensing images. This issue is exactly the intractable part that FSS needs to solve.

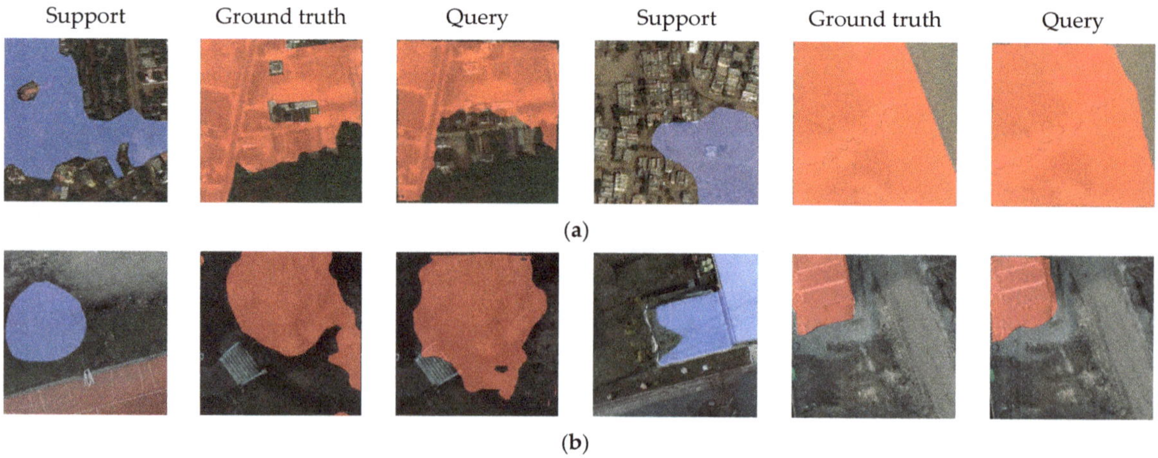

Figure 5. *Cont.*

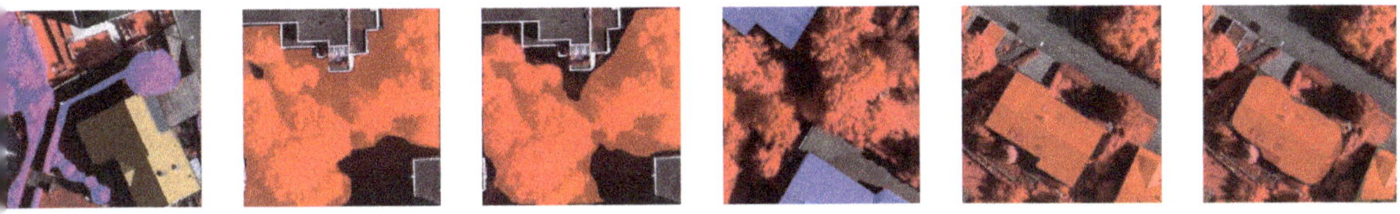

(c)

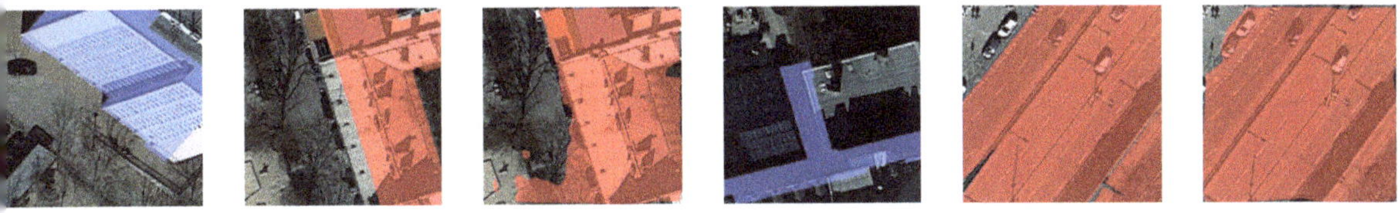

(d)

Figure 5. Qualitative results of the FTNet: (**a**) FSS-RSI-DeepGlobe; (**b**) FSS-RSI-Potsdam; (**c**) FSS-RSI-Vaihingen; and (**d**) FSS-RSI-AISD.

4. Discussion

4.1. Ablation Study

To prove the effectiveness of the FTNet, we carried out an ablation study. The mIoU was selected as the metric. Our baseline was that the architecture removed the HTM and FTM. To further justify the effectiveness of our HTM, we adopted a vanilla transformer module (VTM) for comparison. There were four repeat blocks in the VTM without concatenation; each block was the same as the first in the HTM.

Effects of the HTM and FTM. Table 5 shows the results of the five forms. They were the baseline, adding the FTM, adding the HTM, adding the VTM, and adding both the HTM and FTM. As illustrated, the mIoU of the baseline was similar to that of BAM [19] and the HDMNet [26]. After adding our tailored modules, the FTNet's performance was significantly boosted.

Table 5. The mIoUs (%) of the ablation study. The best results are denoted in bold. The baseline is the architecture that removed the FTM and HTM.

Method	FSS-IRS-DeepGlobe		FSS-IRS-Potsdam		FSS-IRS-Vaihingen		Average	
	1-Shot	5-Shot	1-Shot	5-Shot	1-Shot	5-Shot	1-Shot	5-Shot
Baseline	19.29	30.56	11.56	17.84	21.69	21.72	17.51	23.37
Baseline + FTM	31.62	33.80	22.57	20.86	24.73	**27.61**	26.31	27.42
Baseline + HTM	38.41	43.40	23.03	23.99	23.31	25.41	28.25	30.93
Baseline + VTM	21.23	–	20.27	–	23.16	–	21.55	–
Baseline + HTM + FTM	**41.37**	**44.57**	**24.68**	**25.58**	**24.75**	26.44	**30.27**	**32.20**

Compared to the baseline, the mIoUs in the one-shot and five-shot settings were improved by 50.26% and 17.33%, respectively, after adding the FTM. Furthermore, adding the HTM improved the result by 61.34% and 32.35%, respectively. What surprised us the most is that our complete structure with both the FTM and HTM achieved the highest mIoUs, which were 72.87% and 37.78% higher than the baseline in the one-shot and five-shot settings, respectively. We note that adding the FTM obtained a result comparable to our complete architecture in the one-shot setting and a higher result in the five-shot setting on the FSS-IRS-Vaihingen dataset. These results prove the effectiveness of the proposed modules.

As illustrated in the HDMNet [26], a transformer module follows a backbone similar to ours. Unlike the HDMNet, we added the HTM to fuse low- and high-level information after the prior mask and prototype. Thus, our model's performance was improved. As shown in Table 5, adding the VTM raised the mIoU from 17.51 to 21.55 in the one-shot setting. However, our HTM's result was 31.09% higher than the VTM's result in the one-shot setting. Furthermore, our device was out of memory when we trained the architecture with the VTM in the five-shot setting. Therefore, we could not collect the five-shot result using the VTM. The experimental results justify that our HTM is more effective than the VTM.

The visualization results of the five forms are shown in Figure 6. They are the results based on FSS-RSI-DeepGlobe in the one-shot setting. It is important to note that these qualitative results were unstable across different test rounds, and we consider the quantitative results in Table 5 on the entire dataset to be more reliable.

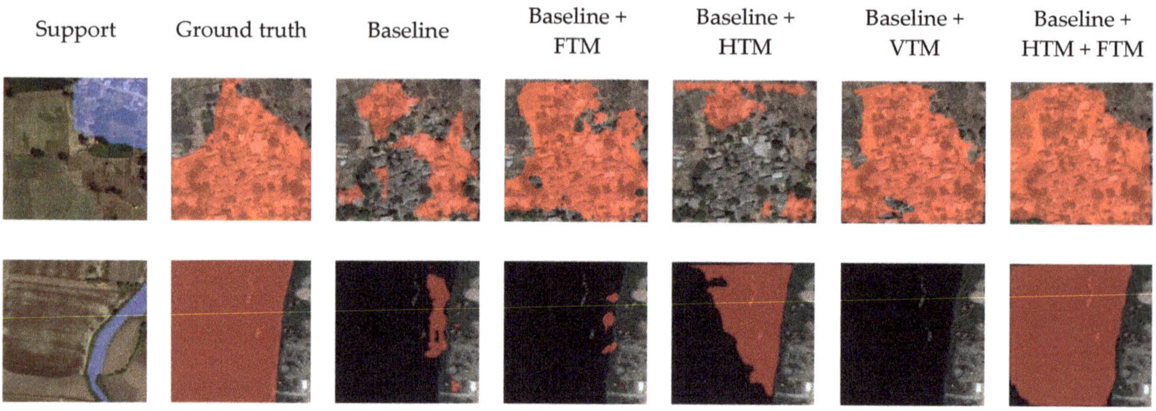

Figure 6. Qualitative results of the ablation study.

Effects of different classes. We counted the mIoUs of each class on the whole benchmark, and the results are shown in Figure 7. To sum up, the FTNet had an unbalanced accuracy for each category. On the FSS-RSI-DeepGlobe dataset, the FTNet had a higher mIoU for agricultural, forested, and barren but a lower mIoU for the remaining three categories. On the FSS-RSI-Potsdam dataset, the category with the highest mIoU was "building", which was 156.40% higher than "clutter". On the FSS-RSI-Vaihingen dataset the best class was "tree", which was 206.92% higher than the categories of "impervious surface". On the FSS-RSI-AISD dataset, the mIoU of "building" was higher. We believe that "building" had a more regular shape relative to "road", which was more conducive to the prediction of the network.

When combined with the pixel distribution in each category in Figure 4, the segmentation results of FSS-RSI-DeepGlobe and FSS-RSI-Potsdam are independent of the number of pixels. On the FSS-RSI-Vaihingen dataset, the smallest pixel ratio was obtained for "impervious surface", and the mIoU in this category was also the smallest. On the FSS-RSI-AISD dataset, the ratio of pixels was the opposite of the ratio of mIoUs. We do not have enough evidence to prove that the accuracy was related to the pixel ratio. Balancing the number of pixels in our benchmark and improving the semantic segmentation accuracy of each category will need to be considered in the future.

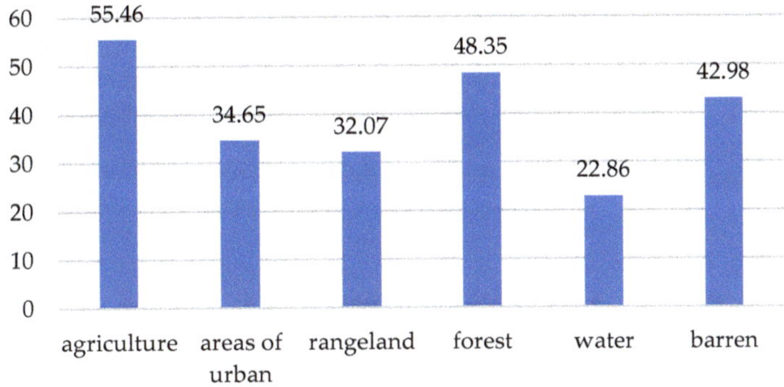

(**a**) Results of FSS-RSI-DeepGlobe

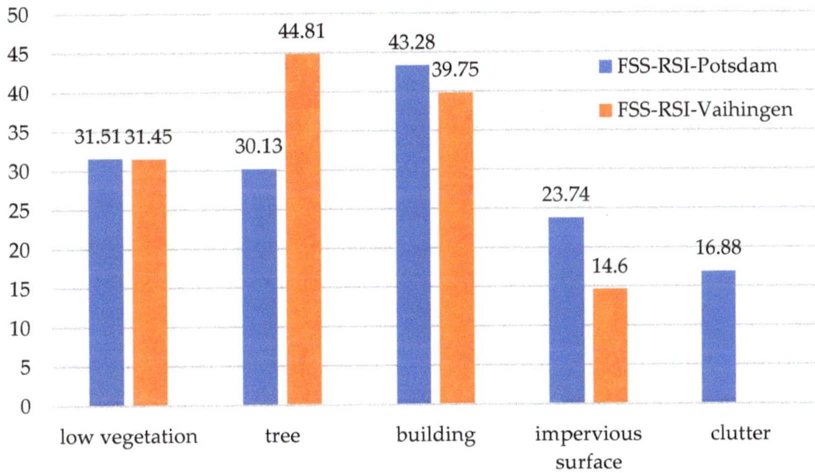

(**b**) Results of FSS-RSI-Potsdam and FSS-RSI-Vaihingen

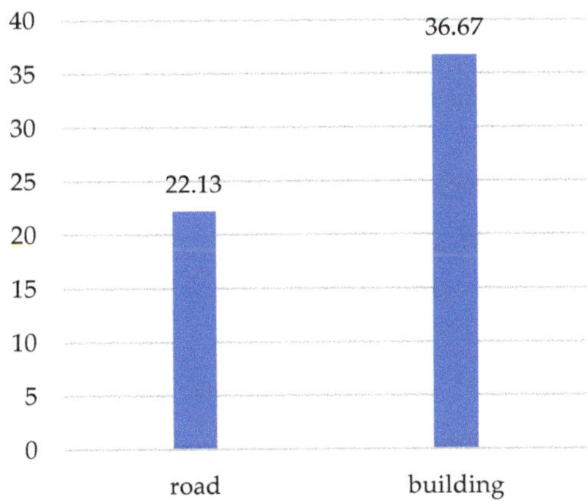

(**c**) Results of FSS-RSI-AISD

Figure 7. The mIoUs (%) of different categories: (**a**) FSS-RSI-DeepGlobe; (**b**) FSS-RSI-Potsdam and FSS-RSI-Vaihingen; and (**c**) FSS-RSI-AISD.

4.2. Limitations

This work introduces FSS into the field of remote sensing image segmentation. Our model achieved an absolute advantage over other SOTA methods, as our experiment shows. The results prove the effectiveness of our approach. However, we need to note that in the FSS-RSI task, the mIoU was only about 30%, which is still far from actual application needs. On the one hand, this phenomenon is attributable to the fact that FSS-RSI is a very challenging task. The training data and the test data were irrelevant. On the other hand, most of the categories in our benchmark did not have fixed shapes, such as low vegetation, impervious surface, and agriculture, which were difficult even for generic semantic segmentation [5,9]. In addition, FSS-RSI did not work well with categories with similar appearances, such as the barren and rangeland areas in the FSS-RSI-DeepGlobe dataset and all categories in the FSS-RSI-AISD dataset.

Some previous works on remote sensing images using FSS have achieved high accuracy [51,52]. Their training and testing data came from the same remote sensing dataset, which was different from our task. And the categories they contained were common categories with fixed shapes, such as airplanes, ships, or cars. In order to extend FSS-RSI to a broader range of applications, some innovative works need to be proposed. For example, tailored models must be designed for categories with no fixed shape to improve their segmentation accuracy. Moreover, FSS-RSI combined with the usual semantic segmentation, which simultaneously segments novel and known categories, would be promising future work.

5. Conclusions

To address the limitations of FSS for remote sensing, we extended the task to a new field called FSS-RSI. Specifically, we established a novel benchmark for evaluating FSS-RSI, which may be useful for other researchers. Moreover, we propose the FTNet with an FTM and an HTM. The FTM transforms the support feature, query feature, and prototype into a domain-agnostic space called the feature anchor. The HTM establishes abundant matching correlations between the support and query patches. In this way, our model can process remote sensing data with data from irrelevant domains.

Experiments were conducted on PASCAL-5^i and our benchmark. The FTNet achieved comparable accuracy to the cutting-edge methods on the in-domain data but obtained an absolute advantage on the FSS-RSI data. The proposed method outperformed the suboptimal model by 25.39% and 21.31% in the one-shot and five-shot settings, respectively. We hope our method will be helpful for few-shot semantic segmentation for remote sensing. For future work, we will focus on two interesting aspects: (1) designing tailored models to improve the accuracy of the FSS-RSI and (2) dealing with FSS problems in some exceptional cases, such as object occlusion, light changes, and similar appearance.

Author Contributions: Conceptualization, Q.S. and J.C.; Methodology, Q.S.; Software, Q.S.; Validation, Z.X. and W.C.; Formal analysis, Z.X.; Investigation, W.L. and N.H.; Writing—original draft, Q.S.; Writing—review & editing, J.C., W.L. and N.H.; Visualization, Z.X. and W.C.; Supervision, J.C.; Funding acquisition, W.L. and N.H. All authors have read and agreed to the published version of the manuscript.

Funding: This research was funded by [Work Enhancement Based on Visual Scene Perception] and [National Key Laboratory Foundation of Human Factors Engineering] grant number [GJSD22007]. The APC was funded by [Work Enhancement Based on Visual Scene Perception].

Data Availability Statement: The PASCAL VOC dataset is available at http://host.robots.ox.ac.uk/pascal/VOC/voc2012 (accessed on 25 June 2012). The Deepglobe dataset is available at https://www.kaggle.com/datasets/balraj98/deepglobe-road-extraction-dataset (accessed on June 2018). The Potsdam dataset is available at https://www.isprs.org/education/benchmarks/UrbanSemLab/2d-sem-label-potsdam.aspx (accessed on February 2015). The Vaihingen dataset is available at https://www.isprs.org/education/benchmarks/UrbanSemLab/2d-sem-labelvaihingen.aspx (accessed on

February 2015). The AISD dataset is available at https://zenodo.org/record/1154821#.XH6HtygzbIU (accessed on July 2017).

Conflicts of Interest: The authors declare no conflict of interest.

Appendix A

We indicated in Section 2.2.2 that the gaps between p^s_{new} and $f^q_{m,\text{ new}}$ and between $f^s_{m,\text{ new}}$ and $f^q_{m,\text{ new}}$ were significantly reduced after using the FTM. To confirm this, we compared the distance between the support and query features before and after using the FTM.

Specifically, for features f^s_m and f^q_m before the use of the FTM, we applied the global averaging pooling operation to change their shape from (B, C, H, W) to $(B, C, 1, 1)$, which is the same shape as p^s. B, C, H, and W indicate the tensor's batch, channel, height, and width, respectively. We defined $B = 1$ for convenience. The same operation was conducted on $f^s_{m,\text{ new}}$ and $f^q_{m,\text{ new}}$. Thus, f^s_m, f^q_m, p^s, $f^s_{m,\text{ new}}$, $f^q_{m,\text{ new}}$, and p^s_{new} were all vectors with the shape of $(1, C, 1, 1)$. The main purpose of the FTM was to transform the features into a domain-agnostic space. We could not directly describe this domain-agnostic space. However, we demonstrated this point by comparing the distance between the support and query features as an alternative. This was plausible because the most important purpose of FSS is to reduce the gap between these two features.

Indeed, the L2 norm was adopted as a metric. We calculated the distance $d_{f^s_m - f^q_m} = \left\| f^s_m - f^q_m \right\|_2$, $d_{p^s - f^q_m} = \left\| p^s - f^q_m \right\|_2$, $d_{f^s_{m,\text{new}} - f^q_{m,\text{new}}} = \left\| f^s_{m,\text{new}} - f^q_{m,\text{new}} \right\|_2$, and $d_{p^s_{\text{new}} - f^q_{m,\text{new}}} = \left\| p^s_{\text{new}} - f^q_{m,\text{new}} \right\|_2$, respectively. We collected 600 samples from the above four distances when testing on the FSS-RSI-DeepGlobe dataset. In Figure A1, we present a visualization of the whole results.

Figure A1a,b presents the distances before the use of the FTM with average distances of 3.40 and 75.85, respectively. Figure A1c,d presents the distances after the use of the FTM, with average distances of 0.47 and 0.88, respectively. The feature distances after the FTM are much smaller than those before the FTM. Thus, we further prove that the FTM is a useful module to reduce the gap between the support and query features.

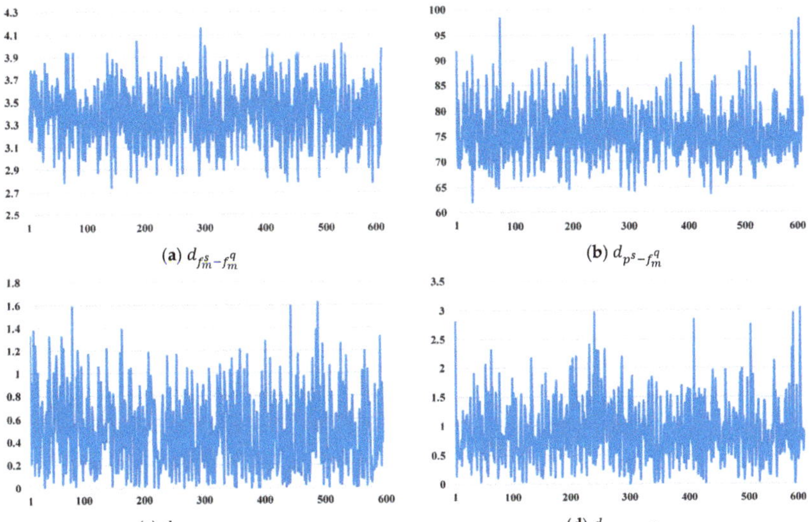

(**a**) $d_{f^s_m - f^q_m}$ (**b**) $d_{p^s - f^q_m}$

(**c**) $d_{f^s_{m,\text{new}} - f^q_{m,\text{new}}}$ (**d**) $d_{p^s_{\text{new}} - f^q_{m,\text{new}}}$

Figure A1. Distances between support and query features: (**a**) f^s_m and f^q_m; (**b**) p^s and f^q_m; (**c**) $f^s_{m,\text{ new}}$ and $f^q_{m,\text{ new}}$; and (**d**) p^s_{new} and $f^q_{m,\text{ new}}$.

Appendix B

To clarify that the HTM hierarchically enhances matching between the support and query features, the Frobenius norm was adopted as a metric. We denoted features before the HTM as f^{s}_{front} and f^{q}_{front}. That is, $f^{q}_{\text{front}} = f^{q}$ and $f^{s}_{\text{front}} = f^{s} \odot M^{s}$, where f^{q}, f^{s}, *and* M^{s} are the features presented in Figure 2. Moreover, we denoted features after each transformer block as $f^{s}_{\text{after},0}$, $f^{q}_{\text{after},0}$, $f^{s}_{\text{after},1}$, $f^{q}_{\text{after},1}$, $f^{s}_{\text{after},2}$, $f^{q}_{\text{after},2}$, $f^{s}_{\text{after},3}$, and $f^{q}_{\text{after},3}$, respectively. The features after merging the four transformer blocks were denoted as f^{s}_{merge} and f^{q}_{merge}. Our purpose was to calculate the Frobenius norms as $F_{\text{front}} = \left\| f^{s}_{\text{front}} - f^{q}_{\text{front}} \right\|_F$, $F_{\text{after},0} = \left\| f^{s}_{\text{after},0} - f^{q}_{\text{after},0} \right\|_F$, $F_{\text{after},1} = \left\| f^{s}_{\text{after},1} - f^{q}_{\text{after},1} \right\|_F$, $F_{\text{after},2} = \left\| f^{s}_{\text{after},2} - f^{q}_{\text{after},2} \right\|_F$, $F_{\text{after},3} = \left\| f^{s}_{\text{after},3} - f^{q}_{\text{after},3} \right\|_F$, and $F_{\text{merge}} = \left\| f^{s}_{\text{merge}} - f^{q}_{\text{merge}} \right\|_F$, respectively. Similar to the information presented in Appendix A, we collected 600 samples from the above six distances when testing on the FSS-RSI-DeepGlobe dataset. In Figure A2, we present a visualization of the results.

As shown in Figure A2a–f, the Frobenius norms decreased gradually, and their average values were reduced from 244.03 to 91.40. Indeed, the transformer is a dense image extraction and matching structure, which is difficult to explain using precise theory. We hope that Figure A2. will justify that the HTM can hierarchically enhance matching between the support and query features. Moreover, our ablation study, as presented in Section 4.1, can further prove the effectiveness of the HTM.

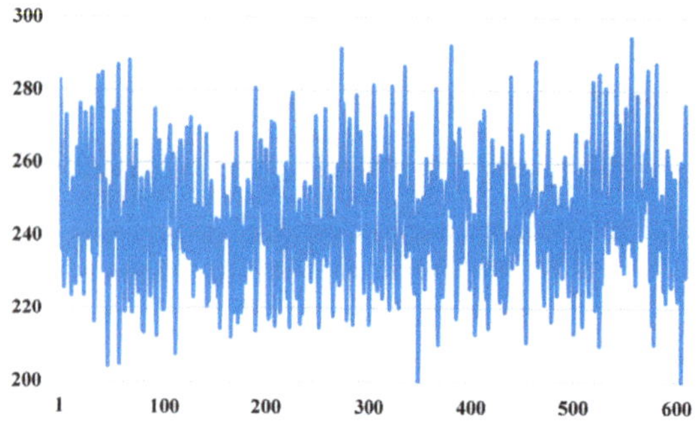

(**a**) F_{front}

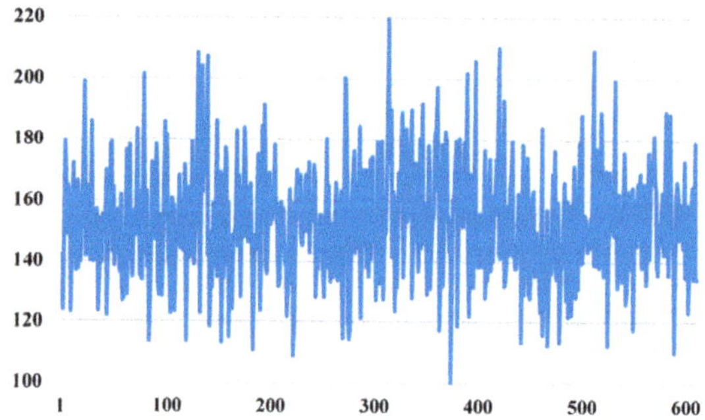

(**b**) $F_{\text{after},0}$

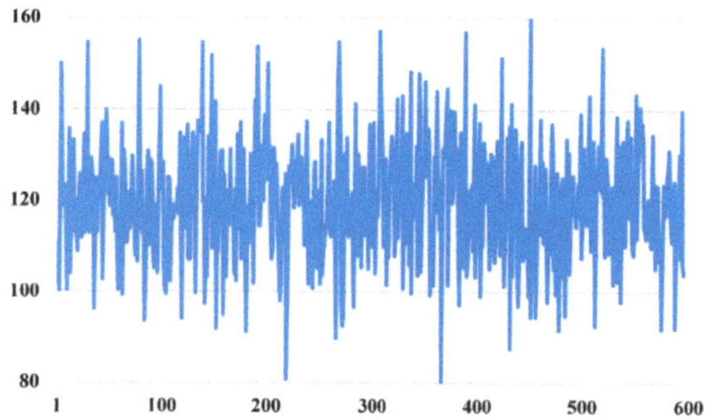

(**c**) $F_{\text{after},1}$

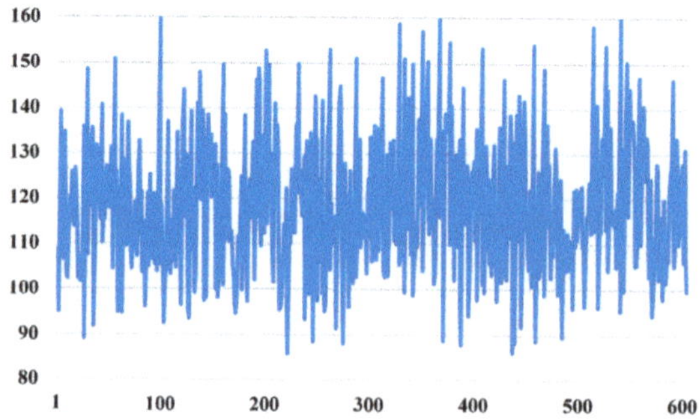

(**d**) $F_{\text{after},2}$

Figure A2. *Cont.*

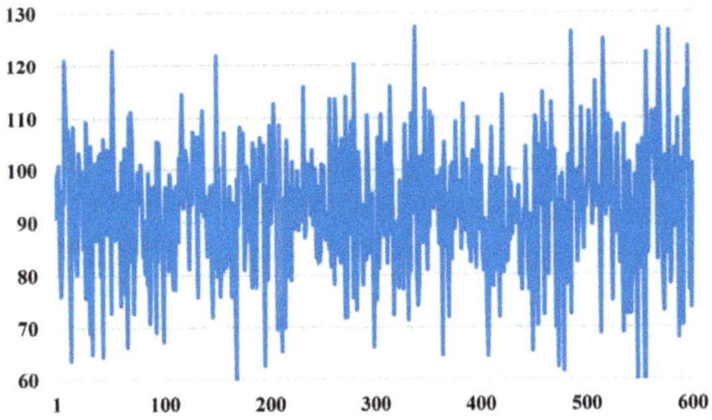

(e) $F_{after,3}$

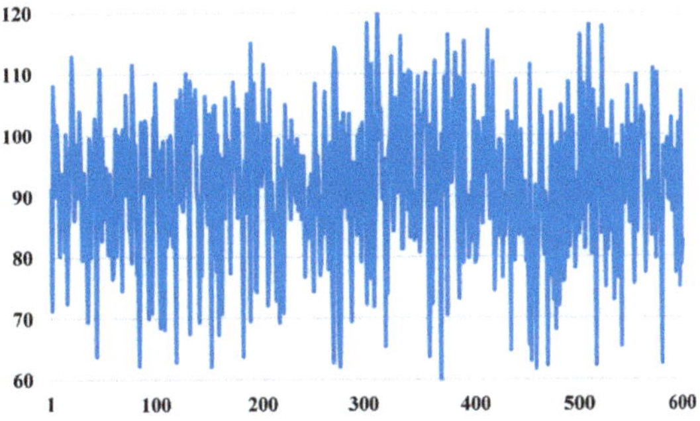

(f) F_{merge}

Figure A2. Frobenius norms between support and query features: (**a**) F^s_{front} and F^q_{front}; (**b**) $F^s_{after,0}$ and $F^q_{after,0}$; (**c**) $F^s_{after,1}$ and $F^q_{after,1}$; (**d**) $F^s_{after,2}$ and $F^q_{after,2}$; (**e**) $F^s_{after,3}$ and $F^q_{after,3}$; and (**f**) F^s_{merge} and F^q_{merge}.

Appendix C

Apart from the FSS-RSI, we further proved the performance of the FTNet on the in-domain dataset. The experiment was performed on PASCAL-5^i, and we followed the commonly used data division of four folds, but we reported only the results in the one-shot setting. The results are shown in Table A1. As already known, a trick is used in the BAM and HDMNet. That is, the image pairs containing novel classes during training are removed. But in other works, novel classes are set as the background. This trick improves the performance of the BAM and HDMNet. We did not use this trick for fairness, i.e., we adopted the same strategy as for the other methods such as the HSNet and PFENet. Thus, we retrained the BAM and HDMNet according to their official settings. The results of their meta branches are shown in Table A1.

Table A1. The mIoUs (%) of different methods on the in-domain dataset. The best results are denoted in bold. Suboptimal results are underlined.

Method	Fold0	Fold1	Fold2	Fold3	Average
PFENet	<u>63.23</u>	70.79	53.28	57.25	61.14
RPMMs	59.50	**71.58**	55.40	51.96	59.61
HSNet	63.03	69.50	59.64	<u>59.88</u>	63.01
BAM	60.94	70.75	<u>61.77</u>	59.45	<u>63.23</u>
HDMNet	**66.92**	75.83	**67.79**	**69.37**	**69.98**
FTNet	62.42	<u>71.06</u>	58.00	58.91	62.60

As can be seen, the HDMNet is the best model, reaching the highest mIoU on three folds. And its average mIoU on all four folds was the highest, reaching 69.98. The PFENet and RPMMs also achieved good results on PASCAL-5^i, reaching 61.14 and 59.61, respectively. The FTNet obtained a suboptimal result on Fold1, which was 10.54% lower than the HDMNet on all folds. However, our primary goal was FSS-RSI. The results in Table A1 further demonstrate the effectiveness of our model on FSS-RSI tasks. Figure A3 illustrates our model's qualitative results on PASCAL-5^i.

Figure A3. Results of the FTNet on PASCAL-5^i.

References

1. Wang, Z.; Wang, B.; Zhang, C.; Liu, Y.; Guo, J. Defending against Poisoning Attacks in Aerial Image Semantic Segmentation with Robust Invariant Feature Enhancement. *Remote Sens.* **2023**, *15*, 3157. [CrossRef]
2. He, Y.; Jia, K.; Wei, Z. Improvements in Forest Segmentation Accuracy Using a New Deep Learning Architecture and Data Augmentation Technique. *Remote Sens.* **2023**, *15*, 2412. [CrossRef]

3. Shelhamer, E.; Long, J.; Darrell, T. Fully Convolutional Networks for Semantic Segmentation. *IEEE Trans. Pattern Anal. Mach. Intell.* **2017**, *39*, 640–651. [CrossRef]
4. Noh, H.; Hong, S.; Han, B. Learning Deconvolution Network for Semantic Segmentation. In Proceedings of the 2015 IEEE International Conference on Computer Vision (ICCV), Piscataway, NJ, USA, 7–13 December 2015; pp. 1520–1528. [CrossRef]
5. Badrinarayanan, V.; Kendall, A.; Cipolla, R. SegNet: A Deep Convolutional Encoder-Decoder Architecture for Image Segmentation. *IEEE Trans. Pattern Anal. Mach. Intell.* **2017**, *39*, 2481–2495. [CrossRef]
6. Lin, G.; Milan, A.; Shen, C.; Reid, I. RefineNet: Multi-path Refinement Networks for High-Resolution Semantic Segmentation. In Proceedings of the 2017 IEEE Conference on Computer Vision and Pattern Recognition (CVPR), Honolulu, HI, USA, 21–26 July 2017; pp. 5168–5177. [CrossRef]
7. Ronneberger, O.; Fischer, P.; Brox, T. U-Net: Convolutional Networks for Biomedical Image Segmentation. *arXiv* **2015**, arXiv:1505.04597.
8. Yu, F.; Koltun, V. Multi-Scale Context Aggregation by Dilated Convolutions. *arXiv* **2016**, arXiv:1511.07122.
9. Chen, L.; Papandreou, G.; Kokkinos, I.; Murphy, K.; Yuille, A.L. DeepLab: Semantic Image Segmentation with Deep Convolutional Nets, Atrous Convolution, and Fully Connected CRFs. *IEEE Trans. Pattern Anal. Mach. Intell.* **2018**, *40*, 834–848. [CrossRef]
10. He, K.; Zhang, X.; Ren, S.; Sun, J. Deep Residual Learning for Image Recognition. In Proceedings of the IEEE Conference on Computer Vision and Pattern Recognition, Las Vegas, NV, USA, 27–30 June 2016; pp. 770–778. [CrossRef]
11. Simonyan, K.; Zisserman, A. Very Deep Convolutional Networks for Large-Scale Image Recognition. *arXiv* **2014**, arXiv:1409.1556.
12. Dosovitskiy, A.; Beyer, L.; Kolesnikov, A.; Weissenborn, D.; Zhai, X.; Unterthiner, T.; Dehghani, M.; Minderer, M.; Heigold, G.; Gelly, S.; et al. An Image is Worth 16x16 Words: Transformers for Image Recognition at Scale. *arXiv* **2020**, arXiv:2010.11929.
13. Zheng, S.; Lu, J.; Zhao, H.; Zhu, X.; Luo, Z.; Wang, Y.; Fu, Y.; Feng, J.; Xiang, T.; Torr, P.H.S.; et al. Rethinking Semantic Segmentation from a Sequence-to-Sequence Perspective with Transformers. In Proceedings of the 2021 IEEE/CVF Conference on Computer Vision and Pattern Recognition (CVPR), Nashville, TN, USA, 20–25 June 2021; pp. 6877–6886. [CrossRef]
14. Xie, E.; Wang, W.; Yu, Z.; Anandkumar, A.; Alvarez, J.M.; Luo, P. SegFormer: Simple and Efficient Design for Semantic Segmentation with Transformers. *arXiv* **2016**, arXiv:2105.15203.
15. Devlin, J.; Chang, M.-W.; Lee, K.; Toutanova, K. BERT: Pre-training of Deep Bidirectional Transformers for Language Understanding. *arXiv* **2016**, arXiv:1810.04805.
16. Vaswani, A.; Shazeer, N.; Parmar, N.; Uszkoreit, J.; Jones, L.; Gomez, A.N.; Kaiser, L.; Polosukhin, I. Attention Is All You Need. *arXiv* **2017**, arXiv:1706.03762.
17. Shaban, A.; Bansal, S.; Liu, Z.; Essa, I.; Boots, B. One-Shot Learning for Semantic Segmentation. *arXiv* **2017**, arXiv:1709.03410.
18. Tian, Z.; Zhao, H.; Shu, M.; Yang, Z.; Li, R.; Jia, J. Prior Guided Feature Enrichment Network for Few-Shot Segmentation. *IEEE Trans. Pattern Anal. Mach. Intell.* **2022**, *44*, 1050–1065. [CrossRef] [PubMed]
19. Lang, C.; Cheng, G.; Tu, B.; Han, J. Learning What Not to Segment: A New Perspective on Few-Shot Segmentation. In Proceedings of the 2022 IEEE/CVF Conference on Computer Vision and Pattern Recognition (CVPR), New Orleans, LA, USA, 18–24 June 2022; pp. 8047–8057. [CrossRef]
20. Vinyals, O.; Blundell, C.; Lillicrap, T.; Kavukcuoglu, K.; Wierstra, D. Matching Networks for One Shot Learning. *arXiv* **2016**, arXiv:1606.04080.
21. Wang, K.; Liew, J.H.; Zou, Y.; Zhou, D.; Feng, J. PANet: Few-Shot Image Semantic Segmentation with Prototype Alignment. In Proceedings of the 2019 IEEE/CVF International Conference on Computer Vision (ICCV), Seoul, Republic of Korea, 27 October–2 November 2019; pp. 9196–9205. [CrossRef]
22. Zhang, C.; Lin, G.; Liu, F.; Yao, R.; Shen, C. CANet: Class-Agnostic Segmentation Networks With Iterative Refinement and Attentive Few-Shot Learning. In Proceedings of the 2019 IEEE/CVF Conference on Computer Vision and Pattern Recognition (CVPR), Long Beach, CA, USA, 15–20 June 2019; pp. 5212–5221. [CrossRef]
23. Yang, B.; Liu, C.; Li, B.; Jiao, J.; Ye, Q. Prototype Mixture Models for Few-shot Semantic Segmentation. *arXiv* **2020**, arXiv:2008.03898.
24. Min, J.; Kang, D.; Cho, M. Hypercorrelation Squeeze for Few-Shot Segmenation. In Proceedings of the 2021 IEEE/CVF International Conference on Computer Vision (ICCV), Montreal, QC, Canada, 10–17 October 2021; pp. 6921–6932. [CrossRef]
25. Siam, M.; Oreshkin, B. Adaptive Masked Weight Imprinting for Few-Shot Segmentation. *arXiv* **2019**, arXiv:1902.11123.
26. Peng, B.; Tian, Z.; Wu, X.; Wang, C.; Liu, S.; Su, J.; Jia, J. Hierarchical Dense Correlation Distillation for Few-Shot Segmentation. *arXiv* **2023**, arXiv:2303.14652.
27. Zhang, G.; Kang, G.; Yang, Y.; Wei, Y. Few-Shot Segmentation via Cycle-Consistent Transformer. *arXiv* **2021**, arXiv:2106.02320.
28. Zhang, J.; Liu, Y.; Wu, P.; Shi, Z.; Pan, B. Mining Cross-Domain Structure Affinity for Refined Building Segmentation in Weakly Supervised Constraints. *Remote Sens.* **2022**, *14*, 1227. [CrossRef]
29. Gao, H.; Zhao, Y.; Guo, P.; Sun, Z.; Chen, X.; Tang, Y. Cycle and Self-Supervised Consistency Training for Adapting Semantic Segmentation of Aerial Images. *Remote Sens.* **2022**, *14*, 1527. [CrossRef]
30. Sun, L.; Cheng, S.; Zheng, Y.; Wu, Z.; Zhang, J. SPANet: Successive Pooling Attention Network for Semantic Segmentation of Remote Sensing Images. *IEEE J. Sel. Top. Appl. Earth Obs. Remote. Sens.* **2022**, *15*, 4045–4057. [CrossRef]
31. Chen, Y.; Wei, C.; Wang, D.; Ji, C.; Li, B. Semi-Supervised Contrastive Learning for Few-Shot Segmentation of Remote Sensing Images. *Remote Sens.* **2022**, *14*, 4254. [CrossRef]
32. Deng, R.; Shen, C.; Liu, S.; Wang, H.; Liu, X. Learning to Predict Crisp Boundaries. In Proceedings of the European Conference on Computer Vision, Munich, Germany, 8–14 September 2018; pp. 570–586. [CrossRef]

33. Demir, I.; Koperski, K.; Lindenbaum, D.; Pang, G.; Huang, J.; Basu, S.; Hughes, F.; Tuia, D.; Raskar, R. DeepGlobe 2018: A Challenge to Parse the Earth through Satellite Images. In Proceedings of the 2018 IEEE/CVF Conference on Computer Vision and Pattern Recognition Workshops (CVPRW), Salt Lake City, UT, USA, 18–22 June 2018; pp. 172–181. [CrossRef]
34. ISPRS. Potsdam. Available online: https://www.isprs.org/education/benchmarks/UrbanSemLab/2d-sem-label-potsdam.aspx (accessed on 20 June 2023).
35. ISPRS Vaihingen. Available online: https://www.isprs.org/education/benchmarks/UrbanSemLab/2d-sem-labelvaihingen.aspx (accessed on 20 June 2023).
36. Kaiser, P.; Wegner, J.D.; Lucchi, A.; Jaggi, M.; Hofmann, T.; Schindler, K. Learning Aerial Image Segmentation from Online Maps. *IEEE Trans. Geosci. Remote Sens.* **2017**, *55*, 6054–6068. [CrossRef]
37. Deng, J.; Dong, W.; Socher, R.; Li, L.J.; Kai, L.; Li, F.-F. ImageNet: A large-scale hierarchical image database. In Proceedings of the 2009 IEEE Conference on Computer Vision and Pattern Recognition, Miami, FL, USA, 20–25 June 2009; pp. 248–255. [CrossRef]
38. Zhao, H.; Shi, J.; Qi, X.; Wang, X.; Jia, J. Pyramid Scene Parsing Network. In Proceedings of the 2017 IEEE Conference on Computer Vision and Pattern Recognition (CVPR), Piscataway, NJ, USA, 21–26 July 2017; pp. 6230–6239. [CrossRef]
39. Ioffe, S.; Szegedy, C. Batch normalization: Accelerating deep network training by reducing internal covariate shift. In Proceedings of the International Conference on International Conference on Machine Learning (ICML), Lille, France, 6–11 July 2015; pp. 448–456.
40. Yu, C.; Wang, J.; Peng, C.; Gao, C.; Yu, G.; Sang, N. BiSeNet: Bilateral Segmentation Network for Real-Time Semantic Segmentation. In Proceedings of the European Conference on Computer Vision, Munich, Germany, 8–14 September 2018; pp. 334–349. [CrossRef]
41. Yu, C.; Gao, C.; Wang, J.; Yu, G.; Shen, C.; Sang, N. BiSeNet V2: Bilateral Network with Guided Aggregation for Real-Time Semantic Segmentation. *Int. J. Comput. Vis.* **2021**, *129*, 3051–3068. [CrossRef]
42. Fan, M.; Lai, S.; Huang, J.; Wei, X.; Chai, Z.; Luo, J.; Wei, X. Rethinking BiSeNet For Real-time Semantic Segmentation. In Proceedings of the IEEE/CVF Conference on Computer Vision and Pattern Recognition, Nashville, TN, USA, 20–25 June 2021; pp. 9711–9720. [CrossRef]
43. Seo, J.; Park, Y.-H.; Yoon, S.W.; Moon, J. Task-Adaptive Feature Transformer with Semantic Enrichment for Few-Shot Segmentation. *arXiv* **2022**, arXiv:2202.06498.
44. Haklay, M.; Weber, P. OpenStreetMap: User-Generated Street Maps. *IEEE Pervasive Comput.* **2008**, *7*, 12–18. [CrossRef]
45. Haklay, M. How good is volunteered geographical information? A comparative study of OpenStreetMap and Ordnance Survey datasets. *Environ. Plan. B-Plan. Des.* **2010**, *37*, 682–703. [CrossRef]
46. Girres, J.-F.; Touya, G. Quality Assessment of the French OpenStreetMap Dataset. *Trans. GIS* **2010**, *14*, 435–459. [CrossRef]
47. Google Maps. Available online: https://support.google.com/mapcontentpartners/answer/144284?hl=en (accessed on 20 September 2023).
48. Heusel, M.; Ramsauer, H.; Unterthiner, T.; Nessler, B.; Hochreiter, S. GANs trained by a two time-scale update rule converge to a local nash equilibrium. *arXiv* **2017**, arXiv:1706.08500.
49. Everingham, M.; Gool, L.V.; Williams, C.K.; Winn, J.; Zisserman, A. The pascal visual object classes (voc) challenge. *Int. J. Comput. Vis.* **2010**, *88*, 303–338. [CrossRef]
50. Hariharan, B.; Arbeláez, P.; Bourdev, L.; Maji, S.; Malik, J. Semantic contours from inverse detectors. In Proceedings of the 2011 International Conference on Computer Vision, Barcelona, Spain, 6–13 November 2011; pp. 991–998. [CrossRef]
51. Lang, C.; Wang, J.; Cheng, G.; Tu, B.; Han, J. Progressive Parsing and Commonality Distillation for Few-Shot Remote Sensing Segmentation. *IEEE Trans. Geosci. Remote Sens.* **2023**, *61*, 5613610. [CrossRef]
52. Li, R.; Li, J.; Gou, S.; Lu, H.; Mao, S.; Guo, Z. Multi-Scale Similarity Guidance Few-Shot Network for Ship Segmentation in SAR Images. *Remote Sens.* **2023**, *15*, 3304. [CrossRef]

remote sensing

MDPI

Article

A New Architecture of a Complex-Valued Convolutional Neural Network for PolSAR Image Classification

Yihui Ren [1,*], Wen Jiang [2] and Ying Liu [1]

1 School of Computer Science and Technology, University of Chinese Academy of Sciences, Beijing 101400, China; yingliu@ucas.ac.cn
2 School of Information Science and Technology, North China University of Technology, Beijing 100144, China; jiangwen19@mails.ucas.ac.cn
* Correspondence: renyihui18@mails.ucas.ac.cn

Abstract: Polarimetric synthetic aperture radar (PolSAR) image classification has been an important area of research due to its wide range of applications. Traditional machine learning methods were insufficient in achieving satisfactory results before the advent of deep learning. Results have significantly improved with the widespread use of deep learning in PolSAR image classification. However, the challenge of reconciling the complex-valued inputs of PolSAR images with the real-valued models of deep learning remains unsolved. Current complex-valued deep learning models treat complex numbers as two distinct real numbers, providing limited assistance in PolSAR image classification results. This paper proposes a novel, complex-valued deep learning approach for PolSAR image classification to address this issue. The approach includes amplitude-based max pooling, complex-valued nonlinear activation, and a cross-entropy loss function based on complex-valued probability. Amplitude-based max pooling reduces computational effort while preserving the most valuable complex-valued features. Complex-valued nonlinear activation maps feature into a high-dimensional complex-domain space, producing the most discriminative features. The complex-valued cross-entropy loss function computes the classification loss using the complex-valued model output and dataset labels, resulting in more accurate and robust classification results. The proposed method was applied to a shallow CNN, deep CNN, FCN, and SegNet, and its effectiveness was verified on three public datasets. The results showed that the method achieved optimal classification results on any model and dataset.

Keywords: polarimetric synthetic aperture radar (PolSAR) image classification; complex-valued convolutional neural network; complex-valued max pooling; complex-valued nonlinear activation; complex-valued cross-entropy

Citation: Ren, Y.; Jiang, W.; Liu, Y. A New Architecture of a Complex-Valued Convolutional Neural Network for PolSAR Image Classification. *Remote Sens.* **2023**, *15*, 4801. https://doi.org/10.3390/rs15194801

Academic Editor: Jocelyn Chanussot

Received: 1 August 2023
Revised: 19 September 2023
Accepted: 27 September 2023
Published: 1 October 2023

1. Introduction

The polarimetric synthetic aperture radar (PolSAR) system was developed from the conventional SAR system, which can provide multidimensional remote sensing information about a target [1]. The PolSAR system is more advanced than the conventional SAR system because it can obtain the target's scattering echo amplitude, phase, and frequency characteristics as well as the polarization characteristics of the target. PolSAR measures the polarization scattering characteristics of the ground target by transmitting and receiving electromagnetic waves with different polarization modes to obtain the target polarization scattering matrix [2]. The polarization of electromagnetic waves is sensitive to physical properties, such as the surface roughness, geometry, and orientation of the target, which means that the polarization scattering matrix contains a wealth of target information. PolSAR technology has been sustained and developed in recent decades, and it has been widely studied and applied in various applications, such as identifying croplands, measuring vegetation height, identifying forest species, describing geological structures, estimating soil humidity and surface roughness, measuring ice thickness, and monitoring coastlines.

PolSAR image classification involves assigning a label to each pixel in an image. As PolSAR systems become more popular, the range and types of ground targets change faster, and the captured target areas are becoming larger and captured more frequently. The traditional pixel-by-pixel manual labeling method is becoming inadequate due to the rapidly expanding PolSAR image data. Machine learning has been introduced to the PolSAR classification task to deal with this issue. PolSAR image classification algorithms can be broadly categorized into traditional machine learning algorithms and deep learning algorithms. Traditional machine learning algorithms can be further classified as unsupervised and supervised algorithms. Unsupervised algorithms include techniques such as Wishart [3–5], Markov random fields (MFRs) [6,7], and objective decomposition [8–11]. Supervised algorithms include supported vector products (SVMs) [12,13], random forests (RFs) [2], and fuzzy clustering [14]. When analyzing PolSAR images, traditional machine learning algorithms usually rely on shallow features of PolSAR images obtained through feature extraction methods. These shallow features include statistical features such as the linear and circular intensities, linear and circular coefficient of variation, and span [13], as well as target decomposition features such as the Pauli decomposition [15], Freeman decomposition [16], and Huynen decomposition [17]. However, this approach has several drawbacks. Firstly, the available features are limited and specific to certain scenes or targets. Secondly, some features, such as target decomposition features, require complex data analysis and computation. Thirdly, manual feature selection is time-consuming and requires many trials. Additionally, machine learning algorithms only utilize the features of a single pixel and ignore contextual information and local dependencies. Lastly, traditional machine learning algorithms do not perform well in nonlinear tasks.

In PolSAR image classification, deep learning has become a popular method for feature extraction. Unlike traditional machine learning, deep learning can automatically extract unlimited features. Deep and high-dimensional features can also be discovered by extracting features layer by layer. Additionally, deep learning can extract contextual information and the local dependency of pixels by inputting a patch containing a pixel to be classified. The feature extractor and classifier are combined into a single model, allowing for adaptive updates to the model parameters from a specific dataset. Deep learning is especially effective in handling nonlinear tasks due to containing a large number of nonlinear modules. Due to its advantages, deep learning has proven to be more accurate and effective than machine learning in PolSAR image classification. De et al. [18] proposed a stacked self-encoder and multi-layer perceptron approach to classify urban buildings in PolSAR images. Zhou et al. [19] designed a convolutional neural network (CNN) with two cascaded layers to extract spatial features with translation invariance in PolSAR images. Bin et al. [20] proposed a semi-supervised deep learning model based on graph convolutional networks for PolSAR image classification. Li et al. [21] developed a method for PolSAR image classification that incorporates a fully convolutional network (FCN) and sparse coding. They called this approach sliding window FCN and sparse coding (SFCN-SC). This approach significantly reduced the computational resources needed. Pham et al. [22] used SegNet to solve the problem of very-high-resolution (VHR) PolSAR image classification. Liu et al. [23] proposed an active ensemble deep learning (AEDL) model that achieved high classification accuracy using only a small amount of training data. Cheng et al. [24] developed a multiscale superpixel-based graph convolutional network (MSSP-GCN) based on a graph convolutional network that fully utilizes the boundary information of superpixels in PolSAR images. Liu et al. [25] used a stacked self-encoder for PolSAR image classification and an evolutionary algorithm to adaptively adjust the weights, activations, and balance factor in the loss function of the stacked self-encoder. Jing et al. [26] designed a method that simultaneously utilizes both the self-attention mechanisms of polarized spatial reconstruction networks for solving the classification of similar objects in PolSAR images. Nie et al. [27] demonstrated that deep reinforcement learning combined with FCN can achieve higher classification accuracy under limited samples. Yang et al. [28] utilized N-clustering generative adversarial networks and deep learning techniques to enhance the accuracy of PolSAR image classification. They achieved this by improving the hard classification accuracy for negative samples. Ren et al. [29] also developed a high-level feature fusion scheme for the multimodal representation of PolSAR images. Their approach was based on a CNN and resulted in the more efficient utilization of different features for the same target.

The PolSAR image classification models mentioned earlier are all based on real-valued CNNs (RV-CNN). This means the models' parameters, inputs, and outputs are all real-valued. However, since the raw data of PolSAR images are complex-valued, it is impossible to input the raw data into the real-valued model directly. Instead, a mapping between the raw data and the input of the real-valued model must be established, and this mapping is selected manually. Although RV-CNN achieves competitive results in PolSAR image classification, this approach still has some issues. First, there are multiple mappings between raw data and real-valued inputs, and it is unclear which is the best. Second, the mapping may cause a significant loss of implicit features in the raw data. Third, complex-to-real mapping often discards phase information, which is useful in PolSAR data [30–32].

Researchers have been investigating the use of a complex-valued neural network to directly process PolSAR data due to the challenges faced in this area. In 1992, Georgiou et al. [33] extended the backpropagation algorithm for neural networks to the complex domain for training complex-valued neural networks. Trabelsi et al. [34] were the first to propose a complex-valued convolutional neural network (CV-CNN), but their complex-valued pooling and loss functions were ineffective, and their proposed complex-valued activation did not work well. Zhang et al. [35] proposed a CV-CNN for PolSAR image classification. Li et al. [36] proposed a model that uses a multiscale contour filter bank and CV-CNN to automatically extract the complex-valued features of PolSAR images using the prior knowledge of the filters. Xiao et al. [37] developed a classification model with a complex-valued encoder and decoder. Additionally, they utilized the complex-valued upsampling module for the first time. Zhao et al. [38] proposed a contrastive-regulated CV-CNN that obtains features from raw back-scatter data. Tan et al. [39] explored the effectiveness of using a 3D complex-valued convolution to extract hierarchical features in both spatial and scattering dimensions. This allowed them to obtain physical features with the polarization resolution of neighboring cells. Zhang et al. [40] investigated the potential of random fields for modeling and complex-valued convolution for representation learning on PolSAR images. They proposed a hybrid conditional random field model based on a complex-valued 3D convolutional neural network. Qin et al. [41] suggested incorporating expert knowledge as input to the CV-CNN model to enhance its performance and make it more robust. Fang et al. [42] proposed a stacked complex-valued convolutional long short-term memory network for PolSAR image classification, which extracts coherence information between different features. Meanwhile, Tan et al. [43] utilized three sets of CV-CNNs to extract coherence information from the PolSAR images. They achieved this by maximizing the inter-class distance and minimizing the intra-class distance to learn the most discriminative features.

Although deep learning models using complex values have made significant breakthroughs in PolSAR classification, they still face major challenges. Firstly, the complex-valued nonlinear module has not received enough attention. The excellent performance of CNNs in PolSAR image classification is due to its strong nonlinear fitting ability, but the nonlinear module has not been optimized in the literature. CNN cannot perform strong nonlinear fitting without an outstanding complex-valued nonlinear module bringing sub-optimal classification results. Secondly, existing CV-CNNs either use only the amplitude of the features while ignoring their phases or treat the real and imaginary parts of the features separately. The first approach generates features that do not contain phase information, and the second approach does not satisfy the complex multiplication theorem. Thirdly, cross-entropy, the most common loss function in classification, can make the probability distribution of the CNN output closer to the real label by minimizing the cross-entropy loss between the label and the CNN output. However, cross-entropy is computed for two real-valued probability distributions, while CV-CNN output is complex.

This paper explores using CV-CNN in PolSAR image classification and suggests a new complex-valued pooling method, nonlinear activation, and cross-entropy approach called CV_CrossEntropy. The nomenclature employed in this study designates our novel approaches as new CV-CNN to mitigate potential ambiguities with previously cited CV-CNN methodologies in the literature. These methodologies are subsequently employed across shallow CNN (SCNN), deep CNN (DCNN), FCN, and SegNet architectures. The results reveal substantial enhancements in the models' classification performance when compared to both real-valued and conventional complex-valued counterparts featuring identical structural configurations and parameters. This paper focuses on four aspects: (1) complex-valued max pooling to reduce computation and expand the receptive field; (2) complex-valued activation to extract high-dimensional nonlinear features; (3) complex-valued probability and labels to calculate loss; and (4) CV_CrossEntropy to train CV-CNN.

To summarize, our contributions can be expressed as follows:

(1) A novel CV-CNN is introduced in this study, featuring complex-valued inputs, outputs, as well as complex-valued weights and biases. Our nonlinear module treats the input as a complex number, respecting the mathematical significance of complex-valued inputs and extracting the most discriminative features, resulting in improved classification ability. Our new complex-valued methods are used in different deep learning models and achieve better results than real-valued or old complex-valued versions with the same structure.
(2) In this research, a novel complex-valued max pooling technique is presented for the downsampling of feature maps. This method is designed to reduce computational demands, accelerate training and inference, and, importantly, retain the most essential features.
(3) A novel complex-valued activation function is employed to acquire high-dimensional nonlinear features. This new activation maps the amplitude and phase of the features into the high-dimensional complex domain space and can make the model more sparse.
(4) A novel complex-valued cross-entropy is applied in the training process of the new CV-CNN. The complex-valued probability principle [44–48] is employed to reallocate one-hot labels within the dataset. This loss function utilizes the complex-valued labels and outputs to compute the classification loss and train a better model by backpropagation.

Three different versions of SCNN, DCNN, FCN, and SegNet were considered: the real-valued version, the old complex-valued version, and the new complex-valued version. In total, 12 models were compared across three publicly available PolSAR datasets. The experimental results demonstrate that the models enhanced by the new complex-valued approach consistently outperform the others, yielding the best results.

The rest of the paper is structured in the following manner. Section 2 provides an in-depth explanation of complex-valued nonlinear modules and CV_CrossEntropy theory. Section 3 presents the experimental results on three public datasets. Furthermore, Section 4 showcases related discussions and ablation experiments. Lastly, Section 5 contains the summary and future work.

2. Materials and Methods

This section presents a new approach called CV-CNN for classifying PolSAR images. The method uses a complex-valued convolutional kernel to extract the features of PolSAR images, which addresses the implicit mapping problem introduced by a real-valued convolutional kernel. The paper also proposes a new complex-valued nonlinear module that processes the input data amplitude and phase to extract better features. The model's training employs a new CV_CrossEntropy loss function, yielding improved accuracy and robustness of the model and guaranteeing unique classification results during inference. Additionally, Section 2.1 describes two deep learning models for PolSAR image classification, which can be either complex-valued or real-valued, depending on the input. It is necessary to design appropriate nonlinear methods for PolSAR data to enhance classification accuracy. Section 2.2 introduces the input format of PolSAR. Sections 2.3 and 2.4 present complex-valued max pooling and complex-valued nonlinear activations. Section 2.5 introduces complex-valued probability, a one-hot label, and cross-entropy for computing the loss during training. Finally, the CV-CNN algorithm for PolSAR image classification is summarized in Section 2.6.

2.1. Two Deep Learning Models for PolSAR Classification

Figure 1 shows two networks that can be used for PolSAR image classification: (a) a CNN and (b) a convolutional encoder–decoder network. The CNN classifies one pixel at a time using a patch of size $h \times w \times C$ as input and produces a prediction of a pixel of a size of 1×1. It consists of a feature extractor (convolutional, pooling, and nonlinear activation layers) and a classifier (fully connected and softmax layers). Two CNNs are used in this paper to test complex-valued methods, with the main difference being the number of convolutional layers. The convolutional encoder–decoder network classifies all pixels of an image at once using a PolSAR image of size $H \times W \times C$ as input and producing a prediction image of size $H \times W$. The encoder and decoder are the feature extractor and classifier, respectively. The encoder has the same structure as the CNN, while the decoder has convolutional, upsampling, nonlinear activation, and softmax layers but no fully connected layer. The experiments use two convolutional encoder–decoder models: FCN and SegNet, with the main difference being the connection between the encoder and decoder. The convolutional encoder–decoder has more parameters but less computational redundancy than the CNN.

Figure 1 shows that convolutional models consist of fundamental modules, including convolution, fully connected, pooling, and activation layers. Convolution and fully connected layers are linear modules while pooling, and activation layers are nonlinear modules. A complex-valued batch normalization layer [34] is commonly inserted between the convolutional and nonlinear activation layers to avoid model overfitting. Linear modules perform addition and multiplication, which can be expressed as Equations (1) and (2) for real and complex numbers.

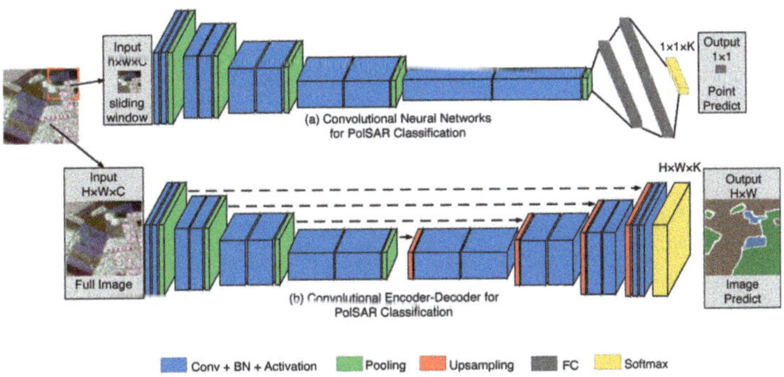

Figure 1. Two types of deep convolutional models for PolSAR image classification. The first model (**a**) is a convolutional neural network with blue and green parts for feature extraction and gray and yellow parts for classification. The second model (**b**) is a convolutional encoder–decoder network with blue and green parts for feature extraction and red, blue, and yellow parts for classification. In both models, 'Conv' refers to the convolutional layer, 'BN' refers to the batch normalization layer, and 'FC' refers to the fully connected layer. The black dotted line indicates that the encoder and decoder feature maps have been fused. Additionally, 'H', 'W', and 'C' represent the input's height, width, and number of channels.

$$(A + j \cdot B) \times (a + j \cdot b) = (Aa - Bb) + j \cdot (Ab + Ba) \quad (1)$$

$$(A + j \cdot B) + (a + j \cdot b) = (A + a) + j \cdot (B + b) \quad (2)$$

By analyzing Equations (1) and (2), it is apparent that complex-valued addition and multiplication are linear computations of real and imaginary parts. As a result two real-valued convolution kernels can replace one complex-valued convolution kernel, and two real-valued fully connected operators can replace one complex-valued fully connected operator.

2.2. Inputs of PolSAR Classification

The inputs with complex and real values have equal width and height but differ in the number of channels. A 2 × 2 complex-valued scattering matrix represents each resolution cell of the PolSAR data, as shown in Equation (3):

$$S = \begin{bmatrix} S_{HH} & S_{HV} \\ S_{VH} & S_{VV} \end{bmatrix} \quad (3)$$

H and V represent the horizontal and vertical polarization bases in this equation, respectively. S_{pg} represents the backscattering coefficient between the polarization scattered and the incident field. It is typically assumed that S_{HV} and S_{VH} are identical due to the reciprocity theorem. This allows the matrix to be simplified and reduced to the scattering vector $\vec{k}$. Using the Pauli decomposition method, the scattering vector $\vec{k}$ can be expressed as shown in Equation (4):

$$\vec{k} = \frac{1}{\sqrt{2}}\left[S_{HH} + S_{VV}, S_{HH} - S_{VV}, 2S_{HV}\right]^T \quad (4)$$

The representation of the consistency matrix for PolSAR data in the multi-look scenario can be found in Equation (5):

$$T = \frac{1}{L}\sum_{i=1}^{L} \vec{k}_i \vec{k}_i^H = \begin{bmatrix} T_{11} & T_{12} & T_{13} \\ T_{21} & T_{22} & T_{23} \\ T_{31} & T_{32} & T_{33} \end{bmatrix} \quad (5)$$

The equation for T, which represents the consistency matrix, includes the number of looks (L) and the conjugate transpose (denoted by H). T is a Hermitian matrix with real-valued elements on the diagonal and complex-valued elements off-diagonal. Only the upper triangular part $[T_{11}, T_{12}, T_{13}, T_{22}, T_{23}, T_{33}]$ is necessary to input T into the deep convolutional model. In the case of the real-valued model, the feature vector is represented by Equation (6) and has nine input channels:

$$[T_{11}, T_{22}, T_{33}, \Re(T_{12}), \Im(T_{12}), \Re(T_{13}), \Im(T_{13}), \Re(T_{23}), \Im(T_{23})] \quad (6)$$

In the model that deals with complex values, there are six input channels, and the feature vector is identified as Equation (7):

$$[T_{11} + 0 \cdot j, T_{22} + 0 \cdot j, T_{33} + 0 \cdot j, T_{12}, T_{13}, T_{23}] \quad (7)$$

2.3. Complex-Valued Amplitude-Based Max Pooling

In machine learning, deep learning is a set of methods requiring much computational power. Unfortunately, a significant portion of this computational power is used redundantly, which can result in slow convergence, poor performance, and overfitting. One technique to address this is pooling, which reduces the amount of data involved by shrinking the feature map. Additionally, pooling also expands the receptive field, allowing the model to extract more meaningful features with contextual and global information. As a result, it is important to use pooling methods that keep the most effective features while reducing the feature map size.

Based on Figure 2a, the complex-valued feature map can be split into real and imaginary feature maps. These two maps can then be combined into an amplitude feature map. The amplitude feature map is then subjected to real-valued max pooling, and the maximum value index is recorded. Afterward, the final pooling result is obtained by utilizing the maximum index and the original feature map. The mathematical expression for amplitude-based max pooling (CVA_Max_Pooling) is shown in Equation (8):

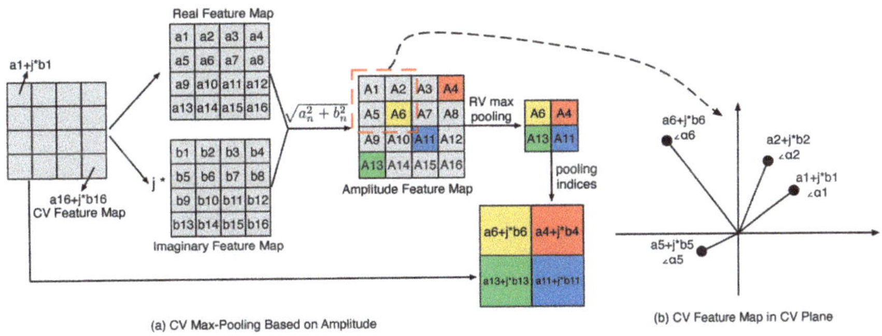

Figure 2. (**a**) displays the process of amplitude-based max pooling, using gray squares to represent the feature map before pooling and colored squares for the feature map after 2 × 2 pooling. (**b**) shows the complex plane representation of the four features in the 2 × 2 feature map identified by the red dashed box in (**a**).

$$CVA\_Max\_Pooling(F) = \{F_{i,j} | i, j = arg\_max(F_{i,j}^2)\} \quad (8)$$

In this equation, F represents the feature map, while w and h refer to the width and height of the pooling kernel. Figure 2b displays a complex plane map of all the data within a pooling kernel. Each feature in this map consists of an amplitude and a phase. The amplitude indicates the strength of the feature, with higher amplitudes indicating greater strength and importance. Meanwhile, the phase of a feature indicates its synchronization relationship with other features. Features with closer phase values are more synchronized. However, comparing the two features' phase sizes is meaningless. In Figure 2b, feature $a6 + j * b6$ has the largest amplitude, so CVA_Max_Pooling will keep that feature in the next layer.

Complex-valued pooling methods, such as max pooling or average pooling, are commonly used on features' real and imaginary parts. However, it is important to note that old complex-valued average pooling can weaken significant features, while old complex-valued max pooling can create "fake" features that could negatively impact the final classification results. On the other hand, CVA_Max_Pooling can efficiently preserve crucial features, thus reducing computational workload, broadening the receptive field, and enhancing classification accuracy.

2.4. Complex-Valued Nonlinear Activation

Using complex-valued nonlinear activation is beneficial in mapping features into a high-dimensional nonlinear space. This greatly enhances the nonlinear fitting ability of

CV-CNNs. In RV-CNN models, the most commonly used nonlinear activations are variant of ReLU. These activations are widely used in real-valued deep learning models and delive outstanding performance due to their computational simplicity, ease of derivation, and ability to sparsify feature maps. A simulated ReLU function is also a preferred design idea for most complex-valued nonlinear activations. Compared to real-valued activation complex-valued activation requires a double nonlinear mapping of the feature amplitude and phase. The three common nonlinear activations in old CV-CNNs are ModReLU CReLU, and ZReLU . Their Equations are (9)–(11), respectively.

$$ModReLU(z) = ReLU(|z| - b)e^{j\theta_z} = \begin{cases} (|z| - b)\frac{z}{|z|} & if\ |z| \geq b \\ 0 + 0 \cdot j & otherwise \end{cases} \quad (9)$$

$$CReLU(z) = ReLU(\Re(z)) + j \cdot ReLU(\Im(z)) \quad (10)$$

$$ZReLU(z) = \begin{cases} z & if\ \theta_z \in [0, \pi/2] \\ 0 + 0 \cdot j & otherwise \end{cases} \quad (11)$$

Based on (9)–(11), it is evident that these three nonlinear activations with complex values imitate ReLU in varying ways. This paper suggests an improved complex-valued nonlinear activation called HReLU, which introduces a new approach and is expressed in Equation (12).

$$HReLU(z) = \begin{cases} z & if\ \theta_z \in [0, \pi] \\ 0 + 0 \cdot j & otherwise \end{cases} \quad (12)$$

To fully comprehend the advantages and disadvantages of complex-valued nonlinear activations, it is crucial to understand why ReLU has succeeded in RV-CNN. ReLU is a segmented mapping with a constant mapping in the range of $(-\infty, 0)$ and a linear mapping in the range of $[0, +\infty)$. This feature makes ReLU convenient for forward inference and for the calculation of derivatives, as its derivatives are 0 in the range of $(-\infty, 0)$ and 1 in the range of $[0, +\infty)$. ReLU maps data in the range of $(-\infty, 0)$ to 0 while keeping data in the range of $[0, +\infty)$ unchanged. This not only sparsifies the feature map and improves the model's generalization ability but also prevents the feature map from being too sparse leading to insufficient model fitting. However, for the complex-valued feature maps, the amplitude and phase ranges are $[0, +\infty)$ and $[0, 2\pi]$, respectively, which makes ReLU unsuitable. To address this, ModReLU, CReLU, ZreLU, and HReLU have been developed to migrate ReLU to the complex domain. If these four complex-valued nonlinear activations are split into the amplitude activation and the phase activation, they can be represented as follows:

ModReLU:

$$ModReLU(|z|) = \begin{cases} |z| - b & if\ |z| \geq b \\ 0 & otherwise \end{cases} \quad (13)$$

$$ModReLU(\theta_z) = \begin{cases} \theta_z & if\ |z| \geq b, \theta_z \in [0, 2\pi) \\ 0 & otherwise, \theta_z \in [0, 2\pi) \end{cases} \quad (14)$$

CReLU:

$$CReLU(|z|) = \begin{cases} |z| & if\ \theta_z \in [0, \pi/2) \\ |\Im(z)| & if\ \theta_z \in [\pi/2, \pi) \\ 0 & if\ \theta_z \in [\pi, 3\pi/2) \\ |\Re(z)| & if\_\theta_z \in [3\pi/2, 2\pi) \end{cases} \quad (15)$$

$$CReLU(\theta_z) = \begin{cases} \theta_z & if\ \theta_z \in [0, \pi/2) \\ \pi/2 & if\ \theta_z \in [\pi/2, \pi) \\ 0 & if\ \theta_z \in [\pi, 3\pi/2) \\ 2\pi & if\ \theta_z \in [3\pi/2, 2\pi) \end{cases} \tag{16}$$

ZReLU:

$$ZReLU(|z|) = \begin{cases} |z| & if\ \theta_z \in [0, \pi/2] \\ 0 & otherwise \end{cases} \tag{17}$$

$$ZReLU(\theta_z) = \begin{cases} \theta_z & if\ \theta_z \in [0, \pi/2] \\ 0 & otherwise \end{cases} \tag{18}$$

HReLU:

$$HReLU(|z|) = \begin{cases} |z| & if\ \theta_z \in [0, \pi] \\ 0 & otherwise \end{cases} \tag{19}$$

$$HReLU(\theta_z) = \begin{cases} \theta_z & if\ \theta_z \in [0, \pi] \\ 0 & otherwise \end{cases} \tag{20}$$

By examining Figure 3 and Equations (13)–(20), it can be observed that only HReLU performs ReLU-like processing on the magnitude and phase of complex-valued feature maps. HReLU is also a segmented function, with the upper half of the complex plane being a linear mapping and the lower half being a constant mapping. Once HReLU is expressed as an amplitude-activated function and a phase-activated function, these two functions also become segmented functions, with half of the data being linear mappings and the other half being constant mappings. HReLU's nonlinear section also maps the data as $0 + 0 \cdot j$, which sparsifies the feature map and improves the model's generalization ability. In contrast, ModReLU's nonlinearization range is too small, making it difficult to extract efficient features. CReLU is not sparse enough, leading to poor generalization. ZReLU is too sparse, resulting in a model prone to underfitting. According to Georgiou et al.'s complex-valued backpropagation algorithm [33], the derivatives of HReLU in the upper and lower halves of the complex plane are simple to compute, being $1 + 1 \cdot j$ and $0 + 0 \cdot j$, respectively.

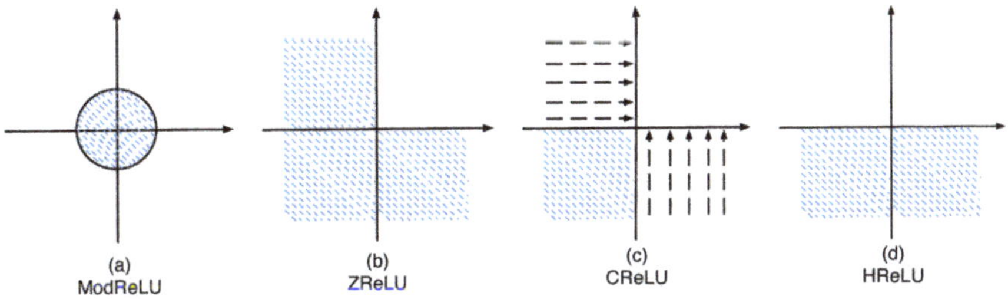

Figure 3. (**a–d**) depict the complex plane mappings for ModReLU, CReLU, ZReLU, and HReLU, respectively. The blue shaded area corresponds to the data set to $0 + 0 \cdot j$, while the dashed region with arrows indicates data mapped to the coordinate axes. Any blank part areas in the data will be preserved for the next layer.

2.5. Complex-Valued Cross-Entropy

CNNs are supervised learning models that rely on the loss between the model output and the label during training. In the case of RV-CNN used for PolSAR image classification, the output is a real-valued probability distribution vector. The labels are a real-valued one-hot vector with dimensions equal to the number of categories. RV-CNN uses real-valued cross-entropy to calculate the loss of PolSAR image classification. However, CV-CNN's output is no longer a real-valued probability distribution vector, which means that real-valued cross-entropy cannot be used to calculate the loss. The old complex-valued classification models only use the real part of the output to calculate the loss value, but this approach loses at least half of the information flow. Thus, this paper proposes a CV_one-hot label, complex-valued probability distribution vector and CV_CrossEntropy to address this issue.

2.5.1. Complex-Valued Probability and CV_one-hot Label

Complex-valued probability is an extension of traditional probability that uses complex numbers to express probability distributions [44–48]. Before delving into this concept, it is important to clarify some related theorems.

Definition 1. $P_m = j \cdot (1 - P_r)$ *represents the probability of a random event A in the imaginary and real fields, where j denotes the imaginary unit.*

Theorem 1. *The norm of a random event in the complex domain is calculated as* $|P_c|$: $|P_c|^2 = P_r^2 + (P_m/j)^2$.

Theorem 2. *The sum of probabilities of a random event's real and imaginary parts in the complex domain is always equal to 1:* $(P_r + P_m/j)^2 = |P_c|^2 - 2jP_rP_m = 1$

From these theorems, it can be inferred that P_r represents the probability of any random event happening, while P_m represents the probability of the associated event in the imaginary domain. P_c is a random event in the complex field given by P_r and P_m. The degree of knowledge and the chaos factor of a random event in the complex domain are denoted by $|P_c|^2$ and $2jP_rP_m$, respectively.

If $P_r = 1$, this means that the random event in the real domain is deterministic, and the degree of knowledge and the chaos factor of the random event in the complex domain are 1 and 0, respectively.

If $P_r = 0$, this means that the random event in the real domain is impossible, and the degree of knowledge and the chaos factor of the random event in the complex domain are 1 and 0, respectively.

When $P_r = 0.5$, the degree of knowledge of the random event in the complex domain is 0.5, and the chaos factor is -0.5.

It is important to note Equations (21) and (22):

$$0.5 \leq |P_c|^2 \leq 1\,,\ -0.5 \leq 2jP_rP_m \leq 0 \tag{21}$$

$$(P_r + P_m/j)^2 = Degree\_of\_knowledge - Chaos\_factor = 1 \tag{22}$$

This means a stochastic system in the complex domain has a constant probability equal to 1, but its degree of knowledge and chaos factor are variable. The more stable the stochastic system is, the greater its degree of knowledge, and the closer the chaos factor is to zero. This can be used to redesign the CV_one-hot label for PolSAR image classification.

In Figure 4a, the real-valued one-hot label can be seen as a probability distribution for an object belonging to a certain category with a 100% probability (the value at the activation point is 1, and the values at the other inactivation points are 0). Figure 4b proposes a K-dimensional complex-valued vector as the CV_one-hot label, with $1 + 0 \cdot j$ values at activation points and the $0 + 1 \cdot j$ values at the inactivation point. Figure 4c shows that when the complex-valued probability is decomposed into P_r and P_m/j probability, the meaning of the CV_one-hot label becomes easier to understand. At the activation point, P_r equals 1, and P_m/j equals 0, while at the inactivation point, P_r equals 0, and P_m/j equals 1. The vector P_r represents the probability of classification and has the same meaning as the vector P, which is used to obtain the unique class of an object via softmax. If a complex-valued probability is treated as a stochastic system, the knowledge degree of any point in the CV_one-hot label equals 1, and the chaos factor equals 0. Therefore, the CV_one-hot label has the highest stability, as well as the largest knowledge degree and the smallest absolute value of the chaos factor.

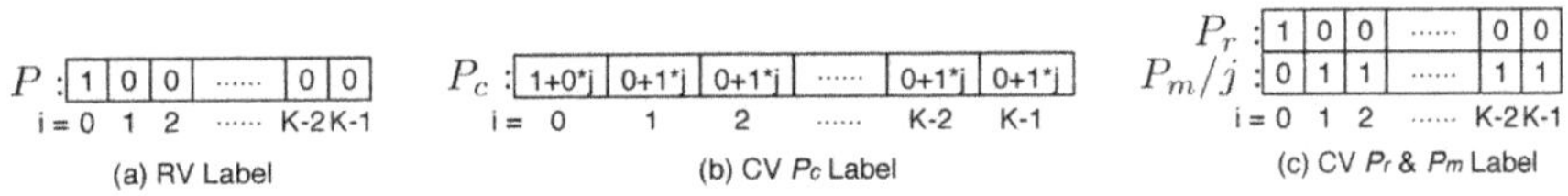

Figure 4. (**a**) shows a real-valued one-hot label, while (**b**,**c**) are CV_one-hot labels. K represents the number of categories. P represents the probability of a random event in the real domain, while P_c represents the probability of a random event in the complex domain. P_r and P_m represent the real and imaginary parts of the random event in the complex domain.

2.5.2. Complex-Valued Cross-Entropy

To effectively train CV-CNNs, it is not sufficient to use CV_one-hot labels. A loss function must also measure the difference between the model's output and the label. RV-CNNs use cross-entropy as their loss function, which calculates the difference between two probability distributions. A high loss value indicates a significant difference between the model's output and the label, while a low value indicates a small difference. Similarly, to train CV-CNNs, this paper proposes a loss function called CV_CrossEntropy, which describes the difference between complex-valued outputs and CV_one-hot labels using the following Equation:

$$\begin{aligned} CV\_Loss &= CrossEntropy(\Re(\hat{y}), \Re(y)) + \sum_{k=0}^{K-1} CrossEntropy([\Re(\hat{y}_k), \Im(\hat{y}_k)], [\Re(y_k), \Im(y_k)]) \\ &= -\frac{1}{N}\sum_{i=1}^{N}\Big(\sum_{k=0}^{K-1}\Re(y_{ik})log\Re(\hat{y}_{ik}) + \sum_{k=0}^{K-1}(\Re(y_{ik})log\Re(\hat{y}_{ik}) + \Im(y_{ik})log\Im(\hat{y}_{ik}))\Big) \end{aligned} \quad (23)$$

In Equation (23), K represents the number of categories, while N represents the number of samples in a mini-batch. In addition, y refers to the ground truth, whereas $\hat{y}$ represents the model's predicted outcome. The initial segment of the loss function only applies to the real part of the labels and the P_r of the outputs. The smaller the value of this part, the more precise the classification outcome of the complex-valued model will be. The second part of the loss function incorporates the labels, P_r, and P_m/j of the outputs. The smaller the value of this part, the more stable the classification system will become.

2.6. Complex-Valued PolSAR Classification Algorithm

Algorithm 1 outlines the PolSAR classification process based on a complex-valued approach proposed in this research. The first step involves constructing a complex-valued convolutional classification network equipped with CVA_Max_Pooling and HReLU in the model. Next, CV_one-hot labels are applied to the training set. Then, the model parameters are updated through iterations using the CV_CrossEntropy loss. Lastly, the trained model is utilized to classify the complete PolSAR dataset.

Algorithm 1: Complex-valued convolutional classification algorithm for PolSAR images

Preprocessing:
1. Construction of complex-valued models for PolSAR image classification with CVA_Max_Pooling and HReLU
2. Assigning CV_one-hot labels to each pixel of the PolSAR dataset
3. Selection of training set from the PolSAR dataset

Input: a training set and corresponding labels, learning rate, batch size, and momentum parameter

4. **Repeat:**
5. Calling CVA_Max_Pooling to obtain the most efficient features
6. Invoking HReLU to map the amplitude and phase of the feature to the nonlinear domain
7. Calling CV_CrossEntropy to compute the loss during training
8. Updating model parameters with loss
9. **Until:** Meeting the conditions for termination
10. Inferring the class of the entire PolSAR image with the trained model

Output: Prediction of the testing set

3. Experimental Results

This section will start by providing a brief description of the three benchmark datasets. Subsequently, the section delves into the specifics of the model inputs, the experimental setup, and the evaluation metrics. Finally, the effectiveness of the proposed complex-valued approach is demonstrated through a comparative analysis of classification model results across the three PolSAR datasets.

3.1. *PolSAR Dataset Description*

3.1.1. Flevoland Dataset 1

On 16 August 1989, the NASA/JPL AIRSAR airborne platform collected a dataset from the Flevoland area in the Netherlands. These data have a size of 750 × 1024, and Figure 5a,b displays the RGB image and corresponding ground truth after Pauli decomposition. The image contains 15 categories: stembeans, peas, forest, lucerne, wheat, beet, potatoes, bare soil, grass, rapeseed, barley, wheat2, wheat3, water, and buildings.

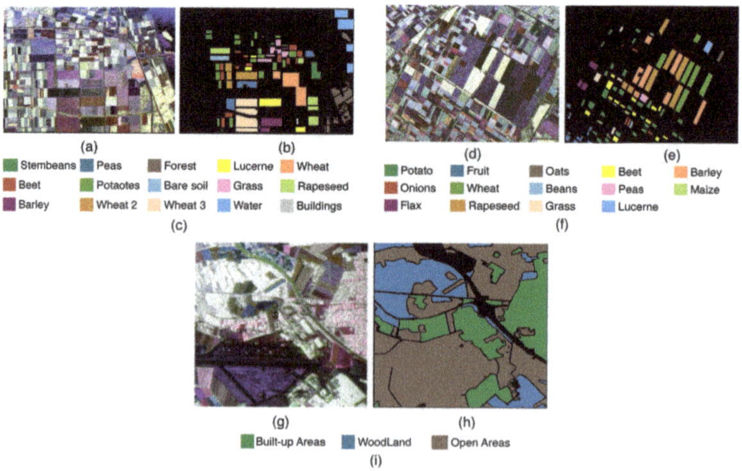

Figure 5. The ground truth and class legends of Flevoland Dataset 1, Flevoland Dataset 2, and Oberpfaffenhofen Dataset. (**a**,**d**,**g**) are RGB images, (**b**,**e**,**h**) are the corresponding ground truth images after Pauli decomposition, and (**c**,**f**,**i**) are class legends.

3.1.2. Flevoland Dataset 2

In 1991, L-band ATRSAR data were collected in the Flevoland area, consisting of a size of 1024 × 1024. Figure 5d displays the RGB image, while Figure 5e shows the ground truth after Pauli decomposition. The image consists of 14 categories: potato, fruit, oats, beet, barley, onions, wheat, beans, peas, maize, flax, rapeseed, grass, and lucerne.

3.1.3. Oberpfaffenhofen Dataset

The German Aerospace Center (DLR) has provided the ESAR data for the Oberpfaffenhofen area in Germany. The dataset is 1300 × 1200, and the RGB image and the corresponding ground truth after Pauli decomposition are displayed in Figure 5g,h. The image depicts three categories: built-up, woodland, and open areas.

3.2. *Parameterization*

Before conducting experiments, it is crucial to establish the appropriate training. For PolSAR image classification, several studies have explored the sampling rate and neural network parameters for PolSAR image classification [20], which renders it unnecessary for this paper to delve into those parameters. Instead, this paper will utilize them directly in the experiments.

Training and testing sets for SCNN, DCNN, FCN, and SegNet needed to be created using PolSAR images and labels. The inputs and outputs for these models were explained in Section 2.1 and will not be repeated here. For the SCNN and DCNN, the input was a 12 × 12 image patch containing a pixel to be classified. For FCN and SegNet, a sliding window of 128 × 128 with a sliding step of 15 generated training and testing sets on PolSAR images and labels. Only labeled pixels in the input of FCN and SegNet were involved in training, and unlabeled pixels could not be used to update model parameters. In all experiments, the model was trained and validated using a ratio of 0.9/0.1 for pixels in the training set.

PyTorch was employed for implementing all codes, and the Adam optimizer with default parameters was utilized. All experiments were conducted on a single workstation with an Intel Core i7-6700K CPU, 32G RAM, an NVIDIA TITAN X GPU, and an Ubuntu 20.04 LTS operating system.

3.3. *Evaluation Metrics*

When evaluating how well PolSAR images are classified, there are three common metrics: overall accuracy (OA), average accuracy (AA), and Kappa coefficient. OA is the ratio of correctly classified samples to the number of test samples. AA is the average accuracy of classification for each category. The kappa coefficient is a metric that measures the effectiveness of classification and consistency testing, especially when the number of samples in different categories varies greatly. The larger these metrics, the better the classification effect.

3.4. *Model Parameters*

The SCNN, DCNN, FCN, and SegNet parameters are listed in Tables 1–3, respectively. For real-valued models, ReLU was used as the activation function, max pooling was used as the pooling function, and cross-entropy was used as the loss function. For old complex-valued models, CReLU was used as the activation function, max pooling was used as the pooling function, and cross-entropy was used as the loss function. For new complex-valued models, HReLU was used as the activation function, CVA_Max_Pooling was used as the pooling function, and CV_CrossEntropy was used as the loss function. To ensure fairness in the experiment, the number of parameters in the different models was kept as equal as possible.

Table 1. Detailed parameters of the RV-SCNN and CV-SCNN. K denotes the total number of categories.

	Module	Dimension		Module	Dimension
RV-SCNN	RV-Convolution RV-Max-Pooling ReLU	$3 \times 3 \times 9 \times 8$ 2×2	CV-SCNN	CV-Convolution CVA_Max_Pooling HReLU	$3 \times 3 \times 6 \times 6$ 2×2
	RV-Convolution RV-Max-Pooling ReLU	$3 \times 3 \times 8 \times 22$ 2×2		CV-Convolution CVA_Max_Pooling HReLU	$3 \times 3 \times 6 \times 12$ 2×2
	RV-Average-Pooling RV-Fully-Connection RV-Fully-Connection	22×180 $180 \times K$		CV-Average-Pooling CV-Fully-Connection CV-Fully-Connection	12×128 $128 \times K$
RV-SCNN Params		FLevoland 1: 9147;	FLevoland 2: 8966;	Oberpfaffenhofen: 6975	
CV-SCNN Params		FLevoland 1: 9214;	FLevoland 2: 8956;	Oberpfaffenhofen: 6118	

Table 2. Detailed parameters of the RV-DCNN and CV-DCNN. K denotes the total number of categories.

	Module	Dimension		Module	Dimension
RV-DCNN	RV-Convolution RV-Max-Pooling ReLU	$3 \times 3 \times 9 \times 18$ 2×2	CV-DCNN	CV-Convolution CVA_Max_Pooling HReLU	$3 \times 3 \times 6 \times 12$ 2×2
	RV-Convolution RV-Max-Pooling ReLU	$3 \times 3 \times 18 \times 36$ 2×2		CV-Convolution CVA_Max_Pooling HReLU	$3 \times 3 \times 12 \times 24$ 2×2
	RV-Convolution RV-Max-Pooling ReLU	$3 \times 3 \times 36 \times 72$ 2×2		CV-Convolution CVA_Max_Pooling HReLU	$3 \times 3 \times 24 \times 48$ 2×2
	RV-Convolution RV-Max-Pooling ReLU	$3 \times 3 \times 72 \times 144$ 2×2		CV-Convolution CVA_Max_Pooling HReLU	$3 \times 3 \times 48 \times 96$ 2×2
	RV-Average-Pooling RV-Fully-Connection RV-Fully-Connection	144×312 $312 \times K$		CV-Average-Pooling CV-Fully-Connection CV-Fully-Connection	96×256 $256 \times K$
RV-DCNN Params		FLevoland 1: 174,405;	FLevoland 2: 174,092;	Oberpfaffenhofen: 170,649	
CV-DCNN Params		FLevoland 1: 168,254;	FLevoland 2: 167,740;	Oberpfaffenhofen: 162,086	

Table 3. Detailed parameters of the RV-(FCN, SegNet) and CV-(FCN, SegNet). K denotes the total number of categories.

	Module	Dimension		Module	Dimension
RV-(FCN, SegNet)	RV-Convolution RV-Max-Pooling ReLU	$3 \times 3 \times 9 \times 17$ 2×2	CV-(FCN, SegNet)	CV-Convolution CVA_Max_Pooling HReLU	$3 \times 3 \times 6 \times 12$ 2×2
	RV-Convolution RV-Max-Pooling ReLU	$3 \times 3 \times 17 \times 34$ 2×2		CV-Convolution CVA_Max_Pooling HReLU	$3 \times 3 \times 12 \times 24$ 2×2
	RV-Convolution RV-Max-Pooling ReLU	$3 \times 3 \times 34 \times 68$ 2×2		CV-Convolution CVA_Max_Pooling HReLU	$3 \times 3 \times 24 \times 48$ 2×2
	RV-Convolution RV-Max-Pooling ReLU	$3 \times 3 \times 68 \times 132$ 2×2		CV-Convolution CVA_Max_Pooling HReLU	$3 \times 3 \times 48 \times 96$ 2×2
	Up-sampling RV-Convolution ReLU	2×2 $3 \times 3 \times 132 \times 68$		Up-sampling CV-Convolution HReLU	2×2 $3 \times 3 \times 96 \times 48$
	Up-sampling RV-Convolution ReLU	2×2 $3 \times 3 \times 68 \times 34$		Up-sampling CV-Convolution HReLU	2×2 $3 \times 3 \times 48 \times 24$
	Up-sampling RV-Convolution ReLU	2×2 $3 \times 3 \times 34 \times 17$		Up-sampling CV-Convolution HReLU	2×2 $3 \times 3 \times 24 \times 12$
	Up-sampling RV-Convolution ReLU	2×2 $3 \times 3 \times 17 \times 9$		Up-sampling CV-Convolution HReLU	2×2 $3 \times 3 \times 12 \times 6$
	Up-sampling RV-Convolution	2×2 $3 \times 3 \times 9 \times K$		Up-sampling CV-Convolution	2×2 $3 \times 3 \times 6 \times K$
RV-(FCN, SegNet) Params	FLevoland 1: 218,345;		FLevoland 2: 218,262;	Oberpfaffenhofen: 217,349	
CV-(FCN, SegNet) Params	FLevoland 1: 223,080;		FLevoland 2: 222,968;	Oberpfaffenhofen: 221,736	

3.5. *Analysis of Experimental Results*

3.5.1. Flevoland Dataset 1 Results

In order to enhance the robustness assessment of the proposed methods, cross-validation was employed to acquire the classification results. Five percent of the labeled samples from each of the 15 dataset categories were randomly selected as the training set, while the remaining samples constituted the testing set. The final result, as depicted in Figure 6 and Table 4, represents the average of ten classification outcomes.

It is evident from the quantitative results that the real-valued version of any classification model has the poorest classification results, while the new complex-valued approach has the best results. This demonstrates the effectiveness of the new complex-valued approach. The complex-valued approach preserves the phase features of the input, thus extracting and retaining more effective features. Moreover, CVA_Max_Pooling preserves the most discriminative features, while HReLU provides sufficient nonlinearity and sparsity. Finally, CV_CrossEntropy enhances the efficiency of feature utilization, leading to the best classification results.

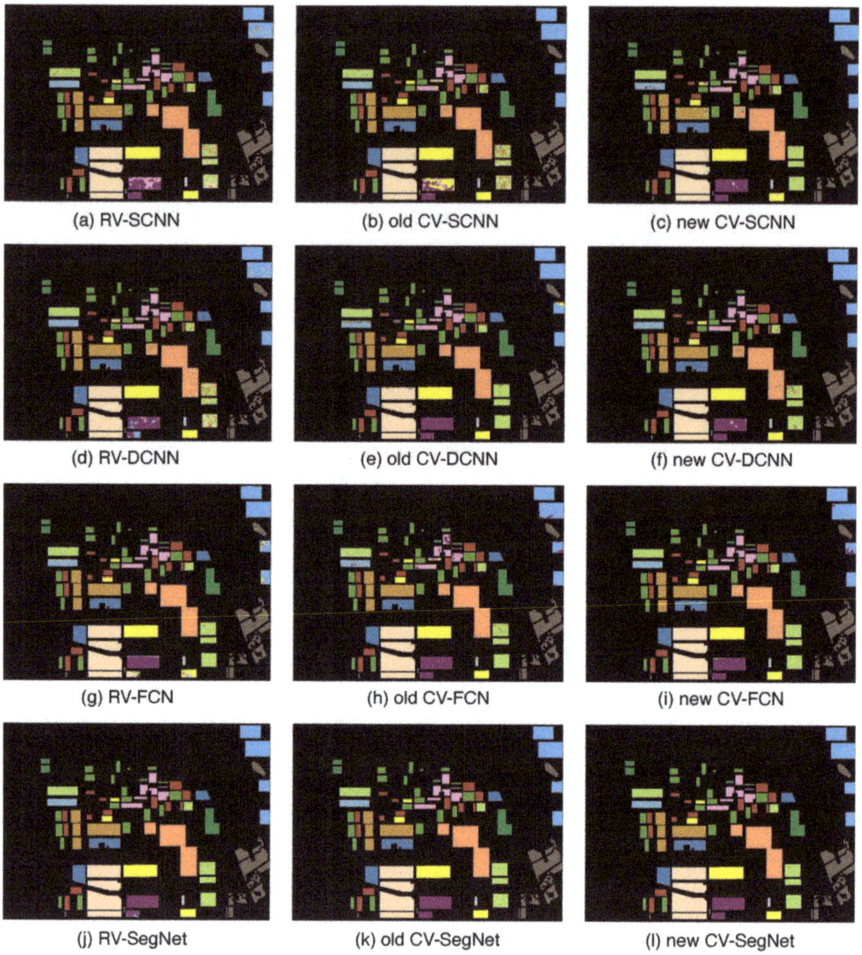

Figure 6. Classification results of Flevoland Dataset 1 with different methods. The classification results of RV-SCNN, RV-DCNN, RV-FCN, and RV-SegNet are represented by (**a**,**d**,**g**,**j**), respectively, while the results of old CV-SCNN, old CV-DCNN, old CV-FCN, and old CV-SegNet are shown by (**b**,**e**,**h**,**k**). The classification results of new CV-SCNN, new CV-DCNN, new CV-FCN, and new CV-SegNet are represented by (**c**,**f**,**i**,**l**).

After analyzing the effects of four classification models, it was observed that SegNet performs the best in achieving classification results under the same version, while SCNN has the poorest classification results, and the encoder–decoder model outperforms the CNN model. The new CV-SCNN, new CV-DCNN, new CV-FCN, and new CV-SegNet have shown an improvement of 4.01%, 4.46%, 3.46%, and 0.45%, respectively, over RV-SCNN, RV-DCNN, RV-FCN, and RV-SegNet. The results indicate that the complex-valued method has a significant improvement effect on CNNs with fewer parameters. This is because the classification results of FCN and SegNet are already satisfactory, and improving them significantly using the complex-valued method is challenging. Therefore, if only CNNs can be selected for PolSAR image classification due to machine performance constraints, the new CV-CNN model would be the best choice. Otherwise, the new CV-SegNet would provide optimal classification results. Figure 6l highlights that the new CV-SegNet's results are almost identical to the ground truth.

Table 4. Overall accuracy (%), average accuracy (%), and Kappa coefficient of all competing methods on the Flevoland Dataset 1. The bolded values represent the highest values among three versions of a model (RV-, old CV-, new CV-).

	RV-SCNN	Old CV-SCNN	New CV-SCNN	RV-DCNN	Old CV-DCNN	New CV-DCNN	RV-FCN	Old CV-FCN	New CV-FCN	RV-SegNet	Old CV-SegNet	New CV-SegNet
Stembeans	99.85	**99.95**	99.33	99.48	**99.67**	99.62	99.72	99.84	**99.92**	99.95	99.98	**100.00**
Peas	95.71	99.57	**99.58**	95.06	99.14	**99.96**	97.56	**99.70**	98.72	98.94	98.76	**99.31**
Forest	98.73	**99.65**	97.12	98.11	99.46	**99.52**	98.65	98.70	**100.00**	99.26	99.48	**99.92**
Lucerne	98.04	**99.91**	96.24	96.39	**96.89**	96.47	88.26	99.88	**99.93**	98.23	**99.95**	99.88
Wheat	97.44	**97.91**	94.21	93.54	**98.89**	95.35	**99.81**	98.35	99.80	99.95	**100.00**	**100.00**
Beet	98.38	**98.60**	98.52	97.84	93.64	**99.79**	96.16	94.78	**98.92**	**99.70**	99.20	99.43
Potaotes	**97.74**	96.76	97.57	**99.44**	95.02	99.35	94.47	**99.80**	98.54	99.25	99.27	**99.88**
Bare soil	**99.97**	94.41	93.01	**100.00**	74.27	98.31	87.69	92.76	**95.58**	**100.00**	99.94	**100.00**
Grass	**94.51**	92.10	92.79	96.47	95.58	**98.68**	98.69	77.51	**99.89**	99.86	99.79	**100.00**
Rapeseed	72.03	69.68	**98.72**	71.44	**94.12**	90.59	97.48	96.42	**99.35**	99.53	**99.92**	99.91
Barley	66.85	45.26	**96.79**	78.30	**99.46**	96.95	77.71	**99.58**	96.03	96.80	**99.83**	99.64
Wheat2	95.52	**99.61**	88.63	97.57	97.55	**99.75**	**98.97**	95.80	98.80	**100.00**	**100.00**	99.92
Wheat3	**99.92**	99.45	97.94	99.90	98.22	**99.97**	99.66	99.65	**99.97**	**99.97**	99.35	99.92
Water	77.20	**99.77**	99.07	87.54	96.99	**99.98**	86.69	93.92	**95.71**	98.66	98.81	**99.46**
Buildings	**98.74**	96.22	93.49	83.82	83.82	**98.53**	85.08	**96.64**	82.98	**85.50**	84.03	82.77
OA	92.65	93.81	**96.66**	93.67	96.79	**98.13**	95.40	97.10	**98.86**	99.31	99.49	**99.76**
AA	92.71	92.59	**96.20**	92.99	94.85	**98.19**	93.77	96.22	**97.61**	98.37	98.55	**98.67**
Kappa	0.9186	0.9315	**0.9634**	0.9300	0.9648	**0.9795**	0.9493	0.9682	**0.9875**	0.9925	0.9944	**0.9974**

3.5.2. Flevoland Dataset 2 Results

In order to enhance the robustness assessment of the proposed methods, cross-validation was employed to acquire the classification results. Five percent of the labeled samples from each of the 14 dataset categories were randomly selected as the training set, while the remaining samples constituted the testing set. The final result, as depicted in Figure 7 and Table 5, represents the average of ten classification outcomes.

According to Table 5, FCN and SegNet can extract more contextual information, resulting in excellent classification results for (RV-, old CV-, new CV-) FCN and SegNet. Although the new CV-FCN and new CV-SegNet perform the best in classification, the improvement is not very noticeable. In comparison, new CV-SCNN and new CV-DCNN show a significant improvement in their classification results compared to RV-SCNN and RV-DCNN. It is worth noting that RV-SCNN and RV-DCNN only achieve 0.09% and 11.18% accuracy, respectively, for the category of beans, while old CV-SCNN and old CV-DCNN only achieve 13.29% and 17.89% accuracy for the onions category. In contrast, the new CV-SCNN and new CV-DCNN show a more balanced performance in these two categories, with no extremely low accuracy. The new CV-SCNN has a classification accuracy of 82.16% and 60% for beans and onions, respectively, while the new CV-DCNN has a classification accuracy of 98.63% and 76.24% for beans and onions, respectively.

From Figure 5e, it is apparent that both beans and onions fall under categories with a limited number of samples. The RV-CNNs and old CV-CNNs have struggled to extract the features of these categories during training. This is because the inputs of these categories have only a few complex-valued features hidden in them. RV-CNNs ignore this part of the features from the input, while old CV-CNNs destroy it during the computation process. However, the new CV-CNNs are designed to retain this part of the features as much as possible during computation. Hence, they can accurately recognize beans and onions.

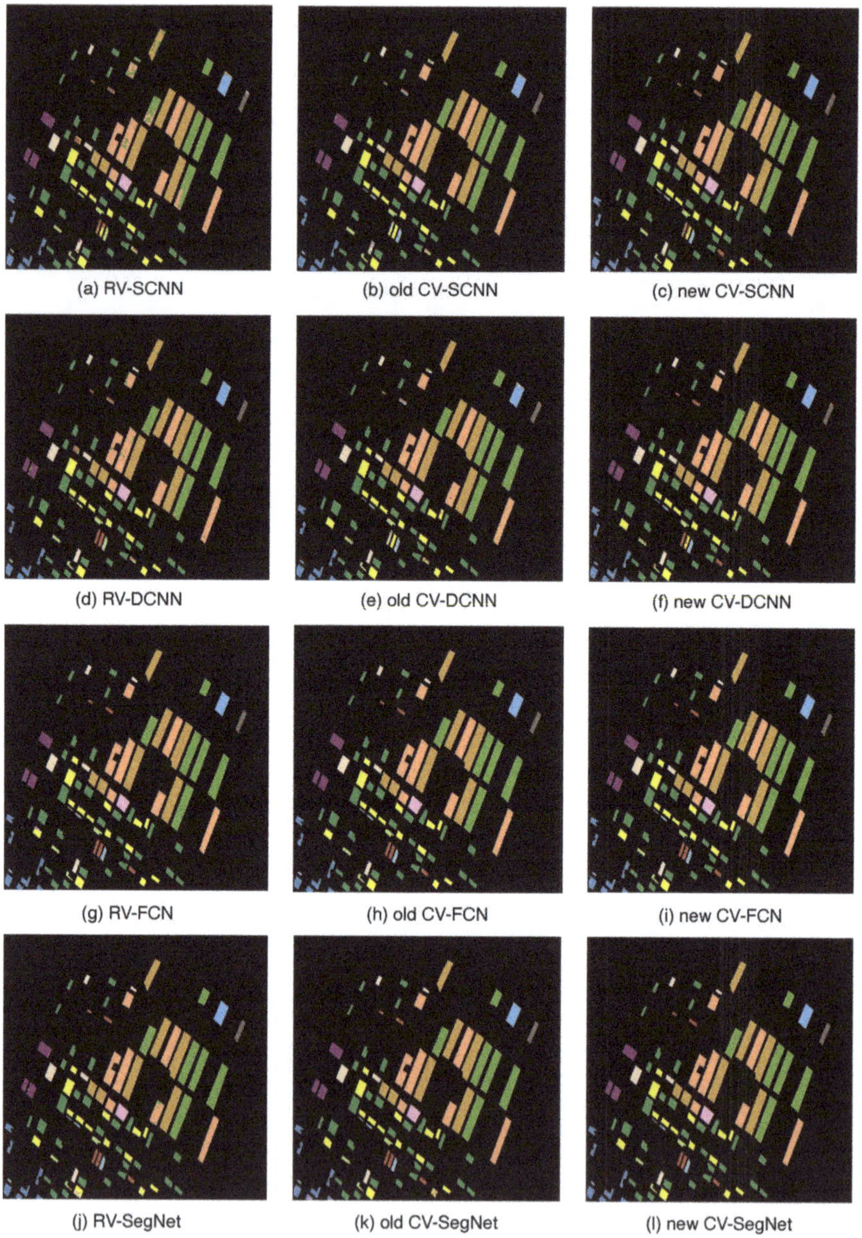

Figure 7. Classification results of Flevoland Dataset 2 with different methods. The classification results of RV-SCNN, RV-DCNN, RV-FCN, and RV-SegNet are represented by (**a**,**d**,**g**,**j**), respectively, while the results of old CV-SCNN, old CV-DCNN, old CV-FCN, and old CV-SegNet are shown by (**b**,**e**,**h**,**k**). The classification results of new CV-SCNN, new CV-DCNN, new CV-FCN, and new CV-SegNet are represented by (**c**,**f**,**i**,**l**).

Table 5. Overall accuracy (%), average accuracy (%), and Kappa coefficient of all competing methods on the Flevoland Dataset 2. The bolded values represent the highest values among three versions of a model (RV-, old CV-, new CV-).

	RV-SCNN	Old CV-SCNN	New CV-SCNN	RV-DCNN	Old CV-DCNN	New CV-DCNN	RV-FCN	Old CV-FCN	New CV-FCN	RV-SegNet	Old CV-SegNet	New CV-SegNet
Potato	99.48	99.50	**99.98**	**99.90**	99.80	99.86	98.72	97.72	**99.97**	99.63	99.46	**99.94**
Fruit	**100.00**	99.70	99.77	99.66	99.66	**99.93**	98.23	**99.98**	99.70	96.97	90.03	**98.51**
Oats	93.62	94.98	**95.62**	**96.41**	92.32	**96.41**	99.93	**100.00**	**100.00**	**100.00**	99.93	99.78
Beet	94.20	**99.06**	98.87	92.75	98.54	**98.87**	94.82	95.21	**97.71**	94.14	95.41	**99.92**
Barley	93.59	99.60	**99.74**	96.26	99.09	**99.99**	98.60	98.92	**99.98**	98.32	**99.98**	**99.98**
Onions	52.77	13.29	**60.00**	**77.75**	17.89	76.24	**100.00**	98.08	98.73	97.18	96.71	**99.39**
Wheat	89.50	**99.80**	99.71	98.54	99.76	**99.95**	99.91	99.45	**100.00**	99.83	99.78	**100.00**
Beans	0.09	**94.27**	82.16	11.18	82.53	**98.43**	84.84	92.42	**95.84**	87.99	97.97	**99.91**
peas	**99.72**	97.69	97.22	**99.91**	99.95	99.44	99.95	**100.00**	**100.00**	**100.00**	**100.00**	**100.00**
Maize	89.61	89.15	**91.86**	**96.28**	81.16	74.11	94.42	98.99	**100.00**	92.56	**97.75**	94.42
Flax	98.72	97.74	**99.28**	97.23	99.95	**99.98**	99.95	**100.00**	**100.00**	98.63	99.98	**100.00**
Rapessed	97.62	99.42	**99.55**	99.29	99.27	**99.95**	99.27	99.58	**99.97**	**99.99**	99.87	**99.99**
Grass	85.94	82.30	**95.15**	97.84	95.20	**99.62**	97.88	98.72	**99.74**	**100.00**	**100.00**	99.95
Lucerne	87.94	92.48	**98.88**	98.17	88.79	**99.80**	99.93	**100.00**	99.97	**100.00**	**100.00**	**100.00**
OA	93.31	97.22	**98.57**	96.95	97.39	**99.17**	98.66	98.72	**99.73**	98.78	99.06	**99.86**
AA	84.49	89.93	**94.13**	90.08	89.56	**95.90**	97.60	98.50	**99.40**	97.52	98.35	**99.41**
Kappa	0.9190	0.9668	**0.9830**	0.9638	0.9689	**0.9902**	0.9841	0.9849	**0.9968**	0.9856	0.9889	**0.9984**

3.5.3. Oberpfaffenhofen Dataset Results

In order to enhance the robustness assessment of the proposed methods, cross-validation was employed to acquire the classification results. Only one percent of the labeled samples from each of the three dataset categories were randomly selected as the training set, while the remaining samples constituted the testing set. The final result, as depicted in Figure 8 and Table 6, represents the average of ten classification outcomes.

Table 6 shows that all models can accurately classify woodland and open areas with classification accuracies above 96%. However, Figure 8a–f shows that CNNs sometimes confuse built-up areas with woodland. Nonetheless, according to Table 6, the classification accuracy of the new CV-SCNN for built-up areas is higher than that of RV-SCNN and old CV-SCNN by 12.6% and 5.14%, respectively. Additionally, the new CV-DCNN has a classification accuracy for built-up areas that is 10.45% and 7% higher than that of RV-DCNN and old CV-DCNN, respectively. These results suggest that the new complex-valued approach significantly improves the classification accuracy of the more challenging categories.

Table 6. Overall accuracy (%), average accuracy (%), and Kappa coefficient of all competing methods on the Oberpfaffenhofen Dataset. The bolded values represent the highest values among three versions of a model (RV-, old CV-, new CV-).

	RV-SCNN	Old CV-SCNN	New CV-SCNN	RV-DCNN	Old CV-DCNN	New CV-DCNN	RV-FCN	Old CV-FCN	New CV-FCN	RV-SegNet	Old CV-SegNet	New CV-SegNet
Built-up areas	79.90	87.36	**92.50**	79.38	82.83	**89.83**	98.55	97.22	**99.46**	96.45	96.24	**98.96**
Wood land	97.36	98.27	**98.65**	98.74	99.11	**99.44**	99.69	99.30	**99.31**	98.74	99.59	**99.20**
Open areas	**96.21**	96.07	96.12	98.71	99.30	**99.78**	97.46	99.05	**99.11**	98.68	99.04	**99.51**
OA	92.35	94.31	**95.69**	93.88	95.14	**97.22**	98.15	98.64	**99.23**	98.13	98.45	**99.31**
AA	91.16	93.90	**95.76**	92.28	93.75	**96.35**	98.57	98.52	**99.29**	97.96	98.29	**99.22**
Kappa	0.8512	0.8941	**0.9221**	0.8831	0.9094	**0.9501**	0.9679	0.9763	**0.9868**	0.9673	0.9729	**0.9882**

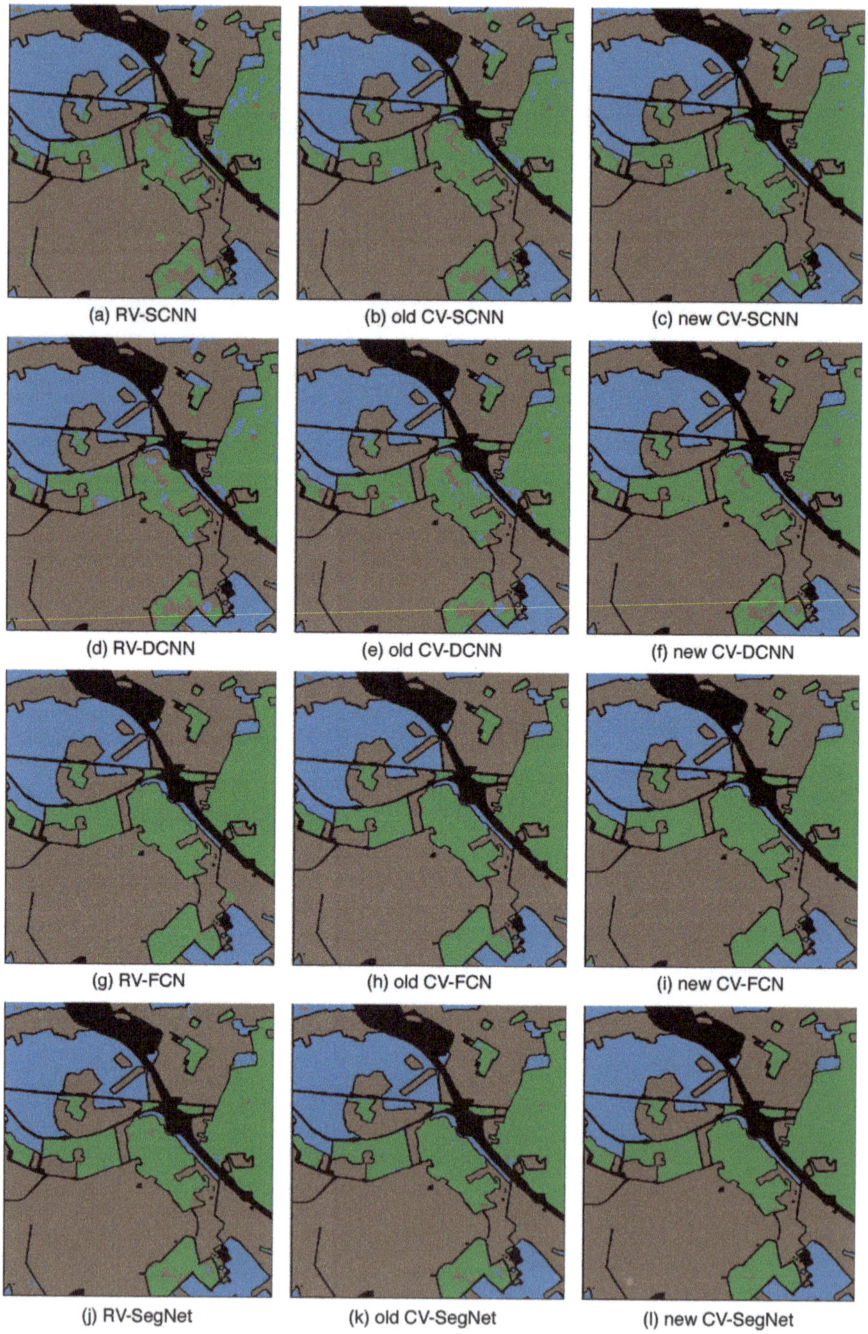

Figure 8. Classification results of Oberpfaffenhofen Dataset with different methods. The classification results of RV-SCNN, RV-DCNN, RV-FCN, and RV-SegNet are represented by (**a**,**d**,**g**,**j**), respectively, while the results of old CV-SCNN, old CV-DCNN, old CV-FCN, and old CV-SegNet are shown by (**b**,**e**,**h**,**k**). The classification results of new CV-SCNN, new CV-DCNN, new CV-FCN, and new CV-SegNet are represented by (**c**,**f**,**i**,**l**).

In summary, it is recommended to use SegNet instead of CNNs to enhance the accuracy of PolSAR image classification without any restrictions on the model size. Although the new CV-SegNet offers better classification outcomes than RV-SegNet, the accuracy improvement is limited. When the model size is limited, the optimal choice is the new CV-CNNs, which can accurately distinguish difficult entries and significantly improve the classification accuracy of small sample categories, thus leading to an overall enhancement in accuracy.

3.5.4. Computational Complexity of CNN

From Figure 9a,b, Tables 1 and 2, it is evident that when the number of convolutional layers in CV-CNN and RV-CNN is the same and the difference in the number of parameters is not substantial, CV-CNN has fewer convolutional kernels in each layer, yet achieves higher classification accuracy. This indicates that, despite extracting fewer feature maps, the new CV-CNN consistently delivers superior classification results.

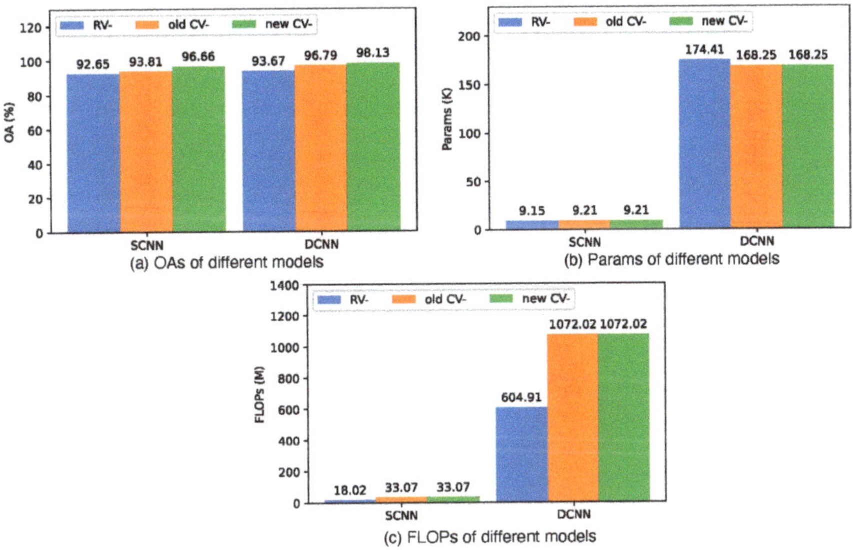

Figure 9. (**a–c**) illustrate the overall accuracy, number of parameters, and FLOPs (floating-point operations per second) for SCNN and DCNN on the Flevoland Dataset 1, respectively. The blue color represents the real-valued version, the red color corresponds to the old complex-valued version, and the green color indicates the new complex-valued version.

Figure 9b,c illustrates that the FLOPs of CV-CNN are significantly larger than those of RV-CNN when they share the same number of convolutional layers and the difference in the number of parameters is not substantial. This discrepancy arises because complex-valued operations can only be approximated by multiple real-valued operations in the PyTorch environment, as depicted in Formulas (1) and (2). For instance, a complex-valued addition operation necessitates two real-valued addition operations, while a complex multiplication operation requires four real-valued multiplication operations and two real-valued addition operations. It is expected that with advancements in complex-valued deep learning techniques, particularly in polarized coordinates, where a real-valued multiplication operation and a real-valued addition operation can replace a complex-valued multiplication operation, this limitation will be mitigated.

Comparing the old CV-CNN and new CV-CNN with the same number of parameters and FLOPs, the new CV-CNN consistently outperforms the old CV-CNN, achieving better classification results. For the Flevoland Dataset 1, the new CV-SCNN yields a 2.99% higher accuracy than RV-DCNN and is only 0.1% lower than the old CV-DCNN. For the Flevoland Dataset 2, the new CV-SCNN achieves results 2% higher than RV-DCNN and 1% higher than the old CV-DCNN. In essence, under the condition of meeting accuracy requirements the new CV-SCNN can effectively replace RV-DCNN and the old CV-DCNN. Moreover the new CV-SCNN boasts approximately half the parameters compared to RV-DCNN and old CV-DCNN, with FLOPs being roughly half of RV-DCNN and about one-third of old CV-DCNN. This trend holds for the Oberpfaffenhofen Dataset as well.

4. Discussion

Three ablation experiments were conducted to validate the performance of various aspects of the new CV-DL models, specifically CVA_Max_Pooling, HReLU, and CV_CrossEntropy. The SCNN classification model was chosen and validated on three different datasets.

4.1. Ablation Experiment 1: Performance of CVA_Max_Pooling

One of the new components in CV-DL models is CVA_Max_Pooling. This component is important because it helps to retain the most important features in the feature map and passes them on to the next convolution layer. This increases the feature utilization efficiency and reduces the computation required. To test the impact of CVA_Max_Pooling on the experimental results, complex-valued classification models were created using both real-valued max pooling and average pooling. These models operate on the real and imaginary parts of the feature map separately. Table 7 shows the results of our experiments, with RMP-CV-SCNN and RAP-CV-SCNN denoting the models using real-valued max pooling and average pooling, respectively.

Table 7. Overall accuracy (%), average accuracy (%), and Kappa coefficient of the different poolings.

Dataset	Methods	OA	AA	Kappa
Flevoland Dataset 1	RMP-CV-SCNN	95.65	95.17	0.9522
	RAP-CV-SCNN	94.10	93.64	0.9349
	new CV-SCNN	96.66	96.20	0.9634
Flevoland Dataset 2	RMP-CV-SCNN	97.94	93.56	0.9756
	RAP-CV-SCNN	96.71	92.30	0.9609
	new CV-SCNN	98.57	94.13	0.9830
Oberpfaffenhofen Dataset	RMP-CV-SCNN	94.78	94.90	0.9043
	RAP-CV-SCNN	94.63	94.64	0.9014
	new CV-SCNN	95.69	95.76	0.9221

Table 7 shows that the new CV-SCNN achieved the best classification results on all three datasets, followed by RMP-CV-SCNN, while RAP-CV-SCNN obtained the worst outcome. Notably, CVA_Max_Pooling is superior to max pooling, as it not only retains the most significant features but also avoids generating "fake" features. Max pooling tends to generate "fake" features by operating on the real and imaginary parts of the feature map separately, resulting in two unrelated features being combined. Although average pooling works on the real and imaginary parts separately, it is a form of complex-valued average pooling. While it does not generate "fake" features, it significantly reduces the weight of the most important features by confusing them with the unimportant ones.

4.2. Ablation Experiment 2: Performance of HReLU

HReLU functions as a complex-domain ReLU by discarding half of the features nonlinearly. To assess its effects on experimental results, complex-valued classification models were created using ModReLU, ZreLU, and CReLU, referred to as Mod-CV-SCNN, ZReLU-CV-SCNN, and CReLU-CV-SCNN, respectively. The outcomes of these experiments can be found in Table 8.

Table 8. Overall accuracy (%), average accuracy (%), and Kappa coefficient of the different activations.

Dataset	Methods	OA	AA	Kappa
Flevoland Dataset 1	CReLU-CV-SCNN	95.95	95.49	0.9554
	ZReLU-CV-SCNN	95.45	95.00	0.9499
	Mod-CV-SCNN	95.53	95.03	0.9508
	new CV-SCNN	96.66	96.20	0.9634
Flevoland Dataset 2	CReLU-CV-SCNN	98.05	93.67	0.9769
	ZReLU-CV-SCNN	97.84	93.42	0.9744
	Mod-CV-SCNN	97.68	93.25	0.9725
	new CV-SCNN	98.57	94.13	0.9830
Oberpfaffenhofen Dataset	CReLU-CV-SCNN	94.97	95.00	0.9082
	ZReLU-CV-SCNN	94.79	94.82	0.9046
	Mod-CV-SCNN	93.77	93.67	0.8844
	new CV-SCNN	95.69	95.76	0.9221

Based on the data presented in Table 8, it is clear that new CV-SCNN outperforms the other models on all three datasets. Additionally, CReLU-CV-SCNN yields better results than ZReLU-CV-SCNN and Mod-CV-SCNN, despite both containing ReLU in their names. However, as explained in Section 2.3, only HReLU produces ReLU-like sparsity and nonlinearity. CReLU is slightly less effective due to a lack of sparsity, while ZReLU underperforms because too many features are dropped, and ModReLU produces the worst results due to its inadequate nonlinearity.

4.3. Ablation Experiment 3: Performance of CV_CrossEntropy

When training deep learning models, the loss function is crucial in driving the model outputs closer to the ground truth and helping the model learn the best classification patterns. For complex domain classification tasks, the CV_CrossEntropy loss function is commonly used in conjunction with complex-valued probabilities to continuously improve the model's accuracy and stability during training. Two common methods for combining cross-entropy and CV-CNN are: (i) calculating cross-entropy loss using only the output's real part and (ii) calculating cross-entropy loss using the real and imaginary parts of the output separately and then summing them up as the final loss. However, the second approach contains a logical error that arises when the model outputs different classification results using real and imaginary parts, leading to confusion.

To assess the influence of CV_CrossEntropy on the experimental outcomes, a complex-valued classification model was developed to utilize real-valued cross-entropy ((i) calculating cross-entropy loss using only the output's real part), called RCE-CV-SCNN. The findings of the experiment are presented in Table 9.

Table 9. Overall accuracy (%), average accuracy (%), and Kappa coefficient of the different loss functions.

Dataset	Methods	OA	AA	Kappa
Flevoland Dataset 1	RCE-CV-SCNN	96.00	95.55	0.9560
	new CV-SCNN	96.66	96.20	0.9634
Flevoland Dataset 2	RCE-CV-SCNN	98.20	93.72	0.9787
	new CV-SCNN	98.57	94.13	0.9830
Oberpfaffenhofen Dataset	RCE-CV-SCNN	95.02	95.16	0.9089
	new CV-SCNN	95.69	95.76	0.9221

According to Table 9, CV-SCNN performs better than RCE-CV-SCNN in all three experiments. This is because CV-SCNN produces complex-valued output, while RCE-CV-SCNN can only calculate the loss using the real part of the output. This results in the loss of half of the information flow. Additionally, the constraints imposed on the model by CV_CrossEntropy are stronger than RV_CrossEntropy, which helps the model learn more accurate classification patterns.

Figure 10 illustrates the training progression of the new CV-SCNN and RCE-CV-SCNN models on Flevoland Dataset 1. Despite the heightened computational complexity and the imposition of more stringent constraints associated with the computation of CV_CrossEntropy, the convergence rate of the models remains unimpeded. Over successive epochs, the new CV-SCNN exhibits a gradual improvement in accuracy over RCE-CV-SCNN. It is pertinent to note that even though the new CV-SCNN consistently manifests higher loss values compared to RCE-CV-SCNN during the convergence phase, this discrepancy arises from the augmented computational elements within CV_CrossEntropy and does not detrimentally impact the overall model performance.

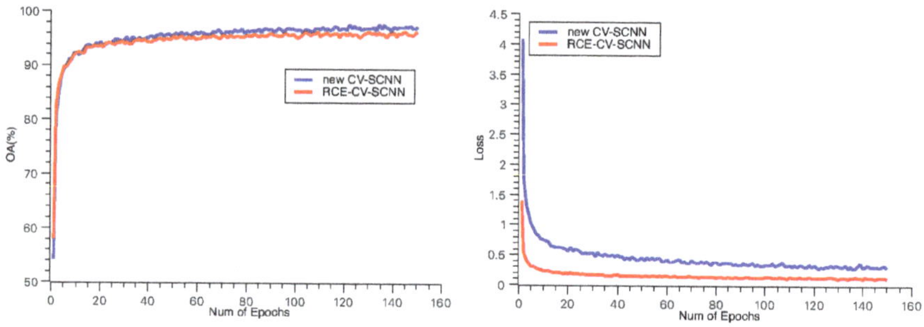

(a) OA curves (b) Loss curves

Figure 10. (**a**,**b**) each represent the variation curves of overall accuracy and loss function during the training process for RCE-CV-SCNN and new CV-SCNN.

4.4. *Comparison with State-of-the-Art Algorithms*

This study also conducted a comparison on Flevoland Dataset 1, evaluating the new CV-SegNet against the state-of-the-art algorithms, with results presented in Table 10. The findings reveal that the new CV-SegNet achieves the highest classification performance. However, it is worth noting that such comparisons may lack full rigor due to the diverse research objectives associated with each algorithm. Consequently, they employ inputs of varying sizes and training datasets with different sampling rates, all of which can influence the final outcomes. For example, RCV-CNN excels in achieving superior accuracy when confronted with limited annotated data, a proficiency that may not confer significant advantages when dealing with relatively large training datasets. The method proposed in this paper is not an independent model but rather an approach aimed at enhancing deep learning models, including CNNs, FCNs, and SegNets. The resulting improved models can lead to performance enhancements or reductions in parameter complexity.

Table 10. Overall accuracy (%), average accuracy (%), and Kappa coefficient of the state-of-the-art algorithms on the Flevoland Dataset 1. The bolded values represent the highest values among all models.

	RCV-CNN [49]	CV-Contourlet-CNN [36]	SF-CNN [50]	AMSE-LSTM [51]	CV-ConvLSTM [42]	New CV-SegNet
Stembeans	98.61	99.81	-	97.16	94.24	**100.00**
Peas	98.56	99.86	99.62	97.62	**99.97**	99.31
Forest	97.81	98.98	-	98.43	99.17	**99.92**
Lucerne	98.22	99.55	**99.93**	97.54	98.56	99.88
Wheat	94.50	99.59	99.46	98.82	97.56	**100.00**
Beet	94.14	99.25	99.22	94.71	99.07	**99.43**
Potaotes	98.90	99.18	99.50	96.40	98.49	**99.88**
Bare soil	98.05	**100.00**	99.72	99.43	99.67	**100.00**
Grass	89.17	99.85	-	98.06	96.73	**100.00**
Rapeseed	97.07	99.00	99.88	96.03	97.68	**99.91**
Barley	98.20	99.77	99.50	99.72	**100.00**	99.64
Wheat2	97.28	99.43	-	98.50	99.88	**99.92**
Wheat3	98.56	99.39	-	99.22	98.32	**99.92**
Water	**99.89**	99.58	-	99.81	99.68	99.46
Buildings	80.88	**99.26**	-	84.90	79.41	82.77
OA	97.22	99.42	99.58	97.09	98.58	**99.76**
AA	-	99.50	**99.61**	-	97.32	98.67
Kappa	0.8930	0.9902	0.9950	0.9683	0.9845	**0.9974**

5. Conclusions

This paper introduced a new method for enhancing deep learning models utilized in PolSAR image classification. The method involves CVA_Max_Pooling, HReLU, and CV_CrossEntropy. CVA_Max_Pooling decreases the computational work and extracts the most important features. HReLU changes the model into a nonlinear sparse model, while CV_CrossEntropy provides a loss computation method for complex-domain classification tasks. The proposed complex-valued deep learning method was applied to improve four PolSAR classification models: SCNN, DCNN, FCN, and SegNet. The models were then validated on three public PolSAR datasets. The experimental results reveal that the method proposed in this paper outperforms the old complex-valued model and is much better than the real-valued model despite having comparable parameters.

In order to continue this work in the future, the following ideas could be explored: (1) While the experiments have shown that the new complex-valued method can significantly improve the performance of shallow CNNs, it is important to note that the inference process of CNNs can be quite time-consuming. On the other hand, FCNs are effective at fast inference but require many model parameters and computation. Therefore, it would be worthwhile to explore the possibility of combining the new complex-valued method with shallow FCNs to improve classification accuracy and reduce inference time simultaneously; (2) The experiments have also demonstrated that the new complex-valued method is suitable for learning with small samples. Further research could be conducted to reduce the sampling rate by utilizing the new complex-valued method.

Author Contributions: Conceptualization, Y.R.; data curation, Y.R.; methodology, Y.R.; supervision, Y.L.; validation, W.J.; writing–original draft, Y.R.; writing–review and editing, W.J. and Y.L. All authors have read and agreed to the published version of the manuscript.

Funding: This work was supported in part by the National Natural Science Foundation of China under Grant #62176247 and research grant #2020-JCJQ-ZD-057-00. It was also supported by the Fundamental Research Funds for the Central Universities.

Data Availability Statement: The datasets utilized in the experiments are accessible to the public and can be accessed from the following website: https://earth.esa.int/eogateway/campaigns/agrisar (accessed on 16 August 2022).

Acknowledgments: The authors wish to extend our thanks to Cheng Wang and the other researchers in Beijing Raying Technologies, Inc. for their great support in radar data analysis.

Conflicts of Interest: The authors declare no conflict of interest.

References

1. Lee, J.S.; Pottier, E. *Polarimetric Radar Imaging: From Basic to Application*; CRC Press: Boca Raton, FL, USA, 2011; pp. 1–22.
2. Hänsch, R.; Hellwich, O. Skipping the real world: Classification of PolSAR images without explicit feature extraction. *ISPRS J Photogramm. Remote Sens.* **2018**, *140*, 122–132. [CrossRef]
3. Lee, J.S.; Grunes, M.R.; Kwok, R. Classification of multi-look polarimetric SAR imagery based on the complex Wishart distribution *Int. J. Remote Sens.* **1994**, *15*, 2299–2311. [CrossRef]
4. Lee, J.S.; Grunes, M.R.; Pottier, E.; Ferro-Famil, L. Unsupervised terrain classification preserving polarimetric scattering characteristics. *IEEE Trans. Geosci. Remote Sens.* **2004**, *42*, 722–731. [CrossRef]
5. Dabboor, M.; Collins, M.; Karathanassi, V.; Braun, A. An unsupervised classification approach for polarimetric SAR data based on the Chernoff distance for the complex Wishart distribution. *IEEE Trans. Geosci. Remote Sens.* **2013**, *51*, 4200–4213. [CrossRef]
6. Wu, Y.; Ji, K.; Yu, W.; Su, Y. Region-based classification of polarimetric SAR images using Wishart MRF. *IEEE Geosci. Remote Sens. Lett.* **2008**, *5*, 668–672. [CrossRef]
7. Song, W.; Li, M.; Zhang, P.; Wu, Y.; Tan, X.; An, L. Mixture WGΓ-MRF model for PolSAR image classification. *IEEE Trans. Geosci. Remote Sens.* **2018**, *56*, 905–920. [CrossRef]
8. Arii, M.; van Zyl, J.J.; Kim, Y. Adaptive model-based decomposition of polarimetric SAR covariance matrices. *IEEE Trans. Geosci. Remote Sens.* **2011**, *49*, 1104–1113. [CrossRef]
9. Clound, S.R.; Pottier, E. An entropy based classification scheme for land applications of polarimetric SAR. *IEEE Trans. Geosci. Remote Sens.* **1997**, *35*, 68–78. [CrossRef]
10. Cloude, S.R.; Pottier, E. A review of target decomposition theorems in radar polarimetry. *IEEE Trans. Geosci. Remote Sens.* **1996**, *34*, 498–518. [CrossRef]
11. An, W.; Cui, Y.; Yang, J. Three-Component Model-Based Decomposition for Polarimetric SAR Data. *IEEE Trans. Geosci. Remote Sens.* **2010**, *48*, 2732–2739. [CrossRef]
12. He, C.; Li, S.; Liao, Z.; Liao, M. Texture Classification of PolSAR Data Based on Sparse Coding of Wavelet Polarization Textons. *IEEE Trans. Geosci. Remote Sens.* **2013**, *51*, 4576–4590. [CrossRef]
13. Lardeux, C.; Frison, P.L.; Tison, C.C.; Souyris, J.C.; Stoll, B.; Fruneau, B.; Rudant, J.P. Support vector machine for multifrequency SAR polarimetric data classification. *IEEE Trans. Geosci. Remote Sens.* **2009**, *47*, 4143–4152. [CrossRef]
14. Melgani, F.; Hashemy, B.A.R.A.; Taha, S.M.R. An explicit fuzzy supervised classification method for multispectral remote sensing images. *IEEE Trans. Geosci. Remote Sens.* **2000**, *38*, 287–295. [CrossRef]
15. Ulaby, F.T.; Elachi, C. Radar polaritnetry for geoscience applications. *Geocarto Int.* **1990**, *5*, 38. [CrossRef]
16. Freeman, A.; Durden, S.L. A three-component scattering model for polarimetric SAR data. *IEEE Trans. Geosci. Remote Sens.* **1998**, *36*, 963–973. [CrossRef]
17. Huynen, J.R. Phenomenological Theory of Radar Targets. Available online: http://resolver.tudelft.nl/uuid:e4a140a0-c175-45a7-ad41-29b28361b426 (accessed on 14 April 2022).
18. De, S.; Bruzzone, L.; Bhattacharya, A.; Bovolo, F.; Chaudhuri, S. A Novel Technique Based on Deep Learning and a Synthetic Target Database for Classification of Urban Areas in PolSAR Data. *IEEE J. Sel. Top. Appl. Earth Obs. Remote Sens.* **2018**, *11*, 154–170. [CrossRef]
19. Zhou, Y.; Wang, H.; Xu, F.; Jin, Y.Q. Polarimetric SAR image classification using deep convolutional neural networks. *IEEE Geosci. Remote Sens. Lett.* **2016**, *13*, 1935–1939. [CrossRef]
20. Bin, H.; Sun, J.; Xu, Z. A Graph-Based Semisupervised Deep Learning Model for PolSAR Image Classification. *IEEE Trans. Geosci. Remote Sens.* **2019**, *57*, 2116–2132. [CrossRef]
21. Li, Y.; Chen, Y.; Liu, G.; Jiao, L. A novel deep fully convolutional network for PolSAR image classification. *Remote Sens.* **2018**, *10*, 1984. [CrossRef]
22. Pham, M.; Lefevre, S. Very high resolution Airborne PolSAR Image Classification using Convolutional Neural Networks. In Proceedings of the 13th European Conference on Synthetic Aperture Radar (EUSAR 2021), Online, 29 March–1 April 2021; pp. 1–4.
23. Liu, S.; Luo, H.; Shi, Q. Active Ensemble Deep Learning for Polarimetric Synthetic Aperture Radar Image Classification. *IEEE Geosci. Remote Sens. Lett.* **2021**, *18*, 1580–1584. [CrossRef]
24. Cheng, J.; Zhang, F.; Xiang, D.; Yin, Q.; Zhou, Y. PolSAR Image Classification with Multiscale Superpixel-Based Graph Convolutional Network. *IEEE Trans. Geosci. Remote Sens.* **2022**, *60*, 1–14. [CrossRef]
25. Liu, G.; Li, Y.; Jiao, L.; Chen, Y.; Shang, R. Multiobjective evolutionary algorithm assisted stacked autoencoder for PolSAR image classification. *Swarm Evol. Comput.* **2021**, *60*, 100794. [CrossRef]
26. Jing, H.; Wang, Z.; Sun, X.; Xiao D.; Fu, K. PSRN: Polarimetric Space Reconstruction Network for PolSAR Image Semantic Segmentation. *IEEE J. Sel. Top. Appl. Earth Obs. Remote Sens.* **2021**, *14*, 10716–10732. [CrossRef]
27. Nie, W.; Huang, K.; Yang, J.; Li, P. A Deep Reinforcement Learning-Based Framework for PolSAR Imagery Classification. *IEEE Trans. Geosci. Remote Sens.* **2022**, *60*, 1–15. [CrossRef]
28. Yang, C.; Hou, B.; Chanussot, J.; Hu, Y.; Ren, B.; Wang, S.; Jiao, L. N-Cluster Loss and Hard Sample Generative Deep Metric Learning for PolSAR Image Classification. *IEEE Trans. Geosci. Remote Sens.* **2022**, *60*, 1–16. [CrossRef]
29. Ren, B.; Zhao, Y.; Hou, B.; Chanussot, J.; Jiao, L. A Mutual Information-Based Self-Supervised Learning Model for PolSAR Land Cover Classification. *IEEE Trans. Geosci. Remote Sens.* **2021**, *59*, 9224–9237. [CrossRef]

30. Lee, J.S.; Hoppel, K.W.; Mango, S.A.; Miller, A.R. Intensity and phase statistics of multilook polarimetric and interferometric SAR imagery. *IEEE Trans. Geosci. Remote Sens.* **1994**, *32*, 1017–1028. [CrossRef]
31. Ainsworth, T.L.; Kelly, J.P.; Lee, J.S. Classification comparisons between dual-pol, compact polarimetric and quad-pol SAR imagery. *ISPRS J. Photogramm. Remote Sens.* **2009**, *64*, 464–471. [CrossRef]
32. Turkar, V.; Deo, R.; Rao, Y.S.; Mohan, S.; Das, A. Classification accuracy of multi-frequency and multi-polarization SAR images for various land covers. *IEEE J. Sel. Top. Appl. Earth Obs. Remote Sens.* **2012**, *5*, 936–941. [CrossRef]
33. Georgiou, G.M.; Koutsougeras, C. Complex domain backpropagation. *IEEE Trans. Circuits Syst. II Analog Digital Signal Process.* **1992**, *39*, 330–334. [CrossRef]
34. Trabelsi, C.; Bilaniuk, O.; Zhang, Y.; Serdyuk, D.; Subramanian, S.; Santos, J.; Mehri, S.; Rostamzadeh, N.; Bengio, Y.; Pal, C. Deep complex networks. In Proceedings of the International Conference on Learning Representations (ICLR2018), Vancouver, BC, Canada, 30 April–3 May 2018.
35. Zhang, Z.; Wang, H.; Xu, F.; Jin, Y. Complex-Valued Convolutional Neural Network and Its Application in Polarimetric SAR Image Classification. *IEEE Trans. Geosci. Remote Sens.* **2017**, *55*, 7177–7188. [CrossRef]
36. Li, L.; Ma, L.; Jiao, L.; Liu, F.; Sun, Q.; Zhao, J. Complex Contourlet-CNN for polarimetric SAR image classification. *Pattern Recognit.* **2020**, *100*, 107110. [CrossRef]
37. Xiao, D.; Liu, C.; Wang, Q.; Wang, C.; Zhang, X. PolSAR Image Classification Based on Dilated Convolution and Pixel-Refining Parallel Mapping network in the Complex Domain. *arXiv* **2020**, arXiv:1909.10783
38. Zhao, J.; Datcu, M.; Zhang, Z.; Xiong, H.; Yu, W. Contrastive-Regulated CNN in the Complex Domain: A Method to Learn Physical Scattering Signatures From Flexible PolSAR Images. *IEEE Trans. Geosci. Remote Sens.* **2019**, *57*, 10116–10135. [CrossRef]
39. Tan, X.; Li, M.; Zhang, P.; Wu, Y.; Song, W. Complex-Valued 3-D Convolutional Neural Network for PolSAR Image Classification. *IEEE Geosci. Remote Sens. Lett.* **2020**, *17*, 1022–1026. [CrossRef]
40. Zhang, P.; Tan, X.; Li, B.; Jiang, Y.; Song, W.; Li, M.; Wu, Y. PolSAR Image Classification Using Hybrid Conditional Random Fields Model Based on Complex-Valued 3-D CNN. *IEEE Trans. Aerosp. Electron. Syst.* **2021**, *57*, 1713–1730. [CrossRef]
41. Qin, X.; Hu, T.; Zou, H.; Yu, W.; Wang, P. Polsar Image Classification via Complex-Valued Convolutional Neural Network Combining Measured Data and Artificial Features. In Proceedings of the 2019 IEEE International Geoscience and Remote Sensing Symposium (IGARSS2019), Yokohama, Japan, 28 July–2 August 2019; pp. 3209–3212.
42. Fang, Z.; Zhang, G.; Dai, Q.; Xue, B. PolSAR Image Classification Based on Complex-Valued Convolutional Long Short-Term Memory Network. *IEEE Geosci. Remote Sens. Lett.* **2022**, *19*, 1–5. [CrossRef]
43. Tan, X.; Li, M.; Zhang, P.; Wu, Y.; Song, W. Deep Triplet Complex-Valued Network for PolSAR Image Classification. *IEEE Trans. Geosci. Remote Sens.* **2021**, *59*, 10179–10196. [CrossRef]
44. Abdo, A.J. The paradigm of complex probability and the Brownian motion. *Syst. Sci. Control Eng.* **2015**, *3*, 478–503. [CrossRef]
45. Abdo, A.J. The paradigm of complex probability and Chebyshev's inequality. *Syst. Sci. Control Eng.* **2016**, *4*, 99–137. [CrossRef]
46. Abdo, A.J. The paradigm of complex probability and Claude Shannon's information theory. *Syst. Sci. Control Eng.* **2017**, *5*, 380–425. [CrossRef]
47. Abdo A.J. The paradigm of complex probability and Ludwig Boltzmann's entropy. *Syst. Sci. Control Eng.* **2018**, *6*, 108–149. [CrossRef]
48. Abdo, A.J. The paradigm of complex probability and Monte Carlo methods. *Syst. Sci. Control Eng.* **2019**, *7*, 407–451. [CrossRef]
49. Xie, W.; Ma, G.; Zhao, F.; Liu, H.; Zhang, L. PolSAR image classification via a novel semi-supervised recurrent complex-valued convolution neural network. *Neurocomputing* **2020**, *388*, 255–268. [CrossRef]
50. Shang, R.; Wang, J.; Jiao, L.; Yang, X.; Li, Y. Spatial feature-based convolutional neural network for PolSAR image classification. *Appl. Soft Comput.* **2022**, *123*, 108922. [CrossRef]
51. Hua, W.; Wang, X.; Zhang, C.; Jin, X. Attention-Based Multiscale Sequential Network for PolSAR Image Classification. *IEEE Geosci. Remote Sens. Lett.* **2022**, *19*, 1–5. [CrossRef]

remote sensing

MDPI

Article

GLF-Net: A Semantic Segmentation Model Fusing Global and Local Features for High-Resolution Remote Sensing Images

Wanying Song [1,*], Xinwei Zhou [1], Shiru Zhang [1], Yan Wu [2] and Peng Zhang [2]

1 Xi'an Key Laboratory of Network Convergence Communication, School of Communication and Information Engineering, Xi'an University of Science and Technology, Xi'an 710054, China; 21207223090@stu.xust.edu.cn (X.Z.); zhangshiru@xust.edu.cn (S.Z.)
2 School of Electronics Engineering, Xidian University, Xi'an 710071, China; ywu@xidian.edu.cn (Y.W.); pzhang@xidian.edu.cn (P.Z.)
* Correspondence: wysong@xust.edu.cn; Tel.: +86-187-9294-7332

Abstract: Semantic segmentation of high-resolution remote sensing images holds paramount importance in the field of remote sensing. To better excavate and fully fuse the features in high-resolution remote sensing images, this paper introduces a novel Global and Local Feature Fusion Network, abbreviated as GLF-Net, by incorporating the extensive contextual information and refined fine-grained features. The proposed GLF-Net, devised as an encoder–decoder network, employs the powerful ResNet50 as its baseline model. It incorporates two pivotal components within the encoder phase: a Covariance Attention Module (CAM) and a Local Fine-Grained Extraction Module (LFM). And an additional wavelet self-attention module (WST) is integrated into the decoder stage. The CAM effectively extracts the features of different scales from various stages of the ResNet and then encodes them with graph convolutions. In this way, the proposed GLF-Net model can well capture the global contextual information with both universality and consistency. Additionally, the local feature extraction module refines the feature map by encoding the semantic and spatial information, thereby capturing the local fine-grained features in images. Furthermore, the WST maximizes the synergy between the high-frequency and the low-frequency information, facilitating the fusion of global and local features for better performance in semantic segmentation. The effectiveness of the proposed GLF-Net model is validated through experiments conducted on the ISPRS Potsdam and Vaihingen datasets. The results verify that it can greatly improve segmentation accuracy.

Keywords: high-resolution remote sensing; semantic segmentation; global context information; fine-grained feature; feature fusion

Citation: Song, W.; Zhou, X.; Zhang, S.; Wu, Y.; Zhang, P. GLF-Net: A Semantic Segmentation Model Fusing Global and Local Features for High-Resolution Remote Sensing Images. *Remote Sens.* **2023**, *15*, 4649. https://doi.org/10.3390/rs15194649

Academic Editors: Jiaojiao Li, Qian Du, Jocelyn Chanussot, Wei Li, Bobo Xi, Rui Song and Yunsong Li

Received: 8 August 2023
Revised: 18 September 2023
Accepted: 19 September 2023
Published: 22 September 2023

1. Introduction

As image processing technology, sensors, and data storage capabilities continue to advance, the acquisition of high-resolution (HR) remote sensing images has become more common and feasible [1]. HR remote sensing images refer to image data with corresponding spatial resolutions acquired by remote sensing platforms, such as satellites, aviation, or unmanned aerial vehicles. These images can provide detailed surface information, including buildings, roads, vegetation, etc. HR remote sensing images are widely used in urban planning, environmental monitoring, and agricultural management [2,3].

Semantic segmentation of HR remote sensing images has always been a difficult challenge in the field of computer vision (CV) [4]. In the early stages, semantic segmentation methods for HR remote sensing images were mainly based on hand-designed features. Researchers scrutinized remote sensing images, dissecting their color, texture, shape, and other distinctive attributes. They harnessed conventional machine learning techniques, like support vector machines and random forests, to execute classification tasks. Davis's method was based on threshold-extracted texture features of images for semantic segmentation [5]. Adams et al. proposed a region-based method to divide an image into regions to realize image segmentation [6]. Kundu et al. proposed an algorithm that could automatically select important edges for human perception [7]. Achanta et al. [8] introduced a novel super-pixel algorithm known as Simple Linear Iterative Clustering, which serves to enhance the performance of semantic segmentation. However, these methods often perform poorly for complex terrain classes and changing environmental conditions.

Compared with traditional methods, CNN possesses the inherent capability to autonomously glean feature representations from raw data, obviating the need for manual design of feature extractors through an end-to-end learning process. And CNN has a more powerful learning ability for image features. The ResNet [9] model was proposed to solve the gradient explosion problem and improve the performance of the model. It is used as the baseline model for many CV tasks and is also suitable for semantic segmentation tasks. The proposal of the fully convolutional network (FCN) [10] extends the traditional convolutional neural network to pixel-level classification and realizes fine semantic segmentation. It used an encoder–decoder structure that produces a layer-hop connection structure to integrate high- and low-dimensional feature maps. To obtain higher segmentation accuracy, researchers have proposed many improved model architectures to further improve the performance of semantic segmentation of HR remote sensing images. Building upon the foundation of FCN, U-Net [11] introduces a streamlined skip connection architecture and optimizes and fuses different feature maps to improve accuracy. Meanwhile, SegNet [12] innovatively captures and utilizes the pooling index during the encoding phase, effectively guiding and standardizing the subsequent decoding procedure. In a similar vein, PSPNet [13] leverages parallel pooling across various scales to extract pivotal features from diverse ground object categories, thereby enhancing the overall segmentation performance of the model. Meanwhile, RS remote sensing images also have the problems of complex labeling and high time consumption, so unsupervised algorithms have also been a hot issue in the semantic segmentation of RS remote sensing images. A method to reduce the prediction uncertainty of target domain data was proposed by Prabhu, S. et al. [14]. Liu, Y. et al. [15] proposed a source-free domain adaptation framework for semantic segmentation, SFDA, in which only well-trained source models and unlabeled target domain datasets are available for adaptation. Chen, J. et al. [16] proposed an unsupervised domain adaptive framework for HRSI semantic segmentation based on adversarial learning. Guan, D. et al. [17] proposed a Scale Variance Minimization (SVMin) technique that uses scale invariance constraints to perform inter-domain alignment while preserving the semantic structure of images in the target domain. Stan, S. et al.'s [18] approach is based on encoding source domain information into the interior for use in guiding the distribution of adaptations in the absence of source samples.

In recent years, attention mechanisms have been widely adopted in the field of computer vision. There are two ways of modeling attention mechanisms: (1) One is to use global information to obtain attentional weights to enhance key local areas or channels without considering the dependencies between global information. SE-Block [19] represents a classical approach to attention, aiming to explicitly establish interdependencies between feature channels. This involves dynamically assigning weights to each channel through model learning, thus boosting relevant features while suppressing irrelevant ones. PSANet [20] proposes the point-wise spatial attention network (PSANet) to relax the local neighborhood constraint. Each position on the feature map is connected to all the other ones through a self-adaptively learned attention mask. (2) The other is to model the dependencies between global as well as local information and enhance the subject information by obtaining the correlation matrix between channels or spatial features. DANet [21] introduces the dual attention (DA) module into the field of semantic segmentation and improves the performance of the model by modeling global information dependencies. Meanwhile, another noteworthy contribution is CBAM-Block [22], an attention module that seamlessly fuses spatial and channel information. In contrast to the singular focus on channel attention exhibited by SE-Block, CBAM combines channel attention and spatial attention, thus enabling the model to focus on both global and local information and to better model global information dependencies when processing images.

However, for the semantic segmentation problem, there are still deficiencies in the existing methods, which can be summarized as follows: (1) Global context information is crucial for the semantic segmentation task. When computing global dependencies, the correlation matrix from a large number of feature maps usually results in high complexity and strong training difficulties. Although some models try to introduce a multi-scale input-output mechanism, how to effectively utilize the information of different scales and how to adequately capture the remote dependency and global context in an image are still difficult problems. (2) RS remotely sensed images contain intricate topographic landscapes that exhibit a wide variety of textures, resulting in both high intra-class diversity and inter-class similarity. As a result, the boundaries in these images can be easily confused with small object features, while some small objects and some regions with unclear boundaries can also be misclassified. This motivates us to mine more distinguishable local fine-grained features for accurate classification. To address the above problems, we propose a covariance attention module (CAM) and a local fine-grained extraction module (LFM) to extract multi-scale global and local fine-grained information, respectively, and a wavelet self-attention module (WST) to fuse global and local features. The main contributions and innovations of this paper include:

1. We designed a CAM that uses the covariance matrix to model the dependencies between the feature map channels, capturing the main contextual information. These features are subsequently encoded by graph convolution, which helps to capture universally applicable and consistent global context information. The covariance matrix can adaptively capture not only the linear relationship between the local context information of the feature map but also the non-local context information of the feature map [23,24]. We model the feature maps of the last three layers of ResNet using covariance matrices to obtain their main context information and fuse them using feature addition. This non-local context information can help GLF-Net understand the relationship between different regions in the image.
2. Building upon the ResNet features, we have introduced a novel approach by integrating the local feature extraction module. This innovative step refines the feature map and yields finely detailed, local-level features. Through a process that involves encoding both spatial and semantic information from the feature map, followed by a comparative analysis against information from global pooling, we successfully capture intricate features that tend to be challenging to discern amidst the complex background of HR remote sensing images. This enhancement improves accuracy when identifying small targets and delineating boundaries, thereby bolstering our model's capacity for feature capture and recognition.
3. We consider the differences and interactions between global features and local features, and simply pursuing maximization or merging class probability maps cannot ensure comprehensive semantic description. Recognizing the intrinsic value of intricate details and texture information residing within an image's high-frequency components, we devised a wavelet self-attention mechanism. This innovation facilitates the fusion of global and local features, harnessing the synergistic interplay between high-frequency and low-frequency information. Importantly, this approach ensures information fusion across varying scales, thereby optimizing the comprehensive utilization of image content.

The subsequent sections of this paper are organized as follows: Section 2 delves into the relevant literature concerning local and global feature extraction. In Section 3, we provide an overview of the materials and methodologies utilized in our study. Moving forward to Section 4, we delve into the presentation of the results stemming from our experimental pursuits. Ultimately, Section 5 encapsulates a concise summary of our concluding insights.

2. Related Work

This section briefly reviews the semantic segmentation methods relevant to this paper, namely, the global feature extraction-based semantic segmentation method and the local feature extraction-based semantic segmentation method.

2.1. Global Context Feature Extraction for Semantic Segmentation

Global context information is crucial in the context of semantic segmentation of HR remote sensing images. It not only helps identify a wide range of objects and distinguish objects and backgrounds, but also captures spatial correlations and enhances the model's ability to understand the overall image semantics. This information holds great importance in enhancing the accuracy and overall effectiveness of semantic segmentation. The Deeplab series [25,26] of networks have established atrous convolution, global average pooling, and atrous spatial pyramid pooling. By employing these techniques, the Deeplab series of networks effectively harness the global context information within images. This enables them to capture semantic details across various scales and facilitates a more profound comprehension of the semantic structure inherent in the images. At the same time, DeeplabV3+ uses the skip connection mechanism to fuse the features in the encoder and the features in the decoder. This allows the decoder to directly access the low-level information from the encoder so that it can better utilize the detailed information of low-level features for segmentation. Zhang, H. et al. [27] introduced a context encoding module based on FCN, which effectively captured and leveraged contextual information, resulting in notable enhancements to the model's segmentation accuracy. Li, R. et al. [28] implemented a feature pyramid network to seamlessly integrate the spatial and contextual features that were extracted. Building upon this foundation, they further refined multi-scale feature acquisition by utilizing attention-guided feature aggregation. Liu, H. et al. [29] introduced additional correspondences between foreground and background, along with incorporating multi-scale contextual semantic features. This strategic augmentation notably aids the encoder in capturing dependable matching patterns.

2.2. Local Fine-Grained Feature Extraction for Semantic Segmentation

In order to handle the classification of small targets and boundaries caused by complex scenes in HR remote sensing images, models usually need to further enhance local information to obtain more subtle fine-grained features. Fine-grained features usually focus on capturing the detailed information in the image, increasing the diversity and discrimination ability of the features. Yang, M. et al. [30] proposed densely connected atrous spatial pyramid pooling, and the features generated by this network can cover the local area in a very dense way to obtain fine-grained local features. Li, R. et al. [31] proposed ABCNet, which uses a bilateral attention network to capture rich spatial details in HR remote sensing images, obtains fine-grained spatial information, and improves the accuracy of the model. Wang, L. et al. [32] proposed the category feature compact module, which solves the problem of feature dispersion in the target domain achieved by cross-domain networks, facilitates the fine-grained alignment of categories, and improves segmentation performance.

3. Materials and Methods

As mentioned above, multi-scale contextual features are crucial for obtaining images in complex scenes. During the down-sampling process, the model inevitably loses important information. Encoding each stage of down-sampling aids in acquiring a broader spectrum of multi-scale contextual and semantic insights. Due to the complexity of HR remote sensing images, some small targets and boundaries are usually confused by global information. Refining the feature map to obtain fine-grained features will help the model recognize these small targets and boundaries. Based on these, we designed GLF-Net.

This section introduces the primary architecture of GLF-Net. As depicted in Figure 1a, an encoder–decoder architecture is employed. The encoder is comprised of a backbone network, alongside a global feature extraction module and a local feature extraction module. We use ResNet50 as the backbone network for feature extraction and down-sample. Figure 1b is a schematic of the ResNet50 residual block. Our CAM is applied to the final three layers of ResNet50, enabling the extraction of comprehensive global context features. The correlation between features is crucial for correctly distinguishing the semantic categories of features. By calculating the covariance matrix of features, we can understand the linear correlation between features, which helps us select the most discriminative combination of features. The extracted multi-scale contextual features can help GLF-Net obtain a wider range of contextual information, including the object's global structure, background information, and contextual relationships, and also enable GLF-Net to better adapt to changes in different images and objects. This contextual information plays a pivotal role in achieving precise object segmentation, comprehending their semantics, and enhancing the overall generalization capability of GLF-Net. The regional fine-grained feature extraction module is used to extract local features, and the fine-grained module can refine the output of ResNet50. Fine-grained features can provide internal details of the object, which helps to distinguish different semantic categories and accurately classify internal regions. It can also capture small changes and edge details of the object to improve the accuracy of boundary recognition and segmentation. This gives better recognition results for small objects in the dataset.

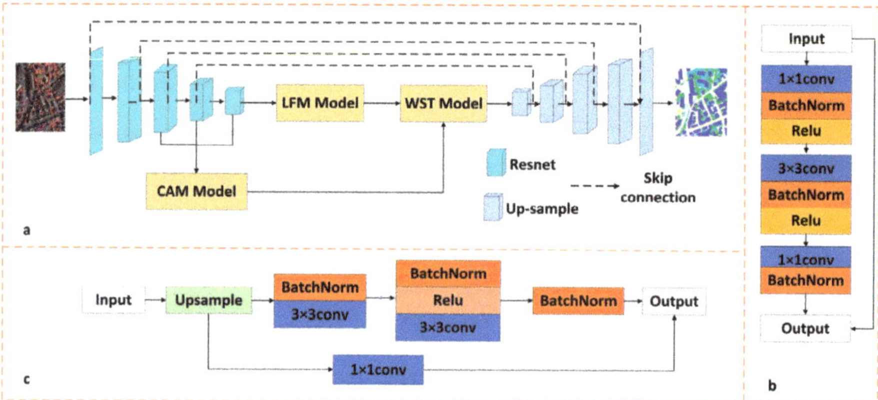

Figure 1. (**a**) Overall structure diagram of GLF-Net. (**b**) ResNet50 residual block. (**c**) Up-sample module.

In the decoder part, we built a WST module that employs the wavelet transform and self-attention to fuse the multi-scale features from the CAM and LFM modules. Then, a sequence of up-convolutions gradually expands the fused output to the original size. The wavelet transform has good sensitivity to edge and texture features. It helps to detect edge and texture information in an image and extract clear boundaries. In semantic segmentation, boundary information helps the model to obtain higher segmentation results. By applying the wavelet transform, the boundary information can be enhanced to improve the ability of GLF-Net to perceive object boundaries. The self-attention mechanism can model the global correlation of different positions in the input features instead of being limited to local regions. By calculating the attentional weights between each location in the input features, the self-attention mechanism can capture the long-range dependencies between different locations. This enables the self-attention mechanism to effectively model global contextual information in feature fusion. In the up-sampling process, as shown in Figure 1c, we designed the up-sampling part based on the ResNet residual block and use the jump connection strategy.

3.1. Global Feature Extraction

In convolutional neural networks, with the transformation of the sensory field and the gradual stacking of features, the semantic information contained in the deeper features of each layer is not exactly the same. Gradually, along with the change of the receptive field and characteristics of the stacked, each layer of ResNet deep features contained in the semantic information is not the same. In this regard, the fusion of multi-scale context information is crucial for the model. By this kind of information fusion, GLF-Net can adapt to different target dimensions effectively and handle the target boundary and complexity so as to improve the flexibility and generalization ability in semantic segmentation tasks.

In the CAM module, we use a covariance matrix (CM) to model the relationship between channels [23], highlight the main channel information while providing a global summary, and then use graph convolution to encode the extracted features to capture the main context information in the last three layers of ResNet50. Figure 2 shows a visualization of the effect of the CM projection, with a1 and b1 showing the original image and a2 and b2 showing the effect of the image covariance projection matrix. It can be seen that the CM has a strong and prominent effect on the main information in the image. Based on this, as shown in Figure 3, we use the covariance matrix to extract the main information of the second-, third-, and fourth-layer features of ResNet50 in an attention mechanism. The first step is to perform the L2 normalization operation on the obtained features and then find the covariance matrix.

$$cov = \frac{1}{H \times W} \sum_{t=1}^{H \times W} \left(F^t - \overline{F^t}\right)^T \left(F^t - \overline{F^t}\right) \tag{1}$$

where C, H, and W are the number of channels, height, and width; $t \in (1, 2, \ldots, H \times W)$; $F^t \in \mathbb{R}^{(H \times W) \times C}$; and $\overline{F^t}$ is the mean of F^{t}. In the dot product process, subject to the effect of the broadcast mechanism, the covariance matrix $cov \in \mathbb{R}^{C \times C}$. Then, we obtain the corresponding covariance attention matrix by the $softmax$ function:

$$S(i) = \frac{exp(cov(i))}{\sum_{i=1}^{C} exp(cov(i))} \tag{2}$$

$$X(i) = F_m(i) \times S(i) \tag{3}$$

where $cov(i)$ represents the middle element of the covariance matrix. The result, $X(i)$, of the covariate attention is obtained by multiplying the original feature, F, with the covariate attention matrix, S. Then, we use covariance attention to extract the main information in this layer. In order to effectively fuse the features of the three layers, we use the dilated convolution strategy to down-sample the features of the second and third layers so that the three-layer features obtain feature maps of the same size. The expanded convolution enables GLF-Net to obtain a larger receptive field, thereby obtaining wider context information. Finally, the three layers of features are added to obtain the fusion feature.

After obtaining the fused multi-scale context features, we use graph convolution [33] to model the global context information of the features. First, our approach involves the projection of the input feature map from the coordinate space onto a graph composed of latent nodes or regions within the interaction space. These latent nodes adeptly aggregate local descriptors using convolutional layers, strategically diminishing the impact of superfluous attributes within the coordinate space. Subsequently, the interrelationships among these nodes are comprehensively deduced through a duo of one-dimensional convolutions.

$$Z = GXw_g \tag{4}$$

where G denotes the adjacency matrix that propagates information across nodes and the adjacency matrix learns edge weights reflecting the relationship between the underlying global pooled features at each node. w_g represents the graph convolution parameters. G and w_g are learned autonomously with gradient descent as the model is continuously trained. During training, the graph's affinity matrix learns the edge weights, thus capturing the nuanced connections between nodes within a fully interconnected graph. This design ensures that each node assimilates information from all the other nodes, constantly updating its state. Upon inference, the output features undergo a transformation back into the original space, yielding the derivation of our global features.

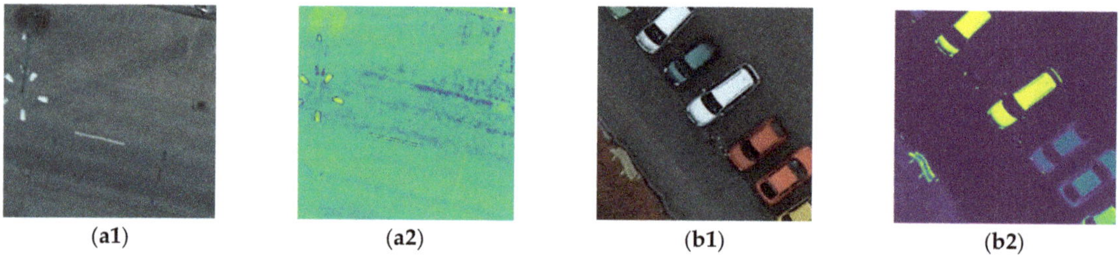

Figure 2. Covariance matrix projection visualization. (**a1**,**b1**) Original image. (**a2**,**b2**) Covariance matrix projection visual effect.

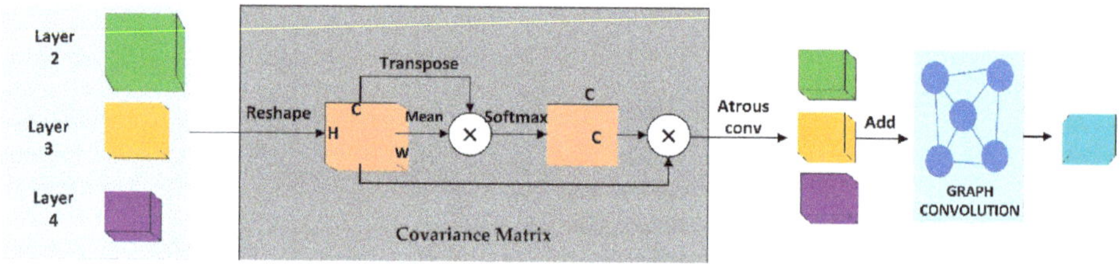

Figure 3. Schematic of the global feature extraction module.

3.2. Local Fine-Grained Feature Extraction

HR remote sensing images have the characteristics of high within-class variance and low within-class variance. In HR remote sensing images, as shown in Figure 4, some small objects present in complex environments are usually misclassified. Therefore, diverging from global features, local features place greater emphasis on recognizing and classifying intricate fine-grained attributes within images. After down-sampling by ResNet, the model eventually extracts a feature map of dimensions 8 × 8; each feature value represents a region of the original image [34]. Through the inference screening of this module, the features of small objects are obtained and highlighted by up-sampling.

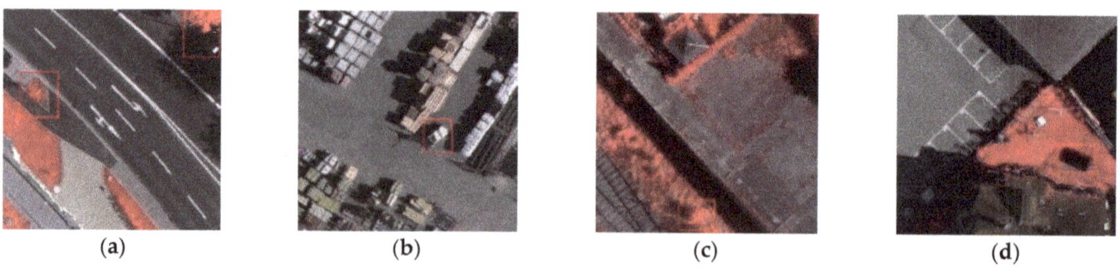

Figure 4. Example of a small object in a complex scene of an HR remote sensing image. The red box selected is the small object that is easy to be ignored. Vegetation in (**a**). Cars in (**b**). Vegetation in (**c**). Cars in (**d**).

Our local feature extraction module is shown in Figure 5. First, we evenly divide our feature map, $V\in R^{(H\times W)}$, into t local areas.

$$V_{hw} = \sum_{t=1}^{D} F_{thw} \tag{5}$$

where V_{hw} represents the information in the dimension (h, w) of V and F_{thw} represents the information in the dimension (t, h, w) of F. We obtain fine-grained local features through semantic and spatial relationships between feature points in each local area. The individual feature points in our local region, V_{hw}, are set to P_j. Specifically, we take the peak point within each local region as the salient point, P_n, and use it as a benchmark to compute semantic and spatial relationships with each point within the local region.

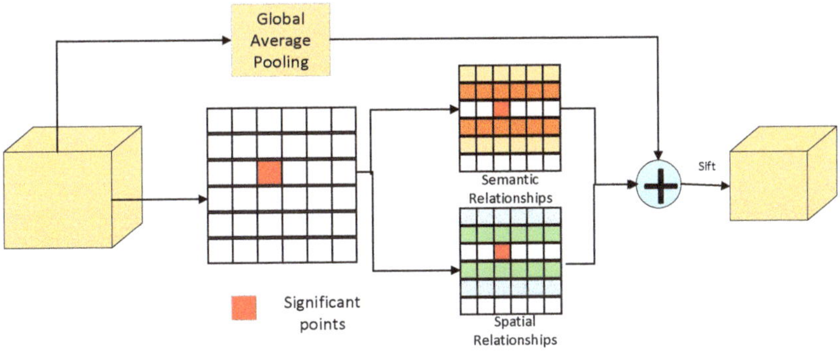

Figure 5. Schematic of local feature extraction.

As mentioned above, the context relationship is particularly important in the task of semantic segmentation, and the simple region division can easily cause the loss of context information in the feature map. To this end, we first calculate the spatial relationship between salient points, P_n, and each feature point, P_j, in each local area based on Euclidean distance, as CR_{nj}:

$$CR_{nj} = \sqrt{(P_n(x) - P_j(x))^2 + (P_n(y) - P_j(y))^2} \tag{6}$$

where $j = 1, \cdots, H \times W$.The smaller the value of CR_{nj}, P_n and P_j get closer. We then use the cosine similarity to calculate the semantic dependency between the salient point, P_n, and the rest of the feature points, P_j:

$$SR_{nj} = \frac{{Q_n}^T Q_j}{||Q_n||||Q_j||} \tag{7}$$

where $Q_n\in R^D$ and $Q_j\in R^D$ are the channel features of point P_n and point P_j in each local area. Considering both spatial relationship and semantic similarity, we define the spatial semantic relationship, R_{nj}, as follows:

$$R_{nj} = \frac{SR_{nj}}{CR_{nj} + 1} \tag{8}$$

The correlation between point P_n and point P_j is proportional to the value of R. Then, we can obtain the local features, F_n^l, of salient points, P_n, by aggregating spatial semantic context information, which is formulated as follows:

$$F_n^l = \sum_{j=1}^{H*W} \frac{exp(R_{nj})}{\sum_{j=1}^{H*W} exp(R_{nj})} \tag{9}$$

After obtaining all the local features, to filter the features of the small target we need from these local features, we first obtain the global features of the original feature, which is denoted as F^G:

$$F^G = \mathrm{GAP(F)} \tag{10}$$

where GAP is the global average pooling. The semantic similarity between each local feature and the global pooling result is then calculated using the cosine similarity and by screening the k groups of local features that are most dissimilar to the global feature, which are the local small target features we need to extract.

3.3. Fusion Module

In CNNs, both convolution and pooling operations inherently entail a certain degree of information loss across different frequencies. However, by incorporating the wavelet transform, the model enables the fusion of various frequency characteristics and the preservation of multi-scale information fusion. This approach optimally exploits the complementarity between high- and low-frequency data.

The deeper convolutional neural network architectures show greater ability to improve the segmentation accuracy of complex image edge contours and details while retaining the multi-frequency attributes. Wavelet transform, employing an array of diverse scale wavelets, decomposes the original function. This process yields coefficients representing the original function under distinct scale wavelets through translation and scale transformations. The translation affords insight into the temporal attributes of the original function, while scale transformation elucidates its frequency characteristics.

Having extracted the global and local features, the subsequent phase revolves around their effective fusion. As depicted in Figure 6, our fusion module harnesses a combination of wavelet transform and self-attention mechanisms to accomplish this fusion task:

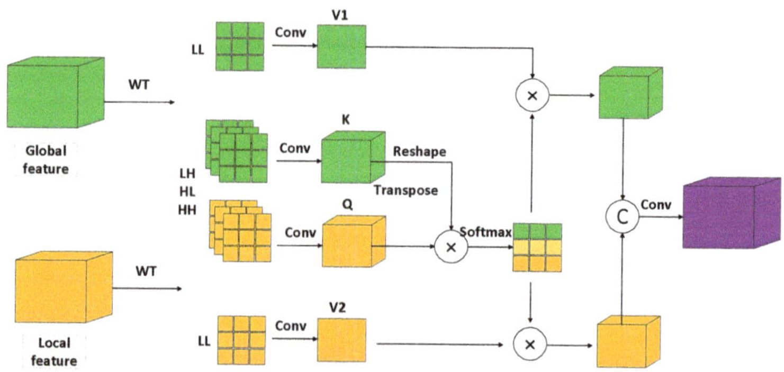

Figure 6. Illustration of the fusion module.

We first use the 2D Haar transform on the global and local features to obtain the low-frequency component, x_{LL}, and three high-frequency components, x_{LH}, x_{HL}, and x_{HH}. The four frequency band components are obtained by Equation (11):

$$\begin{aligned} x_{LL}(i,j) &= x(2i-1,2j-1) + x(2i-1,2j) + x(2i,2j-1) + x(2i,2j) \\ x_{LH}(i,j) &= -x(2i-1,2j-1) - x(2i-1,2j) + x(2i,2j-1) + x(2i,2j) \\ x_{HL}(i,j) &= -x(2i-1,2j-1) + x(2i-1,2j) - x(2i,2j-1) + x(2i,2j) \\ x_{HH}(i,j) &= x(2i-1,2j-1) - x(2i-1,2j) - x(2i,2j-1) + x(2i,2j) \end{aligned} \tag{11}$$

where $i = 1,2,\ldots,H/2, j = 1,2,\ldots,W/2$ and H and W are the height and width of the original feature map, respectively. That is, the width and height of the output component of each level of the DWT will be $1/2$ that of the input image.

V1 and V2 of the self-attention module are obtained by performing a convolution operation on the low-frequency components of the two features. Subsequently, the high-frequency components undergo convolution to yield the Q and K elements of the self-attention module, where $Q, K \in R^{c_k \times H_l \times W_l}$ and C_K is the number of channels in the low-dimensional mapping space. Then, we reshape them into the shape of $C_K \times N$, where $N = H_l \times W_l$ is the number of pixels. Diverging from traditional self-attention mechanisms, our Q and K features establish a mutual interplay to facilitate cross-image information exchange. In light of this, we introduce the concept of two distinct branches tailored to amplify the representation of support and query features. Following this, a matrix multiplication is executed, utilizing the transposed forms of Q and K. This operation culminates in the creation of a novel feature map, which is subsequently transposed once more to derive the feature map for the alternate branch. Lastly, a $softmax$ module is applied to each of these derived maps, individually generating spatial attention maps for the Q and K branches, thereby completing this process [35].

$$A_{ji} = \frac{exp(Q_i \times K_j)}{\sum_{i=1}^{N} exp(Q_i \times K_j)} \tag{12}$$

where A_{ji} measures the impact of querying the ith position on supporting the jth position. The enhanced similarity in feature representations between two locations corresponds to a heightened correlation between them. Then, the final fused features, A_{ji}, are obtained by concatenating them with V1 and V2, respectively.

4. Experimental Results and Analysis

4.1. Data Sets

We validated the performance of GLF-Net using two state-of-the-art airborne image datasets from the City Classification and 3D Building Reconstruction Test projects provided by ISPRS, which are available from the URL Semantic Annotation Benchmark (https://www.isprs.org/education/benchmarks/UrbanSemLab/Default.aspx, accessed on 26 May 2022). The dataset utilizes a Digital Terrain Model (DSM) produced through HR orthogonal photographs and complementary dense image-matching methodologies. Both datasets encompass urban landscapes, capturing diverse urban scenes. Vaihingen portrays a quaint village characterized by numerous individual buildings and multi-story edifices. On the other hand, Potsdam stands as a quintessential historical city replete with expansive building complexes, narrow alleyways, and densely clustered settlement formations. In a meticulous effort, each dataset has been subject to manual classification, resulting in the categorization of land cover into the six most prevalent classes.

(1) Vaihingen dataset: Comprising 33 distinct remote sensing images of varying dimensions, each image is meticulously extracted from a larger-scale orthophoto picture at the top level. A careful image selection process ensures the avoidance of data gaps. The remote sensing images adhere to an 8-bit TIFF file format, encompassing three bands: near-infrared, red, and green. Meanwhile, the DSM is represented as a single-band TIFF file, with its grayscale values (indicative of DSM height) encoded in 32-bit floating point format. The HR remote sensing images and the DSM both share a ground sampling distance of 9 cm. The DSM data are ingeniously derived through dense image matching utilizing the Trimble INPHO 5.3 software. Presented in various channel combinations, HR remote sensing images adopt the form of TIF files, with each channel sporting an 8-bit spectral resolution. Both the HR remote sensing images and label maps take on the form of three-channel images, while DSM data maps are presented as single-channel images. The HR remote sensing images are stored as 8-bit TIF files, each equipped with three frequency bands. These RGB bands correspond to the near-infrared, red, and green bands captured by the camera. Notably, a DSM is encapsulated within a TIFF file, featuring a single frequency band, and its gray levels are encoded as 32-bit floating point values. It is worth mentioning

that HR remote sensing images are spatially defined within the same grid as the DSM, thereby eliminating the necessity to factor in geocoding information during processing.

(2) Potsdam Dataset: Comprising 28 images, all uniformly sized, the spatial resolution of the top image is an impressive 5 cm. Parallel to the Vaihingen dataset, this collection is constructed from remote sensing TIF files characterized by three bands, alongside DSM data, which remain as a single band. It is noteworthy that each remote sensing image within this dataset boasts identical area coverage dimensions.

4.2. Parameter Setting and Evaluation Index

We trained our model within the PyTorch framework, conducting experiments on HR remote sensing image datasets. These experiments were executed on a personal computer featuring an 11th-generation Intel(R) Core(TM) i9-11900F CPU clocked at 2.50GHz(Intel Productions), an NVIDIA GeForce RTX 3090 GPU, and 32 GB of memory (Asus Productions). An initial learning rate of 0.0001 was adopted, spanning a comprehensive training regimen of thirty epochs. The learning rate underwent adjustments every ten epochs, facilitating progressive optimization. For loss computation, the cross-entropy loss function was employed, aiding in the convergence of training. To accommodate the input data within GLF-Net, we meticulously partitioned the HR remote sensing image into smaller 256x256 patches. We introduced image flipping and rotation. These data augmentation techniques effectively expanded the dataset and enhanced its diversity.

The evaluation of GLF-Net's performance was accomplished using metrics such as mean intersection over union (IoU), intersection over union (IoU), overall accuracy, and mean F1-score. IoU is the proportion of the intersection to the union between the predicted outcome and the ground truth value and is calculated for use case segmentation. mIoU is a standard assessment, and it is the mean of all categories of IoU. F1 is a weighted average of the precision and recall of GLF-Net. From the confusion matrix, we can calculate mIoU, IoU, OA, and F1:

$$\mathrm{OA} = \frac{\sum_{K=1}^{K} TP_K}{\sum_{K=1}^{K} TP_K + FP_K + TN_K + FN_K} \tag{13}$$

$$\mathrm{IoU} = \frac{\sum_{K=1}^{K} TP_K}{\sum_{K=1}^{K} TP_K + FP_K + FN_K} \tag{14}$$

$$\mathrm{mIoU} = \frac{1}{K}\frac{\sum_{K=1}^{K} TP_K}{\sum_{K=1}^{K} TP_K + FP_K + FN_K} \tag{15}$$

$$\mathrm{mF1} = \frac{1}{K}\sum_{K=1}^{K} 2 \times \frac{precision_K \times recall_K}{precision_K + recall_K} \tag{16}$$

where TP and TN represent the number of correct and incorrect positive samples, respectively; FP and FN represent the number of negative samples that were correctly and incorrectly judged, respectively; and $precision_K = TP_K/(TP_K + FP_K)$ and $recall_K = TP_K/(TP_K + FN_K)$ are the precision and recall of GLF-Net, respectively.

4.3. Semantic Results and Analysis

This section primarily presents the outcomes attained by GLF-Net. As depicted in Figure 7, the confusion matrix provides a comprehensive overview of our model's performance across these two datasets. Figures 8 and 9 showcase the segmentation results of HR remote sensing images: Figure 8 corresponds to the Potsdam dataset, and Figure 9 pertains to the Vaihingen dataset. Figures 8 and 9 have the same legend. Notably, GLF-Net demonstrates commendable performance on both datasets, substantiating its efficacy in semantic segmentation tasks.

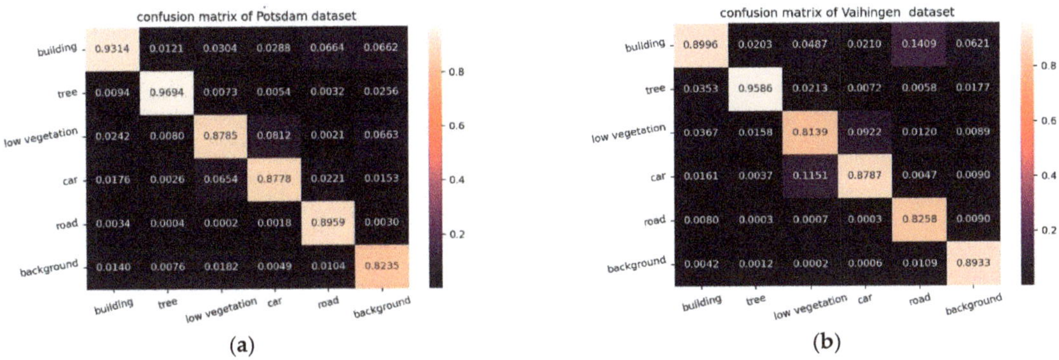

Figure 7. Model confusion matrix. (**a**) The confusion matrix for the Potsdam dataset. (**b**) The confusion matrix for the Vaihingen dataset.

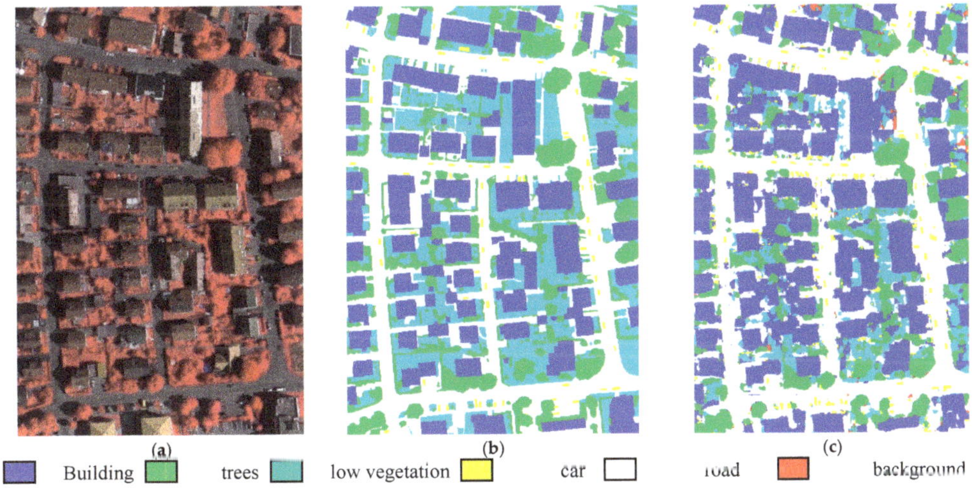

Figure 8. Results of GLF-Net on the Vaihingen dataset. (**a**) Vaihingen dataset image. (**b**) Label image. (**c**) Segmentation result (MioU:0.780).

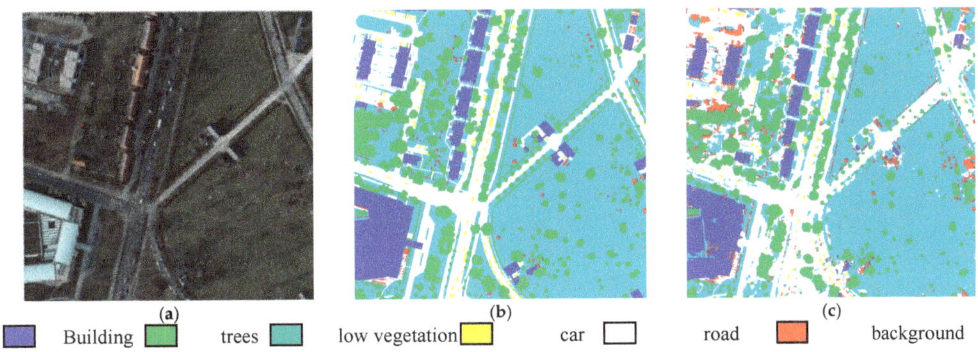

Figure 9. Results of GLF-Net on the Potsdam dataset. (**a**) Potsdam dataset image. (**b**) Label image. (**c**) Segmentation result (MioU:0.811).

To further verify the performance of GLF-Net, we set up a quantitative comparison experiment. We compared GLF-Net with four models: Unet, deeplabV3+, A^2-FPN and BSE-Net [36], and each model consistently uses ResNet50 as the baseline network. DeeplabV3+ employs dilated convolutions to acquire features spanning multiple scales thereby facilitating the extraction of contextual information. A^2-FPN also aggregates global features for image semantic segmentation and derives discriminative features through the accumulation and dissemination of multi-level global contextual attributes. The Bes-Net model is based on boundary information, and incorporating multi-scale context information enhances the precision of the semantic segmentation model.

Tables 1 and 2 present the comparative results from the experimentation conducted on the Vaihingen and Potsdam datasets, respectively. We bold the optimal metrics. Additionally, select outcomes from the test set are showcased in Figures 10 and 11. Notably, GLF-Net demonstrated superior performance across these evaluations. In particular, it stands out for its reduced incidence of misclassified segments and its improved proficiency in discerning certain boundaries and smaller objects. For instance, in the Potsdam dataset, GLF-Net excels at distinguishing the delineation between road and low vegetation. Moreover, the Vaihingen dataset showcases a heightened aptitude for identifying diminutive elements like trees and cars.

Table 1. Comparative experiments on the Vaihingen dataset.

Model	IoU					mIoU	F1	OA
	Building	Low-Veg	Surface	Tree	Car			
Unet	0.795	0.866	0.632	0.744	0.504	0.682	0.804	0.863
DeeplabV3+	0.755	0.826	0.622	0.737	0.513	0.658	0.784	0.846
A^2-FPN	0.817	0.887	0.667	0.771	0.622	0.748	0.853	0.881
Bes-Net	0.830	0.899	**0.698**	**0.789**	0.658	0.774	**0.871**	0.892
OURS	**0.833**	**0.902**	0.692	0.781	**0.668**	**0.780**	0.869	**0.894**

Table 2. Comparison experiments on the Potsdam dataset.

Model	IoU					mIoU	F1	OA
	Building	Low-Veg	Surface	Tree	Car			
Unet	0.814	0.878	0.704	0.774	0.473	0.715	0.827	0.881
DeeplabV3+	0.840	0.924	0.741	0.725	0.777	0.758	0.857	0.890
A^2-FPN	0.869	0.943	0.782	0.759	0.808	0.800	0.886	0.911
Bes-Net	0.871	0.944	0.786	**0.770**	0.825	0.803	0.887	0.913
OURS	**0.876**	**0.946**	**0.791**	**0.770**	**0.827**	**0.811**	**0.893**	**0.916**

4.4. Ablation Experiments

GLF-Net makes full use of the global context information extracted by CAM, LFM extracts fine-grained local features to make GLF-Net better improve the recognition and classification of small targets, and WST effectively integrates the two. To verify that each module can fully play its role, we set up two sets of ablation experiments to verify the performance of our module. Firstly, the ablation strategies of the first group are the baseline network, adding CAM, adding LFM, adding CAM and LFM, and adding three modules (GLF-Net) to verify the performance of our three modules.

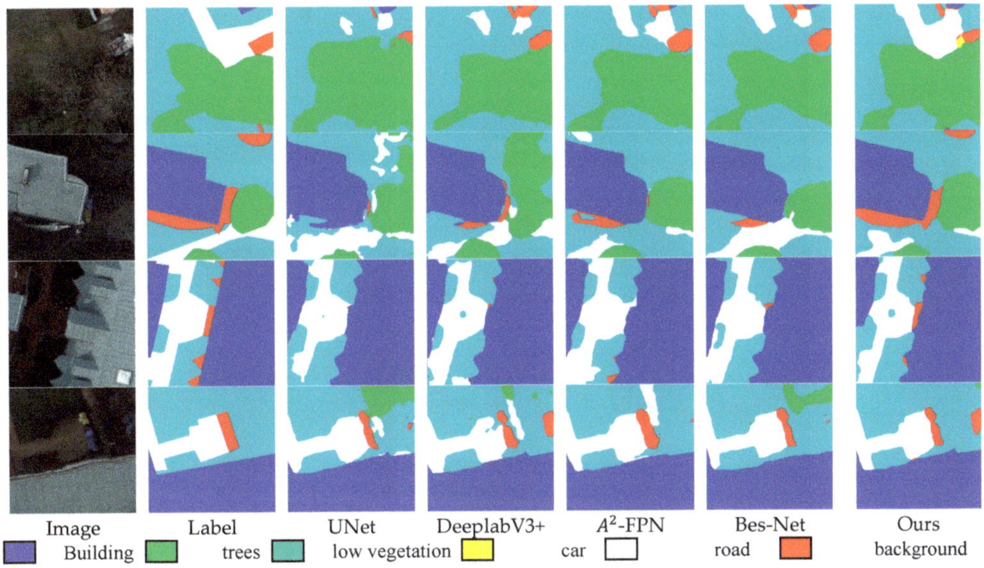

Figure 10. Comparative experimental results on the Potsdam dataset.

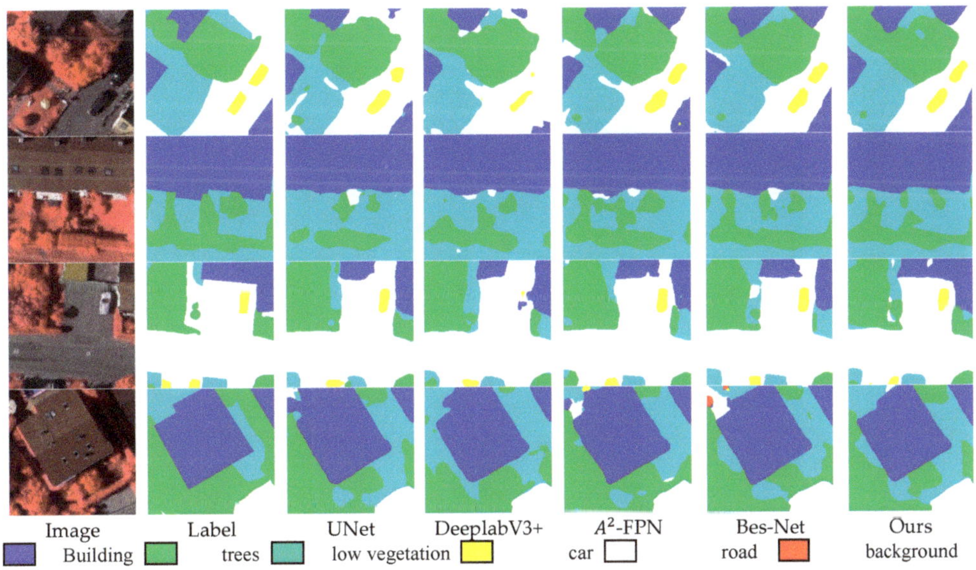

Figure 11. Comparative experimental results on the Vaihingen dataset.

Table 3 showcases the outcomes of ablation experiments conducted on the Vaihingen dataset, while Table 4 presents the results of ablation experiments performed on the Potsdam dataset. We bold the optimal metrics. Moreover, Figures 12 and 13 visually illustrate the findings from ablation experiments on the Vaihingen and Potsdam datasets, respectively. A detailed analysis of the data in these two tables indicates that our modules significantly elevated the performance of GLF-Net when contrasted with the baseline network. And it can be seen from the results that the addition of three modules at the same time is superior to the baseline module and the single use of modules in terms of overall classification and the identification of boundaries and small targets.

Table 3. Ablation experiments on the Vaihingen dataset.

Model	IoU					mIoU	F1	OA
	Building	Low-Veg	Surface	Tree	Car			
Baseline	0.715	0.823	0.584	0.678	0.531	0.592	0.714	0.821
CAM	0.829	0.898	0.687	0.777	0.659	0.764	0.864	0.887
LFM	0.826	0.899	0.685	0.774	0.660	0.761	0.862	0.886
CAM+LFM	0.828	0.900	0.685	0.774	0.652	0.766	0.865	0.888
OURS	**0.833**	**0.902**	**0.692**	**0.781**	**0.668**	**0.780**	**0.869**	**0.894**

Table 4. Ablation experiments on the Potsdam dataset.

Model	IoU					mIoU	F1	OA
	Building	Low-Veg	Surface	Tree	Car			
Baseline	0.788	0.922	0.720	0.728	0.569	0.680	0.795	0.873
CAM	0.869	0.938	0.774	0.769	0.818	0.803	0.887	0.912
LFM	0.865	0.941	0.776	0.763	0.824	0.793	0.880	0.908
CAM+LFM	0.872	0.945	0.787	0.765	**0.827**	0.806	0.889	0.913
OURS	**0.876**	**0.946**	**0.791**	**0.770**	**0.827**	**0.811**	**0.893**	**0.916**

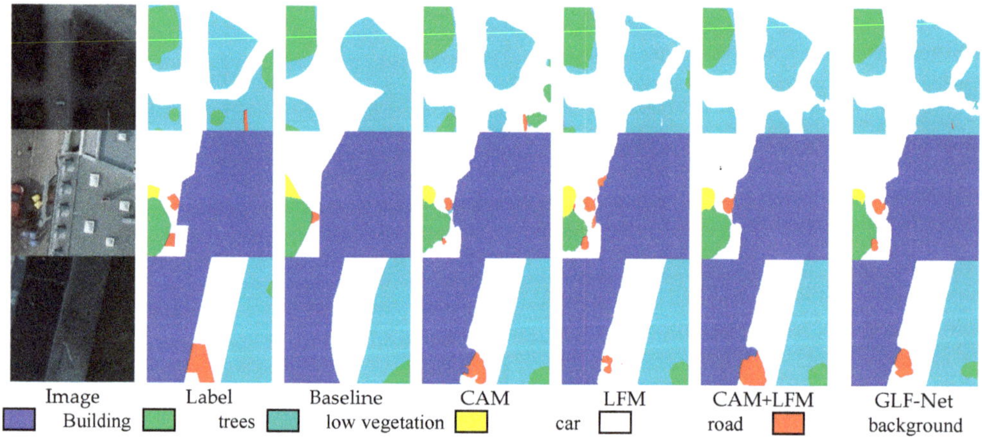

Figure 12. Ablation experimental results on the Potsdam dataset.

To verify which stage of context information of ResNet is most needed for GLF-Net, we set up a second set of ablation experiments to compare the performance of CAM. Our CAM module is used for ResNet stages 123, 124, 134, and 234. Finally, Tables 5 and 6 present the experimental results derived from the Vaihingen dataset and the Potsdam dataset, respectively. The outcomes distinctly highlight the superiority of the CAM module, showcasing its optimal performance when applied to the 234 stages. At the same time, in order to verify the effect of the covariance matrix and graph convolution, we also performed a comparison with two models without using the covariance matrix and without using graph convolution. As shown in Figures 5 and 6, there is a large gap between the performance of the two and CAM. We bold the optimal metrics. Finally, in order to verify the superiority of the CAM module, we also made a comparison with the existing model DANet. The CAM module shows better performance than DANet on both datasets.

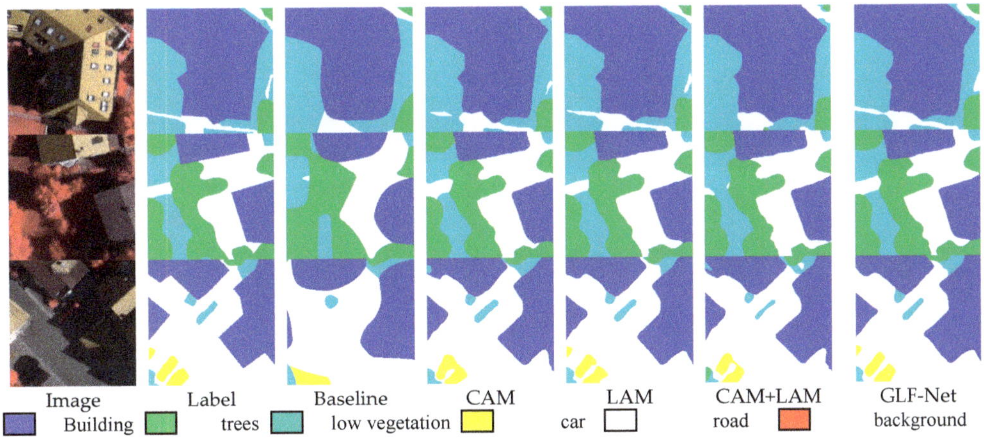

Figure 13. Ablation experimental results on the Vaihingen dataset.

Table 5. CAM ablation experiment on the Vaihingen dataset.

Model	IoU					mIoU	F1	OA
	Building	**Low-Veg**	**Surface**	**Tree**	**Car**			
CAM123	0.827	0.891	0.685	**0.779**	0.647	0.762 ± 0.01	0.862	0.886
CAM124	0.827	0.887	0.685	**0.779**	0.650	0.761 ± 0.01	0.862	0.886
CAM134	0.827	0.887	0.685	**0.779**	0.650	0.761 ± 0.01	0.862	0.886
CAM_nonCM	0.820	0.895	0.675	0.775	0.631	0.755 ± 0.02	0.857 ± 0.01	0.885 ± 0.01
CAM_nonGrapth	0.821	0.893	0.677	0.775	0.608	0.749 ± 0.02	0.853 ± 0.01	0.885 ± 0.01
DANet	0.826	0.885	0.686	0.776	0.643	0.761 ± 0.03	0.862 ± 0.01	0.886 ± 0.01
CAM234	**0.829**	**0.898**	**0.687**	0.777	**0.659**	**0.764 ± 0.03**	**0.864 ± 0.01**	**0.887 ± 0.01**

Table 6. CAM ablation experiment on the Potsdam dataset.

Model	IoU					mIoU	F1	OA
	Building	**Low-Veg**	**Surface**	**Tree**	**Car**			
CAM123	0.870	0.943	**0.783**	0.768	0.818	0.801 ± 0.01	0.886	0.911
CAM124	**0.869**	0.938	0.784	0.767	0.819	0.800 ± 0.01	0.885	0.911
CAM134	**0.869**	0.938	0.784	0.767	0.819	0.800 ± 0.02	0.885	0.911
CAM_nonCM	0.868	0.935	0.776	0.754	0.815	0.798 ± 0.02	0.882 ± 0.02	0.909 ± 0.01
CAM_nonGrapth	0.868	0.936	0.779	0.762	0.812	0.798 ± 0.02	0.883 ± 0.02	0.909 ± 0.01
DANet	0.867	0.935	0.776	0.757	0.810	0.797 ± 0.03	0.882 ± 0.02	0.909 ± 0.01
CAM234	**0.869**	**0.944**	0.777	**0.769**	**0.822**	**0.803 ± 0.02**	**0.887 ± 0.01**	**0.912 ± 0.01**

In particular, to visualize the role of the CAM module in extracting and enhancing contextual features, we visualized ResNet, the CAM module, and the intermediate features of DANet, as shown in Figure 14. The red channel represents a higher degree of responsiveness, while the opposite is true for yellow. Compared to ResNet, DANet does not show significant changes, while CAM extracts channels with primary information.

Finally, in order to verify the performance of the self-attention module in our WST module, we performed ablation experiments on the WST module. Tables 7 and 8 give the results of the ablation experiments. We bold the optimal metrics. It can be seen from the results that the self-attention module has brought significant improvements.

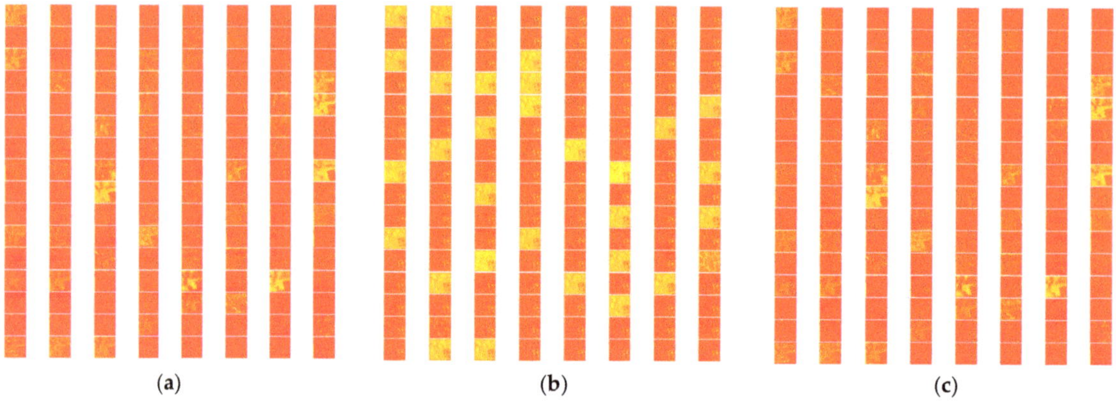

Figure 14. Covariance attention effect. (a) ResNet intermediate features. (b) The effect after using covariance attention. (c) The effect after using DANet.

Table 7. WST ablation experiment on the Potsdam dataset.

Model	IoU					mIoU	F1	OA
	Building	Low-Veg	Surface	Tree	Car			
Non_self attention	0.870	0.945	0.779	0.758	0.820	0.806	0.890	0.912
GLF-Net	**0.876**	**0.946**	**0.791**	**0.770**	**0.827**	**0.811**	**0.893**	**0.916**

Table 8. WST ablation experiment on the Vaihingen dataset.

Model	IoU					mIoU	F1	OA
	Building	Low-Veg	Surface	Tree	Car			
Non_self attention	0.821	0.891	0.677	0.774	0.610	0.750	0.854	0.844
GLF-Net	**0.833**	**0.902**	**0.692**	**0.781**	**0.668**	**0.780**	**0.869**	**0.894**

5. Conclusions

This paper introduces the GLF-Net model for semantic segmentation of HR remote sensing images. This model addresses the complex challenges posed by significant intra-class differences and small inter-class differences in HR remote sensing images. The proposed GLF-Net employs an encoder–decoder architecture with ResNet50 as the base network. The model uses the CAM module to extract global contextual features, uses the LFM module to extract complex local features, and uses WST to effectively integrate these two features. Through the above modules, the proposed GLF-Net simultaneously obtains broader global context information and fine-grained local texture and boundary features, which significantly enhances the model's ability to recognize smaller objects and contributes to the overall enhancement of segmentation performance. The validation of our model on ISPRS's Vaihingen and Potsdam datasets confirms its superior achievement, with GLF-Net outperforming other models when all three modules are effectively integrated.

Although the proposed GLF-Net has achieved good results, it still has a high computational cost, and, next, we will research reducing the complexity of the model while maintaining the existing performance.

Author Contributions: Conceptualization, W.S.; Methodology, W.S.; Software, X.Z.; Validation, W.S. and S.Z; Formal Analysis, Y.W. and P.Z.; Writing—Original Draft Preparation, W.S. and X.Z.; Writing—Review and Editing W.S., X.Z. and S.Z. All authors have read and agreed to the published version of the manuscript.

Funding: This work was supported by the Natural Science Foundation of China (61901358, 62172321, and 61871312), the Outstanding Youth Science Fund of Xi'an University of Science and Technology (2020YQ3-09), the Scientific Research Plan Projects of Shaanxi Education Department (20JK0757), the PhD Scientific Research Foundation (2019QDJ027), the China Postdoctoral Science Foundation (2020M673347), the Natural Science Basic Research Plan in Shaanxi Province of China (2019JZ-14), and the Civil Space Thirteen Five Years Pre-Research Project (D040114).

Data Availability Statement: We employed two publicly available 2D semantic labeling datasets, namely, Vaihingen and Potsdam, graciously provided by the International Society for Photogrammetry and Remote Sensing (ISPRS): https://www.isprs.org/education/benchmarks/UrbanSemLab/Default.aspx, accessed on 26 May 2022.

Acknowledgments: The authors would like to thank the reviewers and the editor for their constructive comments.

Conflicts of Interest: The authors declare no conflict of interest.

References

1. Alganci, U.; Soydas, M.; Sertel, E. Comparative Research on Deep Learning Approaches for Airplane Detection from Very High-Resolution Satellite Images. *Remote Sens.* **2020**, *12*, 458. [CrossRef]
2. Zhu, X.; Tuia, D.; Mou, L.; Xia, G.S.; Zhang, L.; Xu, F.; Fraundorfer, F. Deep Learning in Remote Sensing: A Comprehensive Review and List of Resources. *IEEE Geosci. Remote Sens. Mag.* **2017**, *5*, 8–36. [CrossRef]
3. Ma, L.; Liu, Y.; Liang Zhang, X.; Ye, Y.; Yin, G.; Johnson, B. Deep learning in remote sensing applications: A meta-analysis and review. *ISPRS J. Photogramm. Remote Sens.* **2019**, *152*, 166–177. [CrossRef]
4. Garcia-Garcia, A.; Orts, S.; Opera, S.; Villena-Martinez, V. A Review on Deep Learning Techniques Applied to Semantic Segmentation. *arXiv* **2017**, arXiv:1704.06857.
5. Davis, L.S.; Rosenfeld, A.; Weszka, J.S. Region extraction by averaging and thresholding. *IEEE Trans. Syst. Man Cybern.* **1975**, *SMC-5*, 383–388. [CrossRef]
6. Adams, R.; Bischof, L. Seeded Region Growing. *IEEE Trans. Pattern Anal. Mach. Intell.* **1994**, *16*, 641–647. [CrossRef]
7. Kundu, M.K.; Pal, S.K. Thresholding for edge detection using human psychovisual phenomena. *Pattern Recognit. Lett.* **1986**, *4*, 433–441. [CrossRef]
8. Achanta, R.; Shaji, A.; Smith, K.; Lucchi, A.; Fua, P.; Süsstrunk, S. SLIC superpixels compared to state-of-the-art superpixel methods. *IEEE Trans. Pattern Anal. Mach. Intell.* **2012**, *34*, 2274–2282. [CrossRef]
9. He, K.; Zhang, X.; Ren, S.; Sun, J. Deep residual learning for image recognition. In Proceedings of the IEEE Conference on Computer Vision and Pattern Recognition, Las Vegas, NV, USA, 27–30 June 2016; pp. 770–778.
10. Long, J.; Shelhamer, E.; Darrell, T. Fully convolutional networks for semantic segmentation. In Proceedings of the IEEE Conference on Computer Vision and Pattern Recognition, Boston, MA, USA, 7–12 June 2015; pp. 3431–3440.
11. Ronneberger, O.; Fischer, P.; Brox, T. U-net: Convolutional networks for biomedical image segmentation. In Proceedings of the International Conference on Medical Image Computing and Computer-Assisted Intervention, Munich, Germany, 5–9 October 2015; pp. 234–241.
12. Badrinarayanan, V.; Handa, A.; Cipolla, R. Segnet: A deep convolutional encoder–decoder architecture for robust semantic pixel-wise labeling. *arXiv* **2015**, arXiv:1505.07293.
13. Zhao, H.; Shi, J.; Qi, X.; Wang, X.; Jia, J. Pyramid scene parsing network. In Proceedings of the IEEE Conference on Computer Vision and Pattern Recognition, Honolulu, HI, USA, 21–26 July 2017; pp. 2881–2890.
14. Prabhu, S.; Fleuret, F. Uncertainty Reduction for Model Adaptation in Semantic Segmentation. In Proceedings of the 2021 IEEE/CVF Conference on Computer Vision and Pattern Recognition (CVPR), Nashville, TN, USA, 20–25 June 2021; pp. 9608–9618.
15. Liu, Y.; Zhang, W.; Wang, J. Source-Free Domain Adaptation for Semantic Segmentation. In Proceedings of the 2021 IEEE/CVF Conference on Computer Vision and Pattern Recognition (CVPR), Nashville, TN, USA, 20–25 June 2021; pp. 1215–1224.
16. Chen, J.; Zhu, J.; Guo, Y.; Sun, G.; Zhang, Y.; Deng, M. Unsupervised Domain Adaptation for Semantic Segmentation of High-Resolution Remote Sensing Imagery Driven by Category-Certainty Attention. *IEEE Trans. Geosci. Remote Sens.* **2022**, *60*, 1–15. [CrossRef]
17. Guan, D.; Huang, J.; Lu, S. Scale variance minimization for unsupervised domain adaptation in image segmentation. *Pattern Recognit.* **2020**, *112*, 107764. [CrossRef]
18. Stan, S.; Rostami, M. Domain Adaptation for the Segmentation of Confidential Medical Images. *arXiv* **2021**, arXiv:2101.00522.
19. Hu, J.; Shen, L.; Albanie, S.; Sun, G.; Wu, E. Squeeze-and-Excitation Networks. *IEEE Trans. Pattern Anal. Mach. Intell.* **2020**, *42*, 2011–2023. [CrossRef] [PubMed]
20. Zhao, H.; Zhang, Y.; Liu, S.; Shi, L.; Shi, J.; Loy, C.C.; Lin, D.; Jia, J. PSANet: Point-wise spatial attention network for scene parsing. In Proceedings of the European Conference on Computer Vision, Munich, Germany, 8–14 September 2018; pp. 267–283.
21. Fu, J.; Liu, J.; Tian, H.; Li, Y.; Bao, Y.; Fang, Z.; Lu, H. Dual Attention Network for Scene Segmentation. In Proceedings of the IEEE/CVF Conference on Computer Vision and Pattern Recognition, Long Beach, CA, USA, 15–20 June 2019; pp. 3146–3154.

22. Woo, S.; Park, J.; Lee, J.Y.; Kweon, I.S. CBAM: Convolutional block attention module. In Proceedings of the European Conference on Computer Vision, Munich, Germany, 8–14 September 2018; pp. 3–19.
23. Liu, Y.; Chen, P.; Sun, Q. Covariance Attention for Semantic Segmentation. *IEEE Trans. Pattern Anal. Mach. Intell.* **2022**, *44*, 1805–1818. [CrossRef] [PubMed]
24. He, N.; Fang, L.; Li, S.; Plaza, J.; Plaza, A. Skip-Connected Covariance Network for Remote Sensing Scene Classification. *IEEE Trans. Neural Netw. Learn. Syst.* **2020**, *31*, 1461–1474. [CrossRef]
25. Chen, L.C.; Papandreou, G.; Schroff, F.; Adam, H. Rethinking atrous convolution for semantic image segmentation. *arXiv* **2017**, arXiv:1706.05587.
26. Chen, L.C.; Zhu, Y.; Papandreou, G.; Schroff, F.; Adam, H. Encoder-decoder with atrous separable convolution for semantic image segmentation. *arXiv* **2018**, arXiv:1802.02611.
27. Zhang, H.; Dana, K.; Shi, J.; Zhang, Z.; Wang, X.; Tyagi, A.; Agrawal, A. Context Encoding for Semantic Segmentation. In Proceedings of the IEEE Conference on Computer Vision and Pattern Recognition, Salt Lake City, UT, USA, 18–22 June 2018; pp. 7151–7160.
28. Li, R.; Wang, L.; Zhang, C.; Duan, C.; Wang, L. A^2-FPN for semantic segmentation of fine-resolution remotely sensed images. *Int. J. Remote Sens.* **2022**, *43*, 1131–1155. [CrossRef]
29. Liu, H.; Peng, P.; Chen, T.; Wang, Q.; Yao, Y.; Hua, X.S. FECANet: Boosting Few-Shot Semantic Segmentation with Feature-Enhanced Context-Aware Network. *arXiv* **2023**, arXiv:2301.08160. [CrossRef]
30. Yang, M.; Yu, K.; Zhang, C.; Li, Z.; Yang, K. Denseaspp for semantic segmentation in street scenes. In Proceedings of the IEEE Conference on Computer Vision and Pattern Recognition, Salt Lake City, UT, USA, 18–23 June 2018; pp. 3684–3692.
31. Li, R.; Duan, C. ABCNet: Attentive Bilateral Contextual Network for Efficient Semantic Segmentation of Fine-Resolution Remote Sensing Images. *arXiv* **2021**, arXiv:2102.0253. [CrossRef]
32. Wang, L.; Xiao, P.; Zhang, X.; Chen, X. A Fine-Grained Unsupervised Domain Adaptation Framework for Semantic Segmentation of Remote Sensing Images. *IEEE J. Sel. Top. Appl. Earth Obs. Remote Sens.* **2023**, *16*, 4109–4121. [CrossRef]
33. Chen, Y.; Rohrbach, M.; Yan, Z.; Shui, Y.; Feng, J.; Kalantidis, Y. Graph-Based Global Reasoning Networks. In Proceedings of the 2019 IEEE/CVF Conference on Computer Vision and Pattern Recognition (CVPR), Long Beach, CA, USA, 15–20 June 2019; pp. 433–442.
34. Xiang, X.; Zhang, Y.; Jin, L.; Li, Z.; Tang, J. Sub-Region Localized Hashing for Fine-Grained Image Retrieval. *IEEE Trans. Image Process.* **2022**, *31*, 314–326. [CrossRef] [PubMed]
35. Chen, C.F.; Fan, Q.; Panda, R. CrossViT: Cross-attention multi-scale vision transformer for image classification. In Proceedings of the IEEE/CVF International Conference on Computer Vision (ICCV), Montreal, QC, Canada, 10–17 October 2021; pp. 347–356.
36. Chen, F.; Liu, H.; Zeng, Z.; Zhou, X.; Tan, X. BES-Net: Boundary Enhancing Semantic Context Network for High-Resolution Image Semantic Segmentation. *Remote Sens.* **2022**, *14*, 1638. [CrossRef]

remote sensing MDPI

Article

TCUNet: A Lightweight Dual-Branch Parallel Network for Sea–Land Segmentation in Remote Sensing Images

Xuan Xiong [1], Xiaopeng Wang [1,*], Jiahua Zhang [2], Baoxiang Huang [1] and Runfeng Du [1]

1 Remote Sensing Information and Digital Earth Center, College of Computer Science and Technology, Qingdao University, Qingdao 266071, China; 2021023788@qdu.edu.cn (X.X.); baoxianghuang@qdu.edu.cn (B.H.); 2021023825@qdu.edu.cn (R.D.)
2 Key Laboratory of Digital Earth Science, Aerospace Information Research Institute, Chinese Academy of Sciences, Beijing 100094, China; zhangjh@radi.ac.cn
* Correspondence: wxp@qdu.edu.cn

Abstract: Remote sensing techniques for shoreline extraction are crucial for monitoring changes in erosion rates, surface hydrology, and ecosystem structure. In recent years, Convolutional neural networks (CNNs) have developed as a cutting-edge deep learning technique that has been extensively used in shoreline extraction from remote sensing images, owing to their exceptional feature extraction capabilities. They are progressively replacing traditional methods in this field. However, most CNN models only focus on the features in local receptive fields, and overlook the consideration of global contextual information, which will hamper the model's ability to perform a precise segmentation of boundaries and small objects, consequently leading to unsatisfactory segmentation results. To solve this problem, we propose a parallel semantic segmentation network (TCU-Net) combining CNN and Transformer, to extract shorelines from multispectral remote sensing images, and improve the extraction accuracy. Firstly, TCU-Net imports the Pyramid Vision Transformer V2 (PVT V2) network and ResNet, which serve as backbones for the Transformer branch and CNN branch, respectively, forming a parallel dual-encoder structure for the extraction of both global and local features. Furthermore, a feature interaction module is designed to achieve information exchange, and complementary advantages of features, between the two branches. Secondly, for the decoder part, we propose a cross-scale multi-source feature fusion module to replace the original UNet decoder block, to aggregate multi-scale semantic features more effectively. In addition, a sea–land segmentation dataset covering the Yellow Sea region (GF Dataset) is constructed through the processing of three scenes from Gaofen-6 remote sensing images. We perform a comprehensive experiment with the GF dataset to compare the proposed method with mainstream semantic segmentation models, and the results demonstrate that TCU-Net outperforms the competing models in all three evaluation indices: the PA (pixel accuracy), F1-score, and MIoU (mean intersection over union), while requiring significantly fewer parameters and computational resources compared to other models. These results indicate that the TCU-Net model proposed in this article can extract the shoreline from remote sensing images more effectively, with a shorter time, and lower computational overhead.

Keywords: double-branch; sea–land segmentation; GF-6; CNN; transformer; remote sensing

Citation: Xiong, X.; Wang, X.; Zhang, J.; Huang, B.; Du, R. TCUNet: A Lightweight Dual-Branch Parallel Network for Sea–Land Segmentation in Remote Sensing Images. *Remote Sens.* **2023**, *15*, 4413. https://doi.org/10.3390/rs15184413

Academic Editors: Jiaojiao Li, Qian Du, Jocelyn Chanussot, Wei Li, Bobo Xi, Rui Song and Yunsong Li

Received: 25 July 2023
Revised: 30 August 2023
Accepted: 5 September 2023
Published: 7 September 2023

1. Introduction

A coastline refers to the boundary line or marginal area between the ocean or lake and the land [1]. Different and debated are the definitions of coastline, because defining the sea–land interface is neither conceptually nor physically simple; one of the conceptually simplest is defined as the boundary between the land surface and the ocean surface [2], also known as an instantaneous coastline, in the field of remote sensing application research. The types of coastline are mainly divided into rocky coasts, sandy coasts, silty coasts, biological coasts, and artificial coastlines. Coastline information is an important basis for the implementation of coastal zone protection and disaster management, the basis for the development and use of marine resources, and an important territorial resource for countries bordering the sea, and plays an significant role in the ecological safety of the ocean [3]. However, at the same time, the extraction of the coastline is a very challenging problem, because it is the land–water boundary of the multi-year average high tide, rather than an instantaneous line [4]. Traditional shoreline extraction methods are mainly manual measurements. However, manual surveying and mapping is associated with issues of labor intensiveness and a lengthy surveying and mapping duration, which consequently lead to a reduced efficiency. Additionally, the influence of human factors [5]; for instance, errors introduced during the process of data collection and variations in subjective judgments and drawing styles among different operators when delineating coastlines; results in disparities in the depiction of the same coastline area on different maps. Collectively, these factors will have an impact on the precise depiction of the coastline. In contrast, remote sensing images have the advantages of a wide coverage, fast information acquisition, high data reliability, fewer constraints caused by the weather, geographic environment and other conditions, free access, etc., which can greatly reduce the cost of surveying and mapping and, therefore, have been commonly used in agricultural development, sea monitoring, and other fields [6,7]. Remote sensing technology has become the main technical means of coastline research, and is widely used in the extraction and monitoring of coastlines.

Coastline extraction methods mainly include threshold segmentation methods [8], edge detection algorithms, object-oriented methods, machine learning methods, and deep learning methods [9]. The threshold segmentation method divides the pixels in an image into two or more categories according to the pixel digital number values, so as to divide the image into different regions. In remote sensing images, the spectral water index (SWI) method is often used; i.e., based on the different reflectance properties of water bodies and non-water bodies in the infrared and visible bands, we calculate certain combinations of bands in the remote sensing image, to distinguish between water bodies and non-water bodies. For example, there is the Normalized Difference Water Index (NDWI) [10] and the Modified Normalized Difference Water Index (MNDWI) [11]. However, threshold-based methods often require thresholds to be set manually, but different images often have large differences, and it is likely that different thresholds will need to be set, making threshold selection difficult and, thus, affecting the final shoreline extraction accuracy. In addition, the coastline region has a complex terrain; there are shadows cast by the surrounding terrain, clouds, vegetation, and other factors, meaning that considering only the spectral differences to distinguish between land and water will make the accuracy lower. Image edge detection algorithms, currently commonly used as edge detection models include the Roberts operator [12], Sobel operator [13], Canny operator [14], and so on. However, the coastline detected by such methods is highly affected by noise, and the noise causes distortion in the edge detection results. Thus, the detected edges are not accurate enough. At the same time, these methods are less efficient, and can only detect the significant edges in the image, and the accuracy of the obtained boundaries is not high [4,15,16]. The object-oriented classification method combines pixels into objects, integrating their interrelationships and spatial distributions, and thereby reducing the interference from internal pixel information, and maximizing the utilization of image information. However, due to the complexity of the steps, processing difficulties, and the difficulty of determining the threshold value of image segmentation, it is difficult to use in a wide range of high-resolution images with many features and information. Many machine learning algorithms extract diverse information based on a variety of data, and use traditional machine learning algorithms, such as random forest [17] or support vector machine (SVM) [18], to extract the shoreline. These algorithms are able to extract the shoreline quickly and efficiently compared to traditional methods. Traditional machine learning methods have certain limitations. For instance, when manually extracting image features as input, and selecting features, it is possible that the complex distinctions between the ocean and land cannot be fully captured, thereby restricting the algorithm's generalization ability and robustness.

Furthermore, machine learning algorithms typically focus only on individual pixel features, neglecting the spatial relationships and contextual information among pixels, leading to insufficient smoothness and accuracy in the segmentation results. As a result, these limitations result in a lack of precision in traditional machine learning methods when extracting complex coastlines from high-resolution images [4,16,19].

Advancements in computer technology and artificial intelligence have generated considerable interest in the application of deep learning techniques, particularly in the domain of computer vision, including, but not limited to, semantic segmentation [20] and object detection [21]. In contrast to other approaches, deep learning models, specifically those based on convolutional neural networks (CNNs) [22], have demonstrated a superior capacity to handle intricate image features, and show robust self-learning capabilities. Long et al. [23] proposed the use of a full convolution network (FCN) to solve the problem that a traditional convolutional neural network (CNN) cannot directly handle variable length inputs and outputs. They used convolutional layers instead of fully connected layers, and used methods such as inverse convolution and up-sampling to reduce the feature maps, which provided new ideas for those who came after them. On this basis, U-Net, proposed by Ronneberger O et al. [24], has been extensively employed in the domain of medical image segmentation. Its innovative architecture and the introduction of jump connections bring new methods for image segmentation research. Furthermore, the domain of semantic segmentation encompasses several classical methods, including SegNet [25], PSPNet [26], the Deeplab series [20,27,28], HRNet [29], and so on. In addition, several researchers have endeavored to integrate CNN methods into land and water segmentation in remote sensing images, which has led to substantial enhancements in the accuracy of shoreline extraction. Li et al. [30] proposed a model called DeepUNet, which is deeper than U-Net, and improves the accuracy by 2% compared to U-Net. Shamsolmoali et al. [31] combined the DenseNet [32] and ResNet [33] to develop RDUNet, which has a better classification accuracy than DeepUNet, DenseNet, and other models. He et al. [34] combined the attention mechanism with the classical UNet network to devise a novel segmentation network for extracting glacial lakes in remote sensing images, which enhances the classification accuracy, as well as achieving clearer boundaries compared to the traditional models.

However, traditional CNN methods capture detailed features of an image only from a local scope, and do not determine the target boundaries from the global level, based on the contextual information of the image. In recent years, Transformer [35] has been migrated to computational vision tasks, showing amazing potential and value. By dividing images into image patches, and applying a self-attention mechanism, global contextual information can be utilized for classification, rather than just local features. This global information processing gives Transformer an advantage over other methods when dealing with large-scale images and complex scenes. The Vision Transformer (ViT), proposed by Dosovitskiy et al. [36], is a transformer-based architecture developed for large-scale image recognition tasks. The fundamental concept behind ViT is to divide the input image into a series of image patches, considering each patch as an element in a sequence. These image patches are transformed into corresponding embedding vectors, through a linear mapping layer, and combined with position coding, to form the input to the Transformer model. By processing these input embedding vectors through multiple Transformer encoder layers, ViT is able to capture the global contextual information in the image and, thus, process image tasks efficiently, with a relatively good performance. Several studies have modified the architecture of ViT for dense prediction tasks. The Pyramid Vision Transformer (PVT) [37] was the first transformer-based model to import the feature pyramid of CNNs. With the pyramid structure capturing the multi-scale features, and the Transformer model achieving global context modelling, PVT has shown a good performance in image classification tasks. Later, a hierarchical attention mechanism was proposed in the Swin Transformer [38], which performs attention computation at multiple scales, thus reducing the computational and memory burden. It is able to handle large-size images with a good scalability and efficiency, achieving an excellent image classification performance. Meanwhile, in the domain of semantic segmentation, segmentation transformers (SETR) [39] employ ViT as a backbone to extract features, while the decoder uses progressive up-sampling to mitigate the noise problem. After four up-sampling operations to obtain the segmentation results subsequently, semantic segmentation transformers (SegFormer) [40] achieved some improvements on STER, by removing positional coding, and introducing convolutional operations, while using a hierarchical encoder structure that outputs multi-scale features and, finally, designed a lightweight decoder, to reduce the computational overhead. These changes further improve its segmentation effect.

However, it has been pointed out [41–43] that results based on sheer transformer-based segmentation networks are suboptimal, primarily because transformers are inclined towards global modelling, and lack location awareness. Furthermore, due to the unique self-attention mechanism, and absence of convolutional operations, in Transformer models, they suffer from certain drawbacks in modeling spatial information, expressing local details, preserving image invariance, and maintaining robustness. Consequently, these limitations result in the disruption of image structure, and loss of information. Therefore, many scholars have tried to design methods with better results, by combining the union of CNNs and transformers. TransUNet [41] used a hybrid Vision Transformer structure to stack CNNs and transformers sequentially as an encoder, while the decoder followed the classical UNet, and achieved good results in medical image segmentation. He et al. [44] constructed a novel parallel dual-branch encoder based on TransUNet, using Swin Transformer as a secondary encoder, and the original hybrid Vision Transformer primary encoder, and achieved good segmentation results on hyperspectral images. Chen et al. [45] put forth a dual-branch parallel network for segmentation tasks. In the encoding part, ResNet50 and Swin transformers serve as a dual-branch backbone, to capture the features from the input images, followed by the complete fusion of the extracted information. A new fusion module is proposed during the decoding process for multi-scale feature fusion. The experimental results show that the network maximizes the advantages of both the backbone networks, and improves the accuracy of semantic segmentation tasks related to buildings and water bodies.

Inspired by these works, and in order to solve the problems of complex shoreline extraction and fine water-body identification, in this study, we propose a new two-branch parallel image segmentation network fusing CNN and Transformer, to achieve the accurate segmentation of sea and land in multispectral remote sensing images. The paper primarily contributes via the following four aspects:

- In this paper, we propose TCUNet, a parallel two-branch image segmentation network fusing CNN and Transformer, to achieve a fine segmentation of land and sea in multispectral remote sensing images.
- We design a new lightweight feature interaction module (FIM) to achieve feature exchange and information flow in the dual branch, by embedding it between each coding block in the dual branch, to minimize the semantic gap of the dual branch, enhancing the global representation of the CNN branch, while complementing the local details of the Transformer branch.
- We propose a cross-scale, multi-source feature fusion module (CMFFM) to replace the decoder block in UNet, to solve the issue of feature inconsistency between different scales, and achieve the fusion of multi-source features at different scales.
- Based on three Gaofen-6 satellite images produced in February 2023, we constructed a sea–land semantic segmentation dataset, the GF dataset, covering the entire Yellow Sea region of China, which contains 12,600 sheets, each with a size of 512 pixels × 512 pixels. We have made it available for public use.

2. Methods and Materials

2.1. Overall Network Structure

Most of the existing land and sea segmentation models use a convolutional neural network as an encoder to achieve land and sea feature extraction from remote sensing images. Despite being highly effective in local feature extraction, and significantly enhancing the network's robustness in sea–land segmentation, CNNs extract image features by reusing convolutional and pooling layers, but this results in a limited size of the model's receptive field. When dealing with large images, the convolution kernel needs to become very large, which increases the computational cost and memory consumption. Convolutional neural networks are, after all, only network structures that focus on local information, and this computational mechanism leads to difficulties in capturing and storing global information over long distances. Numerous transformer-based backbone networks have emerged, integrating the self-attention mechanism to effectively capture global contextual information, and address the limitations of CNN in recent years. However, compared to CNN, transformer-based models cannot fully utilize the local features of the image.

To address this limitation, we propose a novel parallel semantic segmentation network based on a transformer and CNN, to extract comprehensive global information and intricate local details between the target and background for sea–land segmentation tasks. The network architecture, as illustrated in Figure 1, comprises a CNN branch, a Transformer branch, a feature interaction module (FIM) and a cross-scale multi-level feature fusion module (CMFFM), which are described in detail below. Considering that the remote sensing image in the GF dataset contains eight bands, in order to facilitate clear viewing, we selected a specific image, and displayed a subset of bands. More specifically, bands 3, 4, and 5 (Red, NIR, and SWIR-1) were selected as illustrative samples, as visually depicted in Figure 1.

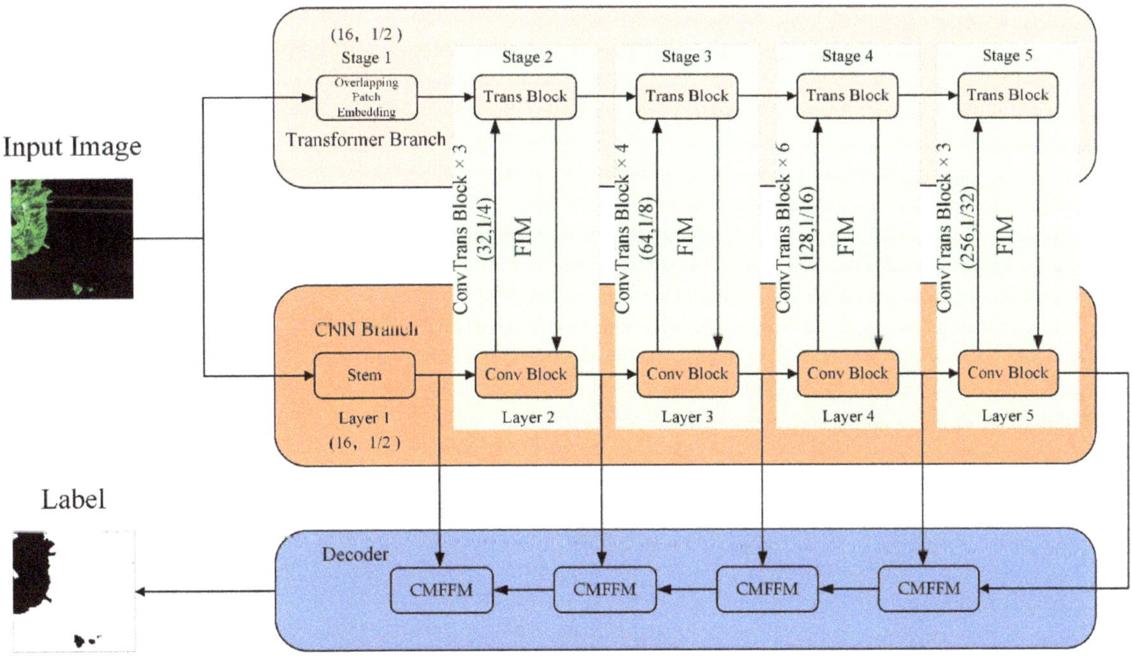

Figure 1. The overall structure of TCUNet.

2.2. CNN Branch

The CNN branch is devised to capture the local contextual information. It is structured in a feature pyramid style, comprising five distinct layers, where the feature map resolution of each subsequent layer is halved compared to the previous layer, and the number of channels is doubled, accordingly. The resolution of the feature map decreases with the increase of the number of network layers, while the number of channels increases. The first layer is the stem module, which consists of a 7 × 7 convolutional kernel with the stride of 2, a batch normalization (BN) layer, and a ReLU activation function. An initial H × W × C remote sensing image is processed by the stem module, to obtain an H × W × 16 feature map, which is used for the image extraction of the initial local features, and the layers 2–5 are all composed of a number of Conv Blocks, as shown in Figure 2. Every Conv Block comprises two bottleneck blocks. Each layer down-samples the input feature map, and inputs it into the next stage and, finally, outputs a feature map with half the resolution, and double the number of channels. Therefore, five hierarchical feature maps with different scales are obtained through these five layers. The shape of the i-th layer feature map is $H/2^i \times W/2^i \times C_i$, where $i \in \{1, 2, 3, 4, 5\}$, and $C_1 = 16$, $C_{i+1} = 2 \times C_i$.

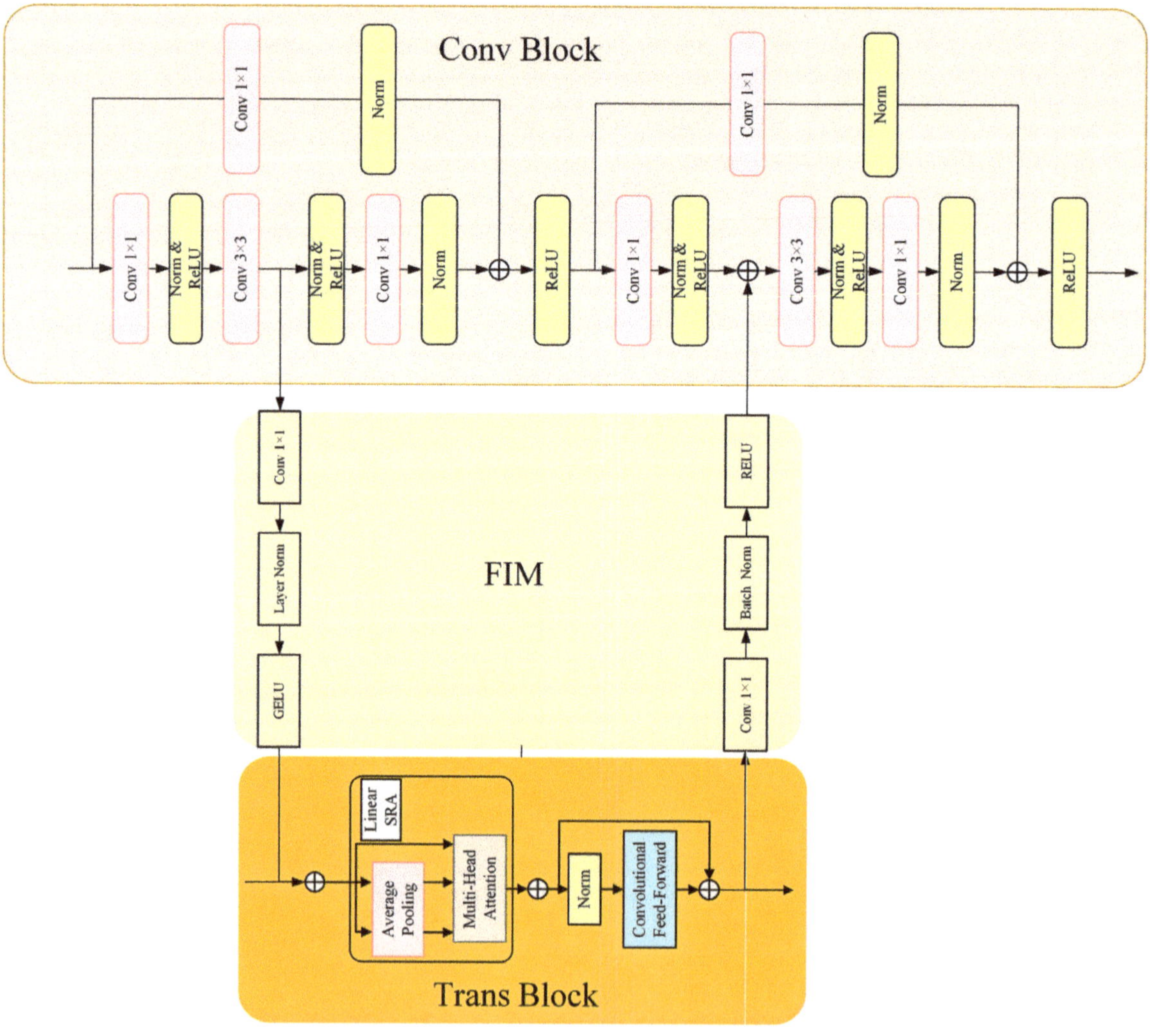

Figure 2. The structure of the Conv Block, FIM, and Tans Block.

2.3. Transformer Branch

The transformer branch is devised to capture the global contextual information from remote sensing images. PVT v2 [46], as the latest transformer backbone network, is designed with overlapping patch embedding to encode the images, removes the fixed-size positional coding in the feed-forward network, introduces zero-filling positional coding, and replaces the spatialreduction attention (SRA) [37]. With these three improvements, PVTv2 can not only ensure the local continuity of image and feature maps, but can also flexibly handle different scales of input signals, and control the computational complexity within the linear range. Therefore, in this paper, PVT v2 is adopted as the encoder of the Transformer branch for feature extraction, and its encoder module is shown in Figure 2. Similarly to the CNN branch, the Transformer branch also employs the feature pyramid structure to divide the whole branch into five layers. In the first level, the input image is initially partitioned into overlapping patches of 7 × 7 dimensions. Subsequently, these patches are fed into the Transformer encoding module, to acquire the first-stage feature maps, which are transmitted to the next stages. The subsequent four-stage feature maps are cut into overlapping 3 × 3-sized patches and, finally, five feature maps with different scales and resolutions are obtained, which are consistent with the size and number of channels of the CNN branch, facilitating interaction between the feature layers of both branches. To mitigate the high computational burden associated with the self-attention mechanism in Transformer encoders, PVT V2 proposes the linear spatial reduction attention (LSRA) as a substitute for the traditional multihead attention (MHA) in Transformer encoders [35]. Similar to the MHA, the LSRA accepts the query Q, key K, and value V as input, and produces refined features as the output. The distinguishing feature of the LSRA is that it reduces the spatial scale of K and V before executing the attention operation, resulting in a significant reduction in the computational and memory overheads. This is described in Equation (1):

$$\mathrm{LSRA}(Q,K,V) = \mathrm{Concat}(\mathrm{head}_0,\ldots,\mathrm{head}_{N_i})W^O \tag{1}$$

$$\mathrm{head}_j = \mathrm{Attention}\left(QW_j^Q, \mathrm{LSR}(K)W_j^K, \mathrm{LSR}(V)W_j^V\right) \tag{2}$$

where Concat (·) represents the channel splicing operation, $W_j^Q \in \mathbb{R}^{C_i \times d_{head}}$, $W_j^K \in \mathbb{R}^{C_i \times d_{head}}$, and $W^O \in \mathbb{R}^{C_i \times C_i}$ are linear projection parameters. In addition, head_i is the attention value of the ith head in Stage_i. LSR(·)represents the operation of reducing the spatial dimensions of K and V, which is written as:

$$\mathrm{LSR}(x) = \mathrm{GELU}(\mathrm{Norm}\left(\mathrm{Reshape}(f(\mathrm{AvgPool}(x,p))W^S\right)). \tag{3}$$

$$\mathrm{Attention}(Q,K,V) = \mathrm{Softmax}\left(\frac{QK^T}{\sqrt{d_{head}}}\right) \tag{4}$$

In contrast to traditional multi-attention operations, the LSRA utilizes average pooling to decrease the spatial dimensions (h × w) to a constant size (p × p). As a result, the LSRA significantly reduces the computational cost, and decreases the model memory footprint. To be specific, when provided with an input of size h × w × c, the computational complexity of the LSRA and the MHA can be expressed as follows:

$$O(\mathrm{LSRA}) = hwp^2c \tag{5}$$

$$O(\mathrm{MHA}) = h^2w^2c \tag{6}$$

Here, p corresponds to the feature map size subsequent to pooling, which is fixed at 7.

2.4. Feature Interaction Module

We considered the problem of the feature differences between the feature maps in the CNN branch and the patch-embedding features in the Transformer branch, as well as aiming to better combine and utilize the global features extracted by the Transformer, and the local features captured by the CNN. Inspired by Conformer [47], starting from stage 2, we embedded a feature interaction module in the middle of each bottleneck block and Transformer encoding block, to realize the feature interaction of the dual branches, as shown in Figure 2. Firstly, the features from the CNN branch undergo a 1×1 convolution, to align with the number of channels in the Transformer branch, while the features are regularised using LayerNorm [48] and, finally, the features from the two branches are summed. In this way, local features extracted from the CNN branch are gradually incorporated into the Transformer block, complementing the local semantic information of the Transformer branch. Similarly, when the features from the Transformer branch are fed back to the CNN branch, the feature maps need to be aligned with the CNN feature maps, in terms of the channel dimensions by 1×1 convolution and, at the same time, the features are regularized using BatchNorm, and the features of the two branches are finally summed, and such a process achieves the advantages of the two-branch feature maps, in such a way that they complement each other.

2.5. Cross-Scale Multi-Level Feature Fusion Module

After five stages of the backbone network, the model extracts multi-layer features with global contextual information. Similar to FPN-like networks, low-level features contain coarse-grained information with a relatively high resolution; high-level features contain fine-grained information, but with a relatively low resolution. While in the decoding stage, traditional UNet models often employ the simple upsampling of high-level features, to match the spatial scale of low-level features, followed by concatenation. However, the simple upsampling only makes the feature size of the high and low layers consistent; it cannot eliminate the corresponding error between the high- and low-layer feature pixels. Consequently, this approach falls short in resolving spatial misalignment between features, resulting in substantial information loss, and adversely affecting the overall performance of the model [49]. In addition, this operation easily generates semantic gaps, which lead to the occurrence of situations such as the omission of small water bodies, and the misclassification of shadow targets. To solve the above problems, we design a cross-scale, multi-source feature fusion module, to replace the decoder block in UNet.

As shown in Figure 3, for two feature maps with different scales and channel numbers as inputs to the module, we assume that the high-level input features are X_h, and the low-level input features are X_l, whose sizes are $2C \times H \times W$ and $C \times 2H \times 2W$, respectively, where C represents the number of channels of the feature map, and H and W are the height and width. To ensure that the high-level features include the same channels as the low-level features, a 1×1 convolution operation is initially applied to X_h. Then, inspired by Li et at. [49] and Huang et al. [50], we put the high-level and low-level features into a designed feature calibration module, so that we could obtain the spatially dimensional aligned high- and low-level features X_{h2} and X_{l1} (both of the sizes $C \times 2H \times 2W$). Subsequently, the high- and low-level features are summed, to obtain the fusion feature X_f. For the fusion feature, we perform the attention mechanism along the spatial and channel dimensions, respectively, to obtain the spatial weight M_s and the channel weight M_c, and then we sum the outputs of the two to obtain the output X_{f1}, and then we obtain the weights via the sigmoid activation function. The output variables X_{h3} and X_{l2} are generated via multiplying the weight coefficients s and $(1 - s)$ with X_{h2} and X_{l1}, respectively, which are then summed to obtain the final feature map X_{out}. The above process can be expressed as a series of equations:

$$X_{l1}, X_{h2} = \mathrm{FAM}\left(X_l, f^{1\times1}(X_h)\right) \tag{7}$$

$$X_f = X_{l1} + X_{h2} \tag{8}$$

$$X_{f1} = \mathrm{CAM}(X_f) + \mathrm{SAM}(X_f) \tag{9}$$

$$s = \mathrm{sigmoid}(X_{f1}) \tag{10}$$

$$X_{out} = X_{l1} \cdot (1 - s) + X_{h2} \cdot s \tag{11}$$

where $f^{1\times1}()$ denotes the 1×1 convolution layer, while the abbreviations FCM, CAM, and SAM, respectively, denote the feature calibration module, channel attention module, and spatial attention module. For further details regarding these modules, please refer to Sections 2.5.1–2.5.3 of this paper.

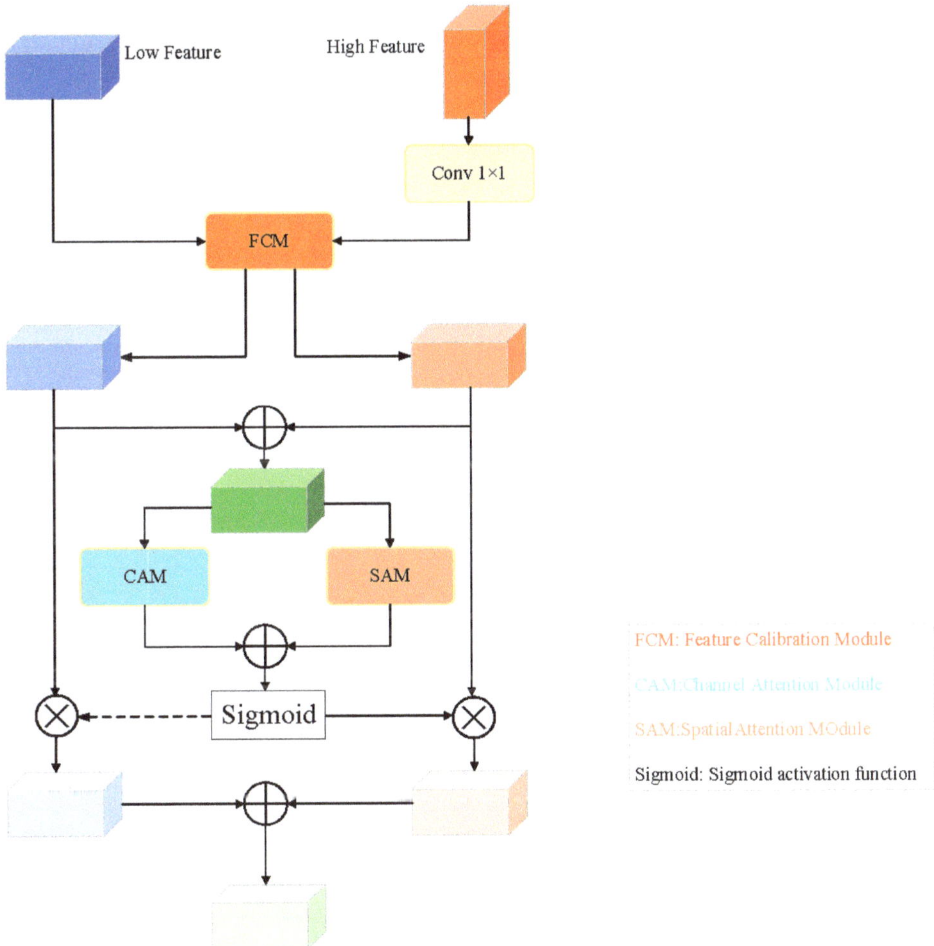

Figure 3. The overall structure of the cross-scale, multi-level feature fusion module.

2.5.1. Feature Calibration Module

In semantic segmentation tasks, low-level features contain abundant spatial information, but are limited in terms of semantic information, while high-level features exhibit the opposite characteristics, being abundant in semantic information, but lacking in contextual and spatial details. In the decoder stage, the challenge lies in how to effectively fuse multi-scale hierarchical semantic features, to obtain rich spatial and semantic information for pixel classification. Previous works have explored this issue [24,26,51,52]. However, many of these works often overlook a crucial problem, which is the feature misalignment issue across different scales.

The problem of feature misalignment refers to the misalignment or mismatch between features caused by differences in the receptive field sizes and resolutions at different scales. This may lead to issues such as blurry boundaries and misclassification of objects in the segmentation results. The main cause of feature misalignment across multiple scales lies in the up-sampling and down-sampling operations used in the models. During scale transformations, the up-sampling and down-sampling operations employed may introduce misalignment in feature maps. For example, the interpolation method used during up-sampling may introduce positional offsets, while down-sampling may result in a loss of information and blurring effects.

To address the issue of the semantic and spatial misalignment of features on different scales, this study proposes a feature calibration module (illustrated in Figure 4). Specifically, the high-level and low-level features are first passed through individual 1 × 1 convolutional layers to adjust their dimensions, followed by up-sampling of the high-level features to align with the low-level features. Subsequently, the concatenated feature maps are processed by a 3 × 3 convolutional layer, to reduce the number of channels to four, which represent the offset maps of the high-level and low-level features in the x and y directions, as shown in Equation (12).

$$\Delta_l, \Delta_h = f^{3\times3}\left(cat\left(f^{1\times1}(F_l), Up(f^{1\times1}(F_h))\right)\right) \quad (12)$$

where $cat(\cdot)$ represents the concatenation operation, and $f^{3\times3}(\cdot)$ is the 3 × 3 convolutional layer, $f^{1\times1}(\cdot)$ is the 1 × 1 convolutional layer, $Up(\cdot)$ denotes the up-sampling operation, and Δ_l, Δ_h represent the offset map (size H × W × 2) of the low- and high-level features.

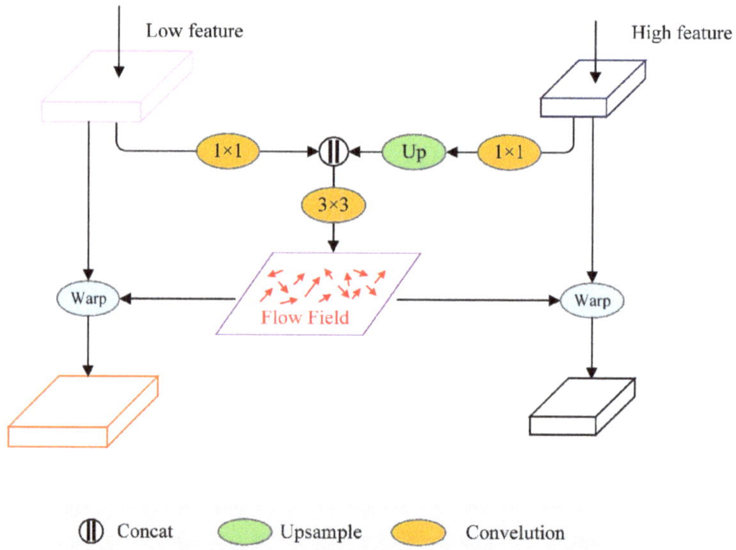

Figure 4. The structure of the Feature Calibration Module.

After obtaining the offset map between the high- and low-level feature maps, we then perform a warp operation (as shown in Figure 5) on the semantic flow field of the two features, which is described in Equation (13):

$$Warp^c_{hw} = \sum_{h'=1}^{H}\sum_{w'=1}^{W} F^c_{h'w'} \cdot \max(0, 1 - |h + \Delta_y - h'|) \cdot \max(0, 1 - |w + \Delta_x - w'|) \quad (13)$$

where $F^c_{h,w,}$ is the value of the position of the original feature at the spatial level (w', h', c), and h, w are the height and width of the output feature map (e.g., for high-level features, $h = 2 \times h'$, and for low-level features, $h = h'$). Δ_y, Δ_x are the offset of the offset map obtained from the feature map in Equation (12) on the y, x axes, i.e., on the height and width.

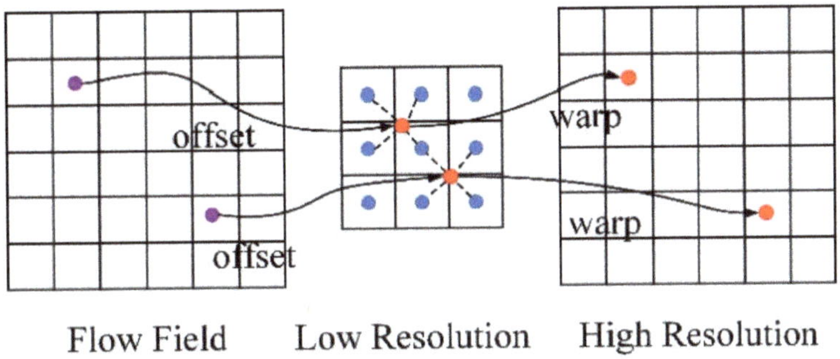

Figure 5. The warp procedure of the feature calibration module.

Finally, two feature maps are obtained after calibration, with consistent height and width dimensions for both (i.e., generating the feature maps with a size of H × W × C for both).

2.5.2. Channel Attention Module

Inspired by the human visual system, attention mechanisms [35] have been introduced into neural networks, to learn more relevant features. In neural networks, attention mechanisms calculate weights for each feature map in a layer, allowing the model to capture critical information more effectively. Building on the work of Liu et al., a channel attention sub-module was designed (as shown in Figure 6) to model the interdependencies between channels in the fused features. To determine the variance between channels, and infer their relative importance, a scaling factor γ was introduced into the calculation of batch normalization (BN) [53], as shown in Equation (14).

$$B_{out} = BN(B_{in}) = \gamma \frac{B_{in} - \mu_B}{\sqrt{\sigma_B^2 + \epsilon}} + \beta \quad (14)$$

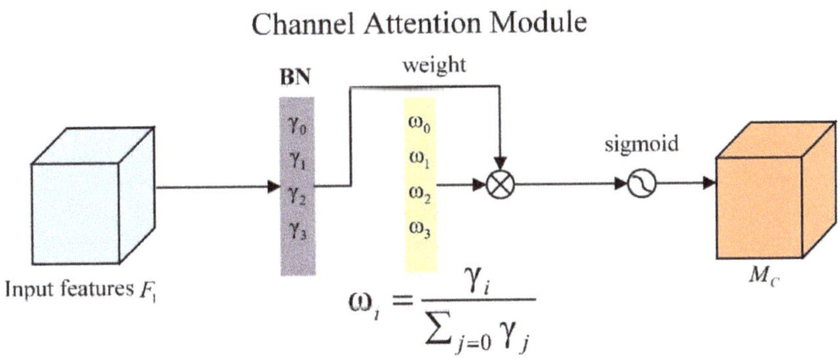

Figure 6. The structure of the channel attention module.

In Equation (14), μ_B and σ_B, respectively, represent the mean and standard deviation of batch B, while β denotes the bias term. The channel attention module weights, denoted as Mc, can be obtained by reversing Equation (14) using Equation (15), where γ is the scaling factor for each channel, and $W\gamma = \frac{\gamma_i}{\sum_{j=0} \gamma_j}$ represents the proportion of the scaling factor for each channel among all the channels. A higher value indicates that the corresponding channel requires more attention, while a lower value suggests that the model should assign less attention to that channel.

$$M_c = sigmoid(W_\gamma(BN(F_1))) \quad (15)$$

2.5.3. Spatial Attention Module

For the spatial attention module, as shown in Figure 7, we directly pass the feature map through three convolutions, followed by BN and ReLU after each convolution, and the first and last of the three convolutions are 1×1 convolutions for channel transformation, similar to the structure in the bottleneck. In the middle is a 3×3 dilation convolution, which is used to enlarge the receptive field without increasing the computational overhead The introduction of dilated convolution and the ability to obtain more context information are of great help in providing spatial modeling. Finally, the spatial weight Ms is obtained through the sigmoid function, as shown in Formula (16).

$$\mathrm{Ms}(\mathrm{F}) = \mathrm{sigmoid}\left(\mathrm{f}_2^{1\times 1}\left(\mathrm{f}_1^{3\times 3}\left(\mathrm{f}_0^{1\times 1}(\mathrm{F})\right)\right)\right) \quad (16)$$

where $\mathrm{f}^{3\times 3}(\cdot)$ denotes a 3×3 two-dimensional dilated convolution, and $\mathrm{f}^{1\times 1}(\cdot)$ denotes a 1×1 two-dimensional convolution.

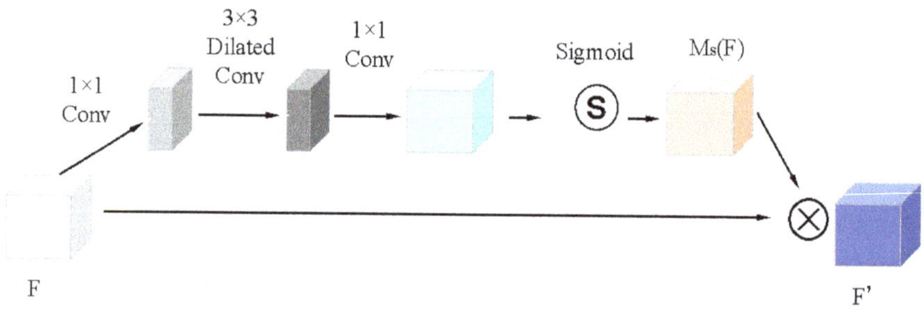

Figure 7. The structure of the spatial attention module.

2.6. Loss Function

Cross-entropy (CE) loss and Dice loss are commonly utilized as the predominant loss function in the semantic segmentation of remote sensing images. However, these loss functions and their variants are based on region similarity, and may lead to a poor performance when dealing with imbalanced classes, as well as small objects and edge details in images. In the task of land–water segmentation, there are often many small segmentation targets in the image, such as lakes, ships, islands, buildings, and clouds. Moreover, the water–land boundary in the image is often jagged and difficult to distinguish. The use of only the Dice loss or CE loss is insufficient to address these issues.

Therefore, in this study, we incorporated the boundary loss function [54] to address the problem of edge detail handling and small object recognition in water–land semantic segmentation. The formula for computing the boundary loss is as follows:

$$\mathrm{L_B} = \frac{1}{\mathrm{N}}\sum_{\mathrm{i}=1}^{\mathrm{N}} \mathrm{d}(\mathcal{B}(\mathrm{y_i}), \mathcal{B}(\hat{\mathrm{y}}_\mathrm{i})) \quad (17)$$

where $\mathrm{L_B}$ denotes the boundary loss function,$\mathrm{y_i}$ is the ground truth label of pixel i, and $\hat{\mathrm{y}}_\mathrm{i}$ is the predicted label of pixel i by the model. The distance function d measures the dissimilarity between two boundaries, with N representing the whole number of pixels. To address the challenges of small object recognition and edge detail handling in land–water semantic segmentation, we propose a hybrid loss function L that integrates the boundary loss function with the CE loss. Specifically, the proposed loss function L is defined as follows:

$$\mathrm{L} = \mathrm{p}\cdot\mathrm{L_{ce}} + (1-\mathrm{p})\cdot\mathrm{L_B} \quad (18)$$

$\mathrm{L_{ce}}$ represents the CE loss function, and p is a weighting coefficient. Through experiments, we set p to 0.8 in this study.

3. Results

3.1. Study Area and Dataset

For this study, the Chinese coastline on the Yellow Sea was chosen as the designated study area. The image data utilized in this study were acquired from the China Center for Resources Satellite Data and Application (CCRSDA; http://www.cresda.cn, accessed on 15 April 2023). Specifically, we acquired three remote sensing images from Gaofen-6 (GF-6), captured in February 2023, with a spatial resolution of 16 m, and eight spectral bands. All the GF-6 images utilized in this study were of the Class 1A product type, characterized by a high quality and an absence of cloud cover, and provided complete coverage of the entire Yellow Sea area (see Figure 8 for further details). A detailed summary of the GF-6 images is presented in Table 1. Subsequently, we preprocessed the original image with radiometric calibration and atmospheric correction and, ultimately, generated remote sensing images suitable for further research purposes.

Figure 8. The geographic location of the research area, on the Yellow Sea.

Table 1. GF6/WFV data.

Project	GF6/WFV Data
Wavelength range/um	B1(Blue): 0.45~0.52
	B2(Green): 0.52~0.59
	B3(Red): 0.63~0.69
	B4(NIR): 0.76~0.90
	B5(SWIR1): 0.69~0.73
	B6(SWIR2): 0.73~0.77
	B7(Purple): 0.40~0.45
	B8(Yellow): 0.59~0.63
Spatial resolution/m	16
Width/km	864.2

In the image of the study area, the part of the sea–land boundary to be used to construct the GF dataset was selected, and sea–land segmentation was carried out. Initially, due to the excessive width of the original GF6 WFV images, and the presence of overlapping regions between the three images, we cropped these three images, while preserving the Yellow Coast as fully as possible, to reduce the difficulty of the task. Please refer to Figure 8 for specific details. Then, the clipped image was divided into two categories: ocean and land. To improve the efficiency of the training, we selected only those cropped images that contained both ocean and land, and obtained 2100 images and 2100 labels, all of which were 512 pixels $\times$ 512 pixels in size.

In cases where the network model requires an insufficient number of training samples, data augmentation becomes a crucial step in enhancing the network's invariance and robustness. In order to increase the data volume of the experimental dataset, five data expansion methods, such as horizontal flip, vertical flip, diagonal mirror, local cropping and magnification, and image sharpening, are used to increase the image quantity of the dataset. Finally, the GF dataset we constructed contained 12,600 images, which were subsequently partitioned into training, validation, and test sets, in a random 7:2:1 ratio.

3.2. Experimental Details and Evaluation Metrics

All experiments were performed on a workstation running Windows 10 with an NVIDIA GeForce RTX 3090 graphics card, and using the deep learning framework Pytorch (2017). All models were trained with an initial learning rate of 0.001, and AdamW [48], with a momentum term of 0.9 and a weight decay of 0.01, was selected as the optimizer to optimize the network model. Additionally, to speed up the training, we set the batch size to 16, and the epoch number to 100. The poly method is used to dynamically adjust the learning rate. The formula is expressed as follows:

$$l_i = l_{base} \times \left(1 - \frac{epoch_{\ i}}{epoch_{\ max}}\right)^{0.9} \quad (19)$$

where l_i is the current learning rate, l_{base} is the base learning rate set to 0.001, $epoch_i$ is the current number of iterations, and $epoch_{max}$ is the maximum epoch set to 100.

In this paper, three metrics normally utilized in semantic segmentation are used to verify the effectiveness of the model, namely the pixel accuracy (PA), mean intersection over union (MIoU), and F1-score. Based on the associated confusion matrix, the PA, MIoU, and F1 are calculated as

$$PA = \frac{\sum_{k=1}^{K} TP_k}{\sum_{k=1}^{K} (TP_k + FP_k + TN_k + FN_k)} \quad (20)$$

$$\text{MIoU} = \frac{1}{\text{K}} \frac{\sum_{\text{K}=1}^{\text{K}} \text{TP}_\text{k}}{(\text{TP}_\text{k} + \text{FP}_\text{k} + \text{FN}_\text{k})} \tag{21}$$

$$\text{F1} = 2 \times \frac{\text{precision}_\text{k} \times \text{recall}_\text{k}}{\text{precision}_\text{k} + \text{recall}_\text{k}} \tag{22}$$

where TP_k, FP_k, TN_k, and FN_k represent the true positive, false positive, true negative, and false negative values for the kth class, respectively. In addition, $\text{precision}_\text{k} = \frac{\text{TP}_\text{k}}{\text{TP}_\text{k}+\text{FP}_\text{k}}$ and $\text{recall}_\text{k} = \frac{\text{TP}_\text{k}}{\text{TP}_\text{k}+\text{FN}_\text{k}}$ are the precision and recall rates for the k classes, respectively.

3.3. *Performance Comparison of Different Band Combinations*

The GF6 image is a typical multispectral remote sensing image. Compared with the traditional RGB image, multispectral images contain a significantly greater amount of information, due to their higher number of bands (the GF6 has eight bands). This paper, firstly, discusses the effectiveness of different band combinations in sea–land segmentation. According to Yu et al. [55] and Mou et al. [56], we selected ten common three-band and all-band combinations, and compared their performance differences on sea–land semantic segmentation. Details of the comparison experiment of band combination are presented in Table 2.

Table 2. Comparison of results of different band combinations on the GF dataset.

Band Combination	PA (%)	MIoU (%)	F1 (%)
B1 + B2 + B3	96.52	91.12	95.30
B1 + B4 + B5	96.81	92.23	95.36
B2 + B3 + B4	96.95	92.64	95.58
B2 + B3 + B5	96.21	91.81	94.88
B3 + B4 + B5	**96.99**	**92.78**	**95.67**
B3 + B4 + B8	96.64	92.07	95.27
B3 + B5 + B7	95.89	89.56	94.32
B4 + B5 + B6	96.69	92.19	95.33
B4 + B6 + B7	96.31	91.28	94.82
B5 + B6 + B7	96.88	92.36	95.43
All-bands	**97.52**	**93.53**	**96.63**

As seen in Table 2, bands 3, 4, and 5 (Red, Nir, and Swir-1) outperformed the other nine bands in the sea–land segmentation task. However, the effect of the all-band combination is better than that of all the three-band combinations. This shows that the eight different bands can contain more spatial and spectral information, and that the complementary information is more advantageous in the task of sea–land semantic segmentation.

3.4. *Ablation Study*

3.4.1. Performance of Feature Interaction Module

To assess the performance of the FIM, this article conducted ablation research to validate the effectiveness of the module design. We divided the experiment into three scenarios: (1) only using CNN branches as encoders; (2) using only transformer branches as encoders; (3) the method proposed in this article, to use dual branches and, simultaneously, use FIM as encoders. For the decoder part, we uniformly used the designed CMFFM. The outcomes of the experiment are presented in Table 3.

Table 3. Results of the module ablation experiments. The best results are in bold.

Method	Encoder	PA (%)	MIoU (%)
TCU-Net	CNN	96.02	92.01
	Transformer	95.89	91.95
	CNN + Transformer + FIM	**97.52**	**93.53**

Table 3 reveals that the segmentation accuracy using only the Transformer branch as the encoder is the worst, with an MIoU of 91.95% and a PA of 95.89%, while the segmentation accuracy using only the CNN branch is not high, with an MIoU of 92.01% and a PA of 96.02. There is a certain gap between the segmentation accuracy of the two branches and the FIM as the encoder. This shows that simply using a CNN or Transformer branch as the encoder has certain defects in feature extraction, and cannot integrate image spatial information, semantic information, and global context information well. After the introduction of the FIM as the information exchange bridge between the two branches, the missing local and global information perception ability between the two branches is enhanced, the information exchange and complementary function are perfectly realized, and the feature extraction ability of the whole model is greatly enhanced.

3.4.2. Performance of Cross-Scale, Multi-Level Feature Fusion Module

In order to evaluate the performance of the cross-scale, multi-level feature fusion module, we verified the effect of the module for small targets and edge extraction in images. In this paper, we use TCU-Net as a baseline to perform ablation experiments on the Yellow Sea sea–land semantic segmentation dataset.

The results of the experiment are shown in Table 4. Through comparing the feature fusion strategies, we find that the simple up-sampling and jump join, as in the original UNet, can not fully fuse semantic features of different scales and levels. Through using the proposed CSMFF module, the PA, MIoU, and F1-score are improved by 0.51%, 0.31%, and 0.36%, respectively, on the test set. Simultaneously, the number of parameters of the CSMFF module designed in this paper is reduced by 0.68 M, compared with the jump connection of the original UNet, which further upgrades the efficiency of the model in processing images.

Table 4. Results of the module ablation experiments.

Method	Decoder	PA (%)	MIoU (%)	F1 (%)	Params (M)
TCU-Net	UNet	96.91	93.01	96.02	2.4 M
	CSMFF	97.52	93.53	96.63	1.72 M

The visualization results from the experiment are shown in Figure 9. In order to show the difference in the prediction results between the two decoders more directly, blue boxes are used to highlight the positions where the model shows differences in the prediction image. It is evident that the TCUNet using the decoder of the original UNet performs poorly on the sea–land boundary and the small water body when predicting the picture, because the shallow feature and the deep feature are spliced only using up-sampling and the jump connection, and semantic gaps are easily generated, resulting in the situation whereby the small water body is missed, and the shadow target is misclassified. Using the CSMFF module designed in this paper as a decoder can effectively improve the detection of small objects and the definition of boundaries in the land–sea segmentation task, so that the classification results of the model are more accurate.

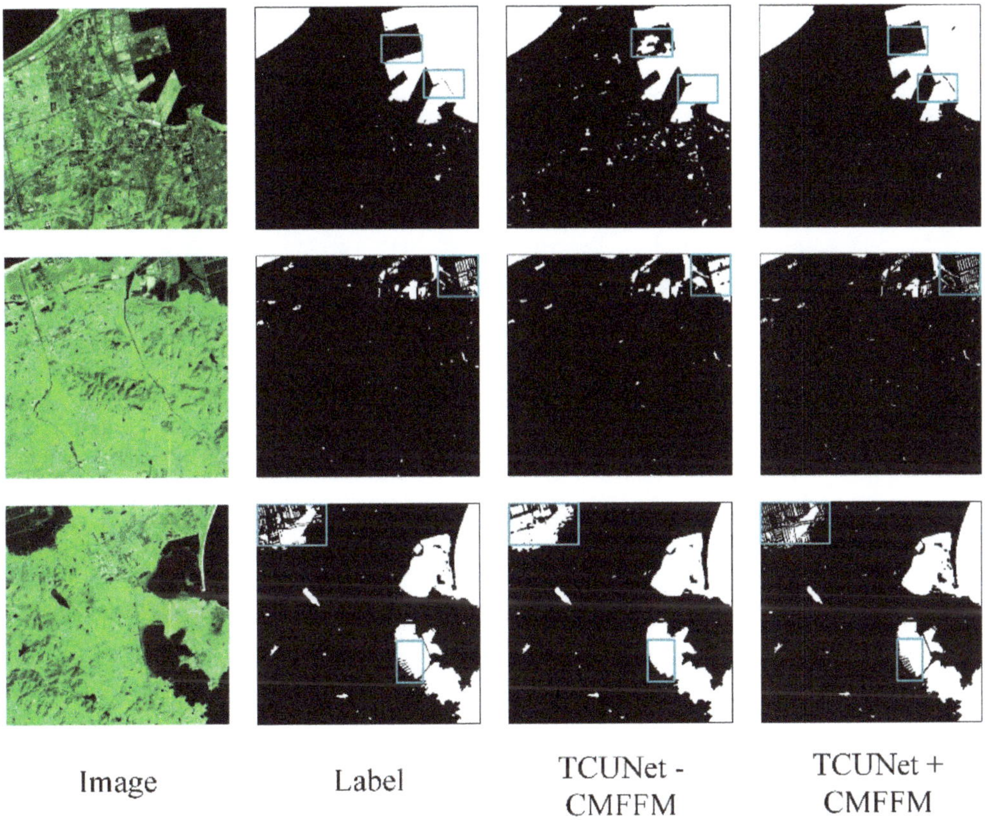

Figure 9. Visualization of CMFFM ablation on the GF dataset. "−" indicates that CMFFM was not used, and "+" indicates that CMFFM was used. The blue boxes highlight where the model differs on the predicted image.

3.5. Contrast Experiment

To more accurately evaluate the performance of the model proposed in this paper, we compare our model with some excellent models commonly found in the field of semantic segmentation, including UNet, Deeplabv3+, DANet [51], Segformer, SwinUNet [57], TransUNet, ST-UNet, and UNetformer [58]. The first three methods are CNN networks, Segformer and SwinUNet are pure vision sensor methods, and TransUNet, ST-UNet, and UNetformer are hybrid models that combine CNNs with sensors. The TransUNet encoder adopts the serial form of standard ViT and ResNet, and the decoder is the same as UNet; ST-UNet improves the encoder part on the basis of TransUNet, using a dual-encoder structure with a Swin transformer and CNN in parallel, while UNetformer uses ResNet18 as the encoder, and develops an efficient global–local attention mechanism to construct transformer blocks in the decoder, as the decoder. In addition, the backbone of Deeplabv3+ and DANet is ResNet50, that of Segformer is MiT-B1, and the backbones of other models are set by the original authors. In addition, according to the experiment in 3.3, for UNetformer, which only accepts three-band image input (its backbone is the officially packaged ResNet18), we chose the band combination of bands 2, 3, and 5 as its input data, while the other models used the full-band combination (8 bands) as their input data. To ensure the fairness of the experiment, no models were pre-trained. The experiments were carried out under the same conditions, and the specific implementation details are shown in Section 3.2. of this paper.

The quantitative analysis results of the GF dataset are shown in Table 5, and the best results of each evaluation index are highlighted in bold.

Table 5. Comparison of all the methods in metrics. Training Time represents the time spent by the model in processing the training and validation sets during training; Inference Time represents the time spent processing the test set when the model makes predictions.

Method	Backbone	PA (%)	MIoU (%)	F1 (%)	Params (M)	FLOPs (GMac)	Training Time (s)	Inference Time (s)
UNet [24]	-	96.95	92.15	95.96	31.04	218.9	695	86.28
Deeplabv3+ [28]	ResNet50	96.87	91.98	95.77	40.36	70.22	385	77.28
DANet [51]	ResNet50	96.68	91.52	95.52	49.61	205.37	680	85.44
Segformer [40]	MiT-B1	97.16	92.71	96.18	13.69	13.49	375	78.48
SwinUNet [57]	Swin-Tiny	96.88	91.95	95.92	27.18	26.56	505	84.36
TransUNet [41]	ViT-R50	97.07	92.41	96.03	100.44	25.5	810	106.26
ST-UNet [44]	-	97.23	92.99	96.34	160.97	95.41	915	135.54
UNetformer [58]	ResNet18	97.15	92.67	96.15	11.72	11.73	**235**	**73.44**
TCUNet	-	**97.52**	**93.53**	**96.63**	**1.72**	**3.24**	445	87.78

The results show that the TCUNet proposed in this paper is superior to the other eight models in its PA, MioU, and F1-score. Overall, the combination of CNN and Transformer worked slightly better than the visual Transformer method, and the CNN-based method showed the worst classification accuracy, but there was no significant difference between the nine methods. This shows that the CNN-based model has some limitations in describing global dependencies. In the CNN method, the effect of UNet is the best, that of Deeplabv3 + is the second best, and the effect of DANet is the worst. This may be due to the fact that UNet adopts a feature pyramid-like structure in the decoder, which fuses the five layers of semantic features extracted by the backbone network through jumping links; it can be applied to the sea–land segmentation of high-level semantic information and detail information. However, Deeplabv3 + uses hole convolution and an ASPP module to integrate multi-layer semantic features, which is too simple, and not good for the fine segmentation of object edges and details. DANet's encoder only uses the high-level semantic features of the backbone network for classification, meaning that it can not make full use of the shallow semantic features, and the classification effect is the worst. Among the Transformer models, TCUNet is the best, ST-UNet is the second, Segformer and UNetformer have the same classification accuracy, slightly better than TransUNet, and SwinUNet is the worst.

Figure 10 shows the segmentation results for all the methods in the six test images. Looking at Figure 10, we can see that TCUNet performed better on segmentation than the other eight models, especially in the blue-rectangular-box-labeled area. As can be seen from these test charts, the proposed method shows the best segmentation effect compared with the other models. Faced with complex types of shorelines (farmed ponds, ports, small rivers), our networks can still clearly delineate boundaries. At the same time, in the edge details and small target recognition, compared with other methods, our network segmentation is better. This shows that our network model can solve the problems of missed detection and misclassification in low contrast areas and small water bodies with a complex background, and effectively improve the problems of pixel classification, small target extraction, and boundary blur, meaning that the effect of classification is more accurate.

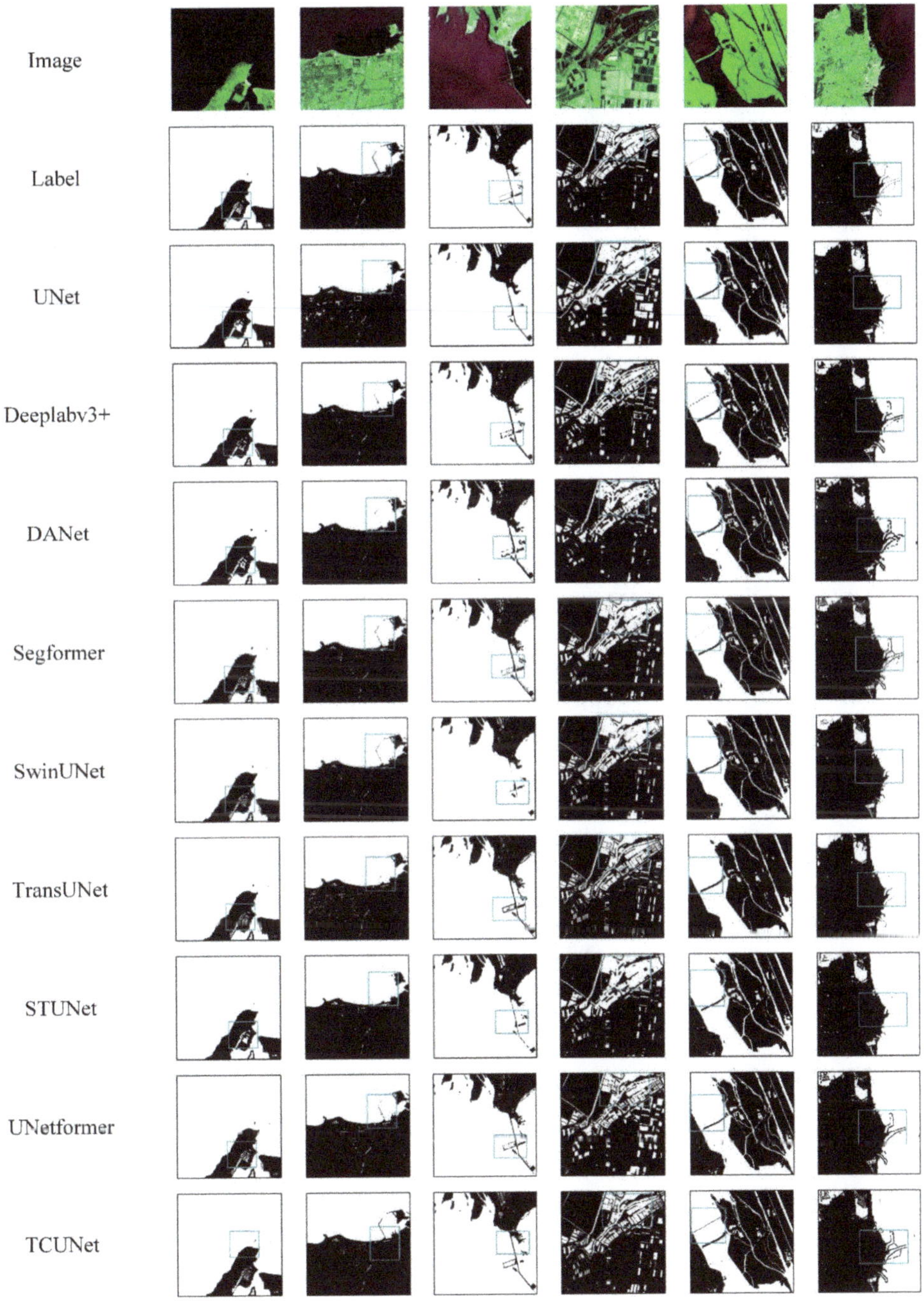

Figure 10. Comparison of different models in the GF dataset.

In order to evaluate the segmentation efficiency of all the models, we also list the number of parameters, the computational complexity, and the time spent on training and reasoning of each model in Table 5, where "M" represents one million parameters, and "GMac" stands for the billion times a model performs a floating-point multiplication and addition operation in the course of a single forward propagation. The model proposed in this paper is only 1.72 M and 3.24 GMac, far lower than the other eight models. This is because the two-branch structure used in this paper greatly enhances the ability of the model to extract semantic features, so we set the number of channels in the first stage of the model to 16 (that is, greatly reducing the width of the network) without affecting the performance of the network. In terms of the training and prediction efficiency, the training time of TCUNet is 89 s per epoch, and the inference time is 14.63 s, which ranks medium among all models. The lightweight model UNetformer runs much faster than other networks. This may be because the hybrid structure of the CNN and transformer in TCUNet slows down the running efficiency of the model, and there are many LN and GELU [59] functions in FIM, which are far less optimized via the graphics card than the convolution and ReLU operations of a traditional CNN. This will cause TCUNet to be slower when processing images. Although the above two issues may limit the application of TCUNet in some scenarios (such as on small mobile devices), TCUNet is still valuable in exploring the role of the transformer and CNN combination in sea–land semantic segmentation in remote sensing images.

4. Discussion

4.1. Comparison of Model Effects on Different Satellite Sensor Images

Various satellite sensors can collect different remote sensing images in the same geographical area. In order to verify the adaptability of our model to different satellite images at different time periods, we selected a Landsat 8/OLI remote sensing image of the Yellow Sea region in October 2019, to verify the portability of the model. The OLI sensor has a total of 9 bands, with bands 1–7 and 9 having a spatial resolution of 30 m, and band 8 having a panchromatic resolution of 15 m. The detailed information is listed in Table 6. In order to ensure that the number of bands in the data is consistent with GF6/WFV, this article will perform image fusion on the first seven bands and panchromatic bands after some preprocessing steps, such as radiation calibration and FLAASH atmospheric correction. Finally, an 8-band image, with a resolution of 15 m, was obtained. As with the GF6 image, we selected part of the sea–land boundary, to construct a Landsat dataset for validation experiments. After cropping and labeling, 112 images and labels were obtained; all images had a size of 512 pixels × 512 pixels. Subsequently, we used five data expansion methods, including horizontal flipping, vertical flipping, diagonal mirroring, local cropping, and zooming in, and image sharpening, to increase the number of images in the dataset, resulting in 672 images and labels.

Table 6. Landsat8/OLI data.

Project	Landsat 8/OLI
Wavelength range/um	B1(Coastal aerosol): 0.43~0.55
	B2(Blue): 0.45–0.51
	B3(Green): 0.53–0.59
	B4(Red):0.64–0.67
	B5(NIR): 0.85–0.88
	B6(SWIR1): 1.57–1.65
	B7(SWIR2): 2.11–2.29
	B8(PAN): 0.50–0.68
Spatial resolution/m	15
Width/km	185

Without training and parameter adjustments, we directly predicted the 672 images using the model weights trained in 3.5, exploring the land and sea segmentation effects of each model on remote sensing images of the same area from different satellites and at different time points. The experimental results are shown in Table 7. The indicators of TCUNet in the PA, F1-Score, and MIoU are 95.46%, 95.19%, and 90.84%, respectively, which are much higher than those of the other nine models.

Table 7. The results of all methods on the Landsat datasets.

Method	PA (%)	MIoU (%)	F1 (%)
UNet	64.63	41.55	61.25
Deeplabv3+	91.75	83.82	91.13
DANet	88.23	76.84	86.72
Segformer	80.88	67.83	80.63
SwinUNet	81.04	68.03	80.96
TransUNet	75.10	60.60	74.92
ST-UNet	84.82	73.41	84.65
UNetformer	90.17	80.20	88.89
TCUNet	95.46	90.84	95.19

In addition, for the segmentation results of Landsat images, as shown in Figure 11, in order to visually verify the segmentation effect of each method, this article uses blue boxes to highlight the positions where the model shows differences in the predicted image. It can be clearly seen that without pre-training, Deeplabv3+, UNeformer, and DANet perform well. However, it can be seen from the graph that these models cannot perform the precise segmentation of water and land, and there is a phenomenon of misclassification and missing segmentation for small targets, such as ships, islands, and ponds. However, the segmentation results of Segformer, TransUNet, and SwinUNet are not satisfactory, and cannot accurately complete land and sea segmentation. They can only roughly distinguish between water and land. U-Net cannot perform land and sea segmentation in Landsat 8 images. This indicates that U-Net struggles to extract water bodies from different remote sensing images across sensors, despite its excellent performance in medical images. The TCUNet method proposed in this article can extract the coastline at different times, across sensors. The accuracy of the extraction results meets the extraction requirements.

4.2. Performance under Different Parameter Settings

In this paper, we continue to explore the sea land segmentation task, by setting different parameters, to test the segmentation performance of the model. The experimental results are shown in Table 8.

Table 8. Performance under different parameter settings. E represents the dimension of the first stage of the Transformer branch. C represents the number of channels in the first layer of the CNN branch, and D represents the number of Conv blocks and Transformer blocks in stages 2–5.

E	C	D	PA (%)	Params
16	16	[2,2,2,2]	96.92	1.04 M
		[3,4,6,3]	97.52	1.72 M
46	32	[2,2,2,2]	97.06	7.07 M
		[3,4,6,3]	97.46	8.50 M
92	64	[2,2,2,2]	97.42	20.43 M
		[3,4,6,3]	97.56	33.51 M

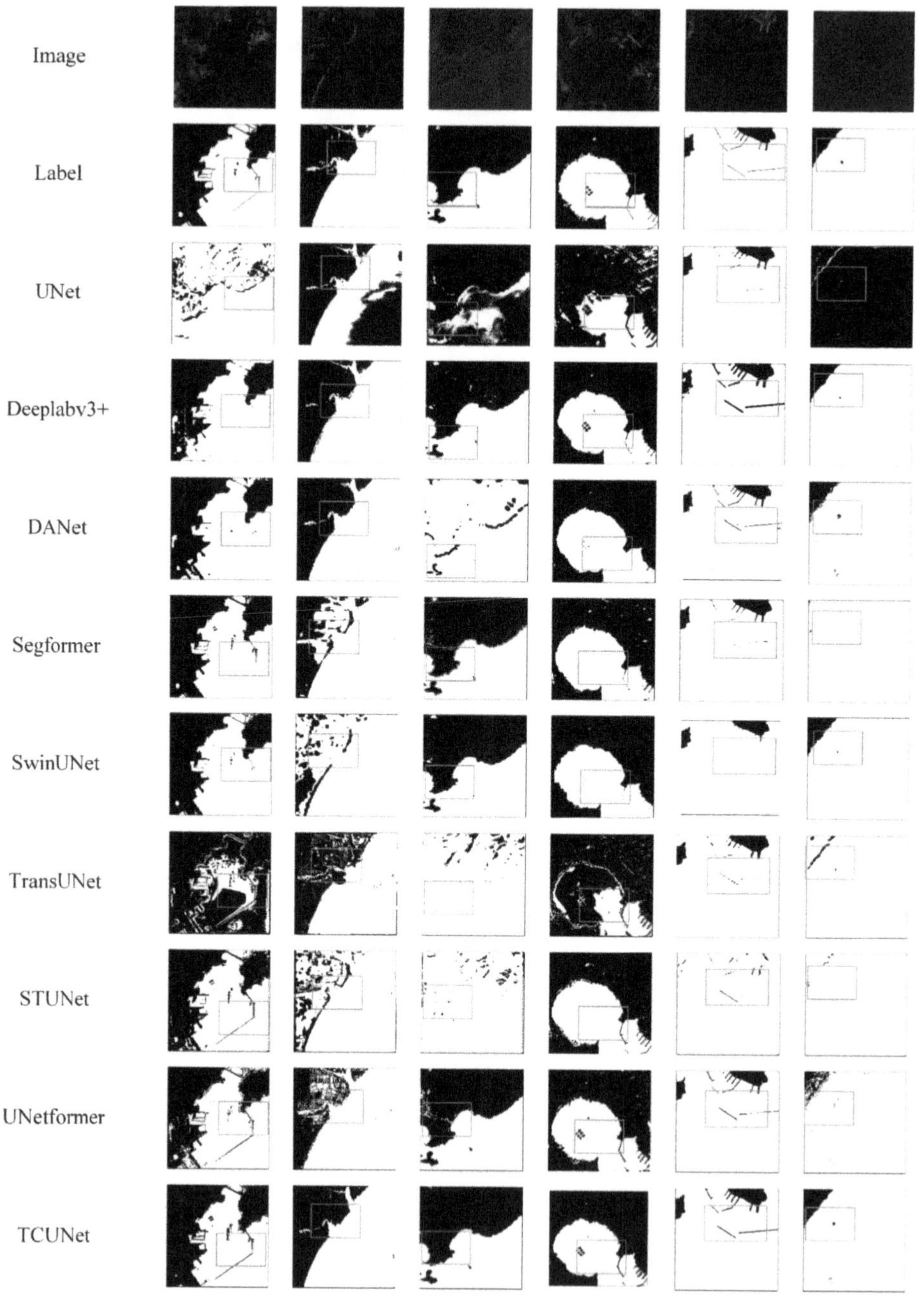

Figure 11. Comparison of the prediction effects of different models on the Landsat dataset.

As can be seen from Table 3, increasing the number of dual-branch channels in the first stage (stages 2–5 have twice as many channels as the previous stage, as described in Section 2.2, i.e., they deepen the width of the network) does not significantly improve the segmentation performance of the model, but the parameters and complexity of the model have increased by tens of times. On the contrary, through keeping the number of model channels constant, and increasing the number of double-branch encoder modules in layers 2–5 (i.e., increasing the depth of the network), the segmentation precision of the model is obviously improved, and the parameters of the model did not increase significantly.

In response to the aforementioned phenomenon, we speculate that our proposed model, which combines a CNN and Transformer, exhibits a considerably enhanced ability in feature extraction compared to conventional CNN networks. Consequently, each layer of the model does not require too many channels (i.e., the width of the network does not need to be too wide) to obtain sufficient rich information. Correspondingly, increasing the number of channels in each layer of the network does not significantly improve the segmentation accuracy. However, network depth enhancement can enable the model to learn deeper feature information, and more complex representations of the image. As a result, enhancing the depth of the model is more effective in improving the accuracy of land and sea segmentation, in comparison with increasing the width of the module, while it also contributes to a reduction in the computational overheads.

Therefore, combining the network complexity and the model segmentation accuracy, we set the channel number of the first stage model to 16, and set the number of encoder blocks in layers 2–5 of the network to 3, 4, 6, and 3, respectively.

4.3. Limitations of the Model and Future Prospects

This paper presents a TCU-Net model specifically designed for the extraction of the shoreline from multispectral remote sensing images. Compared with the latest CNN and Transformer methods, the proposed model achieves a better segmentation accuracy with fewer model parameters and computational resources.

However, due to the inherent computational demands of the parallel dual-branch encoder structure and the Transformer model, despite efforts to reduce the model's computational overhead through narrowing the network width and designing lightweight decoder structures, optimal results in terms of training and inference speed have not been achieved in this study. Future research will focus on further optimizing the model architecture, while ensuring a robust segmentation accuracy. This will involve the design of more efficient model structures and effective training strategies, aiming to alleviate training complexity and difficulty.

5. Conclusions

In order to achieve the high-precision segmentation of sea–land boundaries and coastline extraction from remote sensing images, a lightweight two-branch parallel network model combining CNN and Transformer is designed for sea–land segmentation in remote sensing images.

Specifically, in the encoding process of the algorithm, the CNN branch and the Transformer branch are used to extract the local semantic features and the global spatial features of the multi-spectral remote sensing image. At the same time, we design a feature interaction module (FIM) which is embedded between each corresponding two-branch coding block, serving as a bridge module to fuse the local features from the CNN branch and the global representation from the Transformer branch, to realize information interaction between the twobranches' features. For the decoder part, we designed a cross-scale, multi-source feature fusion module (CMFFM) to replace the original UNet encoder module, achieving the successful integration of low-level semantic and high-level abstract features, and improving the network's ability to capture information flows. For CMFFM, the module is first replaced via up-sampling using a feature calibration module, which can reduce the semantic differences between the "corresponding" pixels of images at different scales. At the same time, a channel attention module and spatial attention module are introduced to obtain channel and spatial attention weights, using two branches with different scales for the fused features, so that the model can capture the spatial and band information of the image, and realize the successful integration of low-level semantics and high-level abstract features. Finally, the fused multi-scale features are obtained. In this study, we generated a dataset, named the GF dataset for sealand segmentation in the Yellow Coastline region using three GF-6 remote sensing satellite images. Subsequently, an extensive series of comprehensive experiments was conducted, to evaluate the segmentation performance and efficiency of TCUNet in comparison to other existing semantic segmentation networks on this dataset. The experimental results demonstrate that TCUNet has a better segmentation effect than other classical semantic segmentation networks, highlighting its superiority and effectiveness. Furthermore, we also discussed the application of the model on different band combinations and different remote sensing sensor images. In summary, this study provides a new method for extracting the coastline from remote sensing images accurately and effectively.

In our future research, we will continue to refine our model, collect multi-spectral satellite remote sensing images taken by different satellites, at different band settings and spatial resolutions, improve the application scope of the model and the accuracy of shoreline extraction, and then expand the research area, to achieve the extraction of shorelines in other sea areas.

Author Contributions: Conceptualization, X.X. and X.W.; methodology, X.X. and X.W.; code, X.X. and R.D.; validation, X.X. and X.W.; formal analysis, B.H. and J.Z.; investigation, X.X. and J.Z.; resources, X.X.; data curation X.X. and R.D.; writing–original draft preparation, X.X.; writing–review and editing, X.X. and X.W.; visualization, X.X. and R.D.; supervision, B.H. and X.W.; funding acquisition, X.W. and J.Z. All authors have read and agreed to the published version of the manuscript.

Funding: This work was jointly supported by the National Natural Science Foundation of China (No. 42276203), Shandong Provincial Natural Science Foundation, China (No. 2020QF067), "Taishan Scholar" Project of Shandong Province (No. TSXZ201712), and Strategic Priority Research Program of the Chinese Academy of Sciences-A (No. XDA19030402).

Data Availability Statement: The code can be found at https://github.com/xx16516/TCUNet, (accessed on 27 August 2023). The datasets in our study are public. The GF dataset can be found at https://aistudio.baidu.com/aistudio/datasetdetail/230558, (accessed on 17 July 2023).

Conflicts of Interest: The authors declare no conflict of interest.

References

1. Zollini, S.; Alicandro, M.; Cuevas-González, M.; Baiocchi, V.; Dominici, D.; Buscema, P.M. Shoreline extraction based on an active connection matrix (ACM) image enhancement strategy. *J. Mar. Sci. Eng.* **2020**, *8*, 9. [CrossRef]
2. Boak, E.H.; Turner, I.L. Shoreline Definition and Detection: A Review. *J. Coast. Res.* **2005**, *21*, 688–703. [CrossRef]
3. Soloy, A.; Turki, I.; Lecoq, N.; Gutiérrez Barceló, Á.D.; Costa, S.; Laignel, B.; Bazin, B.; Soufflet, Y.; Le Louargant, L.; Maquaire, O. A fully automated method for monitoring the intertidal topography using Video Monitoring Systems. *Coast. Eng.* **2021**, *167*, 103894. [CrossRef]
4. Yang, L.; Wang, X.; Zhai, J. Waterline Extraction for Artificial Coast With Vision Transformers. *Front. Environ. Sci.* **2022**, *10*, 16. [CrossRef]
5. Bengoufa, S.; Niculescu, S.; Mihoubi, M.K.; Belkessa, R.; Abbad, K. Rocky Shoreline Extraction Using a Deep Learning Model and Object-Based Image Analysis. Int. Arch. Photogramm. *Remote Sens. Spat. Inf. Sci. ISPRS Arch.* **2021**, *43*, 23–29.
6. Bengoufa, S.; Niculescu, S.; Mihoubi, M.K.; Belkessa, R.; Rami, A.; Rabehi, W.; Abbad, K. Machine Learning and Shoreline Monitoring Using Optical Satellite Images: Case Study of the Mostaganem Shoreline, Algeria. *J. Appl. Remote Sens.* **2021**, *15*, 026509. [CrossRef]
7. Liu, Z.; Chen, X.; Zhou, S.; Yu, H.; Guo, J.; Liu, Y. DUPnet: Water Body Segmentation with Dense Block and Multi-Scale Spatial Pyramid Pooling for Remote Sensing Images. *Remote Sens.* **2022**, *14*, 5567. [CrossRef]
8. Pardo-Pascual, J.E.; Almonacid-Caballer, J.; Ruiz, L.A.; Palomar-Vazquez, J. Automatic extraction of shorelines from Landsat TM and ETM+ multi-temporal images with subpixel precision. *Remote Sens. Environ.* **2012**, *123*, 1–11. [CrossRef]
9. Lecun, Y.; Bengio, Y.; Hinton, G. Deep learning. *Nature* **2015**, *521*, 436. [CrossRef]
10. McFeeters, S.K. The use of the Normalized Difference Water Index (NDWI) in the delineation of open water features. *Int. J. Remote Sens.* **1996**, *17*, 1425–1432. [CrossRef]
11. Xu, H. Modification of normalised difference water index (NDWI) to enhance open water features in remotely sensed imagery. *J. Remote Sens.* **2006**, *27*, 3025–3033. [CrossRef]
12. Yang, C.S.; Park, J.H.; Rashid, H.A. An Improved Method of Land Masking for Synthetic Aperture Radar-based Ship Detection. *J. Navig.* **2018**, *71*, 788–804. [CrossRef]

13. Kanopoulos, N.; Vasanthavada, N.; Baker, R.L. Design of an image edge detection filter using the Sobel operator. *IEEE J. Solid-State Circuits* **1988**, *23*, 358–367. [CrossRef]
14. Liu, H.; Jezek, K.C. Automated extraction of coastline from satellite imagery by integrating Canny edge detection and locally adaptive thresholding methods. *Int. J. Remote Sens.* **2004**, *25*, 937–958. [CrossRef]
15. Toure, S.; Diop, O.; Kpalma, K.; Maiga, A.S. Shoreline detection using optical remote sensing: A review. *ISPRS Int. J. Geo-Inf.* **2019**, *8*, 75. [CrossRef]
16. Wu, Y.; Liu, Z. Research progress on methods of automatic coastline extraction based on remote sensing images. *J. Remote Sens.* **2019**, *23*, 582–602. [CrossRef]
17. Breiman, L. Random forests. *Mach. Learn.* **2001**, *45*, 5–32. [CrossRef]
18. Suykens, J.A.K.; Vandewalle, J. Least Squares Support Vector Machine Classifiers. *Neural Process. Lett.* **1999**, *9*, 293–300. [CrossRef]
19. Cui, B.; Jing, W.; Huang, L.; Li, Z.; Lu, Y. SANet: A Sea–Land Segmentation Network Via Adaptive Multiscale Feature Learning. *IEEE J. Sel. Top. Appl. Earth Obs. Remote Sens.* **2021**, *14*, 116–126. [CrossRef]
20. Chen, L.C.; Papandreou, G.; Kokkinos, I.; Murphy, K.; Yuille, A.L. Semantic image segmentation with deep convolutional nets and fully connected crfs. *arXiv* **2014**, arXiv:1412.7062.
21. Girshick, R.; Donahue, J.; Darrell, T.; Malik, J. Rich Feature Hierarchies for Accurate Object Detection and Semantic Segmentation. In Proceedings of the 2014 IEEE Conference on Computer Vision and Pattern Recognition (CVPR), Columbus, OH, USA, 23–28 June 2014; pp. 580–587.
22. Kalchbrenner, N.; Grefenstette, E.; Blunsom, P. A convolutional neural network for modelling sentences. *arXiv* **2014**, arXiv:14042188.
23. Long, J.; Shelhamer, E.; Darrell, T. Fully Convolutional Networks for Semantic Segmentation. In Proceedings of the IEEE Conference on Computer Vision and Pattern Recognition, Boston, MA, USA, 7–12 June 2015; pp. 3431–3440.
24. Ronneberger, O.; Fischer, P.; Brox, T. U-net: Convolutional networks for biomedical image segmentation. In Proceedings of the International Conference on Medical Image Computing and Computer-Assisted Intervention, Munich, Germany, 5–9 October 2015; pp. 234–241.
25. Badrinarayanan, V.; Kendall, A.; Cipolla, R. SegNet: A Deep Convolutional Encoder-Decoder Architecture for Image Segmentation. *IEEE Trans. Pattern Anal. Mach. Intell.* **2015**, *39*, 2481–2495. [CrossRef] [PubMed]
26. Zhao, H.; Shi, J.; Qi, X.; Wang, X.; Jia, J. Pyramid Scene Parsing Network. In Proceedings of the IEEE Conference on Computer Vision and Pattern Recognition, Honolulu, HI, USA, 21–26 July 2017; pp. 2881–2890.
27. Chen, L.C.; Papandreou, G.; Kokkinos, I.; Murphy, K.; Yuille, A.L. DeepLab: Semantic Image Segmentation with Deep Convolutional Nets, Atrous Convolution, and Fully Connected CRFs. *arXiv* **2017**, arXiv:1606.00915. [CrossRef]
28. Chen, L.C.; Zhu, Y.; Papandreou, G.; Schroff, F.; Adam, H. Encoder-Decoder with Atrous Separable Convolution for Semantic Image Segmentation. In Proceedings of the European Conference on Computer Vision (ECCV), Munich, Germany, 8–14 September 2018; pp. 801–818.
29. Sun, K.; Xiao, B.; Liu, D.; Wang, J. Deep high-resolution representation learning for human pose estimation. In Proceedings of the IEEE/CVF Conference on Computer Vision and Pattern Recognition, Long Beach, CA, USA, 15–20 June 2019; pp. 5693–5703.
30. Li, R.; Liu, W.; Yang, L.; Sun, S.; Hu, W.; Zhang, F.; Li, W. Deepunet: A deep fully convolutional network for pixel-level sea-land segmentation. *IEEE J. Sel. Top Appl. Earth Obs. Remote Sens.* **2018**, *11*, 3954–3962. [CrossRef]
31. Shamsolmoali, P.; Zareapoor, M.; Wang, R.; Zhou, H.; Yang, J. A Novel Deep Structure U-Net for Sea-Land Segmentation in Remote Sensing Images. *IEEE J. Sel. Top Appl. Earth Obs. Remote Sens.* **2019**, *12*, 3219–3232. [CrossRef]
32. Huang, G.; Liu, Z.; van der Maaten, L.; Weinberger, K. Densely Connected Convolutional Networks. In Proceedings of the IEEE Conference on Computer Vision and Pattern Recognition, Honolulu, HI, USA, 21–26 July 2017.
33. He, K.; Zhang, X.; Ren, S.; Sun, J. Deep residual learning for image recognition. In Proceedings of the IEEE Conference on Computer Vision and Pattern Recognition, Las Vegas, NV, USA, 26 June–1 July 2016; pp. 770–778.
34. He, Y.; Yao, S.; Yang, W.; Yan, H.; Zhang, L.; Wen, Z.; Zhang, Y.; Liu, T. An Extraction Method for Glacial Lakes Based on Landsat-8 Imagery Using an Improved U-Net Network. *IEEE J. Sel. Top Appl. Earth Obs. Remote Sens.* **2021**, *14*, 6544–6558. [CrossRef]
35. Vaswani, A.; Shazeer, N.; Parmar, N.; Uszkoreit, J.; Jones, L.; Gomez, A.N.; Kaiser, L.; Polosukhin, I. Attention Is All You Need. *arXiv* **2017**, arXiv:1706.03762.
36. Dosovitskiy, A.; Beyer, L.; Kolesnikov, A.; Weissenborn, D.; Zhai, X.; Unterthiner, T.; Dehghani, M.; Minderer, M.; Heigold, G.; Gelly, S.; et al. An Image Is Worth 16x16 Words: Transformers for Image Recognition at Scale. *arXiv* **2021**, arXiv:2010.11929.
37. Wang, W.; Xie, E.; Li, X.; Fan, D.P.; Song, K.; Liang, D.; Lu, T.; Luo, P.; Shao, L. Pyramid vision transformer: A versatile backbone for dense prediction without convolutions. In Proceedings of the IEEE/CVF International Conference on Computer Vision, Montreal, QC, Canada, 10–17 October 2021; pp. 568–578.
38. Liu, Z.; Lin, Y.; Cao, Y.; Hu, H.; Wei, Y.; Zhang, Z.; Lin, S.; Guo, B. Swin Transformer: Hierarchical Vision Transformer Using Shifted Windows. In Proceedings of the IEEE/CVF International Conference on Computer Vision, Montreal, BC, Canada, 11–17 October 2021; pp. 10012–10022.
39. Zheng, S.; Lu, J.; Zhao, H.; Zhu, X.; Luo, Z.; Wang, Y.; Fu, Y.; Feng, J.; Xiang, T.; Torr, P.H.S.; et al. Rethinking Semantic Segmentation from a Sequence-to-Sequence Perspective with Transformers. *arXiv* **2021**, arXiv:2012.15840.

40. Xie, E.; Wang, W.; Yu, Z.; Anandkumar, A.; Alvarez, J.M.; Luo, P. SegFormer: Simple and Efficient Design for Semantic Segmentation with Transformers. In *Advances in Neural Information Processing Systems, Proceedings of the Conference on Neural Information Processing Systems, Virtual, 6–14 December 2021*; Curran Associates, Inc.: Red Hook, NY, USA, 2021; Volume 34, pp. 12077–12090.
41. Chen, J.; Lu, Y.; Yu, Q.; Luo, X.; Adeli, E.; Wang, Y.; Lu, L.; Yuille, A.L.; Zhou, Y. TransUNet: Transformers Make Strong Encoders for Medical Image Segmentation. *arXiv* **2021**, arXiv:2102.04306.
42. Guo, J.; Han, K.; Wu, H.; Tang, Y.; Chen, X.; Wang, Y.; Xu, C. Cmt: Convolutional neural networks meet vision transformers. In Proceedings of the IEEE/CVF Conference on Computer Vision and Pattern Recognition, New Orleans, LA, USA, 18–24 June 2022; pp. 12175–12185.
43. Zhang, Y.; Liu, H.; Hu, Q. Transfuse: Fusing transformers and cnns for medical image segmentation. In Proceedings of the International Conference on Medical Image Computing and Computer-Assisted Intervention, Strasbourg, France, 27 September–1 October 2021; pp. 14–24.
44. He, X.; Zhou, Y.; Zhao, J.; Zhang, D.; Yao, R.; Xue, Y. Swin transformer embedding UNet for remote sensing image semantic segmentation. *IEEE Trans. Geosci. Remote Sens.* **2022**, *60*, 1–15. [CrossRef]
45. Chen, J.; Xia, M.; Wang, D.; Lin, H. Double Branch Parallel Network for Segmentation of Buildings and Waters in Remote Sensing Images. *Remote Sens.* **2023**, *15*, 1536. [CrossRef]
46. Wang, W.; Xie, E.; Li, X.; Fan, D.P.; Song, K.; Liang, D.; Lu, T.; Luo, P.; Shao, L. Pvtv2: Improved baselines with pyramid vision transformer. *Comput. Vis. Media* **2022**, *8*, 415–424. [CrossRef]
47. Peng, Z.; Huang, W.; Gu, S.; Xie, L.; Wang, Y.; Jiao, J.; Ye, Q. Conformer: Local features coupling global representations for visual recognition. In Proceedings of the IEEE/CVF International Conference on Computer Vision, Montreal, QC, Canada, 10–17 October 2021; pp. 367–376.
48. Ba, J.L.; Kiros, J.R.; Hinton, G.E. Layer normalization. *arXiv* **2016**, arXiv:1607.06450.
49. Li, X.; You, A.; Zhu, Z.; Zhao, H.; Yang, M.; Yang, K.; Tan, S.; Tong, Y. Semantic flow for fast and accurate scene parsing. In *Lecture Notes in Computer Science, Proceedings of the 16th European Conference Computer Vision (ECCV 2020), Glasgow, UK, 23–28 August 2020*; Springer International Publishing: Cham, Switzerland, 2020; pp. 775–793.
50. Huang, Z.; Wei, Y.; Wang, X.; Shi, H.; Liu, W.; Huang, T.S. AlignSeg: Feature-Aligned segmentation networks. *IEEE Trans. Pattern Anal. Mach. Intell.* **2021**, *44*, 550–557. [CrossRef]
51. Fu, J.; Liu, J.; Tian, H.; Li, Y.; Bao, Y.; Fang, Z.; Lu, H. Dual Attention Network for Scene Segmentation. In Proceedings of the 2019 IEEE/CVF Conference On Computer Vision And Pattern Recognition (CVPR 2019), Long Beach, CA, USA, 16–20 June 2019; pp. 3141–3149.
52. Chen, L.C.; Papandreou, G.; Schroff, F.; Adam, H. Rethinking Atrous Convolution for Semantic Image Segmentation. *arXiv* **2017**, arXiv:1706.05587.
53. Szegedy, S.I.a.C. Batch normalization: Accelerating deep network training by reducing internal covariate shift. In Proceedings of the 32nd International Conference on Machine Learning ICML, Lile, France, 6–11 July 2015; pp. 448–456.
54. Kervadec, H.; Bouchtiba, J.; Desrosiers, C.; Granger, E.; Dolz, J.; Ayed, I.B. Boundary loss for highly unbalanced segmentation. *Med. Image Anal.* **2021**, *67*, 101851. [CrossRef]
55. Yu, Z.; Di, L.; Yang, R.; Tang, J.; Lin, L.; Zhang, C.; Rahman, M.S.; Zhao, H.; Gaigalas, J.; Yu, E.G. Selection of landsat 8 OLI band combinations for land use and land cover classification. In Proceedings of the 2019 8th International Conference on Agro-Geoinformatics, Istanbul, Turkey, 16–19 July 2019; pp. 1–5.
56. Mou, H.; Li, H.; Zhou, Y.; Dong, R. Response of different band combinations in Gaofen-6 WFV for estimating of regional maize straw resources based on random forest classification. *Sustainability* **2021**, *13*, 4603. [CrossRef]
57. Cao, H.; Wang, Y.; Chen, J.; Jiang, D.; Zhang, X.; Tian, Q.; Wang, M. Swin-unet: Unet-like pure transformer for medical image segmentation. *arXiv* **2021**, arXiv:2105.05537.
58. Wang, L.; Li, R.; Zhang, C.; Fang, S.; Duan, C.; Meng, X.; Atkinson, P.M. UNetFormer: A UNet-like transformer for efficient semantic segmentation of remote sensing urban scene imagery. *ISPRS J. Photogramm. Remote Sens.* **2022**, *190*, 196–214. [CrossRef]
59. Hendrycks, D.; Gimpel, K. Gaussian error linear units (gelus). *arXiv* **2016**, arXiv:1606.08415.

remote sensing

MDPI

Article

RockSeg: A Novel Semantic Segmentation Network Based on a Hybrid Framework Combining a Convolutional Neural Network and Transformer for Deep Space Rock Images

Lili Fan *, Jiabin Yuan, Xuewei Niu, Keke Zha and Weiqi Ma

College of Computer Science and Technology, Nanjing University of Aeronautics and Astronautics, Nanjing 211106, China; jbyuan@nuaa.edu.cn (J.Y.); xwn@nuaa.edu.cn (X.N.); zhakeke@nuaa.edu.cn (K.Z.); mawqnn@nuaa.edu.cn (W.M.)
* Correspondence: fanlily913@nuaa.edu.cn

Abstract: Rock detection on the surface of celestial bodies is critical in the deep space environment for obstacle avoidance and path planning of space probes. However, in the remote and complex deep environment, rocks have the characteristics of irregular shape, being similar to the background, sparse pixel characteristics, and being easy for light and dust to affect. Most existing methods face significant challenges to attain high accuracy and low computational complexity in rock detection. In this paper, we propose a novel semantic segmentation network based on a hybrid framework combining CNN and transformer for deep space rock images, namely RockSeg. The network includes a multiscale low-level feature fusion (MSF) module and an efficient backbone network for feature extraction to achieve the effective segmentation of the rocks. Firstly, in the network encoder, we propose a new backbone network (Resnet-T) that combines the part of the Resnet backbone and the transformer block with a multi-headed attention mechanism to capture the global context information. Additionally, a simple and efficient multiscale feature fusion module is designed to fuse low-level features at different scales to generate richer and more detailed feature maps. In the network decoder, these feature maps are integrated with the output feature maps to obtain more precise semantic segmentation results. Finally, we conduct experiments on two deep space rock datasets: the MoonData and MarsData datasets. The experimental results demonstrate that the proposed model outperforms state-of-the-art rock detection algorithms under the conditions of low computational complexity and fast inference speed.

Keywords: deep space exploration; planetary rover; rock segmentation; semantic segmentation

Citation: Fan, L.; Yuan, J.; Niu, X.; Zha, K.; Ma, W. RockSeg: A Novel Semantic Segmentation Network Based on a Hybrid Framework Combining a Convolutional Neural Network and Transformer for Deep Space Rock Images. *Remote Sens.* **2023**, *15*, 3935. https://doi.org/10.3390/rs15163935

Academic Editors: Qian Du, Jiaojiao Li, Wei Li, Jocelyn Chanussot, Rui Song, Yunsong Li and Bobo Xi

Received: 19 June 2023
Revised: 28 July 2023
Accepted: 5 August 2023
Published: 9 August 2023

1. Introduction

Obstacle detection is a crucial component of space exploration to assure rover patrol safety of deep space probes. Particularly, on the surface of most celestial bodies, rocks are the main obstacle that interfere with landing probes and rover missions [1–3]. To obtain suitable path planning and ensure the safe driving of planetary rovers, it is important for planetary rovers to perceive and avoid these rock hazards when carrying out a deep space exploration mission. However, the deep space environment is complex and unknown; some rocks have irregular morphology and different size on the surface of the planet. Compared to other nearby targets such as sand, soil, or gravel, they have no distinct distinguishing features, and some rocks may also be affected by changes in illumination, different lighting angles, and the resulting shadow causing a false visual perception. These conditions undoubtedly increase planetary rovers' difficulty in perceiving and understanding the surroundings. Therefore, the exploration of autonomous rock detection on the surface of planets still faces great challenges [4,5].

Recently, autonomous technology has been used for a range of planetary scientific missions, including autonomous landing location [6–8], rover navigation [2,3,9], and autonomous path planning [1,10]. As the distance of deep space exploration increases, autonomous technology becomes the key and necessary technology to support deep space exploration in the future [11]. In deep space environments, edge-based digital image processing methods [12–14] are a common method to achieve rock autonomous detection. Most of them use the local strength gradient operator or the gradient difference in illumination direction to detect the target boundary, which is sensitive to noise and illumination conditions. In order to deal with the influence of sunlight and noise, some studies [15–17] try to classify regional objects by using a super-pixel segmentation region method based on pixel clustering to improve the performance in rock detection. In addition, some machine learning classifiers [18,19] are also used to classify planetary terrain. However, the complexity of super-pixel segmentation increases with the size of the input image, and how to adjust its convergence and detection performance is a challenge. Although most machine-learning techniques are successful at terrain classification, they fall short in accurately identifying rock boundaries and locations.

Convolutional neural network (CNN)-based deep learning technology has achieved great success in the semantic segmentation of 2D images [20,21]. Some efforts towards semantic segmentation-based methods have been made to achieve automatic rock detection. For the deep space autonomous rock segmentation network, when the rover captures an image, it is passed to a semantic segmentation network and the network output is the classification at the pixel level, which is fed back to the detector to sense the surrounding environment information. In order to realize high-precision rock detection in the deep space environment, acquiring multiscale context information of rock images is essential in a semantic segmentation network. Some studies propose convolution pooling, dilated convolution [22], spatial pyramid pooling (SPP) [23], pyramid pooling module (PPM) [24], and atrous spatial pyramid pooling (ASPP) [25] to obtain a larger receptive field and integrate multiscale context information [26]. A U-shape network [27] is a common multiscale semantic segmentation network widely applied to medical image segmentation and analysis, which uses upsampling in the decoder to expand the feature map to the same size as the original image. In addition, there has recently been increased focus on other multiscale semantic segmentation networks, such as FCN [28], PSPNet [24], and DeepLabV3+ [25], for planet rock detection [4,5,29,30].

Convolutional pool operation is a common operation in the encoder of semantic segmentation networks, which is used to obtain the multiscale feature map, expand the field of perception, and reduce the amount of calculation to some extent. However, using convolutional pool operations may cause a loss of information, which causes blurry output results in the process of the network decoder. It is very important to consider how to reduce information loss to restore the clarity output feature mapping for improving the accuracy of rock semantic segmentation. Some works [24,25] use a direct upsampling operation in the network decoder to obtain the output feature map. Although this approach is easy to implement, some details may be lost, resulting in blurred segmentation boundaries. To enhance the clarity of the rock detection boundary, other researchers [5,29–32] recommend fusing low-level feature details and using skip connections and stepwise sampling to generate more rich feature output in the upsampling process. These strategies can improve the clarity of the rock segmentation boundary to a certain extent. However, some overlaps and redundant information may be added to the output feature map in the upsampling process, which affects the accuracy of network segmentation [11]. In addition, the multiple sampling and connection process may increase unnecessary network parameters and computational complexity [33,34]. Most rock detection methods do not consider how to balance accuracy and complexity.

Obtaining local and global context dependencies is the key to extracting the target object [35,36]. CNN can obtain the local context dependencies using multiscale context information in semantic segmentation networks. However, the local feature of the convolution layer of the CNN limits the ability of the network to capture global information. Recently, a transformer network based on a multi-headed attention mechanism has been successful in the field of computer vision. Vision Transformer (ViT) can effectively obtain global information using a self-attention mechanism and enhance the model expression through the multi-head spaces map. Some researchers have applied vision transformers (ViT) in image classification and segmentation [5,29,37]. The VIT model often relies on powerful computing resources and a pre-training model, which limits its use in many tasks. To apply the strong global feature extraction ability of the transformer, some studies propose a new combination of CNN and transformer networks to fuse both advantages for capturing local and global contextual information. Hybrid networks combining CNN and transformer have been attempted in some fields, such as image change detection [38,39], medical image segmentation [35,36], person re-identification [40], and image super-resolution [41].

In previous work, we have proposed [31] an onboard rock detection algorithm based on a spiking neural network to reduce the calculation energy consumption. In this paper, we explore a novel network based on a hybrid framework combining CNN and vision transformer for deep space rock images to improve the efficiency and accuracy of rock detection; the proposed model contains an efficient backbone feature extraction block and a multiscale low-level feature fusion module. Firstly, to efficiently extract rock features, we propose a new backbone (Resnet-T), which utilizes part of the Resnet backbone and combines it with a visual transformer block to capture the global context information of the rock. Secondly, a simple and effective multiscale low-level feature fusion (MSF) module is designed to obtain more rich semantic features, and they are fused into the output feature map in the upsampling process to improve the quality of the output feature map. Last, we use two deep space rock image datasets (MoonData and RockData) to verify the performance of the proposed model. The experimental results show that our model has higher detection accuracy and faster model reasoning speed than other methods when the model parameters and computational complexity are lower.

In summary, our main contributions are as follows.

- We propose a novel semantic segmentation network (RockSeg) based on the combined CNN and transformer framework, which contains an efficient feature extraction backbone and a multiscale low-level feature fusion module to effectively detect rocks on the surface of celestial bodies.
- We combine Resnet blocks and visual transformer blocks to construct an efficient Resnet-T backbone network to extract the global context information. In addition, we design MSF to obtain rich multiscale fusion features and fuse them into the output feature map to improve the segmentation clarity of the target boundary.
- The experiment is conducted on the PyTorch platform with two rock datasets to verify the performance of the RockSeg. The results show that our method outperforms the state-of-the-art rock detection models in terms of detection accuracy and inference speed.

The rest of this paper is organized as follows: Section 2 describes related work. Section 3 describes the proposed network architecture, the design of the feature extraction backbone, and the multiscale low-level feature fusion module. The experimental results and analysis are provided in Section 4. In Section 5, we conclude our work.

2. Related Work

2.1. Deep Learning-Based Obstacle Detection in Space Exploration

Obstacle detection is crucial for rover navigation and path planning of space rovers. Recently, some deep learning-based approaches for improving the accuracy and practicality of obstacle detection have been developed. Craters are a conspicuous and well-preserved feature of star surfaces, with the majority of them being registered. Researchers used CNN to detect the crater pictures obtained during the probe's descent to gain visual global localization [7,8], which helps the lander in locating and selecting a safe landing place. Moreover, other studies concentrate on applying deep learning to terrain classification [42,43], terrain segmentation [33,44], and rock segmentation [4,30] for Mars rovers. (i) Terrain classification. Li et al. [43] suggest using transfer deep learning techniques for autonomous classification of Martian rock images with seven different types of terrain. In order to enhance the clarity of the output feature map texture, Liu et al. [45] also combine a number of modules with generative adversarial networks, attention mechanisms, and a feature pyramid structure to build the detection network. (ii) Terrain segmentation. In order to increase the accuracy of the segmentation result, a hybrid attention semantic segmentation network is proposed [44] for unstructured terrain on Mars, which combines the global and local attention branches to aggregate the contexts for the final segmentation. In addition, Dai et al. [33] propose a lightweight ViT-based terrain segmentation approach with low computational complexity and power consumption for onboard satellites. Furthermore, the semi-supervised learning framework [46,47] is proposed for Mars terrain segmentation to address the lack of training data and training complexity. (iii) Rock segmentation. In previous work, we propose an efficient rock detection algorithm on the surface of the Moon to reduce the complexity of the calculation [31], which uses a spiking neural network with a new brain-like paradigm to achieve onboard rock detection. In Martian rock detection methods, the work [4] employs the Unet convolutional neural network to obtain a segmented rock image by training different sizes, shapes, and textures of rock images in a Mars-like environment. The paper [5] build a U-shaped transformer network that uses a hierarchical encoder–decoder architecture and multiscale features based on an improved vision transformer to capture global dependencies for Martian rock segmentation. In addition, the authors of [30] also design automatic rock segmentation based on deep learning using enhanced Unet-based architecture combined with a visual geometry group and dilated convolutional to improve the accuracy of the rock segmentation.

In general, the above models for deep learning-based obstacle detection have promoted the progress of autonomous technology in deep space exploration to some extent. However, the terrain classification method only divides terrain categories to detect the terrain, which is a coarse-grained recognition and detection of the surroundings. Semantic segmentation methods are fine-grained recognition and detection methods based on pixel classification, which is vital for deep space probes to know the surroundings. Moreover, deep space is far from the Earth, and the probe carrying resource is limited. To achieve autonomous technology in complex and changeable deep space, the deep space spacecraft must meet safety, high recognition accuracy, and low complexity computing requirements. Due to most semantic segmentation methods for planet rock detection only paying attention to detection accuracy or low computational complexity, few of them consider both computational complexity and precision, so most autonomous rock detection methods do not yet have the capability to be used in deep-space environments. In this paper, we propose an effective rock detection network to balance accuracy and computational complexity, and make it more suitable for deep space environments.

2.2. Improved Segmentation Accuracy and Performance

A semantic segmentation network is usually composed of an encoder and a decoder; the encoder is used to extract multiscale features from the input image, and the decoder is used to convert the features into pixel-level segmentation results. In the network encoder, a convolution pool is a common method to enhance the receptive field and reduce the model parameters. However, this may lead to the loss of some information, which has a negative impact on the accuracy of the segmentation results. In order to reduce the loss of information, Yu et al. [22] propose a novel dilated convolution to aggregate multiscale contextual information without losing resolution, which achieves an increase in the receptive field without additional parameters of the network. Inspired by [22], the work [24] utilizes a dilated convolution and pyramid pooling module to integrate contextual information from different regions and embed it in fully convolutional networks. In addition, a stronger encoder–decoder network to refine the result of segmentation is proposed in [25], in which they apply atrous convolution at multiple scales to encode multiscale contextual information in the encoder module and in the decoder module they use spatial information to recover the feature map to refine the object boundary.

Another approach is to improve segmentation accuracy by incorporating more details. When researchers use the simple and direct one-time upsampling methods [24,28] to obtain the output feature image, the edge of the output feature image may be blurred, which may have a bad effect on the segmentation results. In order to obtain a clear segmentation of the boundary, some works [5,7] use skip connections and step-by-step sampling methods to merge more rich fine-grained information and increase the quality of an output feature map. Sun et al. [32] propose the HRNet network using repeated fusion of the high-to-low-resolution representations to obtain rich high-resolution representations. However, multiple upsampling and connection operations may increase unnecessary network parameters and computational complexity. In this paper, a new semantic segmentation network based on a hybrid framework combining CNN and vision transformer is constructed which has an efficient backbone feature extraction module and a multiscale low-level feature fusion module. Similar to [32], we design a more simple and efficient multiscale low-level feature fusion module to fuse more detailed features to the output feature map during upsampling on the network to obtain more fine-grained segmentation results.

To improve the semantic segmentation network's capability in capturing global features, some studies have presented a hybrid framework network combining CNN and transformer to enhance the ability of the network to capture local and global features. In an image change detection task, the authors of [38] construct a new model combining vision transformer and UperNet to effectively transfer the pretrained model. Zhang et al. [39] propose an asymmetric cross-attention hierarchical network by combining CNN and transformer in a series-parallel manner to improve effectiveness in a change detection task. In medical image segmentation, Xiao et al. [35] design a new teacher–student semi-supervised learning optimization strategy fusing CNN and transformer, which improves the utilization of a large number of unlabeled medical images and the effectiveness of model segmentation results. The paper [36] links CNN and a swin transformer as a feature extraction backbone to build a pyramid structure network for improving the quality of breast ultrasound lesion segmentation. To improve the image super-resolution, Fang et al. [41] propose a hybrid network of CNN and transformer for lightweight image super-resolution. In these hybrid networks, most of them embedded the transformer block by image patch in the CNN layer as a new feature extraction block to capture the global context information. However, the CNN and transformer block have their own advantages; the later decision fusion may be more beneficial to the representation of features. In this paper, to fully fuse these advantages, we propose a new hybrid network combining CNN structure and transformer blocks without image patches to apply them to deep rock detection.

3. Methods

In this section, we describe the detail of the novel semantic segmentation network based on a hybrid framework combining CNN and vision transformer, namely RockSeg, the efficient feature extraction backbone, and the multiscale low-level feature fusion module.

3.1. RockSeg

We propose a hybrid framework combining CNN and vision transformer for rock image semantic segmentation in deep space and the whole network includes two parts, an encoder process and a decoder process. Figure 1 depicts the RockSeg network structure. The network input is the rock images on the surface of celestial bodies and the output is the classification results at the pixel level. In the network encoder, the input rock images are first processed through the feature extraction backbone, which contains the two Resnet blocks from the Resnet-34 network and four transformer blocks to extract the important features of the rock. Simultaneously, five different scales of low-level feature maps L_i are obtained from the network encoder, where $L_i = \{l_1, l_2, l_3, l_4, l_5\}$, $1 \leq i \leq 5$, and $i \in \mathbb{N}$. In the network decoder, to improve the quality of the final output feature map, the five low-level feature maps are fused by a simple multiscale feature fusion module; the fused results are denoted as msf_1 and msf_2 shown in Figure 1. Then, the fusing results are added to the output feature map by two upsampling processes, *Decoder*1 and *Decoder*2, to enhance the clarity of semantic segmentation object boundaries.

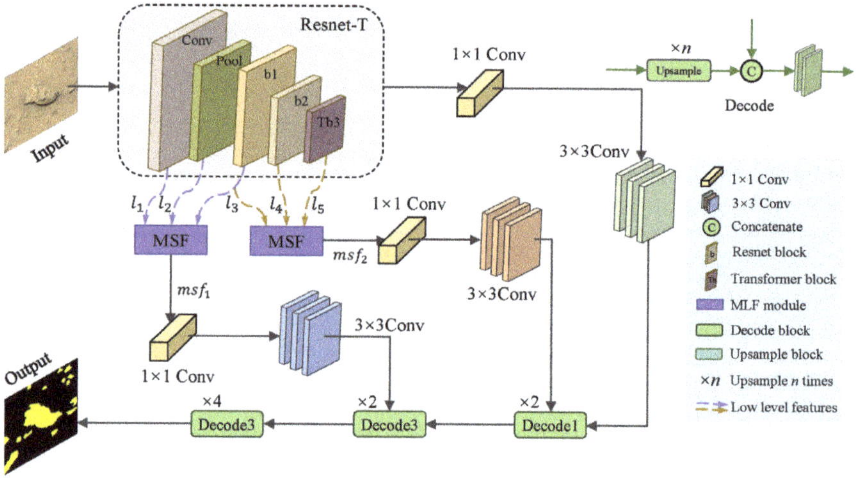

Figure 1. Framework overview of the proposed RockSeg.

3.2. Efficient Backbone Network

In deep space with limited carrying resources, low computing complexity and computational cost are important considerations for the rover to achieve the mission. Deep residual networks [48] have been shown to easily gain accuracy from rapidly increasing depth networks and the results are often superior to those of other networks. However, their network complexity may not apply to deep-space environments with limited resources. To balance the accuracy and complexity of the network model, we design a new efficient backbone network based on a hybrid framework, which combines Resnet-34 blocks and transformer blocks with a multi-head self-attention mechanism to extract the rock's features. The original Resnet-34 backbone and the new proposed backbone structure Resnet-T are shown in Figure 2. Figure 2a shows the original backbone of Resnet-34 with four Resnet blocks. In comparison, Figure 2b is the proposed backbone of Resnet-T with two resnet blocks *b*1, *b*2, and one transformer block *Tb*3. The details of the parameters of Resnet-34 and the Resnet-T are shown in Table 1 and Table 2, separately.

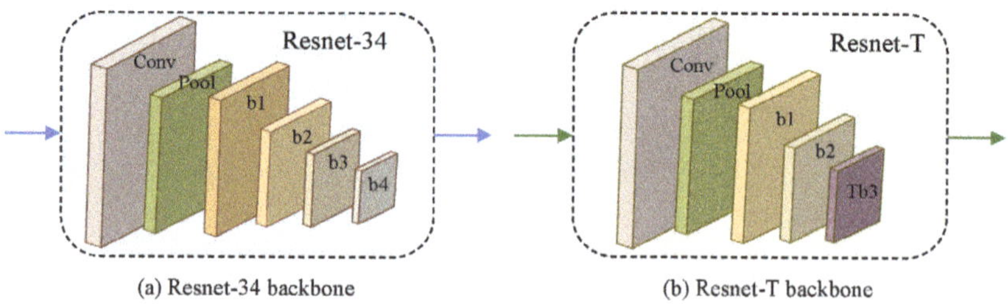

Figure 2. The backbone structure comparison of Resnet-34 and Resnet-T.

In most semantic segmentation networks, full convolution networks without linear fully connected layers are used to extract the object features. So, in Table 1, we remove the linear fully connected layers of the final layer from Resnet-34 as the backbone to extract the object image features. We suppose the input of the network is an RGB image with 256 × 256 pixels, and the output size is obtained by the convolution or pool operation of different blocks. In Table 1, The backbone of Resnet-34 has a 33-layer convolution structure which mainly includes four Resnet blocks, where s is the stride of the convolution operation, k denotes kernel size, B represents the resnet block, $B = \{b1, b2, b3, b4\}$, and n is the number of times repeated for each Resnet block, $n = \{3, 4, 6, 3\}$.

We discovered that using four Resnet blocks to extract rock features is redundant and inefficient in our studies. Feature redundancy may degrade the quality of the output feature map and redundant Resnet extraction blocks also consume additional processing and storage resources. Recently, transformers [37] have achieved significant success in the field of computer vision of 2D image classification. A transformer network is a deep learning mode that uses a self-attention mechanism to better capture long-distance dependencies, compute global dependencies, and more easily interpret predictive results. In particular, some studies have achieved success in the semantic segmentation field [29]; they use the self-attention transformer blocks to build the semantic segmentation networks to improve the performance of object detection. Inspired by the transformer network, in this paper, we design a novel hybrid architecture, which combines Resnet-34 blocks and a transformer block to build a lightweight backbone Resnet-T to effectively extract rock features. In Table 2, we delete the $b3$ and $b4$ blocks from Resnet-34 and replace them with a transformer block $Tb3$ with a multi-headed attention mechanism to create a new backbone Resnet-T for feature extraction.

Table 1. The network parameters of the Resnet-34 backbone.

Layer Name	Output Size	Resnet-34	
Conv	128 × 128	k = 7 × 7, 64, s = 2	
b1	64 × 64	k = 3 × 3 maxpool, s = 2	
		3 × 3, 64 3 × 3, 64	3
b2	32 × 32	3 × 3, 128 3 × 3, 128	4
b3	16 × 16	3 × 3, 256 3 × 3, 256	6
b4	8 × 8	3 × 3, 512 3 × 3, 512	3

Table 2. The network parameters of the Resnet-T backbone.

Layer Name	Output Size	Resnet-T	
Conv	128 × 128	k = 7 × 7, 64, s = 2	
b1	64 × 64	k = 3 × 3 maxpool, s = 2	
		3 × 3, 64 3 × 3, 64	3
b2	32 × 32	3 × 3, 128 3 × 3, 128	4
Tb3	16 × 16	1 × 1, 256, avgpool, s = 2	
		Transfm, 256	4

In Table 2, we can see the Resnet-T network framework is simpler than Resnet-34, where $Conv$, $b1$, and $b2$ are the same as Resnet-34. On the other hand, in order to reduce the computational complexity and obtain good performance, we use $Tb3$ to replace the $b3$ and $b4$ blocks as the enhanced feature extraction block. And we downsample the final output feature map to 1/16 times the input feature map using the Resnet-T backbone. In the proposed Resnet-T backbone, the blocks $b1$ and $b2$ can efficiently extract the basic rock features, and the transformer blocks $TB3$ with multi-headed self-attention mechanisms can weigh features; this hybrid network structure can satisfactorily enhance the feature extraction and reduce the backbone parameters.

In the $Tb3$ block, we first use 1×1 convolution to raise the channel, then, we utilize the average pool to enlarge the receptive field and reduce the size of the feature map, simultaneously. This process can be described as follows:

$$\dot{\boldsymbol{X}} = AvgPool(Conv_{1\times 1}(\boldsymbol{X})) \tag{1}$$

where $\boldsymbol{X}$ is the input of $TB3$, $\dot{\boldsymbol{X}}$ is the output of the raising channel, and $\boldsymbol{X}$ and $\dot{\boldsymbol{X}} \in \mathbb{R}^{C\times H\times W}$. Then, $\dot{\boldsymbol{X}}$ is processed by layer normalization [49] over a mini-batch of inputs, after it is sent to the layer transformer block (Transfm) with multi-headed attention mechanisms (MHead) and multi-layer perceptions (MLP) to obtain the output of the feature map. In the $Transfm$ block, we flatten the feature map to one dimension without the patch and we use the four transformer blocks to extract rock features. The transformer block $Transfm$ can be defined as follows:

$$\begin{aligned} \ddot{\boldsymbol{X}} &= Transfm(Norm(\dot{\boldsymbol{X}})) \\ &= (MLP(MHead(Norm(\dot{\boldsymbol{X}})))) \end{aligned} \tag{2}$$

where $Norm$ represents the layer normalization operation, $MHead$ is the operation of multi-headed attention mechanisms, MLP denotes the operation of the multi-layer perception, and $\ddot{\boldsymbol{X}}$ presents the final feature map output of Resnet-T.

3.3. Multiscale Low-Level Feature Fusion

In CNN networks near the input layer, the network layer becomes shallow and has rich local detail features, the resolution of feature mapping is high, and the receptive field is small [50]. Otherwise, the layer has a large receptive field and high dimension when closer to the output layer, and has abstraction features and global information [51]. In order to keep consistent with the input image, the semantic segmentation network must restore the size of the feature map. The traditional methods of recovering an output feature map are to use upsampling methods once or many times. Although the one-time sampling method is simple and direct, the obtained feature map lacks fine-grained information, which leads to blurring the target boundary. The method of using upsampling multiple times fuses more low-level feature maps by skipping connections and using stepwise sampling to restore the feature size. However, most of these algorithms are complex and inefficient; they need to spend more computation and multiple upsampling to keep the final output feature map clear and detailed.

In this paper, we present a simple and efficient multiscale low-level feature fusion module for fusing more detailed features into the output feature map during the network upsampling process. The diagram of the feature fusing process is shown in Figure 3. We obtain five low-layer feature maps using the feature extraction layers in the network encoder process. The five low level features are denoted L, where L=$\{l_1, l_2, l_3, \ldots, l_i, \ldots\}$, $i = \{1, 2, 3, \ldots\}$, $i \in \mathbb{N}^+$. Due to the closer input layer, the network layer is richer in local detail features, so we use adjacent feature maps to fuse more different detailed information. In our network, $i \in [1,5]$, the two groups of low-level feature maps $\{l_1, l_2, l_3\}$ and $\{l_3, l_4, l_5\}$ are fused to output msf_1 and msf_2 by MSF, respectively.

In Figure 3, we show the fusing process of the three adjacent low-level features $\boldsymbol{X}$ to obtain more detailed information, where $\boldsymbol{X}$=$\{\boldsymbol{X_1}, \boldsymbol{X_2}, \boldsymbol{X_3}\}$ and, for each $\boldsymbol{X_j} \in \boldsymbol{X}$, $\boldsymbol{X_j} \in \mathbb{R}^{B_j\times C_j\times H_j\times W_j}$. The green arrow, yellow arrow, and blue arrow represent the different fusion branches of $\boldsymbol{X}$. In order to describe the fusion process more clearly, we set batch B as 1, so $\boldsymbol{X_j} \in \mathbb{R}^{C_j\times H_j\times W_j}$. Due to $\boldsymbol{X_j}$ being next to each other and obtained from the network encode process, they meet these constraints, $C_1 \le C_2 \le C_3$, $H_1 \ge H_2 \ge H_3$, $W_1 \ge W_2 \ge W_3$, and $H_j \equiv W_j$.

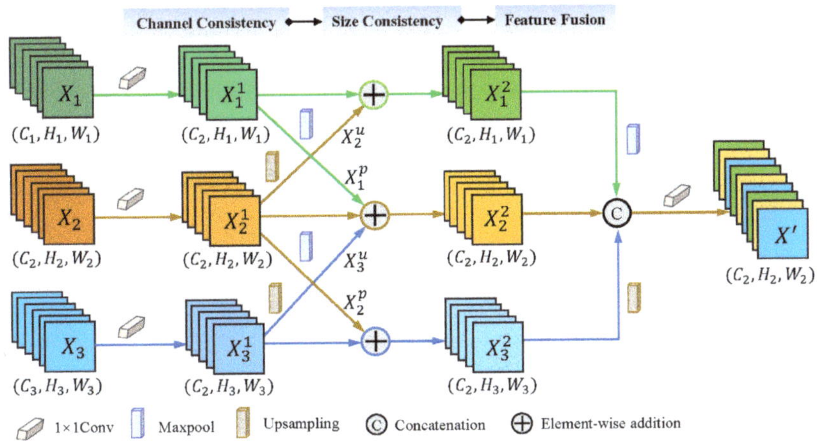

Figure 3. Illustration of the multiscale low-level feature fusion.

In the MSF module, $\mathbf{X}_j$ first is processed by 1×1 convolution to achieve channel consistency; the channel consistency is computed as follows:

$$\mathbf{X}_j^1 = \begin{cases} Conv_{1\times 1}(\mathbf{X}_j), & C_j! = C_2 \\ \mathbf{X}_j, & otherwize \end{cases} \tag{3}$$

where $\mathbf{X}_j^1$ is the output result of the j-th low feature map using the channel consistency process. After, $\mathbf{X}_j^1$ has the same channel C_2 and $\mathbf{X}_j^1 \in \mathbb{R}^{C_2 \times H_j \times W_j}$. Then, the results of the channel consistency are processed by the Maxpool or Upsampling operation to achieve size consistency of the feature map. The size consistency is described as follows:

$$\begin{cases} \mathbf{X}_j^p = Maxpool(\mathbf{X}_j^1), & H_j! = H_{j+1} \\ \mathbf{X}_j^u = Upsampling(\mathbf{X}_j^1), & H_{j+1}! = H_j \\ \mathbf{X}_j^1 = \mathbf{X}_j^1, & otherwise \end{cases} \tag{4}$$

where Maxpool is the maximum pool operation, which is used to reduce the length and width of the feature map to 1/2 of the original size. The Upsampling operation denotes sampling the image to a higher resolution and we use bilinear interpolation to obtain the upsampling results. $\mathbf{X}_j^p$ is the output result of the j-th feature map by Maxpool and $\mathbf{X}_j^u$ denotes the output result of the j-th feature map by Upsampling. After the consistency operation has adjusted the different sizes of the feature map, we use element-wise addition to fuse the neighborhood information of different branches. This simple method can fuse other additional information on the basis of the original information; the fusion process is characterized as follows:

$$\begin{cases} \mathbf{X}_1^2 = \mathbf{X}_1^1 \oplus \mathbf{X}_2^u \\ \mathbf{X}_2^2 = \mathbf{X}_2^1 \oplus \mathbf{X}_1^p \oplus \mathbf{X}_3^u \\ \mathbf{X}_3^2 = \mathbf{X}_3^1 \oplus \mathbf{X}_2^p \end{cases} \tag{5}$$

where the output results of the three branches are $\mathbf{X}_1^2$, $\mathbf{X}_2^2$, and $\mathbf{X}_3^2$, where $\mathbf{X}_j^2 \in \mathbb{R}^{C_2 \times H_j \times W_j}$.

In the feature fusion process, we first use the Maxpool and Upsampling operations to adjust $\mathbf{X}_1^2$, $\mathbf{X}_2^2$, and $\mathbf{X}_3^2$ to the same height H_2 and width W_2 to obtain $\mathbf{X}_j^2$, where $\mathbf{X}_j^2 \in \mathbb{R}^{C_2 \times H_2 \times W_2}$. Then, we connect the three branches in channel dim and use the 1×1

convolution to obtain the final fusing output result $\mathbf{X}'$, $\mathbf{X}' \in \mathbb{R}^{C_2 \times H_2 \times W_2}$; this is computed as follows:

$$\mathbf{X}' = Conv_{1\times1}(Concat(\mathbf{X}_1^2, \mathbf{X}_2^2, \mathbf{X}_3^2)) \tag{6}$$

In our model, we obtain two fusion feature map msf_1 and msf_1 using the MSF module The two fusing feature maps are connected with the upsampling feature map one by one in the decoder process to enhance the clarity of the object boundary of the output segmentation result. The pseudo-code of the multiscale low level feature fusion is described in Algorithm 1. The input parameters of the MSF are $\mathbf{X}, C, W, H$. $\mathbf{X}$ is processed by pre-processing, channel consistency, size consistency, and feature fusion in turn to obtain the final segmentation result $\mathbf{X}'$.

Algorithm 1: Multiscale low-level feature fusion

Input: Input parameters $\mathbf{X}, C, W, H$
A set of feature maps $\mathbf{X} = \{\mathbf{X_1}, \mathbf{X_2}, \ldots, \mathbf{X}_j\}$, $\mathbf{X}_j \in \mathbb{R}^{C_j \times H_j \times W_j}$;
A set of feature channels $C = \{C_1, C_2, \ldots, C_j\}$,$C_1 \leq C_2 \leq C_j$;
The high of feature maps $H = \{H_1, H_2, \ldots, H_j\}$, $H_1 \geq H_2 \geq H_j$;
The wide of feature maps $W = \{W_1, W_2, \ldots, W_j\}$,$W_1 \geq W_2 \geq W_j$;
Constraints: $H_j = W_j, j \in \mathbb{N}^+, j = 1, 2, 3, \ldots$;
Pre-processing: $\mathbf{X}_{sub} = sub(\mathbf{X})$, $\mathbf{X}_{sub} \leftarrow \{\mathbf{X_1}, \mathbf{X_2}, \mathbf{X_3}\}$, and $\mathbf{X}_{sub} \subset \mathbf{X}$;
Output: The output result of the fusion feature $\mathbf{X}'$
begin
    `// step 1:channel consistency`
    $\mathbf{X}_{nsub}$= [];
    for *Each feature map* $\mathbf{X}_j$ *in* $\mathbf{X}_{sub}$ **do**
        if $C_j! = C_2$ **then**
            $\mathbf{X}_j^1 = Conv_{1\times1}(\mathbf{X}_j)$;
        else
            $\mathbf{X}_j^1 = \mathbf{X}_j$;
        end
        $\mathbf{X}_{nsub}$.add($\mathbf{X}_j^1$);
    end
    `// step 2:size consistency`
    for *Each feature map* $\mathbf{X}_j$ *in* $\mathbf{X}_{nsub}$ **do**
        if $H_j! = H_{j+1}$ **then**
            $\mathbf{X}_j^p = Maxpool(\mathbf{X}_j^1)$;
        end
        if $H_{j+1}! = H_j$ **then**
            $\mathbf{X}_j^u = Upsampling(\mathbf{X}_j^1)$;
        else
            $\mathbf{X}_j^1 = \mathbf{X}_j^1$;
        end
    end
    $\mathbf{X}_1^2 = \mathbf{X}_1^1 \oplus \mathbf{X}_2^u$;
    $\mathbf{X}_2^2 = \mathbf{X}_1^p \oplus \mathbf{X}_2^1 \oplus \mathbf{X}_3^u$;
    $\mathbf{X}_3^2 = \mathbf{X}_3^1 \oplus \mathbf{X}_2^p$;
    `// step 3:feature fusion`
    $\mathbf{X}_1^2 \xleftarrow{Maxpool} \mathbf{X}_1^2$, $\mathbf{X}_2^2 \leftarrow \mathbf{X}_2^2$, and $\mathbf{X}_3^2 \xleftarrow{Upsamping} \mathbf{X}_3^2$;
    $\mathbf{X}' = Conv_{1\times1}(Concat(\mathbf{X}_1^2, \mathbf{X}_2^2, \mathbf{X}_3^2))$;
end
Return $\mathbf{X}'$

In the network decoder, we employ the three times upsampling operations to restore the output feature map size to the input size shown in Figure 1. In *Decoder*1 and *Decoder*2, we first upsample the feature map to 2 times scale and fuse the low-level feature (msf_1, msf_2) with detailed information by concatenation in the channel dimension, then use the two 3×3 convolutions to scatter converged information. In *Decoder*3, we upsample the feature map to 4 times the size, utilize the 3 × 3 convolution to reduce the chance to 64 and, lastly, use a 1 × 1 convolution to obtain the segmentation results of N categories.

4. Experiments

In this section, we describe the experimental setup, including the experimental environment and parameter settings, experimental datasets, evaluation measures, comparison algorithms, and experiment results and analysis.

4.1. Experiment Setting

We conducted the experiments on a single GPU (GeForce RTX 3080Ti, 12 GB RAM, 8 CPU/4 core) with Pytorch 1.8.1 + CUDA 11.1. During network training, we set the initial learning rate to 10^{-4}, and used the Adam [52] optimizer and cross-entropy loss function to train the network model. The size of the network training batch was set to 16 and the maximum number of training iterations was 200 epochs. The sign of the end of network training is that the training reaches the maximum number of iterations, or the network is stagnant in 20 epochs. In the experiment, the network input is an RGB image; the image is normalized and processed by a resizing method without distortion to 256 × 256 pixels. All the image label is transformed into gray labels with linear pixel mapping and the output of the network is a grayscale image with different category values.

4.2. Datasets

We used two rock detection datasets in this paper, a lunar rock dataset called MoonData (https://www.kaggle.com/datasets/romainpessia/artificial-lunar-rocky-landscape-dataset (accessed on 9 December 2022)) and a Martian rock dataset called MarsData [17]. The details of the two datasets are as shown in Table 3.

Table 3. Parameter details of two rock datasets.

DataSet	Training	Validation	Testing
MoonData	7812	977	977
MarsData	22,279	5541	3092

MoonData: This lunar rock dataset is a sample of artificial yet realistic lunar landscapes, which was used to train rock detection algorithms. The Moon rock dataset contains 9766 realistic renders of rocky lunar landscapes, which are labeled into four classes: background, sky, smaller rocks, and bigger rocks. MoonData is an RGB image with 480 × 720 pixels and the label is also a three-channel RGB image. In this experiment, we convert the three-channel RGB label to grayscale by linear pixel mapping, and we partition the dataset 8:1:1 into 7812 training images, 977 validation images, and 977 testing images. The details of the Mars dataset are described in Table 3.

MarsData: The Martian rock dataset (https://dominikschmidt.xyz/mars32k/ (accessed on 13 September 2021)) consists of about 32,000 color images collected by the Curiosity rover on Mars with a Mastcam camera between August 2012 and November 2018. All images have been scaled down using linear interpolation to 560 × 500 pixels; unfortunately, they don't have semantic segmentation labels. In previous work, the paper [17] completed a total of 405 labeled rock images of more than 20,000 rocks and the data were augmented to 30,912 images by cropping and rotating. In our experiment, we use the augmented Mars rock dataset to train and evaluate rock segmentation methods. Moreover, we repartitioned the dataset 9:1 according to the train–validation images with 22,279 training images, 5541 validation images, and 3092 testing images.

4.3. Evaluation Criteria

In order to report the research results in the field of semantic segmentation, most researchers used simple and representative measures of pixel accuracy (PA), class pixel accuracy (CPA), mean pixel accuracy (MPA), intersection and union (IoU), and mean intersection and union (MIoU). In this paper, we employ the standard evaluation standards for semantic segmentation to confirm the effectiveness of our model. We computed PA, MPA, Recall, and MIoU based on the corresponding confusion matrix to evaluate the quality of network predictions.

In the confusion matrix, the PA denotes the sum of the true positives and true negatives divided by the total number of queried individuals. The PA is computed as follow:

$$\mathrm{PA} = \frac{\mathrm{TP} + \mathrm{TN}}{\mathrm{TP} + \mathrm{TN} + \mathrm{FP} + \mathrm{FN}} \tag{7}$$

where true positive (TP) represents the number of positive samples that are correctly predicted as positive ones. True negative (TN) denotes the number of negative samples that are correctly determined as negative ones. False positive (FP) represents the number of negative objects that are incorrectly predicted as positive samples and false negative (FN) is the number of positive samples that are incorrectly classified as negative samples.

The class pixel accuracy is the percentage of the total predicted value that is correct for a category and MPA is the mean of CPA; CPA is represented as follow:

$$\mathrm{CPA} = \frac{\mathrm{TP}}{\mathrm{TP} + \mathrm{FP}} \tag{8}$$

where TP is the prediction accuracy of the category and TP + FP is all predictions in this category. $\mathrm{MPA} = \frac{1}{n}\sum_{i=0}^{n-1} CPA_i$, where n denotes the number of categories and CPA_i is the value of CPA in the i-th class. The recall is the probability that a category is predicted correctly, which is calculated by TP divided by TP + FN as follows:

$$\mathrm{Recall} = \frac{\mathrm{TP}}{\mathrm{TP} + \mathrm{FN}} \tag{9}$$

The IoU is the ratio of the intersection and union of the predicted results and the true values for a given classification. The IoU is computed as follows:

$$\mathrm{IoU} = \frac{\mathrm{TP}}{\mathrm{TP} + \mathrm{FN} + \mathrm{FP}} \tag{10}$$

where TP denotes the intersection set and TP + FN + FP is the union set of the predicted results and true values for a category. Moreover, MIoU is the mean of the IoU of the n classes; $\mathrm{MIoU} = \frac{1}{n}\sum_{i=0}^{n-1} IoU_i$, where IoU_i represents the value of IoU in the i-th class.

4.4. Compared Methods

In our experiment, we compared with the six latest semantic segmentation networks for rock detection, DeeplabV3+ [25], FCN [28], CCNet [53], DANet [54], PSPNet [24], and Swin-Unet [29]. Simple descriptions of these compared methods are as follows. FCN [28] is a basic model of classical semantic segmentation with the first full convolution network. PSPNet [24] used a pyramid pooling module (PPM) and dilated convolutions to integrate contextual information from different regions and embed it in FCN. DeeplabV3+ [25] used the ASPP module to obtain multiscale context information. DANet [54] and CCNet [53] employed a dual attention (DA) mechanism and criss-cross attention (CCA) mechanism to improve the accuracy of segmentation. Swin-Unet [29] is a novel vision transformer network-based semantic segmentation used to compare.

The main parameter settings of the compared methods are in Table 4, which contains the network backbone, downsampling multiple (dm), network encoder, and decoder. The *dm* represents the downsampling multiple of the input image in a network encoder; FCN8 denotes using an eight-fold sampling to obtain the output feature map.

The network decoder is divided into three methods to restore the output feature map: (1) the *one_upsamping* method employing upsampling once, (2) the *one_fuse + upsampling* method fusing fine-grained shallow features once and upsampling, and (3) the *muti_fuse + upsampling* method utilizing multiple-fusion and upsampling. Resnet-34-2 is a combination of the proposed model, which consists of two Resnet-34 blocks and a transformer block (T). It utilizes MSF to fuse more shallow features to obtain a finer-grained output.

Table 4. The main parameter settings of the compared methods.

Methods	Backbone	dm	Encoder	Decoder
DeeplabV3+ [25]	Resnet-50	1/16	Resnet-50+ASPP	one_fuse+upsampling
FCN [28]	Resnet-50	1/8	Resnet-50+FCN8	one_upsamping
CCNet [53]	Resnet-50	1/8	Resnet-50+CCA	one_upsamping
DANet [54]	Resnet-50	1/16	Resnet-50+DA	one_upsamping
PSPNet [24]	Resnet-50	1/16	Resnet-50+PPM	one_upsamping
Swin-Unet [29]	Resnet-50	1/16	Vision Transformer	muti_fuse+upsampling
RockSeg (Our)	Resnet-34-2	1/16	Resnet-34-2+MSF+T	one_fuse+upsampling

4.5. Experiment Results

In this section, we compared the state-of-the-art methods for deep space rock detection. All compared networks used the Resnet-50 backbone to extract the feature, and the input image was processed to a uniform size of 256 × 256 pixels by image resize, padding, and scale technology. In experiments, we not only used the evaluation metrics of PA, CPA, MPA, Recall, IoU, and MIoU mentioned in Section 4.3, but we also calculated the network parameters (Params) to evaluate the spatial complexity of the network, evaluated the time complexity of the model by floating-point operations (FLOPs), and computed the inference speed of the network in frames per second (FPS) to evaluate the performance of the networks.

4.5.1. Results on MoonData

The rock detection results on the MoonData dataset are shown in Table 5; the bold data represents the best prediction results. We can see that the proposed RockSeg obtained the best prediction results in the PA, MPA, Recall, and MIoU indicators, and it achieved a faster inference speed with fewer network parameters. Specifically, it improved by about 5.3% and 11.2% on the PSPNet model in MPA and Recall evaluation indicators, respectively. In the MIoU indicator, the proposed RockSeg improved about 2.2%, 6.1%, 1.4%, 6.7%, 10.5%, and 6.1% on DeeplabV3+, FCN, CCNet, DANet, PSPNet, and Swin-Unet, respectively. Moreover, we found that RockSeg not only obtained a high detection precision but the network also had a fast inference speed; the FPS was up to 52.90 HZ. The network parameters of the proposed model were reduced by about seven times compared to the CCNet model.

Table 5. The comparison results with other methods on MoonData.

Methods	PA (%)	MPA (%)	Recall (%)	MIoU (%)	FLOPs (G)	Params (M)	FPS (HZ)
DeeplabV3+	97.01	88.32	83.00	76.71	45.77	40.35	51.15
FCN	96.72	87.83	80.34	72.80	34.72	32.94	**53.31**
CCNet	97.05	89.33	83.26	77.49	59.93	52.27	38.12
DANet	96.29	86.42	77.86	72.18	**14.30**	47.55	51.98
PSPNet	95.86	84.01	73.95	68.37	14.84	46.70	43.43
Swin-Unet	96.54	85.75	79.28	72.78	40.06	17.25	33.26
RockSeg (Ours)	**97.25**	**89.42**	**85.13**	**78.90**	20.29	**7.94**	52.90

Furthermore, we used the CPA and IoU indicators to evaluate the different category detection results shown in Figure 4. The MoonData dataset has four categories including ground, sky, bigger rocks simplified "brocks", and smaller rocks simplified "srocks". In

Figure 4, the horizontal axis represents four different categories and the vertical axis is the value of CPA and IoU, respectively. The legend represents different methods and ranges (R) in two categories of brocks and srocks; R is defined as

$$R = |R_{brocks} - R_{srocks}| \tag{11}$$

where R_{brocks} and R_{srocks} denote the accuracy score in brocks and srocks classes and R represents the difference between the two categories; the larger R, the more difficult it is to distinguish between the two categories; otherwise, the easier it is to distinguish between the two categories.

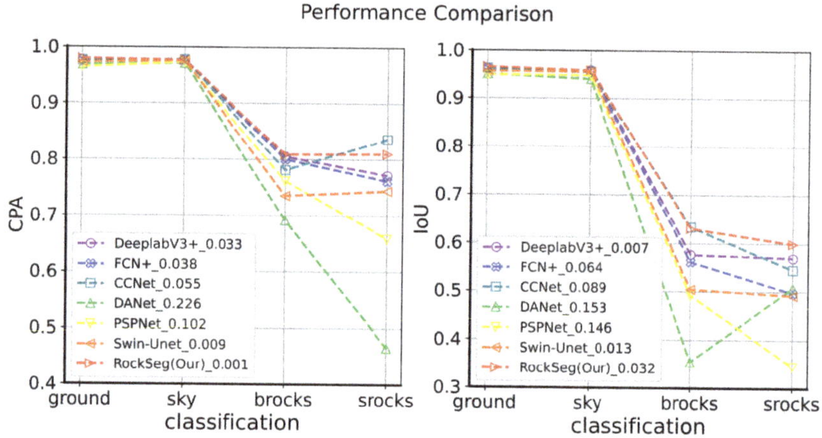

Figure 4. The comparison results of different network models with CPA and IoU on MoonData.

On the whole, we discovered that all compared methods could obtain better detection accuracy in the ground and sky categories, but the detection results of different models have a large gap in the brocks and srocks categories. For an input rock image of the Moon, the pixel ratio of the ground and sky is large, and the pixel ratio of the rocks is relatively small; there is an imbalance of categories in the MoonData data. In semantic segmentation, category objects with different pixel proportions in an image have different detection difficulties [55,56]. Category objects with small proportion pixels are difficult to distinguish, while category objects with multi-proportion pixels are relatively easy to distinguish [7]. So the ground and sky categories have a higher accuracy than the brocks and srocks categories in CPA and IoU evaluation.

From Figure 4, we can see that the DANet model had the worst classification results; the proposed model and the CCNet model had better detection accuracy than other methods. The DANet and PSPNet models obtained a large R between the brocks and srocks classifications; the accuracy range was 0.226 and 0.102 in CPA, and 0.153 and 0.146 in IoU, respectively. In the IoU evaluation, we found that RockSeg obtained the best scores in each classification; in particular, it achieved 63.11% and 59.94% IoU scores in brocks and srocks classifications, respectively. In the CPA evaluation, the RockSeg obtained high CPA values in ground, sky, and brocks classification, in which the brocks and srocks were 63.11% and 59.94%, respectively. The CCNet network also achieved the highest accuracy in the srocks class using the CPA evaluation, in which the brocks and srocks accuracy were 78.17% and 83.65%, respectively. However, RockSeg obtained a smaller R in CPA and IoU than the CCNet model. The accuracy range of RockSeg was only 0.001 compared to 0.055 for CCNet in the CPA evaluation and, in the IoU, the accuracy range of RockSeg was 0.032 and the R was lower than CCNet in the CPA and IoU evaluations. Thus, the proposed RockSeg is more robust than the CCNet model.

In addition, we show the confusion matrix of the probability of different categories being predicted in Figure 5. We can see that most pixels with ground and sky categories can be correctly classified; the probability of brocks being incorrectly classified as the ground category was 0.24 and the probability was only 0.02 of them being incorrectly classified as the srocks category. In the srocks category, there was only a probability of 0.29 and 0.01 of being incorrectly classified as the ground and brocks categories, respectively. Therefore, RockSeg has strong robustness for detecting deep space rocks; both large and small rocks can be detected correctly. Last, we show the visualization segmentation results of different methods on MarsData in Figure 6. There are five visualization segmentation results with different angles of sunlight and shadows in Figure 6. The yellow rectangle represents the contrast of the local details. Figure 6a,d,e denote the vision that follows the sunlight and Figure 6b,c are the visual angle against the sunlight on the surface of the Moon. When the sun's rays shine perpendicular to the surface of the Moon, the rock shadows are small as shown in Figure 6d,e; otherwise, the rock shadows are big as shown in Figure 6a–c. We can see that the proposed RockSeg could accurately obtain segmentation results with different sunlight shadows and angles. Specifically, our model could clearly detect the boundary of the object compared to the other models and some small rock objects could also be accurately detected.

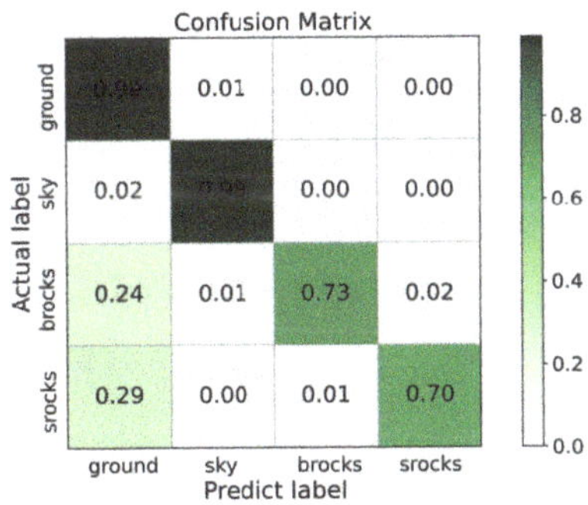

Figure 5. The confusion matrix of the RockSeg model on MoonData.

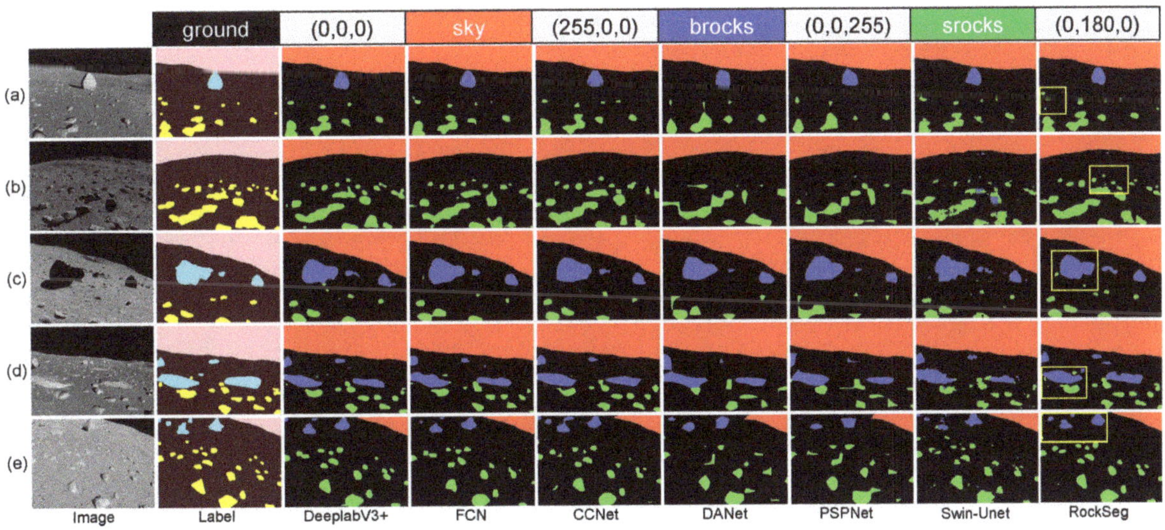

Figure 6. Comparison of the visualization segmentation results for different models on MoonData. (**a–e**) show the different views of the rocks on the lunar surface from different Suns. (**a**,**d**,**e**) denote the vision that follows the sunlight, and (**b**,**c**) represent the visual angle against the sunlight.

4.5.2. Results on MarsData

The comparison results with other methods on MarsData are shown in Table 6; the bold denotes the best prediction accuracy. The MarsData has two categories, the background and rock objects. The pixel ratio of rocks and background is not much different, so it is relatively easy to segment them. We can see that the compared methods are all above 96% accuracy in the PA, Recall, and MIoU indicators. From Table 6, the FCN model obtained the best inference speed compared with other models and the precision of the PSPNet model was relatively low. Our proposed model obtained the best accuracy in each indicator compared to the other methods. Moreover, the proposed model achieved a high inference speed with low network parameters and computation complexity. Furthermore, we evaluated the CPA and IoU of different categories on MarsData; the results of different methods are shown in Table 7. We found that *R* was small in the CPA and IoU evaluations for all compared models. Due to the classes being relatively balanced on MarsData data, they could be very well detected. We can see that the RockSeg model achieved the best score in the IoU evaluation and obtained the best PA value in ground classification compared to the other models. In deep space rock detection, our proposed model had excellent portability and robustness.

Table 6. The comparison results with other methods on MarsData. The best result for each column is in bold.

Methods	PA (%)	MPA (%)	Recall (%)	MIoU (%)	FPS (HZ)
DeeplabV3+	98.72	97.12	98.51	97.12	54.13
FCN	98.52	98.29	98.29	98.29	**55.83**
CCNet	98.74	98.53	98.53	98.53	40.18
DANet	98.03	97.73	97.73	97.73	54.26
PSPNet	97.69	94.85	97.29	96.05	55.10
Swin-Unet	98.39	98.21	98.10	96.39	34.55
RockSeg (Ours)	**98.91**	**98.78**	**98.73**	**97.54**	55.18

Table 7. Comparisons of CPA and IoU for different methods on MarsData. The best result for each column is in bold.

Methods	CPA (%)		IoU (%)	
	Ground	Rocks	Ground	Rocks
DeeplabV3+	99.01	98.12	98.14	96.10
FCN	98.88	97.76	97.84	95.50
CCnet	98.17	**98.32**	98.37	96.59
Danet	99.02	98.17	97.14	94.06
PSPNet	98.19	96.62	96.65	96.18
Swin-Unet	98.71	97.71	97.66	95.11
RockSeg (Ours)	**99.16**	98.23	**98.41**	**96.68**

By comprehensive feature extraction and rich semantic feature fusion, the proposed model could realize high-precision detection. The proposed RockSeg network used combining the CNN and vision transformer to extract the rock features, in which the CNN network is advantageous in obtaining local multiscale context features and the vision transformer block is more suitable for capturing global features. The local and global rock features were fused to achieve a comprehensive feature extraction by the proposed hybrid network, which is beneficial for the detection of objects of different sizes. Moreover, the designed MSF module fused multiscale low-layer features to the output feature map which could improve the accuracy of the segmentation results. Furthermore, we eliminated the feature redundancy and overlap by manually adjusting the network parameters to achieve a lightweight network; see Section 4.6 for details of model parameters. Using the above policies, the proposed model could achieve high accuracy and inference speed under low computation complexity.

The visualization segmentation results of our model and the state-of-the-art methods on MarsData are shown in Figure 7. In the label image, we labeled the object rock as green and the other compared segment results as yellow for visual distinction. In the four image visualization segmentations, we discovered that all of the comparison models could accurately detect large rock objects. But, for some small gravel with burning in the soil and some dense rocks, it is relatively more difficult to distinguish and identify them than big rocks. In terms of accuracy and clarity of the border segmentation, the RockSeg results were finer and closer to the label image than the other model, and we used the red box in our model to show the finer boundary segmentation results. From the visualization segmentation results, we can see that the Swin-Unet, PSPNet, DANet, CCNet, and FCN models had poor detection results in small object detection; their segmentation results show that the target boundary was blurred and rough. In Figure 7b,d, we can see that the proposed model achieved accurate detection in big rocks, and also obtained accurate segmentation in some dense small rocks or small rocks buried in the soil.

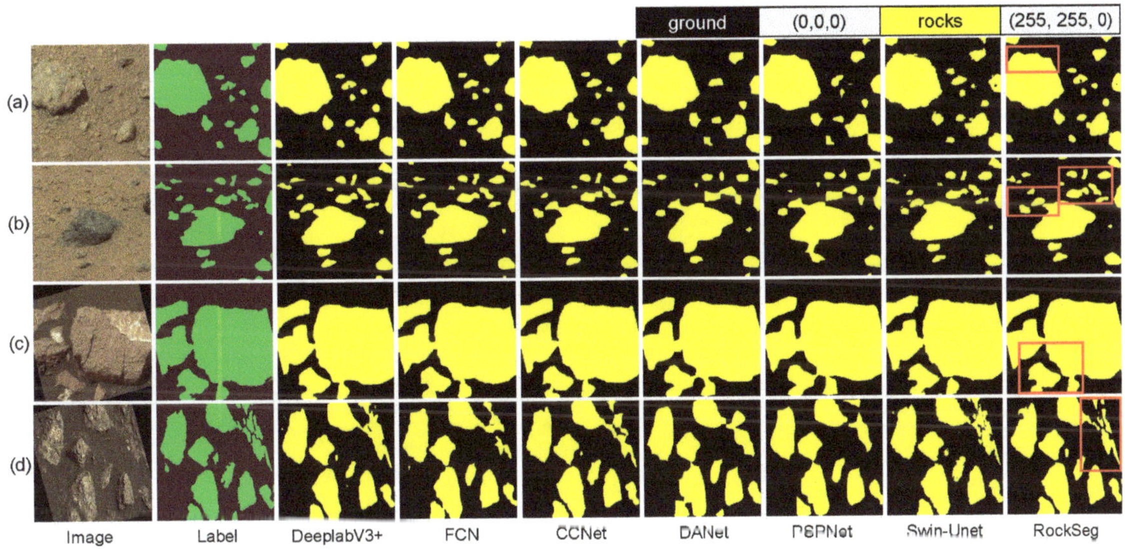

Figure 7. Comparison of the visualization segmentation results for different models on MarsData. (**a–d**) show the different rocky features of the Martian surface. (**a**,**b**) represent the surface of Mars as composed of sparse mudstones and small boulders, and (**c**,**d**) denote dense large rocks and sandstone partially buried in the sand.

4.5.3. Ablation Study

In this section, we ablated our network to validate the performance of the proposed model. The results of the ablation study are shown in Table 8 and the best value in each column is in bold. The MSF represents the multiscale low-level feature fusion module, the transformer block is simplified as T, the ✓ flag represents the module being used, and the – flag denotes the module not being used. In Table 8, we can see that our model obtained the best PA, MPA, and MioU compared to the other ablation models. The T module with a multi-headed attention mechanism could capture the global context information of the rock to improve the rock's object detection accuracy. Thus, we discovered that RockSeg-T and the RockSeg-MSF-T achieved a higher accuracy in PA, MPA, Recall, and MIoU than RockSeg-no-MSF-T. Specifically, RockSeg-T obtained the best accuracy in Recall. The multiscale feature fusion module obtained the rich fusion feature maps msf_1 and msf_2; they were added to the output feature map using the upsampling process to accurately enhance the clarity of the semantic segmentation object boundary and improve the accuracy of segmentation. In Table 8, we found the RockSeg-MSF and RockSeg-MSF-T models also achieved an improvement over the RockSeg-no-MSF-T in the four evaluation indicators. On the whole, our model with T and MSF modules obtained the best performance in rock detection.

Table 8. The ablation results of our model on MoonData.

Model	MSF	T	PA (%)	MPA (%)	Recall (%)	MIoU (%)
RockSeg-no-MSF-T	–	–	97.12	88.70	84.39	77.93
RockSeg-T	–	✓	97.20	88.95	**85.29**	78.72
RockSeg-MSF	✓	–	97.18	88.89	85.10	78.56
RockSeg-MSF-T (Ours)	✓	✓	**97.25**	**89.42**	85.13	**78.90**

Furthermore, we show the visual ablation results of the MSF and T modules with the heatmap output shown in Figure 8. We compared the different channel activation statuses with the different models of RockSeg-no-MSF-T, RockSeg-T, RockSeg-MSF, and RockSeg-MSF-T. We used a blue–red color scheme to show the difference; the smaller the value, the closer it is to blue, the larger the value, the closer it is to red. In Figure 8, the top is the original rock image and label; Figure 8(1–6) show the two low-level feature maps msf_1 and msf_2, where Figure 8(1–3) denote the output results of msf_1 and Figure 8(4–6) are the output results of msf_2 with RockSeg-no-MSF-T, RockSeg-MSF, and RockSeg-MSF-T (our model). Figure 8a–c show the feature map output results of the transformer block using RockSeg-no-MSF-T, RockSeg-T, and RockSeg-MSF-T.

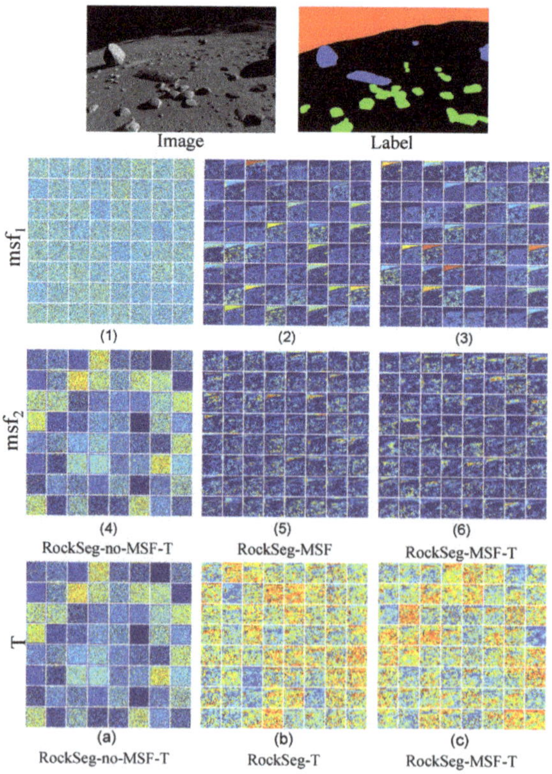

Figure 8. Comparison of the visual results of the ablation study. (**1–6**) are the visual results of the MSF module with different models; (**1–3**) denote the output feature map of the msf_1 module; (**4–6**) are the output feature map of the msf_2 module. (**a–c**) represent the output feature map of the T module with different models.

For the whole network structure, msf_1 is closer to the input network and msf_2 is relatively far from the network input. We can see that most information of the original image was retained in the activation output msf_1. From the activation output msf_1 and msf_2, we discovered that, as the number of layers increased, the activation output became more and more abstract. The density of activation decreased with the deepening of layers and the information about categories was increased. For example, the density of activation contrast, Figure 8(3) $>$ Figure 8(6). In msf_1 and msf_2, we can see that RockSeg-MSF and RockSeg-MSF-T had more channel activation statuses than RockSeg-no-MSF-T; the proposed MSF module obtained more rich semantic information from the context. Due to the T module being far from the network input in the whole network structure, the activation output is very sparse and abstract as shown in Figure 8a–c. The RockSeg-T and RockSeg-MSF-T used multiple attention mechanisms to activate important information by setting different weights of attention. Thus, they had more red feature signatures than the RockSeg-no-MSF-T model in Figure 8. On the whole, from different output feature heatmaps, we found that the proposed semantic segmentation network based on a hybrid framework combining CNN and vision transformer, using an efficient feature extraction backbone and multiscale low-level feature fusion, had an excellent presentation of features to achieve good performance in rock detection.

4.6. Impact of Different Backbones and Parameters on Models

In this section, we discuss the parameter impact on our model and tune them with the MoonData data. The parameters contain the different backbone networks, the number of backbone layers, and the number of layers and heads of the T block. The tuning process is divided into three groups, denoted gps, $gps = \{gp1, gp2, gp3\}$. In the three groups, we kept the same decoder process, normalized the size of the feature map in downsampling to an input size of 1/16 times, and evaluated them by the indicators described in Section 4.3. The tuning results are shown in Table 9. In Table 9, nbs is the number of Resnet blocks. The backbone represents the network encoder with different modules and parameters, where MSF and T denote the multiscale low-level feature fusion module and vision transformer in the backbone, respectively. The T module has two import parameters, the number of heads represented by h and the depth of the transformer layer denoted d. The – represents the process of adjusting their parameters and the ✓ denotes using this module.

Table 9. The impact of different backbones and parameters on models. The best result for each column in gps is in bold.

gps	Backbone	nbs	MSF	T	PA (%)	MPA (%)	Recall (%)	MIoU (%)	FLOPs (M)	Params (G)	FPS (HZ)
gp1	Resnet-50	4	✓	✓	**97.24**	**90.45**	84.11	78.67	27.89	32.01	34.41
	Resnet-34	4	✓	✓	97.22	89.77	84.84	**78.87**	25.40	27.97	41.44
	Resnet-18	4	✓	✓	97.10	88.43	**84.96**	78.16	**22.07**	**17.86**	**51.57**
gp2	Resnet-34-4	4	✓	✓	97.22	**89.77**	84.84	78.87	25.40	27.97	41.44
	Resnet-34-3	3	✓	✓	97.24	88.8	85.96	79.12	22.00	14.72	43.57
	Resnet-34-2	2	✓	✓	**97.25**	89.42	**85.13**	**78.90**	**20.29**	**7.94**	**52.90**
gp3	Resnet-34-2-88	2	✓	–	97.17	88.98	85.29	78.73	21.10	11.09	52.83
	Resnet-34-2-44	2	✓	–	**97.25**	**89.42**	85.13	**78.90**	**20.29**	**7.94**	**52.90**
	Resnet-34-2-14	2	✓	–	97.21	88.97	85.38	78.80	20.29	**7.94**	52.47

In $gp1$, we compared the impact of different Resnet backbones with four Resnet blocks on deep space rock detection. We combined Resnet-50, Resnet-34, and Resnet-18 with the T module as the backbone network separately, and used the same MSF module to decode the network. In $gp1$, we found Resnet-50 obtained the best PA and MPA with maximum parameters and a large amount of computation; Resnet-18 had low parameters, small amounts of computation, and high FPS. Resnet-34 achieved the best MIoU compared to Resnet-50 and Resnet-18; the detection accuracy in PA and MPA indicators was close to Resnet-50, and the model parameters and computations were close to Resnet-18. In order to balance the calculation complexity and accuracy of the rock detection model in a deep space environment with limited resources, we chose Resnet-34 as the backbone for our model. Too many feature extraction layers may cause feature redundancy and overlap. To obtain an efficient and lightweight feature extraction backbone network, after obtaining the Resnet-34 backbone, we tuned the number of Resnet blocks in the backbone to optimize our model. In $gp2$, Resnet-34-n represents the backbone with different numbers of Resnet blocks n, where $n = \{2,3,4\}$. We discovered that Resnet-34-2 with two Resnet blocks achieved better performance than the Resnet-34-4 and Resnet-34-3 models. In $gp2$, the Resnet-34-4 backbone network may have over-representation; the Resnet-34-2 backbone network achieved the appropriate representation for rock feature extraction. The Resnet-34-2 backbone could obtain the best PA, MPA, Recall, and MIoU score under low computation and parameters, and fast inference speed.

Last, in $gp3$, we test the impact on the proposed model by tuning the parameters of h and d in the T block. Resnet-34-2-hd is composed of the Resnet-34-2 backbone network and the T module with h heads and d layers, where h is the number of heads, $h = \{1,4,8\}$; corresponding to the number of transformer layer d denotes $d = \{4,4,8\}$. In $gp3$, we found the parameter of h and d had little effect on the precision of the model, but the complexity of different parameters was different. In deep space, the probe carries limited resources, and onboard computation needs to satisfy not only high precision requirements but also low complexity requirements. We can see that Resnet-34-2-44 achieved a higher PA, MPA, and MIoU than other models with a faster inference speed. Thus, in this paper, in order to create a high accuracy and low complexity rock detection model, we chose the final Resnet-34-244 as the hybrid framework combining CNN and transformer for deep space rock images, which is based on the Resnet-34-2 backbone and the T module containing four heads and transformer layers.

5. Conclusions

In this paper, we proposed an efficient deep space rock detection network, named RockSeg, which is a novel semantic segmentation network based on a hybrid framework combining CNN and vision transformer for deep space rock images. The novel model contains an efficient backbone feature extraction block and a multiscale low-level feature fusion module for deep space rock detection. Firstly, to enhance the feature extraction, we used part of the Resnet-34 backbone and combined it with the visual transformer block as a new backbone network Resnet-T to extract the global context information of the rock. In addition, we proposed a simple and efficient multiscale low-level feature fusion module to obtain more rich detailed feature information. These rich features were fused to the output feature map in the network decoder to obtain a more fine-grained output result and improve the clarity of the semantic segmentation object boundary. Furthermore, the proposed model was applied to two rock segmentation datasets (lunar and Martian rock data) compared with six state-of-the-art segmentation models for deep space rock detection. The results demonstrated that the RockSeg model outperforms the state-of-the-Art rock detection methods; our model achieved good performance in deep space rock detection. In particular, on MoonData data, our model achieved accuracy up to 97.25% in the PA and 78.97% in the MIoU indicators with low parameters, smaller amount of computation, and high inference speeds.

In tuning the network process, we found the deeper network may not be a good choice to achieve the best performance; too many deep network structures may be redundant for feature extraction. The proposed hybrid network combines CNN and transformer; they need to play to their strengths to complement and integrate local and global context information. To obtain the best appropriate network structure, we manually adjust the network backbone structure and optimize the parameter configuration with coarse-grained parameter tuning. We employed a conventional backbone to achieve network feature extraction and used evaluation measures and visual heatmaps simultaneously to decide whether the network feature extraction is insufficient or redundant. Then, the network structure was suitably decreased and increased based on the qualitative and quantitative assessment results to meet the specific detection task. In the future, we need to further study how to integrate CNN and transformer network structures adaptively to remove redundant features and enhance the ability to capture local and global context information. Moreover, we will transplant and expand our work to the detection of deep space multi-category terrain segmentation, further improving the availability of the model in deep space.

Author Contributions: Conceptualization, L.F. and K.Z.; Formal analysis, L.F., K.Z. and W.M.; Funding acquisition, J.Y.; Investigation, J.Y.; Methodology, L.F.; Project administration, J.Y.; Software, L.F., K.Z. and W.M.; Supervision, J.Y.; Visualization, L.F., X.N. and K.Z.; Writing—original draft, L.F.; Writing—review and editing, L.F. and X.N. All authors have read and agreed to the published version of the manuscript.

Funding: This work was funded by the National Natural Science Foundation of China (Grant No. 62076127).

Data Availability Statement: Not applicable.

Conflicts of Interest: The authors declare no conflict of interest.

References

1. Kilic, C.; Martinez, B.; Tatsch, C.A.; Beard, J.; Strader, J.; Das, S.; Ross, D.; Gu, Y.; Pereira, G.A.; Gross, J.N. NASA Space Robotics Challenge 2 Qualification Round: An Approach to Autonomous Lunar Rover Operations. *IEEE Aerosp. Electron. Syst. Mag.* **2021**, *36*, 24–41. [CrossRef]
2. Kuang, B.; Wisniewski, M.; Rana, Z.A.; Zhao, Y. Rock Segmentation in the Navigation Vision of the Planetary Rovers. *Mathematics* **2021**, *9*, 3048. [CrossRef]
3. Turan, E.; Speretta, S.; Gill, E. Autonomous navigation for deep space small satellites: Scientific and technological advances. *Acta Astronaut.* **2022**, *193*, 56–74. [CrossRef]
4. Furlán, F.; Rubio, E.; Sossa, H.; Ponce, V. Rock detection in a Mars-like environment using a CNN. In *Proceedings of the Mexican Conference on Pattern Recognition*; Springer: Queretaro, Mexico, 2019; pp. 149–158.
5. Liu, H.; Yao, M.; Xiao, X.; Xiong, Y. RockFormer: A U-shaped Transformer Network for Martian Rock Segmentation. *IEEE Trans. Geosci. Remote Sens.* **2023**, *61*, 1–16. [CrossRef]
6. Brockers, R.; Delaune, J.; Proença, P.; Schoppmann, P.; Domnik, M.; Kubiak, G.; Tzanetos, T. Autonomous safe landing site detection for a future mars science helicopter. In Proceedings of the 2021 IEEE Aerospace Conference (50100), Big Sky, MT, USA, 6–13 March 2021; pp. 1–8.
7. Fan, L.; Yuan, J.; Zha, K.; Wang, X. ELCD: Efficient Lunar Crater Detection Based on Attention Mechanisms and Multiscale Feature Fusion Networks from Digital Elevation Models. *Remote Sens.* **2022**, *14*, 5225. [CrossRef]
8. Ebadi, K.; Coble, K.; Kogan, D.; Atha, D.; Schwartz, R.; Padgett, C.; Vander Hook, J. Semantic mapping in unstructured environments: Toward autonomous localization of planetary robotic explorers. In Proceedings of the 2022 IEEE Aerospace Conference, Big Sky, MT, USA, 5–12 March 2022.
9. Ugenti, A.; Vulpi, F.; Domínguez, R.; Cordes, F.; Milella, A.; Reina, G. On the role of feature and signal selection for terrain learning in planetary exploration robots. *J. Field Robot.* **2022**, *39*, 355–370. [CrossRef]
10. Jiang, J.; Zeng, X.; Guzzetti, D.; You, Y. Path planning for asteroid hopping rovers with pre-trained deep reinforcement learning architectures. *Acta Astronaut.* **2020**, *171*, 265–279. [CrossRef]
11. Wang, W.; Lin, L.; Fan, Z.; Liu, J. Semi-Supervised Learning for Mars Imagery Classification and Segmentation. *ACM Trans. Multimed. Comput. Commun. Appl.* **2023**, *19*, 1–23. [CrossRef]
12. Gui, C.; Li, Z. An autonomous rock identification method for planetary exploration. In *Emerging Technologies for Information Systems, Computing, and Management*; Springer: Hangzhou, China, 2013; pp. 545–552.
13. Burl, M.C.; Thompson, D.R.; deGranville, C.; Bornstein, B.J. Rockster: Onboard rock segmentation through edge regrouping. *J. Aerosp. Inf. Syst.* **2016**, *13*, 329–342. [CrossRef]
14. Li, Y.; Wu, B. Analysis of rock abundance on lunar surface from orbital and descent images using automatic rock detection. *J. Geophys. Res. Planets* **2018**, *123*, 1061–1088. [CrossRef]
15. Xiao, X.; Cui, H.; Yao, M.; Tian, Y. Autonomous rock detection on mars through region contrast. *Adv. Space Res.* **2017**, *60*, 626–635. [CrossRef]
16. Xiao, X.; Cui, H.; Yao, M.; Fu, Y.; Qi, W. Auto rock detection via sparse-based background modeling for mars rover. In Proceedings of the 2018 IEEE Congress on Evolutionary Computation (CEC), Rio de Janeiro, Brazil, 8–13 July 2018; pp. 1–6.
17. Xiao, X.; Yao, M.; Liu, H.; Wang, J.; Zhang, L.; Fu, Y. A kernel-based multi-featured rock modeling and detection framework for a mars rover. *IEEE Trans. Neural Netw. Learn. Syst.* **2021**, *34*, 3335–3344. [CrossRef] [PubMed]
18. Goh, E.; Ward, I.R.; Vincent, G.; Pak, K.; Chen, J.; Wilson, B. Self-supervised Distillation for Computer Vision Onboard Planetary Robots. In Proceedings of the 2023 IEEE Aerospace Conference, Big Sky, MT, USA, 4–11 March 2023; pp. 1–11.
19. Huang, G.; Yang, L.; Cai, Y.; Zhang, D. Terrain classification-based rover traverse planner with kinematic constraints for Mars exploration. *Planet. Space Sci.* **2021**, *209*, 105371. [CrossRef]

20. Lateef, F.; Ruichek, Y. Survey on semantic segmentation using deep learning techniques. *Neurocomputing* **2019**, *338*, 321–348. [CrossRef]
21. Li, J.; Zi, S.; Song, R.; Li, Y.; Hu, Y.; Du, Q. A stepwise domain adaptive segmentation network with covariate shift alleviation for remote sensing imagery. *IEEE Trans. Geosci. Remote Sens.* **2022**, *60*, 1–15. [CrossRef]
22. Yu, F.; Koltun, V. Multi-scale context aggregation by dilated convolutions. *arXiv* **2015**, arXiv:1511.07122.
23. He, K.; Zhang, X.; Ren, S.; Sun, J. Spatial pyramid pooling in deep convolutional networks for visual recognition. *IEEE Trans. Pattern Anal. Mach. Intell.* **2015**, *37*, 1904–1916. [CrossRef]
24. Zhao, H.; Shi, J.; Qi, X.; Wang, X.; Jia, J. Pyramid scene parsing network. In Proceedings of the IEEE Conference on Computer Vision and Pattern Recognition, Honolulu, HI, USA, 21–26 July 2017; pp. 2881–2890.
25. Chen, L.C.; Zhu, Y.; Papandreou, G.; Schroff, F.; Adam, H. Encoder-decoder with atrous separable convolution for semantic image segmentation. In Proceedings of the European Conference on Computer Vision (ECCV), Munich, Germany, 8–14 September 2018; pp. 833–851.
26. Yuan, Y.; Chen, X.; Wang, J. Object-contextual representations for semantic segmentation. In *Proceedings of the Computer Vision—ECCV 2020: 16th European Conference, Glasgow, UK, 23–28 August 2020*; Proceedings, Part VI 16; Springer: Glasgow, UK, 2020, pp. 173–190.
27. Liu, L.; Cheng, J.; Quan, Q.; Wu, F.X.; Wang, Y.P.; Wang, J. A survey on U-shaped networks in medical image segmentations. *Neurocomputing* **2020**, *409*, 244–258. [CrossRef]
28. Long, J.; Shelhamer, E.; Darrell, T. Fully convolutional networks for semantic segmentation. In Proceedings of the IEEE Conference on Computer Vision and Pattern Recognition, Boston, MA, USA, 7–12 June 2015; pp. 3431–3440.
29. Cao, H.; Wang, Y.; Chen, J.; Jiang, D.; Zhang, X.; Tian, Q.; Wang, M. Swin-unet: Unet-like pure transformer for medical image segmentation. In *Proceedings of the European Conference on Computer Vision*; Springer: Tel Aviv, Israel, 2022; pp. 205–218.
30. Li, H.; Qiu, L.; Li, Z.; Meng, B.; Huang, J.; Zhang, Z. Automatic Rocks Segmentation Based on Deep Learning for Planetary Rover Images. *J. Aerosp. Inf. Syst.* **2021**, *18*, 755–761. [CrossRef]
31. Ma, W.; Jiabin, Y.; Zha, k.; Fan, L. Onboard rock detection algorithm based on spiking neurnal network. In *Computer Science*; China Academic Journal Electronic Publish House: Beijing, China, 2023; pp. 98–104.
32. Sun, K.; Xiao, B.; Liu, D.; Wang, J. Deep high-resolution representation learning for human pose estimation. In Proceedings of the IEEE/CVF Conference on Computer Vision and Pattern Recognition, Long Beach, CA, USA, 15–20 June 2019; pp. 5693–5703.
33. Dai, Y.; Zheng, T.; Xue, C.; Zhou, L. SegMarsViT: Lightweight Mars Terrain Segmentation Network for Autonomous Driving in Planetary Exploration. *Remote Sens.* **2022**, *14*, 6297. [CrossRef]
34. Azkarate, M.; Gerdes, L.; Wiese, T.; Zwick, M.; Pagnamenta, M.; Hidalgo-Carrió, J.; Poulakis, P.; Pérez-del Pulgar, C.J. Design, testing, and evolution of mars rover testbeds: European space agency planetary exploration. *IEEE Robot. Autom. Mag.* **2022**, *29*, 10–23. [CrossRef]
35. Xiao, Z.; Su, Y.; Deng, Z.; Zhang, W. Efficient combination of CNN and transformer for dual-teacher uncertainty-guided semi-supervised medical image segmentation. *Comput. Methods Programs Biomed.* **2022**, *226*, 107099. [CrossRef]
36. Yang, H.; Yang, D. CSwin-PNet: A CNN-Swin Transformer combined pyramid network for breast lesion segmentation in ultrasound images. *Expert Syst. Appl.* **2023**, *213*, 119024. [CrossRef]
37. Dosovitskiy, A.; Beyer, L.; Kolesnikov, A.; Weissenborn, D.; Zhai, X.; Unterthiner, T.; Dehghani, M.; Minderer, M.; Heigold, G.; Gelly, S.; et al. An image is worth 16x16 words: Transformers for image recognition at scale. *arXiv* **2020**, arXiv:2010.11929.
38. Zhang, Y.; Zhao, Y.; Dong, Y.; Du, B. Self-supervised Pre-training via Multi-modality Images with Transformer for Change Detection. *IEEE Trans. Geosci. Remote Sens.* **2023**, *61*, 1–11.
39. Zhang, X.; Cheng, S.; Wang, L.; Li, H. Asymmetric cross-attention hierarchical network based on CNN and transformer for bitemporal remote sensing images change detection. *IEEE Trans. Geosci. Remote Sens.* **2023**, *61*, 1–15. [CrossRef]
40. Luo, H.; Wang, P.; Xu, Y.; Ding, F.; Zhou, Y.; Wang, F.; Li, H.; Jin, R. Self-supervised pre-training for transformer-based person re-identification. *arXiv* **2021**, arXiv:2111.12084.
41. Fang, J.; Lin, H.; Chen, X.; Zeng, K. A hybrid network of cnn and transformer for lightweight image super-resolution. In Proceedings of the IEEE/CVF Conference on Computer Vision and Pattern Recognition, New Orleans, LA, USA, 19–20 June 2022; pp. 1103–1112.
42. Wagstaff, K.; Lu, Y.; Stanboli, A.; Grimes, K.; Gowda, T.; Padams, J. Deep mars: Cnn classification of mars imagery for the pds imaging atlas. In Proceedings of the AAAI Conference on Artificial Intelligence, New Orleans, LA, USA, 2–7 February 2018; Volume 32.
43. Li, J.; Zhang, L.; Wu, Z.; Ling, Z.; Cao, X.; Guo, K.; Yan, F. Autonomous Martian rock image classification based on transfer deep learning methods. *Earth Sci. Inform.* **2020**, *13*, 951–963. [CrossRef]
44. Liu, H.; Yao, M.; Xiao, X.; Cui, H. A hybrid attention semantic segmentation network for unstructured terrain on Mars. *Acta Astronaut.* **2023**, *204*, 492–499. [CrossRef]
45. Liu, M.; Liu, J.; Ma, X. MRISNet: Deep-learning-based Martian instance segmentation against blur. *Earth Sci. Inform.* **2023**, *16*, 965–981. [CrossRef]
46. Panambur, T.; Chakraborty, D.; Meyer, M.; Milliken, R.; Learned-Miller, E.; Parente, M. Self-supervised learning to guide scientifically relevant categorization of martian terrain images. In Proceedings of the IEEE/CVF Conference on Computer Vision and Pattern Recognition, New Orleans, LA, USA, 19–20 June 2022; pp. 1322–1332.

47. Goh, E.; Chen, J.; Wilson, B. Mars terrain segmentation with less labels. In Proceedings of the 2022 IEEE Aerospace Conference (AERO), Big Sky, MT, USA, 5–12 March 2022; pp. 1–10.
48. He, K.; Zhang, X.; Ren, S.; Sun, J. Deep residual learning for image recognition. In Proceedings of the IEEE Conference on Computer Vision and Pattern Recognition, Las Vegas, NV, USA, 26 June–1 July 2016; pp. 770–778.
49. Xiong, R.; Yang, Y.; He, D.; Zheng, K.; Zheng, S.; Xing, C.; Zhang, H.; Lan, Y.; Wang, L.; Liu, T. On layer normalization in the transformer architecture. In Proceedings of the International Conference on Machine Learning, Virtual, 13–18 July 2020; pp. 10524–10533.
50. Zhu, Q.; Liao, C.; Hu, H.; Mei, X.; Li, H. MAP-Net: Multiple attending path neural network for building footprint extraction from remote sensed imagery. *IEEE Trans. Geosci. Remote Sens.* **2020**, *59*, 6169–6181. [CrossRef]
51. Hou, S.; Xiao, S.; Dong, W.; Qu, J. Multi-level features fusion via cross-layer guided attention for hyperspectral pansharpening. *Neurocomputing* **2022**, *506*, 380–392. [CrossRef]
52. Bock, S.; Weiß, M. A proof of local convergence for the Adam optimizer. In Proceedings of the 2019 International Joint Conference on Neural Networks (IJCNN), Budapest, Hungary, 14–19 July 2019; pp. 1–8.
53. Huang, Z.; Wang, X.; Huang, L.; Huang, C.; Wei, Y.; Liu, W. Ccnet: Criss-cross attention for semantic segmentation. In Proceedings of the IEEE/CVF International Conference on Computer Vision, Seoul, Republich of Korea, 27 October–2 November 2019; pp. 603–612.
54. Fu, J.; Liu, J.; Tian, H.; Li, Y.; Bao, Y.; Fang, Z.; Lu, H. Dual attention network for scene segmentation. In Proceedings of the IEEE/CVF Conference on Computer Vision and Pattern Recognition, Long Beach, CA, USA, 15–20 June 2019; pp. 3146–3154.
55. Hou, S.; Liu, Y.; Yang, Q. Real-time prediction of rock mass classification based on TBM operation big data and stacking technique of ensemble learning. *J. Rock Mech. Geotech. Eng.* **2022**, *14*, 123–143. [CrossRef]
56. Zhang, W.; Li, H.; Han, L.; Chen, L.; Wang, L. Slope stability prediction using ensemble learning techniques: A case study in Yunyang County, Chongqing, China. *J. Rock Mech. Geotech. Eng.* **2022**, *14*, 1089–1099. [CrossRef]

remote sensing

MDPI

Article

Neural Network Compression via Low Frequency Preference

Chaoyan Zhang, Cheng Li, Baolong Guo * and Nannan Liao

Institute of Intelligent Control and Image Engineering, Xidian University, Xi'an 710071, China; cyzhang0808@stu.xidian.edu.cn (C.Z.); licheng812@stu.xidian.edu.cn (C.L.); nnliao@stu.xidian.edu.cn (N.L.)
* Correspondence: blguo@xidian.edu.cn; Tel.: +86-130-8896-6638

Abstract: Network pruning has been widely used in model compression techniques, and offers a promising prospect for deploying models on devices with limited resources. Nevertheless, existing pruning methods merely consider the importance of feature maps and filters in the spatial domain. In this paper, we re-consider the model characteristics and propose a novel filter pruning method that corresponds to the human visual system, termed Low Frequency Preference (LFP), in the frequency domain. It is essentially an indicator that determines the importance of a filter based on the relative low-frequency components across channels, which can be intuitively understood as a measurement of the "low-frequency components". When the feature map of a filter has more low-frequency components than the other feature maps, it is considered more crucial and should be preserved during the pruning process. We conduct the proposed LFP on three different scales of datasets through several models and achieve superior performances. The experimental results obtained on the CIFAR datasets and ImageNet dataset demonstrate that our method significantly reduces the model size and FLOPs. The results on the UC Merced dataset show that our approach is also significant for remote sensing image classification.

Keywords: model compression; neural network pruning; frequency domain; lightweight deep neural networks; remote sensing image classification

Citation: Zhang, C.; Li, C.; Guo, B.; Liao, N. Neural Network Compression via Low Frequency Preference. *Remote Sens.* **2023**, *15*, 3144. https://doi.org/10.3390/rs15123144

Academic Editor: Giuseppe Scarpa

Received: 11 May 2023
Revised: 11 June 2023
Accepted: 13 June 2023
Published: 16 June 2023

1. Introduction

Deeper and wider architectures of convolutional neural networks (CNNs) have achieved great success in the field of computer vision and have been widely used in both academia and industry [1–6]. Nevertheless, they also impose high requirements for computing power and memory footprint, resulting in a significant challenge in deploying most state-of-the-art CNNs on mobile or edge devices. Therefore, reducing the parameters and calculations of existing models is still a research hot spot, where an effective technique is model compression. This technique can achieve a balanced trade-off between accuracy and model size.

Conventional compression strategies consist of network pruning [7–11], quantization [12–14], low-rank approximation [15,16], knowledge distillation [17–20] and lightweight neural framework design [21–23]. Network pruning has become the most popular model compression technique. Recent pruning strategies in this category can be roughly divided into weight pruning [8,24,25] and filter pruning [26–28], according to the granularity of pruning. Weight pruning directly removes the selected weights from a filter, resulting in unstructured sparsity. Despite the irregular structure having a high compression ratio, real acceleration cannot be achieved on general hardware platforms or Basic Linear Algebra Subprogram (BLAS) libraries [29]. Filter pruning directly discards the selected filters, leaving a regular network structure, which makes it hardware friendly. CNNs have exerted a great influence on remote sensing classification tasks with their powerful feature representation capability. Zhang et al. [30] and Volpi [31] constructed relatively small networks and trained them using satellite images from scratch. Xia et al. [32] and Marmanis et al. [33]

extracted features from the middle layer of the pre-training network, formed global feature representation and realized remote sensing classification. Nogueira et al. [34] used a remote sensing dataset for fine-tuning and obtained a superior classification performance. Zhu et al. [35] proposed a knowledge-guided land pattern depicting (KGLPD) framework for urban land-use mapping. Ref. [36] constructed a new remote sensing knowledge graph (RSKG) from scratch to support the inference recognition of unseen remote sensing image scenes. Zhang et al. [37] made full use of the advantages of CNNs and CapsNet models to propose an effective framework for remote sensing image scene classification. Ref. [38] proposed a CNN pre-training method guided by the human visual attention mechanism to improve the land-use scene classification accuracy. However, the success of CNNs comes with expensive computing costs and a high memory footprint. However, the classification task of remote sensing images often needs to be carried out on the airborne or satellite-borne equipment with limited computing resources. Insufficient computing resources hinder the application of CNNs in remote sensing imaging. Therefore, model pruning technology can alleviate this resource constraint and enable CNNs to develop in the field of remote sensing. It is worth noting that the scale of public remote sensing image datasets is usually smaller than the scale of natural image datasets, which contain hundreds of thousands or even millions of images. This leads to a lot of parameter redundancy and structural redundancy in the network model, so pruning techniques are needed to reduce these redundancies and avoid overfitting of the model. Therefore, pruning technology has a great application demand and prospect in real-time remote sensing image classification (as shown in Figure 1) for resource-constrained devices such as spaceborne or airborne devices [39,40].

Figure 1. Examples of remote sensing image classification.

To achieve both network speedup (reduction in FLOPs) and a model size reduction (reduction in parameters), we focus on filter pruning aiming to provide a general solution (as shown in Figure 2) for devices with a low computational power.

Inherent Attribute Constraint. The pruning operation on a filter can be regarded as decreasing the constraints generated by different inherent attributes in CNNs. Li et al. [26] calculated the L_1-norm of parameters or features to judge the degree of attribute constraints. The conclusion was that the smaller norm, the less useful the information, which indicates that a smaller norm is a weak constraint for the network and should be pruned first.

Hu et al. [41] measured the constraint of each filter by counting the Average Percentage of Zeros (APoZ) in the activation values output by the filter. The sparser the activation feature map, the weaker the constraints of the feature map. Molchanov et al. [42] used a first-order Taylor expansion to approximate the contribution of feature maps to the network output to estimate the importance of filters. He et al. [43] calculated the geometric median of filters in the same layer, in this case, the filter closest to the geometric median is considered as a weak constraint that should be pruned first. Lin et al. [44] proposed that feature maps with a lower rank have fewer constraints on the network. Therefore, the corresponding filters can be removed first. Sui et al. [45] proposed to estimate the independence of channels by calculating the nuclear norm of the feature map. Channels with a lower independence have weaker constraints and can be deleted first. In brief, these methods follow the principle of "weak constraints are pruned, strong constraints are retained" to achieve fast pruning. Nevertheless, they cannot make up for the loss in the network training process while merely improving the performance by fine-tuning in the later stage.

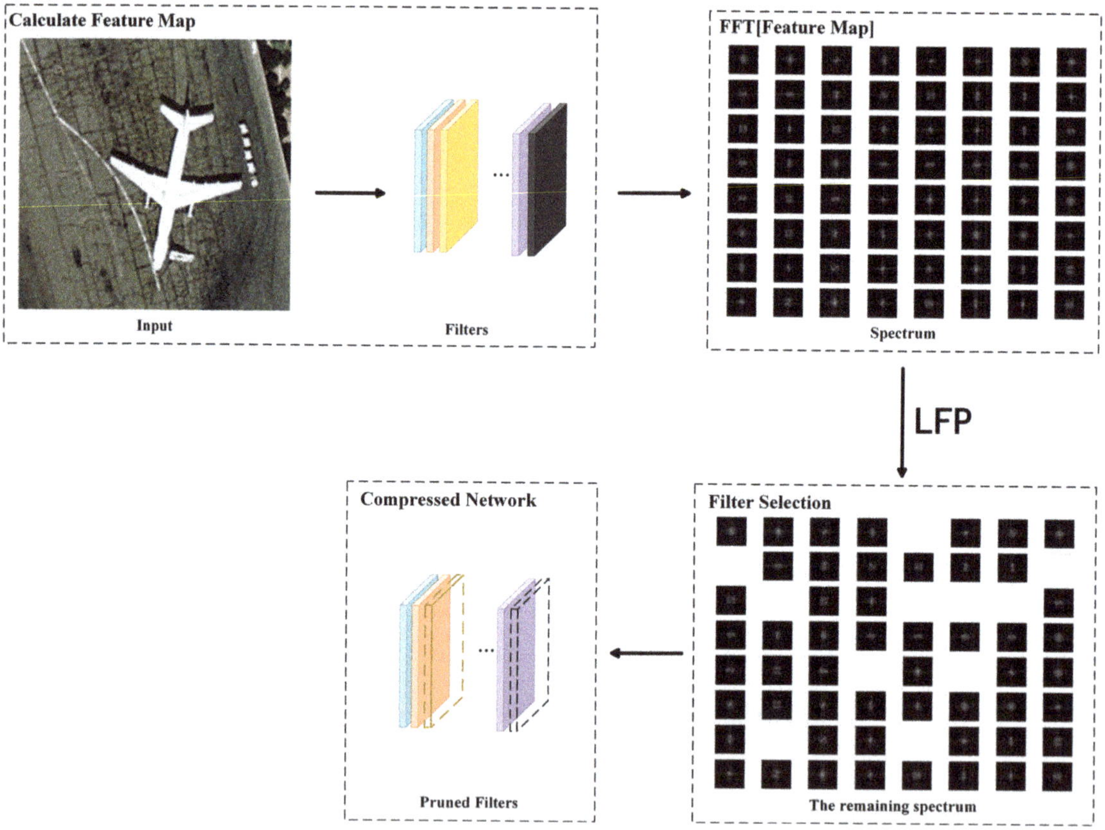

Figure 2. Framework of the proposed LFP. In the left box, we first use images to run through the convolutional layers to obtain the feature maps. The resulting feature map is then calculated by FFT in the second box. In the third box, we then estimate the LFP of each spectrum map, which is used as the criteria for pruning. The last box shows the pruning (the dotted filters) according to LFP calculation results.

Induction of sparsity. These methods learn sparse structure pruning by imposing sparsity constraints on the target function in the network. Wen et al. [28] proposed a compression method based on structured sparse learning, which learns different compact structures by regularizing various network structures. Huang et al. [46] also introduced a new scaling factor, which scales the output of various structures, such as neurons, group convolutions or residual blocks, and then safely removes structures whose corresponding scaling factor is close to zero. In contrast to [46], Liu et al. [47] utilized the scaling parameter in the batch normalization layer to control the output of the corresponding channel without introducing any additional parameters. Zhao et al. [48] further extended the scaling parameters in the batch normalization layer to include bias terms and estimated their probability distributions by variational inference. These methods are not based on deterministic values but on the distribution of the corresponding scaling parameters to prune redundant channels, which makes them more interpretable. Lin et al. [49] studied important filters by incorporating two different regularizations of structural sparsity into the original loss function, achieving a superior performance on a variety of state-of-the-art network frameworks. Chen et al. [50] imposed regularizations on both filter weights and BN scaling factors and then evaluated the filter importance by their combination. Compared with the inherent attribute constraint method, the induction of sparsity achieves better compression and acceleration results. Nevertheless, the sparsity requirement must be embedded into the training process, so it is expensive with regard to training time and manpower.

In general, it is desirable to pursue a higher compression ratio and speedup ratio without losing too much accuracy. In recent years, pruning models according to the constraints provided by different inherent attributes in CNNs has become a popular filter pruning strategy. Instead of directly selecting filters, important feature maps are first determined and then the corresponding channels are retained. As reported in [44,45,51–53], feature maps can inherently reflect rich and important information about the input data and filters. Therefore, calculating the importance of feature maps could provide better pruning guidance for filters/channels. For example, the feature-oriented pruning concept [45] can provide richer knowledge of filter pruning than the intra-channel information when considering the correlation of multiple filter/channel feature information. The importance of a filter that is merely determined by its corresponding feature map could be easily affected by the input data. On the contrary, cross-channel feature information leads to more stable and reliable measurements, as well as a deeper exploitation of the underlying correlations between different feature maps (and corresponding filters). The results in [45] also show that the proposed inter-channel and feature-guided strategy outperforms the state-of-the-art filter-guided methods in terms of task performance (e.g., accuracy) and compression performance (e.g., model size and floating-point operation reduction).

Preference and Frequency Perspective. In previous work, both the feature-map-based strategy and the filter-guided strategy passively formulate the pruning strategy according to the inherent internal structure of CNNs in the spatial domain. Specifically, some theories, such as optimal brain damage [54] and the lottery ticket hypothesis [55,56], propose that there is parameter redundancy inside the model. Therefore, only if the parameters of the filter or feature maps are calculated in the spatial domain can their importance be determined according to experience and mathematical knowledge. Considering the "preference" of the model from the perspective of frequency domain, it can be found that the neural network often learns low-frequency information first, and then slowly learns high-frequency information [57,58] in the process of fitting the data (and some high-frequency information cannot be perfectly fitted). At the same time, the human visual system is sensitive to the representation of low-frequency information [59,60], while the representation of low-frequency information in the spatial domain is not prominent enough. We can observe from Figure 3 that after discarding part of the high-frequency information, the category of the image can still be identified through the retained low-frequency information.

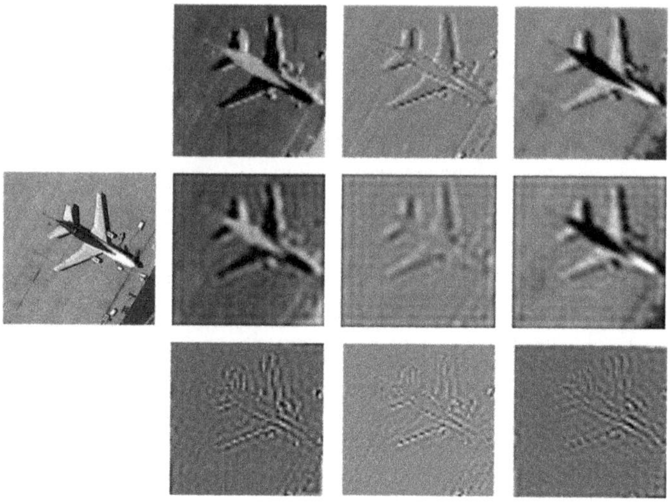

Figure 3. The original image (**left**), three random feature maps (**top**), low-frequency representations of the feature maps (**middle**) and high-frequency representations (**bottom**).

In order to maintain the consistency between the model characteristics and the human visual system, it is necessary to explore new methods in the frequency domain. Experiments in [61] show that, after adding a low-frequency filter in the test image, the robustness of the whole model is enhanced. In addition, adding low-frequency information can efficiently improve the accuracy and gradually achieve a performance similar to the original image. Considering that most real scenario images are predominantly low frequency, the influence of noise is relatively negligible on the low-frequency images but enormous on the high-frequency images, which easily leads to overfitting of the model. Therefore, a better task and compression performance can be obtained by discarding the learning of high-frequency information (the feature maps with more high-frequency components are pruned).

Technical Preview and Contributions. Motivated by these promising potential benefits, in this paper, we exploit the frequency information of cross-channel features for efficient filter pruning. We propose a novel metric termed Low Frequency Preference (LFP) to determine the importance of filters based on the relative frequency components across channels. It can be intuitively understood as a measurement of the "low frequency component". Specifically, if the feature map of a filter is measured with a larger proportion of low-frequency components compared with other feature maps of the layer, the feature map is more important than that in other channels, which needs to be preserved during pruning. On the contrary, feature maps with more high-frequency components are less preferred by the model, which indicates that they contain very limited information or knowledge. Therefore, the corresponding filters are treated as unimportant and can be safely removed without affecting the model capacity.

To sum up, the contributions of this paper can be summarized as follows:

- We analyze the properties of a model from the new perspective of the frequency domain and associate the characteristics of an image with the frequency domain preference characteristics of the model. Similar to the "smaller-norm-less-important" hypothesis, we come up with a novel "lower-frequency-more-important" metric. On this basis, a low-cost, high-robustness, low-frequency component analysis scheme is proposed.
- We propose a novel metric that measures the relative low-frequency components of multiple feature maps to determine the importance of filters, termed LFP. It originates from an inter-channel perspective to determine the importance of filters more globally and precisely, thus providing better guidelines for filter pruning.

- We apply the LFP-based importance determination method to different filter pruning tasks. Extensive experiments show that the proposed method achieves good results while maintaining high precision. Notably, on the CIFAR-10 dataset, our method improves the accuracy by 0.96% and 0.95% over the baseline ResNet-56 and ResNet-110 models, respectively. Meanwhile, the model size and FLOPs are reduced by 44.7% and 48.4% (for ResNet-56) and 39.0% and 47.8% (for ResNet-110), respectively. On the ImageNet dataset, it achieves 40.8% and 46.7% storage and computation reductions, respectively, for ResNet-50 and the accuracy of Top-1 and Top-5 is 1.21% and 1.26% higher than the baseline model, respectively.

2. Proposed Method

2.1. Notation

We formally introduce symbols and notations in this section. Assume a pre-trained convolutional neural network model has L layers. We use C_i and C_{i+1}, to represent the number of input and output channels for the i-th convolutional layer, respectively. $F_{i,j}$ represents the j-th filter of the i-th layer, then the dimension of filter is $F_{i,j}$ is $\mathbb{R}^{C_i \times K \times K}$, where K denotes the kernel size of the network. The i-th layer of the CNN model $\mathcal{W}^{(i)}$ can be represented by $\{F_{i,1}, F_{i,2}, \ldots, F_{i,j}\}$ that contains j filters, where $F_{i,j} \in \mathbb{R}^{C_i \times K \times K}$, $1 \leq j \leq C_{i+1}$. The tensor of connection in the deep CNN network can be parameterized by $\{\mathcal{W}^{(i)} \in \mathbb{R}^{C_{i+1} \times C_i \times K \times K}, 1 \leq i \leq L\}$. The outputs of i-th layer, i.e., i-th feature maps, are denoted as $\mathcal{M}^i = \{M_{i,1}, M_{i,2}, \ldots, M_{i,C_{i+1}}\} \in \mathbb{R}^{C_{i+1} \times h \times w}$. The feature map corresponding to the j-th channel is $M_{i,j} \in \mathbb{R}^h \times w$. The height and width of the feature map are h and w, respectively. In filter pruning, $\mathcal{W}^{(i)}$ can be split into two groups, i.e., a subset I containing n_{i1} filters to be reserved and a subset, with less importance, to be pruned U containing n_{i2} filters. Thus, we have $I \cap U = \varnothing, I \cup U = \mathcal{W}^{(i)}$ and $n_{i1} + n_{i2} = C_{i+1}$.

2.2. Frequency Domain Analysis of Feature Maps

The Fourier transform aims to obtain the signal distribution in the frequency domain, which can also be utilized in digital image processing, since an image is a collection of points sampled in a continuous space (real scenario). It uses a two-dimensional matrix to represent each point in the space, and the image can be represented by $z = f(x,y)$. For the discrete signal of digital image, we choose the discrete Fourier transform (DFT) to obtain its frequency distribution (spectrum). Then, the frequency can be regarded as an indicator of the intensity change in the image, which reveals the gradient of the gray level in the plane space. Specifically, if the gray level changes quickly, the frequency will be high. On the contrary, if the gray level changes slowly, the frequency will be low. In terms of an image, a high-frequency signal usually corresponds to the edge and noise, while a low-frequency signal describes the image contour and background signal. The two-dimensional DFT is defined as follows:

$$\begin{aligned} F(u,v) &= 2D - DFT[f(x,y)] \\ &= \sum_{x=0}^{M-1} \sum_{y=0}^{N-1} f(x,y) e^{-j2\pi(\frac{ux}{M} + \frac{vy}{N})}, \end{aligned} \tag{1}$$

where 2D-DFT $[\cdot]$ stands for the two-dimensional DFT; $f(x,y)$ is a digital image of size $M \times N$; and x and y are spatial variables, which, respectively, represent the specific horizontal and vertical coordinates in the digital image $f(x,y)$. Then, u and v are frequency domain variables, where $u \in \{0,1,2,\ldots,M-1\}$, $v \in \{0,1,2,\ldots,N-1\}$; $e^{-j2\pi(\frac{ux}{M} + \frac{vy}{N})}$ is the transform kernel of the DFT, which has separability.

Therefore, the DFT of the output of i-th layer (i.e., i-th feature map) is denoted as:

$$\begin{aligned} F_{M_{i,C_{i+1}}}(u,v) &= 2D - DFT[M_{i,C_{i+1}}] \\ &= \sum_{x=0}^{h-1}\sum_{y=0}^{w-1} f(x,y)e^{-j2\pi(\frac{ux}{h}+\frac{vy}{w})}, \end{aligned} \tag{2}$$

To further boost the computational efficiency of the DFT, Cooley et al. [62] proposed a special kind of DFT termed a one-dimensional fast Fourier transform (FFT). In this way, the number of multiplications required in the DFT can be greatly reduced. In addition, the more sampling points to be transformed, the more significant the savings of the FFT algorithm computation. Based on the separability of the Fourier transform kernel $e^{-j2\pi(\frac{ux}{h}+\frac{vy}{w})}$, the 2D-DFT can also be computed using the two-step FFT:

$$\begin{aligned} F_{M_{i,C_{i+1}}}(u,v) &= FFT_x\{FFT_y[f(x,y)]\} \\ &= FFT_y\{FFT_x[f(x,y)]\} \\ &= FFT(M_{i,C_{i+1}}), \end{aligned} \tag{3}$$

The spectrum map obtained by the two-dimensional Fourier transform is a distribution of image gradient. The points on the spectrum map do not have a one-to-one correspondence with the points on the image plane, even if the frequency is not shifted. The degree of brightness or darkness on the Fourier spectrum map indicates the intensity difference between the gray value of a point on the image with the neighboring points (i.e., the gradient and the frequency value of a point). Larger differences/gradients indicate higher frequencies and lower energies, which leads to lower values and a darker appearance on the spectrum map. A smaller difference/gradient indicates a lower frequency and a higher energy, resulting in a higher numerical value and a brighter appearance on the spectrum map. In other words, the brighter the frequency spectrum, the higher the energy, the lower the frequency and the smaller the image difference (more flat). Therefore, the result of the FFT on the image is shown in Figure 4c. The low-frequency component of the image is distributed in the four corners of the spectrum map. For better observation, the low-frequency component $F(0,0)$ is translated to the center of the frequency rectangle defined by the interval $[0, M-1]$ and $[0, N-1]$ via the following equation:

$$f(x,y)(-1)^{x+y} \xrightarrow{FFT} F(u-\frac{M}{2}, v-\frac{N}{2}), \tag{4}$$

In the displayed spectrum map, since the dynamic range of other gray values is compressed, the log transformation in Equation (5) is performed once on Figure 4c,d. Therefore, the details can be greatly improved to observe and calculate the spectral law.

$$F'(u,v) = 1 + \log|F(u,v)|, \tag{5}$$

Therefore, the i-th spectrum map (the i-th feature map after FFT) is represented as $\mathcal{M}^i_{FFT} = \{M^F_{i,1}, M^F_{i,2}, \ldots, M^F_{i,C_{i+1}}\}$. To observe the extraction of different frequency features by different filters more apparently, we visualize the feature maps of the model ResNet-50-conv1 as well as the corresponding spectrum map in Figures 5 and 6. The bright areas in the spectrum correspond to the low-frequency components (with higher values), while the dark areas correspond to the high-frequency components (with lower values). In addition, some spectra with fewer low-frequency components and the corresponding feature maps are annotated with red boxes. Therefore, we can prune the filters corresponding to the feature maps with fewer low-frequency components, thus leaving more low-frequency components.

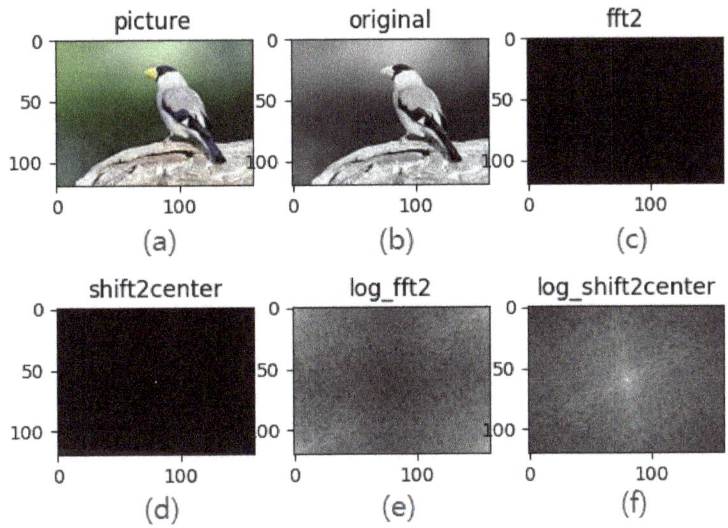

Figure 4. Workflow of the FFT. (**a**) A color bird image; (**b**) the grayscale image of (**a**). The image should be converted into grayscale before the FFT since the frequency is an indicator of the intensity change in the image. (**c**) The result of applying FFT to (**b**); (**d**) the centralized spectrum; (**e**) logarithmic transformation of (**c**) for better observation and calculation of the spectrum; (**f**) the result of (**e**) after centralization.

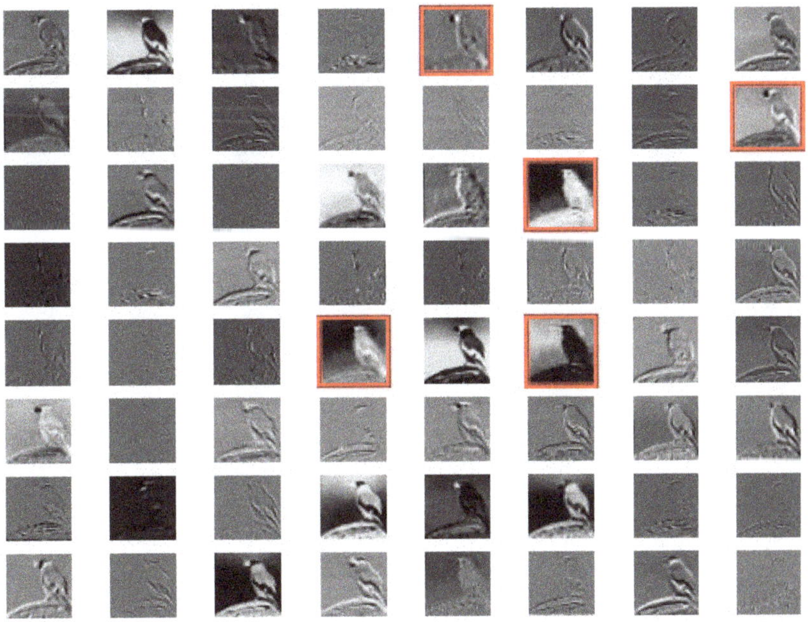

Figure 5. Visualization of feature maps of ResNet-50-conv1.

Figure 6. Spectrum corresponding to the feature map.

2.3. *LFP-Based Model Pruning*

As mentioned above, measuring importance in the frequency domain is a new research approach. Motivated by those promising benefits in Section 1, we propose to explore the filter importance from an inter-channel perspective, and the key idea is to use LFP to measure the importance of each feature map (and its corresponding filter). Specifically, if there are more low-frequency components in a feature map of a channel, the model "prefers" its intrinsic information, that is to say, the Frequency Preference Index of this feature map is higher. The Frequency Preference Index is higher as the filter corresponding to the feature map becomes more important. On the other hand, feature maps with relatively few low-frequency components (i.e., high-frequency components dominate) contain relatively little useful information. Therefore, even if the corresponding filter is excluded, the information and knowledge can still be roughly preserved by feature maps of other filters after the fine-tuning process. In other words, filters that generate low-frequency preference feature maps tend to be more "ignorable", which can be interpreted as having lower importance. Therefore, it would be appropriate to remove those filters that have feature maps with low channel frequency preferences, while still maintaining the high model capacity.

Filter pruning aims to identify and remove the less important filter sets from $\mathcal{W}^{(i)}$. To accurately measure the importance, we design a mathematical metric to quantify the Frequency Preference of a feature map using the Frobenius norm in Equation (6). It was reported in [63,64] that the F-norm can be used to measure the energy and difference of an image. In addition, we have also mentioned that higher frequency locations in the image mean lower energy, lower value, and a darker appearance in the spectrum. On the contrary, lower frequency locations mean higher energy, higher values and a brighter appearance on the spectrum. To this end, we elaborate a mathematical metric to measure Frequency Preference by using the F-norm of the spectrum.

$$\|A\|_F = \sqrt{tr(A^T A)} = \sqrt{\sum_{i=1}^{m}\sum_{j=1}^{n}|a_{ij}|^2} \quad , \tag{6}$$

where A is an $m \times n$ matrix and a_{ij} is each element of matrix A.

If the importance of a filter is merely determined by its corresponding feature map, the results may be sensitive to input data. Cross-channel feature information leads to more stable and reliable measurements, which is suitable for discovering the underlying correlations between different feature maps (and corresponding filters). Thus, in practice, in order to simultaneously remove multiple unimportant filters, a combination of frequency preference on multiple feature maps needs to be calculated. For the i-th layer with output feature maps, $\mathcal{M}^i_{FFT} = \{M^F_{i,1}, M^F_{i,2}, \ldots, M^F_{i,C_{i+1}}\} \in \mathbb{R}^{C_{i+1} \times h \times w}$. Firstly, let $\mathcal{M}^i_{FFT}$ be rewritten as $M^i = [m_{i,1}{}^T, m_{i,2}{}^T, \ldots, m_{i,C_{i+1}}{}^T]^T \in \mathbb{R}^{C_{i+1} \times hw}$, a matrix of C_{i+1} rows and hw columns, $m_{i,C_{i+1}} \in \mathbb{R}^{hw}$. To determine the minimum k-row frequency preference in $\mathcal{M}^i_{FFT}$, we first successively delete row $m_{i,j}$ from M^i and compute the corresponding F-norm change between the remaining $(C_{i+1} - 1)$ row matrix and the original C_{i+1} row matrix M^i. Then, C_{i+1} F-norm change values are obtained after C_{i+1} computations, and the k values with the smallest change are determined by sorting, along with their corresponding feature maps. These selected k feature maps $M_{i,j}$ are interpreted as receiving a lower "preference" from the model compared to other feature maps, so their corresponding filters $F_{i,j}$ are less important and should be pruned. Therefore, computing the change in the global F-norm in the feature map $\mathcal{M}^i$ in i-th layer, that is, the low frequency preference of $\mathcal{M}^i$, can be defined as follows:

$$LFP[\mathcal{M}^i] \triangleq [\left\|M^i\right\|_F - \left\|M^i * Z_j\right\|_F]_{j=1}^{C_{i+1}}, \tag{7}$$

where $\|\cdot\|_F$ is the Frobenius norm, $*$ is the matrix convolution operation and Z_j is the row mask matrix whose j-th row entries are zeros and other entries are ones.

In the set of F-norm changes obtained by LFP$[\mathcal{M}^i]$, the k smallest changes can be determined according to the pruning rate, and the corresponding feature maps and filters are not important and can be pruned. As shown in Figure 7, by randomly extracting the spectra corresponding to five change values, it can be observed that the spectra with more low-frequency components show higher LFP change values.

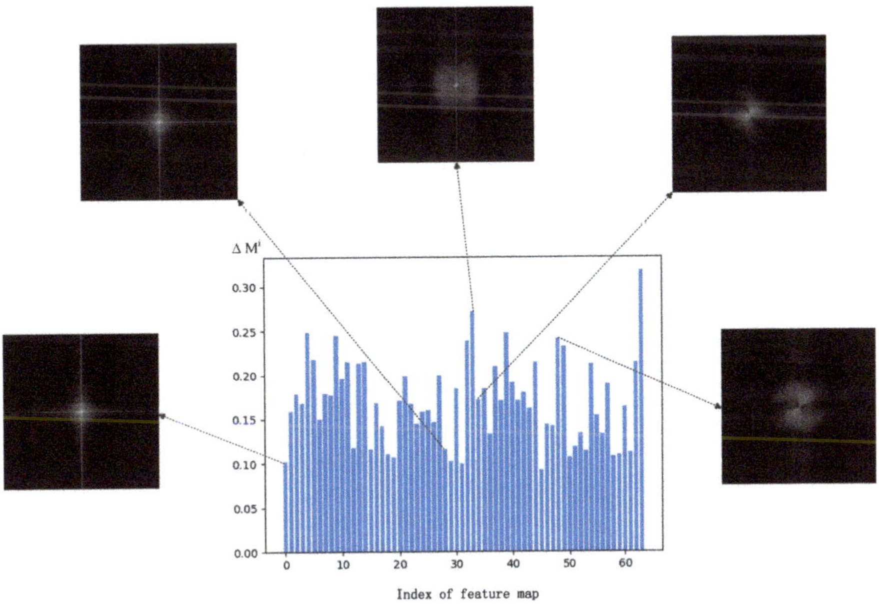

Figure 7. The low frequency preference of feature maps for one layer in ResNet-50. The ordinate is the change value of Mi, while the abscissa is the index of the feature map.

2.4. The Overall Algorithm

Combining the above two steps, the whole filter pruning process is developed from an inter-channel perspective. Figure 8 is a chart of methodology for the proposed method.

The pseudo-code of LFP is provided in Algorithm 1, which gives a lucid description and summary of our proposed filter pruning algorithm. Starting from a pre-trained model $\mathcal{W}^{(i)}$, the feature maps obtained after the image input model are calculated by the FFT to obtain the spectrum. The spectrum is reshaped into a matrix M^i with row C_{i+1} and column hw. Then, an LFP calculation is performed on M^i and the results are sorted. According to the pruning ratio, specific filters can be pruned. After fine-tuning the pruned model, a sub-model $\mathcal{W}^*$ can be obtained.

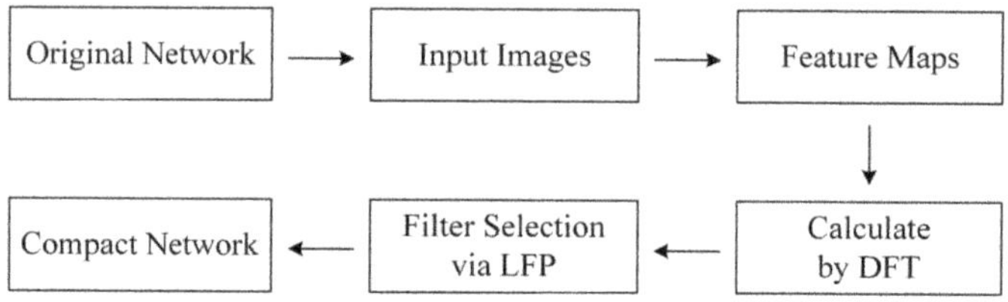

Figure 8. The chart of methodology.

Algorithm 1 Algorithm Description of the LFP method for the i-th layer

Input: An L-layer CNN model with pre-trained weights $\mathcal{W}^{(i)}$; The i-th feature maps $\mathcal{M}^i = \{M_{i,1}, M_{i,2}, \ldots, M_{i,C_{i+1}}\} \in \mathbb{R}^{C_{i+1} \times h \times w}$; target sparsity S; training set D;
Output: A sub-model satisfying the target sparsity S and its optimal weight values $\mathcal{W}^*$;

for Sample a mini-batch from D **do**
  FFT calculation: Calculate $F_{M_i,C_{i+1}}$ by Equation (3)
  Reshape FFT feature maps: $M^i := reshape(\mathcal{M}^i_{FFT}, [C_{i+1}, hw])$
  for $i = 1; i <= C_{i+1}; i++$ **do**
    LFP calculation: LFP$[\mathcal{M}^i]$ by Equation (7)
  end for
end for
Filters Selection: Sort LFP$[\mathcal{M}^i]$;
Pruning: Prune $S \times C_{i+1}$ filters via the $S \times C_{i+1}$ smallest LFP$[\mathcal{M}^i]$;
Fine-tuning;

3. Experiments and Analysis

3.1. Experimental Settings

Baselines Models and Datasets. To demonstrate the effectiveness and generality of the proposed LFP method, we evaluate its pruning performance against various baseline models on three image classification datasets. Specifically, we introduce LFP into three modern CNN models (ResNet-56 [65], ResNet-110 [65] and VGG-16 [66]) on the CIFAR-10 dataset [67] and ResNet-20 [65] on the CIFAR-100 [67] dataset. CIFAR-10 contains 60,000 color images (50,000 for training and 10,000 for testing) with a uniform size of 32×32 and classes of 10, but CIFAR-100 has 100 classes. In addition, we further evaluate and compare the performance with other state-of-the-art pruning methods using the ResNet-50 model [65] on ImageNet [68], which is a large-scale and challenging dataset. In addition, we perform our algorithm on VGG-16 with a publicly available dataset designed for remote sensing image classification, called UC Merced land-use dataset, which consists of images of 21 land-use scene categories [69]. Each class contains 100 images with the size of 256×256 pixels and a one foot spatial resolution. Figure 9 shows some example images randomly selected from the UC Merced dataset.

Figure 9. Remote sensing example images from the UC Merced dataset. (1) Agricultural. (2) Airplane. (3) Baseball diamond. (4) Beach. (5) Building. (6) Chaparral. (7) Dense residential. (8) Forest. (9) Freeway. (10) Golfcourse. (11) Harbor. (12) Intersection. (13) Medium residential. (14) Mobile home park. (15) Overpass. (16) Parking lot. (17) River. (18) Runway. (19) Sparse residential. (20) Storage tank. (21) Tennis court.

Configurations. We use PyTorch 1.6.0, Python 3.7 and CUDA 10.2 for implementation and thop for calculating the parameters and FLOPs. Referring to the experimental design in [44,45], an identical layer-by-layer pruning strategy is adopted in our framework. To determine the LFP of each filter, we randomly sample five batches (total five $\times$ mini-batch input images) to calculate the average LFP of each feature map in all the experiments. After completing filter pruning based on LFP, we perform fine-tuning on the pruned models with stochastic gradient descent (SGD) [70–72] as the optimizer. SGD can more efficiently use information, especially when the information is more redundant [72–74]. In addition, we perform the fine-tuning for 300 epochs on CIFAR and UC Merced datasets with the batch size 256, momentum of 0.9, weight decay of 0.005 and initial learning of 0.01. On the ImageNet dataset, fine-tuning is performed for 150 epochs with the batch size of 128, momentum of 0.99, weight decay of 0.0001 and initial learning rate of 0.1.

3.2. Results on CIFAR Datasets

To prove the feasibility of LFP, we use different pruning ratios (Table 1) to achieve the goal of high accuracy, as well as the goals of model size and FLOP reduction. Tables 2–5 show the evaluation results of the pruned modern CNN models on the CIFAR-10/100 datasets, respectively.

For the ResNet-56 model, our LFP-based method improves the accuracy by 0.96% over the baseline model, and reduces the model size and FLOPs by 44.7% and 48.4%, respectively. When the model size and FLOPs are both reduced by 71.8%, we still achieve a better performance.

Table 1. Pruning ratio of various baseline models on different datasets by LFP.

Model/Dataset	Pruning Ratio Setting of All Layers
ResNet-56/CIFAR-10	[0.0] + [0.15] × 2 + [0.4] × 27 [0.0] + [0.4] × 2 + [0.5] × 9 + [0.6] × 9 + [0.7] × 9
ResNet-110/CIFAR-10	[0.0] + [0.2] × 2 + [0.3] × 18 + [0.35] × 36 [0.0] + [0.4] × 2 + [0.5] × 18 + [0.65] × 36
VGG-16/CIFAR-10	[0.3] × 7 + [0.75] × 5 [0.45] × 7 + [0.78] × 5
ResNet-20/CIFAR-100	[0.0] + [0.1] × 2 + [0.25] × 9 [0.0] + [0.3] × 2 + [0.3] × 3 + [0.4] × 3 + [0.5] × 3
ResNet-50/ImageNet	[0.0] + [0.1] × 3 + [0.35] × 16 [0.0] + [0.5] × 3 + [0.6] × 16

Table 2. Pruning results of ResNet-56 on the CIFAR-10 dataset.

Method	Pruned Top-1%	Δ Top-1	Parameters (↓%)	FLOP (↓%)
ResNet-56 [65]	93.26	0	0.85M (0.0)	125.49M (0.0)
L1-norm [26]	93.06	−0.20	0.73M (14.1)	90.90M (27.6)
NISP [75]	93.01	−0.25	0.49M (42.4)	81.00M (35.5)
GAL-0.6 [76]	92.98	−0.28	0.75M (11.8)	78.30M (37.6)
HRank [44]	93.52	+0.26	0.71M (16.8)	88.72M (29.3)
CHIP [45]	94.16	+0.90	0.48M (43.5)	65.94M (47.5)
RUFP [77]	93.57	+0.52	0.52M (38.8)	79.3M (37.6)
LFP (Ours)	**94.22**	**+0.96**	**0.47M (44.7)**	**64.71M (48.4)**
GAL-0.8 [76]	91.58	−1.68	0.29M (65.9)	49.99M (60.2)
LASSO [78]	91.80	−1.46	N/A	62.00M (50.6)
HRank [44]	90.72	−2.54	0.27M (68.1)	**32.52M (74.1)**
CHIP [45]	92.05	−1.21	**0.24M (71.8)**	34.79M (72.3)
LFP (Ours)	**92.70**	**−0.56**	**0.24M (71.8)**	35.37M (71.8)

Table 3. Pruning results of ResNet-110 on the CIFAR-10 dataset.

Method	Pruned Top-1%	Δ Top-1	Parameters (↓%)	FLOPs (↓%)
ResNet-110 [65]	93.50	0	1.72M (0.0)	252.89M (0.0)
L1-norm [26]	93.30	−0.20	1.16M (32.6)	155.00M (38.7)
HRank [44]	94.23	+0.73	1.04M (39.5)	148.70M (41.2)
CHIP [45]	94.44	+0.94	**0.89M (48.3)**	**121.09M (52.1)**
LFP (Ours)	**94.45**	**+0.95**	1.05M (39.0)	132.08M (47.8)
GAL-0.5 [76]	92.74	−0.76	0.95M (44.8)	130.20M (48.5)
HRank [44]	92.65	−0.85	0.53M (69.2)	79.30M (68.6)
CHIP [45]	93.63	+0.13	**0.53M (69.2)**	**71.69M (71.6)**
LFP (Ours)	**93.72**	**+0.22**	0.54M (68.6)	72.83M (71.2)

Table 4. Pruning results of VGG-16 on the CIFAR-10 dataset.

Method	Pruned Top-1%	Δ Top-1	Parameters (↓%)	FLOPs (↓%)
VGG-16 [66]	93.96	0	15.00M (0.0)	314.00M (0.0)
SSS [46]	93.02	−0.94	3.93M (73.8)	183.13M (41.6)
GAL-0.05 [76]	93.77	−0.19	3.36M (77.6)	189.49M (39.6)
HRank [44]	93.43	−0.53	2.51M (83.3)	145.61M (53.6)
CHIP [45]	93.86	−0.10	2.76M (81.6)	131.17M (58.1)
RUFP [77]	93.81	−0.15	**2.50M (83.3)**	167.00M (46.8)
LFP (Ours)	**93.98**	**+0.02**	**2.51M (83.3)**	**104.96M (66.6)**
GAL-0.1 [76]	93.42	−0.54	2.67M (82.2)	171.89M (45.2)
HRank [44]	91.23	−2.73	**1.78M (92.0)**	73.70M (76.5)
CHIP [45]	93.18	−0.78	1.90M (87.3)	**66.95M (78.7)**
LFP (Ours)	**93.61**	**−0.35**	1.89M (87.4)	67.09M (78.6)

Table 5. Pruning results of ResNet-20 on the CIFAR-100 dataset.

Method	Pruned Top-1%	Δ Top-1	Parameters (↓%)	FLOPs (↓%)
ResNet-20 [65]	68.47	0	278.3k (0.0)	41.20M (0.0)
L1-norm [26]	66.59	−1.88	176.2k (36.7)	20.80M (49.5)
L2-norm [79]	66.61	−1.86	**175.9k (36.8)**	21.00M (49.0)
FPGM-0.4 [43]	66.68	−1.79	183.8k (34.0)	**20.60M (50.0)**
PFP [80]	66.19	−2.28	176.3k (36.7)	21.00M (49.0)
KLNP [81]	66.68	−1.79	187.5k (32.7)	21.20M (48.5)
LFP (Ours)	**67.43**	**−1.04**	**175.8k (36.7)**	**20.62M (50.0)**
IENP [27]	65.76	−2.71	168.8k (39.4)	20.00M (51.5)
LFP (Ours)	**65.82**	**−2.65**	**157.4k (43.4)**	**19.65M (52.3)**

For the ResNet-110 model, the accuracy is improved by 0.95% and the model size and FLOP are reduced by 39.0% and 47.8%, respectively. When the model size and FLOP are reduced by 68.6% and 71.2% for pruning (close to the highest compression ratio of the algorithm), our pruned model can still obtain a 0.22% accuracy improvement over the baseline model.

For the VGG-16 model, our method can reduce the model size and FLOPs by 83.3% and 66.6%, respectively. Meanwhile, it still improves the accuracy by 0.02%. In addition, when the compression ratio of the pruned model is close to [44,45], the storage and computational cost are reduced by 87.4% and 78.6%, respectively, and the accuracy is merely reduced by 0.35%.

For the ResNet-20 model on CIFAR-100, on the premise of little accuracy loss, LFP can reduce the model size and FLOP by 36.7% and 50.0%, respectively. When the model is further compressed, the accuracy of our method is reduced by only 2.65%.

After preliminary pruning on ResNet-56/110 and VGG-16, LFP can be more accurate than the baseline model. This shows that the LFP algorithm can alleviate the overfitting problem of the original model while reducing the model size and calculation costs. Although further pruning on ResNet-56 and VGG-16 will cause a slight drop in accuracy, it is within an acceptable range compared to other algorithms.

3.3. Results on ImageNet

The proposed LFP not only shows good performance on small datasets but works well on large-scale datasets. To verify the effectiveness more comprehensively, we also conducted several experiments on the challenging ImageNet dataset. Table 6 lists the pruning performance of ResNet-50 on the ImageNet dataset via our method. The results indicate that, when targeting a small compression ratio, our method can achieve 40.8% and 46.7% storage and computation reductions, respectively. In addition, the accuracy of top-1 and top-5 is 1.21% and 1.26% higher than the baseline model, respectively. When the compression ratio is further increased, LFP still achieves a superior performance over the state-of-the-art methods. That is, the accuracy can be guaranteed while maintaining a high compression ratio. However, in the case of a small compression ratio, CHEX [82] is slightly more accurate than LFP. At the same time, the reductions in model size and computation are not optimal for LFP. However, in the further compression, LFP shows its superiority in precision, storage and computation reduction.

Table 6. Pruning results of ResNet-50 on the ImageNet dataset.

Method	Pruned Top-1%	Δ Top-1	Pruned Top-5%	Δ Top-5	Parameters (↓%)	FLOPs (↓%)
ResNet-50 [65]	76.15	0	92.87	0	25.50M (0.0)	4.09B (0.0)
ThiNet [83]	72.04	−4.11	90.67	−2.20	16.91M (33.7)	2.58B (36.8)
SFP [84]	74.61	−1.54	92.06	−0.81	N/A	2.38B (41.8)
Auto [85]	74.76	−1.39	92.15	−0.72	N/A	2.10B (48.7)
GAL-0.5 [76]	71.95	−4.20	90.94	−1.93	21.19M (16.9)	2.33B (43.0)
FPGM-0.3 [43]	75.59	−0.56	92.63	−0.24	15.94M (37.5)	2.36B (42.2)
HRank [44]	74.98	−1.17	92.33	−0.54	16.17M (36.6)	2.30B (43.7)
SCOP-0.4 [52]	75.95	−0.20	92.79	−0.08	**14.59M (42.8)**	2.24B (45.3)
CHIP [45]	76.30	+0.15	93.02	+0.15	15.10M (40.8)	2.26B (44.8)
CHEX-0.3 [82]	**77.40**	**+1.25**	N/A	-	N/A	**2.00B (51.1)**
LFP (Ours)	**77.36**	**+1.21**	**94.13**	**+1.26**	15.09M (40.8)	2.18B (46.7)
PFP [80]	75.21	−0.94	92.43	−0.44	17.82M (30.1)	2.29B (44.0)
SCOP-0.5 [52]	75.26	−0.89	92.53	−0.34	12.29M (51.8)	1.86B (54.6)
CHIP [45]	75.26	−0.89	92.53	−0.34	11.04M (56.7)	1.52B (62.8)
CHEX-0.5 [82]	76.00	−0.15	N/A	-	N/A	1.00B (75.6)
LFP (Ours)	**76.07**	**−0.08**	**92.26**	**+0.09**	**8.02M (68.5)**	**0.97B (76.3)**

3.4. Results on the UC Merced Dataset

Table 7 lists the pruning performance of VGG-16 on the UCM dataset via our method. The experimental results show that the proposed LFP also performs well in remote sensing image classification. When targeting a small compression ratio, our method can achieve 78.3% and 40.6% storage and computation reductions, respectively. Meanwhile, the accuracy is 0.23% higher than the baseline model. It can be seen that LFP has a tiny loss in accuracy (it decreases by 0.68) when the compression ratio is further increased. That is, the accuracy can be guaranteed while maintaining a high compression ratio.

Table 7. Pruning results of VGG-16 on the UC Merced land-use dataset.

Method	Pruned Top-1%	Δ Top-1	Parameters (↓%)	FLOPs (↓%)
VGG-16 [66]	93.45	0	15.00M (0.0)	314.00M (0.0)
LFP (Ours)	93.68	+0.23	3.25M (78.3)	186.61M (40.6)
LFP (Ours)	92.77	−0.68	2.04M (86.4)	146.53M (53.3)

4. Discussion

This paper proposes a novel model compression method for frequency domain filtering in accordance with the "smaller-norm-less-important" idea. In contrast to previous algorithms that perform pruning in the spatial domain, we explore the similarity, symmetry and substitutability of feature maps. We re-consider the model characteristics that correspond to the human visual system termed Low Frequency Preference (LFP) in the frequency domain. Based on the new frequency domain perspective and model characteristics, the performance of LFP is even superior to state-of-the-art methods [45,82].

Although our LFP is originally proposed for CNNs, there are few pruning algorithms for recurrent neural networks (RNNs). However, we are working hard to explore this limitation, and hope to extend the pruning algorithm to more diverse network structures in the future. Secondly, although it is effective to utilize F-norm pruning in the pruning process, whether there is a more appropriate and accurate metric for pruning than F-norm will continue to be explored in future work. At the same time, we will also focus on the study of different pruning granularities such as [71] to further compress the model.

5. Conclusions

Convolutional neural networks (CNNs) have been widely used in remote sensing image classification due to their powerful feature representation abilities. However, the accompanying high computational cost is always a problem worth trying to improve. In this paper, we propose a novel pruning method called low frequency preference (LFP) from the new perspective of the frequency domain, which takes into account the model properties (i.e., the preference of the network model) for the data properties. It determines the relative importance of filters by observing and computing the spectrogram of the feature map. We conducted LFP with several modern and popular models on different scale datasets to verify its superiority. The experimental results demonstrate that the LFP pruning method can effectively reduce the computational complexity and model size while maintaining a high classification accuracy.

In future research, we will continue to explore different pruning methods in the frequency domain, as well as combine the spatial domain pruning methods to achieve a higher compression ratio. The goal is to find a method to prune CNNs from scratch for remote sensing image classification. Since the pruned channels are already selected when training the original over-parameterized network, pruning CNNs from scratch can save more computational resources and time. It is also of great significance for resource-constrained remote sensing image classification tasks.

Author Contributions: Conceptualization, C.Z. and C.L.; methodology, C.Z.; software, C.Z. and N.L.; validation, C.Z. and N.L.; formal analysis, C.Z. and C.L.; investigation, C.L.; resources, C.Z.; writing—original draft preparation, C.Z.; writing—review and editing, C.Z., C.L. and N.L.; visualization, C.Z. and N.L.; supervision, B.G.; funding acquisition, B.G. All authors have read and agreed to the published version of the manuscript.

Funding: This work was supported by the National Natural Science Foundation of China (no. 62171341).

Data Availability Statement: The CIFAR dataset and the reference codes in this work are available at: http://www.cs.toronto.edu/kriz/cifar.html (accessed in 2009). The ImageNet dataset and the reference codes in this work are available at: https://image-net.org (accessed in 2020). The UC

Merced dataset and the reference codes in this work are available at: http://weegee.vision.ucmerced edu/datasets/landuse.html (accessed in 2010).

Acknowledgments: We would like to thank the editor and anonymous reviewers for their valuable comments on this paper.

Conflicts of Interest: The authors declare no conflict of interest.

Abbreviations

The following abbreviations are used in this manuscript:

LFP	Low Frequency Preference
CNN	Convolutional Neural Network
RNN	Recurrent Neural Network
BLAS	Basic Linear Algebra Subprograms
FLOPs	Floating Point Operations
APoZ	Average Percentage of Zeros
NISP	Neuron Importance Score Propagation
HRank	High Rank
CHIP	Channel Independence-based Pruning
GAL	Generative Adversarial Learning
RUFP	Reinitializing Unimportant Filters for Soft Pruning
FPGM	Filter Pruning via Geometric Median
SSS	Sparse Structure Selection
FFT	Fast Fourier Transform
DFT	Discrete Fourier Transform
F-norm	Frobenius Norm
ResNet	Residual Network
VGG	Visual Geometry Group
CIFAR	Canadian Institute for Advanced Research
SGD	Stochastic Gradient Descent
ImageNet	A Large-Scale Hierarchical Image Database
UC Merced	University of California, Merced
CUDA	Compute Unified Device Architecture

References

1. Long, J.; Shelhamer, E.; Darrell, T. Fully convolutional networks for semantic segmentation. In Proceedings of the IEEE Conference on Computer Vision and Pattern Recognition (CVPR), Boston, MA, USA, 7–12 June 2015; pp. 3431–3440.
2. Girshick, R.; Donahue, J.; Darrell, T.; Malik, J. Rich feature hierarchies for accurate object detection and semantic segmentation. In Proceedings of the IEEE Conference on Computer Vision and Pattern Recognition, Columbus, OH, USA, 23–28 June 2014; pp. 580–587.
3. Girshick, R. Fast r-cnn. In Proceedings of the IEEE International Conference on Computer Vision (ICCV), Santiago, Chile, 7–13 December 2015; pp. 1440–1448.
4. Ren, S.; He, K.; Girshick, R.; Sun, J. Faster r-cnn: Towards real-time object detection with region proposal networks. *Adv. Neural Inf. Process. Syst.* **2015**, *28*, 91–99. [CrossRef] [PubMed]
5. He, K.; Gkioxari, G.; Dollár, P.; Girshick, R. Mask r-cnn. In Proceedings of the IEEE International Conference on Computer Vision (ICCV), Venice, Italy, 22–29 October 2017; pp. 2961–2969.
6. Chen, L.C.; Papandreou, G.; Kokkinos, I.; Murphy, K.; Yuille, A.L. Deeplab: Semantic image segmentation with deep convolutional nets, atrous convolution, and fully connected crfs. *IEEE Trans. Pattern Anal. Mach. Intell.* **2017**, *40*, 834–848. [CrossRef] [PubMed]
7. Blalock, D.; Gonzalez Ortiz, J.J.; Frankle, J.; Guttag, J. What is the state of neural network pruning? *Proc. Mach. Learn. Syst.* **2020**, *2*, 129–146.
8. Han, S.; Mao, H.; Dally, W.J. Deep compression: Compressing deep neural networks with pruning, trained quantization and huffman coding. *arXiv* **2015**, arXiv:1510.00149.
9. Deng, C.; Liao, S.; Yuan, B. Permcnn: Energy-efficient convolutional neural network hardware architecture with permuted diagonal structure. *IEEE Trans. Comput.* **2020**, *70*, 163–173. [CrossRef]
10. Liang, T.; Glossner, J.; Wang, L.; Shi, S.; Zhang, X. Pruning and quantization for deep neural network acceleration: A survey. *Neurocomputing* **2021**, *461*, 370–403. [CrossRef]
11. Xu, K.; Zhang, D.; An, J.; Liu, L.; Liu, L.; Wang, D. GenExp: Multi-objective pruning for deep neural network based on genetic algorithm. *Neurocomputing* **2021**, *451*, 81–94. [CrossRef]

12. Esser, S.K.; McKinstry, J.L.; Bablani, D.; Appuswamy, R.; Modha, D.S. Learned step size quantization. *arXiv* **2019**, arXiv:1902.08153.
13. Xu, Y.; Wang, Y.; Zhou, A.; Lin, W.; Xiong, H. Deep neural network compression with single and multiple level quantization. In Proceedings of the AAAI Conference on Artificial Intelligence, New Orleans, LO, USA, 2–7 February 2018; Volume 32.
14. Rastegari, M.; Ordonez, V.; Redmon, J.; Farhadi, A. Xnor-net: Imagenet classification using binary convolutional neural networks. In *European Conference on Computer Vision ECCV 2016*; Springer: Berlin/Heidelberg, Germany, 2016; pp. 525–542.
15. Pan, Y.; Xu, J.; Wang, M.; Ye, J.; Wang, F.; Bai, K.; Xu, Z. Compressing recurrent neural networks with tensor ring for action recognition. In Proceedings of the AAAI Conference on Artificial Intelligence, Honolulu, HI, USA, 27 January–1 February 2019; Volume 33, pp. 4683–4690.
16. Yin, M.; Sui, Y.; Liao, S.; Yuan, B. Towards efficient tensor decomposition-based dnn model compression with optimization framework. In Proceedings of the IEEE/CVF Conference on Computer Vision and Pattern Recognition, Nashville, TN, USA, 20–25 June 2021; pp. 10674–10683.
17. Park, W.; Kim, D.; Lu, Y.; Cho, M. Relational knowledge distillation. In Proceedings of the IEEE/CVF Conference on Computer Vision and Pattern Recognition, Long Beach, CA, USA, 15–20 June 2019; pp. 3967–3976.
18. Hinton, G.; Vinyals, O.; Dean, J. Distilling the knowledge in a neural network. *arXiv* **2015**, arXiv:1503.02531.
19. Zhang, L.; Song, J.; Gao, A.; Chen, J.; Bao, C.; Ma, K. Be your own teacher: Improve the performance of convolutional neural networks via self distillation. In Proceedings of the IEEE/CVF International Conference on Computer Vision, Seoul, Republic of Korea, 27 October–2 November 2019; pp. 3713–3722.
20. Li, T.; Li, J.; Liu, Z.; Zhang, C. Few sample knowledge distillation for efficient network compression. In Proceedings of the IEEE/CVF Conference on Computer Vision and Pattern Recognition, Seattle, WA, USA, 13–19 June 2020; pp. 14639–14647.
21. Howard, A.G.; Zhu, M.; Chen, B.; Kalenichenko, D.; Wang, W.; Weyand, T.; Andreetto, M.; Adam, H. Mobilenets: Efficient convolutional neural networks for mobile vision applications. *arXiv* **2017**, arXiv:1704.04861.
22. Howard, A.; Sandler, M.; Chu, G.; Chen, L.C.; Chen, B.; Tan, M.; Wang, W.; Zhu, Y.; Pang, R.; Vasudevan, V.; et al. Searching for mobilenetv3. In Proceedings of the IEEE/CVF International Conference on Computer Vision (ICCV), Seoul, Republic of Korea, 27 October–2 November 2019; pp. 1314–1324.
23. Sandler, M.; Howard, A.; Zhu, M.; Zhmoginov, A.; Chen, L.C. Mobilenetv2: Inverted residuals and linear bottlenecks. In Proceedings of the IEEE Conference on Computer Vision and Pattern Recognition, Salt Lake City, UT, USA, 18–23 June 2018; pp. 4510–4520.
24. Han, S.; Pool, J.; Tran, J.; Dally, W. Learning both weights and connections for efficient neural network. *Adv. Neural Inf. Process. Syst.* **2015**, *28*, 1135–1143.
25. Carreira-Perpinán, M.A.; Idelbayev, Y. "Learning-compression" algorithms for neural net pruning. In Proceedings of the IEEE Conference on Computer Vision and Pattern Recognition, Salt Lake City, UT, USA, 18–23 June 2018; pp. 8532–8541.
26. Li, H.; Kadav, A.; Durdanovic, I.; Samet, H.; Graf, H.P. Pruning filters for efficient convnets. *arXiv* **2016**, arXiv:1608.08710.
27. Molchanov, P.; Mallya, A.; Tyree, S.; Frosio, I.; Kautz, J. Importance estimation for neural network pruning. In Proceedings of the IEEE/CVF Conference on Computer Vision and Pattern Recognition, Long Beach, CA, USA, 15–20 June 2019; pp. 11264–11272.
28. Wen, W.; Wu, C.; Wang, Y.; Chen, Y.; Li, H. Learning structured sparsity in deep neural networks. *Adv. Neural Inf. Process. Syst.* **2016**, *29*, 2082–2090.
29. Chen, Y.; Zheng, B.; Zhang, Z.; Wang, Q.; Shen, C.; Zhang, Q. Deep learning on mobile and embedded devices: State-of-the-art, challenges, and future directions. *ACM Comput. Surv. (CSUR)* **2020**, *53*, 1–37. [CrossRef]
30. Zhang, F.; Du, B.; Zhang, L. Scene classification via a gradient boosting random convolutional network framework. *IEEE Trans. Geosci. Remote Sens.* **2015**, *54*, 1793–1802. [CrossRef]
31. Volpi, M.; Tuia, D. Dense semantic labeling of subdecimeter resolution images with convolutional neural networks. *IEEE Trans. Geosci. Remote Sens.* **2016**, *55*, 881–893. [CrossRef]
32. Xia, G.S.; Hu, J.; Hu, F.; Shi, B.; Bai, X.; Zhong, Y.; Zhang, L.; Lu, X. AID: A benchmark data set for performance evaluation of aerial scene classification. *IEEE Trans. Geosci. Remote Sens.* **2017**, *55*, 3965–3981. [CrossRef]
33. Marmanis, D.; Datcu, M.; Esch, T.; Stilla, U. Deep learning earth observation classification using ImageNet pretrained networks. *IEEE Geosci. Remote Sens. Lett.* **2015**, *13*, 105–109. [CrossRef]
34. Nogueira, K.; Penatti, O.A.; Dos Santos, J.A. Towards better exploiting convolutional neural networks for remote sensing scene classification. *Pattern Recognit.* **2017**, *61*, 539–556. [CrossRef]
35. Zhu, Q.; Lei, Y.; Sun, X.; Guan, Q.; Zhong, Y.; Zhang, L.; Li, D. Knowledge-guided land pattern depiction for urban land use mapping: A case study of Chinese cities. *Remote Sens. Environ.* **2022**, *272*, 112916. [CrossRef]
36. Li, Y.; Kong, D.; Zhang, Y.; Tan, Y.; Chen, L. Robust deep alignment network with remote sensing knowledge graph for zero-shot and generalized zero-shot remote sensing image scene classification. *ISPRS J. Photogramm. Remote Sens.* **2021**, *179*, 145–158. [CrossRef]
37. Zhang, W.; Tang, P.; Zhao, L. Remote sensing image scene classification using CNN-CapsNet. *Remote Sens.* **2019**, *11*, 494. [CrossRef]
38. Chen, J.; Wang, C.; Ma, Z.; Chen, J.; He, D.; Ackland, S. Remote sensing scene classification based on convolutional neural networks pre-trained using attention-guided sparse filters. *Remote Sens.* **2018**, *10*, 290. [CrossRef]
39. Guo, X.; Hou, B.; Ren, B.; Ren, Z.; Jiao, L. Network Pruning for Remote Sensing Images Classification Based on Interpretable CNNs. *IEEE Trans. Geosci. Remote Sens.* **2022**, *60*, 1–15. [CrossRef]

40. Zhu, X.X.; Tuia, D.; Mou, L.; Xia, G.S.; Zhang, L.; Xu, F.; Fraundorfer, F. Deep Learning in Remote Sensing: A Comprehensive Review and List of Resources. *IEEE Geosci. Remote Sens. Mag.* **2017**, *5*, 8–36. [CrossRef]
41. Hu, H.; Peng, R.; Tai, Y.W.; Tang, C.K. Network trimming: A data-driven neuron pruning approach towards efficient deep architectures. *arXiv* **2016**, arXiv:1607.03250.
42. Molchanov, P.; Tyree, S.; Karras, T.; Aila, T.; Kautz, J. Pruning convolutional neural networks for resource efficient inference. *arXiv* **2016**, arXiv:1611.06440.
43. He, Y.; Liu, P.; Wang, Z.; Hu, Z.; Yang, Y. Filter pruning via geometric median for deep convolutional neural networks acceleration. In Proceedings of the IEEE/CVF Conference on Computer Vision and Pattern Recognition (CVPR), Long Beach, CA, USA, 15–20 June 2019; pp. 4340–4349.
44. Lin, M.; Ji, R.; Wang, Y.; Zhang, Y.; Zhang, B.; Tian, Y.; Shao, L. Hrank: Filter pruning using high-rank feature map. In Proceedings of the IEEE/CVF Conference on Computer Vision and Pattern Recognition (CVPR), Seattle, WA, USA, 13–19 June 2020; pp. 1529–1538.
45. Sui, Y.; Yin, M.; Xie, Y.; Phan, H.; Aliari Zonouz, S.; Yuan, B. CHIP: CHannel independence-based pruning for compact neural networks. *Adv. Neural Inf. Process. Syst.* **2021**, *34*, 24604–24616.
46. Huang, Z.; Wang, N. Data-driven sparse structure selection for deep neural networks. In Proceedings of the Computer Vision ECCV—15th European Conference, Munich, Germany, 8–14 September 2018; pp. 304–320.
47. Liu, Z.; Li, J.; Shen, Z.; Huang, G.; Yan, S.; Zhang, C. Learning efficient convolutional networks through network slimming. In Proceedings of the IEEE International Conference on Computer Vision (ICCV), Venice, Italy, 22–29 October 2017; pp. 2736–2744.
48. Zhao, C.; Ni, B.; Zhang, J.; Zhao, Q.; Zhang, W.; Tian, Q. Variational convolutional neural network pruning. In Proceedings of the IEEE/CVF Conference on Computer Vision and Pattern Recognition, Long Beach, CA, USA, 15–20 June 2019; pp. 2780–2789.
49. Lin, S.; Ji, R.; Li, Y.; Deng, C.; Li, X. Toward compact convnets via structure-sparsity regularized filter pruning. *IEEE Trans. Neural Netw. Learn. Syst.* **2019**, *31*, 574–588. [CrossRef]
50. Chen, Y.; Wen, X.; Zhang, Y.; Shi, W. CCPrune: Collaborative channel pruning for learning compact convolutional networks. *Neurocomputing* **2021**, *451*, 35–45. [CrossRef]
51. Wang, Z.; Liu, X.; Huang, L.; Chen, Y.; Zhang, Y.; Lin, Z.; Wang, R. QSFM: Model Pruning Based on Quantified Similarity between Feature Maps for AI on Edge. *IEEE Internet Things J.* **2022**, *9*, 24506–24515. [CrossRef]
52. Tang, Y.; Wang, Y.; Xu, Y.; Tao, D.; Xu, C.; Xu, C.; Xu, C. Scop: Scientific control for reliable neural network pruning. *Adv. Neural Inf. Process. Syst.* **2020**, *33*, 10936–10947.
53. Wang, J.; Jiang, T.; Cui, Z.; Cao, Z. Filter pruning with a feature map entropy importance criterion for convolution neural networks compressing. *Neurocomputing* **2021**, *461*, 41–54. [CrossRef]
54. LeCun, Y.; Denker, J.; Solla, S. Optimal brain damage. *Adv. Neural Inf. Process. Syst.* **1989**, *2*, 598–605.
55. Frankle, J.; Carbin, M. The lottery ticket hypothesis: Finding sparse, trainable neural networks. *arXiv* **2018**, arXiv:1803.03635.
56. Girish, S.; Maiya, S.R.; Gupta, K.; Chen, H.; Davis, L.S.; Shrivastava, A. The lottery ticket hypothesis for object recognition. In Proceedings of the IEEE/CVF Conference on Computer Vision and Pattern Recognition, Nashville, TN, USA, 20–25 June 2021; pp. 762–771.
57. Xu, Z.Q.J.; Zhang, Y.; Xiao, Y. Training behavior of deep neural network in frequency domain. In *Proceedings of the International Conference on Neural Information Processing*; Springer: Berlin/Heidelberg, Germany, 2019; pp. 264–274.
58. Xu, Z.Q.J.; Zhang, Y.; Luo, T.; Xiao, Y.; Ma, Z. Frequency principle: Fourier analysis sheds light on deep neural networks. *arXiv* **2019**, arXiv:1901.06523.
59. Kim, J.; Lee, S. Deep Learning of Human Visual Sensitivity in Image Quality Assessment Framework. In Proceedings of the 2017 IEEE Conference on Computer Vision and Pattern Recognition (CVPR), Honolulu, HI, USA, 21–26 July 2017; pp. 1969–1977.
60. Xu, K.; Qin, M.; Sun, F.; Wang, Y.; Chen, Y.K.; Ren, F. Learning in the frequency domain. In Proceedings of the IEEE/CVF Conference on Computer Vision and Pattern Recognition, Seattle, WA, USA, 13–19 June 2020; pp. 1740–1749.
61. Yin, D.; Gontijo Lopes, R.; Shlens, J.; Cubuk, E.D.; Gilmer, J. A fourier perspective on model robustness in computer vision. *Adv. Neural Inf. Process. Syst.* **2019**, *32*, 13276–13286.
62. Cooley, J.W.; Tukey, J.W. An algorithm for the machine calculation of complex Fourier series. *Math. Comput.* **1965**, *19*, 297–301. [CrossRef]
63. Shan, Y.; Hu, D.; Wang, Z.; Jia, T. Multi-channel Nuclear Norm Minus Frobenius Norm Minimization for Color Image Denoising. *arXiv* **2022**, arXiv:2209.08094.
64. Clerckx, B.; Oestges, C. *MIMO Wireless Networks: Channels, Techniques and Standards for Multi-Antenna, Multi-User and Multi-Cell Systems*; Academic Press: Cambridge, MA, USA, 2013.
65. He, K.; Zhang, X.; Ren, S.; Sun, J. Deep residual learning for image recognition. In Proceedings of the IEEE Conference on Computer Vision and Pattern Recognition (CVPR), Las Vegas, NV, USA, 27–30 June 2016; pp. 770–778.
66. Simonyan, K.; Zisserman, A. Very deep convolutional networks for large-scale image recognition. *arXiv* **2014**, arXiv:1409.1556.
67. Krizhevsky, A.; Hinton, G. Learning Multiple Layers of Features from Tiny Images 2009. Available online: https://www.cs.toronto.edu/kriz/learning-features-2009-TR.pdf (accessed on 8 April 2009).
68. Deng, J.; Dong, W.; Socher, R.; Li, L.J.; Li, K.; Fei-Fei, L. Imagenet: A large-scale hierarchical image database. In Proceedings of the 2009 IEEE Conference on Computer Vision and Pattern Recognition, Miami, FL, USA, 20–25 June 2009; pp. 248–255.

69. Yang, Y.; Newsam, S. Bag-of-visual-words and spatial extensions for land-use classification. In Proceedings of the 18th SIGSPATIAL International Conference on Advances in Geographic Information Systems, San Jose, CA, USA, 2–5 November 2010; pp. 270–279.
70. Li, Y.; Chen, Y.; Liu, G.; Jiao, L. A novel deep fully convolutional network for PolSAR image classification. *Remote Sens.* **2018**, *10*, 1984. [CrossRef]
71. Lin, M.; Zhang, Y.; Li, Y.; Chen, B.; Chao, F.; Wang, M.; Li, S.; Tian, Y.; Ji, R. 1xn pattern for pruning convolutional neural networks. *IEEE Trans. Pattern Anal. Mach. Intell.* **2022**, *45*, 3999–4008. [CrossRef] [PubMed]
72. Bottou, L. Stochastic gradient descent tricks. In *Neural Networks: Tricks of the Trade*, 2nd ed.; Lecture Notes in Computer Science; Springer: Berlin/Heidelberg, Germany, 2012; pp. 421–436.
73. Todorov, V.; Dimov, I. Efficient stochastic approaches for multidimensional integrals in bayesian statistics. In Proceedings of the Large-Scale Scientific Computing: 12th International Conference, LSSC 2019, Sozopol, Bulgaria, 10–14 June 2019; Revised Selected Papers 12; Springer: Berlin/Heidelberg, Germany, 2020; pp. 454–462.
74. Predić, B.; Vukić, U.; Saračević, M.; Karabašević, D.; Stanujkić, D. The possibility of combining and implementing deep neural network compression methods. *Axioms* **2022**, *11*, 229. [CrossRef]
75. Yu, R.; Li, A.; Chen, C.F.; Lai, J.H.; Morariu, V.I.; Han, X.; Gao, M.; Lin, C.Y.; Davis, L.S. Nisp: Pruning networks using neuron importance score propagation. In Proceedings of the IEEE/CVF Conference on Computer Vision and Pattern Recognition, Salt Lake City, UT, USA, 18–23 June 2018; pp. 9194–9203.
76. Lin, S.; Ji, R.; Yan, C.; Zhang, B.; Cao, L.; Ye, Q.; Huang, F.; Doermann, D. Towards optimal structured cnn pruning via generative adversarial learning. In Proceedings of the IEEE/CVF Conference on Computer Vision and Pattern Recognition, Long Beach, CA, USA, 15–20 June 2019; pp. 2790–2799.
77. Zhang, K.; Liu, G.; Lv, M. RUFP: Reinitializing unimportant filters for soft pruning. *Neurocomputing* **2022**, *483*, 311–321. [CrossRef]
78. He, Y.; Zhang, X.; Sun, J. Channel pruning for accelerating very deep neural networks. In Proceedings of the IEEE International Conference on Computer Vision (ICCV), Venice, Italy, 22–29 October 2017; pp. 1389–1397.
79. Ye, J.; Lu, X.; Lin, Z.; Wang, J.Z. Rethinking the smaller-norm-less-informative assumption in channel pruning of convolution layers. *arXiv* **2018**, arXiv:1802.00124.
80. Liebenwein, L.; Baykal, C.; Lang, H.; Feldman, D.; Rus, D. Provable filter pruning for efficient neural networks. *arXiv* **2019**, arXiv:1911.07412.
81. Luo, J.H.; Wu, J. Neural network pruning with residual-connections and limited-data. In Proceedings of the IEEE/CVF Conference on Computer Vision and Pattern Recognition, Seattle, WA, USA, 13–19 June 2020; pp. 1458–1467.
82. Hou, Z.; Qin, M.; Sun, F.; Ma, X.; Yuan, K.; Xu, Y.; Chen, Y.K.; Jin, R.; Xie, Y.; Kung, S.Y. CHEX: CHannel EXploration for CNN Model Compression. In Proceedings of the IEEE/CVF Conference on Computer Vision and Pattern Recognition, New Orleans, LA, USA, 18–24 June 2022; pp. 12287–12298.
83. Luo, J.H.; Wu, J.; Lin, W. Thinet: A filter level pruning method for deep neural network compression. In Proceedings of the IEEE International Conference on Computer Vision (ICCV), Venice, Italy, 22–29 October 2017; pp. 5058–5066.
84. He, Y.; Kang, G.; Dong, X.; Fu, Y.; Yang, Y. Soft filter pruning for accelerating deep convolutional neural networks. *arXiv* **2018**, arXiv:1808.06866.
85. Luo, J.H.; Wu, J. Autopruner: An end-to-end trainable filter pruning method for efficient deep model inference. *Pattern Recognit.* **2020**, *107*, 107461. [CrossRef]

remote sensing

MDPI

Article

Multi-Pooling Context Network for Image Semantic Segmentation

Qing Liu, Yongsheng Dong *, Zhiqiang Jiang, Yuanhua Pei, Boshi Zheng, Lintao Zheng and Zhumu Fu

School of Information Engineering, Henan University of Science and Technology, Luoyang 471023, China
* Correspondence: ysdong@haust.edu.cn

Abstract: With the development of image segmentation technology, image context information plays an increasingly important role in semantic segmentation. However, due to the complexity of context information in different feature maps, simple context capture operations can easily cause context information omission. Rich context information can better classify categories and improve the quality of image segmentation. On the contrary, poor context information will lead to blurred image category segmentation and an incomplete target edge. In order to capture rich context information as completely as possible, we constructed a Multi-Pooling Context Network (MPCNet), which is a multi-pool contextual network for the semantic segmentation of images. Specifically, we first proposed the Pooling Context Aggregation Module to capture the deep context information of the image by processing the information between the space, channel, and pixel of the image. At the same time, the Spatial Context Module was constructed to capture the detailed spatial context of images at different stages of the network. The whole network structure adopted the form of codec to better extract image context. Finally, we performed extensive experiments on three semantic segmentation datasets (Cityscapes, ADE20K, and PASCAL VOC2012 datasets), which fully proved that our proposed network effectively alleviated the lack of context extraction and verified the effectiveness of the network.

Keywords: semantic segmentation; context information; convolutional neural network; attention module

Citation: Liu, Q.; Dong, Y.; Jiang, Z.; Pei, Y.; Zheng, B.; Zheng, L.; Fu, Z. Multi-Pooling Context Network for Image Semantic Segmentation. *Remote Sens.* **2023**, *15*, 2800. https://doi.org/10.3390/rs15112800

Academic Editors: Jiaojiao Li, Qian Du, Wei Li, Bobo Xi, Jocelyn Chanussot, Rui Song and Yunsong Li

Received: 2 April 2023
Revised: 10 May 2023
Accepted: 12 May 2023
Published: 28 May 2023

1. Introduction

Image segmentation is an important part of computer vision, and semantic segmentation is a basic task of image segmentation. Semantic segmentation involves pixel-level semantic image processing, which is mainly utilizes the relationship between pixels and their surroundings. The development of deep learning has led to the widespread use of image semantic segmentation in real-life applications, such as medical imaging [1–3], assisted driving [4–7], and radar image processing [8–10]. Context information usually represents the relationship between its own pixels and surrounding pixels, which is crucial for visually understanding tasks. The main principle of image semantic segmentation is to give corresponding semantic expression to all pixels in the image. This expression not only pays attention to the meaning of its own pixels, but also needs to express the relationship between its own pixels and surrounding pixels. Therefore, context information is an important factor in image semantic segmentation. Contextual information is not only often used in the field of segmentation, but is also a common method of problem solving in other areas [11–13]. We divide the context information into semantic context information and spatial context information according to different image feature maps. Semantic context information is often contained in low-resolution, high-level feature maps, which is mainly used to distinguish pixel categories. The spatial context information is mainly used in a high-resolution, low-level feature map to help the pixel restore the spatial details. The combination of these two context information types greatly improves the quality of image semantic segmentation.

With the development of the convolutional neural network, more and more methods have been used to capture rich semantic context information. For example, Context-Reinforced Semantic Segmentation [14] proposes a context-enhanced semantic segmentation network to explore the advanced semantic context information in a feature graph. It embeds the learned context into the segmentation reasoning based on FCN [15] to further enhance the modern semantic segmentation method. The Co-Occurrent Features Network [16] designs a special module to learn fine-grained spatial information representation and constructs overall contextual feature information by aggregating co-occurrence feature probabilities in co-occurrence contexts. Context Encoding for Semantic Segmentation [17] is used to capture the semantic information in the scene using the encoding and decoding module to selectively filter the information with the same class of features. The Context Deconvolution Network for Semantic Segmentation [18] proposes a context deconvolution network and focuses on the semantic context association in decoding network. The Gated Path Selection Network [19] has developed a gated path selection network. In order to dynamically select the required semantic context, the gate prediction module is further introduced. Unlike previous efforts to capture semantic context information, its network can adaptively capture dense context. LightFGCNet [20] has designed a lightweight global context capture method and combines feature information from different regions during the upsampling phase to enable better global context extraction across the network. BCANet [21] has designed a boundary-guided context aggregation module to capture the correlation context between pixels in the boundary region and other pixels to facilitate the understanding of the semantic information of the overall image. DMAU-Net Network [22] presents an attention-based multiscale maximum pooling dense network, which designs an integrated maximum pool module to improve the image information feature extraction ability in the encoder section, thereby improving the network segmentation efficiency. The Multiscale Progressive Segmentation Network [23] presents a multiscale progressive segmentation network that gradually divides image targets into small, large, and other scales and cascades them into three distinct subnetworks to achieve the final image segmentation result. The Semantic Segmentation Network [24] presents a semantic segmentation network that combines multi-path structure, attention weighting, and multi-scale encoding. It captures spatial information, semantic context information, and semantic map information of images through three parallel structures. The Combining Max-Pooling Network [25] combines the traditional wavelet algorithm with a convolutional neural network pooling operation to propose a new multi-pooling scheme, and it uses this scheme to create two new stream architectures for semantically segmenting images.

There are many ways to use spatial context information. For example, the CBAM [26] aggregates spatially detailed information about pixels through pooling operations and generates different spatial context descriptors through a spatial attention module to capture spatial detail context information. The spatial context is generally found in high-resolution feature maps or in the connection of pixels to other pixels. As a result, they cannot capture spatial context information for objects that reside at different scales. The Feature Pyramid Transformer [27] uses specially designed converters to form feature pyramids in a top-down or opposite interaction to capture high-resolution spatial context. To reduce the computational effort needed to capture more spatial context, the Fast Attention Network [28] captures the same spatial context at a fraction of the computational cost by using different orders of spatial attention. The HRCNet [29] maintains spatial contextual information through a specific network structure, obtains global contextual information during the feature extraction phase, and uses a feature-enhanced feature pyramid structure to fuse contextual information at different scales. The CTNet [30] has designed a spatial context module and a channel context module to capture the semantic and spatial context between different pixel features by exploring inter-pixel correlations.

These methods have excellent performance in extracting semantic context and spatial context information. For better image semantic segmentation, not only rich semantic context, but also sufficient spatial context information is required. We believe that a good combination of these two context information types can better complete the semantic segmentation task and improve the segmentation quality. Therefore, we designed a new network structure: the Multi-Pooling Context Network (MPCNet). The MPCNet captures feature context information in different stages through encoding and decoding structures. Specifically, we designed a Pooling Context Aggregation Module (PCAM), which is composed of multiple pooling operations and dilated convolutions. The application captures rich semantic context information in low-resolution high-level feature map to improve the utilization of semantic-related context in a high-level feature map. In addition, a Spatial Context Module (SCM) was proposed, which is composed of maximum pooling and average pooling. It captures the spatial context in a low-level feature map and provides the output to the encoder in the form of a jump connection to form each decoding stage, so as to better restore the spatial details of pixels. Our MPCNet captures rich semantic context information through the encoder and combines the spatial context information from the decoder that is captured by jump connection to form the encoding and decoding structure of the whole network, which not only improves the information conversion rate of pixels, but also increases the utilization rate of the context information, thus improving the quality of semantic segmentation.

The following are our main contributions:

(1) We constructed a Multi-Pooling Context Network (MPCNet), which captures rich semantic context information through the encoder and restores the spatial context information through the decoder formed by the jump connection. The whole network realizes the effective combination of semantic context and spatial context with the encoding and decoding structure, thus completing the semantic segmentation task.
(2) We designed a Spatial Context Module (SCM), which is composed of different types of pooling layers. It transfers the spatial information in the low-level feature map at the encoding stage to each decoding stage through the jump connection, improves the information utilization of the spatial context, and, thus, increases the pixel location of the semantic category.
(3) We designed a Pooling Context Aggregation Module (PCAM) consisting of a combination of different pooling operations and dilation convolution. It cooperates with the encoder to capture different contexts in the high-level feature graph, thereby creating rich semantic contextual information for pixel classification.

2. Related Work

In this section, we introduce some relevant semantic context and spatial context information capture methods and popular semantic segmentation models.

2.1. Semantic Context Information

An image is composed of several pixels. Semantic segmentation is mainly performed to label several pixels in the image. Successfully partitioning each pixel requires rich semantic context information. Semantic context information can effectively improve the semantic classification of pixel images. In recent years, semantic context has fully verified its effectiveness in semantic segmentation methods. For example, PSPNet [31] collects the feature information of pixels by pooling the pyramids of different sizes to obtain rich semantic context for the semantic segmentation of images. ParseNet [32] uses the average feature of the layer to increase the information of each location, and then adds the semantic context to the full convolutional network to improve the image segmentation quality. DeepLabV3+ [33] is designed to expand the convolutional composition of the atrous spatial pyramid pool module to capture rich semantic contextual information, thereby improving the segmentation performance of the network. DDRNet [34] establishes two parallel depth branches and uses the two-branch structure to search the semantic context in the low-resolution feature map. The Gated Full Fusion for Semantic Segmentation [35] (GFF) uses gates to selectively fuse semantic contexts at all levels in a fully connected way, uses gate control units to control the propagation of useful semantic contexts, and suppresses additional contextual information noise. These methods improve the semantic segmentation performance through their unique network design. They pay more attention to large-scale pixel semantic information. On the contrary, our network method aims to combine the multi-scale semantic context information in the low-resolution feature map and achieve the purpose of multi-scale semantic context information and increase the feature receptive field through different pooling combinations and dilated convolutions, so as to capture more relevant semantic context information.

2.2. Spatial Context Information

Several pixels in the image are closely connected; pixels themselves and between pixels have different meanings. In the process of semantic segmentation, it is necessary to know the different meanings between the pixel itself and other pixels, but these meanings are often contained in the spatial context information. At present, many methods are exploring how to better capture the spatial detail context information of images. One example is the SpaceMeshLab—featuring Spatial Context Memorization. Furthermore, the Meshgrid Through Convergence Consus For Semantic Segmentation [36] proposed a spatial context memo, which preserves the input dimension through the bypass branch of this spatial context and constantly communicates with the backbone network to capture its spatial context information. Context Encoding and Multi-Path Decoding [37] propose a scale selection scheme, thereby selectively fusing information from different scale features, preserving the rich spatial context information fraction in the feature scale, and improving the segmentation performance of pixel spatial details. BiSeNetV2 [38] introduces a new feature fusion module to effectively combine spatial and semantic context information, interactively explore spatial and semantic context information, and find different pixels for semantic segmentation. SGCPNet [39] devises a spatial detail-oriented context propagation strategy that uses shallow spatial detail to guide the global context and also effectively recovers lost spatial detail information. These methods have performed well in completing the capture of spatial context information, and have a good restoration and reconstruction effect on pixel spatial details, whether from the multi-scale or multi-branch. The difference is that our method compensates for the lost spatial context in the down-sampling process by combining pooling operations and transfers it to the corresponding image up-sampling stage in the down-sampling stage, which greatly compensates for the spatial details of image segmentation.

3. Methodology

In this section, we first explain the framework of our Multi-Pooling Context Network (MPCNet) and present the main principles of the two proposed modules—the Pooling Context Aggregation Module (PCAM) and the Spatial Context Module (SCM).

3.1. Overview

The structure of the Multi-Pooling Context Network for semantic segmentation (MPCNet) proposed by us is shown in Figure 1. The network uses codec as its main architecture that uses the pre-training residual network ResNet101 [40] as the encoding stage. Since down-sampling loses the spatial details of the image, we used a 3×3 convolution with a step size of 2 instead of the down-sampling operation of the backbone network. In the last resolution stage, we set the step size to 1 and used a 3×3 dilation convolution with a dilation rate of 2 instead of the convolution. In this way, the image features are retained at resolutions of 1/4, 1/8, 1/16, and 1/16, and the number of channels corresponding to each resolution is 256, 512, 1024 and 2048, respectively. These four feature resolutions also represent four different coding stages. In order to capture more semantic contexts, we applied the Pooling Context Aggregation Module (PCAM) in the last coding stage. At the same time, the Spatial Context Module (SCM) was used to capture the spatial context information of the first three coding stages, and the spatial information of the first three coding stages formed the decoding stage in the form of jump connection with the flow fusion [41] and the output of PCAM module. In this way, the spatial details of the corresponding image encoding phase will exist in the corresponding decoding phase.

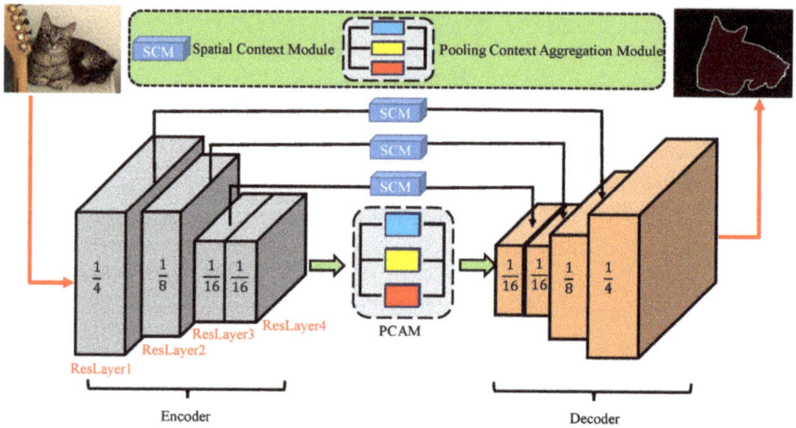

Figure 1. Overview of MPCNet. ResNet is used as the encoder backbone network, and its four different resolution layers such as ResLayer1, ResLayer2, ResLayer3, and ResLayer4 are used as the encoder stage. The PCAM obtains the semantic context of high-level features at the coding stage. The SCM sends the spatial context extracted in each encoding stage to the decoder in the form of jump connection. The whole network uses an encoding and decoding structure for semantic segmentation (Best in color).

Note that our MPCNet aims to capture more context information for semantic segmentation. MPCNet captures three parts of the context in the encoder's high-level feature map by PCAM to form rich semantic context information, divides several categories of image pixels, and then transfers the spatial context of the image pixels to the decoder in the form of skip connection with the spatial context captured by SCM at each stage of encoding, thus restoring the spatial details of the image pixels. In order to better capture the context information, our entire model uses a codec–decode structure, extracts the context information of the image using the backbone network as the encoder to reduce the resolution, captures the semantic context information through PCAM, and combines it with the spatial detail context captured by SCM in the form of jump connection. By sampling step-by-step to form the decoder, each module structure of the whole network is clear, simple, and easy to implement.

3.2. Spatial Context Module

With the continuous down-sampling of the convolutional neural network, the low-resolution pixels of the image will lose the spatial detail information, thus resulting in blurred target boundaries. To reduce the loss of spatial detail, the spatial position of the target pixels was improved. We built the Spatial Context Module (SCM). Figure 2 shows our proposed Spatial Context Module (SCM) structure. It can be seen from Figure 2 that SCM is an integrated design of the whole module, which can be flexibly applied to any network structure. Next, let us introduce SCM in detail.

First, we used high-resolution feature map as input, but because the number of feature map channels in each stage was different, we used common convolution to unify the number of channels, then used maximum pooling and average pooling operations to collect different weight information of feature map, and then fused different weight information. The context weight obtained was calculated by sigmoid function, and then all the weight information output by sigmoid was selected by using the features of the unified channel, filtering out redundant information, and preserving relevant spatial details. To prevent the gradients from disappearing due to the increase in network depth, we initialized the connection of spatial contextual information to ensure smooth transmission of the gradients. For spatial context module output O_{output}, the specific expression is

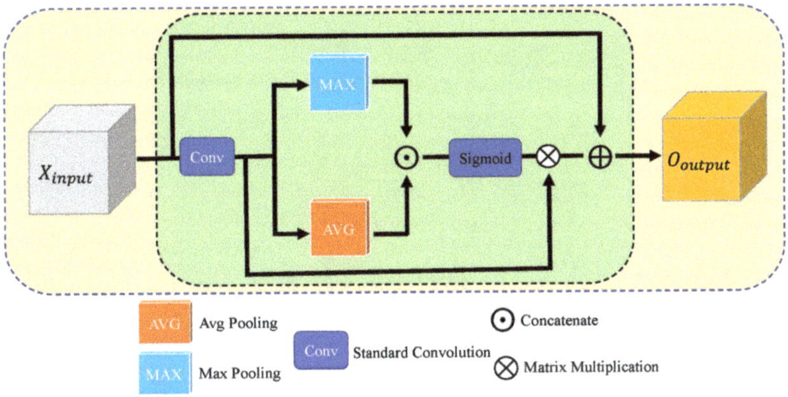

Figure 2. Overview of Spatial Context Module (SCM). It captures the spatial context information in the high-level feature graph through different pooling operations.

$$O_{output} = Sig\left[Max\left(Conv\left(X_{input}\right)\right) \odot Avg\left(Conv\left(X_{input}\right)\right)\right] \otimes Conv\left(X_{input}\right) \oplus X_{input}, \quad (1)$$

where Max represents maximum pooling, Avg represents average pooling, $Conv$ represents standard convolution, Sig represents sigmoid function, X_{input} represents high-resolution input features, $\odot$ represents concat, $\otimes$ represents matrix element multiplication, and $\oplus$ represents element summation.

Our Spatial Context Module aims to capture spatial details in high-resolution feature maps. First, we used the channel number of the convolution uniform feature map and then used the pooling operation to obtain different information weights. Because the maximum pooling can obtain more prominent pixel information weights on the image, and the average pooling can obtain additional target information, we used two parallel poolings to capture the weight information of the image, then used the probability function to effectively select it, and finally filtered out redundant information and output spatial details between image pixels. This preserved effective spatial context information in the high-resolution feature map.

3.3. *Pooling Context Aggregation Module*

Semantic context is crucial for semantic segmentation. Semantic information of dense pixels is generally reserved in low-resolution feature images, so it is necessary to reduce the resolution of the image to extract rich semantic information. However, in an image with complex background, we should not only pay attention to the semantic information of low resolution, but also pay attention to the context information between its own semantic pixels and surrounding pixels. In order to better capture the rich context information with low resolution, we designed the Pooling Context Aggregation Module (PCAM). Figure 3 shows the structure of PCAM. From Figure 3, we can see that PCAM is composed of three parts. Next, we will introduce the Pooling Context Aggregation Module in detail.

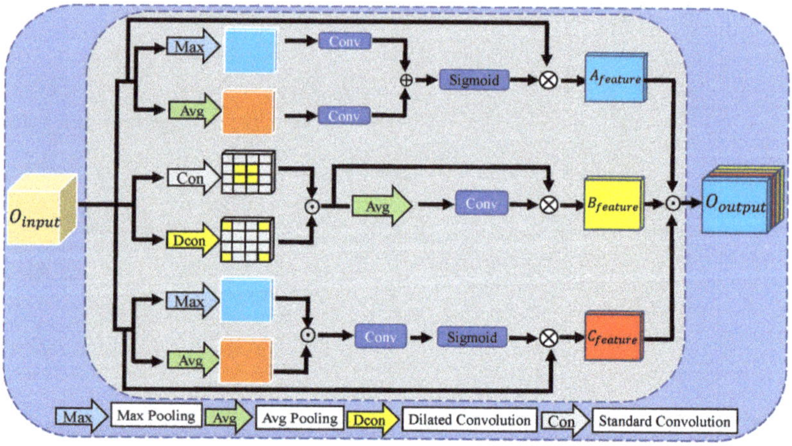

Figure 3. Overview of Pooling Context Aggregation Module(PCAM). It is mainly composed of three parts of context information by capturing the semantic context in the low-resolution feature map.

The Pooling Context Aggregation Module (PCAM) is composed of three different parts, and the corresponding capture $A_{feature}$, B_{feture}, and C_{feture} has three parts of context information. First, the input low-resolution feature O_{input} performs maximum pooling and average pooling operations, and it then uses 1×1 convolution to capture the context information between its channels after each pooling module. The maximum pooling channel information and average pooling channel information are fused to form a complete channel context weight. The weight probability is expressed using the sigmoid function, and then the channel weight is selected with the initial input characteristics to remove redundant channel information, as well as preserve complete and rich channel context information $A_{feature}$. Next, in the second part, we use ordinary and dilation convolution to expand the receptive field of the input features, as well as fuse and retain contextual information between pixels. Then, average pooling and convolution are used to select weights for feature links, remove redundant information, retain useful information between pixels, increase connectivity between pixels, and capture contextual information between pixels B_{feture}. The last part is the spatial context information that is captured by the spatial context module C_{feture}. The captured three-part context information is fused to form a low-resolution semantic context O_{output}. The formal description of output is as follows:

$$O_{output} = A_{feature} \odot B_{feature} \odot C_{feature}, \quad (2)$$

where $A_{feature}$, B_{feture}, and C_{feture} represent channel context information, context information between pixels, and spatial context information, respectively. They are specifically expressed as follows:

$$A_{feature} = Sig\left[Conv(Max(O_{input})) \oplus Conv(Avg(O_{input}))\right] \otimes O_{input}, \quad (3)$$

$$B_{feature} = Conv(Avg(Conv(O_{input}) \odot Dconv(O_{input}))) \otimes (Conv(O_{input}) \odot Dconv(O_{input})), \quad (4)$$

$$C_{feature} = Sig\left[Conv((Max(O_{input}) \odot Avg(O_{input})))\right] \otimes O_{input}, \quad (5)$$

where Max represents maximum pooling, Avg represents average pooling, $Conv$ represents standard convolution, $Dconv$ represents 3×3 dilated convolution. Sig represents sigmoid function, O_{input} represents low-resolution input features, $\odot$ represents concat, $\otimes$ represents matrix element multiplication, and $\oplus$ element summation.

Our proposed Pooling Context Aggregation Module aims to capture rich semantic context information of low-resolution feature maps through different pooling and convolution operations. The channel weight is expressed by probability through maximum pooling and average pooling, and the context information between its channels is obtained; in order to preserve the connection between pixels, we use dilated convolution to capturing the context information between pixels; because the low-resolution feature map also contains spatial details, we use the spatial context module to capture its spatial context. Unlike the high-resolution spatial context module, we remove the unified channel convolution and initialization connection. The whole low-resolution semantic context is composed of these three parts of context information. It not only divides the semantic categories of each pixel, but also distinguishes itself and surrounding pixels by certain pixel categories. It ensures the semantic correctness of different pixels.

4. Experimental Results

In this section, we compare numerical and segmentation results with ten image semantic segmentation methods from recent years on the PASCAL VOC2012 dataset [42], the Cityscape dataset [43], and the ADE20K MIT dataset [44].

4.1. Datasets and Experimental Settings

In this subsection, we first introduce the three semantic segmentation datasets used for network training, and then detail the specific parameter details of the experiments.

4.1.1. PASCAL VOC2012

PASCAL VOC 2012 is a computer vision competition dataset. It is divided into three sections according to the data training requirements: training, evaluation, and test sets. Each set has roughly 1400 images. The categories of these images include not only humans and animals, but also driving tools, indoor scenes, etc. There are 21 categories covering many objects in our lives.

4.1.2. Cityscapes

Cityscapes is a vehicle driving dataset. It has a total of 19 street view category labels, and the dataset is divided into three parts, including a training, evaluation, and test dataset. The corresponding images are 2979, 500, and 1525, respectively, and each image has a high resolution of 2048×1024.

4.1.3. ADE20K MIT

The ADE20K dataset is MIT's open scene understanding dataset. It contains over 20 K images of over 3000 object classes. Because of the complexity of the classes, the samples in the dataset have different resolutions of up to 2400×1800 pixels.

4.1.4. Experimental Settings

We implemented our network on a single GPU using the Python language, which used ResNet101 with a dilated convolution strategy as the backbone of the network. Specifically, we replaced the pooling module with dilated convolution and resolved the size of Resnet's final output feature map to 1/16, thus avoiding 1/8, which would use too much GPU memory, and ensuring sufficient contextual information.

Our experiments generally refer to most previous work [33,45,46] using pixel accuracy (PA), intersection over union (IoU), and mean intersection of union (mIoU) as evaluation metrics [47]. A combination of random gradient descent (SGD) [48] and cross-entropy loss with a small batchsize dataset setup was used to train the network weights. For all datasets, we used a horizontal random flip and random scaling. For the Cityscapes dataset, we used a learning rate of 0.01, set the batchsize to 8, and set the training iterations to 160 K. For the ADE20K and PASCAL VOC2012 datasets, we set the learning rate to 0.007, set the batchsize to 12, and set training iterations to 100 K.

4.2. Ablation Experiments with MPCNet

In this section, we designed ablation experiments on two modules of the MPCNet (Pooling Context Aggregation Module (PCAM) and Spatial Context Module (SCM)) for the Cityscape dataset. In the ablation experiments that follow, we set the training iterations to 100K for the convenience of the experiments.

4.2.1. Ablation Experiment for PCAM

To demonstrate the effectiveness of our proposed PCAM in MPCNet, we performed ablation experiments on its components. Table 1 shows our proposed PCAM ablation experiments on the ResNet101 backbone network for the Cityscapes dataset. We divided PCAM into two parts for ablation experiments—one containing only channel context $A_{feature}$ and spatial context $C_{feature}$ and one containing the context information between pixels $B_{feature}$. From Table 1, we can see that, regardless of whether it contained only channel context $A_{feature}$ and spatial context $C_{feature}$ or context information between pixels $B_{feature}$, the PA and mIOU of the segmentation pixels were greatly reduced, and the results were not as good as those of the three merges.

Table 1. PA and mIoU of our PCAM module for the Cityscapes dataset ($A_{feature}$, $B_{feature}$, and $C_{feature}$ denote the channel context, context information between pixels, and spatial context of our proposed PCA module, respectively). (Note that the bold indicates the best value for that column).

Method	$A_{feature}$	$B_{feature}$	$C_{feature}$	PA (%)	mIOU (%)
ResNet101				90.77	71.25
ResNet101	✓		✓	94.76	77.88
ResNet101		✓		94.27	77.56
ResNet101	✓	✓	✓	**95.81**	**78.05**

To further evaluate the advancement of our PCAM, we compared the results with PCAM using several classic context extraction modules: PPM [31], ASPP [33], and MMP [49]. To increase the fairness of the comparison data, we set consistent training parameters in the comparison experiments. Table 2 shows the results of the module comparison. From Table 2, we can see that our PCAM achieved 97.92% PA and 78.24% mIOU based on the same parameter settings, which outperformed those with the PPM, ASPP and MMP. The main reason is that our proposed PCAM aggregates the channel context, spatial context, and inter-pixel context of the low-resolution feature map, thereby making maximum use of the pixel information of the low-resolution feature map to capture more semantic context information.

Table 2. Comparison of PA and mIoU of our PCAM module with the other three modules (PPM, ASPP, MPM) for the Cityscapes dataset. (Note that the bold indicates the best value for that column).

Method	BaseNet	PPM	ASPP	MPM	PCAM	PA (%)	mIOU (%)
MPCNet	ResNet101	✓				94.56	76.43
MPCNet	ResNet101		✓			95.15	77.68
MPCNet	ResNet101			✓		95.01	77.21
MPCNet	ResNet101				✓	**97.92**	**78.24**

4.2.2. Ablation Experiment for SCM

In order to verify the validity of the SCM module, we conducted an experimental comparative analysis on the backbone network ResNet101 using SPM [49] with the same ability to capture spatial context information. Table 3 shows the experimental analysis for the Cityscapes dataset. From Table 3, we can see that the SCM module was superior to the SPM in the ResNet101 baseline network, and its performance reached 76.74% mIOU. The main reason is that our proposed SCM filters spatial information through different pooling, saves spatial location information in different stages, and transfers spatial details to decoders through skip connections, thereby greatly restoring the pixel location to maintain the consistency of semantic and spatial details.

Table 3. Comparison of PA and mIoU of proposed PCAM module with the SPM modules for the Cityscapes dataset. (Note that the bold indicates the best value for that column).

Method	SPM	SCM	PA (%)	mIOU (%)
ResNet101			90.77	71.25
ResNet101	✓		94.76	75.58
ResNet101		✓	**96.21**	**76.74**

4.3. Segmentation Performances and Comparisons

In this subsection, to demonstrate the segmentation performance of our proposed MPCNet, numerical and visualization results were compared with ten segmentation methods for three image semantic segmentation datasets.

4.3.1. PASCAL VOC2012

To validate the effectiveness of our proposed MPCNet, we conducted a numerical experimental comparison with excellent semantic segmentation algorithms of recent years on the VOC2012 dataset. Table 4 shows comparison of the PA and mIOU for the PASCAL VOC2012 dataset with ten other methods. Since some of the methods did not run on this dataset, we ran the pixel precision (PA) of the FCN [15], PSPNet [31], DeepLab [50], Denseaspp [51], OCNet [52], and DeepLabV3+ [33] on the same device. The results of the OCRNet [53], OCNet [52], and ANN [54] were derived from SA-FFNet [55]. From Table 4, we can see that our method obtained 94.83% PA and 77.48% mIOU. Under ResNet101, our PA was 0.99% to 6.1% higher than other methods. Our MPCNet could achieve an mIOU of 77.48%, which was 1.06% higher than the SA-FFNet [55]. A comparison of different experimental values reveals that our MPCNet maintains good pixel accuracy.

Table 4. Comparison of our proposed MPCNet's PA and mIOU for the PASCAL VOC2012 dataset with ten other methods.

Method	BaseNet	PA (%)	mIOU (%)
FCN [15]	ResNet101	88.73	62.20
DeepLab [50]	ResNet101	92.84	78.51
PSPNet [31]	ResNet101	93.11	82.60
DeepLabv3+ [33]	ResNet101	93.78	80.57
Denseaspp [51]	ResNet101	93.68	75.27
ANN [54]	ResNet101	93.20	72.79
DANet [56]	ResNet101	93.38	80.40
OCRNet [53]	ResNet101	93.47	74.69
OCNet [52]	ResNet101	93.80	75.55
SA-FFNet [55]	ResNet101	93.84	76.42
MPCNet (ours)	ResNet101	94.83	77.48

4.3.2. Cityscapes

In this section, we conducted a comparative experiment for the Cityscapes dataset. Table 5 shows comparison of the PA and mIOU for the Cityscapes dataset. Considering the rigor of the experiment, we also retested the pixel accuracy for DeepLab [50], FCN [15], DeepLabV3+ [33], PSPNet [31], OCNet [52], Denseaspp [51], DANet [56], ANN [54], and OCRNet [53], as well as the mIOU of the FCN [15] for the Cityscapes dataset. From Table 5, we can see that our PA was 97.92%, which was 1.67% higher than other methods. The mIOU was 78.24%, which was 5.11% higher than other methods. Therefore, our MPCNet still has advantages regarding the PA and mIOU.

Table 5. Comparison of our proposed MPCNet's PA and mIOU for the Cityscapes dataset with ten other methods.

Method	BaseNet	PA (%)	mIOU (%)
FCN [15]	ResNet101	94.85	66.61
DeepLab[50]	ResNet101	95.78	79.30
PSPNet [31]	ResNet101	96.49	78.40
DeepLabv3+ [33]	ResNet101	96.66	79.55
Denseaspp [51]	ResNet101	95.85	80.60
ANN [54]	ResNet101	95.16	81.30
DANet [56]	ResNet101	95.45	81.50
OCRNet [53]	ResNet101	95.29	81.80
OCNet [52]	ResNet101	96.53	81.40
SA-FFNet [55]	ResNet101	96.25	73.13
MPCNet (ours)	ResNet101	97.92	78.24

4.3.3. ADE20K

To further validate our proposed MPCNet, we performed experiments on a larger ADE20K dataset. Table 6 shows the mIOU and PA of the MPCNet and ten other methods. It can be seen From Table 6 that the pixel accuracy of the MPCNet was 82.55%, and the mIOU was 38.04%. The results of these two methods still have certain advantages over the other ten methods. The ADE20K dataset has a large number of images and complex pixel types. Our proposed MPCNet extracts different pixel semantic contexts through the PCAM, uses the SCM to compensate for the missing spatial details, and uses codec mode to increase the capture of complex information. Our proposed MPCNet achieved different segmentation performance, so it is effective.

Table 6. Comparison of our proposed MPCNet's PA and mIOU for the ADE20K dataset with ten other methods.

Method	BaseNet	PA (%)	mIOU (%)
FCN [15]	ResNet101	76.32	29.47
SegNet [57]	ResNet101	68.59	21.63
DeepLab[50]	ResNet101	80.26	33.87
PSPNet [31]	ResNet101	81.56	41.68
DeepLabv3+ [33]	ResNet101	82.31	36.42
Denseaspp [51]	ResNet101	81.75	34.55
ANN [54]	ResNet101	81.37	45.24
DANet [56]	ResNet101	82.27	36.33
OCRNet [53]	ResNet101	81.88	45.28
OCNet [52]	ResNet101	82.10	45.04
MPCNet (ours)	ResNet101	82.55	38.04

4.4. Visual Comparison

To demonstrate the proposed visual advantage of the MPCNet, we compared three methods for the Cityscapes dataset in Figure 4, namely, PSPNet, OCNet, and DeepLabv3+. From Figure 4, it can be seen that small targets in a complex background, such as traffic lights, people in the distance, bicycles, etc., were all pixel categories that were difficult to segment. In contrast, our method had a better segmentation result than the other methods and could be successfully segmented. In addition, the loss of spatial detail information for pixels was successfully alleviated, such as the division and positioning of "human contour" and "overlapping vehicle" in line 5. From the perspective of the segmentation effect, our proposed MPCNet can provide the context information needed for segmentation and can accurately segment the image.

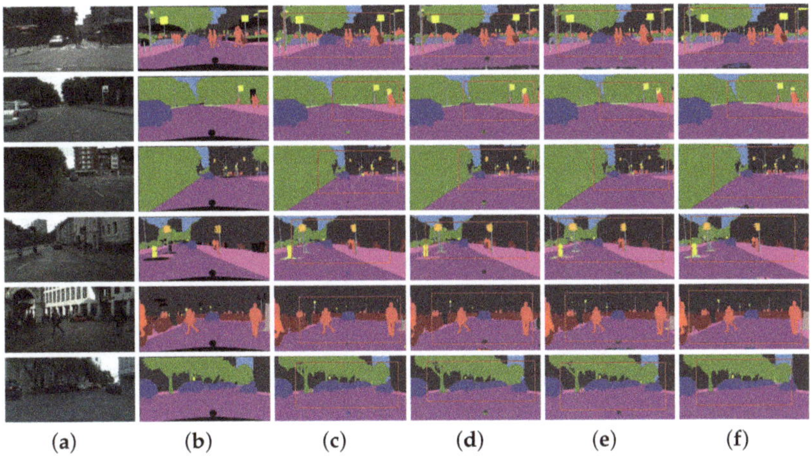

Figure 4. Comparison of the visual segmentation results of our proposed MPCNet with the other three methods for the Cityscapes dataset: (**a**) Image. (**b**) Ground Truth. (**c**) PSPNet [31]. (**d**) OCNet [52]. (**e**) DeepLabv3+ [33]. (**f**) Ours.

To further verify the validity of our method, we compared our proposed MPCNet with three methods for the VOC2012 dataset in Figure 5. From Figure 5, it can be seen that both vehicle and animal MPCNets could result in the semantics being classified correctly and the outline being clear. We propose that PCAM constructs a semantic context by capturing different contextual information and semantically dividing pixels. The SCM improves the spatial positioning ability of each semantic category and ensures that the outline of the category is clear. Therefore, from the perspective of visual analysis, our proposed MPCNet is effective in the application of semantics segmentation.

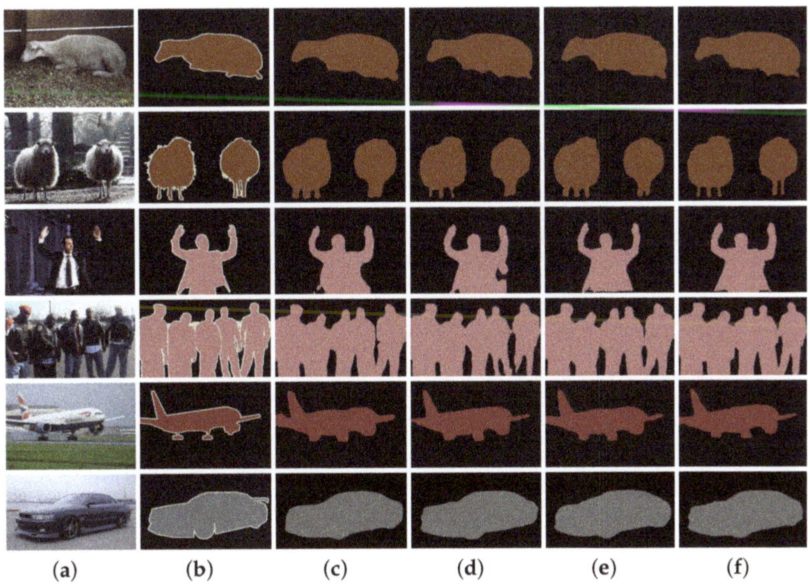

Figure 5. Comparison of the visual segmentation results of our proposed MPCNet with the other three methods for the PASCAL VOC dataset: (**a**) Image. (**b**) Ground Truth. (**c**) PSPNet [31]. (**d**) OCNet [52]. (**e**) DeepLabv3+ [33]. (**f**) Ours.

5. Conclusions

In this paper, we proposed a Multi-Pooling Context Network (MPCNet) for semantic segmentation. Specifically, our proposed PCAM aggregates the semantic context information in the high-level feature graph through three parts of feature information, increases the semantic exploitation of pixels in the low-resolution feature graph, and classifies different pixels in the image into semantic categories. Our proposed SCM captures the spatial contextual information of high-resolution features and passes it to the decoder in the form of a jump connection to enhance the spatial localization of semantic categories. The stable structure of the network using coding and decoding ensures that the contextual information is fully utilized, thus better improving the segmentation results. Experimental results show that our proposed MPCNet is effective.

Our method has initially alleviated the problem of insufficient context information capture in simple images, but the segmentation effect for complex backgrounds and multi-category pixel images still needs to be improved. For different complex background image processing, not only sufficient context information is needed, but also more attention should be paid to the relationships between pixels. For example, overlapping target objects, small target objects, and multi-shape target objects constitute the difficulties of semantic segmentation of complex images, and are also the focus of our research work in the future.

Author Contributions: Q.L.: Methodology, Writing—Original Draft Preparation, Experiments. Y.D.: Conceptualization, Methodology, Writing–Reviewing and Editing. Z.J.: Methodology, Investigation, Experiments. Y.P.: Methodology, Investigation, Experiments. B.Z.: Methodology, Investigation, Experiments. L.Z.: Methodology, Investigation, Experiments. Z.F.: Methodology, Investigation, Writing—Reviewing and Editing. All authors have read and agreed to the published version of the manuscript.

Funding: This work was supported by the Natural Science Foundation of Henan under Grant 232300421023.

Institutional Review Board Statement: Not applicable.

Informed Consent Statement: Not applicable.

Data Availability Statement: The data generated and analyzed during this study are available from the corresponding author by request.

Conflicts of Interest: The authors declare no conflict of interest.

References

1. Li, P.; Liu, Y.; Cui, Z.; Yang, F.; Zhao, Y.; Lian, C.; Gao, C. Semantic graph attention with explicit anatomical association modeling for tooth segmentation from CBCT images. *IEEE Trans. Med. Imaging* **2022**, *41*, 3116–3127. [CrossRef] [PubMed]
2. Song, J.; Chen, X.; Zhu, Q.; Shi, F.; Xiang, D.; Chen, Z.; Fan, Y.; Pan, L.; Zhu, W. Global and local feature reconstruction for medical image segmentation. *IEEE Trans. Med. Imaging* **2022**, *41*, 2273–2284. [CrossRef] [PubMed]
3. Wang, Q.; Du, Y.; Fan, H.; Ma, C. Towards collaborative appearance and semantic adaptation for medical image segmentation. *Neurocomputing* **2022**, *491*, 633–643. [CrossRef]
4. Dai, Y.; Zheng, T.; Xue, C.; Zhou, L. SegMarsViT: Lightweight mars terrain segmentation network for autonomous driving in planetary exploration. *Remote. Sens.* **2022**, *14*, 6297. [CrossRef]
5. Li, X.; Zhao, Z.; Wang, Q. ABSSNet: Attention-based spatial segmentation network for traffic scene understanding. *IEEE Trans. Cybern.* **2021**, *52*, 9352–9362. [CrossRef]
6. Liu, Q.; Dong, Y.; Li, X. Multi-stage context refinement network for semantic segmentation. *Neurocomputing* **2023**, *535*, 53–63. [CrossRef]
7. Wang, H.; Chen, Y.; Cai, Y.; Chen, L.; Li, Y.; Sotelo, M.A.; Li, Z. SFNet-N: An improved SFNet algorithm for semantic segmentation of low-light autonomous driving road scenes. *IEEE Trans. Intell. Transp. Syst.* **2022**, *23*, 21405–21417. [CrossRef]
8. Liu, B.; Hu, J.; Bi, X.; Li, W.; Gao, X. PGNet: Positioning guidance network for semantic segmentation of very-high-resolution remote sensing images. *Remote. Sens.* **2022**, *14*, 4219. [CrossRef]
9. Wang, H.; Chen, X.; Zhang, T.; Xu, Z.; Li, J. CCTNet: Coupled CNN and transformer network for crop segmentation of remote sensing images. *Remote Sens.* **2022**, *14*, 1956. [CrossRef]
10. Nie, J.; Zheng, C.; Wang, C.; Zuo, Z.; Lv, X.; Yu, S.; Wei, Z. Scale–Relation joint decoupling network for remote sensing image semantic segmentation. *IEEE Trans. Geosci. Remote. Sens.* **2022**, *60*, 1–12. [CrossRef]

11. Dong, Y.; Jiang, Z.; Tao, F.; Fu, Z. Multiple spatial residual network for object detection. *Complex Intell. Syst.* **2022**, *9*, 1–16. [CrossRef]
12. Dong, Y.; Tan, W.; Tao, D.; Zheng, L.; Li, X. CartoonLossGAN: Learning surface and coloring of images for cartoonization. *IEEE Trans. Image Process.* **2021**, *31*, 485–498. [CrossRef] [PubMed]
13. Dong, Y.; Yang, H.; Pei, Y.; Shen, L.; Zheng, L.; Li, P. Compact interactive dual-branch network for real-time semantic segmentation. *Complex Intell. Syst.* **2023**, *2023*, 1–11. [CrossRef]
14. Zhou, Y.; Sun, X.; Zha, Z.J.; Zeng, W. Context-reinforced semantic segmentation. In Proceedings of the IEEE Conference on Computer Vision and Pattern Recognition (CVPR), Long Beach, CA, USA, 16–20 June 2019; pp. 4046–4055.
15. Long, J.; Shelhamer, E.; Darrell, T. Fully convolutional networks for semantic segmentation. In Proceedings of the IEEE Conference on Computer Vision and Pattern Recognition (CVPR), Boston, MA, USA, 7–12 June 2015; pp. 3431–3440.
16. Zhang, H.; Zhang, H.; Wang, C.; Xie, J. Co-occurrent features in semantic segmentation. In Proceedings of the IEEE Conference on Computer Vision and Pattern Recognition (CVPR), Long Beach, CA, USA, 16–20 June 2019; pp. 548–557.
17. Zhang, H.; Dana, K.; Shi, J.; Zhang, Z.; Wang, X.; Tyagi, A.; Agrawal, A. Context encoding for semantic segmentation. In Proceedings of the IEEE Conference on Computer Vision and Pattern Recognition (CVPR), Salt Lake City, UT, USA, 18–23 June 2018; pp. 7151–7160.
18. Fu, J.; Liu, J.; Li, Y.; Bao, Y.; Yan, W.; Fang, Z.; Lu, H. Contextual deconvolution network for semantic segmentation. *Pattern Recognit.* **2020**, *101*, 107152. [CrossRef]
19. Geng, Q.; Zhang, H.; Qi, X.; Huang, G.; Yang, R.; Zhou, Z. Gated path selection network for semantic segmentation. *IEEE Trans. Image Process.* **2021**, *30*, 2436–2449. [CrossRef]
20. Chen, Y.; Jiang, W.; Wang, M.; Kang, M.; Weise, T.; Wang, X.; Tan, M.; Xu, L.; Li, X.; Zhang, C. LightFGCNet: A aightweight and focusing on global context information semantic segmentation network for remote sensing imagery. *Remote Sens.* **2022**, *14*, 6193. [CrossRef]
21. Ma, H.; Yang, H.; Huang, D. Boundary guided context aggregation for semantic segmentation. *arXiv* **2021**, arXiv:2110.14587.
22. Yang, Y.; Dong, J.; Wang, Y.; Yu, B.; Yang, Z. DMAU-Net: An Attention-Based Multiscale Max-Pooling Dense Network for the Semantic Segmentation in VHR Remote-Sensing Images. *Remote Sens.* **2023**, *15*, 1328. [CrossRef]
23. Hang, R.; Yang, P.; Zhou, F.; Liu, Q. Multiscale progressive segmentation network for high-resolution remote sensing imagery. *IEEE Trans. Geosci. Remote. Sens.* **2022**, *60*, 1–12. [CrossRef]
24. Lin, Z.; Sun, W.; Tang, B.; Li, J.; Yao, X.; Li, Y. Semantic segmentation network with multi-path structure, attention reweighting and multi-scale encoding. *Vis. Comput.* **2023**, *39*, 597–608. [CrossRef]
25. De Souza Brito, A. Combining max-pooling and wavelet pooling strategies for semantic image segmentation. *Expert Syst. Appl.* **2021**, *183*, 115403. [CrossRef]
26. Woo, S.; Park, J.; Lee, J.Y.; Kweon, I.S. Cbam: Convolutional block attention module. In Proceedings of the European Conference on Computer Vision (ECCV), Munich, Germany, 8–14 September 2018; pp. 3–19.
27. Zhang, D.; Zhang, H.; Tang, J.; Wang, M.; Hua, X.; Sun, Q. Feature pyramid transformer. In *Proceedings of the Computer Vision—ECCV 2020: 16th European Conference, Glasgow, UK, 23–28 August 2020*; Proceedings, Part XXVIII 16; Springer: Berlin/Heidelberg, Germany, 2020; pp. 323–339.
28. Hu, P.; Perazzi, F.; Heilbron, F.C.; Wang, O.; Lin, Z.; Saenko, K.; Sclaroff, S. Real-time semantic segmentation with fast attention. *IEEE Robot. Autom. Lett.* **2020**, *6*, 263–270. [CrossRef]
29. Xu, Z.; Zhang, W.; Zhang, T.; Li, J. HRCNet: High-resolution context extraction network for semantic segmentation of remote sensing images. *Remote Sens.* **2020**, *13*, 71. [CrossRef]
30. Li, Z.; Sun, Y.; Zhang, L.; Tang, J. CTNet: Context-based tandem network for semantic segmentation. *IEEE Trans. Pattern Anal. Mach. Intell.* **2021**, *44*, 9904–9917. [CrossRef] [PubMed]
31. Zhao, H.; Shi, J.; Qi, X.; Wang, X.; Jia, J. Pyramid scene parsing network. In Proceedings of the IEEE Conference on Computer Vision and Pattern Recognition (CVPR), Honolulu, HI, USA, 21–26 July 2017; pp. 2881–2890.
32. Liu, W.; Rabinovich, A.; Berg, A.C. Parsenet: Looking wider to see better. *arXiv* **2015**, arXiv:1506.04579.
33. Chen, L.C.; Zhu, Y.; Papandreou, G.; Schroff, F.; Adam, H. Encoder-decoder with atrous separable convolution for semantic image segmentation. In Proceedings of the European Conference on Computer Vision (ECCV), Munich, Germany, 8–14 March 2018; pp. 801–818.
34. Hong, Y.; Pan, H.; Sun, W.; Jia, Y. Deep dual-resolution networks for real-time and accurate semantic segmentation of road scenes. *arXiv* **2021**, arXiv:2101.06085.
35. Li, X.; Zhao, H.; Han, L.; Tong, Y.; Yang, K. Gff: Gated fully fusion for semantic segmentation. *arXiv* **2019**, arXiv:1904.01803.
36. Kim, T.; Kim, J.; Kim, D. SpaceMeshLab: Spatial context memoization and meshgrid atrous convolution consensus for semantic segmentation. In Proceedings of the 2021 IEEE International Conference on Image Processing (ICIP), Anchorage, AL, USA, 19–22 September 2021; pp. 2259–2263.
37. Ding, H.; Jiang, X.; Shuai, B.; Liu, A.Q.; Wang, G. Semantic segmentation with context encoding and multi-path decoding. *IEEE Trans. Image Process.* **2020**, *29*, 3520–3533. [CrossRef]
38. Yu, C.; Gao, C.; Wang, J.; Yu, G.; Shen, C.; Sang, N. Bisenet v2: Bilateral network with guided aggregation for real-time semantic segmentation. *Int. J. Comput. Vis.* **2021**, *129*, 3051–3068. [CrossRef]

39. Hao, S.; Zhou, Y.; Guo, Y.; Hong, R.; Cheng, J.; Wang, M. Real-Time semantic segmentation via spatial-detail guided context propagation. *IEEE Trans. Neural Netw. Learn. Syst.* **2022**, *2022*, 1–12. [CrossRef]
40. He, K.; Zhang, X.; Ren, S.; Sun, J. Deep residual learning for image recognition. In Proceedings of the IEEE Conference on Computer Vision and Pattern Recognition (CVPR), Las Vegas, NV, USA, 27–30 June 2016; pp. 770–778.
41. Li, X.; You, A.; Zhu, Z.; Zhao, H.; Yang, M.; Yang, K.; Tan, S.; Tong, Y. Semantic flow for fast and accurate scene parsing. In Proceedings of the European Conference on Computer Vision (ECCV), Glasgow, UK, 23–28 August 2020; pp. 775–793.
42. Everingham, M.; Van Gool, L.; Williams, C.K.; Winn, J.; Zisserman, A. The pascal visual object classes (voc) challenge. *Int. J. Comput. Vis.* **2010**, *88*, 303–338. [CrossRef]
43. Cordts, M.; Omran, M.; Ramos, S.; Rehfeld, T.; Enzweiler, M.; Benenson, R.; Franke, U.; Roth, S.; Schiele, B. The cityscapes dataset for semantic urban scene understanding. In Proceedings of the IEEE Conference on Computer Vision and Pattern Recognition (CVPR), Las Vegas, NV, USA, 27–30 June 2016; pp. 3213–3223.
44. Zhou, B.; Zhao, H.; Puig, X.; Fidler, S.; Barriuso, A.; Torralba, A. Scene parsing through ade20k dataset. In Proceedings of the IEEE Conference on Computer Vision and Pattern Recognition (CVPR), Honolulu, HI, USA, 21–26 July 2017; pp. 633–641.
45. Minaee, S.; Boykov, Y.; Porikli, F.; Plaza, A.; Kehtarnavaz, N.; Terzopoulos, D. Image segmentation using deep learning: A survey. *IEEE Trans. Pattern Anal. Mach. Intell.* **2022**, *44*, 3523–3542. [CrossRef]
46. Dong, Y.; Shen, L.; Pei, Y.; Yang, H.; Li, X. Field-matching attention network for object detection. *Neurocomputing* **2023**, *535*, 123–133. [CrossRef]
47. Garcia-Garcia, A.; Orts-Escolano, S.; Oprea, S.; Villena-Martinez, V.; Garcia-Rodriguez, J. A review on deep learning techniques applied to semantic segmentation. *arXiv* **2017**, arXiv:1704.06857.
48. Krizhevsky, A.; Sutskever, I.; Hinton, G.E. Imagenet classification with deep convolutional neural networks. *Commun. ACM* **2017**, *60*, 84–90. [CrossRef]
49. Hou, Q.; Zhang, L.; Cheng, M.M.; Feng, J. Strip Pooling: Rethinking spatial pooling for scene parsing. In Proceedings of the IEEE Conference on Computer Vision and Pattern Recognition (CVPR), Seattle, WA, USA, 18–20 June 2020; pp. 4003–4012.
50. Chen, L.C.; Papandreou, G.; Schroff, F.; Adam, H. Rethinking atrous convolution for semantic image segmentation. *arXiv* **2017**, arXiv:1706.05587.
51. Yang, M.; Yu, K.; Zhang, C.; Li, Z.; Yang, K. Denseaspp for semantic segmentation in street scenes. In Proceedings of the IEEE Conference on Computer Vision and Pattern Recognition (CVPR), Salt Lake City, UT, USA, 18–23 June 2018; pp. 3684–3692.
52. Yuan, Y.; Huang, L.; Guo, J.; Zhang, C.; Chen, X.; Wang, J. Ocnet: Object context network for scene parsing. *arXiv* **2018**, arXiv:1809.00916.
53. Yuan, Y.; Chen, X.; Wang, J. Object-contextual representations for semantic segmentation. In Proceedings of the European Conference on Computer Vision (ECCV), Glasgow, UK, 23–28 August 2020; pp. 173–190.
54. Zhu, Z.; Xu, M.; Bai, S.; Huang, T.; Bai, X. Asymmetric non-local neural networks for semantic segmentation. In Proceedings of the IEEE International Conference on Computer Vision (ICCV), Seoul, Republic of Korea, 27 October–2 November 2019; pp. 593–602.
55. Zhou, Z.; Zhou, Y.; Wang, D.; Mu, J.; Zhou, H. Self-attention feature fusion network for semantic segmentation. *Neurocomputing* **2021**, *453*, 50–59. [CrossRef]
56. Fu, J.; Liu, J.; Tian, H.; Li, Y.; Bao, Y.; Fang, Z.; Lu, H. Dual attention network for scene segmentation. In Proceedings of the IEEE Conference on Computer Vision and Pattern Recognition (CVPR), Long Beach, CA, USA, 16–20 June 2019; pp. 3146–3154.
57. Badrinarayanan, V.; Kendall, A.; Cipolla, R. SegNet: A deep convolutional encoder-decoder architecture for image segmentation. *IEEE Trans. Pattern Anal. Mach. Intell.* **2017**, *39*, 2481–2495. [CrossRef]

remote sensing

MDPI

Article

Unmixing-Guided Convolutional Transformer for Spectral Reconstruction

Shiyao Duan [1], Jiaojiao Li [1,*], Rui Song [1], Yunsong Li [1] and Qian Du [2]

1 The State Key Laboratory of ISN, Xidian University, Xi'an 710071, China; 20012100009@stu.xidian.edu.cn (S.D.); rsong@xidian.edu.cn (R.S.); ysli@mail.xidian.edu.cn (Y.L.)
2 The Department of Electronic and Computer Engineering, Mississippi State University, Starkville, MS 39762, USA; du@ece.msstate.edu
* Correspondence: jjli@xidian.edu.cn

Abstract: Deep learning networks based on CNNs or transformers have made progress in spectral reconstruction (SR). However, many methods focus solely on feature extraction, overlooking the interpretability of network design. Additionally, models exclusively based on CNNs or transformers may lose other prior information, sacrificing reconstruction accuracy and robustness. In this paper, we propose a novel Unmixing-Guided Convolutional Transformer Network (UGCT) for interpretable SR. Specifically, transformer and ResBlock components are embedded in Paralleled-Residual Multi-Head Self-Attention (PMSA) to facilitate fine feature extraction guided by the excellent priors of local and non-local information from CNNs and transformers. Furthermore, the Spectral–Spatial Aggregation Module (S2AM) combines the advantages of geometric invariance and global receptive fields to enhance the reconstruction performance. Finally, we exploit a hyperspectral unmixing (HU) mechanism-driven framework at the end of the model, incorporating detailed features from the spectral library using LMM and employing precise endmember features to achieve a more refined interpretation of mixed pixels in HSI at sub-pixel scales. Experimental results demonstrate the superiority of our proposed UGCT, especially in the *grss_dfc*_2018 dataset, in which UGCT attains an RMSE of 0.0866, outperforming other comparative methods.

Keywords: spectral reconstruction; convolutional transformer; hyperspectral unmixing; multi-head self-attention; hyperspectral image

Citation: Duan, S.; Li, J.; Song, R.; Li, Y.; Du, Q. Unmixing-Guided Convolutional Transformer for Spectral Reconstruction. *Remote Sens.* **2023**, *15*, 2619. https://doi.org/10.3390/rs15102619

Academic Editor: Giuseppe Scarpa

Received: 6 April 2023
Revised: 8 May 2023
Accepted: 15 May 2023
Published: 18 May 2023

1. Introduction

Hyperspectral image (HSI) refers to a three-dimensional data cube generated through the collection and assembly of numerous contiguous electromagnetic spectrums, which are acquired via airborne or spaceborne hyperspectral sensors. Unlike regular RGB or grayscale images, HSI provides more information in the band dimension, which allows subsequent tasks to distinguish materials and molecular components that are difficult to distinguish from normal RGB through their stored explicit or implicit distinctions. As a result, HSI has distinct advantages in a variety of tasks, including object detection [1,2], water quality monitoring [3–5], intelligent agriculture [6–8], geological prospecting [9,10], etc.

However, hyperspectral imaging often requires long exposure times and various costs, making it unaffordable to collect sufficient data using sensors for many tasks with restricted budgets. Instead, acquiring a series of RGB or multispectral images is often a fast and cost-effective alternative. Therefore, using SR methods to inexpensively reconstruct the corresponding HSI from RGB or multispectral images (MSI) is a valuable solution. Currently, there are two main reconstruction approaches: the first involves fusing paired low-resolution hyperspectral (lrHS) and high-resolution multispectral (hrMS) images to produce a high-resolution hyperspectral (HrHs) image [11–13] with both high spatial and spectral resolutions, and the second approach generates the corresponding HSI by learning the inverse mapping from a single RGB image [14–19]. Commonly, image fusion-based methods [11–13] require paired images of the same scene, which can still be overly restrictive. Although reconstruction only from RGB images [14–16,20,21] is an ill-posed task due to the assumptions of inverse mapping, theoretical evidence demonstrates that feasible solutions exist under low-dimensional manifolds [22], and it provides sufficient cost-effectiveness.

Utilizing deep learning to model the inverse mapping in single-image reconstruction problems has been widely studied. Initially, numerous methods leveraged the excellent geometric feature extraction capabilities of CNNs [15–19] to achieve success in SR tasks. However, with the outstanding performance of transformers in various computer vision tasks, many transformer-based approaches [14,23,24] have recently emerged. These approaches take advantage of the transformer's global receptive field and sophisticated feature parsing abilities to achieve more refined HSI reconstruction. Nonetheless, current methods are predominantly limited to single-mechanism-driven frameworks, which often implies that the transformer architecture sacrifices the exceptional geometric invariance prior offered by CNNs. In fact, to ingeniously combine the advantages of both, numerous computer vision tasks have attempted to employ convolutional transformers to enhance the capability of feature extraction in their models, yielding highly impressive results [25–28]. Hence, employing a convolutional transformer to integrate the outstanding characteristics of both approaches is a clearly beneficial solution in SR.

Additionally, to achieve a higher signal-to-noise ratio in hyperspectral imaging, a trade-off between spectral resolution and spatial resolution is inevitable [29]. Most airborne hyperspectral sensors typically have a spatial resolution lower than 1 m/pixel [30,31], while satellite-based sensors, such as the Hyperion dataset of Ahmedabad, only have a 30 m/pixel resolution [32]. This significantly limits the effectiveness of HSI in capturing geographic spatial features. As a result, numerous approaches concentrate on employing mature CNNs or advanced transformer architectures to enhance feature extraction capabilities while overlooking the interpretability of the modeling itself and the pixel-mixing issues that arise during the imaging process.

In recent studies, the HU has been mostly composed of the linear mixing model (LMM) [33], the bilinear mixing model (BMM) [34], and the nonlinear mixing model (NMM) [35]. Among them, LMM has long been a focal point, achieving notable results in balancing time and computational costs, as demonstrated in Figure 1. In real-world environments, it is relatively uncommon for electromagnetic waves to be captured by sensors after only one reflection or refraction, which means NMM often aligns more closely with practical modeling. However, nonlinear unmixing inherently takes into account too numerous complex factors, such as the actual scene distribution, and still faces significant limitations in practical applications. As a result, utilizing the more mature LMM model to obtain the linear abundance distribution and subsequently extract HSI information at the sub-pixel level is a judicious and convenient choice. As one of the most crucial HSI processing tasks, employing a highly interpretable HU architecture enables sub-pixel interpretation of the collected HSIs. In edge regions where pixel mixing is severe and understanding the imagery is critical, the HU mechanism extracts more refined features through unmixing. Consequently, leveraging the HU framework to enhance image understanding and interpretability for the SR network [31,36] would result in notable improvements.

In this paper, we propose a novel hyperspectral reconstruction network that combines the LMM and convolutional transformer blocks. By leveraging the HU mechanism, this network aims to enhance the mathematical interpretability of SR modeling and improve the accuracy of HSI reconstruction at a sub-pixel, fine-grained level. By employing endmembers from a filtered spectral library, the input RGB images are mapped to an HSI with high resolution. Our model capitalizes on the geometric invariance between the original prior of the transformer and the convolutional mechanisms. Our model combines the global receptive field of transformers with the geometric invariance of CNN mechanisms, simultaneously extracting both local and non-local features from the image. Furthermore, to mitigate spectral distortion arising from insufficient channel dimension modeling in

CNNs [37], we embed channel position encoding by mapping transformer features into CNNs. It bolsters the capability of the convolutional transformer, ultimately yielding a precise reconstruction of HSIs. The primary contributions of our work can be summarized as follows:

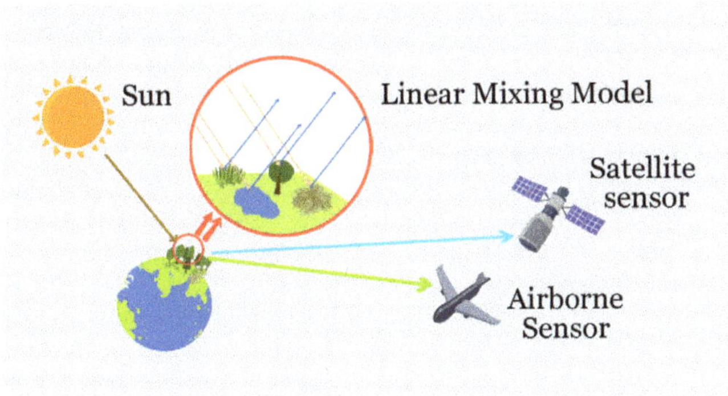

Figure 1. Linear Mixing Model.

1. We introduce an SR network, the UGCT, which tackles HSI recovery from RGB tasks using the LMM as a foundation while employing convolutional transformer to drive fine spectral reconstruction. By employing an unmixing technique and convolutional transformer block, the reconstruction performance of mixed pixels has been notably enhanced. The experiments on two datasets demonstrate that our method's performance is state of the art in the SR task.
2. The Spectral–Spatial Aggregation Module (S2AM) adeptly fuses transformer-based and convolution-based features, thereby enhancing the feature merging capability within the convolutional transformer block. We embed the channel position encoding of the transformer into ResBlock to address positional inaccuracies during the generation of abundance matrices. Notably, such errors can lead to spectral response curve distortions in the reconstructed HSIs.
3. The Paralleled-Residual Multi-Head Self-Attention (PMSA) module generates a more comprehensive spectral feature by synergistically leveraging the transformer's exceptional complex feature extraction capabilities and the CNN's geometric invariance. To the best of our knowledge, we are among the first to incorporate a parallel convolutional transformer block within the single-image SR.

2. Related Work

2.1. Spectral Reconstruction (SR) with Deep Learning

Deep learning technology in SR task encompasses two distinct aspects. The first involves a fusion method based on paired images, while the second entails a direct reconstruction approach that leverages a single image such as those from CASSI or RGB systems. In the first category, a simultaneous capture of lrHS and hrMs images is employed, both possessing the same spectral and spatial resolution as HSIs separately. For example, Yao et al. [11] views hrMS as a degenerate representation of HSI in the spectral dimension and lrHS as a degenerate representation of HSI in the spatial dimension. It is suggested to use cross-attention in coupled unmixing nets based on the complementarities of the two features. Hu et al. [13], on the other hand, employed the Fusformer to obtain the implicit connection between global features and to solve the local neighborhood issue of the finite receptive field of the convolution kernel in the fusion problem using the transformer mechanism. The training process's data load is decreased by learning the spectral and spatial properties, respectively. However, the majority of the models' prior knowledge was created manually, which frequently results in a performance decrease when the domain is changed. Using the HSI denoising iterative spectral reconstruction approach based on deep learning, the MoG-DCN described by Dong et al. [38] has produced outstanding results in numerous datasets.

For the second category, where only single images are input, the model will learn the inverse function of the camera response function of a sensor using a single RGB image as an example. It will separate the RGB image's hidden hyperspectral feature data from it and then combine it with the intact spatial data to reconstruct a fine HSI. Shi et al. [15], for instance, replaced leftover blocks with dense blocks to significantly deepen the network structure and achieved exceptional results in NTIRE 2018 [20]. The pixel-shuffling layer was employed by Zhao et al. [19] to achieve inter-layer interaction, and the self-attention mechanism was used to widen the perceptual field. Cai et al. [14] presented a cascade-based visual transformer model, MST++, to address the numerous issues with convolution networks in SR challenges. Its designed S-MSA and other modules further improved the ability of model to extract spatial and spectral features and achieved outstanding results in a large number of experiments.

The aforementioned analysis reveals that most previous models predominantly focused on enhancing feature extraction capabilities while neglecting the interpretability of physical modeling. This oversight often resulted in diminished performance in practical applications. In response, an SR model with robust interpretability was developed, capitalizing on the autoencoder's prowess in feature extraction and the simplicity of LMM. By harnessing the ability of LMM to extract sub-pixel-level features, ample spatial information is concurrently gathered from RGB images. Subsequently, high-quality HSIs are restored during the reconstruction process.

2.2. Deep Learning-Based Hyperspectral Unmixing

Several deep learning models based on mathematical or physical modeling have been suggested recently and used in real-world tests with positive outcomes due to the growing demand for the interpretability of deep learning models. Among these, HU has made significant progress in tasks such as change detection (CD), SR, and other HSI processing tasks. Guo et al. [39] utilized HU to extract sub-pixel-level characteristics from HSIs to integrate the HU framework into a conventional CD task. In order to obtain the reconstructed HSI, Zou et al. [40] used the designed constraints and numerous residual blocks to obtain the endmember matrix and abundance matrix, respectively. Su et al. [41] used the paired lrHs and hrMs to learn the abundance matrix and endmember from the planned autoencoder network and then rearranged them into HSI using the fundamental LMM presumptions.

Moreover, deep learning-based techniques are frequently used to directly extract the abundance matrix or end endmembers from the HU mechanism. According to Hong et al. [42], EGU-Net can extract a pure-pixel directed abundance matrix extraction model and estimate the abundance of synchronous hyperspectral pictures by using the parameter-sharing mechanism and the two-stream autocoder framework. By utilizing the asymmetric autoencoder network and LSTM to capture spectral information, Zhao et al. [43] were able to address the issue of inadequate spectral and spatial information in the mixed model.

Based on the aforementioned research, utilizing the HU mechanism to drive the SR task evidently improves interpretability. In light of this, our method introduces a parallel feature fusion module that combines the rich geometric invariance present in the residual blocks with the global receptive field of the transformer. This approach ensures the generation of well-defined features and aligns the channel-wise information with the endmembers of the spectral library.

2.3. Convolutional Transformer Module

The transformer-based approach has achieved great success in the field of computer vision, but using it exclusively will frequently negate the benefits of the original CNN structure and add a significant amount of computing burden. Due to this, numerous studies have started fusing the two. Among these, Wu et al. [25] inserted CNN into the conventional vision transformer block, replacing linear projection and other components, and improved the accuracy of various computer vision tasks. Guo et al. [26] linked the two in succession, created the CMT model with both benefits, and created the lightweight visual model. He et al. [27] created the parallel CNN and transformer feature fusion through the developed RAM module and the dual-stream feature extraction component.

The integration of CNN and transformer is inevitable because they are the two most important technologies in the field of image processing. Many performance comparisons between the two have produced their own upsides and downsides [44,45]. Important information will inevitably be lost when using a single module alone. It is crucial to understand how to incorporate the elements that can be derived from both. In order to perform feature fusion for the parallel structure of PMSA, the channel size of the CNN that lacks modeling [37] can be well constrained utilizing the channel information in the transformer.

3. The Proposed Method

In this section, we present an overview of the LMM in the SR network, including the development of an extensive endmember library. We then introduce the UGCT framework, as illustrated in Figure 2, and describe the HSI reconstruction process, comprising the abundance generator framework and LMM architecture. Furthermore, we provide a comprehensive account of the convolutional transformer architecture, driven by the fine abundance generator, as depicted in Figure 3. Subsequently, the PMSA and S2AM are discussed, which are two crucial components of feature extraction. The process of seamlessly integrating transformer and ResBlock features within S2AM will be thoroughly illustrated in Figure 4. Lastly, we explore the loss function and delve into the implementation and configuration of various details.

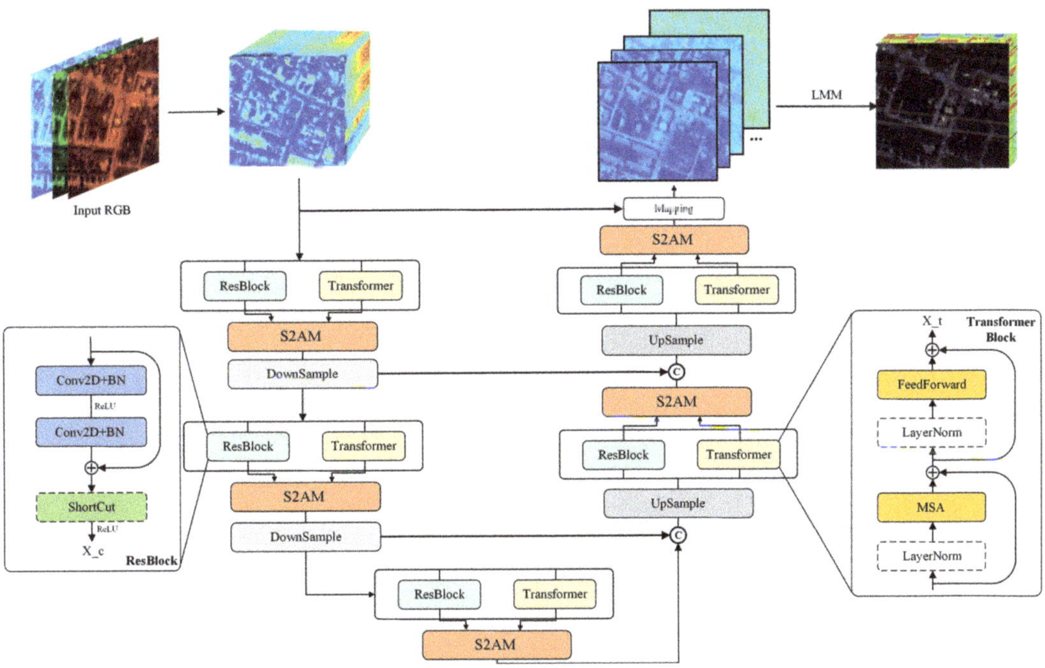

Figure 2. The Struction of Unmixing-Guided Convolutional Transformer Network (UGCT).

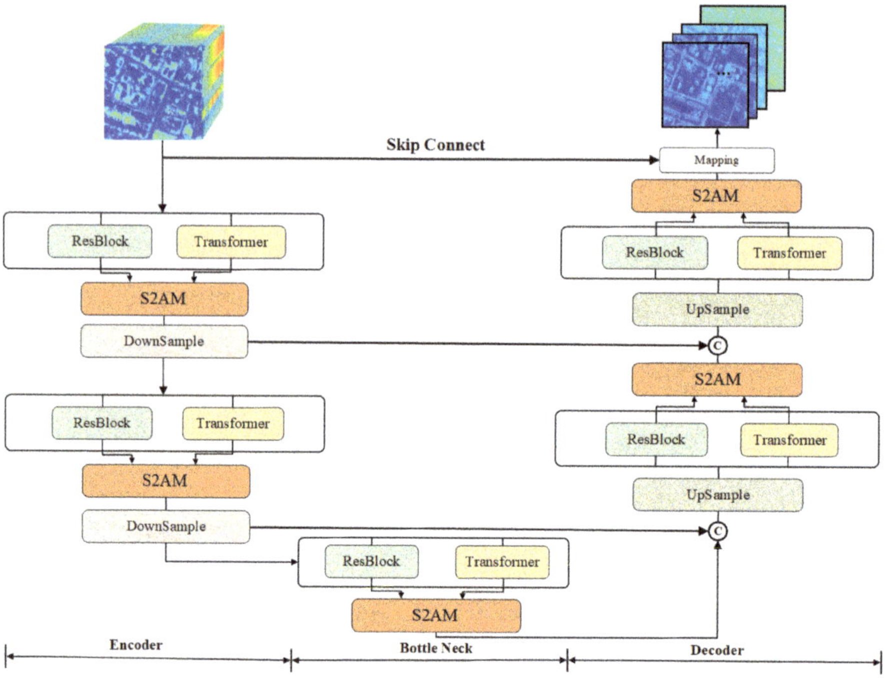

Figure 3. The Struction of Unmixing-Guided Convolutional Transformer Abundance Generator (UGCA).

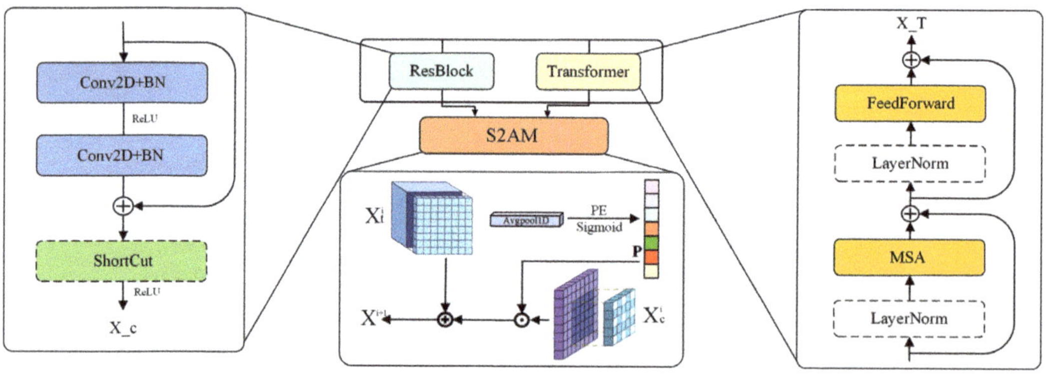

Figure 4. The Paralleled-Residual Multi-Head Self-Attention (PMSA) block and Spectral-Spatial Aggregation Module (S2AM).

3.1. Hu-Based Modeling

During the imaging process of airborne and spaceborne hyperspectral image sensors, a considerable amount of spatial information becomes intermingled within mixed pixels due to factors such as atmospheric absorption, sensor performance complexity, and the actual distribution of ground objects. This substantially reduces the spatial resolution of HSIs. At present, HU is among the most effective algorithms for addressing pixel mixing, with the LMM being one of the most well-developed fundamental modeling algorithms [46]. The HSI, $\boldsymbol{Y} \in \mathbb{R}^{H\times W\times C}$, composed of mixed pixels can be divided into finite pure pixels $\boldsymbol{r} \in \mathbb{R}^{N\times C}$ and corresponding abundance matrices $\boldsymbol{A} \in \mathbb{R}^{H\times W\times N}$ in the classic LMM model.

$$\boldsymbol{Y} = \boldsymbol{A}\boldsymbol{r} + \boldsymbol{b} \tag{1}$$

in which $b \in \mathbb{R}^{H \times W \times C}$ means the noise matrices, H and W represent the spatial scale and N is the number of the endmembers.

With the help of Equation (1), we can create HSIs with high spatial resolution at sub-pixel scales by obtaining a complete endmember library $\mathbb{L}$ of HSIs and their corresponding fine abundance matrices. Because of the low spatial resolution of hyperspectral imaging, multiple ground objects are quite common in the same pixel. Within a mixed pixel, the abundance matrix describes the pure pixel content ratio. According to the basic assumption in the LMM [47], only one reflection and refraction of light occurs between emitting and being captured by the sensor.

$$y_n = \sum_{i=1}^{N} \alpha_i r_i + \beta \tag{2}$$

where $\alpha_i \in A$ and $r_i \in r$ and y_n represent the n-th pixel in the mixing HSI. It should be noted that r_i is the i-th endmember vector from a well-known complete spectral library, which represents continuous spectral data obtained by sensors under pure light from certain pure ground objects such as bushes and gravels. Furthermore, α_i is the spectral abundance of the i-th endmember corresponding to the nth mixed pixel. The β denotes noise disturbances, which include complex atmospheric noise as well as environmental disturbances. It is simply modeled as a bias matrix due to difficulties in accurate modeling or being eliminated in the preprocessing section.

The abundance matrix has practical physical significance, and during calculation, LMM specifies two constraints for it: a sum-to-one constraint and a non-negative abundance constraint [31]. Because the information content of the mixed pixel cannot exceed that of the pure pixel itself in the actual imaging process and because the proportion of a pure pixel included in the pixel cannot be negative, the following constraints will be used:

$$\alpha_i \geq 0; \quad \boldsymbol{\alpha}^{\top}\mathbf{1} = 1 \tag{3}$$

The entire spectrum library $\mathbb{L}$ is already available which was obtained in the laboratory and during onboard practical testing [48]. As a result, obtaining a fine abundance matrix from a single RGB image input is central to improving the performance of spectral reconstruction tasks based on the HU mechanism. This does not imply that we will only use the weak spectral information in RGB to reconstruct a complete HSI. On the contrary, the highly effective, complete, and pure pixels collected will be used as a key reference index to guide model training. In fact, for a high level of a priori comprehensiveness, a deep layer-by-layer autoencoder network utilizing a convolutional transformer will be used.

3.2. The Struction of UGCT

In our network, we employ the Unmixing-Guided Convolutional Transformer Abundance Generator (UGCA) in Figure 3, denoted as $\mathcal{F}$, which is specifically designed for the generation of fine abundance matrices. By providing an accurate remote sensing RGB and a complete spectral set [48] of endmembers from the relevant band, the created network will recover all of its abundance values pixel by pixel using learnable parameters $\boldsymbol{\theta}_l$ and then combine them into a complete spectral abundance matrix.

$$A = \boldsymbol{Soft}(\overline{A}) = \mathcal{F}(\overline{\mathbf{X}}|\boldsymbol{\theta}_l) \tag{4}$$

in which $\overline{\mathbf{X}}$ represents the upsampled RGB input and $\boldsymbol{Soft}(\cdot)$ stands for the softmax operator in order to fit the sum-to-one constraint in Formula (3).

$$\overline{\mathbf{X}} = Upsampling(\mathbf{X}) \tag{5}$$

In an effort to emulate the complex mixing process of light propagation, an autoencoder approach is employed to obtain the full abundance. In this method, the input RGB

X must first undergo a predefined spectral upsampling to map it to the initial spectral features $\overline{X}$. As illustrated in Figure 3, the abundance matrix A is processed through an encoding–decoding procedure where upsampling and downsampling modules are modeled as conv4 and deconv layers, respectively, to facilitate the spatial feature transformation while accommodating the corresponding channel dimension changes.

It is worth noting that this may lead to redundant features and parameters if upsampling and downsampling operations are not incorporated in an autoencoder framework [14], which inevitably leads to redundant features and parameters. To alleviate the pressure from excessive parameters and invalid repetitive features on the training process, they are widely employed in such frameworks. Specifically, as the encoder progresses deeper, the channel dimension will gradually undergo upsampling, while the spatial dimension will experience downsampling. Subsequently, in the decoder section, the spatial dimension is incrementally upsampled in accordance with the input feature scale. Concurrently, the processing of spatial dimensions facilitates the model in acquiring features at different scales. Overall, the model is designed with a symmetric architecture and employs a Conv2D (mapping) layer after the original skip connection to map the features to the desired abundance matrix.

$$\overline{A} = Map(\boldsymbol{PMSA}^{(n)}(\overline{X}) + \overline{X}|\theta_{map}) \tag{6}$$

The input hyperspectral features undergo processing through an n-layer PMSA $\boldsymbol{PMSA}^{(n)}$ module, which encodes them into abundance features using trainable parameters. A skip connection is then employed to project these features into refined abundance representations that fulfill the specified requirements. The n-layer PMSA module can be dissected into three primary components: encoder, bottleneck, and decoder.

$$\boldsymbol{PMSA}^{i}_{encoder} = [f_T^{i-1} \otimes f_C^{i-1}]^{\downarrow} \tag{7}$$

During the encoder phase, the original features are partitioned into two separate streams, which are subsequently processed by transformer blocks and residual blocks (ResBlock). Distinct from conventional transformer blocks, the PMSA module harnesses the combined power of convolutional and transformer networks' prior knowledge to execute accurate abundance extraction driven by local and non-local information.

The i-th encoder module, denoted as $\boldsymbol{PMSA}^{i}_{encoder}$, employs a S2AM $\otimes$ to integrate the two acquired features, thereby maximizing their exceptional extraction capabilities in both spatial and channel dimensions. Following this, a downsampling operation $\downarrow$ is utilized to guarantee that no erroneous features impede the learning process while expanding band dimensions. Within the S2AM module in the encoder, image features undergo upsampling (doubling) in the channel dimension. To prevent the generation of an excessive number of redundant features, spatial downsampling operations $\downarrow$ prove to be highly advantageous. To avert irregularities during model training, a finer feature representation is either recommended for subsequent computation or utilized in a skip connection, ensuring a more stable and accurate learning process.

$$\boldsymbol{PMSA}^{j}_{decoder} = Concat(\boldsymbol{PMSA}^{i}_{encoder}, \boldsymbol{PMSA}^{j-1}_{decoder})^{\uparrow} \tag{8}$$

The encoder process maps hyperspectral features to abundance matrix features within the bottleneck section while maintaining consistent feature spatial and spectral scales. In the subsequent decoder step, spectral features and abundance features from the prior decoder section are amalgamated in the channel dimension using the concatenation operation.

Contrasting the previously described encoder module, the decoder section $\boldsymbol{PMSA}_{decoder}$ upsamples features in the spatial dimension to augment the spatial information of the abundance matrix features while simultaneously compressing channel characteristics. Spatial upsampling $\uparrow$ and channel downsampling operations are implemented within the same deconvolution layer in order to maintain the symmetry of the autoencoder structure. This method ensures an effective balance between spatial and spectral information in the final abundance matrix feature.

Finally, we will discuss in detail the issue of setting the number of blocks in the Discussion section.

3.3. Paralleled-Residual Multi-Head Self-Attention

A Paralleled-Residual Multi-Head Self-Attention (PMSA) block is composed of four key components: two parallel convolutional transformer blocks, an S2AM, and a sampling module (either upsampling or downsampling, excluding the bottleneck layer). In this architecture, the input features are explicitly divided into two separate parts, which are then fed independently into the CNN and transformer blocks.

$$\begin{aligned} \hat{\mathbf{X}}^{i} &= MSA(\mathbf{X}^{i-1}) + \mathbf{X}^{i-1} \\ \mathbf{X}_{t}^{i} &= FFN(\hat{\mathbf{X}}^{i}) + \hat{\mathbf{X}}^{i} \end{aligned} \tag{9}$$

in which MSA means the multi-head self-attention module, and FFN consists of three Conv2D and two GELU operations.

In the ResBlock, as illustrated in Figure 4, the input must first undergo two consecutive 2D convolution and batch normalization layers (Conv2D+BN). The inclusion of a residual connection assists the model in training and converging more effectively. In the encoder and decoder part, the final ShortCut operation becomes a 2D convolution with a convolution kernel of one, while in the bottleneck section, this part is set as an empty layer.

The PMSA block leverages the strengths of both the CNN and transformer architectures to process multi-scale features effectively. The block can capture both local and global contextual information simultaneously. The parallel transformer and CNN outputs are combined in the feature fusion S2AM module to further improve the model's capacity for pattern recognition. Finally, the sampling module adjusts the spatial resolution of the features as required, depending on the specific layer in the network.

$$\mathbf{X}^{i} = [\mathbf{X}_{t}^{i} \otimes \mathbf{X}_{c}^{i}]^{\downarrow} \tag{10}$$

The main distinction between features $\mathbf{X}_t^i$ and $\mathbf{X}_c^i$ lies in their methods for handling scale within their respective blocks. Feature $\mathbf{X}_t^i$ implements channel upsampling within the resblock, which results in an increase in the number of channels while preserving spatial dimensions. On the other hand, Feature $\mathbf{X}_c^i$ maintains the same scale within the transformer block, retaining both the spatial dimensions and the number of channels. The S2AM is then employed to fuse the features from both $\mathbf{X}_t^i$ and $\mathbf{X}_c^i$, even though they have different scales. This fusion process enables the model to combine the information from various scales effectively, capturing diverse contextual information and improving the overall performance of the network.

Specifically, as depicted in Figure 2, within the encoder section, we first take input $\mathbf{X}^{i-1} \in \mathbb{R}^{h,w,c}$ and feed it into the parallel convolutional transformer section. Following this, it passes through a channel upsampling module with a convolution with one kernal size in ShortCut(), and $\mathbf{X}_c^i \in \mathbb{R}^{h,w,2c}$ is output after ResBlock. Subsequently, within the built-in upsampling module of S2AM, features $\mathbf{X}_c^i$ and $\mathbf{X}_t^i \in \mathbb{R}^{h,w,c}$ are fused to produce output $\bar{\mathbf{X}}^i \in \mathbb{R}^{h,w,2c}$. To reduce feature redundancy and prevent additional complexity, spatial downsampling is applied to $\bar{\mathbf{X}}^i$, ultimately yielding $\mathbf{X}^i \in \mathbb{R}^{\frac{h}{2},\frac{w}{2},2c}$. In a similar manner, the decoder section will exhibit symmetry with the encoder.

3.4. Spectral–Spatial Aggregation Module

Transformer and CNN models use distinctly different priors and feature extraction techniques. We suggest the S2AM in Figure 4, which addresses ResBlock's inaccurate assumption of channel dimensions brought on by convolutional kernel constraints [37] in order to significantly increase the benefits of both models. This module utilizes the transformer block to encode the weights of features along the channel dimension. These encoded weights are then embedded into the ResBlock to assist in aligning features along the channel dimension. This integration results in the reconstruction of a more detailed HSI.

Enhancing each feature separately and achieving feature scale alignment along the channel dimension are prerequisites for efficiently processing features X_t^i and X_c^i in simultaneous transmission. Both features must go through a careful preprocessing stage to achieve this.

$$\hat{X}_t^i = \tau(X_t^i)$$
$$\hat{X}_c^i = \delta(X_c^i) \tag{11}$$

in which δ represents a 3×3 dilation convolution, and τ represents a group convolution. Utilizing the δ's expansion factor gives features a larger spatial receptive field, which aids in capturing more contextual information from the input. Group convolution, on the other hand, helps reduce the redundant parameters introduced by the transformer during channel dimension alignment. These enhanced features can then be effectively fused and processed in subsequent layers of the network. Next, the feature $\hat{X}_t^i$ will be encoded as a one-dimensional position code along the channel dimension.

$$T^i = \hat{X}_c^i \odot sig\left(£\left(Avgpool\left(\hat{X}_t^i\right)\right)\right) \tag{12}$$

where £ stands for the fully connected layer and $sig(\cdot)$ is the sigmoid operator to map the feature with 0–1.

It becomes difficult to model the distribution of many ground objects and their relationships when pixels are mixed. This complexity significantly affects the generation of abundance matrices, which are crucial for understanding the composition of mixed pixels in remote sensing and hyperspectral imaging applications. In the position encoder component of the S2AM, three cascaded, fully connected layers £ are employed to simulate the complex relationships between ground objects.

$$X^{i+1} = T^i + \hat{X}_t^i \tag{13}$$

In conclusion, the aligned transformer features and the position-encoded embedded ResBlock information are carefully combined through element-wise addition. This process achieves information aggregation for the transformer, enabling the model to effectively fuse the strengths of them. By integrating the position-encoded information and leveraging the S2AM module, the model is better equipped to handle the challenges of spectral reconstruction.

3.5. Loss Function and Details

Our model is specifically designed to address the single-image SR task. It begins by taking a three-channel image as input, and through model mapping, it produces a reconstructed HSI Y. To ensure that it closely resembles the ground-truth HSI $\hat{Y}$, it is essential to constrain the model to learn the inverse function of the camera response function. Designing a superior loss function is a key component of achieving this objective. We primarily use the mean relative absolute error (MRAE) loss as the loss function for this purpose. By using MRAE loss, the model is encouraged to learn a more accurate mapping between the input three-channel image and the corresponding HSI, resulting in improved reconstruction quality.

$$\mathcal{L}oss(Y, \hat{Y}) = \frac{1}{N} \sum_{i=1}^{N} \frac{|Y_i - \hat{Y}_i|}{\hat{Y}_i} \tag{14}$$

It is important to note that due to the presence of a significant number of zero values (minimum values) in some datasets (AVIRIS [49]), the MRAE loss calculation may fail. For all comparative experiments involving such datasets, we use the L1 loss as a substitute for the previously mentioned loss function.

In order to generate a more sufficient abundance matrix and subsequently reconstruct the HSI, we have adopted a dual-stream PMSA architecture to process features. This design choice enables the model to leverage the strengths of both convolutional and transformer-based methods, resulting in improved feature representation and fusion. During the design process, the number of blocks in the backbone network is set to 7, including two symmetric encoder and decoder blocks in Figure 3, with one serving as the bottleneck layer. This configuration allows for a more efficient flow of information through the network while maintaining an appropriate balance between the model's complexity and performance.

Additionally, the spectral dimension is designed with a reference point of 32 in $\overline{\mathbf{X}}$ to ensure the stability of parameter quantities and model performance. This choice helps to keep the number of model parameters at a manageable level while still achieving high-quality SR.

4. Experiments and Results

4.1. Spectral Library

The success of incorporating LMM into the SR task depends on the a priori integration of the accurate spectral library. The quality and completeness of this endmember library directly influence the model's effectiveness in practical applications. To ensure a comprehensive and accurate data source, we have chosen the United States Geological Survey (USGS) [50] Spectral Library Version 7. This library offers an extensive collection of well-characterized reference spectra, enhancing the reliability of our model. To maximize compatibility with various hyperspectral datasets, we selected the 2014 version of the Airborne Visible/Infrared Imaging Spectrometer (AVIRIS) [49] sensor measurements, owing to its wide spectral range (0.4–2.5 μm) and a fine spectral resolution of 10 nm. This choice ensures that our model can accommodate the widest possible range of hyperspectral cubes.

However, the USGS v7 includes a large number of spectra that cannot be detected by airborne or satellite-based sensors, such as those of laboratory-made substances. Including these redundant spectra not only increases the number of parameters but also potentially impacts the recontruction HSIs performance. Therefore, it is crucial to carefully curate the spectral library by eliminating irrelevant spectra and retaining only those pertinent to the specific remote sensing application.

To improve the spectral library's precision, we first undertook a rigorous data-cleaning process. This involved the removal of officially calibrated invalid spectral locations, resulting in the elimination of 914 targets containing invalid channels. After that, we concentrated on identifying ground objects that are typically difficult to detect in remote sensing images, such as minerals and lab-created organic compounds, in their pure pixels. Through this process, we identified 1019 pure pixels that met our criteria for further analysis. In order to optimize our results, we conducted additional screening to isolate pure pixels that were not needed, as in Refs. [31,36]. This comprehensive screening process ultimately yielded 345 calibrated endmembers, which are expected to significantly improve the quality and precision of our spectral analysis.

4.2. Datasets and Training Setup

We experiment with the UGCT on the *grss_dfc*_2018 [31] and AVIRIS [51] datasets. The IEEE *grss_dfc*_2018 dataset is a remote sensing dataset for change detection analysis. It was collected on 16 February 2017 by the National Center for Airborne Laser Mapping (NCALM) from Houston University. The dataset includes hyperspectral data acquired by an ITRES CASI 1500 sensor with a spectral range of 380–1050 nm and 48 bands. It covers two urban areas, Las Vegas and Paris, with a total of 180 image pairs. The original dataset consisted of 27, 512 × 512 pixel hyperspectral image patches. We randomly selected 24 of these patches for training and 3 for testing. Since the original dataset did not provide corresponding RGB channels, we chose to superimpose the features of channels 23, 12, and 5 to create RGB input.

The AVIRIS [49] dataset is a collection of high-spectral-resolution images captured by the AVIRIS sensor, which has 224 contiguous spectral bands between 0.4 and 2.5 μm and a spatial resolution of 10–20 m. Its large imaging coverage is a major advantage. After preprocessing, we extracted 48 spectral features in the 380–1050 range to form the hyperspectral image (HSI) and selected three channels similar to those in the $grss\_dfc\_2018$ dataset as RGB inputs. In total, 3768 patches of size 64 × 64 were used as the training set, and a large image of size 500 × 1000 was used as the validation set.

The proposed UGCT model was trained on an RTX2080Ti GPU for approximately 6 h. The training data for the model input were divided into patches of size 64 × 64. The batch size was set to 20, and the optimizer used was Adam [52] with β_i = 0.9 and β_2 = 0.999. The learning rate was initialized at 0.0004, and a cosine annealing [53] learning rate strategy was used for 100 epochs. Due to the limited size of the training set, random rotation and flipping augmentation methods were used to enhance the data [54].

We selected several SR methods for comparison to demonstrate the superiority of our method, including AWAN [16], HRNet [19], HSCNN+ [15], MST++ [14], and Restormer [55]. Additionally, Ours− was introduced, representing the UGCT model without LMM modeling. To ensure a fair comparison, each method was fully optimized and retrained in the same scene.

$$MRAE = \frac{1}{N}\sum_{i=1}^{N}\frac{|\mathbf{Y}_i - \widehat{\mathbf{Y}}_i|}{\widehat{\mathbf{Y}}_i} \tag{15}$$

$$RMSE = \sqrt{\frac{1}{N}\sum_{i=1}^{N}(\mathbf{Y}_i - \widehat{\mathbf{Y}}_i)^2} \tag{16}$$

To quantitatively compare the results, we used several parameters, including Root Mean Square Error (RMSE) [14,15], Mean Relative Absolute Error (MRAE), Structural SIMmilarity (SSIM) [17], Peak Signal-to-Noise Ratio (PSNR) and Spectral Angle Mapper (SAM) [56]. The RMSE, MRAE, and SAM are metrics for evaluating the accuracy of the reconstructed results, where lower values indicate better reconstruction. Meanwhile, higher SSIM and PSNR values indicate better performance.

4.3. Comparision with Other Networks

Figure 5 showcases the performance results of different methods on the $grss\_dfc\_2018$ dataset. Five channels were selected as examples to demonstrate the MRAE loss error of the comparison model on the validation set. It should be noted that if the reconstructed result performs poorly in terms of MRAE, the pixel will appear brighter. Conversely, if the reconstruction is similar to HSI, the image will appear darker as a whole.

Due to its large number of parameters, HRNet tends to overfit when faced with small sample datasets, resulting in widespread errors in the spectral response curve of a patch in Figure 6. Although HSCNN+, MST++, and Restormer generally maintain alignment in spatial features when compared to HRNet, displaying only minor and consistent distortions at the fine edges, they still exhibit more severe reconstruction errors in comparison to UGCT.

The Ours−, which removes the LMM, achieves results that are comparable to the aforementioned models. However, by incorporating spectral library priors, our method clearly provides more accurate reconstruction results. For the 830 nm feature, other approaches exhibit distortions on the streets, whereas our method, due to the inclusion of priors, demonstrates a significant advantage in maintaining the accuracy of the reconstructed HSI. Based on the data presented in Table 1, our proposed method achieves competitive results across multiple metrics. In terms of RMSE, our approach outperforms the second-best result by 0.0048, while for MRAE, our method and the UGCT variant without LMM obtain the best and second-best results, respectively. These outcomes collectively demonstrate the effectiveness of our method in comparison to the competing algorithms.

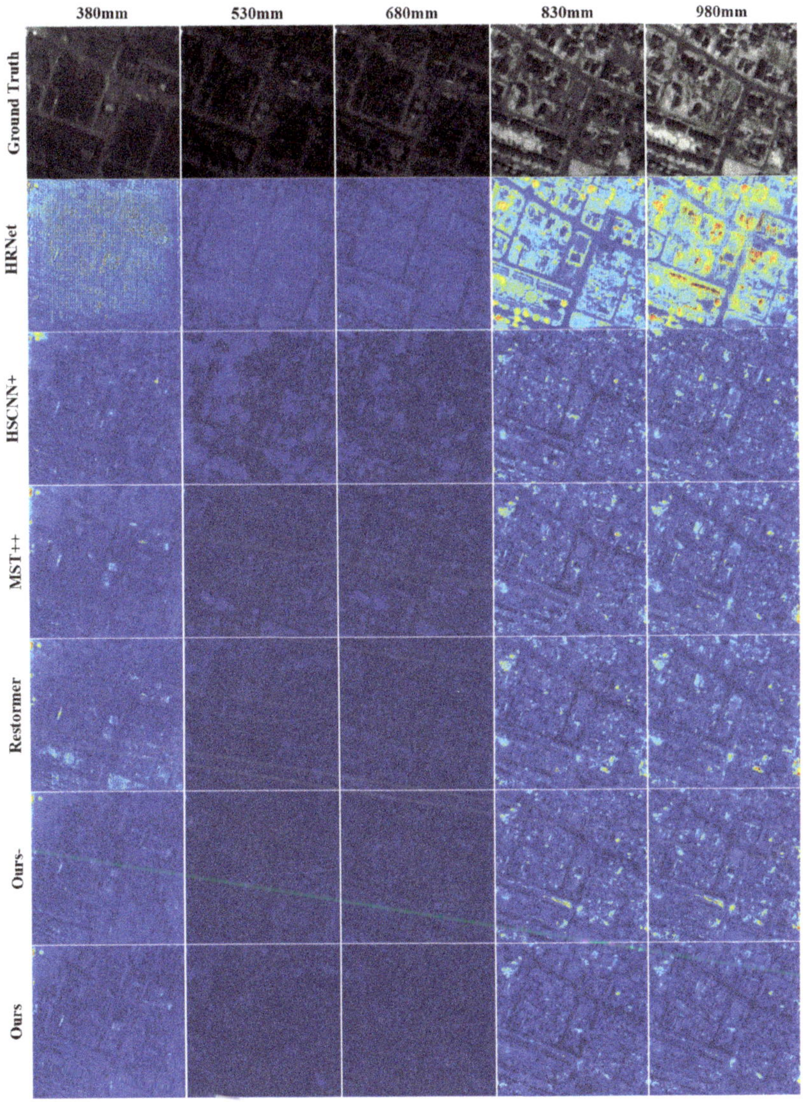

Figure 5. Visual error map of five selected bands on the *grss_dfc*_2018 validation dataset.

Table 1. The quantitative results of the *grss_dfc*_2018 validation dataset. The best and second-best methods are **bolded** and underlined.

Method	RMSE ↓	MRAE ↓	SSIM ↑	SAM ↓
HRNet [19]	0.2020	0.1630	0.882	8.53
AWAN [16]	0.1027	0.0757	0.970	4.64
HSCNN+ [15]	0.1001	0.0724	0.967	4.09
MST++ [14]	<u>0.0914</u>	0.0649	0.972	4.17
Restormer [55]	0.0973	0.0668	0.971	3.96
Ours−	0.0954	<u>0.0614</u>	<u>0.977</u>	**3.89**
Ours	**0.0866**	**0.0587**	**0.979**	<u>3.91</u>

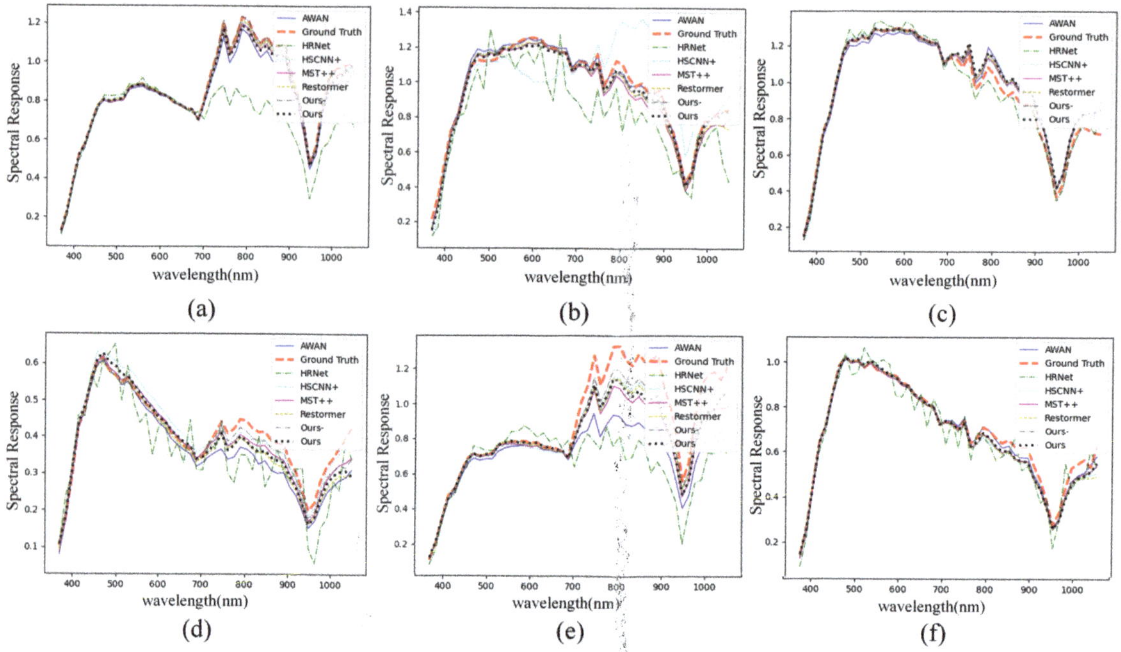

Figure 6. Spectral response curve of the patch (**a–f**) of the validation set for *grss_dfc_*2018.

Due to the validation images in the AVIRIS dataset being large, with dimensions of 1010 × 662, we have reduced computational costs by dividing the images into three overlapping 515 × 512 patches. To demonstrate our results in comparison with other models, we have displayed the MRAE error maps for five selected channels in Figure 7 and the spectral response curves for two selected regions in Figure 8. The closer the curve is to the ground truth, the better the reconstruction performance, and vice versa.

As the results of Table 2 demonstrate, our method achieves the best performance in all four metrics and exhibits the highest similarity to the ground truth curve in the spectral response curves. Notably, HRNet and HSCNN+ appear unable to obtain adequate training or extract sufficient features, leading to substantial distortion in the results, as depicted in Figure 7, which implies that the AVIRIS dataset, characterized by its limited data volume and elevated image noise, demands a more robust feature extraction capability from the network. In contrast, the more lightweight MST++ achieves comparatively improved results, demonstrating a markedly better fit of the spectral response curve in Figure 8 when compared to the previously mentioned methods. While the UGCT exhibits a marginally lower performance than Ours− in SAM metrics, it is evident that both methods substantially outperform other comparison techniques, which indicates the superiority of the convolutional transformer in feature extraction. It is worth noting that the removal of LMM from UGCT results in a significant decline in the performance of the three indexes, which can be attributed to the loss of prior knowledge from the spectral library. When faced with the smaller, noisier AVIRIS dataset, this approach encounters considerable challenges. However, it still manages to produce satisfactory reconstruction results, ranking near the top overall.

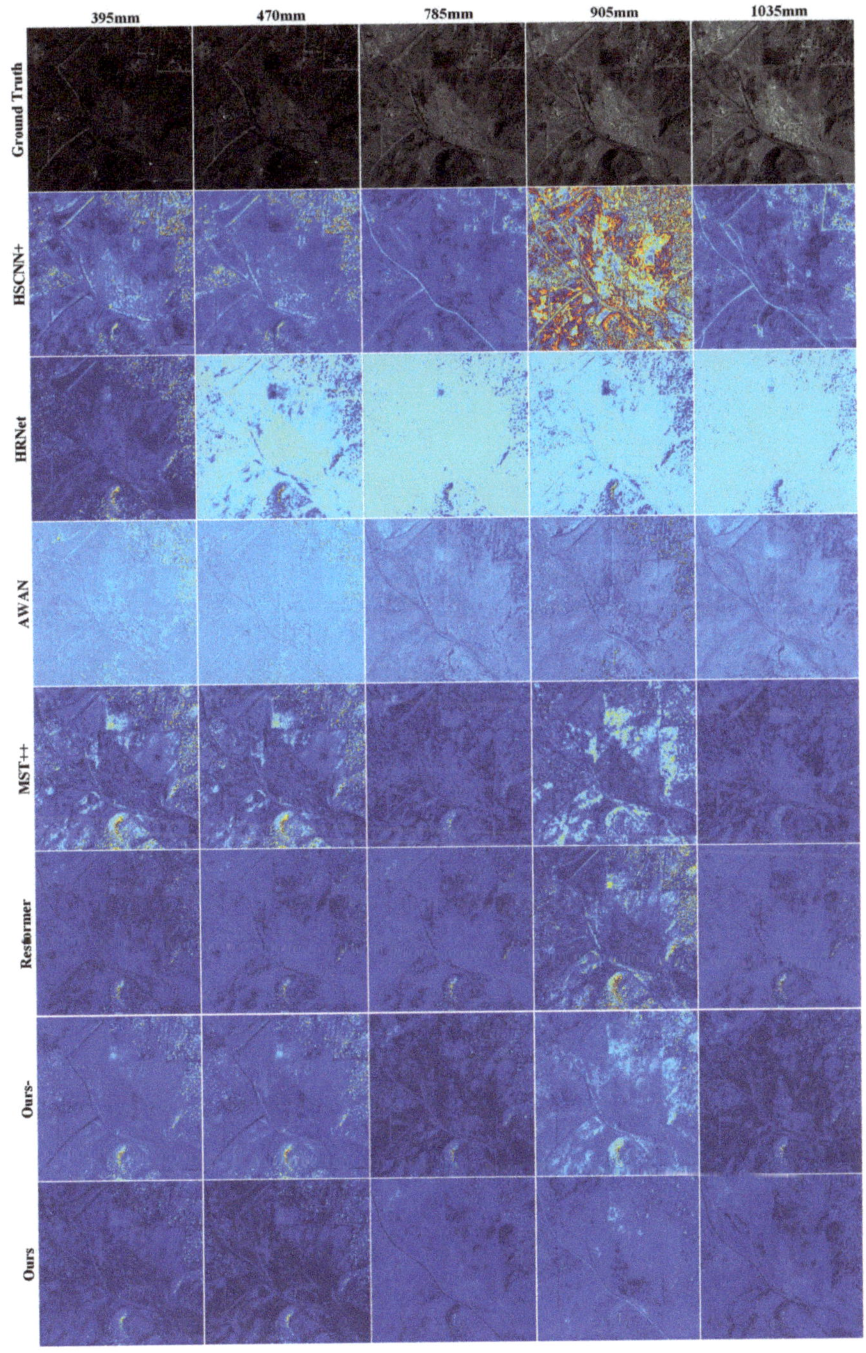

Figure 7. Visual error map of five selected bands on the AVIRIS validation dataset.

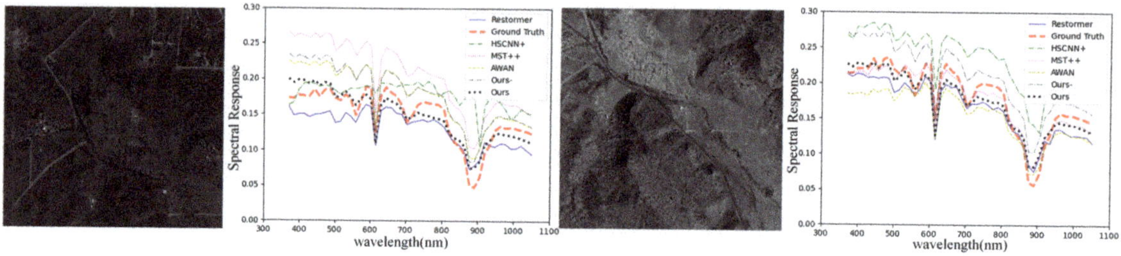

Figure 8. Spectral response curve of the patch of the validation set for AVIRIS.

Table 2. The quantitative results of the AVIRIS validation dataset. The best and second-best methods are **bolded** and underlined.

Method	RMSE ↓	MRAE ↓	SSIM ↑	SAM ↓
HRNet [19]	0.1400	0.8158	0.105	59.63
AWAN [16]	0.0408	0.2141	0.779	12.30
HSCNN+ [15]	0.0775	0.4744	0.716	9.08
MST++ [14]	0.0446	0.2806	0.748	12.61
Restormer [55]	0.0324	0.1883	0.846	8.38
Ours−	0.0357	0.2424	0.875	7.71
Ours	**0.0271**	**0.1451**	**0.886**	**6.80**

The superior performance of our method on the two small-sample remote sensing datasets demonstrates its enhanced reconstruction capabilities for scenes with low spatial resolution, limited sample size, and high noise when compared to alternative approaches. This improvement stems from the integration of the exceptional feature extraction capabilities in the convolutional transformer with the sub-pixel information interpretation offered by the LMM. This combination enables a more effective extraction of mixed pixel information and refined HSI reconstruction.

Specifically, we showcase the superiority of our method on the dataset through Tables 1 and 2. Moreover, to observe the reconstruction ability of our method on remote sensing datasets from the channel dimension, we randomly selected five channel visualization error maps from two datasets, 380 mm, 530 mm, 680 mm, 830 mm and 980 mm in the *grss_dfc*_2018 dataset and 395 mm, 470 mm, 785 mm, 905 mm and 1035 mm in the AVIRIS dataset. It is evident that our method achieved better results/lower error (indicated by darker colors) in both complex scene regions and simple, consistent regions. This demonstrates that the local and non-local features extracted by the convolutional transformer are effectively utilized in the task. Furthermore, spectral response curves serve as a valuable method for visualizing reconstruction tasks. By observing the degree of curve fitting in the selected area, we can clearly see that our method has achieved the best results in multiple comparisons.

In summary, based on the comprehensive comparison results, we found that the Unmixing Guided Convolutional Transformer (UGCT) driven by the LMM model outperforms the model without the unmixing module Our-, indicating that the unmixing-driven model excels in spectral reconstruction tasks. Furthermore, employing the Spectral–Spatial Aggregation Module to combine the benefits of CNN and transformer models surpasses those models that use either convolution or transformer alone. Lastly, our initial attempt at utilizing the self-encoder structured convolutional transformer for SR tasks demonstrated a state-of-the-art performance.

5. Discussion

We further discuss and analyze the impact of the modules and hyperparameter settings on the results through ablation experiments. The ablation study was divided into two parts. The first part compared the performance of different parameter settings, including spectral dimension and block number. The second part focused on the internal modules of the UGCT model, including the LMM module and the PMSA module, etc.

5.1. Network Details

In the first part, we compared the performance of different parameter settings to determine the optimal configuration for spectral dimension and block number in the *grss_dfc_2018* dataset. We modified the spectral dimension while keeping other parameters constant, and we evaluated the results by measuring the corresponding indicators. The results showed that when the initial spectral dimension of the $\hat{X}$ channel was set to 32, the model achieved higher performance, as shown in Table 3.

Table 3. Ablation study about the setting of spectral dim and block number.

Spectral Dim	RMSE	MRAE	SSIM	PNSR
8	0.0924	0.0624	0.976	25.39
16	0.0943	0.0667	0.973	25.34
32	**0.0865**	**0.0587**	**0.979**	**25.69**
48	0.0877	0.0602	0.978	25.60
Block Number	**Params**	**RMSE**	**MRAE**	**SSIM**
5	**2.41M**	0.0882	0.0618	0.977
7	9.56M	**0.0865**	**0.0587**	**0.979**
9	38.12M	0.0975	0.0678	0.969

In summary, for the hyperparameter design of the model, setting the spectral dimension to 32 and the block number to 7 is the optimal choice. All subsequent experiments will be conducted under these settings.

On the other hand, we also examined the effect of block number on the performance of the model while keeping the spectral dimension at 32. It should be noted that the block number significantly affects the model's parameter count due to channel expansion, so we only conducted experiments on three block number values: 5, 7, and 9. According to the table above, although the optimal value 7 has a larger parameter compared to 5, this is a trade-off. As the block number further increases, the parameter count will sharply increase, and the performance may decrease. Therefore, 7 is a relatively better choice.

5.2. Module Ablation Analysis

In this section, we will investigate three aspects of the model: the S2AM feature fusion component, the dual-stream parallel convolutional transformer part, and the LMM module in Table 4.

Table 4. The module ablation analysis in the *grss_dfc_2018* validation dataset.

Description	R_a	R_b	R_c	R_d	R_e	Ours
LMM	✔	✔	✔	✗	✗	✔
S2AM	✗	✗	✗	✗	✔	✔
Resblock	✔	✗	✔	✔	✔	✔
Transformer	✔	✔	✗	✔	✔	✔
MRAE ↓	0.0638	0.0642	0.0712	0.0674	0.0614	**0.0587**

Firstly, in the comparison between $\boldsymbol{R_a}$ and **Ours**, we find that the removal of the S2AM module results in a significant decrease in the reconstruction capability in terms of MRAE. This is because although the PMSA block can effectively extract two excellent features, the lack of a suitable combination method may cause the features to interfere with or mask each other. The results of $\boldsymbol{R_a}$ are similar to those of $\boldsymbol{R_b}$, which also demonstrates the masking effect of the transformer on the ResBlock features.

Secondly, in R_b and R_c, we tested the reconstruction effects of retaining only one part of the dual-stream model to demonstrate its working principle. Both experiments showed a decline in performance, but it is evident that the transformer plays a leading role in feature extraction, while ResBlock also has a crucial function when the S2AM module is present.

Lastly, in R_e, we demonstrated the crucial role of the LMM mechanism, as the loss of the excellent prior knowledge from the spectral library led to a significant decline in the results. To illustrate the impact of the implicit relationship between the spectral position encoding embedded in the S2AM module and the endmember positions in the spectral library on reconstruction accuracy, we compared Experiment R_d with Experiment R_e. The results highlight the importance of the position encoder in S2AM.

6. Conclusions

In this study, we present a novel SR network, UGCT, which is based on the LMM. Specifically, the backbone of the UGCT model consists of several dual-stream PMSA blocks, divided into encoder, bottleneck, and decoder sections. The convolutional transformer block PMSA is a combination of the transformer model and the CNN with various levels. Additionally, considering that CNN does not explicitly model the band dimension, we propose S2AM to fuse the dual-stream features and obtain globally refined image features. To enhance the model's interpretability and incorporate the clear prior knowledge from the spectral library, we propose an HU-based model framework. Finally, comparative experiments conducted on two small and noisy datasets demonstrate the superiority of UGCT in reconstruction accuracy and spectral response curve fitting.

Author Contributions: S.D. and J.L. conceived and designed the original idea; R.S. performed the experiments and shared part of the experiment data; J.L. and Y.L. analyzed the data and conceptualization; S.D. and J.L. wrote the paper; R.S. and Q.D. reviewed and edited the manuscript; Y.L. and Q.D. formal analysis. All authors have read and agreed to the published version of the manuscript.

Funding: This work was supported in part by the Fundamental Research Funds for the Central Universities under Grant JBF220101, in part by the state Key Laboratory of Geo-Information Engineering (No. SKLGIE2020-M-3-1), in part by the science and technology on space intelligent control laboratory ZDSYS-2019-03, in part by the Open Research Fund of CAS Key Laboratory of Spectral Imaging Technology (No. LSIT201924W), in part by the Wuhu and Xidian University special fund for industry-university-research cooperation (No. XWYCXY-012021002), in part by the 111 Project (B08038), and in part by the Youth Innovation Team of Shaanxi Universities.

Data Availability Statement: The data presented in this study are available in the article.

Conflicts of Interest: The authors declare no conflict of interest.

References

1. Li, Y.; Shi, Y.; Wang, K.; Xi, B.; Li, J.; Gamba, P. Target detection with unconstrained linear mixture model and hierarchical denoising autoencoder in hyperspectral imagery. *IEEE Trans. Image Process.* **2022**, *31*, 1418–1432. [CrossRef] [PubMed]
2. Chhapariya, K.; Buddhiraju, K.M.; Kumar, A. CNN-Based Salient Object Detection on Hyperspectral Images Using Extended Morphology. *IEEE Geosci. Remote Sens. Lett.* **2022**, *19*, 6015705. [CrossRef]
3. Liu, H.; Yu, T.; Hu, B.; Hou, X.; Zhang, Z.; Liu, X.; Liu, J.; Wang, X.; Zhong, J.; Tan, Z.; et al. Uav-borne hyperspectral imaging remote sensing system based on acousto-optic tunable filter for water quality monitoring. *Remote Sens.* **2021**, *13*, 4069. [CrossRef]
4. Niroumand-Jadidi, M.; Bovolo, F.; Bruzzone, L. Water quality retrieval from PRISMA hyperspectral images: First experience in a turbid lake and comparison with sentinel-2. *Remote Sens.* **2020**, *12*, 3984. [CrossRef]
5. Niu, C.; Tan, K.; Jia, X.; Wang, X. Deep learning based regression for optically inactive inland water quality parameter estimation using airborne hyperspectral imagery. *Environ. Pollut.* **2021**, *286*, 117534. [CrossRef] [PubMed]
6. Li, K.Y.; Sampaio de Lima, R.; Burnside, N.G.; Vahtmäe, E.; Kutser, T.; Sepp, K.; Cabral Pinheiro, V.H.; Yang, M.D.; Vain, A.; Sepp, K. Toward automated machine learning-based hyperspectral image analysis in crop yield and biomass estimation. *Remote Sens.* **2022**, *14*, 1114. [CrossRef]
7. Arias, F.; Zambrano, M.; Broce, K.; Medina, C.; Pacheco, H.; Nunez, Y. Hyperspectral imaging for rice cultivation: Applications, methods and challenges. *AIMS Agric. Food* **2021**, *6*, 273–307. [CrossRef]
8. Khan, A.; Vibhute, A.D.; Mali, S.; Patil, C. A systematic review on hyperspectral imaging technology with a machine and deep learning methodology for agricultural applications. *Ecol. Inform.* **2022**, *69*, 101678. [CrossRef]

9. Chakraborty, R.; Kereszturi, G.; Pullanagari, R.; Durance, P.; Ashraf, S.; Anderson, C. Mineral prospecting from biogeochemical and geological information using hyperspectral remote sensing-Feasibility and challenges. *J. Geochem. Explor.* **2022**, *232*, 106900. [CrossRef]
10. Pan, Z.; Liu, J.; Ma, L.; Chen, F.; Zhu, G.; Qin, F.; Zhang, H.; Huang, J.; Li, Y.; Wang, J. Research on hyperspectral identification of altered minerals in Yemaquan West Gold Field, Xinjiang. *Sustainability* **2019**, *11*, 428. [CrossRef]
11. Yao, J.; Hong, D.; Chanussot, J.; Meng, D.; Zhu, X.; Xu, Z. Cross-attention in coupled unmixing nets for unsupervised hyperspectral super-resolution. In Proceedings of the Computer Vision—ECCV 2020: 16th European Conference (Part XXIX 16), Glasgow, UK, 23–28 August 2020; pp. 208–224.
12. Hu, J.F.; Huang, T.Z.; Deng, L.J.; Jiang, T.X.; Vivone, G.; Chanussot, J. Hyperspectral image super-resolution via deep spatiospectral attention convolutional neural networks. *IEEE Trans. Neural Netw. Learn. Syst.* **2021**, *33*, 7251–7265. [CrossRef] [PubMed]
13. Hu, J.F.; Huang, T.Z.; Deng, L.J.; Dou, H.X.; Hong, D.; Vivone, G. Fusformer: A transformer-based fusion network for hyperspectral image super-resolution. *IEEE Geosci. Remote Sens. Lett.* **2022**, *19*, 6012305. [CrossRef]
14. Cai, Y.; Lin, J.; Lin, Z.; Wang, H.; Zhang, Y.; Pfister, H.; Timofte, R.; Van Gool, L. Mst++: Multi-stage spectral-wise transformer for efficient spectral reconstruction. In Proceedings of the IEEE/CVF Conference on Computer Vision and Pattern Recognition, New Orleans, LA, USA, 19–20 June 2022; pp. 745–755.
15. Shi, Z.; Chen, C.; Xiong, Z.; Liu, D.; Wu, F. Hscnn+: Advanced cnn-based hyperspectral recovery from rgb images. In Proceedings of the IEEE Conference on Computer Vision and Pattern Recognition Workshops, Salt Lake City, UT, USA, 18–22 June 2018; pp. 939–947.
16. Li, J.; Wu, C.; Song, R.; Li, Y.; Liu, F. Adaptive weighted attention network with camera spectral sensitivity prior for spectral reconstruction from RGB images. In Proceedings of the IEEE/CVF Conference on Computer Vision and Pattern Recognition Workshops, Seattle, WA, USA, 14–19 June 2020; pp. 462–463.
17. Hu, X.; Cai, Y.; Lin, J.; Wang, H.; Yuan, X.; Zhang, Y.; Timofte, R.; Van Gool, L. Hdnet: High-resolution dual-domain learning for spectral compressive imaging. In Proceedings of the IEEE/CVF Conference on Computer Vision and Pattern Recognition, New Orleans, LA, USA, 18–24 June 2022; pp. 17542–17551.
18. Koundinya, S.; Sharma, H.; Sharma, M.; Upadhyay, A.; Manekar, R.; Mukhopadhyay, R.; Karmakar, A.; Chaudhury, S. 2D-3D CNN based architectures for spectral reconstruction from RGB images. In Proceedings of the IEEE Conference on Computer Vision and Pattern Recognition Workshops, Salt Lake City, UT, USA, 18–22 June 2018; pp. 844–851.
19. Zhao, Y.; Po, L.M.; Yan, Q.; Liu, W.; Lin, T. Hierarchical regression network for spectral reconstruction from RGB images. In Proceedings of the IEEE/CVF Conference on Computer Vision and Pattern Recognition Workshops, Seattle, WA, USA, 14–19 June 2020; pp. 422–423.
20. Arad, B.; Ben-Shahar, O.; Timofte, R.N.; Van Gool, L.; Zhang, L.; Yang, M.N. Challenge on spectral reconstruction from RGB images. In Proceedings of the Conference on Computer Vision and Pattern Recognition Workshops, Salt Lake City, UT, USA, 18–23 June 2018; pp. 18–22.
21. Arad, B.; Timofte, R.; Ben-Shahar, O.; Lin, Y.T.; Finlayson, G.D. Ntire 2020 challenge on spectral reconstruction from an rgb image. In Proceedings of the IEEE/CVF Conference on Computer Vision and Pattern Recognition Workshops, Seattle, WA, USA, 14–19 June 2020; pp. 446–447.
22. Arad, B.; Ben-Shahar, O. Sparse recovery of hyperspectral signal from natural RGB images. In Proceedings of the Computer Vision—ECCV 2016: 14th European Conference (Part VII 14), Amsterdam, The Netherlands, 11–14 October 2016; pp. 19–34.
23. He, J.; Yuan, Q.; Li, J.; Xiao, Y.; Liu, X.; Zou, Y. DsTer: A dense spectral transformer for remote sensing spectral super-resolution. *Int. J. Appl. Earth Obs. Geoinf.* **2022**, *109*, 102773. [CrossRef]
24. Yuan, D.; Wu, L.; Jiang, H.; Zhang, B.; Li, J. LSTNet: A Reference-Based Learning Spectral Transformer Network for Spectral Super-Resolution. *Sensors* **2022**, *22*, 1978. [CrossRef]
25. Wu, H.; Xiao, B.; Codella, N.; Liu, M.; Dai, X.; Yuan, L.; Zhang, L. Cvt: Introducing convolutions to vision transformers. In Proceedings of the IEEE/CVF International Conference on Computer Vision, Montreal, QC, Canada, 10–17 October 2021; pp. 22–31.
26. Guo, J.; Han, K.; Wu, H.; Tang, Y.; Chen, X.; Wang, Y.; Xu, C. Cmt: Convolutional neural networks meet vision transformers. In Proceedings of the IEEE/CVF Conference on Computer Vision and Pattern Recognition, New Orleans, LA, USA, 18–24 June 2022; pp. 12175–12185.
27. He, X.; Zhou, Y.; Zhao, J.; Zhang, D.; Yao, R.; Xue, Y. Swin transformer embedding UNet for remote sensing image semantic segmentation. *IEEE Trans. Geosci. Remote Sens.* **2022**, *60*, 4408715. [CrossRef]
28. Liu, Z.; Luo, S.; Li, W.; Lu, J.; Wu, Y.; Sun, S.; Li, C.; Yang, L. Convtransformer: A convolutional transformer network for video frame synthesis. *arXiv* **2020**, arXiv:2011.10185.
29. He, J.; Yuan, Q.; Li, J.; Zhang, L. PoNet: A universal physical optimization-based spectral super-resolution network for arbitrary multispectral images. *Inf. Fusion* **2022**, *80*, 205–225. [CrossRef]
30. Xu, Y.; Du, B.; Zhang, L.; Cerra, D.; Pato, M.; Carmona, E.; Prasad, S.; Yokoya, N.; Hänsch, R.; Le Saux, B. Advanced multi-sensor optical remote sensing for urban land use and land cover classification: Outcome of the 2018 IEEE GRSS data fusion contest. *IEEE J. Sel. Top. Appl. Earth Obs. Remote Sens.* **2019**, *12*, 1709–1724. [CrossRef]
31. Liu, L.; Li, W.; Shi, Z.; Zou, Z. Physics-informed hyperspectral remote sensing image synthesis with deep conditional generative adversarial networks. *IEEE Trans. Geosci. Remote Sens.* **2022**, *60*, 5528215. [CrossRef]

32. Mishra, K.; Garg, R.D. Single-Frame Super-Resolution of Real-World Spaceborne Hyperspectral Data. In Proceedings of the 2022 12th Workshop on Hyperspectral Imaging and Signal Processing: Evolution in Remote Sensing (WHISPERS), Rome, Italy, 13–16 September 2022; pp. 1–5.
33. West, B.T.; Welch, K.B.; Galecki, A.T. *Linear Mixed Models: A Practical Guide Using Statistical Software*; CRC Press: Boca Raton, FL, USA, 2022.
34. Luo, W.; Gao, L.; Zhang, R.; Marinoni, A.; Zhang, B. Bilinear normal mixing model for spectral unmixing. *IET Image Process.* **2019**, *13*, 344–354. [CrossRef]
35. Wang, M.; Zhao, M.; Chen, J.; Rahardja, S. Nonlinear unmixing of hyperspectral data via deep autoencoder networks. *IEEE Geosci. Remote Sens. Lett.* **2019**, *16*, 1467–1471. [CrossRef]
36. Liu, L.; Zou, Z.; Shi, Z. Hyperspectral Remote Sensing Image Synthesis based on Implicit Neural Spectral Mixing Models. *IEEE Trans. Geosci. Remote Sens.* **2022**, *61*, 5500514. [CrossRef]
37. Hu, J.; Shen, L.; Sun, G. Squeeze-and-excitation networks. In Proceedings of the IEEE Conference on Computer Vision and Pattern Recognition, Washington, DC, USA, 18–23 June 2018; pp. 7132–7141.
38. Dong, W.; Zhou, C.; Wu, F.; Wu, J.; Shi, G.; Li, X. Model-guided deep hyperspectral image super-resolution. *IEEE Trans. Image Process.* **2021**, *30*, 5754–5768. [CrossRef]
39. Guo, Q.; Zhang, J.; Zhong, C.; Zhang, Y. Change detection for hyperspectral images via convolutional sparse analysis and temporal spectral unmixing. *IEEE J. Sel. Top. Appl. Earth Obs. Remote Sens.* **2021**, *14*, 4417–4426. [CrossRef]
40. Zou, C.; Huang, X. Hyperspectral image super-resolution combining with deep learning and spectral unmixing. *Signal Process. Image Commun.* **2020**, *84*, 115833. [CrossRef]
41. Su, L.; Sui, Y.; Yuan, Y. An Unmixing-Based Multi-Attention GAN for Unsupervised Hyperspectral and Multispectral Image Fusion. *Remote Sens.* **2023**, *15*, 936. [CrossRef]
42. Hong, D.; Gao, L.; Yao, J.; Yokoya, N.; Chanussot, J.; Heiden, U.; Zhang, B. Endmember-guided unmixing network (EGU-Net): A general deep learning framework for self-supervised hyperspectral unmixing. *IEEE Trans. Neural Netw. Learn. Syst.* **2021**, *33*, 6518–6531. [CrossRef]
43. Zhao, M.; Yan, L.; Chen, J. LSTM-DNN based autoencoder network for nonlinear hyperspectral image unmixing. *IEEE J. Sel. Top. Signal Process.* **2021**, *15*, 295–309. [CrossRef]
44. Zhou, H.Y.; Lu, C.; Yang, S.; Yu, Y. ConvNets vs. Transformers: Whose visual representations are more transferable? In Proceedings of the IEEE/CVF International Conference on Computer Vision, Montreal, BC, Canada, 11–17 October 2021; pp. 2230–2238.
45. Heo, B.; Yun, S.; Han, D.; Chun, S.; Choe, J.; Oh, S.J. Rethinking spatial dimensions of vision transformers. In Proceedings of the IEEE/CVF International Conference on Computer Vision, Montreal, QC, Canada, 10–17 October 2021; pp. 11936–11945.
46. Manolakis, D.; Siracusa, C.; Shaw, G. Hyperspectral subpixel target detection using the linear mixing model. *IEEE Trans. Geosci. Remote Sens.* **2001**, *39*, 1392–1409. [CrossRef]
47. Xu, X.; Shi, Z.; Pan, B. $\ell 0$-based sparse hyperspectral unmixing using spectral information and a multi-objectives formulation. *ISPRS J. Photogramm. Remote Sens.* **2018**, *141*, 46–58. [CrossRef]
48. Clark, R.N.; Swayze, G.A.; Wise, R.A.; Livo, K.E.; Hoefen, T.M.; Kokaly, R.F.; Sutley, S.J. *USGS Digital Spectral Library Splib06a*; Technical Report; US Geological Survey: Reston, VA, USA, 2007.
49. Green, R.O.; Eastwood, M.L.; Sarture, C.M.; Chrien, T.G.; Aronsson, M.; Chippendale, B.J.; Faust, J.A.; Pavri, B.E.; Chovit, C.J.; Solis, M.; et al. Imaging spectroscopy and the airborne visible/infrared imaging spectrometer (AVIRIS). *Remote Sens. Environ.* **1998**, *65*, 227–248. [CrossRef]
50. Kokaly, R.; Clark, R.; Swayze, G.; Livo, K.; Hoefen, T.; Pearson, N.; Wise, R.; Benzel, W.; Lowers, H.; Driscoll, R.; et al. *Usgs Spectral Library Version 7 Data: Us Geological Survey Data Release*; United States Geological Survey (USGS): Reston, VA, USA, 2017.
51. AVIRIS Homepage. Available online: https://aviris.jpl.nasa.gov/ (accessed on 22 March 2023).
52. Kingma, D.P.; Ba, J. Adam: A method for stochastic optimization. *arXiv* **2014**, arXiv:1412.6980.
53. Loshchilov, I.; Hutter, F. Decoupled weight decay regularization. *arXiv* **2017**, arXiv:1711.05101.
54. Cai, Y.; Lin, J.; Hu, X.; Wang, H.; Yuan, X.; Zhang, Y.; Timofte, R.; Van Gool, L. Mask-guided spectral-wise transformer for efficient hyperspectral image reconstruction. In Proceedings of the IEEE/CVF Conference on Computer Vision and Pattern Recognition, New Orleans, LA, USA, 18–24 June 2022; pp. 17502–17511.
55. Zamir, S.W.; Arora, A.; Khan, S.; Hayat, M.; Khan, F.S.; Yang, M.H. Restormer: Efficient transformer for high-resolution image restoration. In Proceedings of the IEEE/CVF Conference on Computer Vision and Pattern Recognition, New Orleans, LA, USA, 18–24 June 2022; pp. 5728–5739.
56. Zamir, S.W.; Arora, A.; Khan, S.; Hayat, M.; Khan, F.S.; Yang, M.H.; Shao, L. Multi-stage progressive image restoration. In Proceedings of the IEEE/CVF Conference on Computer Vision and Pattern Recognition, Online, 19–25 June 2021; pp. 14821–14831.

remote sensing

MDPI

Article

Application of Machine Learning to Tree Species Classification Using Active and Passive Remote Sensing: A Case Study of the Duraer Forestry Zone

Su Rina [1,2], Hong Ying [1,2], Yu Shan [1,2,*], Wala Du [3,4], Yang Liu [1,2], Rong Li [1,2] and Dingzhu Deng [5]

1 College of Geographic Science, Inner Mongolia Normal University, Hohhot 010022, China; 20214019033@mails.imnu.edu.cn (S.R.); hongy864@nenu.edu.cn (H.Y.); 20204019011@mails.imnu.edu.cn (Y.L.); 20224016019@mails.imnu.edu.cn (R.L.)
2 Inner Mongolia Key Laboratory of Remote Sensing and Geographic Information Systems, Inner Mongolia Normal University, Hohhot 010022, China
3 Chinese Academy of Agricultural Sciences Grassland Research Institute, Hohhot 010022, China; duwala@caas.cn
4 Arxan Forest and Grassland Disaster Prevention and Mitigation Research Station of Inner Mongolia Autonomous Region, Alxan 137400, China
5 Inner Mongolia Autonomous Region Surveying, Mapping and Geographic Information Center, Hohhot 010022, China; nmchddz@126.com
* Correspondence: yushan@imnu.edu.cn; Tel.: +86-187-4799-9666

Abstract: The technology of remote sensing-assisted tree species classification is increasingly developing, but the rapid refinement of tree species classification on a large scale is still challenging. As one of the treasures of ecological resources in China, Arxan has 80% forest cover, and tree species classification surveys guarantee ecological environment management and sustainable development. In this study, we identified tree species in three samples within the Arxan Duraer Forestry Zone based on the spectral, textural, and topographic features of unmanned aerial vehicle (UAV) multispectral remote sensing imagery and light detection and ranging (LiDAR) point cloud data as classification variables to distinguish among birch, larch, and nonforest areas. The best extracted classification variables were combined to compare the accuracy of the random forest (RF), support vector machine (SVM), and classification and regression tree (CART) methodologies for classifying species into three sample strips in the Arxan Duraer Forestry Zone. Furthermore, the effect on the overall classification results of adding a canopy height model (CHM) was investigated based on spectral and texture feature classification combined with field measurement data to improve the accuracy. The results showed that the overall accuracy of the RF was 79%, and the kappa coefficient was 0.63. After adding the CHM extracted from the point cloud data, the overall accuracy was improved by 7%, and the kappa coefficient increased to 0.75. The overall accuracy of the CART model was 78%, and the kappa coefficient was 0.63; the overall accuracy of the SVM was 81%, and the kappa coefficient was 0.67; and the overall accuracy of the RF was 86%, and the kappa coefficient was 0.75. To verify whether the above results can be applied to a large area, Google Earth Engine was used to write code to extract the features required for classification from Sentinel-2 multispectral and radar topographic data (create equivalent conditions), and six tree species and one nonforest in the study area were classified using RF, with an overall accuracy of 0.98, and a kappa coefficient of 0.97. In this paper, we mainly integrate active and passive remote sensing data for forest surveying and add vertical data to a two-dimensional image to form a three-dimensional scene. The main goal of the research is not only to find schemes to improve the accuracy of tree species classification, but also to apply the results to large-scale areas. This is necessary to improve the time-consuming and labor-intensive traditional forest survey methods and to ensure the accuracy and reliability of survey data.

Keywords: active–passive remote sensing; canopy height model (CHM); classification; random forest (RF)

Citation: Rina, S.; Ying, H.; Shan, Y.; Du, W.; Liu, Y.; Li, R.; Deng, D. Application of Machine Learning to Tree Species Classification Using Active and Passive Remote Sensing: A Case Study of the Duraer Forestry Zone. *Remote Sens.* **2023**, *15*, 2596. https://doi.org/10.3390/rs15102596

Academic Editor: Emanuel Peres

Received: 21 March 2023
Revised: 9 May 2023
Accepted: 12 May 2023
Published: 16 May 2023

1. Introduction

Forest resources are a major component of terrestrial ecosystems and play an increasingly important role in regulating the global carbon balance and mitigating climate change [1–3]. The quantity and quality of forest areas are, therefore, of great importance, as is monitoring forests to ensure the stability of forest ecosystems [4]. However, traditional manual monitoring methods are not only time-consuming and labor-intensive but also subject to human error [5]. Remote sensing monitoring provides a rich source of data, and the applied remote sensing methods are constantly being updated [6]; thus, such methods have played an increasingly important operational role in the implementation of national forest inventories (NFIs).

Research using remotely sensed data to classify and map tree species dates back several decades. Several studies of tree species classification based on data sources to improve accuracy have shown that classifiers that combine image pixels with spectra outperform pure spectral classifiers [7–9]. Although optical remote sensing is sufficiently mature, in many cases, it is difficult to identify small differences (e.g., similar species) in land cover classification due to the similar spectral characteristics [10]. However, the accuracy of stand identification based only on single features is very limited [11]. Combining textural features and vertical structure information can improve the accuracy of the classification results obtained with optical remote sensing techniques [12]. In some research based on the optimization of classification methods, classification methods based on remotely sensed data have advantages and disadvantages; usually, different classification methods are better for different regional features [13]. The CART methodology assesses the nonparametric discriminative statistical relationships among multiple data layers and generates a binary tree [14,15]. However, the limitations of the decision tree approach are its potential for overfitting and underfitting [16]. SVMs are machine learning methods with powerful generalization capabilities [17,18]; they have been shown to be powerful for local feature recognition in images [19,20]. The RF methodology is another approach for identifying local features in images. It is an integrated learning technique that builds multiple classification trees based on random bootstrap samples of training data [21,22]. In RFs, redundant variables can be removed automatically using the best classification tree [23]. In recent years, RF has been widely used in land cover and forest classification. Ke et al. integrated spectral and LiDAR data and used machine learning decision trees to construct classification rule sets. The results of a quantitative segmentation quality assessment and the classification accuracy showed improved forest classification accuracy in image segmentation and object-based classification [24].

Drones can carry a variety of sensors that can acquire a variety of different data types and resolutions. Because UAV remote sensing data acquisition requires considerable money and has various limitations, such as flight altitude, the application of satellite active–passive remote sensing data is needed to classify the entire Duraer Forestry Zone, which contains a large range of tree species. Satellite-based studies are becoming more common due to the increasing availability of satellite data, image resolution and time series datasets, and time and computational costs [25]. Researchers reported an overall accuracy of 83.2% for a model constructed using only Sentinel-2 data and an improvement in overall accuracy (OA) for combined Sentinel-1 broadleaf and conifer groups, with significant improvements in producer accuracy (PA) and user accuracy (UA) for all species and relatively good separation of the two species, which could not be separately classified using Sentinel-1 data alone [26]. This difference was because of the time-consuming satellite data search and download activities of traditional methods and the huge storage space required for aerial remote sensing data. In addition, the increased number of classified areas and tree species affects the difficulty and workload of the classification process, requiring strong computational processing power to manage all the data and run different algorithms. Therefore, cloud-based platforms, also known as virtualized supercomputer infrastructures, provide a more user-friendly approach [27]. In this respect, Google Earth Engine (GEE) has been successful because it is a cloud-based platform used for geospatial analysis that allows users to efficiently solve the main problems related to managing large quantities of data and their storage, integration, processing, and analysis [28].

The forest resources in the Arxan region cover 80% of the area, affecting the local ecosystem and representing a national reserve forest resource and a treasure trove of ecological resources [3]. The topography of the Duraer Forestry Zone is complex and mountainous, and its slope orientation has a direct impact on the growth of forest stands. Therefore, integrating multiple data sources [29] and optimal classification features [30] and selecting the best classification method are key to the classification of tree species. The aim of this study is to provide a logical basis for forest management measures to better support the monitoring and conservation of forests and their sustainable development [31,32].

2. Materials and Methods

2.1. Study Area

The study area of this paper is in Duraer National Forest in Arxan, northwest of Xing'an League, Inner Mongolia Autonomous Region (119°28′–120°01′E, 47°15′–47°35′N), at the southwest foothills of the Greater Khingan Mountains, bordering Mongolia in the west and Xin Barag Right Banner in Hulun Buir, Inner Mongolia in the north (Figure 1). The total area of forestry operation is 49,812 hectares, with 33,466 hectares of forestry land, including 14,603 hectares of forested land; a total timber accumulation of 900,000 cubic meters; and a forest coverage rate of 40%. The area has a cold-temperate continental monsoon climate, with long and severe winters, hot summers with short periods of precipitation, and large daily and annual temperature differences. First, the Duraer Forestry Zone is a comprehensive management forestry plantation with natural forests (the main species is birch), planted forests (the main species is larch), farms, breeding, gathering, and wood processing. We classified three sample strips of birch and larch in the Duraer Forest with the same size from aerial photographs: sample a, 950 m × 2150 m; sample b, 910 m × 1970 m; and sample c, 450 m × 4250 m (Figure 1). The number of small classes covered by the three sample strips reached 62, with the number of forest classes being 13 and 2 major tree species being present (birch and larch). Satellite data were then used to create equivalent conditions to classify six species of trees throughout the forest site: willow, poplar, spruce, camphor pine, birch, larch, and nonforest.

2.2. Data

2.2.1. Field Survey Data

Data collected in the field included UAV multispectral data, airborne LIDAR data, UAV orthophotos, and forest sample survey data. Due to the border location of the study area, UAV flight work required multiple applications for permission. All aerial photography was completed between 10 July 2021 and 19 July 2021. Field tree survey work was performed from 10 July 2021 to 19 July 2021 and 16 to 25 July 2022. The forest survey mainly included sample coordinates, tree coordinates, community structure, woodland status, origin, slope orientation, and tree species height.

The orthophotos played an auxiliary role in building the prediction model. The main operation of the orthophoto shooting used Pegasus V10 large-load vertical takeoff and a landing unmanned aerial system (UAS) (Figure 2), and for the complex terrain of the survey area, a variable accuracy model of 8 cm for low flat areas and 13 cm for high steep standing areas was adopted for the route; the regional flight height was approximately 500 to 800 m from the ground. To ensure the accuracy of the model edge, the route exceeded the national border line by 100 to 1000 m.

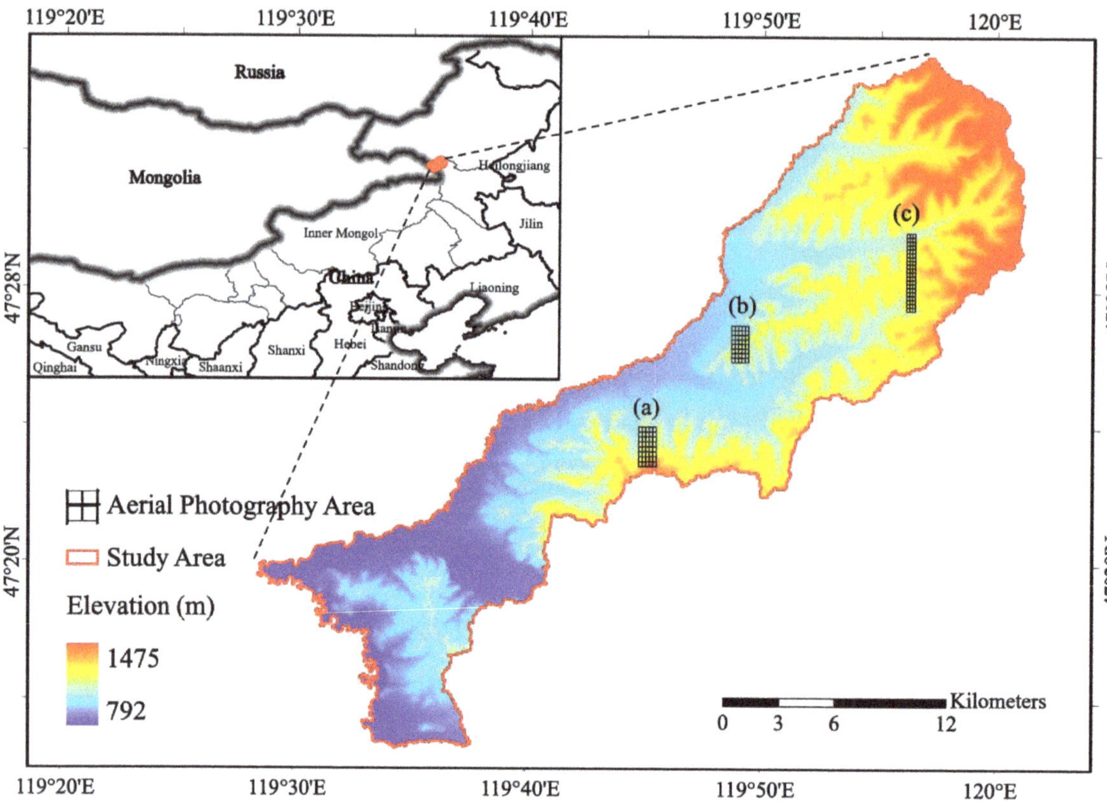

Figure 1. Location of the Duraer Forestry Zone and spatial distribution of the aerial photography areas; (a), (b) and (c) in the figure represent different experimental sample areas acquired by aerial photography, respectively.

Figure 2. Pegasus V10 large-load vertical takeoff and landing UAV.

2.2.2. Drone Multispectral Data

UAV multispectral image data acquisition was performed using the Pegasus V300 product equipped with a camera model Mica Sense Red Edge-MX aerial survey (Figure 3). This product was equipped with an all-in-one multispectral imaging system, using five multispectral cameras (blue, green, red, red edge, and NIR) to form a multispectral image. There were no clouds during the aerial photography, the resolution was adjusted to 10–20 cm for the complex terrain in the survey area, the starting flight altitude was 220 m, and there was no altitude change throughout the survey; the airspeed was 16 m/s, the heading overlap was 80%, and the side overlap was 60%; the camera characteristics are shown in Table 1; and the radiation calibration was performed using a whiteboard.

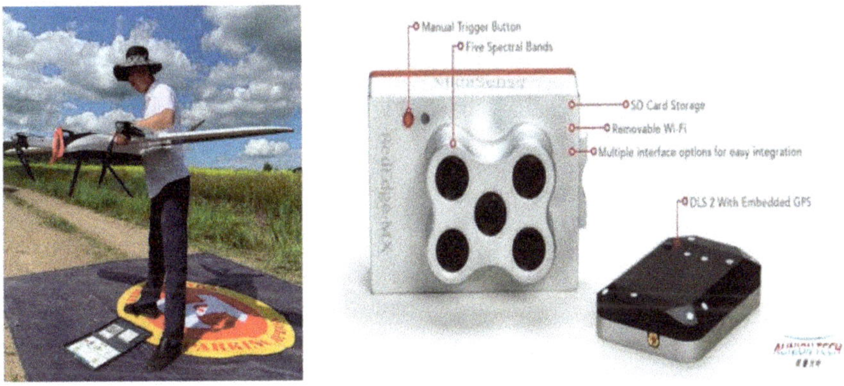

Figure 3. Pegasus V300 UAV and camera Mica Sense Red Edge-MX introduction.

Table 1. Multispectral band information.

Band	Band Name	Wavelength	Wave Width
Band 1	Blue (B)	475	20
Band 2	Green (G)	560	20
Band 3	Red (R)	668	10
Band 4	Near-infrared (NIR)	840	40
Band 5	Red edge (RE)	717	10

The processing of the raw data was performed by the fully automated and fast UAV data processing software Pix4Dmapper from the Swiss company Pix4D. The software is based on the principle of photogrammetry and multivision reconstruction and can be used to quickly obtain point cloud data from aerial footage and process it in postprocessing. We loaded the acquired image into the software to automatically identify the coordinate information and added the image control points to obtain the stitched multispectral image.

2.2.3. UAV Lidar Data

The UAV LiDAR data were collected from a Hurtigruten six-rotor UAV Long-120 equipped with the Hurtigruten ARS-1000 L long-range LiDAR measurement system (Hurtigruten, Wuhan, China) (Figure 4); the core parameters are shown in Table 2. LiDAR data were collected between 12 July 2021, and 17 July 2021, covering a total area of 21.8 km^2. The platform flew at altitudes between 200 and 400 m, with flight speeds of 6 m/s to 10 m/s and an overlap of 60% in the side direction and 70% in the heading direction. The LiDAR sensor beam divergence fraction was 0.5 rmad, so the acquired data footprint diameter was between 0.1 m and 0.2 m.

Figure 4. Hurtigruten six-rotor UAV Long-120 equipped with the Hurtigruten ARS-1000 L long-range LiDAR measurement system.

Table 2. Lidar sensor core parameters.

Core Parameters ARS-1000 L	
Maximum flight height	1350 m
Range resolution	±5 cm
Scanning angle	±330°
Angle resolution	0.001°
Pulse frequency	820 KHZ
Laser wavelength	Near-infrared
Beam divergence	0.5mrad

The processing of raw data was handled using Inertial Explorer (IE) postprocessing software, an open-source software developed by NovAtel's Waypoint product group, and by UAV Butler, a one-stop commercial software for intelligent geographic information systems (GIS) launched by Pegasus Robotics. IE is powerful and highly configurable postprocessing software for processing all available GNSS IE and processing all available GNSS data for decomposition and export to the SBET (OUT) format, which is recognized by common commercial software and can provide high-precision combined navigation information, including position, velocity, and attitude information. The SBET (OUT) format is then converted to the LAS (las) format common to general geoprocessing software using the Drone Butler Smart Laser.

2.2.4. Satellite Data

The Sentinel-2 satellite carries a multispectral imager (MSI) with an altitude of 786 km; it covers 13 spectral bands with an amplitude of 290 km. The ground resolutions are 10 m, 20 m, and 60 m, and the revisit period is 10 days for one satellite and 5 days for two complementary satellites. With different spatial resolutions, from visible and near-infrared to shortwave infrared, the Sentinel-2 data are the only data with three bands in the red-edge range among the available optical data; thus, Sentinel-2 products are very effective for monitoring vegetation health information (Table 3).

Table 3. Spectral bands of the Sentinel-2 sensors (S2A).

Band Number	Band Name	Band Length (nm)	Bandwidth (nm)	Resolution (m)
1	Coastal Aerosol	443.9	27	60
2	Blue	496.6	98	10
3	Green	560.0	45	10
4	Red	664.5	38	10
5	Vegetation red edge (RE)	703.9	19	20
6	Vegetation red edge (RE)	740.2	18	20
7	Vegetation red edge (RE)	782.5	28	20
8	Near-infrared (NIR)	835.1	145	10
8a	Vegetation red edge (RE)	864.8	33	20
9	Water Vapour	945.0	26	60
10	SWIR_Cirrus	1373.5	75	60
11	SWIR	1613.7	143	20
12	SWIR	2202.4	242	20

NASA SRTM Digital Elevation 30 m (SRTM DEM) is a joint effort between NASA and the Department of Defense's National Mapping Agency (NIMA), as well as German and Italian space agencies, and was completed by the U.S.-launched Space Shuttle Endeavour with the SRTM system on board. The SRTM system was used to obtain a near-global DEM. This SRTM V3 product (SRTM Plus) was provided by NASA JPL and has a resolution of 1 arc second (~30 m). This dataset underwent a void-filling process using open-source data (ASTER GDEM2, GMTED2010, and NED), while other versions contained voids or were filled with voids from commercial sources.

ALOS DSM: Global 30 m v3.2 (AW3D30) is a global digital surface model (DSM) dataset with a horizontal resolution of approximately 30 m (1 arc second grid). The dataset is a DSM dataset based on the world's 3D topographic data (5 m grid version). Version 3.2, released in January 2021, is an improved version created by reconsidering the format, ancillary data, and processing methods at high latitudes. The elevations of the AW3D DSM are calculated via an image-matching process that uses pairs of stereo-optical images. Clouds, snow, and ice are automatically identified during processing and mask information is applied.

Data processing is performed using the Google Earth Engine (GEE) code editor, an interactive environment for developing Earth Engine applications, with a central panel that provides a JavaScript code editor. The application programming interface (API) is the core functionality of GEE and is the platform that GEE users are most concerned about. Compared to the graphical user interface (GUI), the API can call all the data and functions in the GEE platform.

2.3. Methods

2.3.1. Extraction of Spectral Features and Texture Features

The vegetation index is very suitable for discriminating vegetation over large areas, where the deviation of the general reflectance curve of vegetation between red and near-infrared constitutes a variable that is sensitive to the presence of green vegetation [33]. For example, depreciation of the NDVI can distinguish unvegetated areas [34], and the EVI belongs to atmospheric impedance [35]. The RVI can assess and monitor vegetation cover [33], and the GRVI is sensitive to subtle disturbances and differences in ecosystem types due to visible red-green band reflectance [36], both of which are sensitive in densely vegetated areas. The VDVI was proposed because chlorophyll absorbs red and blue light and reflects green light, so the classification principle in the study is to determine whether the average value of red and blue light is greater than that of green light and also to distinguish between soil and plants [37]. Other vegetation indices such as the DVI and simple ratios in the NIR and blue bands are more sensitive to the spectral response of green plants [38]. The OSAVI is an optimized index of the Soil Adjusted Vegetation Index (SAVI) which can reduce soil background effects during classification [39]. The IPVI is a linear extension of the NDVI, which can avoid negative numbers during classification [40]. Details of these vegetation indices are shown in (Table 4). In remote sensing, texture describes the variation between light intensity values reflected to the sensor to distinguish valuable data associated with different objects [41]. The red-edge band is valuable in measuring plant health and helping in vegetation classification [42]. The difference in reflectance between birch and larch in the images of the study area in this analysis was more obvious in the red-edge band and the near-infrared band (Figure 5), so these two bands were used in the selection of texture features. The band operation equation is given as follows.

$$\text{Band} = (\text{NIR} + \text{RE})/2 \tag{1}$$

* NIR: near-infrared band; RE: red-edge band

Table 4. Feature information in this research.

Features	Abbreviation	Formula	Reference
Normalized Difference Vegetation Index	NDVI	$\text{NDVI} = \frac{\text{NIR} - \text{R}}{\text{NIR} + \text{R}}$	[34]
Ratio Vegetation Index	RVI	$\text{RVI} = \frac{\text{NIR}}{\text{R}}$	[33]
Enhanced Vegetation Index	EVI	$\text{EVI} = \frac{2.5 * (\text{NIR} - \text{R})}{\text{NIR} + 6 * \text{R} - 7.5 * \text{B} + 1}$	[35]
Difference Vegetation Index	DVI	$\text{DVI} = \text{NIR} - \text{R}$	[38]
Green-Red Vegetation Index	GRVI	$\text{GRVI} = \frac{\text{G} - \text{R}}{\text{G} + \text{R}}$	[36]
Infrared Percentage Vegetation Index	IPVI	$\text{IPVI} = \frac{\text{NIR}}{\text{NIR} + \text{R}}$	[40]
Near infrared and Blue Band Ratios	-	$\frac{\text{NIR}}{\text{B}}$	[38]
Renormalized Difference Vegetation Index	RDVI	$\text{RDVI} = \frac{\text{NIR} - \text{R}}{\sqrt{\text{NIR} + \text{R}}}$	[43]
Visible-band Difference Vegetation Index	VDVI	$\text{VDVI} = \frac{(\text{G} - \text{R}) + (\text{G} - \text{B})}{\text{G} + \text{R} + \text{G} + \text{B}}$	[37]
Optimized Soil Adjusted Vegetation Index	OSAVI	$\text{OSAVI} = \frac{\text{NIR} - \text{R}}{\text{NIR} + \text{R} + 0.16}$	[39]
Grayscale Symbiosis Matrix	GLCM	Mean Variance Contrast Homogeneity Dissimilarity Correlation Angular Second Moment Entropy	
Edge Enhancement	-	Median Sobel Roberts User-defined	
Statistical Filter	-	Data range Mean Variance Entropy Skewness	

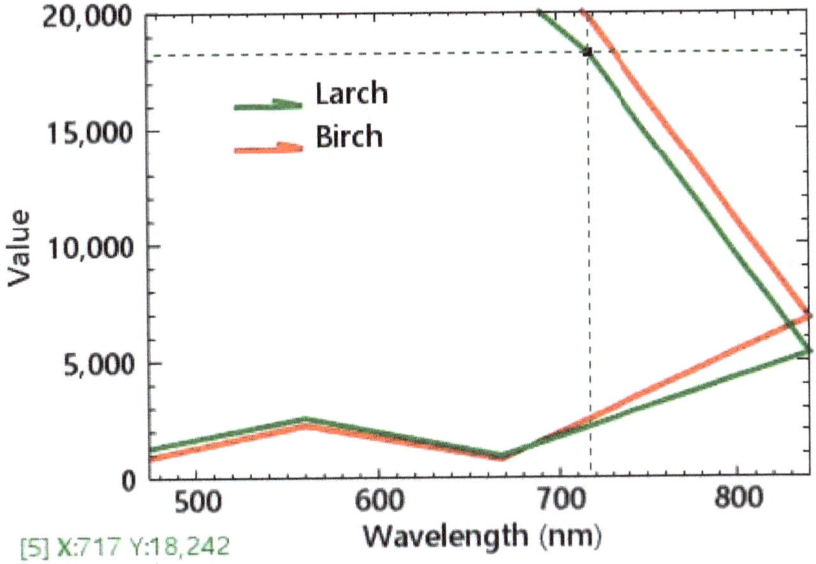

Figure 5. Differences in reflectance of different tree species: the red line represents the spectral reflectance of white birch; the green line represents the spectral reflectance of larch.

2.3.2. Extraction of Vertical Features

The digital elevation model (DEM), digital surface model (DSM), and CHM were obtained from LiDAR360 software developed by Digital Green Earth. This software can preprocess point cloud data with functions such as noise removal, ground point normalization, and extraction of various parameters.

First, the point cloud data were smoothed, resampled, and denoised to ensure that the abnormal point clouds were removed. Then, the ground points were classified, and subsequently, the DEM and DSM were extracted. After obtaining the DEM and DSM, the CHM was extracted and used to segment the airborne point cloud into single trees. Finally, the number of single tree species in a small class was estimated based on the classification results obtained. When classifying by forest landscape (coniferous, broadleaf, and mixed coniferous), we determined whether the ratio of single species in a small class reached 7:3. Simply put, if the percentage of the dominant species was 70% or less, it was considered a mixed forest. When classifying by tree species (birch, larch, mountain poplar, etc.), the specific location of each tree was verified. However, due to the limitations of airborne data, it was not possible to achieve 100% accuracy with the single-wood segmentation.

$$CHM = DSM - DEM \tag{2}$$

2.3.3. Classification Technique

A CART decision tree is a binary tree that can be "pruned" after it is generated [44]. That is, each nonleaf node can only lead to two branches, so when a nonleaf node is a discrete variable with multiple levels (more than 2), the variable has the potential to be used multiple times. CART can be used not only for classification but also for regression. SVMs represent a class of supervised learning that performs binary classification of data [45]. The SVM classification method separates samples belonging to different classes by tracking the maximum-edge hyperplane in the kernel space of the sample mapping [46]. An RF is an integrated classifier consisting of multiple decision trees, where the strength of individual trees and the correlation between trees can be used to generalize the error [21]. RF methodology is an augmentation of traditional decision trees that classifies new data by taking a majority vote among the classification results of all constructed decision trees [47]. In an RF, each node is split using the best combination in a randomly selected subset of feature variables at that node [31].

2.3.4. Confusion Matrix

A confusion matrix summarizes the classification results from a machine learning method in the form of a matrix that classifies the records in a dataset according to two features: the true category and the category predicted by the classification model. In this study, the results of the classification by the machine learning method were considered the predicted category, and the classification results derived from secondary forest inventory data and orthophotos were considered the true category. We analyzed the comparison matrix summarizing the number of image elements and ground tests in every category [48]. The confusion matrix can provide three descriptive accuracy metrics: overall accuracy (OA), producer accuracy (PA), and user accuracy (UA). The OA is equal to the sum of correctly classified pixels divided by the total number of pixels and directly reflects the proportion of correctly classified pixels. PA is the ratio of the number of images that the classifier correctly classifies into a category to the total number of true references in that category. UA is the ratio of the number of pixels correctly classified into a class to the total number of pixels classified into the same class by the classifier. The kappa coefficient is based on the confusion matrix and is used to assess the classification accuracy, and the higher the kappa value is, the greater the classification accuracy of remote sensing images. The value of the OA varies for each category, and the kappa value decreases once the classification result of a category is poor.

2.3.5. GEE Workflow

In the following section, we only describe the conditions created in the GEE to verify the applicability of our proposed scheme to a larger area for the equivalent of Scheme II. Our workflow in the GEE is divided into the following main parts.

(1) Data query and display based on the study area boundary, where the study area vector boundary (feature collection: ao) is imported and the retrieved data are cropped based on the boundary.
(2) Extraction of the best classification elements, which include the best spectral bands, vegetation indices, and texture features (glcm), as well as the CHM derived from the DEM and DSM.
(3) Importation of training sample data based on feature combination, for which the extracted elements are combined and imported into the region of interest (ROI).
(4) Comparison of classification methods and accuracy check, for which the classification accuracy of three classifiers in the ROI are combined to obtain the confusion matrix. Finally, the classification results, accuracy, and kappa of each classifier are calculated, as are the PA and UA of individual tree species.

3. Results

3.1. Comparison of Tree Species Classification Schemes

When classifying image information, one should focus on how to define a meaningful set of features to describe the entire image. Once the best combination of features for classification is selected, the images can be classified using RF in machine learning methods. We designed two schemes based on the above classification features. Scheme I is a combination of the six bands of multispectral reflectance and the extracted vegetation index and texture features, while Scheme II is an additional CHM based on the combination of the six bands of multispectral reflectance and the extracted vegetation index and texture features. RF was used to assess the accuracy of the above two schemes for enhancing tree species classification. As seen in Table 5, the overall accuracy of Scheme I was 79%, and the kappa coefficient was 0.63. The overall accuracy of Scheme II with one more CHM vertical feature was improved by 7%, and the kappa coefficient was 0.75 compared with that of Scheme I.

Table 5. Comparison of the accuracy of Scheme I and Scheme II.

		Birch	Larch	Nonforest
Scheme I	PA	80%	48%	85%
	UA	87%	51%	76%
	OA: 79%		Kappa: 0.63	
Scheme II	PA	90%	70%	84%
	UA	91%	83%	87%
	OA: 86%		Kappa: 0.75	

The results of classification Schemes I and II demonstrated that tree species classification can significantly improve the classification accuracy by increasing its vertical structure on top of the two-dimensional image. This also indicates that canopy height is effective in distinguishing forest from nonforest areas and in classifying tree species. The addition of CHM not only significantly improved the classification accuracy of birch and larch but also significantly improved the misclassification between species, and CHM had no effect on the misclassification of nonforest areas. Therefore, we believe that the hypothesis that the participation of vertical features will improve the classification accuracy of tree species is valid, i.e., Scheme II is the best classification scheme. Among the species, birch was classified most accurately by both schemes and with fewer misclassifications; larch was classified with low accuracy and with more misclassifications relative to other categories. With the addition of CHM, the classification accuracy of both birch and larch improved significantly, and the misclassification rate also decreased significantly. In particular, the classification accuracy of larch was significantly improved, and the misclassification decreased significantly; the classification accuracy of nonforest areas was also significantly improved, but the misclassification was not decreased.

As shown in (Figure 6), group (a) images show the spectral features of tree species, and group (b) includes the CHM features extracted from the point cloud data for the corresponding locations in group (a). The CHM can distinguish tree species from tree height and can compensate for misclassification caused by the shadowed part in the spectral images. Individual trees can also be classified accurately in mixed forests, and low trees do not affect the interpretation of the classifier even if they are blocked by the shadows of taller trees. Small clearings in large woods cannot be discerned spectrally, but the CHM fills this gap well. This is the advantage and notable contribution of active remote sensing in classification.

3.2. Comparison of Tree Species Classification Methods

Based on classification Scheme II, the comparison of tree classification by applying the CART, SVM, and RF is shown in Table 5. The overall accuracy of RF was higher than SVMs and the CART, with 5% and 8% improvement, respectively, and the kappa coefficient was also the highest, indicating that RF has the best classification performance. In addition, we found (Table 6) that RF not only had higher classification accuracy for birch than for other categories, but also led to the lowest misclassification rate for all three categories, and the distinction between birch and larch was more accurate. Although we found that the SVMs and CART classified larch and nonforest areas slightly better than RF, they led to higher misclassification rates. Compared to the SVMs and CART, RF had the least misclassification of larch and nonforest, with 18% and 26% lower misclassification rates for larch and 12% and 14% lower misclassification rates for nonforest areas, respectively. The overall average height of birch was higher than that of the other categories, so each classification method generally classified birch higher than larch and nonforest. The overall accuracy of tree classification improved by 7% with the addition of vertical features; 4% for birch; 32% for larch; and 9% for nonforest areas; with a 10% reduction in misclassification for birch;

22% for larch; and no effect on misclassification for nonforest areas. The improvement of misclassification for larch using RF was significant compared to that for birch and nonforest areas. Overall, RF was the best tree classification method for the data source and the selected scheme of this study, as well as for the Duraer Forestry Zone site.

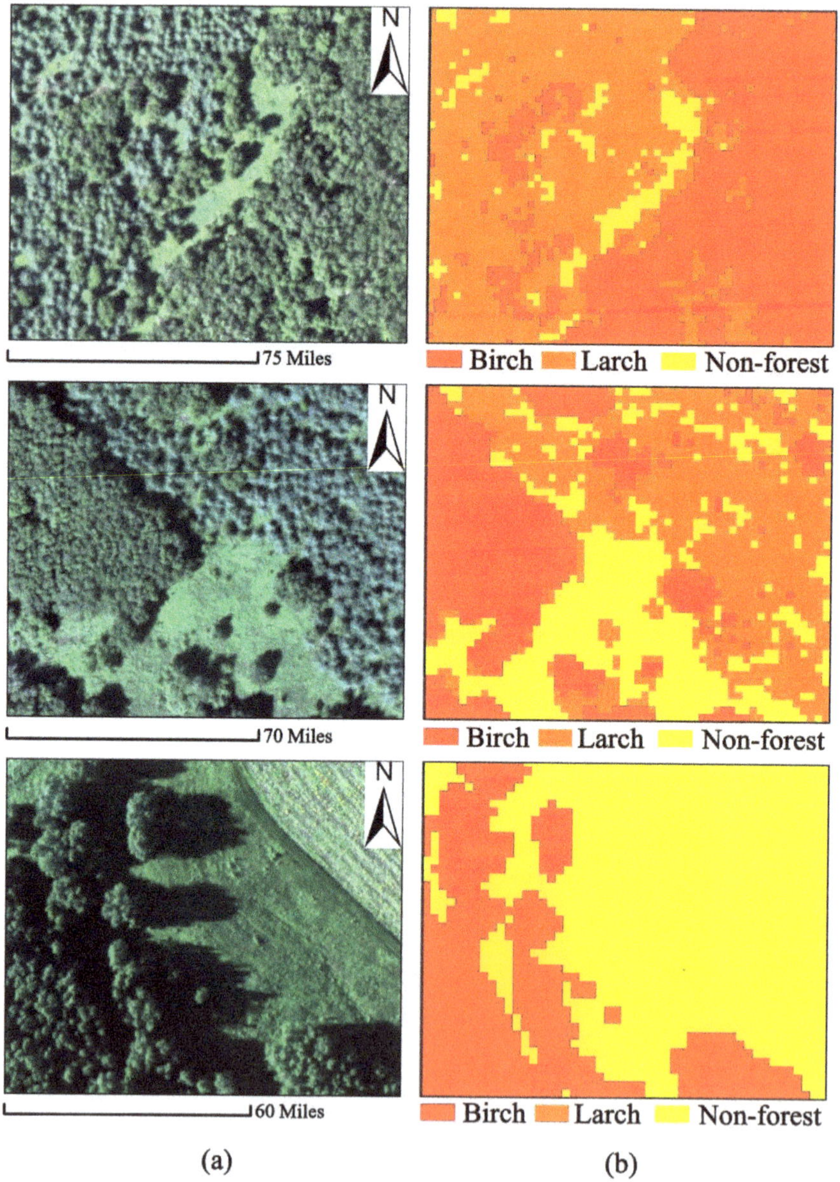

Figure 6. (**a**) Spectral image (RGB) difference among tree species, (**b**) height (CHM) difference among tree species.

Table 6. Comparison of the classification accuracy of machine learning methods.

		RF	SVM	CART
Birch	PA	90%	93%	95%
	UA	91%	77%	75%
Larch	PA	70%	52%	44%
	UA	63%	65%	62%
Nonforest	PA	84%	72%	70%
	UA	87%	93%	90%
OA		86%	81%	78%
kappa		0.75	0.67	0.63

3.3. Spatial Distribution of the Tree Species Classification Based on RF

(Figure 7) shows the spatial distribution of tree species (birch, larch, and nonforest) areas covered by the three sample strips within the Duraer Forestry Zone in Arxan. From left to right in the figure are sample strips (a), sample strips (b), and sample strips (c). The difference image is highlighted in yellow (representing birch in the tree species classification image), orange (representing larch in the tree species classification image), and RGB (representing nonforest areas in the tree species classification image) to show the difference between the three types. The tree species classification in the figure was the result of the RF with the best accuracy. The nonforest RGB image shows that very few tree species were not classified and that small clearings in the forest were accurately classified as nonforest areas.

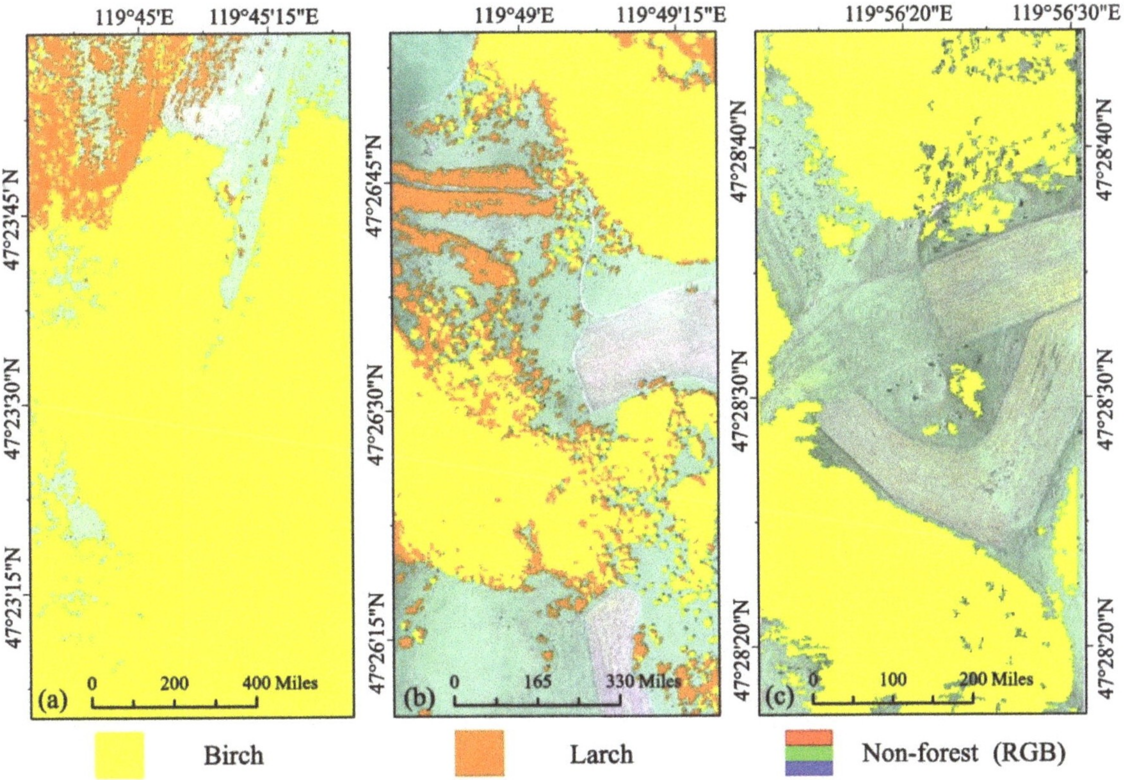

Figure 7. The classification results of sample strips (**a**–**c**).

3.4. Spatial Distribution of the Tree Species Classification Based on GEE

Due to the different data sources, classification schemes, and classification methods used for different data products, the suitability and accuracy of data in some specific areas are often uncertain. Therefore, it is crucial to produce more precise and accurate classification products for a given region [49]. We tested the applicability of the UAV-based study protocol and the various classification features by fusing active and passive satellite data over a large study area. Regions of interest (ROIs) were established based on field-sampled data, and the accuracy of the overall classification results was assessed.

The OA of the CART decision tree classification was 0.96, and the kappa coefficient was 0.94. The OA of the SVM classifier was 0.96, and the kappa coefficient was 0.95. The results of the RF classifier with the highest accuracy are shown in (Figure 8). The OA of the RF reached 0.98, and the kappa coefficient was 0.97. The most common forest type in the Duraer Forestry Zone is natural forest (most of the coniferous forests are planted forests, which are arranged in a regular way; larch is the most planted; and spruce (landscape forest) is mostly planted along the roadside), and the UA reaches 0.98. The UA of nonforest reaches 0.99 because the addition of the CHM contributes greatly to nonforest classification. Therefore, under the same conditions the satellite data are suitable not only for large areas, but also for specific terrain areas.

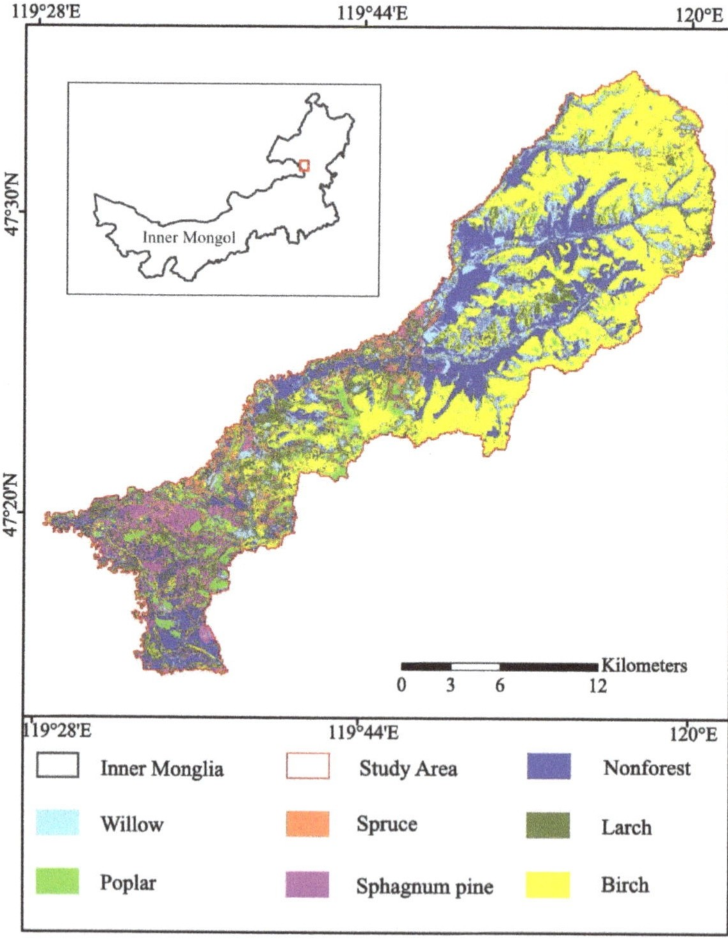

Figure 8. The classification results for the Duraer Forestry Zone.

(Table 7) shows the results of the confusion matrix analysis using ground truth ROIs for the case of applying RF in GEE. To visualize the significance of the variables of the training sample and the test dataset, the rows of the matrix represent the actual categories, while the columns represent the predicted categories. Because the distribution of nonforest and birch is the widest within the Duraer Forestry Zone, the addition of CHM makes the difference in nonforest and birch height obvious. Therefore, the judgments for nonforest and birch are more accurate, and less misjudged. The confusion between birch and larch is most frequent. Most of the larch in the forest is planted, with different planting years, and some early planted larch do not differ significantly in height from immature birch, so the contribution of the CHM to the classification of these two categories is reduced. Because the differences between pine species are more obvious in the leaves, Sphagnum pine is more often misclassified as larch.

Table 7. Confusion matrix using ground truth ROIs.

	Swamp Willow (ROI)	Poplar (ROI)	Spruce (ROI)	Sphagnum Pine (ROI)	Birch (ROI)	Larch (ROI)	Nonforest (ROI)	Total
Swamp Willow	2125	2	6	2	0	16	4	2155
Poplar	1	2259	0	1	9	23	1	2294
Spruce	7	0	2004	6	0	2	14	2033
Sphagnum pine	12	4	12	8742	16	70	7	8863
Birch	28	33	7	21	31,750	100	58	31,997
Larch	37	30	29	41	24	11,866	38	12,065
Nonforest	19	0	10	26	25	26	16,561	16,667
Total	2229	2328	2068	8839	31,824	12,103	16,683	76,074

4. Discussion

Optical sensors have been widely used in classification for a long time, but they are sensitive only to the upper layers of the canopy and have low intercategory separation and high intracategory variability [50]. The quality of a sensor's work is influenced by many environmental factors, and data need to be collected at midday when the sun is shining without cloud cover. In alpine woodland areas, the difficulty of the work can be challenging and data quality can be low due to the terrain and the forest landscape [51,52], making the data can be difficult to separate spectrally. LiDAR radar systems can identify forest canopy structures very well [53] and provide information on understory vegetation [54]. Airborne laser scanning is an active remote sensing data acquisition technique that can provide high-quality vertical structure details [55]; however, its application to forest surveys is limited by the inherent complexity of the canopy structure, and the quality of point clouds collected in naturally dense stands is usually not as good as that in sparse, evenly distributed stands [56]. In conclusion, the remote sensing data obtained from different sensors complement each other [57]. Although many data sources or region-specific methods have been proposed regarding the application of remote sensing data in tree species classification in recent decades, the application to tree species inventoried at large geographic scales remains one of the greatest challenges in this research area [58]. Z. Xie and others found that RF and SVM classification methods performed particularly well when using multisource data and that adding canopy height features to multisource data improved the classification accuracy for some tree species [39]. Researchers verified the classification of individual tree species by combining laser-scanned point clouds and spectral reflectance data and mapping the LiDAR-generated canopy features to the corresponding pixels in multispectral images, resulting in a significant improvement in the overall classification accuracy of all the classified species groups. The results of this study were also consistent with the findings of the above study, concluding that canopy height contributes to tree species classification and significantly influences the classification results among tree species.

Data redundancy may occur when machine learning methods are used to process complex categorical variables, and methods should be chosen considering whether they positively affect classification accuracy. Machine learning methods are efficient and accurate automated techniques but are prone to overfitting when processing large amounts of complex data [15,50]. RFs are integrated models with many classification trees and classifiers and work internally based on a tree pruning mechanism by automatically filtering the input classification features and then voting on the classification results to generalize the classification error [21,59]. In this study, the classification method automatically generated multiple classification trees internally. These classification trees consisted of multiple decision trees related to the reflectance of multispectral bands; the trees were used to extract multiple vegetation indices, textural signatures, and vertical structures. The design of the classification scheme was determined based on the response of the classification accuracy of the RF method to different combinations of the above indices and features. The final combination with the highest classification accuracy constituted a runnable decision tree. Until now, most of the studies on forest classification optimization based on RFs achieved greater than 90% classification accuracy. Part of the reason why the accuracy was not as high in this study was the effect of the predictive classification model when calculating the confusion matrix. The prediction model was based on the most recent forest Scheme II inventory data using the dominant tree species and established species in small classes with the aid of UAV orthophotos and field survey data. However, based on the tree species classification method, the gaps in the forest within small groups were classified as nonforest areas, and there were some large gaps or single trees in nonforest areas that differed from the predicted classification result; therefore, the classification accuracy was affected when calculating the confusion matrix. Although the classification accuracy in this paper was not as good as that of the previous classification optimization study, the objectives of this study were to investigate whether the use of the CHM could improve the classification accuracy of tree species and to compare three machine learning methods to identify the most suitable classification method for the selected study area. Therefore, the classification accuracy we observed was sufficient given the nature of the study. GEE is currently used in various fields, such as agriculture, forestry, ecology, economics, and medicine, with forest and vegetation being the most frequently applied disciplines, followed by land use and land cover [60]. Its development environment supports popular coding languages, and these core features enable users to discover, analyze, and visualize geospatial big data in a powerful way without the use of supercomputers or specialized coding knowledge [61]. In the field of remote sensing and geospatial data science, GEE has become a new method and a key tool for researchers. However, during our research, we found that the accuracy was insufficient if the training sample was too large or complex. Therefore, we relied on the training samples obtained from field surveys for accuracy testing. However, using forest type II survey data to verify accuracy may result in metrics that indicate lower modeling performance than if other less-accurate verification data are used.

In the context of ecologically sustainable development, the United Nations, to ensure the sustainable development of forest ecosystems and woodlands, established measures for different forest types to protect biodiversity and functions [62,63]. Tree species diversity is a key parameter for describing forest ecosystems [47]. The classification of tree species also plays an important role in sustainable forest management. Most of the current research on tree species classification tends to focus on how to optimize the classification results, with few targeted applications. The single-wood segmentation mentioned in this paper can extract information such as absolute coordinates, tree height, and crown width of a single wood. Combining these data with the classification results can solve the time-consuming and labor-intensive problem of traditional forest two-class inventory operations. Although airborne multispectral data and airborne LiDAR data can be effective for tree species surveys in small groups, they are also difficult to implement in forest surveys due to their relatively expensive acquisition costs. Due to the geographical environment of the Duraer Forestry Zone and the natural dense birch forest, the difficulty of airborne LiDAR scanning and data quality cannot be guaranteed. Moreover, the extraction of canopy height requires overlapping points to complement the integrity of forest canopy data, and the contribution of the CHM is affected by the small number of overlapping points in the edge of the scanned area. In addition, the larch in the study area was of an immature plantation forest, so the classification may be confused with taller shrubs (marsh willow, mountain wattle, hoodia, etc.) in terms of height, thus affecting the classification accuracy. In this regard, time series data may improve the identification of larch and thus the classification accuracy of tree species in the entire forest. Recently, some researchers have proposed alpha integrals that can integrate multi-class classifiers, which can combine the best scores to each class by all classifiers separately, thus breaking the limitations of individual classifiers and optimizing the classification results [64,65]. Therefore, the most critical factor to optimize tree species classification is to find the best classification method for individual tree species. Eventually, multiple classification models are fused to obtain the best tree species classification results.

5. Conclusions

The main findings of this paper can be summarized with the following points.

(1) When the classification features were selected, we found that the addition of the CHM to the combination of spectral and textural features for classification improved the overall classification results, indicating that the CHM is an important indicator for improving the classification accuracy of tree species and is important in distinguishing forest from nonforest and white birch from larch.
(2) Comparing the accuracy of machine learning methods under the conditions of choosing equal classification elements, we observed the clear advantage of the random forest among a group of machine learning methods when classifying tree species. This also indicated that RF was the best tree classification method applicable to the data source and the selected scheme of this paper and to the Duraer Forestry Zone.
(3) Our study showed that combining the spectral features, textural features, and vertical features of multisource data (UAV multispectral, LiDAR data, and auxiliary data) and using RF could effectively improve the forest species classification accuracy in the three sample strips within the Duraer Forestry Zone in Arxan.
(4) When applied to a large area following the above research process, the use of the GEE program combined with the required satellite data can support accurate, complex, and rapid tree species classification. The classification results are not limited to specific environments or in cases with data-limited conditions.

Author Contributions: Conceptualization, S.R.; methodology, Y.S. and H.Y.; software, S.R., D.D. and R.L.; validation, S.R.; investigation, S.R., Y.S., Y.L. and H.Y.; resources, Y.S.; writing—original draft preparation, S.R.; writing—review and editing, S.R., Y.S. and H.Y.; visualization, S.R.; supervision, W.D., Y.S., H.Y. and D.D.; All authors have read and agreed to the published version of the manuscript.

Funding: The research was supported by the "14th Five-Year Plan" Social Public Welfare Key R&D and Achievement Transformation Project of Inner Mongolia Autonomous Region [Approval No. 2022YFSH0027], the Key Special Project of Inner Mongolia "Science and Technology Xing Inner Mongolia" Action [Approval No. 2020ZD0028], the National Natural Science Foundation of China [Approval No. 42201374), the Inner Mongolia Natural Science Foundation [Approval No. 2022LHQN04001], the project of "Forest and Grassland Fire Monitoring and Early Warning and Emergency Management System" of the autonomous region [Approval No. 022YFSH0027], the central leading local science and technology development funds "Integrated Demonstration of Ecological Protection and Comprehensive Utilization of Resources in Arxan City", the project of introduction of high-level talents in Inner Mongolia Autonomous Region in 2021 "Key Technology Research on Forest and Grassland Fire Risk Assessment", and the Project for the introduction of high-level talents of Inner Mongolia Normal University [Approval No. 2020YJRC050].

Institutional Review Board Statement: Not applicable.

Informed Consent Statement: Not applicable.

Data Availability Statement: Not applicable.

Acknowledgments: We are grateful to the fieldwork support from the Inner Mongolia Key Laboratory of Remote Sensing and Geographic Information Systems. The authors are very grateful for the support of the Field Scientific Observation and Research Institute for Disaster Prevention and Mitigation of Arxan Forest and Grassland in Inner Mongolia Autonomous Region.

Conflicts of Interest: The authors declare that they have no conflict of interest.

References

1. Dixon, R.K.K.; Solomon, A.; Brown, S.; Houghton, R.; Trexier, M.; Wisniewski, J. Carbon Pools and Flux of Global Forest Ecosystems. *Science* **1994**, *263*, 185–190. [CrossRef] [PubMed]
2. Hansen, M.C.; Potapov, P.V.; Moore, R.M.; Hancher, M.; Turubanova, S.; Tyukavina, A.; Thau, D.; Stehman, S.; Goetz, S.; Loveland, T.R.; et al. High-Resolution Global Maps of 21st-Century Forest Cover Change. *Science* **2013**, *342*, 850–853. [CrossRef] [PubMed]
3. Zhao, J.; Xie, H.; Ma, J.; Wang, K. Integrated remote sensing and model approach for impact assessment of future climate change on the carbon budget of global forest ecosystems. *Glob. Planet. Change* **2021**, *203*, 103542. [CrossRef]
4. Grabska, E.; Hostert, P.; Pflugmacher, D.; Ostapowicz, K. Forest Stand Species Mapping Using the Sentinel-2 Time Series. *Remote Sens.* **2019**, *11*, 1197. [CrossRef]
5. Bouvier, M.; Durrieu, S.; Fournier, R.; Renaud, J. Generalizing predictive models of forest inventory attributes using an area-based approach with airborne LiDAR data. *Remote Sens. Environ.* **2015**, *156*, 322–334. [CrossRef]
6. Barrett, F.; McRoberts, R.E.; Tomppo, E.; Cienciala, E.; Waser, L.T. A questionnaire-based review of the operational use of remotely sensed data by national forest inventories. *Remote Sens. Environ.* **2016**, *174*, 279–289. [CrossRef]
7. Franklin, S.E.; Peddle, D.R. Classification of SPOT HRV imagery and texture features. *Int. J. Remote Sens.* **1990**, *11*, 551–556. [CrossRef]
8. Soh, L.-K.; Tsatsoulis, C. Texture analysis of SAR sea ice imagery using gray level co-occurrence matrices. *IEEE Trans. Geosci. Remote Sens.* **1999**, *37*, 780–795. [CrossRef]
9. Zou, X.; Li, D. Application of image texture analysis to improve land cover classification. *WSEAS Trans. Comput. Arch.* **2009**, *8*, 449–458.
10. Xie, Y.; Sha, Z.; Yu, M. Remote sensing imagery in vegetation mapping: A review. *J. Plant Ecol.* **2008**, *1*, 9–23. [CrossRef]
11. Brovkina, O.V.; Cienciala, E.; Surový, P.; Janata, P. Unmanned aerial vehicles (UAV) for assessment of qualitative classification of Norway spruce in temperate forest stands. *Geo-Spat. Inf. Sci.* **2018**, *21*, 12–20. [CrossRef]
12. Deur, M.; Ga\vsparović, M.; Balenovic, I. Tree Species Classification in Mixed Deciduous Forests Using Very High Spatial Resolution Satellite Imagery and Machine Learning Methods. *Remote Sens.* **2020**, *12*, 3926. [CrossRef]
13. Chen, W.; Xie, X.; Wang, J.; Pradhan, B.; Hong, H.; Bui, D.; Duan, Z.; Ma, J. A comparative study of logistic model tree, random forest, and classification and regression tree models for spatial prediction of landslide susceptibility. *Catena* **2017**, *151*, 147–160. [CrossRef]
14. Hssina, B.; Merbouha, A.; Ezzikouri, H.; Erritali, M. A comparative study of decision tree ID3 and C4.5. *Int. J. Adv. Comput. Sci. Appl.* **2014**, *4*, 13–19. [CrossRef]
15. Loh, W.-Y. Classification and regression trees. *WIREs Data Min. Knowl. Discov.* **2011**, *1*, 14–23. [CrossRef]
16. Song, Y.; Lu, Y. Decision tree methods: Applications for classification and prediction. *Shanghai Arch. Psychiatry* **2015**, *27*, 130–135.
17. Vapnik, V.N. *The Nature of Statistical Learning Theory*; Springer Science & Business Media: Berlin/Heidelberg, Germany, 2000.
18. Wallraven, C.; Caputo, B.; Graf, A.B.A. Recognition with local features: The kernel recipe. In Proceedings of the Ninth IEEE International Conference on Computer Vision, Nice, France, 13–16 October 2003. [CrossRef]
19. Pontil, M.; Verri, A. Support Vector Machines for 3D Object Recognition. *IEEE Trans. Pattern Anal. Mach. Intell.* **1998**, *20*, 637–646. [CrossRef]
20. Schüldt, C.; Laptev, I.; Caputo, B. Recognizing human actions: A local SVM approach. In Proceedings of the International Conference on Pattern Recognition, Cambridge, UK, 23–26 August 2004; Volume 3, pp. 32–36.
21. Breiman, L. Random forests. *Mach. Learn.* **2001**, *45*, 5–32. [CrossRef]
22. Sonobe, R.; Tani, H.; Wang, X.; Kobayashi, N.; Shimamura, H. Random forest classification of crop type using multi-temporal TerraSAR-X dual-polarimetric data. *Sens. Lett.* **2014**, *5*, 157–164. [CrossRef]
23. Dobrinic, D.; Gašparović, M.; Medak, D. Sentinel-1 and 2 Time-Series for Vegetation Mapping Using Random Forest Classification: A Case Study of Northern Croatia. *Remote Sens.* **2021**, *13*, 2321. [CrossRef]
24. Ke, Y.; Quackenbush, L.J.; Im, J. Synergistic use of QuickBird multispectral imagery and LIDAR data for object-based forest species classification. *Remote Sens. Environ.* **2010**, *114*, 1141–1154. [CrossRef]
25. Tassi, A.; Vizzari, M. Object-Oriented LULC Classification in Google Earth Engine Combining SNIC, GLCM, and Machine Learning Algorithms. *Remote Sens.* **2020**, *12*, 3776. [CrossRef]
26. Lechner, M.; Dostálová, A.; Hollaus, M.; Atzberger, C.; Immitzer, M. Combination of Sentinel-1 and Sentinel-2 Data for Tree Species Classification in a Central European Biosphere Reserve. *Remote Sens.* **2022**, *14*, 2687. [CrossRef]
27. Praticò, S.; Solano, F.; Di Fazio, S.; Modica, G. Machine Learning Classification of Mediterranean Forest Habitats in Google Earth Engine Based on Seasonal Sentinel-2 Time-Series and Input Image Composition Optimisation. *Remote Sens.* **2021**, *13*, 586. [CrossRef]

28. Gorelick, N.; Hancher, M.; Dixon, M.; Ilyushchenko, S.; Thau, D.; Moore, R. Google Earth Engine: Planetary-scale geospatial analysis for everyone. *Remote Sens. Environ.* **2017**, *202*, 18–27. [CrossRef]
29. Holmgren, J.; Persson, Å.; Söderman, U. Species identification of individual trees by combining high resolution LiDAR data with multi—Spectral images. *Int. J. Remote Sens.* **2008**, *29*, 1537–1552. [CrossRef]
30. Eisavi, V.; Homayouni, S.; Yazdi, A.M.; Alimohammadi, A. Land cover mapping based on random forest classification of multitemporal spectral and thermal images. *Environ. Monit. Assess.* **2015**, *187*, 291. [CrossRef]
31. Dalponte, M.; Bruzzone, L.; Gianelle, D. Tree species classification in the Southern Alps based on the fusion of very high geometrical resolution multispectral/hyperspectral images and LiDAR data. *Remote Sens. Environ.* **2012**, *123*, 258–270. [CrossRef]
32. Heinzel, J.C.; Koch, B. Investigating multiple data sources for tree species classification in temperate forest and use for single tree delineation. *Int. J. Appl. Earth Obs. Geoinf.* **2012**, *18*, 101–110. [CrossRef]
33. Bannari, A.; Morin, D.; Bonn, F.J.; Huete, A.R. A review of vegetation indices. *Remote Sens. Rev.* **1995**, *13*, 95–120. [CrossRef]
34. Trier, Ø.D.; Salberg, A.-B.; Kermit, M.; Rudjord, Ø.; Gobakken, T.; Næsset, E.; Aarsten, D. Tree species classification in Norway from airborne hyperspectral and airborne laser scanning data. *Eur. J. Remote Sens.* **2018**, *51*, 336–351. [CrossRef]
35. Baugh, W.; Groeneveld, D. Broadband vegetation index performance evaluated for a low—Cover environment. *Int. J. Remote Sens.* **2006**, *27*, 4715–4730. [CrossRef]
36. Motohka, T.; Nasahara, K.N.; Oguma, H.; Tsuchida, S. Applicability of Green-Red Vegetation Index for Remote Sensing of Vegetation Phenology. *Remote Sens.* **2010**, *2*, 2369–2387. [CrossRef]
37. Louhaichi, M.; Borman, M.M.; Johnson, D.E. Spatially Located Platform and Aerial Photography for Documentation of Grazing Impacts on Wheat. *Geocarto Int.* **2001**, *16*, 65–70. [CrossRef]
38. Wang, Y.; Lu, D. Mapping Torreya grandis Spatial Distribution Using High Spatial Resolution Satellite Imagery with the Expert Rules-Based Approach. *Remote Sens.* **2017**, *9*, 564. [CrossRef]
39. Xie, Z.; Chen, Y.; Lu, D.; Li, G.; Chen, E. Classification of Land Cover, Forest, and Tree Species Classes with ZiYuan-3 Multispectral and Stereo Data. *Remote Sens.* **2019**, *11*, 164. [CrossRef]
40. Crippen, R. Calculating the vegetation index faster. *Remote Sens. Environ.* **1990**, *34*, 71–73. [CrossRef]
41. Hernandez, I.E.R.; Shi, W. A Random Forests classification method for urban land-use mapping integrating spatial metrics and texture analysis. *Int. J. Remote Sens.* **2018**, *39*, 1175–1198. [CrossRef]
42. DigitalGlobe. *The Benefits of the Eight Spectral Bands of Worldview-2*; DigitalGlobe: Westminster, CO, USA, 2011.
43. Feng, Y.; Lu, D.; Chen, Q.; Keller, M.; Moran, E.; dos-Santos, M.; Bolfe, É.L.; Batistella, M. Examining effective use of data sources and modeling algorithms for improving biomass estimation in a moist tropical forest of the Brazilian Amazon. *Int. J. Digit. Earth* **2017**, *10*, 996–1016. [CrossRef]
44. Lewis-Beck, M.; Bryman, A.; Futing Liao, T. *CART (Classification and Regression Trees)*; Sage Publications, Inc.: Thousand Oaks, CA, USA, 2004.
45. Dalponte, M.; Bruzzone, L.; Vescovo, L.; Gianelle, D. The role of spectral resolution and classifier complexity in the analysis of hyperspectral images of forest areas. *Remote Sens. Environ.* **2009**, *113*, 2345–2355. [CrossRef]
46. Camps-Valls, G.; Bruzzone, L. Kernel-based methods for hyperspectral image classification. *IEEE Trans. Geosci. Remote Sens.* **2005**, *43*, 1351–1362. [CrossRef]
47. Immitzer, M.; Atzberger, C.; Koukal, T. Tree Species Classification with Random Forest Using Very High Spatial Resolution 8-Band WorldView-2 Satellite Data. *Remote Sens.* **2012**, *4*, 2661–2693. [CrossRef]
48. Niu, Z.G.; Shan, Y.X.; Gong, P. Accuracy evaluation of two global land cover data sets over wetlands of China. *Int. Arch. Photogramm. Remote Sens. Spat. Inf. Sci.* **2012**, *XXXIX-B7*, 223–228. [CrossRef]
49. Yang, Y.; Yang, D.; Wang, X.; Zhang, Z.; Nawaz, Z. Testing Accuracy of Land Cover Classification Algorithms in the Qilian Mountains Based on GEE Cloud Platform. *Remote Sens.* **2021**, *13*, 5064. [CrossRef]
50. Rodriguez-Galiano, V.F.; Ghimire, B.; Rogan, J.; Chica-Olmo, M.; Rigol-Sánchez, J.P. An assessment of the effectiveness of a random forest classifier for land-cover classification. *ISPRS J. Photogramm. Remote Sens.* **2012**, *67*, 93–104. [CrossRef]
51. Guyot, G.; Guyon, D.; Riom, J. Factors affecting the spectral response of forest canopies: A review. *Geocarto Int.* **1989**, *4*, 3–18. [CrossRef]
52. Lapini, A.; Pettinato, S.; Santi, E.; Paloscia, S.; Fontanelli, G.; Garzelli, A. Comparison of Machine Learning Methods Applied to SAR Images for Forest Classification in Mediterranean Areas. *Remote Sens.* **2020**, *12*, 369. [CrossRef]
53. Bortolot, Z.J.; Wynne, R. Estimating forest biomass using small footprint LiDAR data: An individual tree-based approach that incorporates training data. *ISPRS J. Photogramm. Remote Sens.* **2005**, *59*, 342–360. [CrossRef]
54. Frazer, G.; Magnussen, S.; Wulder, M.; Niemann, K.O. Simulated impact of sample plot size and co-registration error on the accuracy and uncertainty of LiDAR-derived estimates of forest stand biomass. *Remote Sens. Environ.* **2011**, *115*, 636–649. [CrossRef]
55. Ussyshkin, V.; Theriault, L. Airborne Lidar: Advances in Discrete Return Technology for 3D Vegetation Mapping. *Remote Sens.* **2011**, *3*, 416–434. [CrossRef]
56. Li, M.; Im, J.; Quackenbush, L.J.; Liu, T. Forest Biomass and Carbon Stock Quantification Using Airborne LiDAR Data: A Case Study Over Huntington Wildlife Forest in the Adirondack Park. *IEEE J. Sel. Top. Appl. Earth Obs. Remote Sens.* **2014**, *7*, 3143–3156. [CrossRef]
57. Lu, D.; Weng, Q. A survey of image classification methods and techniques for improving classification performance. *Int. J. Remote Sens.* **2007**, *28*, 823–870. [CrossRef]

58. Fassnacht, F.E.; Latifi, H.; Stereńczak, K.; Modzelewska, A.; Lefsky, M.A.; Waser, L.T.; Straub, C.; Ghosh, A. Review of studies on tree species classification from remotely sensed data. *Remote Sens. Environ.* **2016**, *186*, 64–87. [CrossRef]
59. Puttonen, E.; Suomalainen, J.M.; Hakala, T.; Räikkönen, E.; Kaartinen, H.; Kaasalainen, S.; Litkey, P. Tree species classification from fused active hyperspectral reflectance and LIDAR measurements. *For. Ecol. Manag.* **2010**, *260*, 1843–1852. [CrossRef]
60. Tian, S.; Zhang, X.; Tian, J.; Sun, Q. Random Forest Classification of Wetland Landcovers from Multi-Sensor Data in the Arid Region of Xinjiang, China. *Remote Sens.* **2016**, *81*, 954. [CrossRef]
61. Phan, T.N.; Kuch, V.; Lehnert, L.W. Land Cover Classification using Google Earth Engine and Random Forest Classifier—The Role of Image Composition. *Remote Sens.* **2020**, *12*, 2411. [CrossRef]
62. Tamiminia, H.; Salehi, B.; Mahdianpari, M.; Quackenbush, L.; Adeli, S.; Brisco, B. Google Earth Engine for geo-big data applications: A meta-analysis and systematic review. *ISPRS J. Photogramm. Remote Sens.* **2020**, *164*, 152–170. [CrossRef]
63. Lindenmayer, D.; Margules, C.; Botkin, D. Indicators of Biodiversity for Ecologically Sustainable Forest Management. *Conserv. Biol.* **2000**, *14*, 941–950. [CrossRef]
64. McCammon, A.L.T. United Nations Conference on Environment and Development, held in Rio de Janeiro, Brazil, during 3–14 June 1992, and the '92 Global Forum, Rio de Janeiro, Brazil, 1–14 June 1992. *Environ. Conserv.* **1992**, *19*, 372–373. [CrossRef]
65. Salazar, A.; Safont, G.; Vergara, L.; Vidal, E. Pattern recognition techniques for provenance classification of archaeological ceramics using ultrasounds. *Pattern Recognit. Lett.* **2020**, *135*, 441–450. [CrossRef]

remote sensing

MDPI

Article

ACTNet: A Dual-Attention Adapter with a CNN-Transformer Network for the Semantic Segmentation of Remote Sensing Imagery

Zheng Zhang, Fanchen Liu, Changan Liu, Qing Tian * and Hongquan Qu

School of Information, North China University of Technology, Beijing 100144, China
* Correspondence: tianqing@ncut.edu.cn

Abstract: In recent years, the application of semantic segmentation methods based on the remote sensing of images has become increasingly prevalent across a diverse range of domains, including but not limited to forest detection, water body detection, urban rail transportation planning, and building extraction. With the incorporation of the Transformer model into computer vision, the efficacy and accuracy of these algorithms have been significantly enhanced. Nevertheless, the Transformer model's high computational complexity and dependence on a pre-training weight of large datasets leads to a slow convergence during the training for remote sensing segmentation tasks. Motivated by the success of the adapter module in the field of natural language processing, this paper presents a novel adapter module (ResAttn) for improving the model training speed for remote sensing segmentation. The ResAttn adopts a dual-attention structure in order to capture the interdependencies between sets of features, thereby improving its global modeling capabilities, and introduces a Swin Transformer-like down-sampling method to reduce information loss and retain the original architecture while reducing the resolution. In addition, the existing Transformer model is limited in its ability to capture local high-frequency information, which can lead to an inadequate extraction of edge and texture features. To address these issues, this paper proposes a Local Feature Extractor (LFE) module, which is based on a convolutional neural network (CNN), and incorporates multi-scale feature extraction and residual structure to effectively overcome this limitation. Further, a mask-based segmentation method is employed and a residual-enhanced deformable attention block (Deformer Block) is incorporated to improve the small target segmentation accuracy. Finally, a sufficient number of experiments were performed on the ISPRS Potsdam datasets. The experimental results demonstrate the superior performance of the model described in this paper.

Keywords: remote sensing; semantic segmentation; transformer; adapter

Citation: Zhang, Z.; Liu, F.; Liu, C.; Tian, Q.; Qu, H. ACTNet: A Dual-Attention Adapter with a CNN-Transformer Network for the Semantic Segmentation of Remote Sensing Imagery. *Remote Sens.* **2023**, *15*, 2363. https://doi.org/10.3390/rs15092363

Academic Editors: Qian Du, Jiaojiao Li, Wei Li, Jocelyn Chanussot, Rui Song, Yunsong Li and Bobo Xi

Received: 27 February 2023
Revised: 18 April 2023
Accepted: 26 April 2023
Published: 29 April 2023

1. Introduction

With the development of modern remote sensing technology and the launch of a series of important high-resolution remote sensing satellites, high-resolution remote sensing (HRRS) images are increasingly captured and applied to research. They contain a rich amount of information on the texture, shape, structure, and neighborhood relationship of various features. The traditional mathematical theory-based semantic segmentation methods [1–3] for the remote sensing of images can be used for relatively simple contents, but are often not suitable for images with complex features. With the excellent image feature extraction capability shown by CNN in recent years, an end-to-end network structure has been established for use in image classification, semantic segmentation, object detection, and other fields, and is effectively used for remote sensing applications [4–6].

Transformer is an architecture proposed in 2017 in the field of NLP, and is a structure for learning global features through a self-attention mechanism. It has achieved extraordinary results in the field of NLP and was quickly introduced by researchers into the field of CV. The Vision Transformer (ViT) [7] cuts images into patches and maps them onto one-dimensional vectors for processing so that they can be converted into sequences for input into the self-attention module, which better captures long-range features and global information. ViT are slightly more accurate than CNN structures after pre-training on large-scale datasets, which demonstrates the powerful potential of Transformer in the imaging domain. Subsequently, more and more CV tasks use Transformer-based models, including semantic segmentation [8–10], target detection [11–13], pose estimation [14,15], etc.

However, the existing models still have serious shortcomings for remote sensing using multi-objective segmentation. First of all, using weights pre-trained o a large dataset to initialize the parameters leads to a better model performance [16], but Transformer lacks inductive biases in CNN, such as translation invariance and local relations, resulting in a poor performance of Transformer-based networks on small datasets. Secondly, based on the square linear relationship between the computational complexity and the image size in the ViT model, it cannot achieve better convergence in training. Thirdly, the increase in the resolution of remote sensing images brings greater intra-class differences and inter-class similarity, and while Transformer can construct a global semantic representation of the images, it loses much detailed information in the process of patching, which is particularly significant for HRRS images. In addition, the flattening process also destroys the structural information of the images, resulting in small targets or multi-branch targets with obvious texture features in HRRS images that cannot be well segmented.

To solve the above problems, this paper proposes a multi-objective segmentation network (ACTNet) with a hybrid CNN and Transformer, which is based on the Swin Transformer and uses a shifted window-based attention algorithm, so that the computational complexity is linear with the image size. In order to not change the structure of the Swin Transformer, the ResAttn module is designed as an adapter in this paper. Its dual attention mechanism ensures that sufficient global information is obtained during the training for remote sensing segmentation tasks and does not lead to excessive computation. Meanwhile, for small and multi-branch targets, we also propose a CNN-based multi-scale feature extraction module (LFE), which refers to the ResNet [17] and mainly consists of a series of convolution and pooling layers to extract as many local details of different targets as possible. In addition, a residual structure is added to the Mask2Former [18] algorithm, so that the mask feature can incorporate more information on deep-level features to improve the segmentation performance of the multi-target.

The main contributions of the article are summarized as follows:

In order to solve the problem of excessive computational complexity in the training phase of HRRS image semantic segmentation, we propose an adapter module (ResAttn) capable of remote sensing semantic segmentation. It uses a dual-attention mechanism to ensure that sufficient global information can be obtained from the feature map. For better integration into the Swin Transformer structure, we use the same patch merging method for down-sampling.

In order to enhance small target segmentation, we explore a CNN-based multi-scale feature extraction module (LFE), which aims to fully extract the texture, color, and other shallow features according to the convolutional filter weights. Meanwhile, local correlation and kernel weight sharing help to keep the parameters relatively small, which also compensates for the lack of local information extraction in Transformer.

We use a mask-based segmentation method with enhanced residual structure. The segmentation accuracy of the model on the occluded targets is improved by using residual connections to process the feature maps before and after through the multi-scale deformable attention layer.

The remainder of this paper is organized as follows. Section 2 introduces the related work. Section 3 presents the design details of our proposed network. Section 4 provides the relevant experiments and setups, and Section 5 summarizes our approach and presents the outlook for future research.

2. Related Work

This section describes the related work in CNN-based remote sensing semantic segmentation methods, Vision Transformer, and adapters. Table 1 and Figure 1 show the context of this section.

Table 1. Summary of related work.

HRRS Image Segmentation Methods			Adapter
Transformer Based	**CNN-Based**	**CNN-Transformer**	
Swin [19], ST-UNet [20]	FCN [21]	CCTNet [22]	K-Adapter [23]
DC-Swin [24]	U-Net [25]	Swin + SASPP + SE [26]	Clip-Adapter [27]
TransRoadNet [28]	DeepLab [29–31]		AdapterFusion [32]
SwinSUNet [33]	MC-FCN [34]	CTNet [35]	ViT-Adapter [36]

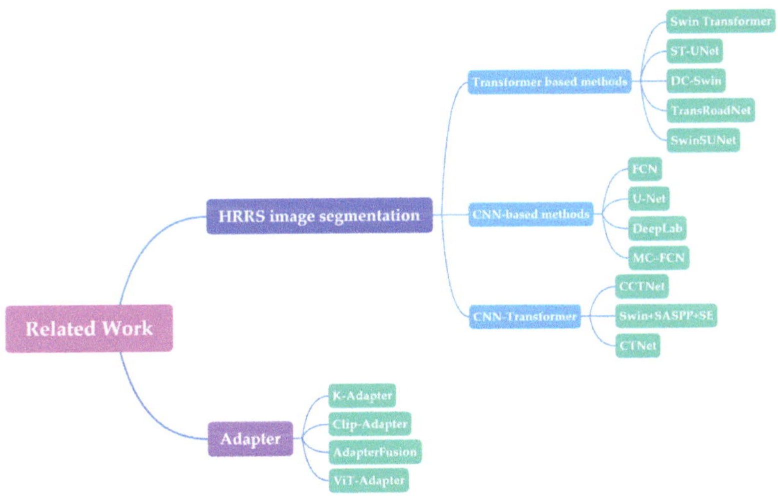

Figure 1. The overall context of related work.

2.1. CNN-Based Semantic Segmentation Methods on Remote Sensing Images

With the development of remote sensing technology and the outstanding performance of CNN in deep learning, the research related to remote sensing semantic segmentation has received wide attention. Since the introduction of the fully convolutional network (FCN), an encoder–decoder architecture has been widely used. The encoder performs convolution and down-sampling on the image to extract the image features, while the decoder recovers the spatial resolution by upsampling the small-size feature map. Based on FCN, Ronneberger et al. [25] developed the U-Net network with a symmetric encoder–decoder structure (i.e., contracting path and expansive path), where the encoder features are introduced in the decoding stage to gather more spatial information. The MC-FCN network proposed by Wu et al. [34] added a residual structure and multi-scale subconstraints based on the U-Net to improve performance in building segmentation.

Despite the successful application of Deep Convolutional Neural Networks (DCNNs) to various tasks, they lack an effective method to acquire global information, which is a critical limitation for understanding complex scenes. To address this issue, PSPNet, proposed by Zhao et al. [37], invokes the spatial pyramid pooling (SPP) method to obtain multi-scale features by pooling layers of different sizes, and then performs feature fusion and upsampling to improve the network's ability to acquire global information. Furthermore, the DeepLab model (DeepLab v2, DeepLab v3, and DeepLab v3+) proposed by Chen et al. [29–31] replaces the pooling layer in SPP with inflated convolution, allowing for the learning of more feature information from the previous input. Although the above methods have improved the performance of CNN model segmentation, they still lack the capability to effectively extract dense target segmentation and fine-branch segmentation.

2.2. Vision Transformer

Due to the excellent performance of Transformer in the field of NLP, it was soon adopted by CV and presented in the Vision Transformer architecture, which relies on its attention mechanism to learn the long-distance information in images. The Swin Transformer proposed by Liu et al. [19] uses shifted window-based attention mechanisms, whose computational complexity is squarely related to the window size and linearly related to the image size. The shifted windows scheme ensures the information interaction among the windows, which enables the Transformer-based model to further explore the features in HRRS images. He et al. [20] introduced the Swin Transformer module into the U-Net shape model, which enhances the spatial feature analysis and small-scale object extraction to improve the global modeling capability. Wang et al. [24] designed the DCFAM module based on an attention mechanism and inflated convolution in the decoder to strengthen the relationship between spatial-wise and channel-wise. To improve the road extraction, Yang et al. [28] performed contextual modeling on high-level features to enhance the foreground information learning capability in order to combat similarity and occlusion problems. Zhang et al. [33] designed a Siamese U-shaped network using Swin Transformer blocks; the encoder generates multiscale features by using a hierarchical Swin Transformer.

The Transformer structure used for the extraction of global information can effectively compensate for the lack of CNN models; therefore, many researches have begun to explore suitable methods to fuse these two components. Wang et al. [22] proposed LAFM and CAFM to efficiently fuse the dual-branch features of the CNN and Transformer models. Zhang et al. [26] used depthwise-separable, convolution-based, atrous spatial pyramid-pooling modules to connect the Swin Transformer-based backbone and CNN-based decoder to capture multi-scale contextual information. The CTNet proposed by Deng et al. [35] uses a dual-stream structure to combine the Transformer and CNN models in its overall architecture, and uses concatenated semantic features and structural features to predict the scene categories.

The Introduction of the Transformer module made the remote sensing segmentation task pay full attention to the information of the target context, resulting in both improved continuity and noise immunity. To solve the problem of the high computational complexity of the attention algorithm, some attention-limited networks, such as cSwin Transformer [38], have been proposed to further reduce the computational effort, but this has led to the loss of the extraction of global features. Moreover, the networks which have Transformer as the backbone still require a large number of computational resources for transfer learning, which has a great deal of room for improvement for the training of remote sensing segmentation as a downstream task.

2.3. Adapters

CV tasks, such as classification, target detection, and semantic segmentation, have been significantly improved with better architectural design and large-scale high-quality datasets. However, collecting datasets for each task is too costly for the scale. To address this problem, the "pretrain-finetune" paradigm, in which large-scale datasets such as ImageNet are pre-trained to obtain weights and then applied to various downstream tasks and finetune, has been widely adopted in the CV field [16,39].

The Adapter was first proposed in NLP (Houlsby et al. [40]), and has been widely used in both the NLP and CV fields [27,32]. Its core idea is to update only the parameters in the adapter module while keeping the other parameters unchanged, so that it can achieve the same effect as finetuning. The K-Adapter structure proposed by Wang et al. [23] makes the adapter more modular for knowledge-intensive tasks. Recently, the ViT-Adapter model proposed by Chen et al. [36] successfully applied this idea to image dense prediction, where the missing local continuity information from the ViT is supplemented by the adapter, allowing it to perform well in dense segmentation. However, it is still a challenge to design an effective adapter module to cope with multi-scale targets and dense targets in remote sensing segmentation.

3. Methodology

In this paper, we propose a new semantic segmentation scheme for remote sensing images. First, we will introduce the various modules contained in ACTNet's encoder and decoder and the general flow. Then we propose an adapter module (ResAttn) based on a dual attention structure to fully extract global information without excessively increasing the parameters. To enhance the model for the extraction of shallow features, such as texture, color, etc., we propose a CNN-based LFE module. Finally, we propose the Deform Block with residual enhancement to improve the segmentation of occluded targets.

3.1. Overall Architecture

The general overview structure of ACTNet is shown in Figure 2. The network is divided into encoder and decoder parts. In the encoder part, a stem block is used to preprocess the H × W size input image first. It consists of four convolutional layers and one pooling layer, each followed by a batch normalization and a ReLU activation function, with an output size of 1/44 of the original image. The output of the stem block is used as input for the LFE module, ResAttn module, and Swin Transformer backbone.

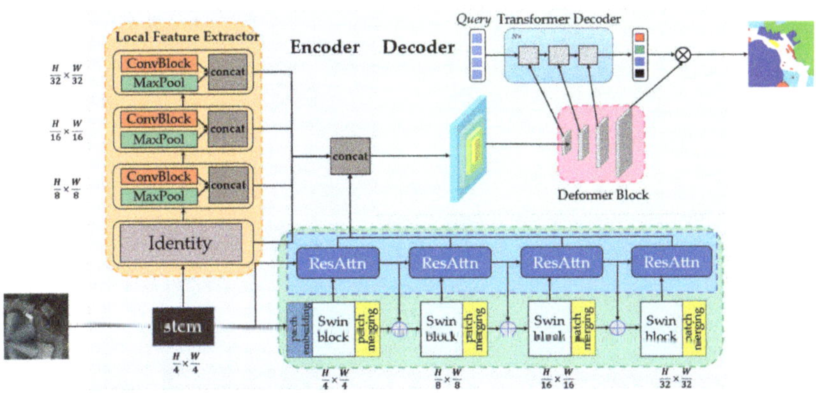

Figure 2. Overview structure of ACTNet.

As shown in the green area in Figure 2, due to the high resolution of the HRRS images, global modeling at large imaging sizes is required. Therefore, the Swin Transformer backbone is used as the main global modeling method, which significantly reduces the computational effort with the help of the shifted window attention algorithm. It consists of four Swin Transformer blocks, each of which contains several MSA and SW-MSA blocks in a series to form a structure.

As shown in the blue-dashed part of the green area in Figure 2, a lightweight ResAttn module is applied behind each Swin Transformer block. It has an input consisting of the output of the current Swin Transformer block and the output of the previous ResAttn module. After generating tokens and fusing them with each other, the global dependency between the features of the two levels can be derived by using the self-attention mechanism to minimize feature loss during down-sampling. In order to keep the structure of the Swin Transformer, element-wise additions are made between the output and the original Swin Transformer block output, so that the pre-trained weight information can be fully utilized during migration training. We use the same patch merging method as the Swin

Transformer uses for down-sampling the output features of the first three blocks to form a multi-scale feature extraction.

Meanwhile, we use several sets of CNN structures to obtain the low-level features of the image, as shown in the orange area of Figure 2. The feature maps obtained from the stem are fed into the LFE module; here the design pattern of the ResNet is invoked, which consists of three stages with a series of convolutions and poolings. The Identity module will not perform any processing, so that the module will output feature maps of 1/4, 1/8, 1/16, and 1/32 of the original image size, finally concatenating with the output of the 4 ResAttn modules as the encoder part of the output.

In the decoder section, we use Mask2Former as a base structure. To improve the segmentation of small objects, a multi-scale decoder structure is used, where the encoder output will be sent into the deformer block first to generate pyramid-like features. As shown in the pink area in Figure 2, here we calculate the correlation of each pixel in the feature map with the surrounding sampled points using 3 N deformable attention for 1/8-, 1/16-, and 1/32-size feature maps. We then upsample the 1/16- and 1/32-size feature maps with the 1/4-size feature map using a bilinear interpolation method for element-wise addition, in order to enhance the effect of small target segmentation while preventing network degradation. These features are next sent to the Transformer decoder module, where N-length queries with random initialization parameters will be learned to obtain global information from masked attention. After that, the mask result and the classification result are calculated with the feature map of 1/4 the original image size. Finally, the mask output and classification output are combined to obtain the network output.

3.2. *ResAttn*

As shown in Figure 2, the first ResAttn module begins with the output of the stem block. The stem block consists of 4 convolutional layers and 1 pooling layer and, as shown in Figure 3, it performs simple feature extraction and down-sampling operations on the input image, which is used to initially reduce the image size and decrease the network complexity.

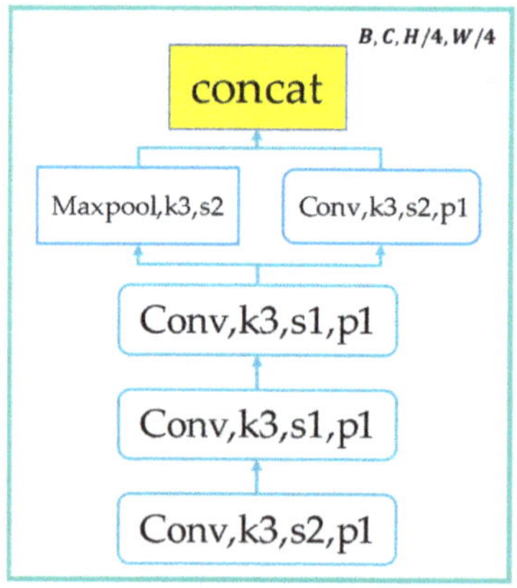

Figure 3. Illustration of stem block.

The existing models still have a risk of gradient disappearance as the network layer deepens, and the deep feature map loses a large number of small object features. Therefore, we propose the ResAttn module, as shown in Figure 4, which is based on an attention structure and incorporates the idea of residual structure. Specifically, it uses the output of the current Swin Transformer block and the previous ResAttn module, then generates 1×1-size tokens and fuses the features together for input into the self-attention module, which uses a multi-head self-attention algorithm. It then concatenates the two parts of the output. Finally, the result is passed through the FFN module. It contains 2 linear layers and 2 activation functions. For this reason, the computation performed in the self-attention mechanism is mainly matrix multiplication, i.e., it is a linear transformation; therefore, its learning ability is still not as strong as the nonlinear transformation, so the expression ability of the query is enhanced by means of activation functions. The features from this step are collected as part of the encoder output. In order to keep the structure of the Swin Transformer, the same method of down-sampling is used to process the features as they are, doubling the number of channels and halving the size. The output features perform element-wise addition with the Swin Transformer block output.

Assuming that the input feature size is (c, h, w), the two inputs are passed through the convolution layer to generate a token of size $(c, 1, 1)$ and a query of size c, $h \times w$, respectively. Then they are fused with each other and sent to the self-attention layer after the addition of position encoding to calculate the weight between the elements in the query sequence. For this we use the self-attention algorithm from ViT, which essentially uses a matrix multiplication to calculate the relationship between each patch and the other patches in the query, and to update the weight matrix by back propagation, whose specific formulas are as follows:

$$Attention(Q, K, V) = Softmax(\frac{QK^T}{\sqrt{d_k}})V \quad (1)$$

$$Q = X \times W_q \quad (2)$$

$$K = X \times W_k \quad (3)$$

$$V = X \times W_v \quad (4)$$

where X is the query, W_q, W_k, and W_v are the learnable weight matrices, and the association between the previous layer features and the current features is constructed by self-attention. The output query is then restored to its original size and concatenated, for which we use a 1×1 convolutional layer to adjust the length to $(c, h \times w)$ and an FFN module to enhance its nonlinear representation. Finally, the image size and number of channels are adjusted by patch merging.

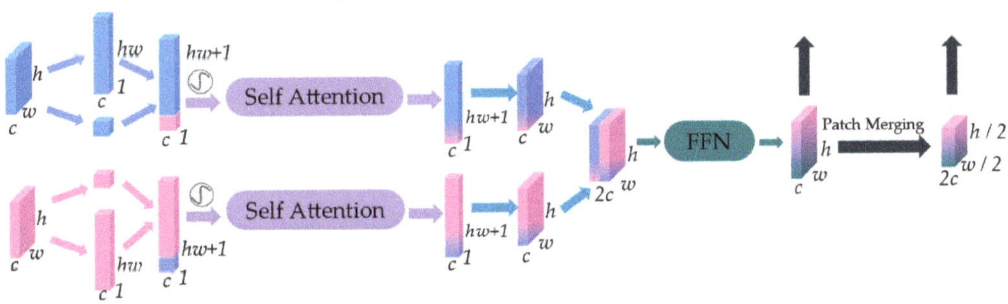

Figure 4. Illustration of ResAttn module.

In ACTNet we add a ResAttn module after each Swin Transformer block, and the final feature sizes obtained are 1/4, 1/8, 1/16, and 1/32 of the original image. This method achieves a result very similar to that of the feature pyramid of the SPP network.

3.3. LFE

Previous studies have shown that the overuse of the Transformer model in the encoder part causes the network to become less capable of extracting shallow features. This indicates there are difficulties in the extraction of most objects with distinct boundaries for multi-target semantic segmentation in HRRS images. CNN-based networks, on the other hand, can obtain local features with relatively small numbers of parameters by gradually increasing the perceptual field through layer-by-layer convolutions, which have distinct geometric properties and are often concerned with consistency or covariance under transformations such as translation, rotation, etc. For example, a CNN convolution filter detects key points, object boundaries, and other basic units that constitute visual elements and that should be transformed simultaneously under spatial transformations, such as translation. CNN networks are a natural choice for dealing with such covariance, so that positional transformations under the same objects have little effect. Therefore, a multi-scale CNN-based LFE module is proposed to enhance the extraction and analysis of high-frequency information in images and to improve the segmentation accuracy of small and multi-branch targets to compensate for the shortcomings of the Transformer networks.

The purple area shown in Figure 5 is the LFE module, which borrows the design pattern of ResNet. We take the original image as the input, and an initial feature block of 1/4 the size of the original image is generated by stem for initial processing. Then a 3-stage convolution block is used to extract the image features. Each stage contains one maxpooling layer and one ConvBlock. Each ConvBlock has N residual convolutional blocks, as shown on the right side of Figure 5. The small cell composed of convolutional layers and residual structures ensures feature extraction while preventing network degradation. After concatenating the output of the maxpool and ConvBlock as the next input, the LFE module finally extracts the features from the original image sizes of 1/4, 1/8, 1/16, and 1/32, as the complement of the Transformer structure. The number of residual convolutional blocks of each ConvBlock in ACTNet is 3, 4, and 3, respectively, so that the number of parameters are small.

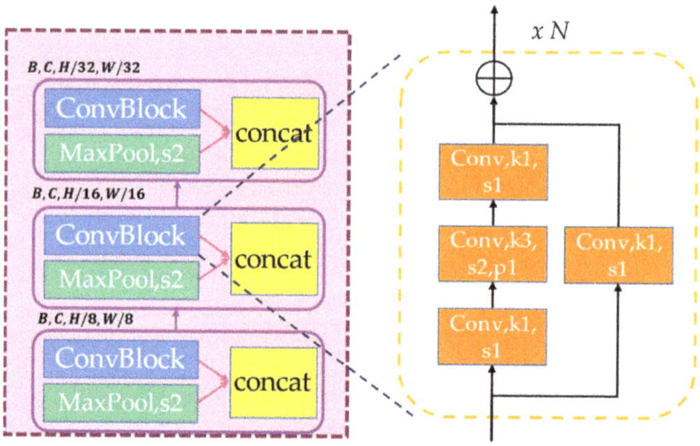

Figure 5. Illustration of LFE module.

3.4. Deformer Block and Loss Function

The decoder section can be seen in the upper right corner of Figure 2, which consists mainly of a deformer block and a Transformer decoder. After the output from the encoder module, a mask-based classification method is used for segmentation instead of the per-pixel classification that we had been using. Many objects in remote sensing images have occlusions, such as houses occluded by tree branches and cars occluded by leaves, which leads to the wrong classification of pixels. Mask segmentation predicts the class of an object using a binary mask, which works better in cases where per-pixel classification fails due to background noise or image complexity, and requires fewer parameters and computations [41].

As shown in the Figure 6, the 4-scale feature maps output by the encoder are first fed into the deformer block module. We calculate the weights using 3 N multi-scale deformable Transformers for the offset of the reference points, which are generated by each query in the feature map sizes of 1/8, 1/16, and 1/32, where N represents 2. The deformable attention formula is as follows:

$$Deformable\ Attention\ (z_q, p_q, x) = \sum_{m=1}^{M} W_m [\sum_{k=1}^{K} A_{mqk} \cdot W'_m x(p_q + \Delta p_{mqk})] \quad (5)$$

where z_q is obtained from the input x by linear transformation, p_q is a 2D vector representing the coordinates, M represents the attention head, K represents the number of positions sampled by 1 query in 1 head, A_{mqk} represents the normalized attention weight, $W'_m x$ is the transformation matrix of value, and Δp_{mqk} is the sampling offset.

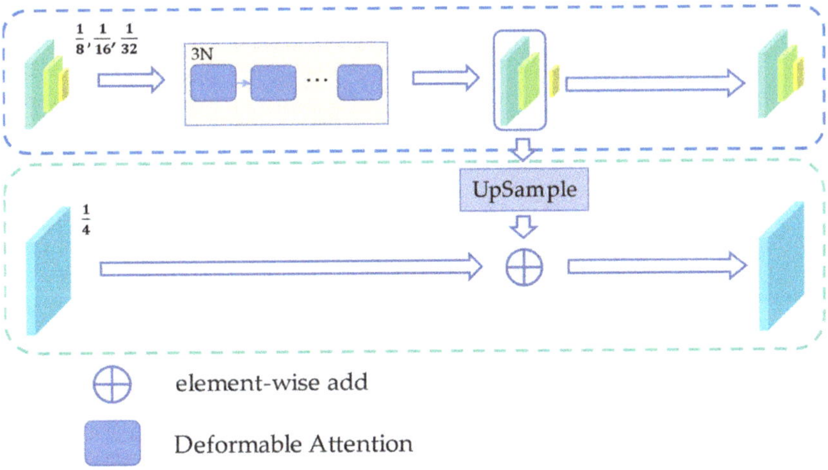

Figure 6. Illustration of our Deformer Block module.

Then the output of the 1/16- and 1/8-size features are added to the 1/4 feature map using bilinear interpolation upsampling to obtain the masked attention. After that, the 1/8-, 1/16-, and 1/32-size features are fed into the Transformer decoder with 3L attention blocks. Finally, the binary mask of each feature map and its corresponding classification result are calculated by query.

In order to accurately calculate the deviation between the result and the ground truth value, the loss function we use combines Cross Entropy Loss (*CELoss*), *FocalLoss*, and *DiceLoss*, each of which has its own role in improving the overall performance. The function can be expressed as follows:

$$Loss = (CELoss + FocalLoss) + DiceLoss \quad (6)$$

CELoss is used to calculate the category probability loss, which is suitable for multi-category tasks and is good for remote sensing multi-target segmentation. The formula is as follows, where M represents the number of categories, y_c is the ground truth value, and p_c is the predicted value:

$$CELoss = -\sum_{c=1}^{M} y_c \log(p_c) \quad (7)$$

FocalLoss is used to calculate the loss value of a mask. Since the ratio between categories in a remote sensing dataset is very unbalanced, using cross entropy loss will cause the training process to be skewed towards the side with more categories. *FocalLoss* adds a modulating factor, γ, to overcome this drawback based on *CELoss*. The formula is as follows, where p is the predicted value, y is the ground truth, and in this paper $\alpha = 0.25$, and $\gamma = 2$:

$$FocalLoss = \begin{cases} -\alpha(1-\alpha)^{\gamma}\log(p), & y = 1 \\ -(1-\alpha)p^{\gamma}\log(1-p), & y = 0 \end{cases} \tag{8}$$

DiceLoss [42] is derived from the dice coefficient, which is an ensemble similar to the measure function. *DiceLoss* is used as a measure function to evaluate the similarity between two samples and is designed to cope with a scenario of a strong imbalance between positive and negative samples in semantic segmentation. It is defined in the formula below, where ε is used to prevent the extreme case where the denominator is 0. In this paper, $\varepsilon = 1$.

$$DiceLoss = 1 - \frac{2yp + \varepsilon}{y^2 + p^2 + \varepsilon} \tag{9}$$

4. Experiment

4.1. Dataset

In this article, we use the ISPRS Potsdam dataset to evaluate the performance of ACTNet. The ISPRS Potsdam dataset is extracted from the Potsdam region and contains 38 true radiographic images of 6000 × 6000 size. Each remote sensing image area covers the same size. Categories include Impervious surfaces, Buildings, Low vegetation, Tree, Car, and Clutter/background. The Clutter/background class includes bodies of water and other objects that look very different from everything else. Considering the size of the HRRS images and the limitations of GPU memory, we cut the images and corresponding labels into 600 × 600 pixel-sized images and then randomly divided them, with 80% going into a training set and 20% going into a validation set in disorder.

4.2. Evaluation Metrics

The semantic segmentation evaluation metrics used in this experiment contain two main categories. One is the metrics used to evaluate the accuracy of the model, including mean Intersection over Union ($mIoU$) and mean (class) accuracy ($mAcc$). The other category is a metric used to evaluate the inference speed and training speed of the module. Consider $mIoU$ as the primary metric, which calculates the intersection ratio of two sets and is widely used in semantic segmentation model evaluation. The formulas for the evaluation metrics are as follows:

$$mIoU = \frac{1}{k+1}\sum_{i=0}^{k}\frac{TP}{FN + FP + TP} \tag{10}$$

$$mAcc = \frac{1}{k+1}\sum_{i=0}^{k}\frac{TP}{FP + TP} \tag{11}$$

4.3. Implementation Details

We built our model using the MMSegmentation framework with Python 3.8. MMSegmentation is a deep-learning framework based on Pytorch, but is easier to scale and build complex networks with than the latter. To initialize our network parameters, the weights pre-trained by the BEiT [43] model on the ADE20K dataset were used. ResAttn and LFE modules use random initialization methods for the initial parameters, while the deformer block and Transformer decoder modules use the Kaiming initialization method [44] for the initial parameters.

For the hyperparameter setting we used a batch size of two and an initial learning rate of 1×10^{-4}. A warmup training strategy was used to avoid instability during training and to optimize the overall training effect. We used AdamW as the parameter update algorithm and Poly as the learning-rate adjustment strategy. All the experiments were trained in parallel on an NVIDIA Geforce RTX2080Ti with an 11-GB memory GPU and a maximum Epoch of 100. In addition, we used random crop, random flip, and other measures to enhance the training data.

4.4. Comparative Experiments

The ACTNet model was compared to other mainstream remote sensing semantic segmentation networks, namely the CNN-based FCN [21], U-Net [25], DeepLabV3+ [31], the Transformer-based Swin-ViT [19], ST-UNet [20], and SwinSUNet [33], respectively, on the Potsdam dataset using the same experimental settings, and the experimental results are shown in Table 2.

Table 2. Comparative experimental result on the Potsdam dataset.

Method	Evaluation Metrics		Inference Time (ms)	Training Time (min/epoch)
	mIoU (%)	*mAcc* (%)		
FCN	75.85	86.33	5.7	4.04
U-Net	77.23	87.45	8.5	7.31
DeepLabV3+	78.47	88.12	13.83	9.87
Swin-ViT	79.63	87.73	27.93	15.60
ST-UNet	75.84	85.26	30.39	16.21
SwinSUNet	82.36	91.94	51.04	27.83
ACTNet	**82.15**	**90.28**	**46.29**	**19.02**

Our proposed ACTNet achieved an 82.15% in $mIoU$ score and a 90.28% in $mAcc$ score. The experiments showed that our model performed better than the CNN-based models or the Transformer-based models and required less training time and inference time compared to other CNN and Transformer-combined models

The visualization results from the comparison experiments are presented in Figure 7, where row (a) is the randomly selected image for the experiment and row (b) is the image corresponding to the ground truth value. From rows (c–e) of the figure, it can be seen that the traditional CNN-based model could not depict the specific outline of the object well due to too many details being lost during down-sampling, which resulted in less detailed results when performing multi-branching objectives. In column (1), the classifications of "Low Vegetation" and "Clutter/background" were incorrectly mixed due to the similarity of their colors. In column (2), the DeepLabV3+ model incorrectly split "Tree" into "Low Vegetation" and "Clutter/Background." The Swin-ViT model correctly classified these, but the area was incomplete. From the black box of columns (3) and (4), the ST-UNet and SwinSUNet were less effective in segmenting the "Clutter/Background" and "Building" objects due to foreground occlusion. The ACTNet achieved better results than the Transformer-based model due to the LFE module's ability to extract local features and its use of the mask-based segmentation method. ACTNet also outperformed DeepLabV3+, U-Net, and other CNN-based networks due to the global modeling capability of the attention mechanism. Furthermore, ACTNet also demonstrated better results on fragmented targets such as "Car" when compared to the CNN and Swin-ViT models.

Although the overall performance of ACTNet was superior to that of the other models, there is still potential for improvement regarding the segmentation effect. We analyzed the test results and visualized the confusion matrix, as shown in Figure 8. In the confusion matrix, we found that "Tree" was misclassified as "Building" or "Low Vegetation" in a large number of cases, which led to a decrease in the overall $mIoU$ and $mAcc$ values.

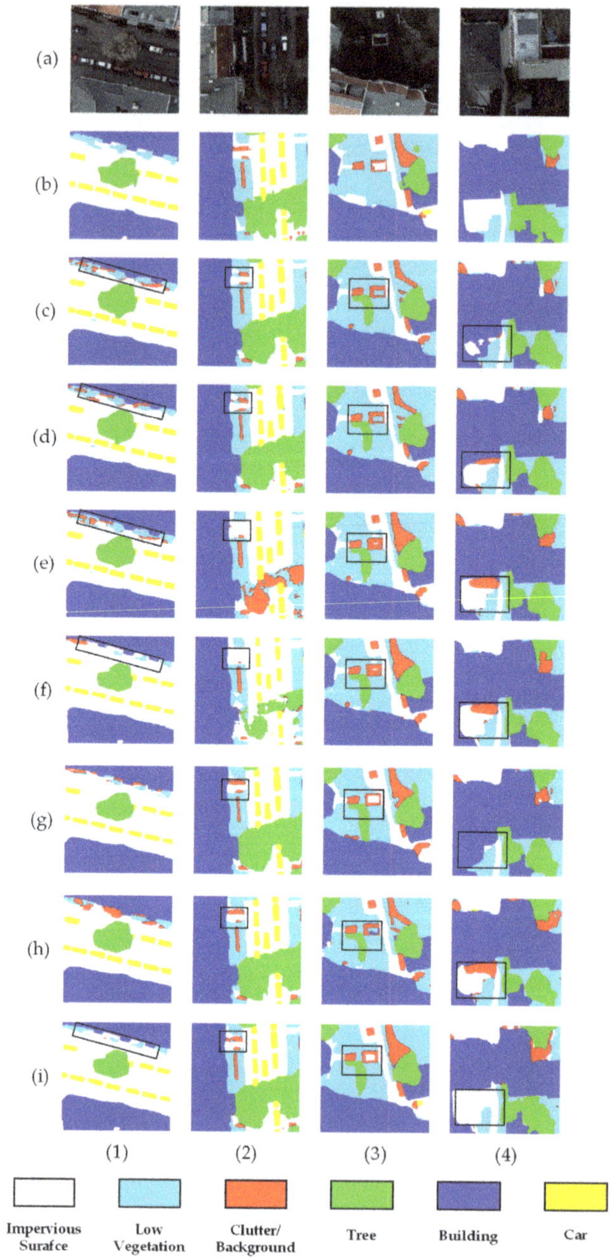

Figure 7. Comparative experimental results from the different models. The black boxes mark the areas with significant differences. Column (1–4) represents the segmentation results of four different test images. Row (**a**) represents the randomly selected image, row (**b**) represents the ground truth corresponding to the image, and rows (**c–i**) represent the experimental results from the FCN, U-Net, DeepLabV3+, Swin-ViT, ST-UNet, SwinSUNet, and ACTNet methods, respectively.

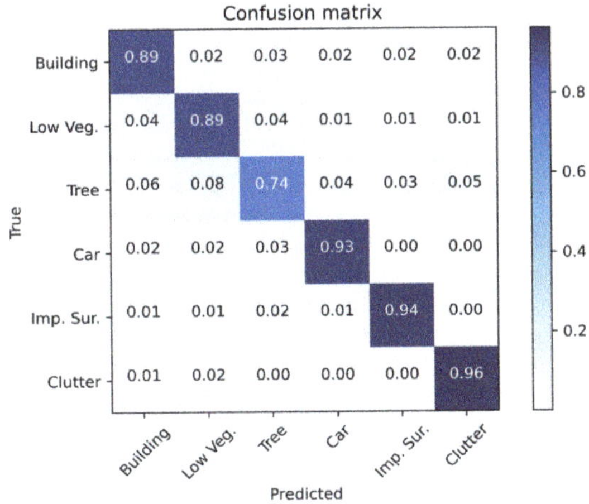

Figure 8. Confusion matrix for the ACTNet segmentation results.

4.5. Ablation Study

To further investigate the performance of the ResAttn, LFE module and the improved mask-based segmentation method in this paper, we conducted a series of ablation experiments on the Potsdam dataset. In these experiments, the baseline Swin Transformer model used a Swin-T configuration with layer numbers = {2, 2, 6, 2} and window size $M = 7$; all the models except ACTNet uniformly used Mask2Former as the segmentation decoder. Our image enhancement methods on these two datasets used random cropping and a 50% probability random flip, and the image resolution was uniformly scaled to 512×512 pixels. The overall experimental results are shown in Table 2, the specific values for each classification are shown in Table 3, and the visualization results of the ablation experiments are shown in Figure 9.

In Table 3, with the addition of the LFE and ResAttn modules leading to an increase in model's computations, the training time per epoch increased by 2.25 min and 3.13 min, respectively. In the experiments where both LFE and ResAttn were added, we used the pretrained parameters from the experiments with only the LFE module added. We froze the weights of the Swin Transformer backbone and LFE modules and updated the weights of the ResAttn module and decoder part. The experimental results showed only a small increase in the training time and an improvement in segmentation performance, which proves the effectiveness of an adapter in HRRS image segmentation. Meanwhile, the inference time increased by 18.36 ms compared to the baseline; however, this is acceptable in consideration of the improvement in classification accuracy.

Table 3. Overview of the results from the ablation experiments.

Method	Evaluation Metrics		Inference Time (ms)	Training Time (min/epoch)
	mIoU (%)	*mAcc* (%)		
Swin-ViT	79.63	87.73	27.93	15.60
+LFE	80.73	88.88	33.44	17.85
+ResAttn	80.38	88.32	36.05	18.73
+LFE, ResAttn	81.52	89.57	46.08	19.02
ACTNet	**82.15**	**90.28**	**46.29**	**19.02**

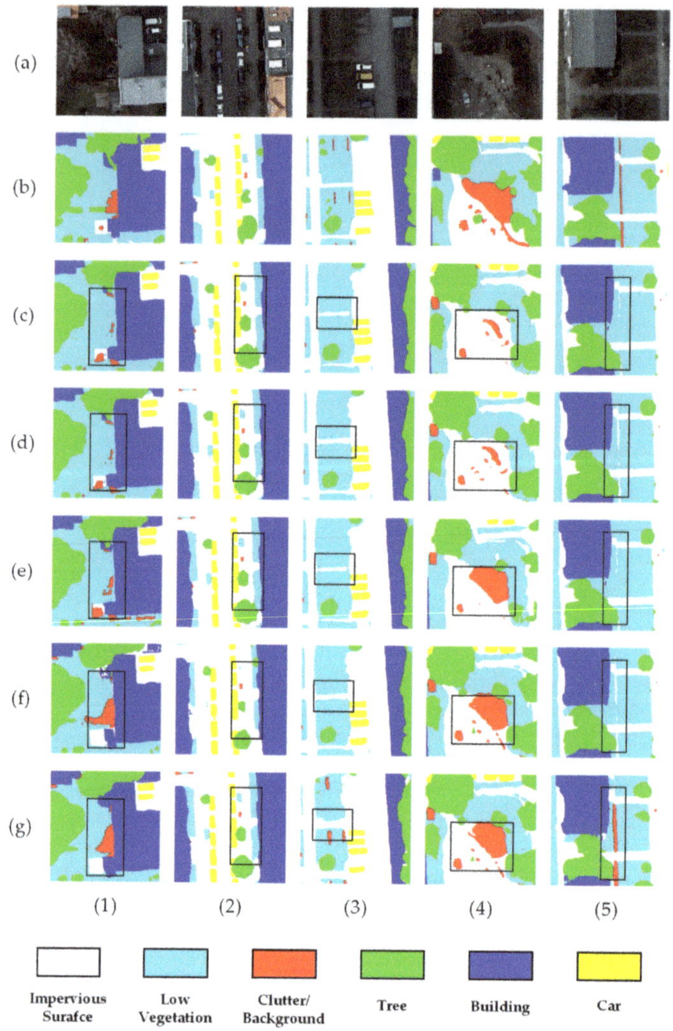

Figure 9. Visualization results from the ablation experiment. The black boxes mark the areas with significant differences. Columns (1–5) represent the segmentation results of five different test images. Row (**a**) represents the randomly selected image, row (**b**) represents the ground truth corresponding to the image, and rows (**c**–**g**) represent the experimental results from the Swin-ViT, Swin-ViT with LFE, Swin-ViT with ResAttn, Swin-ViT with LFE and ResAttn, and ACTNet models, respectively.

From Table 4, it can be seen that adding the LFE module to Swin-ViT increased the $mIou$ value by 1.1% and the $mAcc$ value by 1.15%. The IoU value for the classes "Low Vegetation", "Tree", "Car", and "Impervious Surface" were significantly improved. From Figure 9, row (d) shows a significant increase in IoU for the classes "Low Vegetation" and "Car". This demonstrates the effectiveness of CNN and a multi-scale structure in LFE modules for target edge analysis and small target segmentation. The relatively small increase in inference time relative to the improved segmentation effect is shown in Table 2, which proves the efficiency of the LFE module; these effectively compensate for the shortcomings of the Transformer model in this regard.

Table 4. Specific results from the ablation experiments.

Method	*IoU*						Evaluation Metrics	
	Building	Low Vegetation	Tree	Car	Impervious Surface	Clutter/ Background	*mIoU*	*mAcc*
Swin-ViT(baseline)	77.06	76.40	60.16	83.89	86.86	93.41	79.63	87.73
+LFE	77.74	77.77	61.20	85.00	87.78	94.89	80.73	88.88
+ResAttn	77.22	77.48	61.02	84.81	87.08	90.83	79.74	88.32
+LFE, ResAttn	78.13	78.33	63.79	85.20	88.26	95.41	81.52	89.57
ACTNet	**78.19**	**78.54**	**65.38**	**86.09**	**88.89**	**95.81**	**82.15**	**90.28**

Meanwhile, Table 4 shows that the *IoU* values of "Low Vegetation", "Tree", and "Car" obviously improved after adding the ResAttn module to Swin-ViT. This shows that the double attention mechanism in the ResAttn module and the fusion of token and query between the different features improved the segmentation effect for multiple identical targets in a certain region. However, the *IoU* value of "Background" decreased, and it can be seen from the black box of row (e) in Figure 9 that the "Impervious Surface" area had some incorrect segmentation as "Background". This indicates that the overuse of the attention mechanism caused the model to forcibly associate a target with other targets of different categories in a certain region, leading to segmentation errors. However, when the LFE module was used in combination with the ResAttn module, it could suppress some of the over-association effects of global modeling. As shown in Table 4, the use of both LFE and ResAttn improved $mIoU$ by 1.89% and $mAcc$ by 1.15%, with a significant improvements in all categories.

From the black box of row (f) in Figure 9, we can see that the misclassification of "Impervious Surface" and "Low Vegetation" was suppressed; the boundary between different targets was more clearly segmented. The segmentation results of the "Background" category were also closer to the ground truth. This is because after concatenating the output of LFE and ResAttn in the encoder part, the feature map set contained rich global modeling information and local feature information simultaneously, which further improved the model's ability to discriminate between object features. Finally, after the addition of our modified mask2former-based decoder under the above conditions, the model performance was further improved, which demonstrates the importance of fusing more high-level feature maps into the feature maps in multi-target segmentation.

5. Conclusions

In this paper, a high-performance HRRS image semantic segmentation method ACTNet was proposed. To address the problem of the high computational complexity of the existing Transformer models for training downstream tasks and its dependence on a pre-training weight of large datasets, we proposed a Transformer-based adapter module for HRRS image semantic segmentation (ResAttn). This module uses a dual-attention mechanism to ensure the acquisition of global information while the structure of Swin-ViT remains unchanged. To enhance the extraction of edge and texture features, we designed a CNN-based LFE module and used a pyramid-like structure to fit multi-scale objects. Moreover, we used a mask-based segmentation method with a residual-enhanced deformable attention block to further improve the extraction of small objects. Our series of experiments on the Potsdam dataset demonstrated the excellent performance of ACTNet. In the future, we hope to further reduce the overall training parameters and computational resources used by ACTNet. We will try to find a unified semantic segmentation network based on the structure of ACTNet to support more HRRS image datasets. Furthermore, we will explore its role in urban rail transportation planning, and to demonstrate the generality of the ACTNet structure.

Author Contributions: Conceptualization, Z.Z. and F.L.; methodology, Z.Z. and F.L.; software, F.L.; validation, Z.Z.; formal analysis, C.L.; investigation, Q.T. and H.Q.; resources, Q.T.; data curation, F.L.; writing, Z.Z. and F.L.; original draft preparation, F.L.; visualization, Z.Z.; supervision, F.L. and Q.T.; project administration, C.L. and H.Q. All authors have read and agreed to the published version of the manuscript.

Funding: This research was funded by the National Key Research and Development Program under Ministry of Science and Technology of the People's Republic of China (2020YFB1600702).

Data Availability Statement: Not applicable.

Conflicts of Interest: The authors declare there are no conflict of interest.

References

1. Ke, L.; Xiong, Y.; Gang, W. Remote Sensing Image Classification Method Based on Superpixel Segmentation and Adaptive Weighting K-Means. In Proceedings of the 2015 International Conference on Virtual Reality and Visualization (ICVRV), Xiamen, China, 17–18 October 2015; pp. 40–45.
2. Wu, T.; Xia, L.; Luo, J.; Zhou, X.; Hu, X.; Ma, J.; Song, X. Computationally efficient mean-shift parallel segmentation algorithm for high-resolution remote sensing images. *J. Indian Soc. Remote Sens.* **2018**, *46*, 1805–1814. [CrossRef]
3. Moser, G.; Serpico, S.B. Classification of High-Resolution Images Based on MRF Fusion and Multiscale Segmentation. In Proceedings of the IGARSS 2008-2008 IEEE International Geoscience and Remote Sensing Symposium, Boston, MA, USA, 8–11 July 2008; pp. II–277–II–280.
4. Zhang, Z.; Miao, C.; Liu, C.A.; Tian, Q. DCS-TransUperNet: Road segmentation network based on CSwin transformer with dual resolution. *Appl. Sci.* **2022**, *12*, 3511. [CrossRef]
5. Lu, X.; Zhong, Y.; Zheng, Z.; Liu, Y.; Zhao, J.; Ma, A.; Yang, J. Multi-scale and multi-task deep learning framework for automatic road extraction. *IEEE Trans. Geosci. Remote Sens.* **2019**, *57*, 9362–9377. [CrossRef]
6. Zhang, Z.; Xu, Z.; Liu, C.A.; Tian, Q.; Wang, Y. Cloudformer: Supplementary aggregation feature and mask-classification network for cloud detection. *Appl. Sci.* **2022**, *12*, 3221. [CrossRef]
7. Dosovitskiy, A.; Beyer, L.; Kolesnikov, A.; Weissenborn, D.; Zhai, X.; Unterthiner, T.; Dehghani, M.; Minderer, M.; Heigold, G.; Gelly, S. An image is worth 16x16 words: Transformers for image recognition at scale. *arXiv* **2020**, arXiv:2010.11929.
8. Xie, E.; Wang, W.; Yu, Z.; Anandkumar, A.; Alvarez, J.M.; Luo, P. SegFormer: Simple and efficient design for semantic segmentation with transformers. *Adv. Neural Inf. Process. Syst.* **2021**, *34*, 12077–12090.
9. Zhang, Z.; Miao, C.; Liu, C.; Tian, Q.; Zhou, Y. HA-RoadFormer: Hybrid attention transformer with multi-branch for large-scale high-resolution dense road segmentation. *Mathematics* **2022**, *10*, 1915. [CrossRef]
10. Sertel, E.; Ekim, B.; Osgouei, P.E.; Kabadayi, M.E. Land Use and Land Cover Mapping Using Deep Learning Based Segmentation Approaches and VHR Worldview-3 Images. *Remote Sens.* **2022**, *14*, 4558. [CrossRef]
11. Carion, N.; Massa, F.; Synnaeve, G.; Usunier, N.; Kirillov, A.; Zagoruyko, S. End-to-end Object Detection with Transformers. In Proceedings of the Computer Vision–ECCV 2020: 16th European Conference, Glasgow, UK, 23–28 August 2020; pp. 213–229.
12. Zhang, Z.; Lu, X.; Cao, G.; Yang, Y.; Jiao, L.; Liu, F. ViT-YOLO: Transformer-Based YOLO for Object Detection. In Proceedings of the IEEE/CVF International Conference on Computer Vision, Montreal, QC, Canada, 10–17 October 2021; pp. 2799–2808.
13. Zhang, Z.; Xu, Z.; Liu, C.A.; Tian, Q.; Zhou, Y. Cloudformer V2: Set Prior Prediction and Binary Mask Weighted Network for Cloud Detection. *Mathematics* **2022**, *10*, 2710. [CrossRef]
14. Yang, S.; Quan, Z.; Nie, M.; Yang, W. Transpose: Keypoint Localization via Transformer. In Proceedings of the IEEE/CVF International Conference on Computer Vision, Montreal, QC, Canada, 10–17 October 2021; pp. 11802–11812.
15. He, Y.; Yan, R.; Fragkiadaki, K.; Yu, S.-I. Epipolar Transformers. In Proceedings of the IEEE/CVF Conference on Computer Vision and Pattern Recognition, Seattle, WA, USA, 14–19 June 2020; pp. 7779–7788.
16. He, K.; Girshick, R.; Dollár, P. Rethinking Imagenet Pre-Training. In Proceedings of the IEEE/CVF International Conference on Computer Vision, Seoul, Republic of Korea, 27 October–2 November 2019; pp. 4918–4927.
17. He, K.; Zhang, X.; Ren, S.; Sun, J. Deep Residual Learning for Image Recognition. In Proceedings of the IEEE Conference on Computer Vision and Pattern Recognition, Las Vegas, NV, USA, 27–30 June 2016; pp. 770–778.
18. Cheng, B.; Misra, I.; Schwing, A.G.; Kirillov, A.; Girdhar, R. Masked-Attention Mask Transformer for Universal Image Segmentation. In Proceedings of the IEEE/CVF Conference on Computer Vision and Pattern Recognition, New Orleans, LA, USA, 18–24 June 2022; pp. 1290–1299.
19. Liu, Z.; Lin, Y.; Cao, Y.; Hu, H.; Wei, Y.; Zhang, Z.; Lin, S.; Guo, B. Swin Transformer: Hierarchical Vision Transformer Using Shifted Windows. In Proceedings of the IEEE/CVF International Conference on Computer Vision, Montreal, BC, Canada, 11–17 October 2021; pp. 10012–10022.
20. He, X.; Zhou, Y.; Zhao, J.; Zhang, D.; Yao, R.; Xue, Y. Swin transformer embedding UNet for remote sensing image semantic segmentation. *IEEE Trans. Geosci. Remote Sens.* **2022**, *60*, 1–15. [CrossRef]
21. Long, J.; Shelhamer, E.; Darrell, T. Fully Convolutional Networks for Semantic Segmentation. In Proceedings of the IEEE Conference on Computer Vision and Pattern Recognition, Boston, MA, USA, 7–12 June 2015; pp. 3431–3440.

22. Wang, H.; Chen, X.; Zhang, T.; Xu, Z.; Li, J. CCTNet: Coupled CNN and transformer network for crop segmentation of remote sensing images. *Remote Sens.* **2022**, *14*, 1956. [CrossRef]
23. Wang, R.; Tang, D.; Duan, N.; Wei, Z.; Huang, X.; Cao, G.; Jiang, D.; Zhou, M. K-adapter: Infusing knowledge into pre-trained models with adapters. *arXiv* **2020**, arXiv:2002.01808.
24. Wang, L.; Li, R.; Duan, C.; Zhang, C.; Meng, X.; Fang, S. A novel transformer based semantic segmentation scheme for fine-resolution remote sensing images. *IEEE Geosci. Remote Sens. Lett.* **2022**, *19*, 1–5. [CrossRef]
25. Ronneberger, O.; Fischer, P.; Brox, T. U-net: Convolutional Networks for Biomedical Image Segmentation. In Proceedings of the Medical Image Computing and Computer-Assisted Intervention–MICCAI 2015: 18th International Conference, Munich, Germany, 5–9 October 2015; pp. 234–241.
26. Zhang, C.; Jiang, W.; Zhang, Y.; Wang, W.; Zhao, Q.; Wang, C. Transformer and CNN hybrid deep neural network for semantic segmentation of very-high-resolution remote sensing imagery. *IEEE Trans. Geosci. Remote Sens.* **2022**, *60*, 1–20. [CrossRef]
27. Gao, P.; Geng, S.; Zhang, R.; Ma, T.; Fang, R.; Zhang, Y.; Li, H.; Qiao, Y. Clip-adapter: Better vision-language models with feature adapters. *arXiv* **2021**, arXiv:2110.04544.
28. Yang, Z.; Zhou, D.; Yang, Y.; Zhang, J.; Chen, Z. TransRoadNet: A novel road extraction method for remote sensing images via combining high-level semantic feature and context. *IEEE Geosci. Remote Sens. Lett.* **2022**, *19*, 1–5. [CrossRef]
29. Chen, L.-C.; Papandreou, G.; Kokkinos, I.; Murphy, K.; Yuille, A.L. Deeplab: Semantic image segmentation with deep convolutional nets, atrous convolution, and fully connected crfs. *IEEE Trans. Pattern Anal. Mach. Intell.* **2017**, *40*, 834–848. [CrossRef]
30. Chen, L.-C.; Papandreou, G.; Schroff, F.; Adam, H. Rethinking atrous convolution for semantic image segmentation. *arXiv* **2017**, arXiv:1706.05587.
31. Chen, L.-C.; Zhu, Y.; Papandreou, G.; Schroff, F.; Adam, H. Encoder-Decoder with Atrous Separable Convolution for Semantic Image Segmentation. In Proceedings of the European Conference on Computer Vision (ECCV), Munich, Germany, 8–14 September 2018; pp. 801–818.
32. Pfeiffer, J.; Kamath, A.; Rücklé, A.; Cho, K.; Gurevych, I. AdapterFusion: Non-destructive task composition for transfer learning. *arXiv* **2020**, arXiv:2005.00247.
33. Zhang, C.; Wang, L.; Cheng, S.; Li, Y. SwinSUNet: Pure transformer network for remote sensing image change detection. *IEEE Trans. Geosci. Remote Sens.* **2022**, *60*, 1–13. [CrossRef]
34. Wu, G.; Shao, X.; Guo, Z.; Chen, Q.; Yuan, W.; Shi, X.; Xu, Y.; Shibasaki, R. Automatic building segmentation of aerial imagery using multi-constraint fully convolutional networks. *Remote Sens.* **2018**, *10*, 407. [CrossRef]
35. Deng, P.; Xu, K.; Huang, H. When CNNs meet vision transformer: A joint framework for remote sensing scene classification. *IEEE Geosci. Remote Sens. Lett.* **2021**, *19*, 1–5. [CrossRef]
36. Chen, Z.; Duan, Y.; Wang, W.; He, J.; Lu, T.; Dai, J.; Qiao, Y. Vision transformer adapter for dense predictions. *arXiv* **2022**, arXiv:2205.08534.
37. Zhao, H.; Shi, J.; Qi, X.; Wang, X.; Jia, J. Pyramid Scene Parsing Network. In Proceedings of the IEEE Conference on Computer Vision and Pattern Recognition, Honolulu, HI, USA, 21–26 July 2017; pp. 2881–2890.
38. Dong, X.; Bao, J.; Chen, D.; Zhang, W.; Yu, N.; Yuan, L.; Chen, D.; Guo, B. Cswin Transformer: A General Vision Transformer Backbone with Cross-Shaped Windows. In Proceedings of the IEEE/CVF Conference on Computer Vision and Pattern Recognition, New Orleans, LA, USA, 18–24 June 2022; pp. 12124–12134.
39. Zhu, X.X.; Tuia, D.; Mou, L.; Xia, G.-S.; Zhang, L.; Xu, F.; Fraundorfer, F. Deep learning in remote sensing: A comprehensive review and list of resources. *IEEE Geosci. Remote Sens. Mag.* **2017**, *5*, 8–36. [CrossRef]
40. Houlsby, N.; Giurgiu, A.; Jastrzebski, S.; Morrone, B.; De Laroussilhe, Q.; Gesmundo, A.; Attariyan, M.; Gelly, S. Parameter-Efficient Transfer Learning for NLP. In Proceedings of the International Conference on Machine Learning, Long Beach, CA, USA, 10–15 June 2019; pp. 2790–2799.
41. Cheng, B.; Schwing, A.; Kirillov, A. Per-pixel classification is not all you need for semantic segmentation. *Adv. Neural Inf. Process. Syst.* **2021**, *34*, 17864–17875.
42. Milletari, F.; Navab, N.; Ahmadi, S.-A. V-Net: Fully Convolutional Neural Networks for Volumetric Medical Image Segmentation. In Proceedings of the 2016 Fourth International Conference on 3D Vision (3DV), Stanford, CA, USA, 25–28 October 2016; pp. 565–571.
43. Bao, H.; Dong, L.; Piao, S.; Wei, F. Beit: Bert pre-training of image transformers. *arXiv* **2021**, arXiv:2106.08254.
44. He, K.; Zhang, X.; Ren, S.; Sun, J. Delving Deep into Rectifiers: Surpassing Human-Level Performance on Imagenet Classification. In Proceedings of the IEEE International Conference on Computer Vision, Santiago, Chile, 7–13 December 2015; pp. 1026–1034.

remote sensing

MDPI

Article

A Semantic Segmentation Framework for Hyperspectral Imagery Based on Tucker Decomposition and 3DCNN Tested with Simulated Noisy Scenarios

Efrain Padilla-Zepeda †, Deni Torres-Roman *,† and Andres Mendez-Vazquez †

Center for Research and Advanced Studies of the National Polytechnic Institute, Telecommunications Group, Av del Bosque 1145, Zapopan 45017, Mexico
* Correspondence: deni.torres@cinvestav.mx
† These authors contributed equally to this work.

Abstract: The present work, unlike others, does not try to reduce the noise in hyperspectral images to increase the semantic segmentation performance metrics; rather, we present a classification framework for noisy Hyperspectral Images (HSI), studying the classification performance metrics for different SNR levels and where the inputs are compressed. This framework consists of a 3D Convolutional Neural Network (3DCNN) that uses as input data a spectrally compressed version of the HSI, obtained from the Tucker Decomposition (TKD). The advantage of this classifier is the ability to handle spatial and spectral features from the core tensor, exploiting the spatial correlation of remotely sensed images of the earth surface. To test the performance of this framework, signal-independent thermal noise and signal-dependent photonic noise generators are implemented to simulate an extensive collection of tests, from 60 dB to −20 dB of Signal-to-Noise Ratio (SNR) over three datasets: Indian Pines (IP), University of Pavia (UP), and Salinas (SAL). For comparison purposes, we have included tests with Support Vector Machine (SVM), Random Forest (RF), 1DCNN, and 2DCNN. For the test cases, the datasets were compressed to only 40 tensor bands for a relative reconstruction error less than 1%. This framework allows us to classify the noisy data with better accuracy and significantly reduces the computational complexity of the Deep Learning (DL) model. The framework exhibits an excellent performance from 60 dB to 0 dB of SNR for 2DCNN and 3DCNN, achieving a Kappa coefficient from 0.90 to 1.0 in all the noisy data scenarios for a representative set of labeled samples of each class for training, from 5% to 10% for the datasets used in this work. The source code and log files of the experiments used for this paper are publicly available for research purposes.

Keywords: semantic segmentation; 3D convolutional neural network; noisy hyperspectral image; Tucker tensor decomposition; spectral–spatial feature extraction

Citation: Padilla-Zepeda, E.; Torres-Roman, D.; Mendez-Vazquez, A. A Semantic Segmentation Framework for Hyperspectral Imagery Based on Tucker Decomposition and 3DCNN Tested with Simulated Noisy Scenarios. *Remote Sens.* **2023**, *15*, 1399. https://doi.org/10.3390/rs15051399

Academic Editors: Jiaojiao Li, Qian Du, Wei Li, Bobo Xi, Jocelyn Chanussot, Rui Song and Yunsong Li

Received: 20 January 2023
Revised: 17 February 2023
Accepted: 22 February 2023
Published: 1 March 2023

1. Introduction

Hyperspectral imaging studies the interactions between observed scenes and the electromagnetic spectrum [1]. For example, it allows for measuring the amount of light reflected into a spectral sensor. From these measurements, it is possible to obtain a distinctive spectral signature composed of different wavelength channels [2]. If it is assigned a label corresponding to the ground truth, with the help of a human expert or a clustering algorithm, a Machine Learning classifier can be trained using supervised learning [2]. The hyperspectral image capturing process is far from ideal [3]; it is well known that every signal will be prone to being corrupted by different kinds of noise depending on the electronics' quality, environment of capture, and many others factors. For example, hyperspectral sensors are mounted on airplanes, drones, or satellites causing the capture of data cubes to be highly expensive and noisy. Thus, the need for new methods that are robust to noisy environments for expanding the possible range of applications. To address this task, this work tests the performance of a 3DCNN for noisy hyperspectral images, which is a semantic segmentation algorithm based on the spatial–spectral feature extraction pixel-by-pixel. Given the high computational complexity of the 3DCNN with the original HSI, a dimensionality reduction method based on Tucker Decomposition compresses the spectral dimension of the input, independent of the low signal-to-noise ratio.

1.1. Related Work

It is usual to consider data denoising as a preprocessing step for classification. As is well known, there are many denoising algorithms for hyperspectral imagery [3–9], which aims to recover the clean signal from the noisy one. These algorithms are particularly useful when the posterior tasks analyze the spectral signature for qualitatively studies. On the other hand, the pixel-based semantic segmentation (classification) could be based on a wide range of algorithms, artificial neural networks architectures, or feature extraction techniques; for example, Convolutional Neural Networks (CNNs) have demonstrated outstanding results performing spatial and spectral feature extraction [2,10–15]. Semantic segmentation techniques for RGB images, such as transfer learning [16–19], have been applied to spectral imagery [20] in combination with fully convolutional models, such as the well-known U-net [21–24]. There are other techniques specifically designed for noise–robust classification, e.g., based on band fusion [25,26] or feature extraction as a pre-processing step for a classification algorithm [27–31]. In some applications of optical remote sensing satellites in orbit, using atmospheric correction as an example [32,33], for hyperspectral [34,35] earth surface monitoring missions, the first task is to perform semantic segmentation to obtain a classification mask, from which the atmospherically corrected image is estimated.

This framework aims to classify the noisy data pixel by pixel. For example, 3DCNN can extract spatial and spectral features despite the low Signal to Noise Ratio (SNR). However, there is a drawback of using these models caused by the computational complexity of having such huge datasets. Thus, we propose using Tucker Decomposition with a 3DCNN to combine and improve the properties of DL and Decomposition in a single noise robust framework. TKD shows excellent compression ratios with minimal or no effects in the segmentation performance of DL models given that it found a lower-rank representation of the original tensor, capturing the high spatial–spectral correlation of the data [36,37]. Not only that, in this work, we have shown that TKD helps to improve the classification performance where there are a representative number of samples of each class for training. Finally, in Table 1, we have a summary of the major papers consulted for the proposed semantic segmentation framework.

Table 1. Main papers used for the proposed framework and its contribution.

	Author	Contribution
Tensor	Kolda and Bader [38]	Tensor theory
	López et al. [36]	Use of the TKD for semantic segmentation tasks
Noise	Bourennane et al. [4]	Noise theory and noise model
	Liu et al. [39]	Noise model and noise generation
	Rasti et al. [3]	Noise theory and classification test methodology
Classification	Paoletti et al. [2]	Classifiers code, architectures, and theory
	Chen et al. [10]	Spatial–Spectral feature extraction theory
	Li et al. [40]	3DCNN architecture
	Fu et al. [25]	Noisy-robust classification
Metrics	Grandini et al. [41]	Metrics used for multi-class classification evaluation
	Luque et al. [42]	Impact of class unbalance for classification performance metrics

1.2. Contributions

Our main contributions are three-fold:

- This work provides the remote sensing community with a framework based on a 3DCNN and Tucker Decomposition, performing semantic segmentation of noisy hyperspectral images, from an SNR of from 60 dB to 0 dB, outperforming other classical classifiers such as RF and SVM.
- Taking advantage of the spectral correlation of the data, we perform the Tucker Decomposition compressing only in the spectral domain; for example, for the three data sets studied to 40 new tensor bands and achieving a relative reconstruction error of less than 1%. This compression of the spectral domain of the input space reduces the computational complexity, consequently reducing the training time ratio by up to 29 times with respect to the original input space, depending on the case.
- TKD not only reduces the computational complexity but also increases the classification performance. This improvement was most significant for training set sizes on the order of from 5% to 3%. Furthermore, the behavior of TKD under different SNR are studied for the three used datasets.

The remainder of this work explains the basic concepts of tensor algebra, the noise model used, and the architecture of the DL model in Section 2. Section 3 describes the proposed framework. Section 4 analyzes the experiments. Finally, Sections 5 and 6 present the discussion and conclusions, respectively.

2. Mathematical Background

This section presents a short description of the theoretical concepts used in each stage of the proposed semantic segmentation framework. First, tensor theory for HSI representation and compression based on the TKD is used [38]. Second, a noise model for hyperspectral imagery is used in this paper for noise generation [3,4,39]. Finally, DL spectral and spatial feature extraction models [2,10,40] with metrics are used to compare ground truth labels and predicted ones for unbalanced training scenarios [41–43].

2.1. Tensor Algebra

Nowadays, research in tensor data analysis is finding new novel properties and applications on spectral images [4,8,9,36,44–48]. For this reason, this section presents an overview of the tensors theory, and the representation of an HSI as a tensor.

For tensor algebra, the work of Kolda and Bader [38] is our main reference. Using its notation, a scalar is denoted by x, and the vectors and matrices are denoted by $\mathbf{x}$ and $\mathbf{X}$, respectively, which can also be seen as tensors. For example, a first-order tensor is a vector, and a second-order tensor is a matrix. A third- or higher-order tensor is denoted by $\mathcal{X}$ with elements $x_{i,j,\ldots,n}$. Thus, naturally, a third-order tensor may be represented as a cube of elements. Besides, $\mathbf{x}_i$ is the ith column of a matrix $\mathbf{X}$, and $\mathbf{a}^{(n)}, \mathbf{A}^{(n)}$ are the nth vector or matrix in a sequence of vectors or matrices, respectively. Now, a mode-n fiber is a vector obtained by fixing all indices, except the one corresponding to the nth-dimension [38] (pp. 457–460). $\mathbf{X}_{(n)}$ is a mode-n matricization of an Nth-order tensor $\mathcal{X}$, where the mode-n fibers are arranged to be the matrix columns. Lastly, $\|\mathcal{X}\|$ denotes the tensor norm of $\mathcal{X} \in \mathbb{R}^{I_1 \times I_2 \times \cdots \times I_N}$, as described by Equation (1); this is analogous to the matrix Frobenius norm.

$$\|\mathcal{X}\| = \sqrt{\sum_{i_1=1}^{I_1} \sum_{i_2=1}^{I_2} \cdots \sum_{i_N=1}^{I_N} x_{i_1 i_2 \cdots i_N}^2}. \tag{1}$$

An element of the new vector space generated by the outer product of the N vector spaces on the same field $\mathbb{R}$ is an Nth-order tensor $\mathcal{X}$. Thus, a tensor can be seen as a multi-dimensional array of N dimensions. The order of a tensor is also called the ways or modes [38]. An hyperspectral image $\mathcal{H}$ in Figure 1 can be represented as a third-order tensor $\mathcal{H} \in \mathbb{R}^{I_1 \times I_2 \times I_3}$, where I_1 and I_2 are image height and width, respectively. I_3 is the number of bands in which the electromagnetic spectrum is captured. Then, the element x_{i_1,i_2,i_3} is the pixel value at position (i_1, i_2) at band i_3.

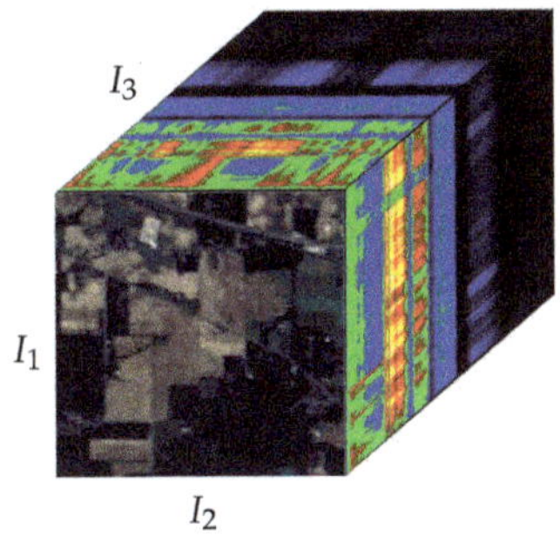

Figure 1. AVIRIS HSI of Indian Pines, NW Indiana. NASA/JPL [49].

In order to introduce the concept of TKD, the concepts of rank-one, vector outer product, *n-mode* product, and n-rank are required. A rank-one tensor of the Nth-order $\mathcal{X} \in \mathbb{R}^{I_1 \times I_2 \times \cdots \times I_N}$ can be represented as the outer product of vectors [38],

$$\mathcal{X} = \mathbf{a}^{(1)} \circ \mathbf{a}^{(2)} \circ \cdots \circ \mathbf{a}^{(N)}, \tag{2}$$

where the symbol "$\circ$", at Equation (2), represents the vector outer product. Thus, each element $x_{i_1,i_2,\ldots,i_n}$ of $\mathcal{X}$ is obtained from the corresponding vector elements [38]:

$$x_{i_1,i_2,\ldots,i_n} = a^{(1)}_{i_1} a^{(2)}_{i_2} \cdots a^{(N)}_{i_N} \quad \text{for all} \quad 1 \leq i_n \leq I_n. \tag{3}$$

The n-rank, $\text{rank}_n(\mathcal{X})$, is the column rank of the matrix $\mathbf{X}_{(n)}$. For easy-reading reasons, it is defined $\text{rank}_n(\mathcal{X})$ as R_n of $\mathcal{X}$ for $n = 1, \ldots, N$. For an HSI represented as a third-order tensor, $\mathcal{H} \in \mathbb{R}^{I_1 \times I_2 \times I_3}$, $\text{rank}_1(\mathcal{H})$, and $\text{rank}_2(\mathcal{H})$ correspond to the spatial domain of the image, such that $1 \leq \text{rank}_1(\mathcal{H}) \leq I_1$ and $1 \leq \text{rank}_2(\mathcal{H}) \leq I_2$. In the same way, $\text{rank}_3(\mathcal{H})$ and $1 \leq \text{rank}_3(\mathcal{H}) \leq I_3$ corresponds to the spectral domain. Finally, for the n-mode product of a tensor with a matrix, $\mathcal{X} \times_n \mathbf{U}$, with $\mathbf{U} \in \mathbb{R}^{J \times I_n}$, the resultant tensor will have dimensions $I_1 \times \cdots \times I_{n-1} \times J \times I_{n+1} \times \cdots \times I_N$ [38] (pp. 460–461):

$$(\mathcal{X} \times_n \mathbf{U})_{i_1 \cdots i_{n-1} j i_{n+1} \cdots i_N} = \sum_{i_n=1}^{I_n} x_{i_1 i_2 \cdots i_N} u_{j i_n}. \tag{4}$$

Tucker in 1966 introduced the now-called Tucker Decomposition [50]. It is a form of higher-order PCA and there are several tensor decompositions derived from this one. The Tucker Decomposition (TKD) decomposes a tensor of Nth-order into a core tensor of the same order but could have different dimensions, multiplied by a transformation matrix along each mode [38]. The Tucker Decomposition for a third-order tensor, e.g., see Figure 2, for an HSI representation, $\mathcal{X} \in \mathbb{R}^{I \times J \times K}$, is defined as:

$$\mathcal{X} \approx \mathcal{G} \times_1 \mathbf{A} \times_2 \mathbf{B} \times_3 \mathbf{C} = \sum_{p=1}^{P} \sum_{q=1}^{Q} \sum_{r=1}^{R} g_{pqr} \, \mathbf{a}_p \circ \mathbf{b}_q \circ \mathbf{c}_r = [\![\mathcal{G}; \mathbf{A}, \mathbf{B}, \mathbf{C}]\!], \tag{5}$$

where $P \leq I$, $Q \leq J$ and $R \leq K$.

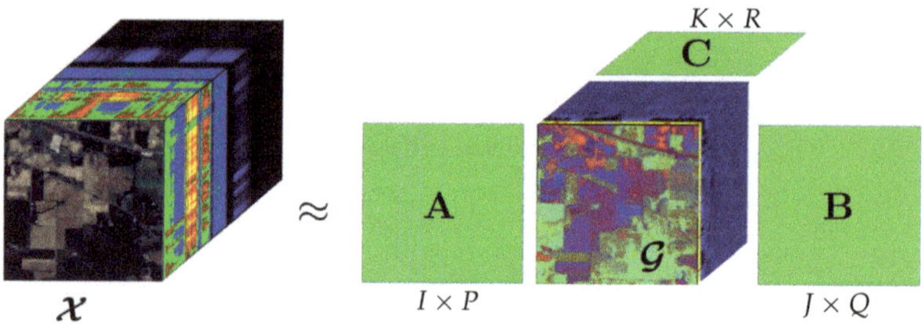

Figure 2. Tucker Decomposition of Indian Pines HSI, a third-order tensor.

Element-wise, the Tucker Decomposition in Equation (5) is provided by:

$$x_{ijk} \approx \sum_{p=1}^{P} \sum_{q=1}^{Q} \sum_{r=1}^{R} g_{pqr}\, a_{ip} b_{jq} c_{kr} \text{ for } i = 1, \ldots, I,\ j = 1, \ldots, J,\ k = 1, \ldots, K. \tag{6}$$

The entries of the core tensor $\mathcal{G} \in \mathbb{R}^{P \times Q \times R}$ "show the level of interaction between the different components". P, Q, and R are the new dimensions of the factor matrices $\mathbf{A}$, $\mathbf{B}$, and $\mathbf{C}$, respectively, which can be seen as the principal components in each mode. "If P, Q, R are smaller than I,J,K, the core tensor $\mathcal{G}$ is a compressed version of $\mathcal{X}$".

According with Kolda and Bader [38], for $R_n = \text{rank}_n(\mathcal{X})$, an exact Tucker Decomposition of rank $(R_1, R_2, \ldots, R_N)$ can be computed. Another case is "the truncated Tucker Decomposition of rank $(R_1, R_2, \ldots, R_N)$, when $R_n < \text{rank}_n(\mathcal{X})$ for one or more n, then it will be necessarily inexact and more difficult to compute." Tucker Decomposition can compress along selected dimensions. To do this, we use the Tucker1 Decomposition, defined in [38], which fixes two factor matrices to be the identity matrix,

$$\mathcal{X} \approx \mathcal{G} \times_1 \mathbf{A} = [\![\mathcal{G}; \mathbf{A}, \mathbf{I}, \mathbf{I}]\!]. \tag{7}$$

2.2. Noise Model

Noise is intrinsic to any signal and hyperspectral imaging is not an exception. There are many sources and kinds of noise present on HSIs mentioned in this section. In general, the noise can be distinguished into two classes [4]: the fixed pattern noise and the random noise. The first one is due to calibration errors, and it is not of interest in this work. Instead, random noise, due to its stochastic nature, can be studied and generated from a suitable noise model. For new-generation imaging spectrometers used in hyperspectral imagery, the random noise mainly comes from two aspects: Signal-Dependent (SD) photonic noise and Signal-Independent (SI) electronic noise, also known as thermal (Johnson) noise [51].

Although, the noise model used in this paper is due to the work of Bourennane et al. [4], we are not addressing the signal denoising process; rather, we use it for the simulation of the noisy data scenarios. For this, we carefully study the calculation of variances focused on the implementation of a noise generator in Section 3.2. Using the tensor theory, the noisy HSI is represented as a sum of the clean signal and additive noise [3].

$$\mathcal{H} = \mathcal{X} + \mathcal{N}, \tag{8}$$

where $\mathcal{H}, \mathcal{X}, \mathcal{N} \in \mathbb{R}^{I_1 \times I_2 \times I_3}$. $\mathcal{H}$ is the noisy HSI, $\mathcal{X}$ is the clean signal, and $\mathcal{N}$ is the noise for both photon and thermal noise [4]. Note that $\mathcal{H}$ is quantized depending on the capturing sensor; this process is described in Section 3.2 by Equation (26). The noise model in Equation (8) is valid under the assumption of high-SNR of $\mathcal{X}$. The variance of the noise depends of each pixel value x_{i_1,i_2,i_3} in the clean signal $\mathcal{X}$. The tensor $\mathcal{N}$ is composed of the sum of two tensors, the photonic noise tensor $\mathcal{P} \in \mathbb{R}^{I_1 \times I_2 \times I_3}$, and the thermal noise tensor $\mathcal{T} \in \mathbb{R}^{I_1 \times I_2 \times I_3}$. Thus:

$$\mathcal{N} = \mathcal{P} + \mathcal{T}, \tag{9}$$

where $\mathcal{P}$ is dependent of the clean signal $\mathcal{X}$, and $\mathcal{T}$ is signal independent.

The improvement of the Charged Couple Device (CCD) sensors for new generation instruments exhibited a tendency to increase the spatial resolution. Therefore, the number of photons that reach a pixel per unit time becomes smaller, causing the random fluctuation of photons arriving at the sensor. Consequently, the photonic noise is now more relevant than before [4]. Photonic noise follows Poisson distribution [52], but it can be approximated by a Gaussian distribution [53]. A single photon noise element p_{i_1,i_2,i_3} of tensor $\mathcal{P} \in \mathbb{R}^{I_1 \times I_2 \times I_3}$ can be expressed in terms of its corresponding element x_{i_1,i_2,i_3} of the clean signal $\mathcal{X}$ as follows [53]:

$$p_{i_1,i_2,i_3} = \left(x_{i_1,i_2,i_3}\right)^{\gamma} \cdot u_{i_1,i_2,i_3}, \tag{10}$$

where u_{i_1,i_2,i_3} is a stationary, zero-mean uncorrelated random process independent of x_{i_1,i_2,i_3} with variance σ_u^2". "In the case for earth remote sensing images captured by instruments mounted in airborne or spaceborne platforms, the exponent γ is equal to 0.5" [51]. Thus:

$$p_{i_1,i_2,i_3} = \sqrt{x_{i_1,i_2,i_3}} \cdot u_{i_1,i_2,i_3}. \tag{11}$$

The thermal agitation of the charge carriers inside the electronics of the instruments used for hyperspectral images causes the thermal noise. A single thermal noise element of the noise tensor $\mathcal{T}$ is denoted by t_{i_1,i_2,i_3}; this random process can be modeled as an additive zero-mean white Gaussian noise with variance σ_t^2 [4].

From Equations (8) and (9), the noise model used in this paper is:

$$\mathcal{H} = \mathcal{X} + \mathcal{P} + \mathcal{T}. \tag{12}$$

Element-wise, using Equations (10) and (11), the noise model is:

$$h_{i_1,i_2,i_3} = x_{i_1,i_2,i_3} + \sqrt{x_{i_1,i_2,i_3}} \cdot u_{i_1,i_2,i_3} + t_{i_1,i_2,i_3}. \tag{13}$$

To highlight the dependency, another useful notation for Equation (9) is [39]:

$$\mathcal{N}(\mathcal{X}) = \mathcal{N}_{SD}(\mathcal{X}) + \mathcal{N}_{SI}. \tag{14}$$

$$n_{i_1,i_2,i_3}(\mathcal{X}) = \sqrt{x_{i_1,i_2,i_3}} \cdot u_{i_1,i_2,i_3} + t_{i_1,i_2,i_3}. \tag{15}$$

Given this, Equation (12) can be rewritten as:

$$\mathcal{H}(\mathcal{X}) = \mathcal{X} + \mathcal{N}_{SD}(\mathcal{X}) + \mathcal{N}_{SI}. \tag{16}$$

This SD and SI noise model is used in the framework of this paper, because it considers two of the main sources of random noise for new generation sensors.

2.3. *Spectral–Spatial Deep Learning Models*

Generally, spectral signatures of equally labeled pixels are highly correlated between them, and this is a feature that most of the classification algorithms take advantage of for class separation. Spatial correlation is present when a group of neighbor pixels belongs to the same class, which is a common case for remote sensing optical images of the earth surface. Convolutional Neural Networks (CNN) are DL models designed to extract features of neighbor pixels and bands, based on this, the architecture depends on the feature analysis they perform. CNNs for HSI classification are divided in three kinds: spectral, spatial, and spectral–spatial [2].

For the Spectral DL model (1DCNN) in Figure 3, the spectral pixels $\mathbf{x}_i \in \mathbb{N}^{n_{bands}}$ are the input data, where n_{bands} is the number of bands of the image with or without compression. On each convolutional layer (CONV), 1D-kernels are applied, such that $K^{(l)} \times q^{(l)}$, obtaining as a result an output $\mathbf{X}^{(l)}$ composed of $K^{(l)}$ feature vectors [2].

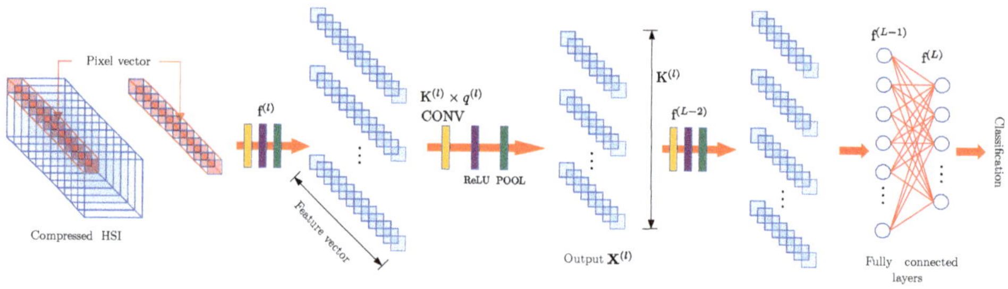

Figure 3. Traditional architecture of spectral convolutional model employed using 1DCNN [2].

Spatial DL models (2DCNN) consider the spatial information obtained from the neighbor pixels of the HSI, see Figure 4. For that reason, the input will be spatial patches of $d \times d$ pixels cropped from the complete HSI with the pixel of interest at the center. To extract spatial features, on each CONV layer, 2D-kernels are applied over the input data, such that $K^{(l)} \times k^{(l)} \times q^{(l)}$, obtaining $K^{(l)}$ feature maps as a result [2].

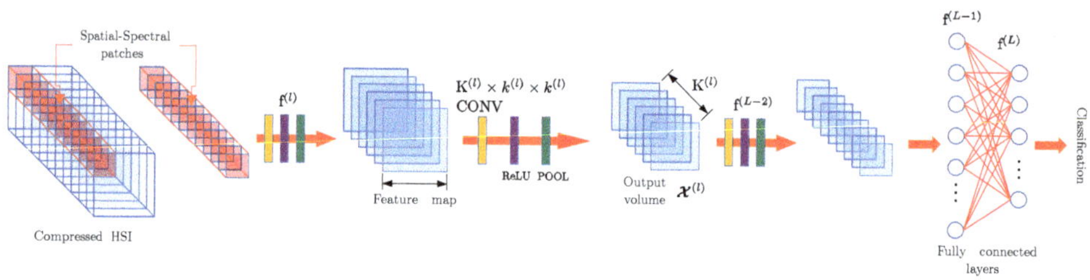

Figure 4. Traditional architecture of spatial convolutional model employed using 2DCNN [2].

Spectral–spatial DL models (3DCNN) extract spectral and spatial features at the same time, see Figure 5. Similarly, as with 2DCNN, the features are extracted from spatial patches of $d \times d$, associated with a single pixel of the HSI. This model uses 3D-kernels, such that $K^{(l)} \times k^{(l)} \times k^{(l)} \times q^{(l)}$, extracting $K^{(l)}$ feature volumes as output [2].

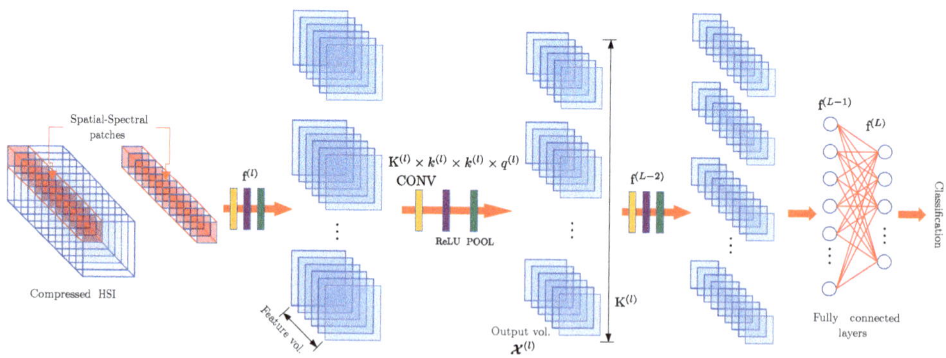

Figure 5. Traditional architecture of spectral–spatial convolutional model employed using 3DCNN [2].

All these spectral feature extraction DL methods basically infer the ground truths based on the spectral signature. Moreover, remote sensing images exhibit an obvious correlation between neighbor pixels, causing the spatial feature extraction to be a good candidate for this task. Spectral–Spatial feature extraction adopts both characteristics, creating features volumes from a pixel of interest, and, additionally, contains information from its neighbor pixels. For this reason, we use 3DCNN, which is used for the classification of hyperspectral remote sensing noisy images.

2.4. Unbalanced Classification Performance Metrics

Unbalanced classification performance metrics are a key piece for this framework, because we cannot guarantee the same number of labeled samples per each class for HSIs to be tested. Each image will have different spatial Ground Truth (GT) distribution, and there is a need to highlight that, when identifying targets with a low sample count, classical metrics in multi-class classification may show biases. For example, the Overall Accuracy is defined as the number of correct classified samples divided by the overall number of samples. This metric is not reliable when the classification problem is imbalanced. The Average Accuracy metric is essentially an average of the accuracies per each class. If the classification problem shows an unbalanced distribution of classes, this metric takes into account the accuracy per each class as equal, independent of the number of samples. On the other hand, Cohen in 1960 evaluated the classification of two raters (prediction of the model and the actual GT) in order to measure the agreement between them [41]. Cohen's Kappa coefficient (K) is widely used for the performance evaluation of remote sensing image classification. For this reason, all the results in this work are presented with this metric.

Given a predicted classification map $\hat{\mathbf{Y}}$, obtained from the trained classifier, and the ground truth $\mathbf{Y}$, of the HSI, a multi-class confusion matrix $\mathbf{M} = \left(m_{i,j}\right) \in \mathbb{Z}^{c \times c}$ is computed, where c is the number of classes in $\mathbf{Y}$.

For the case of a binary confusion matrix, Cohen´s Kappa coefficient is defined as follows:

$$K = \frac{P_o - P_e}{1 - P_e}. \tag{17}$$

where P_o is the accuracy achieved by the model, P_e is the level of accuracy to obtain by chance. For a multi-class confusion matrix, K is defined as:

$$K = \frac{\sum_{k=1}^{c} m_{k,k} \sum_{i=1}^{c} \sum_{j=1}^{c} m_{i,j} - \sum_{k=1}^{c}\left(\sum_{i=1}^{c} m_{i,k} \sum_{j=1}^{c} m_{k,j}\right)}{\left(\sum_{i=1}^{c} \sum_{j=1}^{c} m_{i,j}\right)^2 - \sum_{k=1}^{c}\left(\sum_{i=1}^{c} m_{i,k} \sum_{j=1}^{c} m_{k,j}\right)}. \tag{18}$$

The case when $K = 1$ shows a perfect agreement between the GT and predicted labels. $K = 0$ means that there is a chance of agreement, but if K is negative, it is a clear disagreement. Each class classification must have importance; for that reason, all the results are presented with Cohen's Kappa coefficient, but the other two metrics (OA and AA) can be consulted in the log files available in the public repository of this paper [54].

3. Proposed Framework

The proposed framework consists of the following three blocks:

- Noise Generation and Quantization: Having as input the clean signal power, the variances for signal-dependent and independent noise processes are calculated for a specified SNR. In order to follow the non-negative integer values of a digital image, a quantization is performed.
- Tucker Decomposition: Transforms $\mathcal{H}$ into a new input space through a core-tensor $\mathcal{G}$ and factor matrices $\mathbf{I_A}$, $\mathbf{I_B}$ and $\mathbf{C}$, where $\mathcal{G}$ is a spectrally compressed version of $\mathcal{H}$.
- Deep Learning Model: The model is fitted in terms of the Softmax loss with $\mathcal{G}$ and the class labels present in the ground truth $\mathbf{Y}$, evaluating the prediction $\hat{\mathbf{Y}}$ of the trained model with metrics that consider a possible unbalanced class scenario.

In Figure 6, a block diagram of the proposed framework is shown.

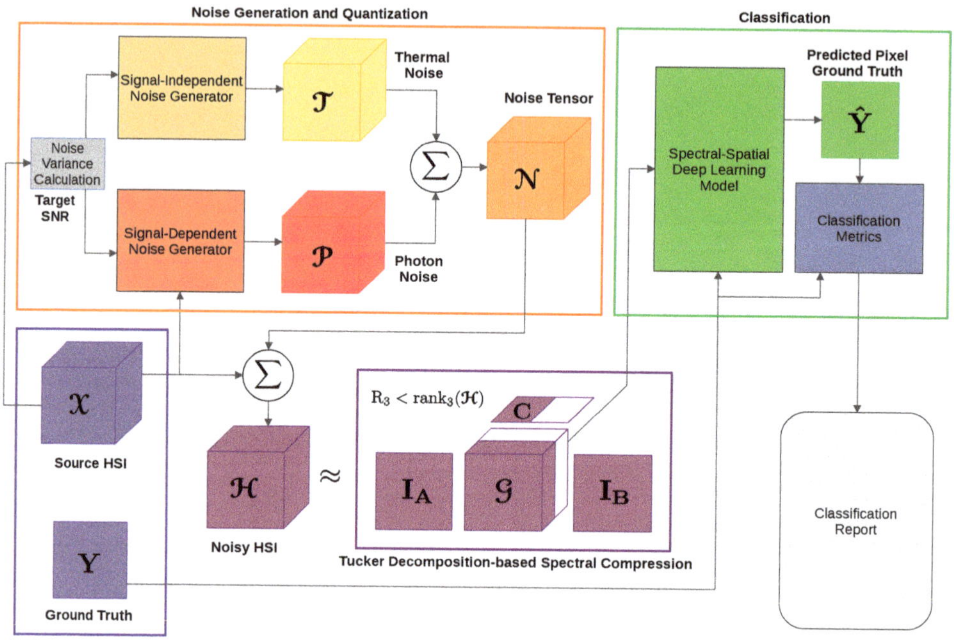

Figure 6. Proposed framework.

3.1. Problem Statement

Given $\mathcal{X} \in \mathbb{R}^{I_1 \times I_2 \times I_3}$, which is a source HSI in a third-order tensor form, assuming high-SNR, and $\mathbf{Y} \in \mathbb{N}^{I_1 \times I_2}$, and given the corresponding pixel ground truth matrix, a noise tensor $\mathcal{N}$ must be generated with the same size of $\mathcal{X}$, $\mathcal{N} \in \mathbb{R}^{I_1 \times I_2 \times I_3}$. $\mathcal{N}$ is the sum of two third-order tensors, the signal-dependent photon noise $\mathcal{P}$, and the signal-independent thermal noise $\mathcal{T}$; hence, $\mathcal{N} = \mathcal{P} + \mathcal{T}$. $\mathcal{N}$ must be generated in such a way that the SNR between $\mathcal{X}$ and $\mathcal{N}$ is at a desired target and the power of both noise tensors are at different proportions. Thus, the new noisy HSI $\mathcal{H}$ is obtained from the sum of the original HSI $\mathcal{X}$, and the generated noise tensor $\mathcal{N}$; therefore, $\mathcal{H} = \mathcal{X} + \mathcal{N}, \mathcal{H} \in \mathbb{R}^{I_1 \times I_2 \times I_3}$.

The purpose of $\mathcal{H}$ is to evaluate the classification performance of the Spectral–Spatial DL models when the input HSI is noisy and no denoising method is applied, but the training complexity of these models is too high compared with some classical classifiers. Hence, there is a need to reduce the size of the input to decrease the computational complexity of the DL model. This task could be performed using a Tucker Decomposition-based Spectral Compression, setting a suitable compression ratio. Thus, it is necessary to find a core tensor $\mathcal{G} \in \mathbb{R}^{I_1 \times I_2 \times R_3}$, which will be a spectrally compressed version of $\mathcal{H}$, such that $R_3 \leq \text{rank}_3(\mathcal{H})$.

With the pair $(\mathcal{G}, \mathbf{Y})$, divide the ground truth available pixels in training and testing sets, taking into account a possible imbalanced classification case. Train the DL Spectral–Spatial Model and predict with it a $\hat{\mathbf{Y}}$. Finally, evaluate the performance of the DL model with multi-class classification metrics between $\mathbf{Y}$ and $\hat{\mathbf{Y}}$.

3.2. Noise Generation and Quantization

For experiment purposes of this paper, under the assumption of high-SNR, each available HSI obtained from space agencies is considered as the clean signal $\mathcal{X}$ from Equation (16). From $\mathcal{X}$, the noise variances σ^2_{u,i_3} and σ^2_{t,i_3} are calculated to generate samples of the random processes u_{i_1,i_2,i_3} and t_{i_1,i_2,i_3}, which correspond to generate the noise tensors $\mathcal{N}_{SD}(\mathcal{X})$ and $\mathcal{N}_{SI}$ at a specified SNR in dB [39]. If the variance of the signal is calculated on homogeneous pixels, this is $\sigma^2_{x_{i_1,i_2,i_3}} = 0$ by definition [51]; assuming that x, u, and t are independent, and both u and t are zero mean and stationary, the noise variance of each element $n(\mathcal{X})$ of the noise tensor $\mathcal{N}(\mathcal{X})$ can be written as [39]:

$$\sigma^2_{n_{i_1,i_2,i_3}}(\mathcal{X}) = x_{i_1,i_2,i_3} \cdot \sigma^2_{u,i_3} + \sigma^2_{t,i_3}; \tag{19}$$

"in practice, homogeneous pixels with $\sigma^2_{x_{i_1,i_2,i_3}} = 0$ may be extremely rare and theoretical expectation are approximated with local averages" [51]. The mean variance of the noise tensor $\mathcal{N}(\mathcal{X})$ is composed of the sum of the SD and SI noise variances:

$$\sigma^2_{\mathcal{N}(\mathcal{X})} = \sigma^2_{\mathcal{N}_{SD}(\mathcal{X})} + \sigma^2_{\mathcal{N}_{SI}}. \tag{20}$$

Besides, it can be expressed in terms of the mean power of the signal $\mathcal{X}$ and the SNR (dB):

$$\sigma^2_{\mathcal{N}(\mathcal{X})} = \overline{P}_{\mathcal{X}} \cdot 10^{-\left(\frac{\mathrm{SNR}}{10}\right)}, \tag{21}$$

where $\overline{P}_{\mathcal{X}} = \frac{\|\mathcal{X}\|^2}{I_1 I_2 I_3}$. Assuming a parameter α, which controls the contribution of both noise processes to the noise tensor $\mathcal{N}(\mathcal{X})$, such that:

$$\alpha = \frac{\sigma^2_{\mathcal{N}_{SD}(\mathcal{X})}}{\sigma^2_{\mathcal{N}_{SI}}}. \tag{22}$$

From Equations (20) and (22), the mean SI and SD noise variances are expressed in terms of α as follows:

$$\begin{aligned} \sigma^2_{\mathcal{N}_{SD}(\mathcal{X})} &= \frac{\sigma^2_{\mathcal{N}(\mathcal{X})} \cdot \alpha}{\alpha + 1}, \\ \sigma^2_{\mathcal{N}_{SI}} &= \frac{\sigma^2_{\mathcal{N}(\mathcal{X})}}{\alpha + 1}. \end{aligned} \tag{23}$$

Furthermore, the noise variances to draw samples are:

$$\begin{aligned} \sigma^2_{u,i_3} &= \frac{\sigma^2_{\mathcal{N}_{SD}(\mathcal{X})}}{\mu_{i_3}}, \\ \sigma^2_{t,i_3} &= \sigma^2_{\mathcal{N}_{SI}}. \end{aligned} \tag{24}$$

Some mathematical details of the noise model can be consulted in Appendix A.1.

Once obtained, the noise variances are drawn as random samples from a normal continuous random variable to obtain the tensor $\mathcal{N}$. As seen in Equation (8), $\mathcal{H}$ is obtained by the addition of $\mathcal{X}$ and $\mathcal{N}$. The elements h_{i_1,i_2,i_3} of such tensors are integers, in the range $0 \le h_{i_1,i_2,i_3} \le L$, where L is the number of quantization levels, provided by Equation (25), which depends on Q, the number of bits of the sensor. Then,

$$L = 2^Q - 1, \tag{25}$$

and where a uniform quantization was performed according to the following rule:

$$h_{i_1,i_2,i_3} = \begin{cases} 0 & \text{if } h_{i_1,i_2,i_3} \le 0 \\ L & \text{if } h_{i_1,i_2,i_3} \ge L \\ \lfloor h_{i_1,i_2,i_3} \rfloor & \text{if } (h_{i_1,i_2,i_3} - \lfloor h_{i_1,i_2,i_3} \rfloor) < \frac{1}{2} \\ \lceil h_{i_1,i_2,i_3} \rceil & \text{otherwise.} \end{cases} \tag{26}$$

In Figure 7, the behavior of both different kinds of noise is observed, where a specific case of Pavia University is displayed. The SD noise is clearly more present in the high-reflectance pixels. On the other hand, the SI noise is uniformly distributed along the spatial domain.

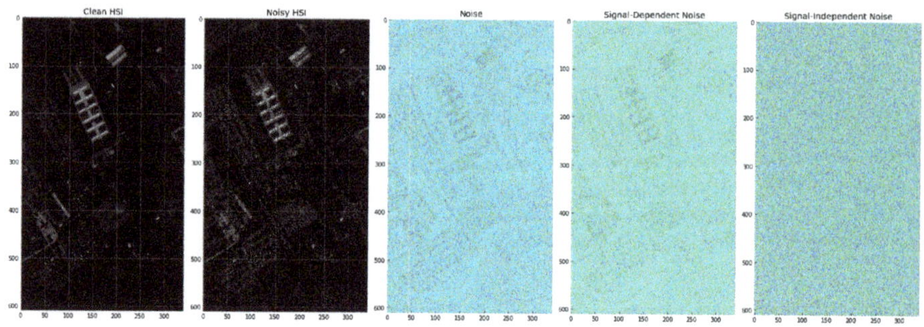

Figure 7. 30th band of Pavia University HSI. $SNR = -15$, $\alpha = 5$.

3.3. Tucker Decomposition-Based Spectral Compression

For our particular case of HSI, we need to reduce the dimensionality in the spectral domain only. If the factor matrices **A** and **B** are the identity matrices, which correspond to the spatial components, then:

$$\mathcal{H} = \mathcal{G} \times_3 \mathbf{C} = [\![\mathcal{G}; \mathbf{I}, \mathbf{I}, \mathbf{C}]\!], \tag{27}$$

$$\mathbf{H}_{(3)} = \mathbf{C}\mathbf{G}_{(3)}. \tag{28}$$

Thus, the core tensor keeps the first two dimensions or spatial domain but reduces the third dimension or spectral domain, causing $\mathcal{G}$ to be a spectrally compressed version of $\mathcal{H}$. To perform this computation, using the truncated Tucker Decomposition, set the n-ranks to: $R_1 = \text{rank}_1(\mathcal{H})$, $R_2 = \text{rank}_2(\mathcal{H})$, and $R_3 < \text{rank}_3(\mathcal{H})$, and reduce the spectral domain from I_3 spectral bands to R_3 new tensor bands.

3.4. Deep Learning Model Architecture

In this paper, the experiments were performed using a 3DCNN model, given that, generally, the spectral and spatial domains of HSIs are highly correlated. In Table 2, it is shown that the architecture used is [2,10,40]. The fundamentals of the 3DCNN model were explained in Section 2.3.

Table 2. Architecture of the 3DCNN model.

Main Layer	Normalization	Activation Function	Downsampling
Linear input ($19 \times 19 \times n_{bands}$)	-	-	-
CONV($32 \times 5 \times 5 \times 24$)	BN	ReLU	-
CONV($64 \times 5 \times 5 \times 16$)	BN	ReLU	POOL ($2 \times 2 \times 1$)
FC(300)	BN	ReLU	-
FC(n_{class})	-	Softmax	-

4. Dataset Experiments and Results

The following section explains the setup and details for the experiments performed to test the framework. The source code and log files with the obtained results are available in the following GitHub repository [54]: github.com/EfrainPadilla, https://github.com/EfrainPadilla/Noisy-Hyperspectral-Semantic-Segmentation-Framework-based-on-Tucker-Decomposition-and-3D-CNN (accessed on 10 March 2022).

4.1. Hardware

The experiments were performed using SSH in a High-Performance Computing (HPC) server installed at the Cinvestav Guadalajara Campus. The hardware is described in Table 3. The implementation ran in Python 3.8.5 and the neural network was implemented in Keras-Tensorflow 2.3.0 with CUDA 10.1. Google Collab was used for developing and testing.

Table 3. Hardware of HPC server for experiments.

Hardware	Cinvestav Guadalajara HPC Server
CPU	×2 Intel Xeon 2.20 GHz 13.75 MB Cache L3
Cores per CPU	10
Threads	40
RAM	6×16 GB 96 GB DDR4 HPE Smart 2666 MHz ECC
GPU	×1 NVIDIA Tesla V100 PCIe 3.0
GPU Memory	16 GB HBM2
CUDA cores	5120

4.2. Datasets Description

4.2.1. Indian Pines

The Indian Pines (IP) dataset was captured using the AVIRIS sensor [55] in 1992, an agricultural area in NW Indiana, characterized by its crops of regular geometry and its irregular forest regions. The spatial resolution is 20 m per pixel with dimensions 145 × 145. From 224 bands in a wavelength range of 0.4 to 2.5 μm, 24 were removed for being null or water absorption bands (in particular 104–108, 150–163, and 220), considering the remaining 200 bands for the experiments [2]. The ground truth described in Table 4 has 10,249 labeled samples divided into 16 classes and true color (RGB) is composed from bands 28, 16, and 9 as red, green, and blue, respectively.

Table 4. Indian Pines ground truth description and true RGB visualization, from [2].

Class Number-Name	Samples	Color	Ground Truth	True RGB
"0-Background	10,776			
1-Alfalfa	46			
2-Corn-notill	1428			
3-Corn-mintill	830			
4-Corn	237			
5-Grass-pasture	483			
6-Grass-trees	730			
7-Grass-pasture-mowed	28			
8-Hay-windrowed	478			
9-Oats	20			
10-Soybean-notill	972			
11-Soybean-mintill	2455			
12-Soybean-clean	593			
13-Wheat	205			
14-Woods	1265			
15-Buildings-Grass-Trees-Drives	386			
16-Stone-Steel-Towers"	93			

4.2.2. University of Pavia

The campus of the University of Pavia (UP) was captured using the ROSIS sensor [56] in 2002, an urban environment in the North of Italy with multiple solid structures, natural objects, and shadows. The spatial resolution is 1.3 m per pixel with dimensions 610 × 340 and 103 bands in a wavelength range of from 0.43 to 8.6 μm. The ground truth described in Table 5 has 42,776 labeled samples divided into nine classes and true color (RGB) is composed from bands 48, 24, and 9 as red, green, and blue, respectively.

Table 5. University of Pavia ground truth description and true RGB visualization, from [2].

Class Number-Name	Samples	Color	Ground Truth	True RGB
"0-Background	164,624		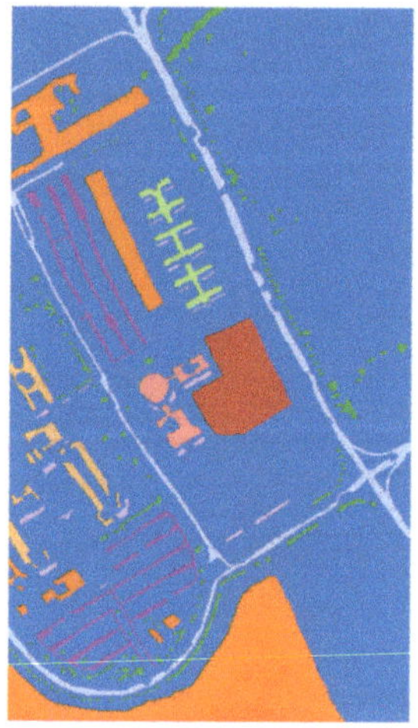	
1-Asphalt	6631			
2-Meadows	18,649			
3-Gravel	2099			
4-Trees	3064			
5-Painted metal sheets	1345			
6-Bare Soil	5029			
7-Bitumen	1330			
8-Self-Blocking Bricks	3682			
9-Shadows"	947			

4.2.3. Salinas

The Salinas (SAL) HSI was captured using the AVIRIS sensor [55] in 2001 over several agricultural fields in the Salinas Valley, CA, USA. The spatial resolution is 3.7 m per pixel with dimensions 512 × 217. As in the case of IP, from 224 bands in a wavelength range of from 0.43 to 8.6 μm, 20 were discarded due to water absorption and noise [2]. The ground truth described in Table 6 has 54,129 labeled samples divided into 16 classes and true color (RGB) is composed from bands 28, 16, and 9 as red, green, and blue, respectively.

Table 6. Salinas ground truth description and true RGB visualization, from [2].

Class number-name	Samples	Color	Ground Truth	True RGB
"0-Background	56,975			
1-Brocoli-green-weeds-1	2009			
2-Brocoli-green-weeds-2	3726			
3-Fallow	1976			
4-Fallow-rough-plow	1394			
5-Fallow-smooth	2678			
6-Stubble	3959			
7-Celery	3579			
8-Grapes-untrained	11,271			
9-Soil-vinyard-develop	6203			
10-Corn-senesced-green-weeds	3278			
11-Lettuce-romaine-4wk	1068			
12-Lettuce-romaine-5wk	1927			
13-Lettuce-romaine-6wk	916			
14-Lettuce-romaine-7wk	1070			
15-Vinyard-untrained	7268			
16-Vinyard-vertical-trellis"	1807			

4.3. Data Pre-Processing for Reduction of Number of Bands

The noise generation is implemented using the random.normal package of Numpy [57], where we draw samples with the computed variances, seen in Section 3.2. Tucker Decomposition is implemented using Tensorly [58], but the code was modified to set the spatial projection matrices to be the identity matrix, as seen in Equation (27). To select the spectral compression ratio, the reconstruction error of the original $\mathcal{H}$ using TKD for different number of tensor bands $R_3 = 1, 2, ..., I_3$ was calculated. The relative reconstruction error ξ is obtained using Equation (29).

$$\xi = \frac{\|\mathcal{H} - \hat{\mathcal{H}}_{R_3}\|^2}{\|\mathcal{H}\|^2}. \quad (29)$$

In Figure 8, the relative reconstruction error is shown for each compression case of IP, UP, and SAL. We have selected $R_3 = 40$ for all the experiments of this paper, given its low relative reconstruction error (below 1%) for the three images. Table 7 shows the reconstruction error ξ and the running time of TKD with $R_3 = 40$, for each HSI used in this paper.

Table 7. Reconstruction error and running time of TKD compressing to 40 tensorial bands.

HSI	ξ	TKD Running Time (s)
IP	0.994 %	2.89
UP	0.019 %	16.83
SAL	0.027 %	12.41

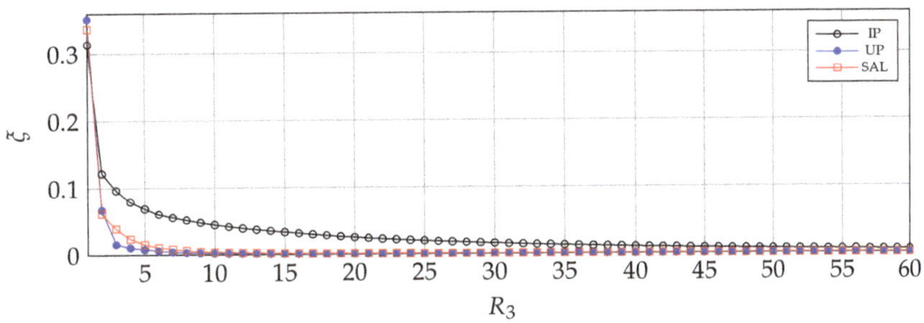

Figure 8. Relative error between original data and the reconstruction from core tensor.

TKD Behavior for Low-SNR HSI Analysis

The aim of the TKD is to find a vector space of a smaller dimension to represent the same information of the original space, taking advantage of the spatial and spectral correlation of pixels. As explained in Section 2, the core tensor $\mathcal{G}$ defines the contribution by the weights on the ith frontal slice $\mathbf{G}_{::i}$. A simple way to visualize the contribution of each tensorial band is to compute the power per pixel p_i of the ith frontal slice using $p_i = \frac{\|\mathbf{G}_{::i}\|^2}{I_1 I_2}$. In Figure 9, the behaviors of the contribution for the first tensorial bands from 60 dB to −20 dB are shown for the three datasets used in this paper. The last bands show very low power compared to the first ones; for this reason, they are not included in Figure 9. It is clear that the main contribution is on the first two tensorial bands of $\mathcal{G}$ for the three datasets, but the contribution becomes smaller for low-SNR scenarios. This is explained because the data becomes uncorrelated by the random noise processes with higher variances, causing it to be harder to find a projection matrix that defines the direction of the data. For IP in Figure 9a, the power per pixel of the first tensorial bands start to decay approximately at 20 dB, for UP in Figure 9b at 0 dB, and SAL in Figure 9c at 10 dB; note that SAL decays less than IP and UP. The higher power per pixel values corresponds to high spatial correlated scenes, such as IP and SAL, which are composed of agricultural crops, different to UP, which is an urban scenario. This could explain the power value differences.

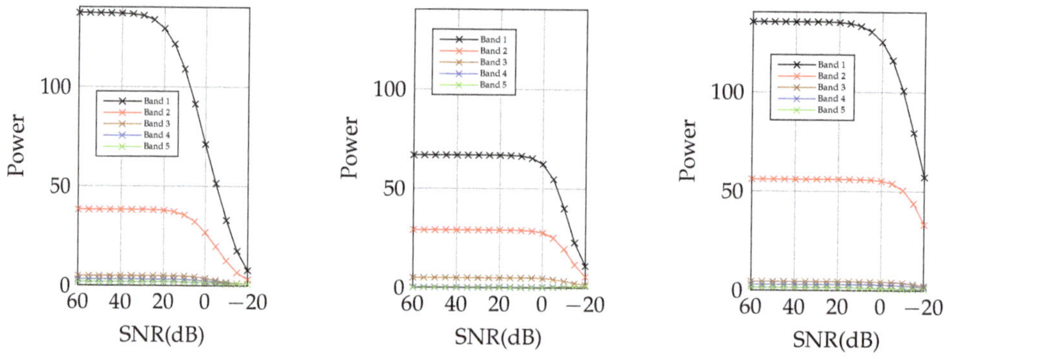

(**a**) IP (**b**) UP (**c**) SAL

Figure 9. First five tensorial bands power from $\mathcal{G}$, after performing TKD from 60 dB to −20 dB, $\alpha = 1$.

To take advantage of the spatial correlation feature extraction of the 3DCNN model, it is necessary to generate a patch with the pixel of interest and its neighbors to train and test the DL model. Each patch $\mathcal{P}$ is composed of 361 pixels ($\mathcal{P} \in \mathbb{R}^{19\times 19\times bands}$), with the pixel of interest in the center ($p_{9,9,i_3}$); if this pixel is associated with a background label, the patch is discarded. The patch is padded with zeros if the pixel is at the edge of the image. Note that the labeled patches contain unlabeled pixels. In Table 8, the running time for patch generation is shown.

Table 8. Running time for patch generation.

Compressed HSI (40 Bands)	Patches Generation Running Time (s)
IP	0.71
UP	18.51
SAL	10.53

We have used scikit-learn [59] for the split of patches in the training and testing sets. The samples are randomly chosen in each experiment with a stratification strategy based on *k*-folds, which returns the stratified folds, where we ensure that the train and test sets have approximately the same percentage of samples of each class available in the ground truth **Y**. In this paper, the size for the training dataset is called the Train Size (TS), and is represented in a percentage from the total labeled samples available. The size for the testing data will be the remaining labels. In Table 9, the number of samples for each case are shown. For the IP case, a smaller set than 5% for the training implies selecting less than one sample for the "Oats" class; for this reason, we do not perform experiments with smaller TS for IP. On the other hand, a bigger TS than 20% for IP or 15% for UP and SAL obtain redundant results and are not aggregated for easy-reading reasons.

Table 9. Number of samples for training and testing sets.

HSI	Set	20%	15%	10%	5%	3%	1%
IP	Training	2049	1537	1024	512	-	-
	Testing	8200	8712	9225	9737	-	-
UP	Training	-	6416	4277	2138	1283	427
	Testing	-	36,360	38,499	40,638	41,493	42,349
SAL	Training	-	8118	5412	2706	1623	541
	Testing	-	46,011	48,717	51,423	52,506	53,588

4.4. 3DCNN Prediction of Data with Variable SNR from 60 to −20 dB

To test the 3DCNN DL model robustness against noise, this experiment considers the model trained with the original data "clean signal", but the prediction is performed for noisy data with different SNR levels and α values. The model was trained for IP, UP, and SAL with a TS of 15%; the results are shown in Figure 10.

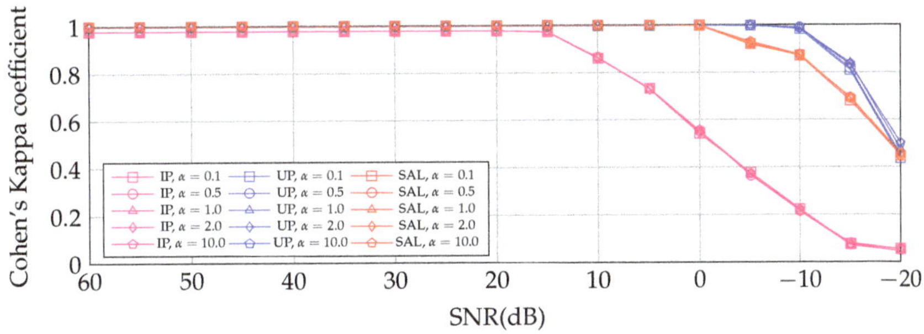

Figure 10. Cohen's Kappa coefficient obtained by the prediction of the 3DCNN DL model over raw noisy data, trained with 15% of the noise-free data of IP, UP, and SAL, without TKD. UP shows the highest Cohen's Kappa coefficient values until −20 dB; after that SAL and IP are the lowest.

This experiment shows that the prediction made by 3DCNN does not change significantly when the signal power is greater than the noise power. For IP, noise affects the prediction for SNR $\leq$ 15 dB. For UP and SAL, noise affects the classifier when the SNR is below 0 dB. The different tested α values do not seem to affect the prediction performance in any particular way.

4.5. Comparison between 3DCNN and Other Classical Algorithms

In this section, Figure 11–13 show a comparison of the performance robustness to high-level noisy data scenarios of spectral (1DCNN), spatial (2DCNN), and spatial–spectral (3DCNN) DL models; additionally, Random Forest (RF) and Support Vector Machine (SVM), which are widely used classifiers, are tested in the same scenario. The training and testing were individually performed for each SNR level with $\alpha = 1$. The results are the average of 10 runs, showing very low variability. Table A1 in Appendix A.2 shows the average Kappa Coefficient with the standard deviation for each experiment.

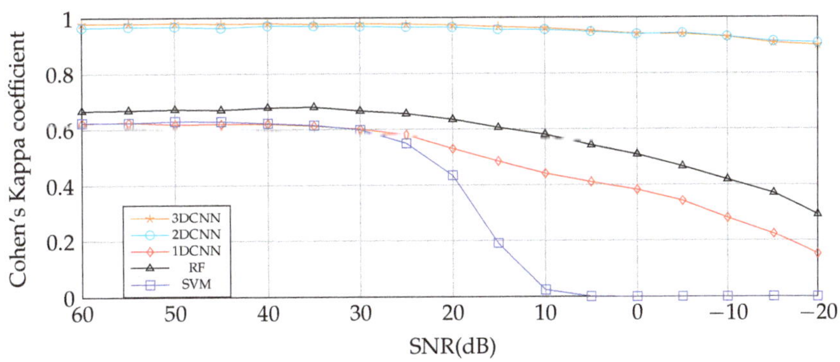

Figure 11. Indian Pines. Cohen's Kappa coefficient obtained by different classifiers for IP HSI compressed with TKD to 40 tensorial bands training with 10% of TS. The best performance is achieved by 3DCNN and 2DCNN, then RF, 1DCNN, and SVM in descending order.

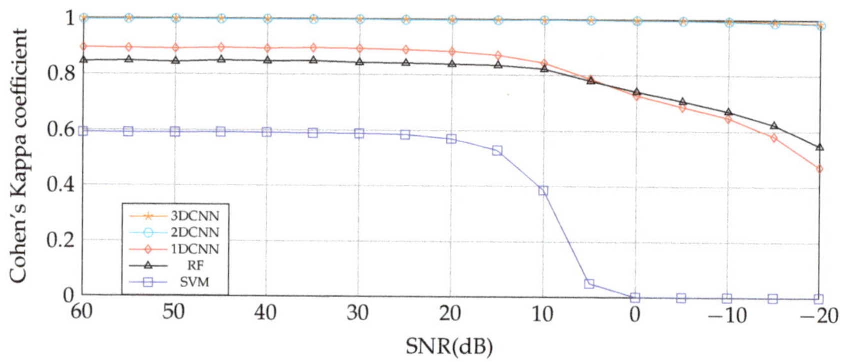

Figure 12. University of Pavia. Cohen's Kappa coefficient obtained by different classifiers for UP HSI compressed with TKD to 40 tensorial bands training with 10% of TS. The best performance is achieved by 3DCNN and 2DCNN, then 1DCNN, RF, and SVM in descending order.

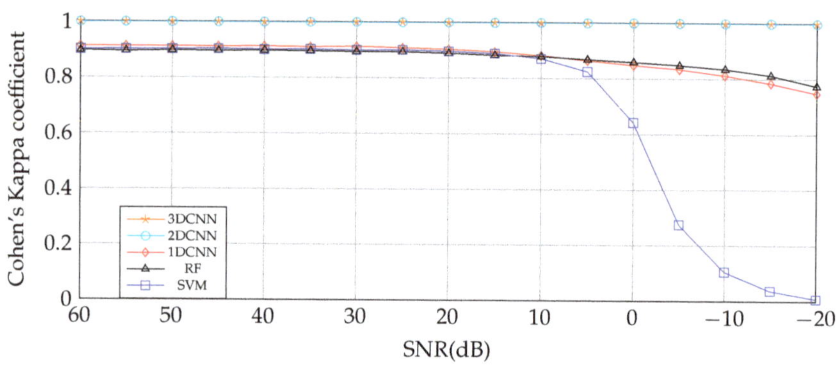

Figure 13. Salinas. Cohen's Kappa coefficient obtained by different classifiers for SAL HSI compressed with TKD to 40 tensorial bands training with 10% of TS. The best performance is achieved by 3DCNN and 2DCNN, then 1DCNN, RF, and SVM show aproximately the same performance above 10 dB, but, below that, SVM was shown to be the worst.

Given the high spatial correlation of the agricultural crops present in IP, the classifiers based on spatial feature extraction achieved better results in all the noise level scenarios. Besides, 3DCNN performs slightly better than 2DCNN at low-level noisy data scenarios; on the other hand, at highly noisy scenarios, 2DCNN performs better. The performance of the spectral-based feature extraction classifiers, 1DCNN, RF, and SVM, is considerably lower than 2DCNN and 3DCNN in all cases, and they are severely affected for SNR ≤ 0 dB.

4.6. *Performance, Computational Complexity, and Training Time Comparison between Original and Compressed Data Using TKD for 3DCNN Model*

The purpose of this section is to show how TKD improves the performance and reduces computational complexity. We have tested the 3DCNN DL model for different noise levels with an equivalent presence of SD and SI noise ($\alpha = 1$). The compression is performed reducing from 200 (IP), 103 (UP), and 204 (SAL) bands, to 40 new tensor bands in the three cases and for a relative reconstruction error less than 1%. The DL model is trained for 40 epochs.

For the Indian Pines HSI with 10249 labeled samples available, we show three training scenarios with 15%, 10%, and 5% for the TS of them in Figure 14.

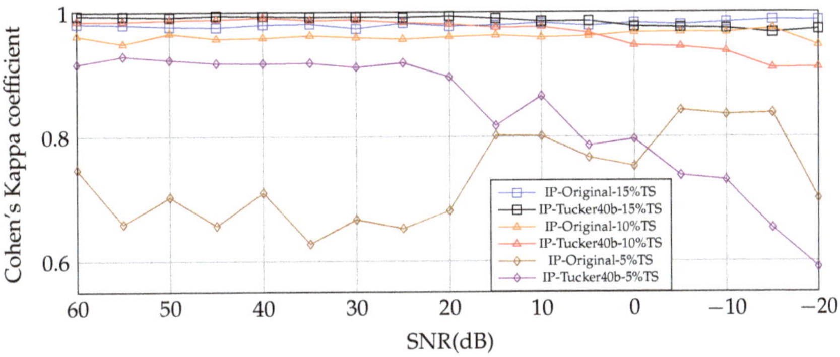

Figure 14. Indian Pines. Cohen's Kappa coefficient obtained using 3DCNN DL model trained for 40 epochs from the original IP HSI and compressed with TKD to 40 tensorial bands. From 60 to −5 dB, Tucker improved the performance for Cohen's Kappa coefficient.

It can be seen that the achieved performance of the DL model training with 15% and 10% of the available labels is always high, even in the high-level noisy scenarios. The Tucker Decomposition improves the classification performance in low- and mid-level noisy scenarios in all the cases. For from 5% to 15% of TS, and from SNR 60 to 0 dB, TKD improves the prediction and is more significant for a TS of 5% than 10% or 15%. The performance achieved, fpr the training with 5% of the labels, is significantly lower than the other two cases, but the TKD remarkably improves the classification performance for low- and mid-level noisy scenarios, while reducing the training time.

For the University of Pavia, HSI with 42,773 labeled samples were available; we show two training scenarios with 5% of TS and 1% in Figure 15.

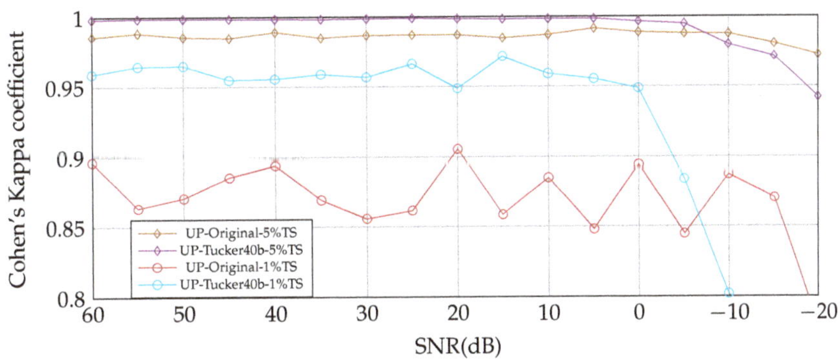

Figure 15. University of Pavia. Cohen's Kappa coefficient obtained using 3DCNN DL model trained for 40 epochs from the original UP HSI and compressed with TKD to 40 tensorial bands. From 60 to −5 dB, Tucker improved the performance for Cohen's Kappa coefficient.

In this case, a consistent improvement of the training is observed with the data compressed by TKD. The improvement increases as the number of available samples for training decreases.

For the Salinas HSI with 54,129 labeled samples available, we show two training scenarios with 5% and 1% of TS in Figure 16.

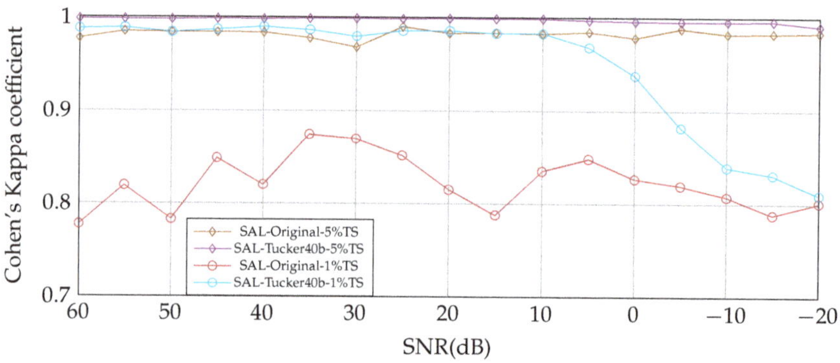

Figure 16. Salinas. Cohen's Kappa coefficient obtained using the 3DCNN DL model trained for 40 epochs from the original SAL HSI and compressed with TKD to 40 tensorial bands. From 60 to −20 dB, Tucker improved the performance for Cohen's Kappa coefficient.

The same behavior is observed in the SAL case, where TKD remarkably improves the performance in all noisy scenarios. Furthermore, the improvement is more notorious with a lower quantity of samples for training.

From Figures 14–16, for 5% of TS until 0 dB of SNR, the lowest score is achieved by IP and the highest by SAL, while UP is in between these two. In these three cases, TKD improves the classification performance in terms of the Cohen's Kappa coefficient. The classification performance of the DL model trained with the compressed HSI by TKD achieves slightly better results at high-SNR levels compared with the original HSI, while, at low-SNR values (close to 0 dB), the performance decreases as the noise power increases. It is important to note that the classification results for the input data compressed by TKD follows the trends in Figure 9, where, for low-SNR, the weight of the contribution of the first tensorial bands of SAL is greater than that corresponding to IP and UP. For IP, the improvement of TKD decreased from SNR ≤ -5 dB; for UP, when SNR ≤ -10 dB, and for SAL, it is always superior. Generally, the aim of compression methods is to reduce the input data size for decreased computational complexity of post-processing algorithms. In this case, TKD not only reduces that complexity but it also improves the classification performance in some cases. The training times of the above experiments are shown in Table 10. Some of them are not displayed in the graphics because of easy-reading reasons (all the log files are available at the public repository for this paper [54]), but all cases follow the same behavior. TKD reduces the original number of bands of each HSI to 40 tensor bands in all the experiments, with a low relative reconstruction error ($\xi \leq 1\%$). The times shown in Table 10 are approximately the average of the experiments presented above, the variation of the training time is insignificant in all the experiments with the same TS.

Table 10. Training time comparison for IP, UP, and SAL datasets.

		15% TS	10% TS	5% TS	1% TS
Indian Pines	Original-200 bands (s)	1889.73	1702.36	1469.47	1284.38
	TKD-40 bands (s)	67.56	63.54	58.43	55.32
	Time reduction ratio	27.97	26.79	25.49	22.56
University of Pavia	Original-103 bands (s)	2987.26	2660.66	2321.35	2041.95
	TKD-40 bands (s)	255.90	239.80	228.65	208.53
	Time reduction ratio	11.67	11.09	10.15	9.79
Salinas	Original-204 bands (s)	9762.00	8783.62	7714.61	6692.29
	TKD-40 bands (s)	328.07	302.23	279.74	266.59
	Time reduction ratio	29.75	29.06	25.57	25.10

4.7. Framework Testing with Datasets for Different α-Values and TS Percentage

The aim of the experiments of this subsection is to show the 3DCNN model performance for different levels and kinds of noise, simulating an extensive set of scenarios for the framework testing. The parameter α controls the dominance of signal-dependent over signal-independent noise (see Equation (22)). The following experiments in Figures 17–19 were performed compressing the HSIs to 40 tensor bands and training the 3DCNN DL model for 40 epochs.

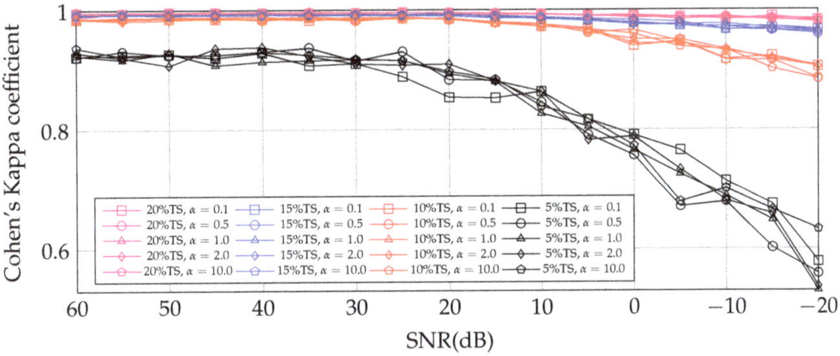

Figure 17. Indian Pines. Cohen's Kappa coefficient obtained using 3DCNN DL model trained with 20%, 15%, 10%, and 5% of the samples of compressed IP HSI for different α values. α-values do not seem to influence Cohen's Kappa coefficient.

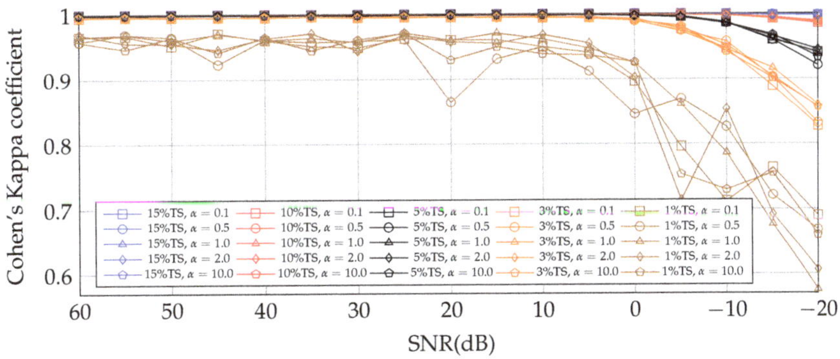

Figure 18. University of Pavia. Cohen's Kappa coefficient obtained using 3DCNN DL model trained with 15%, 10%, 5%, 3%, and 1% of the samples of compressed UP HSI for different α values. α-values do not seem to influence on Cohen's Kappa coefficient.

These experiments show the capabilities of the DL model to extract representative features for all datasets employed in this work. These results have shown that a representative number of samples of each class for training is key for the consistent performance for SNR $\geq$ 0 dB. In terms of TS, for IP TS $\geq$ 10%, UP TS $\geq$ 3%, and SAL TS$\geq$ 3%. The changes in the α values tested, from 0.1 to 10, do not seem to influence the classification performance.

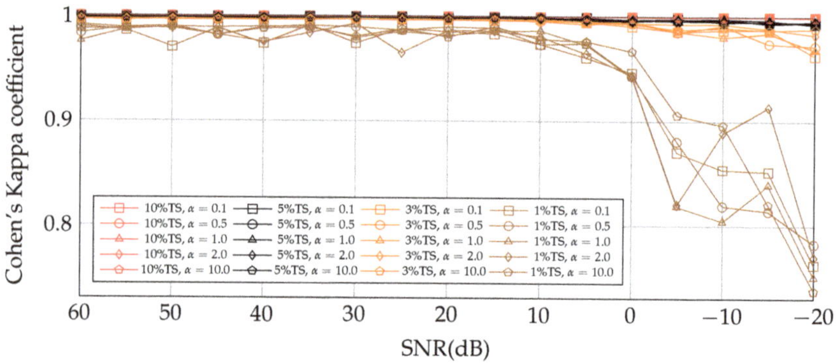

Figure 19. Salinas. Cohen's Kappa coefficient obtained using 3DCNN DL model trained with 10%, 5%, 3%, and 1% of the samples of compressed SAL HSI for different α values. α-values do not seem to influence Cohen's Kappa coefficient.

5. Discussion

In this paper, we provide a classification framework for remote sensing hyperspectral imagery, which allows for adding simulated signal-dependent and signal-independent noise to test their robustness for different SNR values. This kind of framework allows for performing semantic segmentation for noisy hyperspectral images with different SNR values.

The framework is based on a 3DCNN, which is a spectral–spatial deep learning feature extraction model. This method proved to be robust against the training for low-signal-to-noise ratio cases (even when the noise power is greater than the signal power), see Figures 11–13. It is also possible when predicting noisy data from training with noise-free data, such that prediction is affected until the noise power is of the same magnitude as the signal power (SNR close to 0 dB), see Figure 10. However, the computational complexity and resource requirements are higher, compared to other classification algorithms. For that reason, we have implemented spectral compression based on Tucker tensor decomposition, resulting in shorter training times and less hardware resources for implementation.

Tucker Decomposition performs compression correctly until the noise power is of the same magnitude as the signal power, which is a borderline noise case. In most cases, compression, based on TKD, improves the performance of the classifier, see Figures 14–16. This improvement is most noticeable when the model is trained with a set of samples in the order from 5% to 3%.

Since remote sensing images present in nature an unbalanced classification problem, all the results were analyzed primarily using the multi-class unbalanced classification metric, the Cohen's Kappa coefficient, which provides us with a summary of the confusion matrix between the predicted labels and the ground truth of the original image. Three unbalanced hyperspectral images widely studied in the state of the art (described in Tables 4–6) were used to generate the noise and to test the framework: University of Pavia, Salinas, and Indian Pines.

In this way, the presented framework can effectively classify images directly from raw data, with high- and low-signal-to-noise ratios. In the state-of-the-art context, this article includes a detailed analysis for different noisy cases and training with low availability of labeled samples. Our current experiments have demonstrated outstanding results. Although, some related papers use the same datasets, a direct comparison is not fair because a different noise model or SNR are used, with a different number of samples for the classifier training as well as different objectives. The work closest to us is [25], but it is only comparable with the original datasets or high-SNR cases of Figures 14–16, where our approach obtains slightly higher results in terms of the Kappa coefficient for UP (from 0.954 to 0.958) and SAL (from 0.965 to 0.988), but slightly lower results for IP, (from 0.94 to 0.91). For the same original datasets, our approach is competitive with other approaches such as [26–30]. As some of the references present their classification results with the overall accuracy metric only, we prefer to present the results in terms of the Kappa coefficient, because this metric does not hide the imbalanced classification problem.

6. Conclusions

All the results and behaviors can be summarized in the following four conclusions:

- This framework, based on a 3DCNN spectral–spatial deep learning feature extraction model and Tucker Decomposition, proved to be robust in most cases for different combinations and levels of simulated signal-dependent and signal-independent noises, even when the SNR is close to 0 dB.
- Tucker Decomposition reduces from 103 to 224 bands to 40 new tensor bands with $\xi < 1\%$, reducing the computational complexity for the classifier. Different to other compression algorithms, Tucker Decomposition does not affect the performance of the deep learning model; conversely, it improves the classification performance of the 3DCNN deep learning model in the three studied datasets. This improvement is more noticeable for the training set size in the order of from 5% to 3% for the three datasets tested.
- Tucker Decomposition performs well until SNR is close to 0 dB; for SNR ≤ 0 dB, TKD cannot represent the useful information in the core tensor, resulting in an obvious loss of performance.
- With a representative number of labeled samples of each class (depending on the hyperspectral image and accuracy we want), for an SNR $\geq$ of 0 dB, our proposal is not affected by different α-values; in other words, different noisy scenarios of signal-dependent and signal-independent noise.

Open Issues

- To test the spatial–spectral feature extraction of 3DCNN in other types of applications for hyperspectral imagery.
- An algorithm is needed to find the minimum n-rank that fully represents the data into the core tensor, reducing the computational and spatial complexity for posterior stages in the framework.
- To test the framework with a larger number of hyperspectral images, considering distributions of ground truth with less spatial correlation, and for RGB and multispectral imagery.
- Mathematical and statistical analysis of Tucker Decomposition for noisy data.

Author Contributions: Conceptualization, D.T.-R.; methodology, E.P.-Z., D.T.-R. and A.M.-V.; software, E.P.-Z.; validation, E.P.-Z., D.T.-R. and A.M.-V.; formal analysis, E.P.-Z.; investigation, E.P.-Z.; resources, D.T.-R. and A.M.-V.; writing—original draft preparation, E.P.-Z.; writing—review and editing, D.T.-R. and A.M.-V.; visualization, E.P.-Z.; supervision, D.T.-R. and A.M.-V.; project administration, E.P.-Z., D.T.-R. and A.M.-V.; funding acquisition, D.T.-R. and A.M.-V. All authors have read and agreed to the published version of the manuscript.

Funding: This research was funded by Consejo Nacional de Ciencia y Tecnología grant numbers 717754 and 789304.

Data Availability Statement: Restrictions apply to the availability of these data. Data was obtained from [2] and are available in github.com/mhaut, https://github.com/mhaut/hyperspectral_deeplearning_review(accessed on 10 March 2022).

Acknowledgments: We thank Cinvestav-Guadalajara, Mexico, for the master's studies of E. Padilla, first author of this work, and personally thank J. López for the help provided for the implementation of TKD for this work.

Conflicts of Interest: The authors declare no conflict of interest.

Abbreviations

The following abbreviations are used in this manuscript:

CNN	Convolutional Neural Network.
DL	Deep Learning.
GT	Ground Truth.
HPC	High Performance Computing.
HSI	Hyperspectral Image.
IP	Indian Pines.
PCA	Principal Component Analysis.
SAL	Salinas.
SD	Signal Dependent.
SI	Signal Independent.
SNR	Signal-to-Noise Ratio.
TKD	Tucker Decomposition.
TS	Train Size.
UP	University of Pavia.

Appendix A

Appendix A.1

In this appendix, the variance calculations for noise generation used in this paper are explained, which was formulated on [4,39]. First of all, to obtain the noise variances of the random processes σ^2_{u,i_3} and σ^2_{t,i_3}, the mean variance of the noise tensors $\mathcal{N}_{SD}(\mathcal{X})$ and $\mathcal{N}_{SI}$ [39] are required. For a signal-dependent mean noise variance tensor:

$$\sigma^2_{\mathcal{N}_{SD}(\mathcal{X})} = \frac{1}{I_1 I_2 I_3} \sum_{i_1=1}^{I_1} \sum_{i_2=1}^{I_2} \sum_{i_3=1}^{I_3} \sigma^2_{u,i_3} \cdot x_{i_1,i_2,i_3}. \tag{A1}$$

Let μ_{i_3} be the mean of the clean signal at band i_3:

$$\mu_{i_3} = \frac{1}{I_1 I_2} \sum_{i_1=1}^{I_1} \sum_{i_2=1}^{I_2} x_{i_1,i_2,i_3}, \tag{A2}$$

from (A2), Equation (A1) can be rewritten as:

$$\sigma^2_{\mathcal{N}_{SD}(\mathcal{X})} = \frac{1}{I_3} \sum_{i_3=1}^{I_3} \sigma^2_{u,i_3} \cdot \mu_{i_3}; \tag{A3}$$

additionally, the signal-independent noise has constant variance $\sigma^2_{\mathcal{N}_{SI}}$ in all bands. The signal-independent mean variance noise tensor is:

$$\sigma^2_{\mathcal{N}_{SI}} = \frac{1}{I_1 I_2 I_3} \sum_{i_3=1}^{I_3} \sigma^2_{t,i_3}; \tag{A4}$$

thus, using Equation (19), the mean variance of the noise tensor $\mathcal{N}(\mathcal{X})$ is:

$$\sigma^2_{\mathcal{N}(\mathcal{X})} = \sigma^2_{\mathcal{N}_{SD}(\mathcal{X})} + \sigma^2_{\mathcal{N}_{SI}}, \tag{A5}$$

$$\sigma^2_{\mathcal{N}(\mathcal{X})} = \frac{1}{I_1 I_2 I_3} \sum_{i_1=1}^{I_1} \sum_{i_2=1}^{I_2} \sum_{i_3=1}^{I_3} \left(\sigma^2_{u,i_3} \cdot x_{i_1,i_2,i_3} + \sigma^2_{t,i_3} \right). \tag{A6}$$

From the SNR (dB) formula:

$$\mathrm{SNR} = 10 \cdot \log_{10} \frac{\|\mathcal{X}\|^2}{\|\mathcal{N}(\mathcal{X})\|^2}, \tag{A7}$$

$\|\mathcal{N}(\mathcal{X})\|^2$ in terms of $\mathcal{X}$ and a specified SNR is expressed as:

$$\|\mathcal{N}(\mathcal{X})\|^2 = \|\mathcal{X}\|^2 \cdot 10^{-\left(\frac{\text{SNR}}{10}\right)}. \tag{A8}$$

If Equation (A8) is divided by the total number of pixels $I_1 I_2 I_3$, note that $\sigma^2_{\mathcal{N}(\mathcal{X})} = \frac{\|\mathcal{N}(\mathcal{X})\|^2}{I_1 I_2 I_3}$ (see Equation (A6)). If $\overline{P}_{\mathcal{X}} = \frac{\|\mathcal{X}\|^2}{I_1 I_2 I_3}$ is the mean power of tensor $\mathcal{X}$, then:

$$\sigma^2_{\mathcal{N}(\mathcal{X})} = \overline{P}_{\mathcal{X}} \cdot 10^{-\left(\frac{\text{SNR}}{10}\right)}. \tag{A9}$$

Assuming a parameter α, which controls the dominance of the signal-dependent noise variance over the signal-independent noise variance, such that:

$$\alpha = \frac{\sigma^2_{\mathcal{N}_{SD}(\mathcal{X})}}{\sigma^2_{\mathcal{N}_{SI}}}, \tag{A10}$$

Then, from Equations (A10) and (A5), follows:

$$\sigma^2_{\mathcal{N}_{SD}(\mathcal{X})} = \frac{\sigma^2_{\mathcal{N}(\mathcal{X})} \cdot \alpha}{\alpha + 1}, \tag{A11}$$

and:

$$\sigma^2_{\mathcal{N}_{SI}} = \frac{\sigma^2_{\mathcal{N}(\mathcal{X})}}{\alpha + 1}. \tag{A12}$$

Note that both results depend only on α and $\sigma^2_{\mathcal{N}(\mathcal{X})}$, which are already available in Equations (A9) and (A10). Finally, solving for the noise variance of the random process σ^2_{u,i_3} from Equation (A1):

$$\sigma^2_{u,i_3} = \frac{\sigma^2_{\mathcal{N}_{SD}(\mathcal{X})}}{\mu_{i_3}}; \tag{A13}$$

as well, the noise variance of the random process σ^2_{t,i_3}, from Equation (A3):

$$\sigma^2_{t,i_3} = \sigma^2_{\mathcal{N}_{SI}}. \tag{A14}$$

Appendix A.2

Table A1. Average Cohen's Kappa coefficient of 10 runs obtained using different classifiers for IP, UP, and SAL compressed with TKD to 40 tensorial bands training with 10% of TS. Standard deviation shows very low variability for high-SNR cases, growing as the noise variance increases.

	SNR	SVM	RF	1DCNN	2DCNN	3DCNN
Indian Pines	**60 dB**	0.6244 ± 0.0068	0.6664 ± 0.0111	0.6231 ± 0.0113	0.9654 ± 0.0059	**0.9802** ± 0.0029
	55 dB	0.6241 ± 0.0087	0.6682 ± 0.0068	0.6250 ± 0.0099	0.9683 ± 0.0024	**0.9803** ± 0.0024
	50 dB	0.6303 ± 0.0066	0.6716 ± 0.0105	0.6186 ± 0.0104	0.9690 ± 0.0078	**0.9817** ± 0.0040
	45 dB	0.6289 ± 0.0090	0.6697 ± 0.0068	0.6193 ± 0.0086	0.9641 ± 0.0059	**0.9793** ± 0.0056
	40 dB	0.6223 ± 0.0029	0.6780 ± 0.0112	0.6203 ± 0.0139	0.9720 ± 0.0045	**0.9799** ± 0.0050
	35 dB	0.6148 ± 0.0103	0.6795 ± 0.0075	0.6133 ± 0.0103	0.9712 ± 0.0045	**0.9781** ± 0.0029
	30 dB	0.5985 ± 0.0105	0.6660 ± 0.0111	0.6008 ± 0.0061	0.9692 ± 0.0052	**0.9793** ± 0.0035
	25 dB	0.5501 ± 0.0100	0.6553 ± 0.0107	0.5803 ± 0.0093	0.9665 ± 0.0043	**0.9773** ± 0.0033
	20 dB	0.4354 ± 0.0066	0.6346 ± 0.0092	0.5303 ± 0.0126	0.9660 ± 0.0049	**0.9737** ± 0.0059
	15 dB	0.1919 ± 0.0093	0.6054 ± 0.0094	0.4856 ± 0.0108	0.9574 ± 0.0060	**0.9673** ± 0.0043
	10 dB	0.0248 ± 0.0037	0.5808 ± 0.0063	0.4419 ± 0.0131	0.9573 ± 0.0068	**0.9628** ± 0.0035
	5 dB	0.0006 ± 0.0003	0.5424 ± 0.0069	0.4106 ± 0.0063	0.9493 ± 0.0060	**0.9522** ± 0.0061
	0 dB	0.0000 ± 0.0001	0.5086 ± 0.0069	0.3831 ± 0.0128	0.9406 ± 0.0065	**0.9425** ± 0.0081
	−5 dB	0.0000 ± 0.0000	0.4664 ± 0.0083	0.3433 ± 0.0120	**0.9443** ± 0.0092	0.9404 ± 0.0068
	−10 dB	0.0000 ± 0.0000	0.4191 ± 0.0039	0.2826 ± 0.0130	**0.9318** ± 0.0096	0.9299 ± 0.0042
	−15 dB	0.0000 ± 0.0000	0.3700 ± 0.0081	0.2246 ± 0.0085	**0.9143** ± 0.0099	0.9083 ± 0.0079
	−20 dB	0.0000 ± 0.0000	0.2938 ± 0.0069	0.1534 ± 0.0099	**0.9091** ± 0.0164	0.9002 ± 0.0079
University of Pavia	**60 dB**	0.5909 ± 0.0048	0.8461 ± 0.0036	0.8946 ± 0.0045	0.9973 ± 0.0009	**0.9994** ± 0.0003
	55 dB	0.5900 ± 0.0043	0.8484 ± 0.0037	0.8935 ± 0.0044	0.9972 ± 0.0004	**0.9994** ± 0.0003
	50 dB	0.5904 ± 0.0041	0.8444 ± 0.0076	0.8920 ± 0.0061	0.9977 ± 0.0009	**0.9995** ± 0.0005
	45 dB	0.5911 ± 0.0033	0.8502 ± 0.0050	0.8948 ± 0.0052	0.9977 ± 0.0004	**0.9995** ± 0.0004
	40 dB	0.5906 ± 0.0040	0.8475 ± 0.0089	0.8924 ± 0.0042	0.9978 ± 0.0006	**0.9994** ± 0.0003
	35 dB	0.5890 ± 0.0050	0.8489 ± 0.0041	0.8947 ± 0.0069	0.9976 ± 0.0008	**0.9996** ± 0.0002
	30 dB	0.5880 ± 0.0034	0.8431 ± 0.0061	0.8931 ± 0.0050	0.9977 ± 0.0005	**0.9994** ± 0.0002
	25 dB	0.5841 ± 0.0028	0.8415 ± 0.0040	0.8897 ± 0.0048	0.9967 ± 0.0011	**0.9992** ± 0.0004
	20 dB	0.5692 ± 0.0041	0.8380 ± 0.0051	0.8840 ± 0.0083	0.9972 ± 0.0013	**0.9995** ± 0.0003
	15 dB	0.5281 ± 0.0016	0.8350 ± 0.0037	0.8713 ± 0.0035	0.9972 ± 0.0008	**0.9996** ± 0.0002
	10 dB	0.3853 ± 0.0049	0.8224 ± 0.0032	0.8438 ± 0.0049	0.9977 ± 0.0008	**0.9993** ± 0.0003
	5 dB	0.0492 ± 0.0021	0.7800 ± 0.0036	0.7870 ± 0.0045	0.9973 ± 0.0004	**0.9993** ± 0.0003
	0 dB	0.0000 ± 0.0000	0.7421 ± 0.0035	0.7274 ± 0.0056	0.9961 ± 0.0011	**0.9989** ± 0.0005
	−5 dB	0.0000 ± 0.0000	0.7077 ± 0.0028	0.6873 ± 0.0030	0.9954 ± 0.0011	**0.9977** ± 0.0012
	−10 dB	0.0000 ± 0.0000	0.6716 ± 0.0032	0.6478 ± 0.0039	0.9934 ± 0.0018	**0.9966** ± 0.0009
	−15 dB	0.0000 ± 0.0000	0.6241 ± 0.0030	0.5806 ± 0.0069	0.9894 ± 0.0013	**0.9930** ± 0.0012
	−20 dB	0.0000 ± 0.0000	0.5463 ± 0.0026	0.4703 ± 0.0084	0.9853 ± 0.0028	**0.9874** ± 0.0017
Salinas	**60 dB**	0.9010 ± 0.0021	0.8971 ± 0.0023	0.9131 ± 0.0036	0.9996 ± 0.0004	**0.9999** ± 0.0001
	55 dB	0.9017 ± 0.0018	0.8954 ± 0.0026	0.9125 ± 0.0030	0.9995 ± 0.0002	**0.9999** ± 0.0001
	50 dB	0.9016 ± 0.0025	0.8977 ± 0.0040	0.9126 ± 0.0025	0.9994 ± 0.0004	**0.9999** ± 0.0001
	45 dB	0.9029 ± 0.0021	0.8967 ± 0.0019	0.9114 ± 0.0031	0.9995 ± 0.0002	**0.9999** ± 0.0001
	40 dB	0.9006 ± 0.0019	0.8975 ± 0.0020	0.9129 ± 0.0025	0.9995 ± 0.0002	**0.9998** ± 0.0001
	35 dB	0.9020 ± 0.0017	0.8968 ± 0.0021	0.9106 ± 0.0041	0.9995 ± 0.0003	**0.9999** ± 0.0001
	30 dB	0.8993 ± 0.0019	0.8944 ± 0.0019	0.9122 ± 0.0023	0.9995 ± 0.0003	**0.9999** ± 0.0001
	25 dB	0.9003 ± 0.0019	0.8940 ± 0.0028	0.9076 ± 0.0038	0.9995 ± 0.0001	**0.9999** ± 0.0001
	20 dB	0.8961 ± 0.0016	0.8892 ± 0.0017	0.9039 ± 0.0026	0.9995 ± 0.0003	**0.9999** ± 0.0001
	15 dB	0.8881 ± 0.0012	0.8828 ± 0.0026	0.8947 ± 0.0029	0.9994 ± 0.0003	**0.9998** ± 0.0001
	10 dB	0.8699 ± 0.0015	0.8777 ± 0.0015	0.8815 ± 0.0034	0.9996 ± 0.0002	**0.9998** ± 0.0001
	5 dB	0.8234 ± 0.0025	0.8677 ± 0.0022	0.8634 ± 0.0041	0.9993 ± 0.0003	**0.9998** ± 0.0001
	0 dB	0.6413 ± 0.0035	0.8589 ± 0.0020	0.8468 ± 0.0025	0.9994 ± 0.0003	**0.9997** ± 0.0001
	−5 dB	0.2741 ± 0.0100	0.8474 ± 0.0025	0.8324 ± 0.0031	0.9992 ± 0.0003	**0.9998** ± 0.0001
	−10 dB	0.1056 ± 0.0013	0.8344 ± 0.0020	0.8114 ± 0.0029	0.9993 ± 0.0003	**0.9995** ± 0.0003
	−15 dB	0.0377 ± 0.0005	0.8121 ± 0.0015	0.7837 ± 0.0039	0.9994 ± 0.0003	**0.9997** ± 0.0001
	−20 dB	0.0079 ± 0.0005	0.7743 ± 0.0027	0.7478 ± 0.0046	0.9989 ± 0.0004	**0.9996** ± 0.0002

References

1. Borengasser, M.; Hungate, W.S.; Watkins, R.L. *Hyperspectral Remote Sensing: Principles and Applications*; CRC Press: Boca Raton, FL, USA, 2008; p. 119.
2. Paoletti, M.E.; Haut, J.M.; Plaza, J.; Plaza, A. Deep learning classifiers for hyperspectral imaging: A review. *ISPRS J. Photogramm. Remote Sens.* **2019**, *158, 279–317.* [CrossRef]
3. Rasti, B.; Scheunders, P.; Ghamisi, P.; Licciardi, G.; Chanussot, J. Noise Reduction in Hyperspectral Imagery: Overview and Application. *Remote Sens.* **2018**, *10*, 482. [CrossRef]
4. Bourennane, S.; Fossati, C.; Lin, T. Noise Removal Based on Tensor Modelling for Hyperspectral Image Classification. *Remote Sens.* **2018**, *10*, 1330. [CrossRef]
5. Gu, S.; Zhang, L.; Zuo, W.; Feng, X. Weighted nuclear norm minimization with application to image denoising. In Proceedings of the IEEE Computer Society Conference on Computer Vision and Pattern Recognition, Columbus, OH, USA, 23–28 June 2014; IEEE Computer Society: Washington, DC, USA, 2014; pp. 2862–2869. [CrossRef]
6. Karami, A.; Yazdi, M.; Zolghadre Asli, A. Noise reduction of hyperspectral images using kernel non-negative tucker decomposition. *IEEE J. Sel. Top. Signal Process.* **2011**, *5*, 487–493. [CrossRef]
7. Yuan, Q.; Zhang, Q.; Li, J.; Shen, H.; Zhang, L. Hyperspectral image denoising employing a spatial-spectral deep residual convolutional neural network. *IEEE Trans. Geosci. Remote Sens.* **2019**, *57*, 1205–1218. [CrossRef]
8. Fan, H.; Li, C.; Guo, Y.; Kuang, G.; Ma, J. Spatial-Spectral Total Variation Regularized Low-Rank Tensor Decomposition for Hyperspectral Image Denoising. *IEEE Trans. Geosci. Remote Sens.* **2018**, *56*, 6196–6213. [CrossRef]
9. Huang, Z.; Li, S.; Fang, L.; Li, H.; Benediktsson, J.A. Hyperspectral Image Denoising with Group Sparse and Low-Rank Tensor Decomposition. *IEEE Access* **2017**, *6*, 1380–1390. [CrossRef]
10. Chen, Y.; Jiang, H.; Li, C.; Jia, X.; Ghamisi, P. Deep Feature Extraction and Classification of Hyperspectral Images Based on Convolutional Neural Networks. *IEEE Trans. Geosci. Remote Sens.* **2016**, *54*, 6232–6251. [CrossRef]
11. Hu, J.; Li, L.; Lin, Y.; Wu, F.; Zhao, J. A Comparison and Strategy of Semantic Segmentation on Remote Sensing Images. *Adv. Intell. Syst. Comput.* **2019**, *1074*, 21–29. [CrossRef]
12. Niu, Z.; Liu, W.; Zhao, J.; Jiang, G. DeepLab-Based Spatial Feature Extraction for Hyperspectral Image Classification. *IEEE Geosci. Remote Sens. Lett.* **2019**, *16*, 251–255. [CrossRef]
13. Zhong, Z.; Li, J.; Ma, L.; Jiang, H.; Zhao, H. Deep residual networks for hyperspectral image classification. In Proceedings of the International Geoscience and Remote Sensing Symposium (IGARSS), Fort Worth, TX, USA, 23–28 July 2017; Institute of Electrical and Electronics Engineers Inc.: Piscataway, NJ, USA, 2017; pp. 1824–1827. [CrossRef]
14. Feng, J.; Yu, H.; Wang, L.; Cao, X.; Zhang, X.; Jiao, L. Classification of Hyperspectral Images Based on Multiclass Spatial-Spectral Generative Adversarial Networks. *IEEE Trans. Geosci. Remote Sens.* **2019**, *57*, 5329–5343. [CrossRef]
15. Hu, W.; Huang, Y.; Wei, L.; Zhang, F.; Li, H. Deep convolutional neural networks for hyperspectral image classification. *J. Sens.* **2015**, *2015*, 258619. [CrossRef]
16. Xiao, H.; Wei, Y.; Liu, Y.; Zhang, M.; Feng, J. Transferable Semi-Supervised Semantic Segmentation. *Proc. AAAI Conf. Artif. Intell.* **2018**, *32*, 7420–7427. [CrossRef]
17. Sun, R.; Zhu, X.; Wu, C.; Huang, C.; Shi, J.; Ma, L. Not all areas are equal: Transfer learning for semantic segmentation via hierarchical region selection. In Proceedings of the IEEE Computer Society Conference on Computer Vision and Pattern Recognition, Long Beach, CA, USA, 15–19 June 2019; pp. 4355–4364. [CrossRef]
18. Stan, S.; Rostami, M. Unsupervised Model Adaptation for Continual Semantic Segmentation. *Proc. AAAI Conf. Artif. Intell.* **2021**, *35*, 2593–2601. [CrossRef]
19. Sun, J.; Wei, D.; Ma, K.; Wang, L.; Zheng, Y. Boost Supervised Pretraining for Visual Transfer Learning: Implications of Self-Supervised Contrastive Representation Learning. *Proc. AAAI Conf. Artif. Intell.* **2022**, *36*, 2307–2315. [CrossRef]
20. Cui, B.; Chen, X.; Lu, Y. Semantic Segmentation of Remote Sensing Images Using Transfer Learning and Deep Convolutional Neural Network with Dense Connection. *IEEE Access* **2020**, *8*, 116744–116755. [CrossRef]
21. Ronneberger, O.; Fischer, P.; Brox, T. U-Net: Convolutional Networks for Biomedical Image Segmentation. In *Medical Image Computing and Computer-Assisted Intervention—MICCAI 2015*; Lecture Notes in Computer Science (Including Subseries Lecture Notes in Artificial Intelligence and Lecture Notes in Bioinformatics); Springer: Cham, Switzerland, 2015; Volume 9351, pp. 234–241. [CrossRef]
22. Pasquali, G.; Iannelli, G.C.; Dell'Acqua, F. Building Footprint Extraction from Multispectral, Spaceborne Earth Observation Datasets Using a Structurally Optimized U-Net Convolutional Neural Network. *Remote Sens.* **2019**, *11*, 2803. [CrossRef]
23. Wieland, M.; Li, Y.; Martinis, S. Multi-sensor cloud and cloud shadow segmentation with a convolutional neural network. *Remote Sens. Environ.* **2019**, *230*, 111203. [CrossRef]
24. Giang, T.L.; Dang, K.B.; Le, Q.T.; Nguyen, V.G.; Tong, S.S.; Pham, V.M. U-net convolutional networks for mining land cover classification based on high-resolution UAV imagery. *IEEE Access* **2020**, *8*, 186257–186273. [CrossRef]
25. Fu, H.; Zhang, A.; Sun, G.; Ren, J.; Jia, X.; Pan, Z.; Ma, H. A Novel Band Selection and Spatial Noise Reduction Method for Hyperspectral Image Classification. *IEEE Trans. Geosci. Remote Sens.* **2022**, *60*, 1–13. [CrossRef]
26. Prasad, S.; Li, W.; Fowler, J.E.; Bruce, L.M. Information fusion in the redundant-wavelet-transform domain for noise-robust hyperspectral classification. *IEEE Trans. Geosci. Remote Sens.* **2012**, *50*, 3474–3486. [CrossRef]

27. Duan, P.; Kang, X.; Li, S.; Ghamisi, P. Noise-robust hyperspectral image classification via multi-scale total variation. *IEEE J. Sel. Top. Appl. Earth Obs. Remote Sens.* **2019**, *12*, 1948–1962. [CrossRef]
28. Gong, Z.; Zhong, P.; Yao, W.; Zhou, W.; Qi, J.; Hu, P. A CNN with noise inclined module and denoise framework for hyperspectral image classification. *IET Image Process.* **2022**. [CrossRef]
29. Chen, C.; Li, W.; Tramel, E.W.; Cui, M.; Prasad, S.; Fowler, J.E. Spectral–Spatial Preprocessing Using Multihypothesis Prediction for Noise-Robust Hyperspectral Image Classification. *IEEE J. Sel. Top. Appl. Earth Obs. Remote Sens.* **2014**, *7*, 1047–1059. [CrossRef]
30. Gao, L.; Zhao, B.; Jia, X.; Liao, W.; Zhang, B.; Wang, Q.; Younan, N.H.; López-Martínez, C.; Thenkabail, P.S. Optimized Kernel Minimum Noise Fraction Transformation for Hyperspectral Image Classification. *Remote Sens.* **2017**, *9*, 548. [CrossRef]
31. Fu, P.; Sun, X.; Sun, Q. Hyperspectral Image Segmentation via Frequency-Based Similarity for Mixed Noise Estimation. *Remote Sens.* **2017**, *9*, 1237. [CrossRef]
32. de Los Reyes, R.; Langheinrich, M.; Schwind, P.; Richter, R.; Pflug, B.; Bachmann, M.; Müller, R.; Carmona, E.; Zekoll, V.; Reinartz, P. PACO: Python-Based Atmospheric COrrection. *Sensors* **2020**, *20*, 1428. [CrossRef]
33. Zekoll, V.; Main-Knorn, M.; Alonso, K.; Louis, J.; Frantz, D.; Richter, R.; Pflug, B. Comparison of Masking Algorithms for Sentinel-2 Imagery. *Remote Sens.* **2021**, *13*, 137. [CrossRef]
34. Guanter, L.; Kaufmann, H.; Segl, K.; Foerster, S.; Rogass, C.; Chabrillat, S.; Kuester, T.; Hollstein, A.; Rossner, G.; Chlebek, C.; et al. The EnMAP Spaceborne Imaging Spectroscopy Mission for Earth Observation. *Remote Sens.* **2015**, *7*, 8830–8857. [CrossRef]
35. Alonso, K.; Bachmann, M.; Burch, K.; Carmona, E.; Cerra, D.; de los Reyes, R.; Dietrich, D.; Heiden, U.; Hölderlin, A.; Ickes, J.; et al. Data Products, Quality and Validation of the DLR Earth Sensing Imaging Spectrometer (DESIS). *Sensors* **2019**, *19*, 4471. [CrossRef]
36. López, J.; Torres, D.; Santos, S.; Atzberger, C. Spectral Imagery Tensor Decomposition for Semantic Segmentation of Remote Sensing Data through Fully Convolutional Networks. *Remote Sens.* **2020**, *12*, 517. [CrossRef]
37. Padilla-Zepeda, E.; Torres-Roman, D.; Mendez-Vazquez, A. Noise analysis using Tucker decomposition and PCA on spectral images. *ECORFAN J.-Boliv.* **2020**, *7*, 10–16. [CrossRef]
38. Kolda, T.G.; Bader, B.W. Tensor decompositions and applications. *SIAM Rev.* **2009**, *51*, 455–500. [CrossRef]
39. Liu, X.; Bourennane, S.; Fossati, C. Reduction of signal-dependent noise from hyperspectral images for target detection. *IEEE Trans. Geosci. Remote Sens.* **2014**, *52*, 5396–5411. [CrossRef]
40. Li, Y.; Zhang, H.; Shen, Q. Spectral–Spatial Classification of Hyperspectral Imagery with 3D Convolutional Neural Network. *Remote Sens.* **2017**, *9*, 67. [CrossRef]
41. Grandini, M.; Bagli, E.; Visani, G. Metrics for Multi-Class Classification: an Overview. *arXiv* **2020**, arXiv:2008.05756.
42. Luque, A.; Carrasco, A.; Martín, A.; de las Heras, A. The impact of class imbalance in classification performance metrics based on the binary confusion matrix. *Pattern Recognit.* **2019**, *91*, 216–231. [CrossRef]
43. Gonzalez-Ramirez, A.; Lopez, J.; Torres-Roman, D.; Yañez-Vargas, I. Analysis of multi-class classification performance metrics for remote sensing imagery imbalanced datasets. *ECORFAN J. Quant. Stat. Anal.* **2021**, *8*, 11–17. [CrossRef]
44. Makantasis, K.; Doulamis, A.D.; Doulamis, N.D.; Nikitakis, A. Tensor-based classification models for hyperspectral data analysis. *IEEE Trans. Geosci. Remote Sens.* **2018**, *56*, 6884–6898. [CrossRef]
45. Sidiropoulos, N.D.; De Lathauwer, L.; Fu, X.; Huang, K.; Papalexakis, E.E.; Faloutsos, C. Tensor Decomposition for Signal Processing and Machine Learning. *IEEE Trans. Signal Process.* **2017**, *65*, 3551–3582. [CrossRef]
46. An, J.; Lei, J.; Song, Y.; Zhang, X.; Guo, J. Tensor Based Multiscale Low Rank Decomposition for Hyperspectral Images Dimensionality Reduction. *Remote Sens.* **2019**, *11*, 1485. [CrossRef]
47. Kong, X.; Zhao, Y.; Xue, J.; Chan, J.C.W. Hyperspectral Image Denoising Using Global Weighted Tensor Norm Minimum and Nonlocal Low-Rank Approximation. *Remote Sens.* **2019**, *11*, 2281. [CrossRef]
48. Lu, H.; Plataniotis, K.N.; Venetsanopoulos, A.N. MPCA: Multilinear principal component analysis of tensor objects. *IEEE Trans. Neural Netw.* **2008**, *19*, 18–39. [CrossRef] [PubMed]
49. AVIRIS—eoPortal Directory—Airborne Sensors. Available online: https://aviris.jpl.nasa.gov/ (accessed on 15 June 2020).
50. Tucker, L.R. Some mathematical notes on three-mode factor analysis. *Psychometrika* **1966**, *31*, 279–311. [CrossRef] [PubMed]
51. Alparone, L.; Selva, M.; Aiazzi, B.; Baronti, S.; Butera, F.; Chiarantini, L. Signal-dependent noise modelling and estimation of new-generation imaging spectrometers. In Proceedings of the WHISPERS '09—1st Workshop on Hyperspectral Image and Signal Processing: Evolution in Remote Sensing, Grenoble, France, 26–28 August 2009. [CrossRef]
52. Faraji, H.; MacLean, W.J. CCD noise removal in digital images. *IEEE Trans. Image Process.* **2006**, *15*, 2676–2685. [CrossRef]
53. Jain, A.K. *Fundamentals of Digital Image Processing*; Prentice-Hall, Inc.: Hoboken, NJ, USA, 1989.
54. Padilla-Zepeda, E. Noisy-Hyperspectral-Semantic-Segmentation-Framework-Based-on-Tucker-Decomposition-and-3D-CNN. 2022. Available online: https://github.com/EfrainPadilla/Noisy-Hyperspectral-Semantic-Segmentation-Framework-based-on-Tucker-Decomposition-and-3D-CNN (accessed on 10 March 2022).
55. Green, R.O.; Eastwood, M.L.; Sarture, C.M.; Chrien, T.G.; Aronsson, M.; Chippendale, B.J.; Faust, J.A.; Pavri, B.E.; Chovit, C.J.; Solis, M.; et al. Imaging spectroscopy and the Airborne Visible/Infrared Imaging Spectrometer (AVIRIS). *Remote Sens. Environ.* **1998**, *65*, 227–248. [CrossRef]
56. Kunkel, B.; Blechinger, F.; Lutz, R.; Doerffer, R.; van der Piepen, H.; Schroder, M. ROSIS (Reflective Optics System Imaging Spectrometer) - A Candidate Instrument For Polar Platform Missions. *Optoelectron. Technol. Remote Sens. Space SPIE* **1988**, *868*, 134. [CrossRef]

57. Harris, C.R.; Millman, K.J.; van der Walt, S.J.; Gommers, R.; Virtanen, P.; Cournapeau, D.; Wieser, E.; Taylor, J.; Berg, S.; Smith, N.J.; et al. Array programming with NumPy. *Nature* **2020**, *585*, 357–362. [CrossRef]
58. Kossaifi, J.; Panagakis, Y.; Anandkumar, A.; Pantic, M. TensorLy: Tensor Learning in Python. *J. Mach. Learn. Res.* **2019**, *20*, 1–6.
59. Pedregosa, F.; Varoquaux, G.; Gramfort, A.; Michel, V.; Thirion, B.; Grisel, O.; Blondel, M.; Prettenhofer, P.; Weiss, R.; Dubourg, V.; et al. Scikit-learn: Machine Learning in Python. *J. Mach. Learn. Res.* **2011**, *12*, 2825–2830.

remote sensing

MDPI

Article

MCAFNet: A Multiscale Channel Attention Fusion Network for Semantic Segmentation of Remote Sensing Images

Min Yuan [1,*], Dingbang Ren [1], Qisheng Feng [2], Zhaobin Wang [1], Yongkang Dong [1], Fuxiang Lu [1] and Xiaolin Wu [1]

1 School of Information Science and Engineering, Lanzhou University, Lanzhou 730000, China
2 College of Pastoral Agriculture Science and Technology, Lanzhou University, Lanzhou 730000, China
* Correspondence: yuanm@lzu.edu.cn

Abstract: Semantic segmentation for urban remote sensing images is one of the most-crucial tasks in the field of remote sensing. Remote sensing images contain rich information on ground objects, such as shape, location, and boundary and can be found in high-resolution remote sensing images. It is exceedingly challenging to identify remote sensing images because of the large intraclass variance and low interclass variance caused by these objects. In this article, we propose a multiscale hierarchical channel attention fusion network model based on a transformer and CNN, which we name the multiscale channel attention fusion network (MCAFNet). MCAFNet uses ResNet-50 and Vit-B/16 to learn the global–local context, and this strengthens the semantic feature representation. Specifically, a global–local transformer block (GLTB) is deployed in the encoder stage. This design handles image details at low resolution and extracts global image features better than previous methods. In the decoder module, a channel attention optimization module and a fusion module are added to better integrate high- and low-dimensional feature maps, which enhances the network's ability to obtain small-scale semantic information. The proposed method is conducted on the ISPRS Vaihingen and Potsdam datasets. Both quantitative and qualitative evaluations show the competitive performance of MCAFNet in comparison to the performance of the mainstream methods. In addition, we performed extensive ablation experiments on the Vaihingen dataset in order to test the effectiveness of multiple network components.

Keywords: semantic segmentation; transformer; channel attention module; hybrid structure

Citation: Yuan, M.; Ren, D.; Feng, Q.; Wang, Z.; Dong, Y.; Lu, F.; Wu, X. MCAFNet: A Multiscale Channel Attention Fusion Network for Semantic Segmentation of Remote Sensing Images. *Remote Sens.* **2023**, *15*, 361. https://doi.org/10.3390/rs15020361

Academic Editors: Jiaojiao Li, Qian Du, Wei Li, Bobo Xi, Jocelyn Chanussot, Rui Song and Yunsong Li

Received: 23 October 2022
Revised: 21 December 2022
Accepted: 4 January 2023
Published: 6 January 2023

1. Introduction

Semantic segmentation assigns semantic labels to each pixel of the image [1]. In the field of remote sensing, high-resolution remote sensing images provide corresponding data and information support for the construction of smart cities. However, high-resolution urban remote sensing imagery contains rich information about ground objects, which leads to the common phenomenon of large intraclass variance and small interclass variance. Figure 1 shows the local remote sensing image of Potsdam; the orange boxes display the importance of capturing multiscale semantic information, and the black boxes illustrate the difference between small-scale objects. Therefore, extracting useful relevant information from remote sensing images has become a key issue.

In recent years, remote sensing images have developed great potential for application in the field of smart city construction. However, traditional image semantic segmentation of remote sensing images is typically performed by extracting low-level features from the image. When establishing the corresponding semantic segmentation model, there is a gap between the artificially designed features and the high-level semantic features, so the generalizability of the established semantic segmentation model is poor. The interpreted results of deep-learning-based semantic segmentation algorithms in remote sensing city images often present lump-like fuzzy boundaries, which do not sufficiently preserve the feature information of objects. This leads to the confusion of semantic classifiers and brings great challenges to the task of semantic segmentation. Effectively segmenting small objects and improving the interpretation accuracy are still extremely challenging tasks.

Figure 1. The challenge of urban remote sensing image interpretation.

The transformer is a concept proposed by Google in the literature [2]. Since its birth in 2017, the transformer has made rapid breakthroughs in the NLP field. With the advancement of research, it also shows great potential in the field of computer vision, providing novel solutions and achieving good results. Image Transformer [3], released in 2018, was the first to migrate the transformer architecture to the field of computer vision. Since 2019, transformer-based visual models have developed rapidly, and many new attention achievements have appeared. For example, the segmentation transformer (SETR) [4] uses a transformer encoder to completely replace the CNN backbones, discards the convolution and downsampling processes, and uses split tasks as sequence-to-sequence prediction tasks. The detection transformer (DETR) [5] applies the advantages of the transformer in the field of target detection. In July 2020, Chen et al. [6] proposed the iGPT model in order to explore the performance of the approach on images, as well as the performance of unsupervised accuracy. In October 2020, Dosovitskiy et al. [7] proposed the Vision Transformer model, an image classification scheme that is based entirely on the mechanism of self-attention, which was the first work using a transformer to substitute a standard convolution. In January 2021, Esser et al. [8] constructed the vector quantized generative adversarial network (VQGAN), which combines a transformer and CNN, and it is the first transformer architecture to generate megapixel images by semantic guidance. It is worth noting that researchers from Facebook and Berkeley [9] rechecked the design space and tested the limits that pure ConvNet can reach, indicating that the performance of a convolutional neural network is no less than that of a visual transformer, while maintaining the simplicity and effectiveness of the standard ConvNet.

Therefore, given the difficulty of identifying different scales in remote sensing images, we propose a hybrid network structure based on a transformer and CNN, which makes full use of semantic feature information at different scales. In the decoder module, a variety of effective blocks is applied to study an urban scene with complex surface features based on a CNN. The following are this paper's main contributions:

(1) We combine the ResNet-50 and a transformer hybrid model to improve the current mainstream semantic segmentation network structure, and the proposed global–local transformer block models the spatial distance correlation in the image while maintaining the hierarchical characteristics.
(2) We propose a channel attention module decoder (CAMD). In the module, a pooling fusion module is designed to enrich the feature expression of the network. We evaluated the efficiency of each part of the decoder module through ablation research.
(3) We added a fusion module to optimize the structure of the hybrid model, merge feature maps from different scales, and improve semantic representation of the underlying features.

2. Related work

2.1. Methods for Semantic Segmentation Based on Deep Learning

Deep learning [10] is a new research direction in machine learning in recent years. The field of remote sensing image interpretation has gradually implemented deep learning algorithms to deal with problems that were difficult to solve by traditional machine learning methods. At present, the mainstream deep semantic segmentation networks include three forms: a network based on a spatial pyramid structure, a multibranch network, and an encoder–decoder network. These three networks can handle problems in multiscale semantic information extraction and output resolution degradation.

The network based on the spatial pyramid structure uses the pyramid structure to capture the scale semantic information. This kind of network introduces numerous branches to reference the network's end, and each branch corresponds to a fixed scale. For example, PSPNet [11] generates input with different resolutions through an adaptive average pooling operation. PANet [12] changes the input feature map's resolution through a convolutional operation with various core sizes and step sizes. DeepLab [13,14] proposes the atrous spatial pyramid structure (ASPP), which fixes the input resolution of each branch of the pyramid structure and introduces convolutional layers with different expansion rates to expand the network's receptive field. DensAspp [15] improves the receptive field of the pyramid structure in DeepLab by introducing dense connections, making the structure suitable for large-resolution pictures.

The multibranch network sends the input image into multiple branches, and each branch has a different output resolution. For example, icnet [16] uses two spatial branches to capture small-scale targets. Reference [17] uses the branches with higher output resolution to generate a proportional fraction map, which optimized the spatial information of low-resolution branches. Bisenet [18] proposes a lightweight branch with high output resolution and introduces the attention mechanism into the fusion process of different branches, which greatly improves the network speed while maintaining the network accuracy. By combining the shallow characteristics of many branches, Fast SCNN [19] creates a multibranch network, which considerably reduces the amount of calculation consumed on the high-resolution output branches.

The encoder–decoder network gradually integrates the high-dimensional feature map into the low-dimensional feature map to improve the resolution of the output. At the same time, different levels of feature maps have different resolutions. Integrating them can enable the network to capture semantic information of different scales. Therefore, the encoder–decoder network is an effective method to address resolution degradation and multiscale complications. The first deep semantic segmentation network, FCN [20] is a famous encoder–decoder network. It generates a layer-hopping connection structure to integrate high-dimensional and low-dimensional feature maps. On this basis, U-Net [21] proposes a more efficient layer-hopping connection structure, which realizes the fusion of different feature maps with higher accuracy. SegNet [22] records the pooled index in the encoding process and uses the pooled index to supervise the decoding process, making the decoding process more standardized. Refinenet [23] introduces a large number of optimization modules to optimize the feature map fusion results, which could increase the information capture ability of the fused feature map.

Compared with the above two networks, the encoder–decoder structure does not need to change the reference framework and draw additional branches to obtain small-scale semantic information. It needs only to properly optimize the semantic feature maps at different levels on the basis of the reference network. Therefore, this approach is best suited to the domain of semantic segmentation in remote sensing.

Although the encoder–decoder network has great advantages in the field of semantic segmentation, the existing studies have not found an accurate optimization and fusion method for high- and low-dimensional feature maps. Therefore, improving the optimal fusion efficiency of high- and low-dimensional feature maps is a bottleneck in the application of encoder–decoder networks.

2.2. Methods for Semantic Segmentation Based on Transformers

For image problems, convolution has a natural inherent bias translation of equivalence and locality. The transformer obviously does not have these advantages, but its core self-attention operation can obtain a large range of global information, which has obvious advantages for the information extraction range of images. The reasons for the rapid development of the transformer can be attributed to its strong ability to learn long-distance dependencies, multimodal fusion ability, and more interpretable models.

Therefore, many segmentation algorithms take ViT as the backbone network, with Segmenter [24], Segformer [7], and Swin Transformer [25] as typical representatives. Strudel et al. [24] proposed a converter model for semantic segmentation based on the research results from ViT. Segment adopts the ViT model structure in the coding stage, divides the image into blocks, performs a linear mapping, and outputs the embedded sequence after being processed by the encoder. In the decoding stage, learnable category embedding is introduced, and the output of the encoder and category embedding are sent to the decoder, which obtains the class label. Xie et al. [7] proposed SegFormer, a simple, effective, yet powerful semantic segmentation framework. SegFormer uses a hierarchical feature representation method that combines a transformer with a light multilayer perceptron (MLP). Swin Transformer [25] uses a multi-stage design similar to the convolutional neural network, and each stage has a different resolution of the feature map. This mechanism of using a local window attention fully proves that convolution, a method of extracting local feature information, can play its role.

In summary, the transformer has proven to be more powerful than a CNN in feature extraction in semantic segmentation. However, during the semantic segmentation test, the resolution of the picture is not fixed. Its requirements for pixel classification and contour details are meticulous. Transformer-based semantic segmentation methods have poor effects in processing image details. Therefore, more research is needed to build an effective transformer structure and combine it with the current CNN model.

3. The Proposed Method

In this section, we elaborate the method for the semantic segmentation of high-resolution images that combines the hybrid transformer and CNN encoder model. In Section 3.1, we describe the principle and network structures of the hybrid model based on the transformer and ResNet-50. In Section 3.2, we describe the structure of the CAMD module in the CNN-based decoder. Finally, the overall design of our network is described in Section 3.3.

We propose a multiscale channel attention fusion network (MCAFNet) in the context of the semantic segmentation of remotely sensed images. The framework of the MCAFNet is shown in Figure 2. The overall structure of the MCAFNet follows an encoder–decoder structure. A remotely sensed image of an urban area has rich spectral information and texture structure and irregular ground object boundaries, and these characteristics require a higher feature extractor. Therefore, in the encoder part, we used the hybrid model of a CNN and a transformer to extract the multilayer features of the image and optimize the structure of the transformer block. When facing the decoder part of the MCAFNet, the channel attention decoder module is introduced to learn the complex relationship between high- and low-dimensional semantic features. The fusion module is used to improve the fusion efficiency.

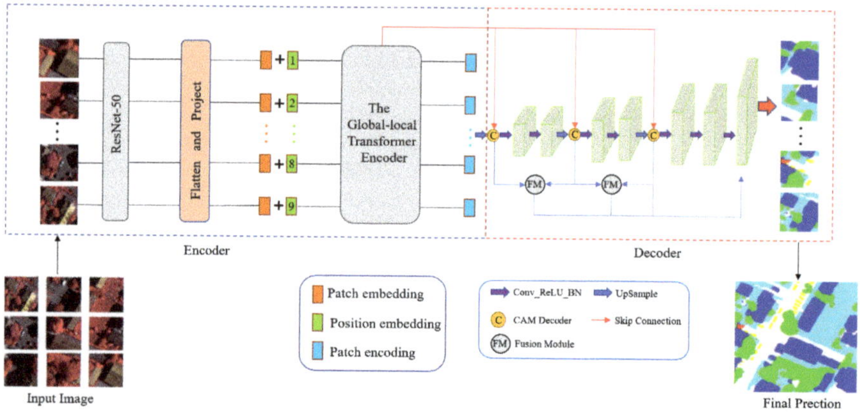

Figure 2. Overall structure diagram of the MCAFNet.

3.1. CNN-Transformer Hybrid as Encoder

The encoder module, which is presented in Figure 3, is designed as a hybrid network model of ResNet-50 and Vit-B/16 [7]. The ResNet-50 convolutional layer is used to enhance the expression of the local context information. The linear multihead self-attention of the transformer module is used to capture the global context information of urban remote sensing images.

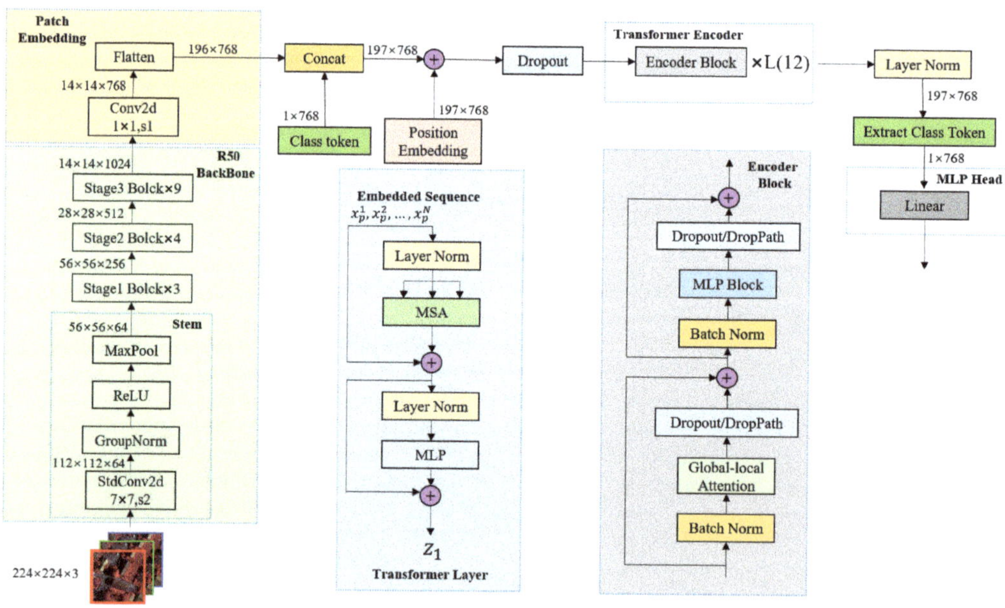

Figure 3. The encoding part structure of the MCAFNet.

We first cut the input remote sensing image x into fixed size patches $\{x = [x_1, \cdots, x_N] \in \mathbb{R}^{(N \times P^2 \times C)}\}$ for feature extraction, where $N = H \times W/P^2$ is the number of image patches and C is the number of slice channels. Then, we used ResNet-50 to perform preliminary semantic feature extraction on the patch. We flattened each patch into a one-dimensional vector $\{X_0 = [E_{X_1}, \cdots, E_{X_N}] \in \mathbb{R}^{N \times D}, E \in \mathbb{R}^{P^2C}\}$ and, then, performed a linear projection to produce a series of patch embeddings to retain low-dimensional semantic feature information. Finally, we modeled the global image context information based on position embedding in the transformer, which perfectly removes the dependence on convolution.

Specifically, inspired by Wang et al. [26], we introduced convolutional groups to extend the multihead attention module. The specific structure is shown in Figure 4b; it adds a convolutional group branch to extract the local features of the image, while retaining the transformer's self-attention mechanism as the global feature extraction branch. In addition, we used batch normalization to solve the variable shift problem in transformer training, accelerate the convergence rate of the model, and resolve the overfitting problem.

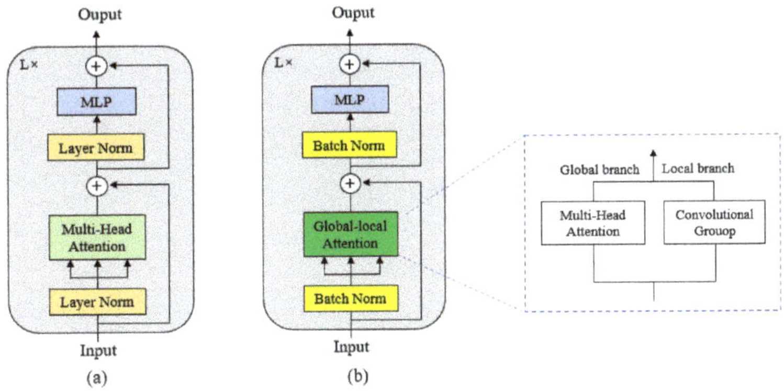

Figure 4. Optimization example of the transformer block. (**a**) is the basic block, and (**b**) is the optimized block.

Figure 5 shows the combination design idea of the module's convolution and self-attention. In order to obtain high- and low-level semantic information at the same time, we processed one part of the input image as a one-dimensional sequence based on the QKV mechanism, and the other part recovers from the sequence the two-dimensional feature map for convolution processing and, finally, splices according to the channel dimensions to output feature vectors with rich semantic information.

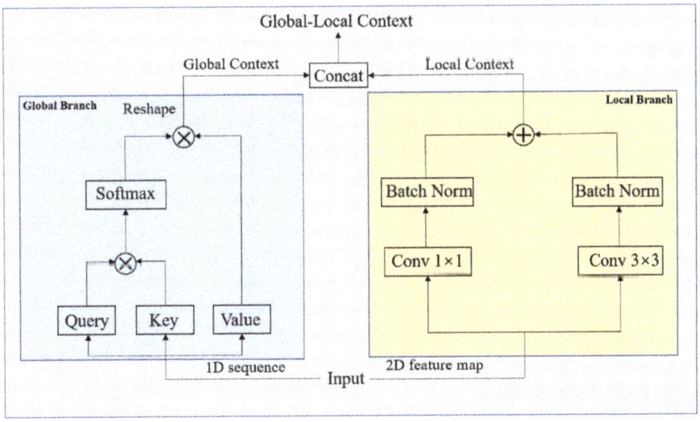

Figure 5. Specific structure design of the global–local attention module.

The global branch deploys the multihead self-attention to capture the global context, and the local branch uses two parallel convolutional layers with core sizes of 3 and 1 to extract the local context. In the transformer, the self-attention mechanism is represented by a linear layer of three points mapped to an intermediate layer. The QKV mechanism is calculated as follows:

$$Attention(Q,K,V) = softmax(\frac{QK^T}{\sqrt{d_K}})V \quad (1)$$

Based on the QKV mechanism of self-attention, in this paper, we combined the deep and shallow semantic features of the input semantic features through weighted fusion, which calculates the correlation between the deep semantic features and the other shallow features in the transformer encoder.

We used the encoded feature vector corresponding to the deep feature map as the query and the value of the multihead attention mechanism and used the encoded feature vector corresponding to the shallow feature map as the key to perform attention fusion. Then, we multiplied the fusion attention map by the encoded feature vector corresponding to the deep feature map, which obtains $Attention(Q,K,V)$ through residual connections and layer normalization. Finally, more precise semantic features are output through the feed-forward network.

3.2. CNN-Based Decoder

As is shown in Figure 6, the process of the MCAFNet interpretation of urban remote sensing images can be summarized as follows: information encoding, information optimization, and information fusion. In addition, a channel attention module is added to adjust the weight of semantic features. Inspired by Ma et al. [27], we added a pooling fusion module (PFM) to enrich the feature expression of remote sensing image semantic information. Its specific structure is shown in Figure 7.

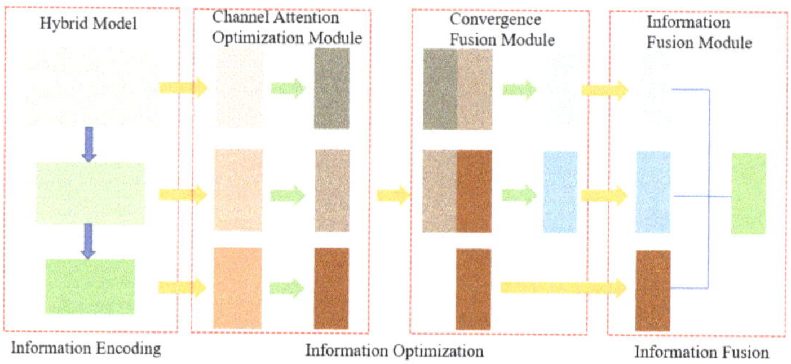

Figure 6. Three steps to decode multiscale semantic information.

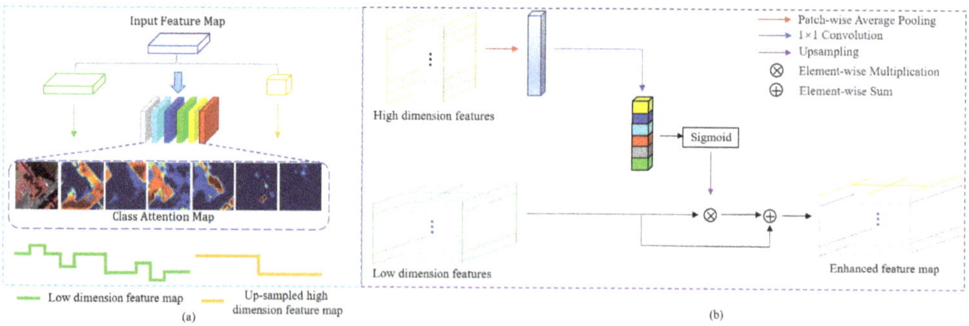

Figure 7. Design motivation and detailed structure of the PFM. (**a**) is the design motivation of PFM, and (**b**) is the specific structure of PFM.

Figure 7a shows the design motivation for the PFM. We materialized the semantic feature maps of six categories of objects and found that the requirements for high-resolution and high-level semantic segmentation contradict the design of convolutional networks. The broken lines with different fluctuations are used to represent the low-dimensional feature map and the high-dimensional feature map after upsampling. Intraclass differences can be shown by discounting small fluctuations, and the differences between classes can be represented by large discounted fluctuations. When they are merged along the spliceosome of the channel dimension, the high- and low-dimensional feature maps of each location are not equally effective, and there are uneven spatial dimensions. Simple upsampling cannot solve the semantic gap problem.

Figure 7b shows the specific structure of the PFM. The operation core of the pooling fusion module is to embed the local attention information in the high-dimensional semantic features of remote sensing images into the low-dimensional semantic features. In this manner, low-dimensional features can be fused to sense the field context information, while this original spatial information will not be lost. First, the average pooling layer is used to optimize the high-dimensional feature map, retain the background information, and obtain the channel attention vector. Then, we reused a 1×1 convolutional layer encoder of each channel weight vector that unifies the number of high- and low-dimensional feature map channels. Finally, we extracted the local attention feature map Z_c from high-dimensional semantic features. The calculation formula is as follows:

$$Z_c = \frac{1}{h_p w_p} \sum_{i=1}^{h_p} \sum_{j=1}^{w_p} x_c(i,j) \tag{2}$$

where $h_p w_p$ denotes the split window size of the average pooling operation and x_c represents a pixel from the c channel.

On this basis, we set the extracted high-dimensional feature map as $Z_H \in \mathbb{R}^{C_h \times H_h \times W_h}$, and set the original low-dimensional feature map as $X_l \in \mathbb{R}^{C_l \times H_l \times W_l}$. Based on the move flip bottleneck convolutional operation, we generated attention maps for the low-dimensional features M_l [28] by transformation. The calculation formula is as follows:

$$M_l = F_u\{\sigma[H_l \delta(H_r Z_h)]\} \tag{3}$$

where σ and δ stand for the sigmoid and ReLU functions. A dimension-reduction convolution of 1×1 with the reduction ratio r is represented by H_r [28]; H_l adjusts the number of channels to match X_l; F_u is the upsampling operation. In addition, we added a residual design to emphasize the importance of low-dimensional features. The augmented features are computed as follows:

$$B_l = X_l + X_l M_l \tag{4}$$

Finally, the PFM outputs feature maps with both precise semantic and spatial information.

The attention mechanism can greatly improve the information capture ability of feature maps. Through the observation of urban remote sensing images, it was found that there are not too many irregular boundaries between adjacent objects in the image, and there was less detail information in the image, making the spatial information of high- and low-dimensional feature maps more accurate. Therefore, this paper used the channel attention mechanism in the soft attention mechanism and did not introduce the spatial attention branch to optimize the spatial information of low-dimensional feature maps. Therefore, we designed the channel attention decoder by combining the channel attention module and the PFM. Figure 8 shows the specific structure of the channel attention module decoder (CAMD).

First, we improved the semantic and spatial information acquisition ability of the MCAFNet based on the channel attention module (CAM). Then, we used upsampling and a 3×3 convolutional operation the further optimize and unify the resolution and channel number of the redefined feature map. Finally, the PFM is used to emphasize the underlying feature information in the key high-dimensional features and filter the background information, restore the pixel position of the target category, and output the enhanced feature information.

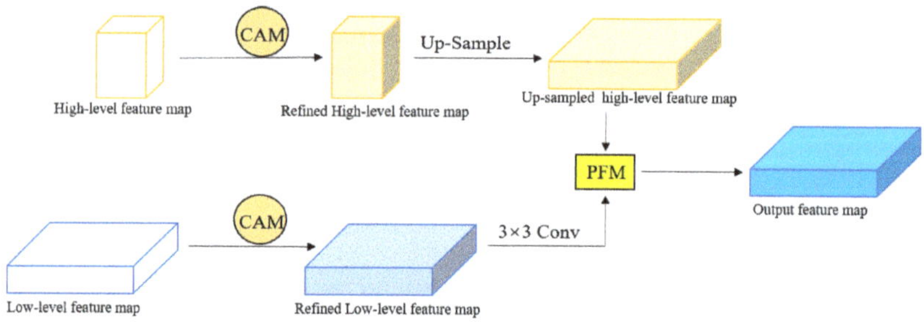

Figure 8. Specific structure of channel attention module decoder.

3.3. Network Architecture

Inspired by Peng et al. [29], we took the output feature maps of the hybrid model as the optimization targets, in order for the network to be able to capture semantic information for three different scales. The characteristic images of Output 1 and Output 2 have a large resolution, which makes them more sensitive to small-scale targets in urban remote sensing images. Therefore, they are the most-important optimization objectives of the network. The overall network structure of the hierarchical encoding and decoding network is shown in Figure 9.

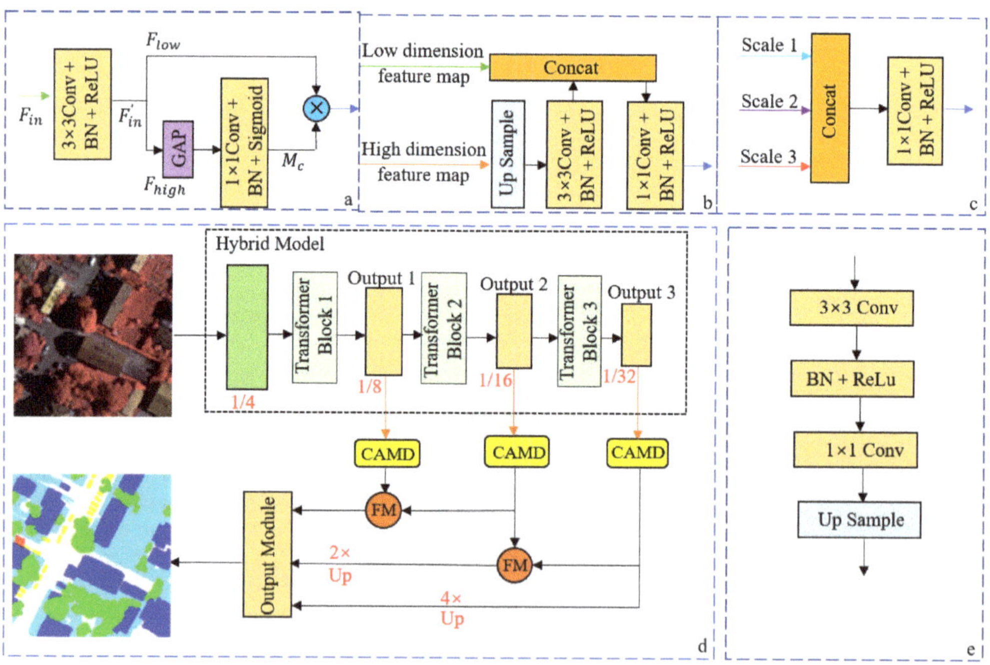

Figure 9. Each module of the network and its overall network structure. (**a**) Channel attention optimization module network structure, (**b**) convergence module network structure, (**c**) information fusion module network structure, (**d**) overall network structure, and (**e**) output module network structure.

A transformer is more suitable for extracting global image information and has a limited ability to capture local semantic information. Directly fusing its intermediate feature map to obtain multiscale semantic information leads to poor segmentation results. Therefore, before fusing the multilevel feature maps, we added a 3 × 3 convolutional layer combined with BN and ReLU to the front-end of the channel attention branch so that it has the function of unifying the number of output channels, which optimizes the feature maps of each dimension, especially low-dimensional feature maps. We assumed that the feature map $F_{in} \in R^{C\times h\times w}$ is the input of the channel attention mechanism and $M_c \in R^{C\times 1\times 1}$ is the channel attention mask. The calculation process of the output feature map F_{cout} after the channel attention mechanism optimization is as follows. The improved channel attention branch is shown in Figure 9a.

$$F_{cout} = F_{low} \bigotimes M_C \quad (5)$$

where ⊗ denotes multiplication by elements. However, the network is still unable to extract accurate small-scale semantic information based on the improved channel attention branch. Therefore, after using the channel attention branch to optimize the feature maps of each dimension, we continued to use the pooling fusion module of the CAMD to optimize the low-dimensional feature maps. The specific fusion module structure used in this article is illustrated in Figure 9b.

The fusion module that we adopted has two input feature maps F_{cout} and F_{high}. The resolution of the high-dimensional feature map is half of the resolution of the low-dimensional feature map. This setting limits the information spread between the two feature maps, which is convenient for information fusion. The fusion module first uses the bilinear interpolation operation UP(·) so that the resolution of the high-dimensional feature map is consistent with the low-dimensional feature map. Thereafter, we adopted a 3 × 3 convolutional layer $CV_{3\times3}$ (·) combining BN and ReLU to optimize the high-dimensional feature maps after upsampling. Finally, the optimized high- and low-dimensional feature maps are aggregated through the concat operation C (· , ·), and the aggregation is optimized using a 1 × 1 convolutional layer $CV_{1\times1}$ (·) that combines BN and ReLU to generate the output. The process of realizing the entire fusion module can be expressed as follows:

$$F_{cfout} = CV_{1\times1}(C(F_{low}, CV_{3\times3}(UP(F_{high})))) \quad (6)$$

Through the fusion module, the low-dimensional feature map obtains more abstract information, and its ability to capture semantic information is also significantly improved. In addition, the fusion module has a simple structure, which quickly improves the ability of low-dimensional feature maps.

We introduce the channel attention module and fusion module to optimize the feature maps of each dimension, the feature maps of each dimension are simultaneously sent to the information fusion module shown in Figure 9c. This module uses cascade operations to fuse the characteristics of different scales, optimizes the fusion result with a 1 × 1 convolutional layer combining BN and ReLU, and finally, outputs a feature map that captures multiscale semantic information.

The specific MCAFNet architecture is illustrated in Figure 9d. First, we used a hybrid network model to extract global and local features from remotely sensed imagery. Then, the channel attention branches were used to optimize the feature maps from different levels within the network to improve their ability to capture semantic information. Next, the feature maps of Output 1 and Output 2 are sent to the same fusion module, and the feature maps of Output 2 and Output 3 are sent to another fusion module to greatly improve the ability of the low-dimensional feature map to capture small-scale semantic information. Unlike encoder–decoder networks, our network does not introduce the information of Output 3 to optimize Output 1 in the decoding process, which avoids the impact of the information gap on the optimization efficiency of Output 1. After optimizing the feature maps of each dimension, they are upsampled at the same time to make them have the same resolution as Output 1, and they are input into the information fusion module, which can improve the ability of capturing semantic information by output feature maps. Finally, the output result of the information fusion module is processed by the output module to generate the final network output feature map, as shown in Figure 9e.

4. Experiment Setup

4.1. Datasets and Evaluation Metrics

All the data needed for repeating the experiments described in the paper are available at https://www.isprs.org (accessed on 10 April 2022). The dataset includes two sub-datasets: the ISPRS Vaihingen and Potsdam 2D semantic segmentation datasets, which correspond to two high-resolution urban remote sensing images of Vaihingen and Potsdam in Germany. The ground objects in the image are marked and distinguished with bright colors, rich ground structures, and representative categories and are suitable for verifying the generalization and robustness of semantic segmentation models. According to the experimental results officially given by ISPRS, generally, only the classification accuracy of the first five categories is evaluated. Therefore, we conducted ablation and interpretation experiments based on them.

Since the image resolution of the experimental dataset is very high, it cannot be directly sent to the GPU for training on the network. We cut the tif image into nonoverlapping blocks to produce 10,000 images, each with a size of 256×256 as the training set and test set. The image as rotated, flipped, tilted, translated, elastically transformed, perspective, cropped, and zoomed to complete the expansion of the dataset. Figure 10 shows some examples of the data augmentation.

We calculated the confusion matrix of these datasets and extracted the overall accuracy (OA), the mean F_1-score, and the mean intersection over union (MIoU) [30] of each class in order to assess the semantic segmentation results. The confusion matrix was obtained by comparing the segmentation result of the predicted output with the labeled image. The formula for the calculation is:

$$CM = \begin{bmatrix} c_{11} & c_{12} & \cdots & c_{1n} \\ c_{21} & c_{22} & \cdots & c_{2n} \\ \vdots & \vdots & \ddots & \vdots \\ c_{n1} & c_{n2} & \cdots & c_{nn} \end{bmatrix}$$

The definitions of the relevant evaluation indicators are shown in the following formulas:

$$Precision = \frac{TP}{TP + FP} \tag{7}$$

$$Recall = \frac{TP}{TP + FN} \tag{8}$$

$$OA = \frac{\sum_{i=1}^{n} c_{ii}}{\sum_{i=1}^{n} \sum_{j=1}^{n} c_{ij}} \tag{9}$$

$$F1 = 2 \cdot \frac{Precision \cdot Recall}{Precision + Recall} \tag{10}$$

$$MIoU = \frac{1}{n} \sum_{i=1}^{n} \frac{c_{ii}}{\sum_{i=1}^{n} c_{ij} + \sum_{j=1}^{n} c_{ji} + c_{ii}} \tag{11}$$

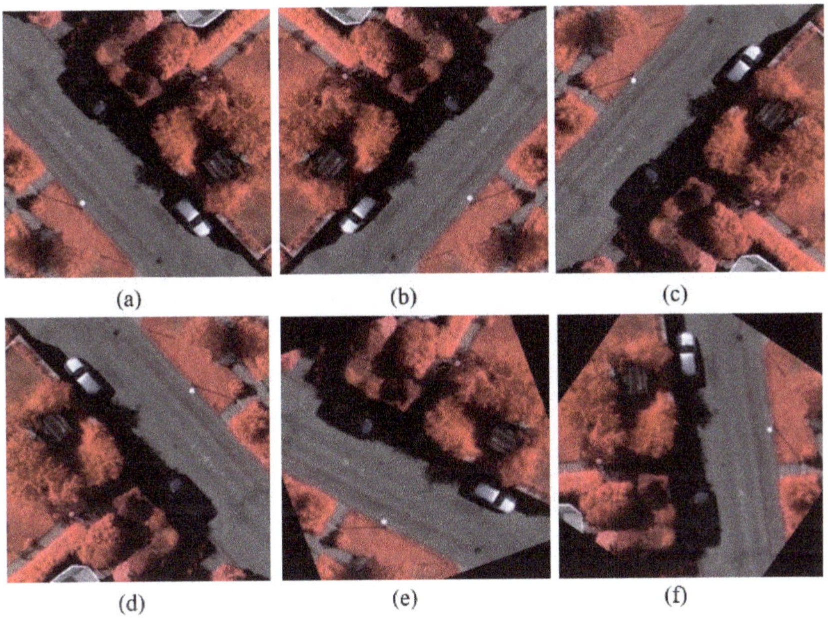

Figure 10. Examples of experimental dataset enhancements. (**a**) is the reference image, and (**b–f**) is examples of multiple enhanced operations based on opencv.

4.2. Implementation Details

We used Python 3.6 and the open-source deep learning framework PyTorch for the experiments in this paper. The optimization algorithm we adopted is the random gradient descent algorithm with momentum equal to 0.9 and weight attenuation equal to 0.0005. In order to train the proposed model on both datasets, the number of iterations was fixed at 15,000. Some experimental parameter settings were as follows: the batch size was set to six; the initial learning rate was set to 0.01; the initial learning rate of the encoder in the network was 0.005. The image was randomly scaled during the training process to be between 0.5- and 2-times higher than the original resolution. At the same time, the image was randomly flipped horizontally to increase the robustness of the model. On this basis, the input image was cropped and padded to a 224 × 224 resolution rate to unify the resolution in each batch of training data.

The skewed distribution of ground objects in remote sensing image sample sets leads to class imbalance. Inspired by D. Eigen et al. [31], we introduced a focal loss function to make the model focus on complex and difficult samples. The definition of the loss function is:

$$MFB\_Focal_{loss} = -\frac{1}{N}\sum_{n=1}^{N}\sum_{c=1}^{C} w_c \cdot l_c^{(n)}(1-p_c^{(n)})^2 \cdot log(p_c^{(n)}) \quad (12)$$

where N represents the number of samples of a minibatch, C represents the number of categories, w_c represents the weight corresponding to category c, $l_c^{(n)}$ represents the true label corresponding to sample n, and $p_c^{(n)}$ represents the probability of sample n for category c.

5. Experiments And Results

5.1. Result Display

To confirm the efficiency of our approach, we visualize the results of the model segmentation on the ISPRS dataset, as shown in Figures 11 and 12. The segmentation results of our proposed network model are almost close to the label values, and the interpretation effect is outstanding in the high-resolution urban remote sensing scene with a dense distribution of ground objects. The boundary of the segmentation results of different types of ground objects is smooth and accurate, and the confusion of classification rarely occurred.

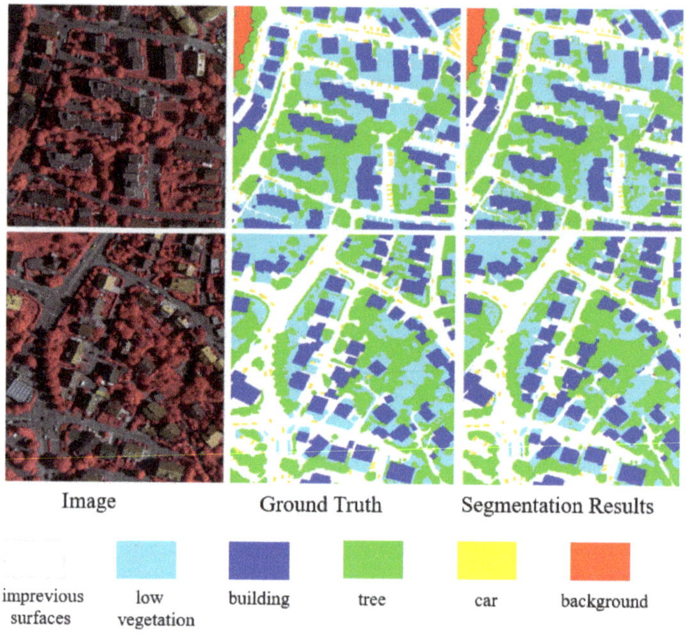

Figure 11. The segmentation results of the MCAFNet on the Vaihingen dataset.

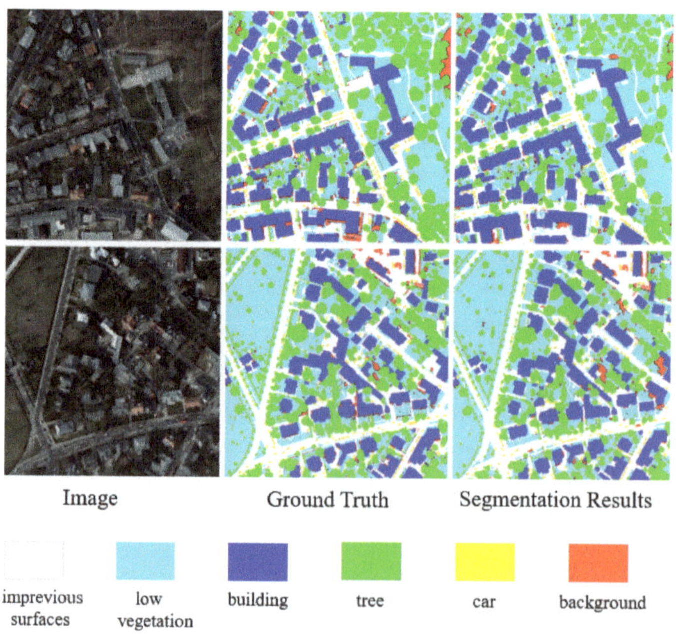

Figure 12. The segmentation results of the MCAFNet on the Potsdam dataset.

5.2. CAM Visualization Analysis

We took the output feature map of the three stages in the MCAFNet encoding part as the optimization goal. According to the change of the resolution of the output feature map, we dynamically adjusted the weight of each channel, so that the CAMD can better optimize the small-scale target. To make the effect more intuitive, we visualize the heat map of the attention mechanism of three stage, as shown in Figure 13. The weight of small target features increases with the deepening of the process, which makes the model more sensitive to small-scale targets.

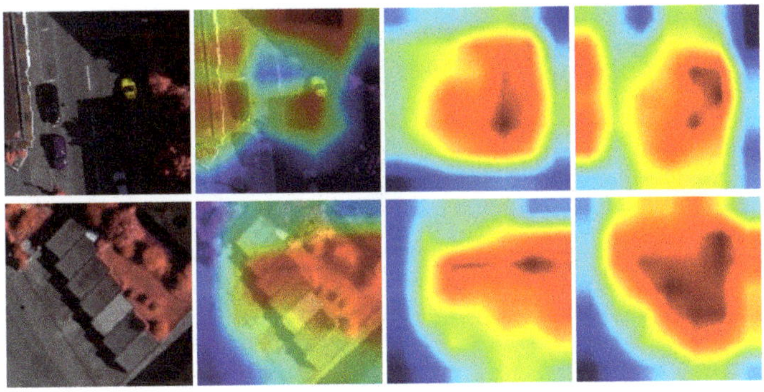

Figure 13. Heat map of small target objects in different stages.

5.3. Architecture Ablation Study

Comprehensive ablation experiments were designed to analyze the efficiency of every part in the MCAFNet. The influence of each module of the network is shown in Table 1, and the improvement effect of the overall and local remote sensing images is shown in Figure 14. It can be seen from Table 1 that different modules of the MCAFNet improved the segmentation performance. However, the performance gain of the transformer and FM is relatively marginal. This is because transformer applications in computer vision are less compatible with urban remote sensing scenes, and the improvement of semantic segmentation accuracy in multi-category scenes is limited. The effect will be obvious if it is combined with the CAMD modules. FM has a simple structure and a small amount of calculation, which can increase the network depth of low-dimensional feature maps, so as to improve the ability of low-dimensional feature maps to capture small-scale targets. It is mainly for the optimized high- and low-dimensional feature maps, which are placed after the CAMD. If a feature map does not go through the channel attention optimization branch, even after FM optimization, this limits the accuracy of the segmentation due to the sufficient depth of the baseline network.

Table 1. The influence of each module on the MCAFNet's performance.

	Transformer	**CAMD**	**FM**	**Mean F_1 (%)**
MCAFNet	×	×	×	81.24
	√	×	×	83.25
	√	√	×	85.46
	√	×	√	83.78
	√	√	√	88.41

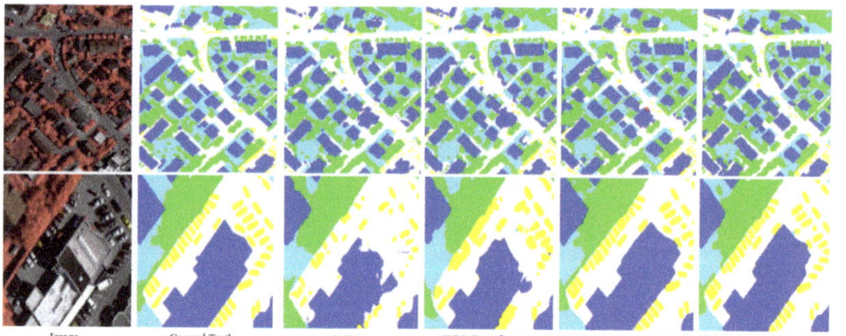

Figure 14. The improvement effect of the overall and local remote sensing images.

(1) Baseline network:

In order to evaluate the performance improvement brought by the architecture in the CNN transformer hybrid as the encoder, we carried out visual analysis on the attention map of different ground object categories for the models before and after the transformer was removed, as shown in Figure 15. The upper row represents the attention map of the figure category of the CNN transformer hybrid model, and the lower row represents the attention map of the category after the transformer is removed. Through comparison, it can be clearly seen that the CNN transformer structure plays an important role in distinguishing the semantic features of different kinds of ground objects when interpreting ground objects. In the process of feature reconstruction, more attention is paid to the pixels of the same category. After the transformer is removed, the model is relatively seriously interfered with by the features of other categories when interpreting a single category of ground objects, which effectively proves that the CNN transformer structure can extract global–local context feature information and improve the segmentation accuracy of multiple types of ground objects.

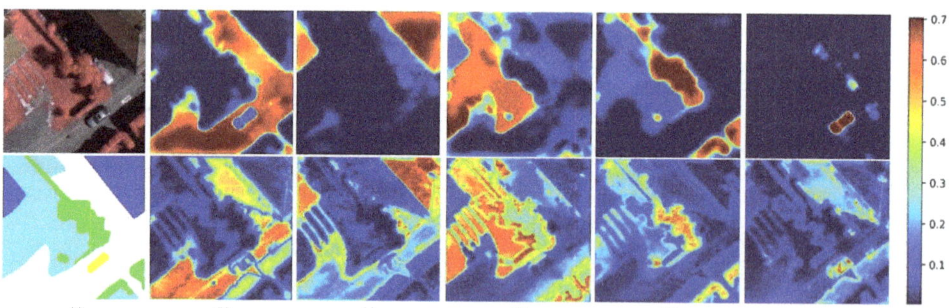

Figure 15. The effect of baseline network on the remote sensing semantic segmentation results. (**a**) is the input image and ground truth, and (**b–f**) are different types of surface feature attention maps.

(2) Channel attention optimization module:

In the ablation experiment, the channel attention decoder was removed, and only the front-end convolutional layers were kept. At this time, the mean F_1-score of the network dropped from 88.41 to 83.78, indicating that the channel attention decoder module plays a very large role in remote sensing scene segmentation. Figure 16 shows the effect of the CAMD on the remote sensing semantic segmentation results.

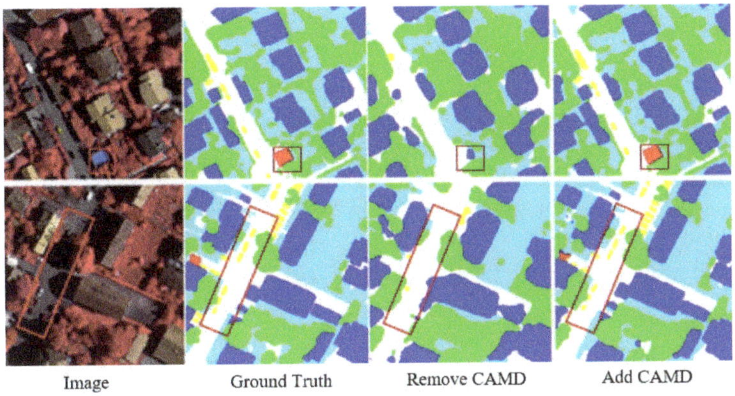

Figure 16. The effect of the CAMD on the remote sensing semantic segmentation results.

(3) Fusion module:

To test the gains in performance brought by the fusion module to the network, we removed it from the network and tested the change in network performance. After removing it, the mean F_1-score of the network decreased from 88.41 to 85.46, which fully proved the importance of the fusion module. From the visualization results shown in the figure, the fusion module increased the network depth of the low-dimensional feature map, which improved the ability of the low-dimensional feature map to capture small-scale targets, and it can efficiently optimize the segmentation results of small-scale targets. Figure 17 shows the effect of the FM on the remote sensing semantic segmentation results.

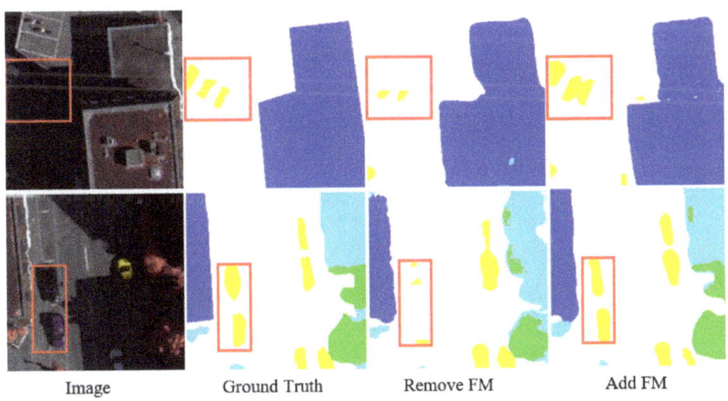

Figure 17. The effect of the FM on the remote sensing semantic segmentation results.

(4) Advanced contrast:

Compared to the performance of U-Net, SegNet, PSPNet, HRNetV2 [32], DeepLab V3+ [33], TransUnet [34], SegFormer, Inception-ResNetV2 [35], and Swin Transformer, the performance of our MCAFNet model in urban remote sensing image interpretation was significantly improved. In Table 2, we provide four test-set-based assessment indices for various models. Compared with the mainstream semantic segmentation model, our model achieved greater improvement in various indicators.

Table 2. The metrics (%) of the semantic segmentation models in the testing phase.

Method	Overall Accuracy	Recall	Mean F_1	MIoU
U-Net	87.5	83.6	82.7	81.2
SegNet	89.4	86.9	86.7	83.6
PSPNet	89.7	87.1	86.9	84.6
HRNetV2	87.2	84.1	83.2	81.4
DeepLab V3+	89.8	87.0	86.7	85.2
TransUnet	90.1	87.2	87.3	86.2
SegFormer	89.5	86.8	87.1	85.9
Inception-ResNetV2	88.1	86.5	86.4	85.5
Swin Transformer	90.2	87.3	87.9	87.3
MCAFNet	**90.8**	**87.9**	**88.4**	**88.2**

To prove the superiority of the method in remote sensing scenes, based on the same experimental conditions, comparison experiments were carried out from local small-scale object segmentation and whole remote sensing scene interpretation. Some visualization results on Vaihingen and Potsdam are shown in Figures 18 and 19. The mean F_1-score of the different models is shown in Tables 3 and 4.

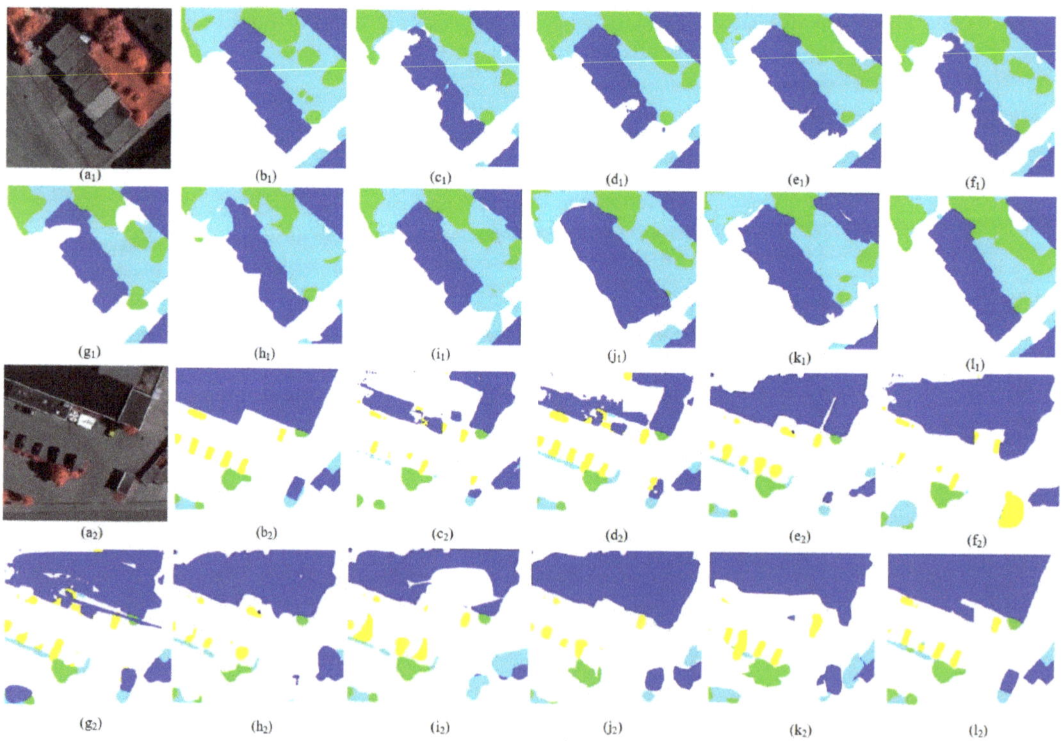

Figure 18. Semantic segmentation results on Vaihingen. (**a1**,**a2**) Original image, (**b1**,**b2**) ground truth, (**c1**,**c2**) U-Net, (**d1**,**d2**) SegNet, (**e1**,**e2**) PSPNet, (**f1**,**f2**) HRNetV2, (**g1**,**g2**) DeepLab V3+, (**h1**,**h2**) TransUnet, (**i1**,**i2**) SegFormer, (**j1**,**j2**) Inception-ResNetV2, (**k1**,**k2**) Swin Transformer, and (**l1**,**l2**) MCAFNet.

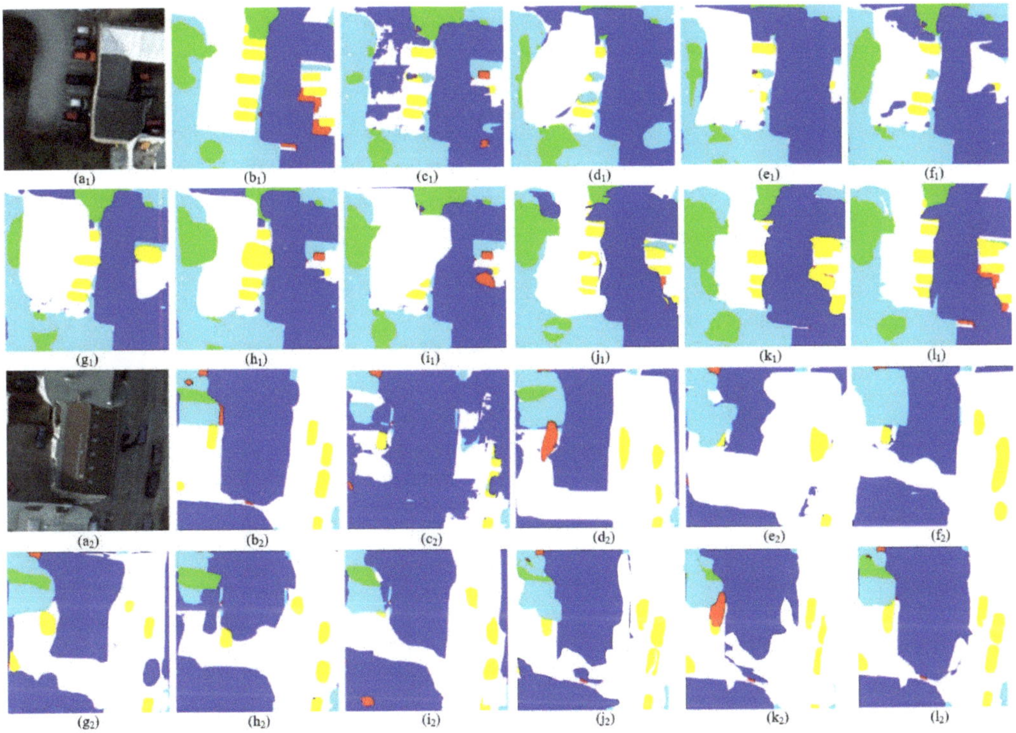

Figure 19. Semantic segmentation results on Potsdam. (**a1**,**a2**) Original image, (**b1**,**b2**) ground truth, (**c1**,**c2**) U-Net, (**d1**,**d2**) SegNet, (**e1**,**e2**) PSPNet, (**f1**,**f2**) HRNetV2, (**g1**,**g2**) DeepLab V3+, (**h1**,**h2**) TransUnet, (**i1**,**i2**) SegFormer, (**j1**,**j2**) Inception-ResNetV2, (**k1**,**k2**) Swin Transformer, and (**l1**,**l2**) MCAFNet.

The performance of these mainstream semantic segmentation networks may be related to their own structure. There are many convolutional layers and pooling layers with steps in the networks, but the convolution lacks an overall understanding of the image itself; moreover, it cannot model feature dependence and does not dynamically adapt to changes in the input. The networks performed poorly in capturing the semantic information of different scales and processing the spatial output resolution of the network. However, we applied a hybrid model network architecture that can reduce the influence of the convolutional operation. It also better integrates semantic information and spatial information through the attention module to optimize the segmentation accuracy of urban features.

Table 3. Performance of ground objects interpreted by the different models on the Vaihingen dataset.

Method	#Param	Building	Car	Low_veg	Imp	Tree	GFLOPs
U-Net	118 M	88.2	75.2	80.2	86.9	85.4	135.4
SegNet	104 M	90.1	84.2	81.7	90.5	86.8	82.9
PSPNet	121 M	91.2	85.7	82.9	91.1	86.4	20.5
HRNetV2	40 M	90.8	76.8	80.4	87.5	86.5	51.5
DeepLab V3+	223 M	91.7	81.4	82.1	88.9	87.6	72.3
TransUnet	257 M	92.3	85.1	83.2	89.7	87.1	112.4
SegFormer	246 M	91.1	81.3	81.5	86.9	86.9	88.7
Inception-ResNetV2	153 M	90.7	84.9	82.5	89.1	86.7	98.5
Swin Transformer	238 M	92.4	85.5	84.1	91.3	87.2	131.4
MCAFNet	334 M	**93.6**	**86.4**	**84.9**	**92.6**	**88.1**	**164.2**

Table 4. Performance of ground objects interpreted by the different models on the Potsdam dataset.

Method	#Param	Building	Car	Low_veg	Imp	Tree	GFLOPs
U-Net	114M	87.2	76.2	80.8	86.2	84.6	123.5
SegNet	97M	89.4	83.6	82.3	90.1	85.9	80.5
PSPNet	114M	90.3	84.7	81.9	89.6	85.4	16.1
HRNetV2	38M	91.2	77.8	80.1	86.9	85.6	43.8
DeepLab V3+	207M	91.4	80.9	81.8	88.2	87.2	62.7
TransUnet	231M	91.8	84.6	82.8	89.1	86.7	98.7
SegFormer	220M	90.7	81.1	81.1	86.4	86.3	81.2
Inception-ResNetV2	141M	90.1	83.9	80.9	87.7	85.5	87.3
Swin Transformer	217M	91.6	85.1	83.1	90.4	87.4	108.7
MCAFNet	320M	**92.4**	**86.1**	**83.9**	**91.3**	**88.3**	**153.3**

6. Discussion

The MCAFNet model we proposed effectively reduces the probability of the misclassification of ground objects in the interpretation of urban remote sensing images and improves the accuracy by integrating low-level semantic features, such as the shape and boundary of ground objects and the high-level semantic information of ground object categories. Two factors ensure the superiority of the model. First, the MCAFNet model realizes the structural innovation in the encoder part and fully combines the advantages of the CNN and transformer when processing semantic segmentation tasks. Second, the proposed network adopts a pooling fusion module in the decoder section. This elaborate design alleviates the information gap and improves the utilization of low-dimensional feature maps. However, the parameters of our model are relatively large, and how to better simplify the transformer structure and combine the advantages of CNNs requires further exploration.

7. Conclusions

We proposed the MCAFNet to realize fast and high-precision semantic segmentation of remote sensing images. The designed network structure successfully integrates the advantages of the transformer and CNNs. Furthermore, we used a channel attention decoder to emphasize the key areas, especially the small-scale target semantic information. The research of our method realizes the robustness and generalization of the model. However, in exchange for high accuracy, the proposed model relies on a large amount of computation, and the CAMD is mainly aimed at the feature extraction of small-scale objects. Its performance is not good when dealing with large-scale target objects. At the same time, there are far more than three bands available in the remote sensing task of image segmentation, and the DSM in the remote sensing image can be used for auxiliary segmentation. This information was not used in this paper. Therefore, future research needs to consider how to reduce the amount of computation of the encoding part, further improve the proposed CAMD to focus on multi-category and large-scale feature extraction, and utilize multiple band information of urban remote sensing images.

Author Contributions: Methodology and writing—original draft preparation, D.R.; writing—review and editing and funding acquisition, M.Y.; resources, Q.F.; data curation, Y.D. and X.W.; supervision, Z.W.; project administration, F.L.; funding acquisition, M.Y. All authors have read and agreed to the published version of this manuscript.

Funding: This work was funded by the Fundamental Research Funds for the Central Universities under Grant No. lzujbky-2021-ct09, the Science and Technology support program of Gansu Province of China under Grant No. 21JR7RA457, the Science and Technology innovation Project of Forestry and Grassland Bureau of Gansu Province (kjcx2022010) and the National Natural Science Foundation of China under Grant No. 62176108.

Data Availability Statement: All the data needed for experiments described in our paper are available at https://www.isprs.org (accessed on 10 April 2022).

Conflicts of Interest: The authors declare no conflict of interest.

References

1. Tapasvi, B.; Udaya Kumar, N.; Gnanamanoharan, E. A Survey on Semantic Segmentation using Deep Learning Techniques. *Int. J. Eng. Res. Technol.* **2021**, *9*, 50–56.
2. Vaswani, A.; Shazeer, N.; Parmar, N.; Uszkoreit, J.; Jones, L.; Gomez, A.N.; Kaiser, Ł.; Polosukhin, I. Attention is all you need. *Adv. Neural Inf. Process. Syst.* **2017**, *30*, 5998–6008.
3. Parmar, N.; Vaswani, A.; Uszkoreit, J.; Kaiser, L.; Shazeer, N.; Ku, A.; Tran, D. Image transformer. In Proceedings of the International Conference on Machine Learning, Stockholm, Sweden, 10–15 July 2018; pp. 4055–4064.
4. Zheng, S.; Lu, J.; Zhao, H.; Zhu, X.; Luo, Z.; Wang, Y.; Fu, Y.; Feng, J.; Xiang, T.; Torr, P.H.; et al. Rethinking semantic segmentation from a sequence-to-sequence perspective with transformers. In Proceedings of the IEEE/CVF Conference on Computer Vision and Pattern Recognition, Nashville, TN, USA, 20–25 June 2021; pp. 6881–6890.
5. Wang, H.; Zhu, Y.; Adam, H.; Yuille, A.; Chen, L.C. Max-deeplab: End-to-end panoptic segmentation with mask transformers. In Proceedings of the IEEE/CVF Conference on Computer Vision and Pattern Recognition, Nashville, TN, USA, 20–25 June 2021; pp. 5463–5474.
6. He, L.; Zhou, Q.; Li, X.; Niu, L.; Cheng, G.; Li, X.; Liu, W.; Tong, Y.; Ma, L.; Zhang, L. End-to-end video object detection with spatial-temporal transformers. In Proceedings of the 29th ACM International Conference on Multimedia, Virtual Event, 20–24 October 2021; pp. 1507–1516.
7. Dosovitskiy, A.; Beyer, L.; Kolesnikov, A.; Weissenborn, D.; Zhai, X.; Unterthiner, T.; Dehghani, M.; Minderer, M.; Heigold, G.; Gelly, S.; et al. An image is worth 16 × 16 words: Transformers for image recognition at scale. *arXiv* **2020**, arXiv:2010.11929.
8. Esser, P.; Rombach, R.; Ommer, B. Taming transformers for high-resolution image synthesis. In Proceedings of the IEEE/CVF Conference on Computer Vision and Pattern Recognition, Nashville, TN, USA, 20–25 June 2021; pp. 12873–12883.
9. Liu, Z.; Mao, H.; Wu, C.Y.; Feichtenhofer, C.; Darrell, T.; Xie, S. A ConvNet for the 2020s. In Proceedings of the IEEE/CVF Conference on Computer Vision and Pattern Recognition (CVPR), New Orleans, LA, USA, 19–20 June 2022; pp. 11976–11986.
10. LeCun, Y.; Bengio, Y.; Hinton, G. Deep learning. *Nature* **2015**, *521*, 436–444. [CrossRef] [PubMed]
11. Zhao, H.; Shi, J.; Qi, X.; Wang, X.; Jia, J. Pyramid scene parsing network. In Proceedings of the IEEE Conference on Computer Vision and Pattern Recognition, Honolulu, HI, USA, 21–26 July 2017; pp. 2881–2890.
12. Li, H.; Xiong, P.; An, J.; Wang, L. Pyramid attention network for semantic segmentation. *arXiv* **2018**, arXiv:1805.10180.
13. Chen, L.C.; Papandreou, G.; Schroff, F.; Adam, H. Rethinking atrous convolution for semantic image segmentation. *arXiv* **2017**, arXiv:1706.05587.
14. Yu, F.; Koltun, V.; Funkhouser, T. Dilated residual networks. In Proceedings of the IEEE Conference on Computer Vision and Pattern Recognition, Honolulu, HI, USA, 21–26 July 2017; pp. 472–480.
15. Yang, M.; Yu, K.; Zhang, C.; Li, Z.; Yang, K. Denseaspp for semantic segmentation in street scenes. In Proceedings of the IEEE Conference on Computer Vision and Pattern Recognition, Salt Lake City, UT, USA, 18–23 June 2018; pp. 3684–3692.
16. Borjigin, S.; Sahoo, P.K. Color image segmentation based on multi-level Tsallis–Havrda–Charvát entropy and 2D histogram using PSO algorithms. *Pattern Recognit.* **2019**, *92*, 107–118. [CrossRef]
17. Wu, Z.; Shen, C.; Hengel, A.V.d. Real-time semantic image segmentation via spatial sparsity. *arXiv* **2017**, arXiv:1712.00213.
18. Xu, Q.; Ma, Y.; Wu, J.; Long, C. Faster BiSeNet: A Faster Bilateral Segmentation Network for Real-time Semantic Segmentation. In Proceedings of the 2021 International Joint Conference on Neural Networks (IJCNN), Shenzhen, China, 18–22 July 2021; pp. 1–8.
19. Poudel, R.P.; Liwicki, S.; Cipolla, R. Fast-scnn: Fast semantic segmentation network. *arXiv* **2019**, arXiv:1902.04502.
20. Long, J.; Shelhamer, E.; Darrell, T. Fully convolutional networks for semantic segmentation. In Proceedings of the IEEE Conference on Computer Vision and Pattern Recognition, Boston, MA, USA, 7–12 June 2015; pp. 3431–3440.
21. Ronneberger, O.; Fischer, P.; Brox, T. U-net: Convolutional networks for biomedical image segmentation. In Proceedings of the International Conference on Medical Image Computing and Computer-Assisted Intervention, Munich, Germany, 5–9 October 2015; pp. 234–241.
22. Badrinarayanan, V.; Handa, A.; Cipolla, R. Segnet: A deep convolutional encoder–decoder architecture for robust semantic pixel-wise labelling. *arXiv* **2015**, arXiv:1505.07293.
23. Lin, G.; Milan, A.; Shen, C.; Reid, I. Refinenet: Multi-path refinement networks for high-resolution semantic segmentation. In Proceedings of the IEEE Conference on Computer Vision and Pattern Recognition, Honolulu, HI, USA, 21–26 July 2017; pp. 1925–1934.
24. Xie, E.; Wang, W.; Yu, Z.; Anandkumar, A.; Alvarez, J.M.; Luo, P. SegFormer: Simple and efficient design for semantic segmentation with transformers. *Adv. Neural Inf. Process. Syst.* **2021**, *34*, 12077–12090.
25. Liu, Z.; Lin, Y.; Cao, Y.; Hu, H.; Wei, Y.; Zhang, Z.; Lin, S.; Guo, B. Swin Transformer: Hierarchical Vision Transformer using Shifted Windows. In Proceedings of the IEEE/CVF International Conference on Computer Vision, Montreal, QC, Canada, 10–17 October 2021.

26. Wang, L.; Fang, S.; Zhang, C.; Li, R.; Duan, C. Efficient Hybrid Transformer: Learning Global-local Context for Urban Scene Segmentation. *arXiv* **2021**, arXiv:2109.08937.
27. Peng, C.; Tian, T.; Chen, C.; Guo, X.; Ma, J. Bilateral attention decoder: A lightweight decoder for real-time semantic segmentation. *Neural Netw.* **2021**, *137*, 188–199. [CrossRef] [PubMed]
28. Ding, L.; Tang, H.; Bruzzone, L. LANet: Local attention embedding to improve the semantic segmentation of remote sensing images. *IEEE Trans. Geosci. Remote Sens.* **2020**, *59*, 426–435. [CrossRef]
29. Peng, C.; Zhang, K.; Ma, Y.; Ma, J. Cross fusion net: A fast semantic segmentation network for small-scale semantic information capturing in aerial scenes. *IEEE Trans. Geosci. Remote Sens.* **2021**, *60*, 5601313. [CrossRef]
30. Yang, H.; Wu, P.; Yao, X.; Wu, Y.; Wang, B.; Xu, Y. Building extraction in very high resolution imagery by dense-attention networks. *Remote Sens.* **2018**, *10*, 1768. [CrossRef]
31. Eigen, D.; Fergus, R. Predicting depth, surface normals and semantic labels with a common multiscale convolutional architecture. In Proceedings of the IEEE International Conference on Computer Vision, Santiago, Chile, 7–13 December 2015; pp. 2650–2658.
32. Sun, K.; Zhao, Y.; Jiang, B.; Cheng, T.; Xiao, B.; Liu, D.; Mu, Y.; Wang, X.; Liu, W.; Wang, J. High-resolution representations for labeling pixels and regions. *arXiv* **2019**, arXiv:1904.04514.
33. Chen, L.C.; Papandreou, G.; Kokkinos, I.; Murphy, K.; Yuille, A.L. Deeplab: Semantic image segmentation with deep convolutional nets, atrous convolution, and fully connected crfs. *IEEE Trans. Pattern Anal. Mach. Intell.* **2017**, *40*, 834–848. [CrossRef] [PubMed]
34. Chen, J.; Lu, Y.; Yu, Q.; Luo, X.; Adeli, E.; Wang, Y.; Lu, L.; Yuille, A.L.; Zhou, Y. Transunet: Transformers make strong encoders for medical image segmentation. *arXiv* **2021**, arXiv:2102.04306.
35. Szegedy, C.; Ioffe, S.; Vanhoucke, V.; Alemi, A. Inception-v4, Inception-ResNet and the Impact of Residual Connections on Learning. In Proceedings of the Thirty-First AAAI Conference on Artificial Intelligence (AAAI-17), San Francisco, CA, USA, 4–9 February 2017.

remote sensing

MDPI

Article

SegMarsViT: Lightweight Mars Terrain Segmentation Network for Autonomous Driving in Planetary Exploration

Yuqi Dai [1,2], Tie Zheng [1,2], Changbin Xue [1,*] and Li Zhou [1]

[1] National Space Science Center, Chinese Academy of Sciences, Beijing 100190, China
[2] University of Chinese Academy of Sciences, Beijing 100049, China
* Correspondence: xuechangbin@nssc.ac.cn

Abstract: Planetary rover systems need to perform terrain segmentation to identify feasible driving areas and surround obstacles, which falls into the research area of semantic segmentation. Recently, deep learning (DL)-based methods were proposed and achieved great performance for semantic segmentation. However, due to the on-board processor platform's strict comstraints on computational complexity and power consumption, existing DL approaches are almost impossible to be deployed on satellites under the burden of extensive computation and large model size. To fill this gap, this paper targeted studying effective and efficient Martian terrain segmentation solutions that are suitable for on-board satellites. In this article, we propose a lightweight ViT-based terrain segmentation method, namely, SegMarsViT. In the encoder part, the mobile vision transformer (MViT) block in the backbone extracts local–global spatial and captures multiscale contextual information concurrently. In the decoder part, the cross-scale feature fusion modules (CFF) further integrate hierarchical context information and the compact feature aggregation module (CFA) combines multi-level feature representation. Moreover, we evaluate the proposed method on three public datasets: AI4Mars, MSL-Seg, and S5Mars. Extensive experiments demonstrate that the proposed SegMarsViT was able to achieve 68.4%, 78.22%, and 67.28% mIoU on the AI4Mars-MSL, MSL-Seg, and S^5Mars, respectively, under the speed of 69.52 FPS.

Keywords: Mars terrain segmentation; semantic segmentation; planetary exploration

Citation: Dai, Y.; Zheng, T.; Xue, C.; Zhou, L. SegMarsViT: Lightweight Mars Terrain Segmentation Network for Autonomous Driving in Planetary Exploration. *Remote Sens.* **2022**, *14*, 6297. https://doi.org/10.3390/rs14246297

Academic Editors: Jiaojiao Li, Qian Du, Wei Li, Bobo Xi, Jocelyn Chanussot, Rui Song and Yunsong Li

Received: 17 November 2022
Accepted: 9 December 2022
Published: 12 December 2022

1. Introduction

Intelligent environmental perception is a necessity for planetary rovers toward autonomous driving, which provides crucial semantic information, e.g., identifying feasible driving areas and surrounding obstacles. For such a panoptic perception mission, terrain segmentation is the most critical procedure, which also can be viewed as a semantic segmentation task. Semantic segmentation is a widely used perception method for self-driving vehicles on earth that can assign a separate predefined class label to each pixel of an image [1] it is the foundation of many high-level tasks that need to infer relevant semantic information from images for subsequent processing. This applies on self-driving vehicles on Mars as well. Therefore, this study explored the task of terrain segmentation on the Martian surface, aiming to characterize semantic information from rover images. As shown, the Figure 1a shows the Tianwen-1 Zhurong rover, China's first Mars rover, which is undergoing its fantastic exploration on the red planet. RGB sample images of the Mars surface and the corresponding terrain segmentation annotation are depicted in Figure 1b,c, respectively. It can be observed that semantic segmentation is a pixel-level dense prediction task, which requires an in-depth understanding of the semantics of the entire scene and is in some ways more challenging than those image-level prediction tasks.

Early image segmentation approaches dedicated to divide images into regions based on little more than basic color and low-level textual information [2,3]. With the rapid development of deep learning techniques in the 2010s, deep convolutional neural networks

(CNNs) became dominant in automatic semantic segmentation technology due to their tremendous modeling and learning capabilities, which strive to boost algorithm accuracy on the strength of massively parallel GPUs and large labelled datasets [4,5]. Long et al. [6] first proposed a fully convolutional network (FCNet), which is a revolutionary work and the majority of following state-of-the-art (SOTA) studies are extensions of the FCN architecture. One of the most pioneering works is UNet presented by Ronneberger et al. [7] for biomedical image segmentation, which adopts the influential encoder–decoder architecture and proved to be very useful for other types of image data [8–11]. Meanwhile, inspired by the high precision that CNNs achieved in semantic segmentation, many CNNs-based approaches were proposed for the Martian terrain segmentation (MTS) task. Rothrock et al. [12] proposed a soil property and object classification (SPOC) system based on DeepLab for visually identifying terrain types as well as terrain features (e.g., scarps, ridges) on a planetary surface. They also presented two successful applications to Mars rover missions, including the landing site traversability analysis and slip prediction. Iwashita et al. [13] proposed TU-Net and TDeelLab robust to illumination changes via data fusion from visible and thermal images. Liu et al. [14] proposed a hybrid attention-based terrain segmentation network called HASS for unstructured Martian images. Claudet et al. [15] employed advanced semantic segmentation algorithms to generate binary safety maps for the spacecraft safe planetary landing problem. Furthermore, several existing studies attempted to resolve the terrain segmentation issue by using wear-supervised techniques. Wang et al. [16] adopted the element-wise contrastive learning technique and proposed a semi-supervised learning framework for Mars imagery classification and segmentation through introducing online pseudo labels on the unlabeled areas. Goh et al. [17] proposed another semi-supervised Mars terrain segmentation algorithm with contrastive pretraining techniques. Zhang et al. [18] proposed a novel hybrid representation learning-based framework, which consists of a self-supervised pre-training stage and a semi-supervised learning phase for sparse data. Li et al. [19] introduced a stepwise domain adaptation Martian terrain segmentation network, which effectively alleviates covariate shift through unifying the color mapping space to further enhance the segmentation performance.

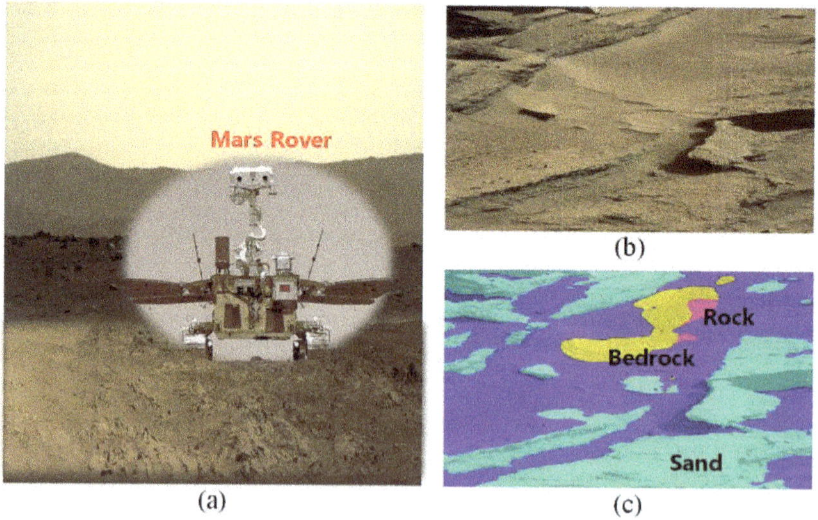

Figure 1. Planetary rover on Mars (**a**) and a sample image (**b**) along with its segmentation annotation for terrain types (**c**).

Furthermore, data-driven deep learning generally refers to learning directly through sufficient experience data. The level of success for deep learning applications is to a great extent determined by the quality and the depth of the data being used for training. In this respect, Mars terrain segmentation is currently attracting more and more attention, and scientific interest for deep learning-based segmentation datasets is growing rapidly. Several large-scale 2D image sets were established for the Mars terrain segmentation problem, the relevant information of which is listed in Table 1. Swan et al. [20] built the first large-scale dataset, AI4Mars, for the task of Mars terrain classification and traversability assessment, of which labels were obtained through a crowdsourcing approach and consisted four classes: soil, bedrock, sand, and big rock. Li et al. [19] extensively released a Mars terrain dataset annotated finely with nine classes, named Mars-Seg. Liu et al. [14] established a panorama semantic segmentation dataset for Mars rover images, named MarsScapes, which provides pixel-wise annotations for eight fined-grained categories. Zhang et al. [18] presented a high-resolution Mars terrain segmentation dataset, S^5Mars, annotated with pixel-level sparse labels for nine categories. The Martian surface condition is complicated and the corresponding annotation process is challenging. Hence, we thank all the above dataset creators that enabled us to conduct the research for this paper.

Table 1. To the best of our knowledge, there are already four public datasets established for MTS task up to now.

Dataset	Year	Classes	RGB
AI4Mars	2021	4	1.6 k
Mars-Seg	2021	8	~4.1 k
S^5Mars	2022	9	6 k
MarsScapes	2022	8	~18.5 k

In comparison to natural scene images, the Martian images have their particular characteristics. Objects on the surface of Mars exhibit unstructured characteristics with rich textures, ambiguous boundaries and diverse sizes, such as rocks and gravel [21]. Understanding unstructured scenes quite heavily depend on modeling the connection between the target pixel and its relevant surrounding content to a certain extent. Therefore, a limited receptive field is hard-pressed to meet demand, and several acquired rare target instances available for training are in small numbers. Class imbalance remains a problem in the MTS task. The above difficulties make it unreliable to directly apply the semantic segmentation methods designed for natural images on Martian terrain segmentation tasks.

On the other hand, CNN-based semantic segmentation methods always made brilliant achievements at the expense of high computational costs, large model size, and inference latency. This situation prevented recent state-of-the-art methods from being applied to real-world applications. Real on-board applications have a strong demand of semantic segmentation algorithms to run on resource-constrained edge devices in a timely manner. Therefore, deep models for the Mars terrain segmentation task should be efficient and accurate. Considering the performance limitations of spacecraft equipment, it is essential to develop efficient networks for accurate Mars terrain segmentation.

Toward this end, this paper proposes a novel lightweight Martian terrain segmentation model, named SegMarsViT. In the encoder part, the mobile vision transformer (MobileViT) backbone is leveraged to extract local–global spatial and capture high-level multiscale contextual information concurrently. An effective layer aggregation decoder (ELAD) is designed to further integrate hierarchical feature context information and generate powerful representations. Moreover, we evaluate the proposed method on three public datasets: AI4Mars, MSL-Seg, and S^5Mars. Extensive experiments demonstrate that the proposed SegMarsViT achieves comparable accuracy as the state-of-the-art semantic segmentation method. In the meantime, SegMarsViT has much less computation burden with a smaller model size.

The main contributions of this work can be summarized as follows:

(1) To the best of our knowledge, this is the first effort toward introducing the lightweight semantic segmentation model into the field of Martian terrain segmentation. We evaluate several representative semantic segmentation models and conduct enough comparable experiments. This is expected to facilitate the development and benchmarking of terrain segmentation algorithms in Martian images.
(2) We investigate a novel vision transformer-based deep neural network SegMarsViT for real-time and accurate Martian terrain segmentation. In the encoder, we employ a lightweight MobileViT backbone to capture a hierarchical feature. Notably, the proposed SegMarsViT is the first transformer-based network for the Martian terrain segmentation task. In the decoder part, a cross-scale feature fusion (CFF) module and a compact feature aggregation (CFA) technique are designed to strengthen and merge the multi-scale context feature.
(3) We conduct extensive experiments on AI4Mars, S5Mars, and MSL-Seg datasets. The results validate the effectiveness and efficiency of the proposed model, which can obtain competitive performance with 68.4%, 78.22%, and 67.28% mIoU, respectively. In the meantime, SegMarsViT has much less computation burden with smaller model size.

The remainder of this article is organized as follows: In Section 2, we will briefly introduce some previous work related to lightweight semantic segmentation and vision transformer. Section 3 describes the proposed method in detail. Section 4 provides overall performance and comparison results of the proposed method with analysis and discussion, and Section 5 concludes this study.

2. Related Work

2.1. Lightweight Semantic Segmentation

In real-world applications, such as robotics [22] and land resource monitoring [23], it is hard to deploy high-precision, high-complexity, and time-consuming semantic segmentation models for real-time inference speed in need. Hence, lightweight semantic segmentation networks came into being. Several research works were proposed to address the challenge of real-time semantic segmentation.

The standard convolution layer is the basic building layer in CNNs, which is computationally expensive. Real-time semantic segmentation pursues the fast data processing capability of the network. In order to meet the requirements of real-time inference performance and ensure high-quality prediction as much as possible, efficient convolution operations are generally used. For example, DABNet [24] introduced the depth-wise asymmetric bottleneck module, which increases efficiency through the combination of depth-wise separable and asymmetric factorized convolutions. ESPNet [25] proposed an efficient spatial pyramid module utilizing 1×1 grouped convolution to reduce dimension complexity and parallel dilation convolution modules to increase the effective receptive field, which results in a very compact and significant network. In addition, many other segmentation models, e.g., RTSeg [26] and EACNet [27], straightly employ the lightweight backbone networks designed for classification tasks as the feature extractor to improve the inference speed.

In addition to commonly used techniques for decreasing the latency and model size, designing novel and lightweight architectures is another effective solution. BiseNetV1 [28] is a two-branch architecture to reserve spatial feature information and enlarge the receptive field, which consists of a context path based on Xception architecture and a spatial branch based on strided convolution layers. Attention refinement modules (ARM) are applied to encode global context. The improved version, BiseNetV2 [29], further simplifies the architecture through utilizing the inverted bottleneck blocks of MobileNetv2 and efficient convolutions and obtains more favorable performance. The real-time general purpose semantic segmentation network (RGPNet) introduces a novel adapter module and a lightweight asymmetric encoder–decoder architecture. The adaptor module intermediates between encoder and decoder through the combination of features of three different levels. The strategy of integrating multi-scale context information results in excellent segmentation performance and the optimized progressive resizing training scheme makes RGPNet achieve an effective balance between speed and accuracy.

2.2. Vision Transformer-Based Semantic Segmentation

In spite of the exceptional representational power, CNN-based approaches generally exhibit limitations for modeling explicit long-range relations, due to the intrinsic local connectivity mechanism of convolution operations. Recently, transformer became a "hotspot" in the computer vision community, which was initially designed for sequence-to-sequence prediction and was powerful at modeling global contexts [30]. To overcome the limitation of the local receptive field of CNN, the latest efforts were focused on adapting transformer models into the computer vision sector [31,32], named vision transformer (ViT). Many scholars introduced the ViT mechanism into the semantic segmentation task. The two most common ways to do this are applying ViTs in conjunction with CNNs and developing pure ViTs. Wang et al. [33] proposed PVT, a pyramid vision transformer for dense prediction tasks, which is a natural extension of ViT with pyramid structures. Zheng et al. [34] proposed SETR, which is a hierarchical transformer from a sequence-to-sequence learning perspective, and it shows that good results can still be obtained without relying on the convolution operation. Huang et al. [35] designed a scale-wise intra-scale transformer, named ScaleFormer, of which the elaborate hybrid CNN-transformer backbone can effectively extract intra-scale local features and global information. Shi et al. [36] took the idea of the SwinTransformer [37] and presented the hierarchical SSformer with an elaborate and simple MLP decoder for semantic segmentation. Xie et al. [38] proposed SegFormer, which comprises a novel hierarchically structured transformer encoder and a lightweight all-MLP decoder, yielding great results. Hatamizadeh et al. proposed the UNetFormer [39] with a 3D SwinTransformer [40]-based encoder and a hybrid CNN-transformer decoder, which can achieve a trade-off performance between efficiency and accuracy for medical image segmentation. Similarly, there are UNETR [41] and nnFormer [42] in the same vein. Motivated by the astounding achievements of ViT, this paper presents the first study to explore the potential of ViT and fulfill the local–global semantics research gap in the context of Martian terrain segmentation.

3. Methodology

In this section, we first provide an overview of our method in the Section 3.1. Then, we introduce the lightweight encoder and effective decoder in the Sections 3.2 and 3.3, respectively. Finally, we present the loss function in the Section 3.4.

3.1. Framework Overview

The overall structure of the proposed SegMarsViT is illustrated in Figure 2. This paper is dedicated to the encoder–decoder segmentation architecture through ViT modules. The whole SegMarsViT is a novel combination of CNN and transformers to some extent, which has the local advantage of CNN and the long-range dependency merit of a transformer. The proposed network utilizes MobileViT backbone to extract corresponding features of five stages (stage1~stage5), whose outputs are denoted as F_1, F_2, F_3, F_4 and F_5, with scales of $\frac{1}{2}$, $\frac{1}{4}$, $\frac{1}{8}$, $\frac{1}{16}$ and $\frac{1}{32}$, respectively. In other words, the output feature maps F_i after each stage are down-sampled with strides of 2^i. After the backbone, we perform an efficient stage-wise layer aggregation decoder, named ELAD, to generate segmentation outputs. The novel ELAD is designed to make multiscale features more distinguishable to learn representative features for SegMarsViT. In ELAD, a series of cross-scale feature fusion (CFF) modules are proposed to further enhance the context modeling and boost the cross-scale communication, which are built upon the top-down pathway. After obtained, we introduce a compact feature aggregation (CFA) module to ensure that feature maps extracted from different stages can be well merged. As shown in Figure 2, the proposed SegMarsViT is asymmetric and the contracting path is deeper than the expansion path. In what follows, we describe all the structures of the above-mentioned modules in detail.

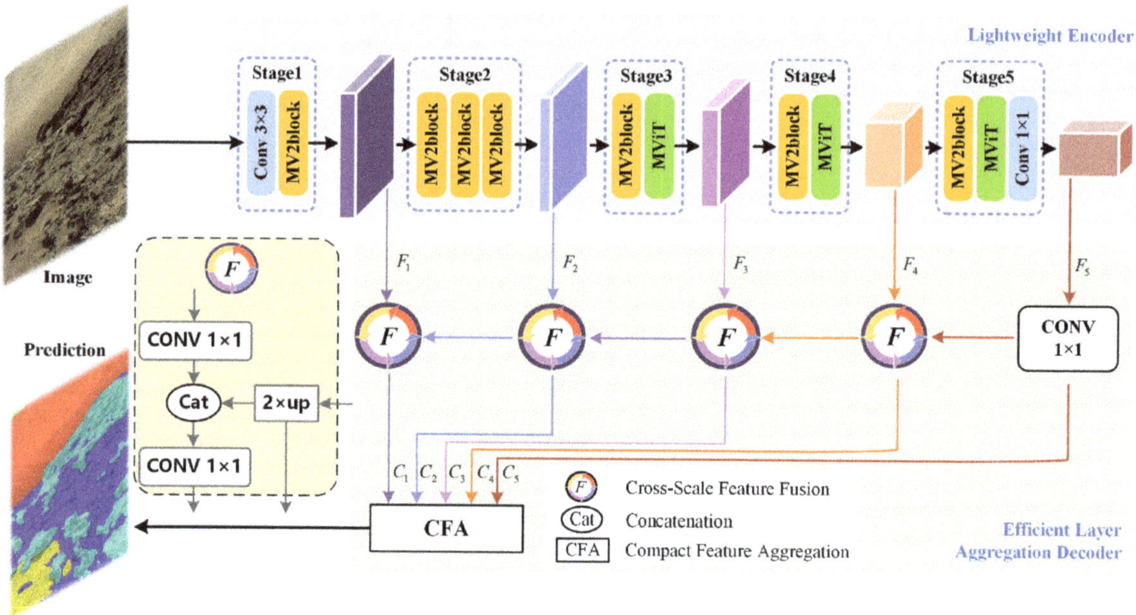

Figure 2. Framework overview of the proposed SegMarsViT.

3.2. Lightweight MViT-Based Encoder

Context modeling in not yet proven to be critical for segmentation and the encoder progressively reduces the spatial resolution and learns more abstract visual concepts with larger receptive fields. However, the encoder is always the most vital part of the whole framework and accounts for the dominant proportion of model size and computational budget.

Considering the strict complexity limitations on the spaceborne payload hardware, we use MobileViT as the backbone to accelerate feature extraction and improve the real-time performance of the proposed method. MobileViT is a lightweight and general-purpose neural network architecture introduced by Apple ML researchers. We removed the last pooling layers and all fully connected layers for image-to-image semantic segmentation prediction. With a special perspective to encode both local and global representations effectively, MobileViT is a hybrid network with both CNN and ViT-like properties. MobileViT improves its stability and performance through incorporating spatial inductive biases of CNN in ViT. As can be seen in Figure 2, the architecture of MobileViT contains the initial fully convolution layer, followed by several MV2 blocks and MViT blocks. Figure 3 visually depicts the design of the two main modules. The MV2 blocks (Figure 3a) come from MobileNetv2 [43] and are mainly responsible for down-sampling in the backbone. Even more to the point, unlike conventional ViTs, the elaborate MViT block (Figure 3b) can learn local and global information with an effective receptive field of $H \times W$.

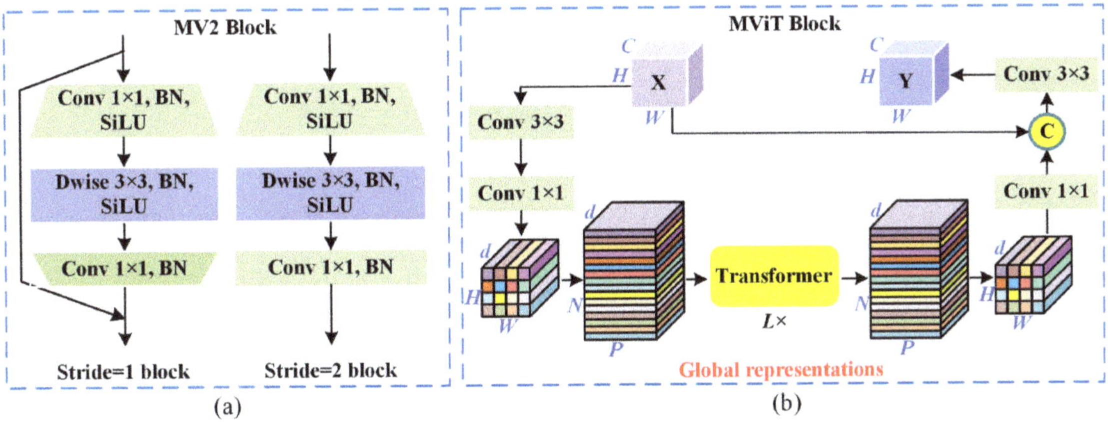

Figure 3. Two types of building blocks in MobileViT backbone: (**a**) MV2 block; (**b**) MViT block.

The first two layers are a standard 3 × 3 convolution layer and a 1 × 1 point-wise expansion layer, where the given input tensor $\mathbf{X} \in \mathbf{R}^{H\times W\times C}$ is projected to $\mathbf{X}_L \in \mathbf{R}^{H\times W\times d}(d > C)$. As is the first step of all the ViTs, $\mathbf{X}_L$ is then split into N non-overlapping patches $\mathbf{X}_U \in \mathbf{R}^{P\times N\times d}$. Next, the standard transformer blocks of multi-headed self-attention (MHA) [44] is applied to model long-range non-local dependencies as:

$$\mathbf{X}_G(p) = \text{Transformer}(\mathbf{X}_U(p)), 1 \leq p \leq P. \quad (1)$$

Then $\mathbf{X}_G \in \mathbf{R}^{P\times N\times d}$ will be folded to obtain $\mathbf{X}_F \in \mathbf{R}^{H\times W\times d}$ as the order of unfolding process. In the end, $\mathbf{X}_F$ will be projected to low C-dimensional space with a point-wise contraction layer and integrate with the raw input tensor $\mathbf{X}$ via concatenation and convolution operations.

The detailed configurations of the MobileViT model are shown in Table 2. The MobileViT models provide three different network sizes (s: small, xs: extral small, and xxs: extra extra small). To obtain multi-scale terrain information, the hierarchical output of five stages will be forwarded into the following decoder module.

Table 2. Detailed architecture of the lightweight backbone used in our SegMarsViT.

	Layer	Output Size	Repeat	Channel		
				xxs	xs	s
Stage 1	Conv 3 × 3	$\frac{H}{2} \times \frac{W}{2}$	1	16	16	16
	MV2 block	$\frac{H}{2} \times \frac{W}{2}$	1	16	32	32
Stage 2	MV2 block	$\frac{H}{2} \times \frac{W}{2}$	1	24	48	64
	MV2 block	$\frac{H}{4} \times \frac{W}{4}$	2	24	48	64
Stage 3	MV2 block	$\frac{H}{4} \times \frac{W}{4}$	1	48	64	96
	MViT block	$\frac{H}{8} \times \frac{W}{8}$	1	48	64	96
Stage 4	MV2 block	$\frac{H}{8} \times \frac{W}{8}$	1	64	80	128
	MViT block	$\frac{H}{16} \times \frac{W}{16}$	1	64	80	128
Stage 5	MV2 block	$\frac{H}{16} \times \frac{W}{16}$	1	80	96	160
	MViT block	$\frac{H}{16} \times \frac{W}{16}$	1	80	96	160
	Conv 1 × 1	$\frac{H}{32} \times \frac{W}{32}$	1	320	384	512

Through leveraging the vision transformer to focus on modeling the global context at all stages, the proposed SegMarsViT can better establish long-range semantic relationships between feature representation. The capacity to model local–global context relationships of images would benefit to learn more abstract semantic visual concepts through enlarging the receptive field. Moreover, the mobile vision transformer is a lightweight and low-latency architecture, which can meet the requirements of accuracy and model complexity in practical satellite missions. The lightweight encoder, therefore, makes the network suitable for real-time applications, as it provides rich semantic information.

3.3. Efficient Layer Aggregation Decoder

In order to further model and fuse multi-level information from the feature encoder, we design an efficient layer aggregation decoder (ELAD) consisting of two primary elements: cross-scale feature fusion (CFF) module and compact feature aggregation (CFA) module in SegMarsViT, as shown in Figure 4. In ELAD, CFF modules are designed to interact multiscale information and strengthen the feature representation learning of lightweight backbone network, and the CFA module is conducted to efficiently aggregation multi-scale deep features and obtain the final segmentation results.

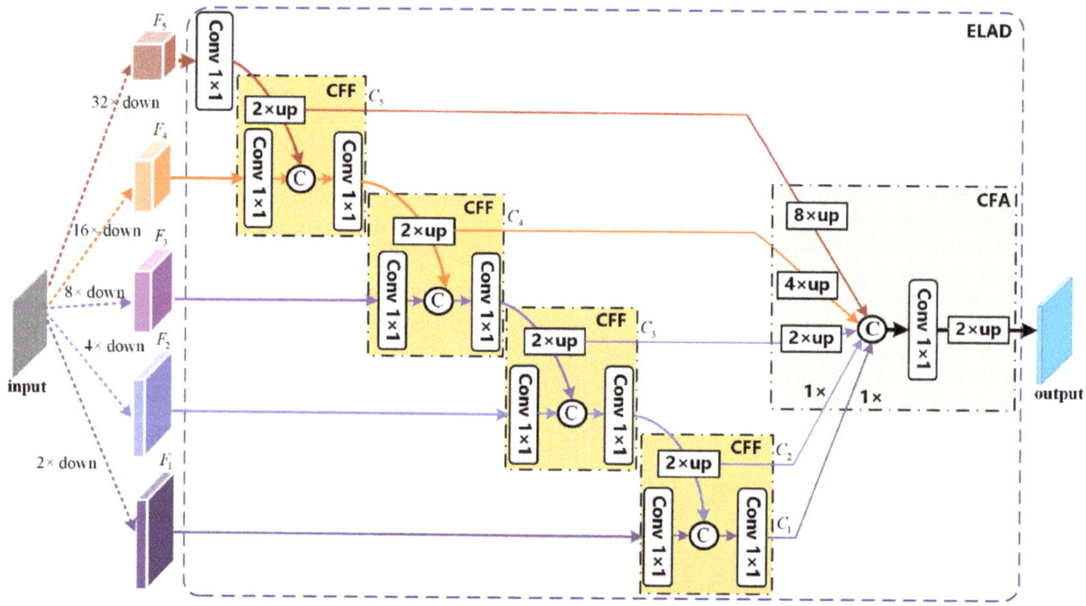

Figure 4. Illustration of the proposed Efficient Layer Aggregation Decoder.

- Cross-Scale Feature Fusion: The utilization of our CFF modules allows high-level context information to be delivered to multi-scale feature maps at different pyramid levels, each of which contains four sub-branches. As can be seen, following the top-down pathway, the input feature maps with coarser resolutions are firstly up-sampled by a factor of 2 to obtain C_i. Meanwhile, we utilize the 1×1 convolution layer for the feature maps in the lower level F_i to unify the channel dimension. Then the up-sampled feature maps C_i are concatenated and fused to the F_i. A 1×1 convolution layer is attached after fusion. Specifically, we have $C_1 = \mathrm{Conv}_{1\times1}(\mathrm{Conv}_{1\times1}(F_1) \oplus C_2)$, where $\mathrm{Conv}_{1\times1}(\cdot)$ represents a 1×1 convolution and $\oplus$ denotes the concatenation operation. In this way, the proposed CFF modules assist our model to enlarge the receptive field through enabling each spatial location to view the local context in different scale spaces.

- Compact Feature Aggregation: After CFF, we perform multi-level feature integration for predicting segmentation maps with fine details. To accomplish multi-level feature fusion, we construct the compact feature aggregation (CFA) module. The output of CFF consists of five fusion maps. We first reshape the high-level feature maps $\{C_1, C_2, C_3\}$ to the same size as C_4 and C_5. Then, all these feature maps in the same spatial resolution are concatenated and followed by a 1×1 convolution for feature fusion. By this means, our lightweight decoder merges multi-level features from top to bottom till the segmentation map of size equal to input image is reconstructed.

3.4. Loss Function

We continue by introducing our loss function for optimizing the proposed SegMarsViT. Our loss function combines the weighted intersection over union (IoU) loss and the weighted binary cross-entropy (BCE) loss:

$$\mathcal{L}oss = \mathcal{L}^{\omega}_{\text{IoU}} + \mathcal{L}^{\omega}_{\text{BCE}} \tag{2}$$

where $\mathcal{L}^{\omega}_{\text{IoU}}$ and $\mathcal{L}^{\omega}_{\text{BCE}}$ represent the weighted IoU loss and BCE loss for the global restriction and local (pixel-level) restriction. Different from the standard IoU loss, which was widely adopted in segmentation tasks, $\mathcal{L}^{\omega}_{\text{IoU}}$ increases the weights of hard pixels to highlight their importance. In addition, compared with the standard BCE loss, $\mathcal{L}^{\omega}_{\text{BCE}}$ pays more attention to hard pixels rather than assigning all the pixels equal weights. The definitions of these losses are the same as in [45,46], and their effectiveness was validated in the field of semantic segmentation.

4. Results and Analysis

In this section, we first provide the experimental setup in the Section 4.1. Then the Section 4.2 presents the results achieved with our model and a comparison made with other segmentation models. In Section 4.3, we conduct comprehensive ablation studies.

4.1. Experimental Settings

4.1.1. Datasets

In order to demonstrate the proposed network's performance, we extensively evaluate our SegMarsViT on three publicly available MTS datasets, including AI4Mars-MSL, MSL-Seg, and S^{5}Mars. These three datasets consist of 17,030, 4155, and 6000 real images of Mars with corresponding pixel-level labels. We offer a brief view in Table 3.

- AI4Mars-MSL: AI4Mars is the first large-scale semantic segmentation dataset build for terrain-aware autonomy on Mars contains 17,000 images with a spatial resolution of 1024×1024, which consists of 3-band RGB images taken by NASA's Mars Science Laboratory (MSL). It contains four classes: Soil, Bedrock, Sand and Big Rock.
- MSL-Seg: The MSL-Seg dataset contains 4184 images with the size of 560×500, which consists of 3-band RGB images from the mars32k dataset (available at https://dominikschmidt.xyz/mars32k/ (accessed on 20 February 2022)). It contains eight categories: Martian soil, Sands, Gravel, Bedrock, Rocks, Tracks, Shadows, and Background.
- S^{5}Mars: The S^{5}Mars dataset contains 6000 images with a spatial resolution of 1200×1200, which are collected by the color mast camera (Mastcam) from the Curiosity rover on Mars. Different from AI4Mars-MSL and MSL-Seg, the overall annotations in S^{5}Mars are employed in a sparse style, which only the pixels with enough human confidence are labeled. It contains nine classes: Sky, Ridge, Soil, Sand, Bedrock, Rock, Rover, Trace, and Hole.

Table 3. Statistics of experimental datasets in this research.

Dataset	Classes	Annotated Images	Image Size	Split
AI4Mars-MSL	4	17,030	1024 × 1024	16,064:322:322
MSL-Seg	8	4155	560 × 500	2893:827:414
S^5Mars	9	6000	1200 × 1200	5000:200:800

4.1.2. Implementation Details

We implement our experiments with the MMSegmentation [47] open source toolbox and Pytorch [48] accelerate training via NVIDIA GPUs. During training, we applied data augmentation operations through random mirror, random resize with ratio 0.5–2.0, random horizontal flipping, random rotation between −10 and 10 degrees and random Gaussian blur for all datasets. Particularly, we random crop to 512 × 512 for AI4Mars, S^5Mars, and MSL-Seg datasets. The proposed model was trained for 400 epochs with a mini-batch size of 16 over 4 GPUs RTX2080Ti. We use the SGD optimizer with the initial learning rate (LR) $1e^{-2}$. The polynomial LR policy [49] was used to update the learning rate and help the model in faster convergence for improving performance.

4.1.3. Evaluation Metrics

For all experiments, we run the same training recipe three times and report several widely used metrics, such as the mean of class-wise intersection over union (mIoU), pixel-wise accuracy (pixelACC), the mean of F1 score (mFscore), and the mean of precision value (mPrecision).

4.2. Comparison with SOTA Methods

In this paper, we compared the proposed SegMarsViT with existing lightweight semantic segmentation methods. We evaluate SegMarsViT against eight SOTA natural image semantic segmentation methods, including FCN [10], DeepLabV3+ [50], Segmenter [51], PSPNet [52], PSANet [53], SegFormer [38], and FPN-PoolFormer [54].

4.2.1. Results on AI4Mars-MSL

Table 4 summarizes our results including parameters, FLOPS and other accuracy metrics of different lightweight semantic segmentation methods on the AI4Mars-MSL dataset. Red, blue, and green denote the best, the second-best, and the third-best results, respectively. For AI4Mars-MSL, there is a relatively small amount of labeled terrain types. With the computing power constraint of available GPUs, we mainly report the results trained with a lightweight backbone. From the results, in comparison with several SOTA approaches, our proposed SegMarsViT outperforms most of them. As shown, on AI4Mars, SegMarsViT yields 68.4% mIoU using only 8.54 M parameters and 5.6 G FLOPs, achieving competitive results in contrast to all other real-time counterparts in terms of parameters and efficiency comprehensively. For instance, compared to SegFormer (MIT-B0), SegMarsViT keeps 0.66% better mIoU.

Table 4. Comparison with state-of-the-art methods on AI4Mars-MSL.

Method (PubYear)	Encoder	pixelACC	mIoU	FLOPs (G)	Params (M)
Segmenter (2021)	ViT-s	92.04	66.85	17.93	26.03
SegFormer (2021)	MIT-B0	92.76	67.74	6.39	3.72
FPN-PoolFormer (2022)	S12	92.72	67.79	30.69	15.64
FCN (2016)	MobileNetv2	92.27	67.12	39.6	9.8
PSPNet (2018)	MobileNetv2	92.41	66.58	52.94	13.72
DeepLabV3+ (2018)	MobileNetv2	91.17	62.04	69.4	15.35
PSANet (2018)	ResNet50	86.83	54.6	194.8	54.07
SegMarsViT (Ours)	MobileViT-s	92.46	68.4	8.54	5.61

4.2.2. Results on S^5Mars

Here, we show both quantitative and qualitative results on S^5Mars. Table 5 shows the comparative results on the test set of S^5Mars. We achieve 78.22% in terms of mIoU, with the standard MobileViT structure as the backbone. The depicted results demonstrate that our model outperforms most of current real-time semantic segmentation works. Figure 5 shows the visual comparison of Martian terrain segmentation methods on five examples from the S^5Mars dataset. The examples include a diverse scene context and backgrounds. The proposed methodology can achieve better or comparable performance in Martian terrain segmentation. What should be noted is that the samples Figure 5a,b are of scenarios in which rough and scattered terrains coexist. From the visual results, the proposed SegMarsViT has less false-positive detection. As for the samples (c) and (d), unstructured scene properties particularly stand out in the images. The proposed method can model contextual information well under the circumstance of unstructured scenes on Mars, which benefit from that ViT-based self-attention technique is applicable to explore spatial correlations. While other competitors may not detect the whole semantic objects or even not find some semantic objects in difficult scenarios, SegMarsViT can segment semantic regions with more accurate results. Especially in the boundary part, the loss of spatial details leads to the loss of accuracy. However, when the difference among foreground objects is relatively small, such as the Figure 5e, some missed detections occur in the results and there is still room for improvement.

Table 5. Comparison with state-of-the-art methods on S^5Mars.

Method	pixelACC	mIoU	mFscore	mPrecision	FLOPs(G)	Params (M)	FPS
Segmenter—ViT-s	90.99	77.15	84.21	85.4	17.93	26.03	48.44
SegFormer—MIT-B0	91.99	79.05	85.74	85.61	6.39	3.72	59.21
FPN—PoolFormer-s12	91.74	76.82	83.3	84.28	30.69	15.64	37.35
FCN—MobileNetv2	86.53	56.57	64.46	72.7	39.6	9.8	58.17
PSPNet—MobileNetv2	90.68	74.64	82.32	82.26	52.94	13.72	50.21
DeepLabV3+—MobileNetv2	89.64	69.5	78.22	80.94	69.4	15.35	37.64
FCN—HRNetv2-w18s	87.71	63.68	73.41	79.71	9.6	3.94	49.53
PSANet—ResNet50	89.11	72.18	80.78	81.96	194.8	54.07	17.64
SegMarsViT—MobileViT-s (Ours)	92.15	78.22	84.74	85.86	8.54	5.61	69.52

4.2.3. Results on MSL-Seg

Table 6 summarizes our results including FLOPS, frame per seconds (FPS), and other four metrics to evaluate the segmentation accuracy on the MSL-Seg dataset. Compared with other latest methods, the proposed SegMarsViT exhibited significant improvement of 2.96% and 5.17% in terms of the pixelACC and mIoU, respectively. We further analyze the classwise segmentation performance of the proposed SegMarsViT on eight classes, we obtain classwise IoU on the test dataset, and is shown in Table 7. It can be observed that IoU for few classes is low, e.g., the Martian soil and bedrocks. This is because the notion of these classes is ambiguous in MSL-Seg. Their low IoU score is due to the low pixel count of these objects in the training data.

Figure 6 shows the ground truth segmentation maps of five sample images along with their predicted segmentation maps. It can be observed that the proposed SegMarsViT has much better comparative results in scenes. As shown in the last two rows of Figure 6, our method can work well on several kinds of complex scenarios with noisy information, while others may fail in such scenarios. It can be seen from the overall detection effect that the main Martian terrain feature can be extracted. However, missed detections and error detections of some objects existed, and the segmentation accuracy needs to be further improved.

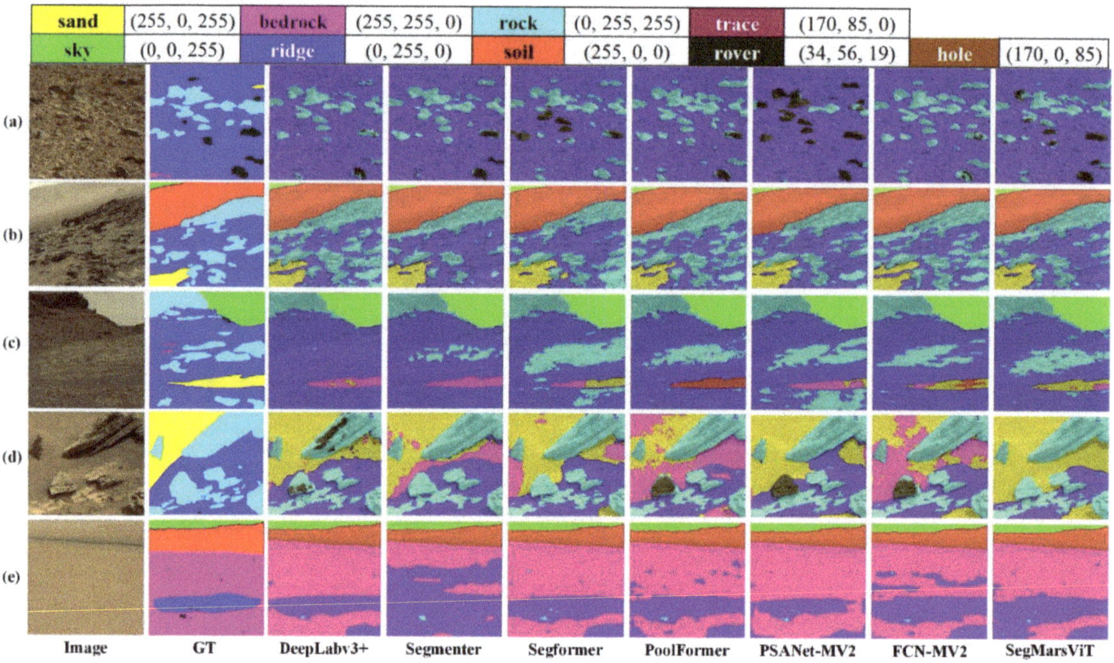

Figure 5. Qualitative comparison on S^5Mars test set. (**a**–**e**) show five experimental samples.

Table 6. Comparison with state-of-the-art methods on MSL-Seg.

Method	pixelACC	mIoU	mFscore	mPrecision	FLOPs (G)	Params (M)	FPS
Segmenter—ViT-s	84.39	66.2	78.32	76.4	17.93	26.03	48.44
SegFormer—MIT-B0	83.84	64.37	76.99	74.74	6.39	3.72	59.21
FPN—PoolFormer-s12	83.9	63.41	76.56	76.03	30.69	15.64	37.35
FCN—MobileNetv2	81.67	54.96	68.32	77.72	39.6	9.8	58.17
PSPNet—MobileNetv2	82.32	60.62	74.2	71.1	52.94	13.72	50.21
DeepLabV3+—MobileNetv2	82.47	59.08	72.78	70.56	69.4	15.35	37.64
FCN—HRNetv2-w18s	82.59	58.57	71.98	75.44	9.6	3.94	49.53
PSANet—ResNet50	83.23	62.43	75.41	73.47	194.8	54.07	17.64
SegMarsViT—MobileViT-s (Ours)	86.05	67.28	78.69	78.75	8.54	5.61	69.52

Table 7. Classwise IoU of the proposed SegMarsViT on MSL-Seg dataset.

Martian Soil	Sands	Gravel	Bedrock	Rocks	Tracks	Shadows	Unknown	mIoU
41.3	77	82.43	47.91	74.39	54.22	89.23	71.78	67.28

To further analyze the model efficiency, we summarize the efficiency-related metrics on the three datasets mentioned above and state them in Figure 7. As shown, the proposed SegMarsViT has the fewest parameters among all the models. These metrics are crucial for Martian terrain segmentation on satellite, which has limited storage. Here, frames per second (FPS) is an average speed that per second with size 512 × 512. Data and parameters load time is not considered, and the employed single GPU is NVIDIA 3070Ti with 8-G storage. The time spent per image of our SegMarsViT is less than other semantic segmentation methods. In conclusion, our method achieves the state-of-the-art performance in Martian terrain segmentation and meanwhile is much more efficient than methods with comparable accuracy.

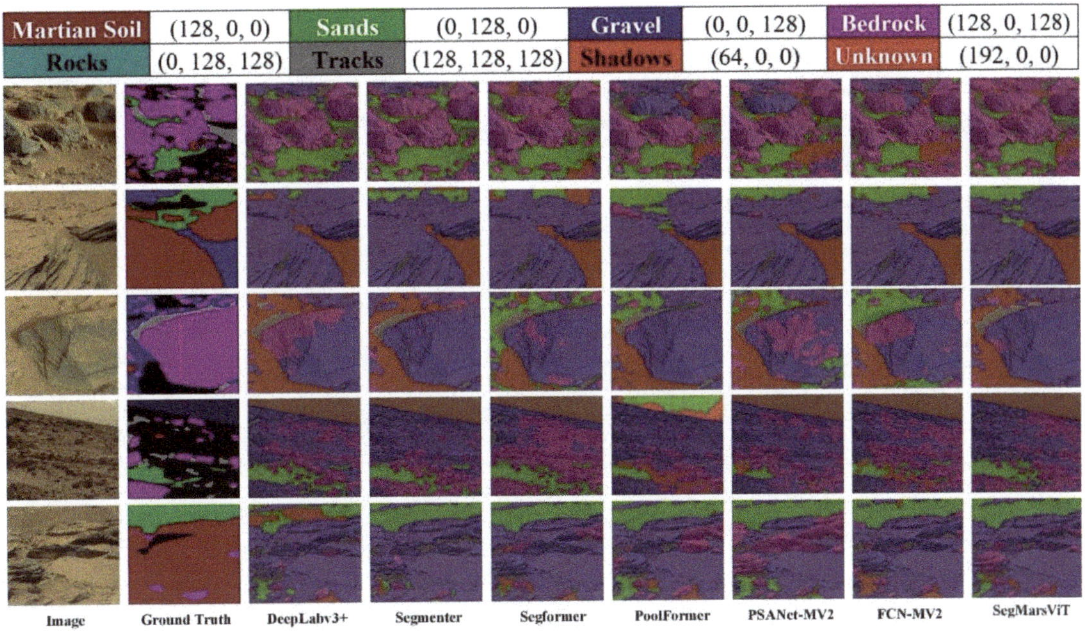

Figure 6. Visualization examples on MSL-Seg test set. (**a**–**e**) represent five experimental samples.

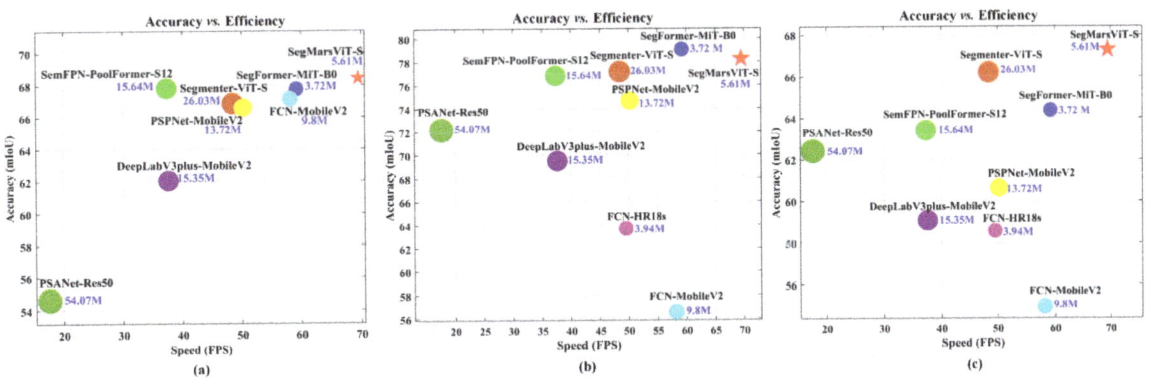

Figure 7. Speed and accuracy comparison on three datasets, (**a**) AI4Mars; (**b**) S^5Mars; (**c**) MSL-Seg. Compared with other regular semantic segmentation methods, the proposed SegMarsViT is competitive.

4.3. Ablation Studies

4.3.1. Effect of Backbone

We first analyze the effect of increasing the size of the encoder on the performance and model efficiency. MobileViT-xxs, MobileViT-xs, and MobileViT-s are the series of mobile transformer encoders with the same architecture but different sizes (as illustrated in Table 2 of Section 3.2). Table 8 summarizes the comparison results for three datasets. It can be observed that both the largest model SegMarsViT-s and the super small model SegMarsViT-xxs achieve close to or exceeding SOTA performance. Furthermore, the super small model SegMarsViT-xxs has good performance, and the number of parameters and FLOPs are 1.84 M and 1.16 G, along with 66.81%, 74.83%, and 65.80% mIoU on the three datasets, respectively. Because the model parameters of the backbone network are smaller and the structure is compact, the pressure on computing resources is smaller. Hence, the proposed model can be better applied to engineering.

Table 8. Evaluation of encoder with different model sizes for SegMarsViT.

Encoder	Complexity			AI4Mars		S^5Mars		MSL-Seg	
	FLOPs	Params	FPS	pixelACC	mIoU	pixelACC	mIoU	pixelACC	mIoU
MobileViT-xxs	1.16 G	1.84 M	110.1	91.92	66.81	89.34	74.83	83.17	65.80
MobileViT-xs	2.23 G	4.61 M	80.3	91.99	67.73	91.58	76.32	84.35	66.83
MobileViT-s	8.54 G	5.61 M	69.5	92.46	68.4	92.15	78.22	86.05	67.28

4.3.2. Effect of ELAD

In this subsection, we test the proposed ELAD with different decoders. As mentioned earlier, the proposed efficient layer aggregation decoder (ELAD) consists of stagewise CFF modules and one CFA module in a nutshell, which are constructed in the way shown in Figure 8. In addition, we select two representative decoders (Figure 9a,b) for test: the All-MLP decoder, termed AMD, first proposed in SegFormer [38], and the classic decoder of the U-shaped network [11], termed UNetD. In practice, we use the official code provided by the authors to implement our experiments. From Table 9, with the same encoder, e.g., MobileViT-s encoder, we find that the proposed ELAD produces higher performance.

Compared with AMD, the proposed ELAD achieves through introducing the CFF modules to build internal communications between the adjacent feature stages. The experimental results in Table 9 verify the significant effect of our CFFs on better fusing feature maps at different scales from another perspective, and this is exactly the common point of ELAD and UNetD. Both consist of an information fusion path for modeling a more representative and robust context. The key difference is the way they implement feature fusion across adjacent stages. The comparison results on Table 9 show that our ELAD has the least FLOPs with comparable parameters, which are the vital part of the construction for the lightweight segmentation network.

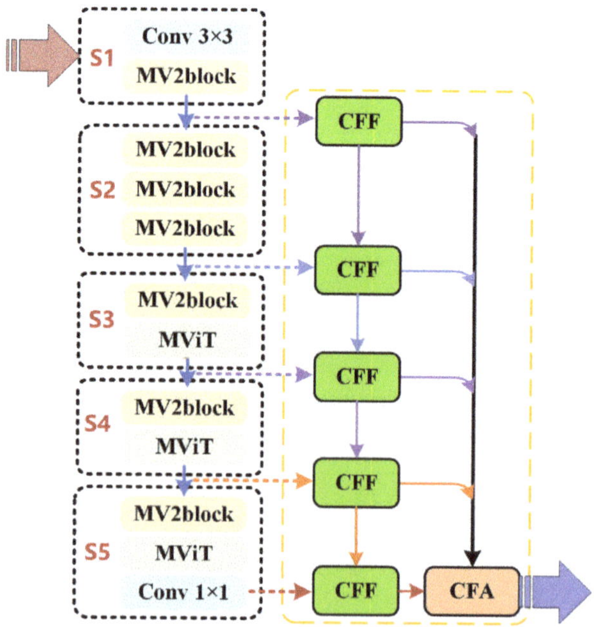

Figure 8. Illustrating the ELAD architecture of the proposed SegMarsViT.

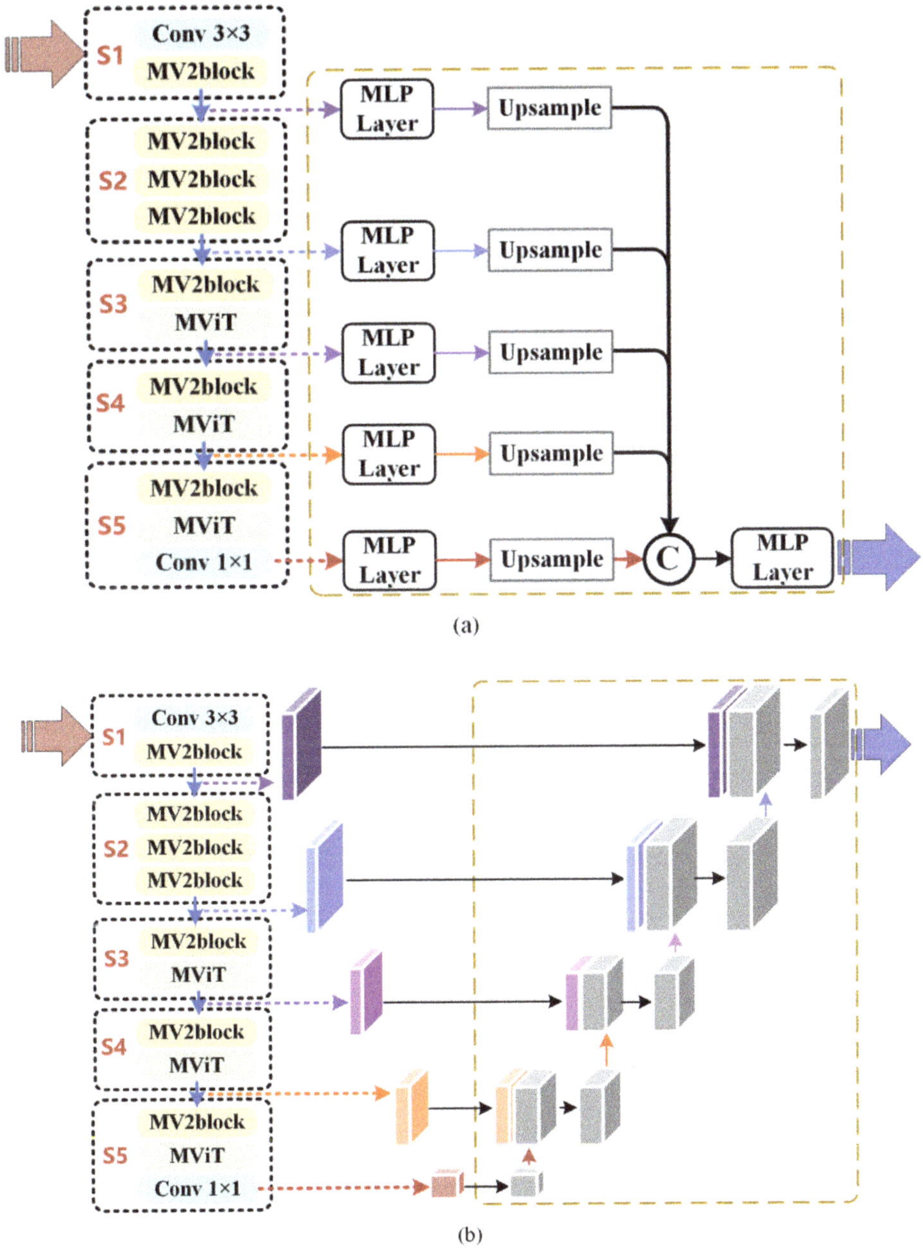

Figure 9. Illustration of different decoder architectures: (**a**) All-MLP Decoder, (**b**) UNet Decoder.

Table 9. Ablation studies for decoder on MSL-Seg dataset.

Encoder	Decoder	pixelACC	mIoU	FLOPs(G)	Params (M)
MobileViT-s	AMD	85.66	66.69	13.14	5.09
MobileViT-s	UNetD	85.48	66.83	12.85	6.23
MobileViT-s	ELAD	86.05	67.28	8.54	5.61

4.3.3. Effect of Components in Decoder

In this subsection, we design ablation experiments on SegMarsViT to examine the validity of CFF and CFA modules.

(1) Baseline: MobileViT-s + ELAD (Without CFA).
(2) SegMarsViT: MobileViT-s + ELAD (CFF + CFA).

Table 10 reports the ablation studies of the baseline and our model on the MSL-Seg test set. We can see that incorporating both the CFF and CFA modules results in consistent and significant increase over the baseline. In particular, when compared with the baseline model, mIoU and PixelACC of the SegMarsViT with both CFF and CFA blocks integrated are improved by 4.06% and 2.62%, respectively. The great improvement of SegMarsViT proves the gain effect of their combination.

Table 10. Ablation results on MSL-Seg dataset.

Method	Modules		pixelACC	mIoU	FLOPs(G)	Params (M)
	CFF	CFA				
Baseline	-	✓	83.43	63.22	7.54	5.0
SegMarsViT	✓	✓	86.05	67.28	8.54	5.61

5. Conclusions

In this paper, we propose SegMarsViT, a lightweight network for the real-time Martian terrain segmentation task. We adopt a deployment-friendly MobileViT backbone to extract discriminative local–global context information from multi-scale feature space. Further, an effective cross-scale feature fusion module was designed to encode context information in the multi-level features, with a cross-scale feature fusion mechanism applied to help further aggregate feature representations. In the end, a compact prediction head is used to aggregate hierarchical features and help enhance feature learning, yielding run-time efficiency. Empirical results validate the superiority of the proposed SegMarsViT over mainstream semantic segmentation methods. The ablation study verifies the effectiveness of each module. More specifically, MobileViT helps obtain the semantic properties of terrain objects in terms of morphology and distribution, while the compact decoder can lead to both high efficiency and performance. Through the comparison of parameters, FLOPs and FPS, the SegMarsViT further demonstrates its advantages in terms of space and computation complexity. All of the results fully demonstrate the capability of the SegMarsViT in efficient and effective Martian terrain segmentation, which provides significant potential for the further development of MTS task.

One potential limitation is that there's an enormous gap between high-end GPU and a low-memory spacecraft device. Our future work will experiment on a realistic hardware platform to evaluate the model efficiency. Energy consumption and practical performance will be our primary focus. Moreover, we will proceed to refine our approach and be committed to investigate MTS methods for more challenging cases, e.g., multi-source heterogeneous data and a multi-task perception system.

Author Contributions: Conceptualization, Y.D., C.X. and L.Z.; funding acquisition, C.X.; investigation, Y.D., T.Z. and C.X.; methodology, Y.D.; writing—original draft, Y.D.; writing—review and editing, T.Z., C.X. and L.Z. All authors have read and agreed to the published version of the manuscript.

Funding: This research was funded by the Chinese Academy of Sciences Project, grant number: CXJJ-20S017.

Data Availability Statement: Not applicable.

Conflicts of Interest: The authors declare no conflict of interest.

References

1. Cakir, S.; Gauß, M.; Häppeler, K.; Ounajjar, Y.; Heinle, F.; Marchthaler, R. Semantic Segmentation for Autonomous Driving: Model Evaluation, Dataset Generation, Perspective Comparison, and Real-Time Capability. *arXiv* **2022**, arXiv:2207.12939.
2. Csurka, G.; Perronnin, F. A Simple High Performance Approach to Semantic Segmentation. In Proceedings of the BMVC, Leeds, UK, 1 September 2008; pp. 1–10.
3. Corso, J.J.; Yuille, A.; Tu, Z. Graph-Shifts: Natural Image Labeling by Dynamic Hierarchical Computing. In Proceedings of the 2008 IEEE Conference on Computer Vision and Pattern Recognition, Anchorage, AK, USA, 23–28 June 2008; pp. 1–8.
4. Holder, C.J.; Shafique, M. On Efficient Real-Time Semantic Segmentation: A Survey. 19. *arXiv* **2022**, arXiv:2206.08605.
5. Garcia-Garcia, A.; Orts-Escolano, S.; Oprea, S.; Villena-Martinez, V.; Garcia-Rodriguez, J. A Review on Deep Learning Techniques Applied to Semantic Segmentation 2017. *arXiv* **2017**, arXiv:1704.06857.
6. Long, J.; Shelhamer, E.; Darrell, T. Fully Convolutional Networks for Semantic Segmentation. In Proceedings of the IEEE Conference on Computer Vision and Pattern Recognition, Boston, MA, USA, 7–12 June 2015; pp. 3431–3440.
7. Ronneberger, O.; Fischer, P.; Brox, T. *U-Net: Convolutional Networks for Biomedical Image Segmentation*; Springer: Cham, Switzerland, 2015.
8. McGlinchy, J.; Johnson, B.; Muller, B.; Joseph, M.; Diaz, J. Application of UNet Fully Convolutional Neural Network to Impervious Surface Segmentation in Urban Environment from High Resolution Satellite Imagery. In Proceedings of the IGARSS 2019—2019 IEEE International Geoscience and Remote Sensing Symposium, Yokohama, Japan, 28 July–2 August 2019; pp. 3915–3918.
9. Sun, J.; Shen, J.; Wang, X.; Mao, Z.; Ren, J. Bi-Unet: A Dual Stream Network for Real-Time Highway Surface Segmentation. *IEEE Trans. Intell. Veh.* **2022**, *15*. [CrossRef]
10. Chattopadhyay, S.; Basak, H. Multi-Scale Attention u-Net (Msaunet): A Modified u-Net Architecture for Scene Segmentation. *arXiv* **2020**, arXiv:2009.06911.
11. Chu, Z.; Tian, T.; Feng, R.; Wang, L. Sea-Land Segmentation with Res-UNet and Fully Connected CRF. In Proceedings of the IGARSS 2019—2019 IEEE International Geoscience and Remote Sensing Symposium, Yokohama, Japan, 28 July–2 August 2019; pp. 3840–3843.
12. Rothrock, B.; Kennedy, R.; Cunningham, C.; Papon, J.; Heverly, M.; Ono, M. SPOC: Deep Learning-Based Terrain Classification for Mars Rover Missions. In Proceedings of the AIAA SPACE 2016, American Institute of Aeronautics and Astronautics, Long Beach, CA, USA, 13–16 September 2016.
13. Iwashita, Y.; Nakashima, K.; Stoica, A.; Kurazume, R. Tu-Net and Tdeeplab: Deep Learning-Based Terrain Classification Robust to Illumination Changes, Combining Visible and Thermal Imagery. In Proceedings of the 2019 IEEE Conference on Multimedia Information Processing and Retrieval (MIPR), San Jose, CA, USA, 28–30 March 2019; pp. 280–285.
14. Liu, H.; Yao, M.; Xiao, X.; Cui, H. A Hybrid Attention Semantic Segmentation Network for Unstructured Terrain on Mars. *Acta Astronaut.* **2022**, *in press*. [CrossRef]
15. Claudet, T.; Tomita, K.; Ho, K. Benchmark Analysis of Semantic Segmentation Algorithms for Safe Planetary Landing Site Selection. *IEEE Access* **2022**, *10*, 41766–41775. [CrossRef]
16. Wang, W.; Lin, L.; Fan, Z.; Liu, J. Semi-Supervised Learning for Mars Imagery Classification and Segmentation. *arXiv* **2022**, arXiv:2206.02180. [CrossRef]
17. Goh, E.; Chen, J.; Wilson, B. Mars Terrain Segmentation with Less Labels. *arXiv* **2022**, arXiv:2202.00791.
18. Zhang, J.; Lin, L.; Fan, Z.; Wang, W.; Liu, J. S^5Mars: Self-Supervised and Semi-Supervised Learning for Mars Segmentation. *arXiv* **2022**, arXiv:2207.01200.
19. Li, J.; Zi, S.; Song, R.; Li, Y.; Hu, Y.; Du, Q. A Stepwise Domain Adaptive Segmentation Network with Covariate Shift Alleviation for Remote Sensing Imagery. *IEEE Trans. Geosci. Remote Sens.* **2022**, *60*, 3152587. [CrossRef]
20. Swan, R.M.; Atha, D.; Leopold, H.A.; Gildner, M.; Oij, S.; Chiu, C.; Ono, M. AI4MARS: A Dataset for Terrain-Aware Autonomous Driving on Mars. In Proceedings of the 2021 IEEE/CVF Conference on Computer Vision and Pattern Recognition Workshops (CVPRW), Nashville, TN, USA, 19–25 June 2021; pp. 1982–1991.
21. Dai, Y.; Xue, C.; Zhou, L. Visual Saliency Guided Perceptual Adaptive Quantization Based on HEVC Intra-Coding for Planetary Images. *PLoS ONE* **2022**, *19*, e0263729. [CrossRef]
22. Tian, Y.; Chen, F.; Wang, H.; Zhang, S. Real-Time Semantic Segmentation Network Based on Lite Reduced Atrous Spatial Pyramid Pooling Module Group. In Proceedings of the 2020 5th International Conference on Control, Robotics and Cybernetics (CRC), Wuhan, China, 16 October 2020; pp. 139–143.
23. Liu, R.; Tao, F.; Liu, X.; Na, J.; Leng, H.; Wu, J.; Zhou, T. RAANet: A Residual ASPP with Attention Framework for Semantic Segmentation of High-Resolution Remote Sensing Images. *Remote Sens.* **2022**, *14*, 3109. [CrossRef]
24. Li, G.; Yun, I.; Kim, J.; Kim, J. DABNet: Depth-Wise Asymmetric Bottleneck for Real-Time Semantic Segmentation. *arXiv* **2019**, arXiv:1907.11357.
25. Mehta, S.; Rastegari, M.; Caspi, A.; Shapiro, L.; Hajishirzi, H. Espnet: Efficient Spatial Pyramid of Dilated Convolutions for Semantic Segmentation. In Proceedings of the European Conference on Computer Vision (ECCV), Munich, Germany, 8–14 September 2018; pp. 552–568.
26. Siam, M.; Gamal, M.; Abdel-Razek, M.; Yogamani, S.; Jagersand, M. Rtseg: Real-Time Semantic Segmentation Comparative Study. In Proceedings of the 2018 25th IEEE International Conference on Image Processing (ICIP), Athens, Greece, 7–10 October 2018; pp. 1603–1607.

27. Li, Y.; Li, X.; Xiao, C.; Li, H.; Zhang, W. EACNet: Enhanced Asymmetric Convolution for Real-Time Semantic Segmentation. *IEEE Signal Process. Lett.* **2021**, *28*, 234–238. [CrossRef]
28. Yu, C.; Wang, J.; Peng, C.; Gao, C.; Yu, G.; Sang, N. Bisenet: Bilateral Segmentation Network for Real-Time Semantic Segmentation. In Proceedings of the European Conference on Computer Vision (ECCV), Munich, Germany, 8–14 September 2018; pp. 325–341.
29. Yu, C.; Gao, C.; Wang, J.; Yu, G.; Shen, C.; Sang, N. BiSeNet V2: Bilateral Network with Guided Aggregation for Real-Time Semantic Segmentation. *Int. J. Comput. Vis.* **2021**, *129*, 3051–3068. [CrossRef]
30. Yang, Y.; Jiao, L.; Liu, X.; Liu, F.; Yang, S.; Feng, Z.; Tang, X. Transformers Meet Visual Learning Understanding: A Comprehensive Review. *arXiv* **2022**, arXiv:2203.12944.
31. Ye, L.; Rochan, M.; Liu, Z.; Wang, Y. Cross-Modal Self-Attention Network for Referring Image Segmentation. In Proceedings of the IEEE/CVF Conference on Computer Vision and Pattern Recognition, Long Beach, CA, USA, 15–20 June 2019; pp. 10502–10511.
32. Zhu, X.; Su, W.; Lu, L.; Li, B.; Wang, X.; Dai, J. Deformable DETR: Deformable Transformers for End-to-End Object Detection. *arXiv* **2021**, arXiv:2010.04159.
33. Wang, W.; Xie, E.; Li, X.; Fan, D.-P.; Song, K.; Liang, D.; Lu, T.; Luo, P.; Shao, L. Pyramid Vision Transformer: A Versatile Backbone for Dense Prediction without Convolutions. *arXiv* **2021**, arXiv:2102.12122.
34. Zheng, S.; Lu, J.; Zhao, H.; Zhu, X.; Luo, Z.; Wang, Y.; Fu, Y.; Feng, J.; Xiang, T.; Torr, P.H.S.; et al. Rethinking Semantic Segmentation from a Sequence-to-Sequence Perspective with Transformers. *arXiv* **2021**, arXiv:2012.15840.
35. Huang, H.; Xie, S.; Lin, L.; Iwamoto, Y.; Han, X.-H.; Chen, Y.-W.; Tong, R. ScaleFormer: Revisiting the Transformer-Based Backbones from a Scale-Wise Perspective for Medical Image Segmentation. In Proceedings of the Thirty-First International Joint Conference on Artificial Intelligence, International Joint Conferences on Artificial Intelligence Organization, Vienna, Austria, 23–29 July 2022; pp. 964–971.
36. Shi, W.; Xu, J.; Gao, P. SSformer: A Lightweight Transformer for Semantic Segmentation. *arXiv* **2022**, arXiv:2208.02034.
37. Liu, Z.; Lin, Y.; Cao, Y.; Hu, H.; Wei, Y.; Zhang, Z.; Lin, S.; Guo, B. Swin Transformer: Hierarchical Vision Transformer Using Shifted Windows. In Proceedings of the IEEE/CVF International Conference on Computer Vision, Nashville, TN, USA, 19–25 June 2021; pp. 10012–10022.
38. Xie, E.; Wang, W.; Yu, Z.; Anandkumar, A.; Alvarez, J.M.; Luo, P. SegFormer: Simple and Efficient Design for Semantic Segmentation with Transformers. *Adv. Neural Inf. Process. Syst.* **2021**, *14*, 12077–12090.
39. Wang, L.; Li, R.; Zhang, C.; Fang, S.; Duan, C.; Meng, X.; Atkinson, P.M. UNetFormer: An UNet-like Transformer for Efficient Semantic Segmentation of Remotely Sensed Urban Scene Imagery. *ISPRS J. Photogramm. Remote Sens.* **2022**, *190*, 196–214. [CrossRef]
40. Liu, Z.; Ning, J.; Cao, Y.; Wei, Y.; Zhang, Z.; Lin, S.; Hu, H. Video Swin Transformer. In Proceedings of the IEEE/CVF Conference on Computer Vision and Pattern Recognition, New Orleans, LA, USA, 18–24 June 2022; pp. 3202–3211.
41. Hatamizadeh, A.; Tang, Y.; Nath, V.; Yang, D.; Myronenko, A.; Landman, B.; Roth, H.R.; Xu, D. Unetr: Transformers for 3d Medical Image Segmentation. In Proceedings of the IEEE/CVF Winter Conference on Applications of Computer Vision, New Orleans, LA, USA, 18–24 June 2022; pp. 574–584.
42. Zhou, H.-Y.; Guo, J.; Zhang, Y.; Yu, L.; Wang, L.; Yu, Y. Nnformer: Interleaved Transformer for Volumetric Segmentation. *arXiv* **2021**, arXiv:2109.03201.
43. Sandler, M.; Howard, A.; Zhu, M.; Zhmoginov, A.; Chen, L.-C. Mobilenetv2: Inverted Residuals and Linear Bottlenecks. In Proceedings of the IEEE Conference on Computer Vision and Pattern Recognition, Salt Lake City, UT, USA, 18–23 June 2018; pp. 4510–4520.
44. Dosovitskiy, A.; Beyer, L.; Kolesnikov, A.; Weissenborn, D.; Zhai, X.; Unterthiner, T.; Dehghani, M.; Minderer, M.; Heigold, G.; Gelly, S.; et al. An Image Is Worth 16x16 Words: Transformers for Image Recognition at Scale. *arXiv* **2021**, arXiv:2010.11929.
45. Wu, Y.-H.; Liu, Y.; Xu, J.; Bian, J.-W.; Gu, Y.-C.; Cheng, M.-M. MobileSal: Extremely Efficient RGB-D Salient Object Detection. *IEEE Trans. Pattern Anal. Mach. Intell.* **2022**, *44*, 10261–10269. [CrossRef]
46. Wu, Y.-H.; Liu, Y.; Zhang, L.; Cheng, M.-M.; Ren, B. EDN: Salient Object Detection via Extremely-Downsampled Network. *IEEE Trans. Image Process.* **2022**, *31*, 3125–3136. [CrossRef]
47. Contributors, Mms. MMSegmentation: Openmmlab Semantic Segmentation Toolbox and Benchmark. 2020. Available online: https://github.com/open-mmlab/mmsegmentation (accessed on 18 May 2022).
48. Paszke, A.; Gross, S.; Massa, F.; Lerer, A.; Bradbury, J.; Chanan, G.; Killeen, T.; Lin, Z.; Gimelshein, N.; Antiga, L.; et al. Pytorch: An Imperative Style, High-Performance Deep Learning Library. *Adv. Neural Inf. Process. Syst.* **2019**, *32*, 1–12.
49. Mishra, P.; Sarawadekar, K. Polynomial Learning Rate Policy with Warm Restart for Deep Neural Network. In Proceedings of the TENCON 2019—2019 IEEE Region 10 Conference (TENCON), Kochi, India, 17–20 October 2019; pp. 2087–2092.
50. Chen, L.-C.; Zhu, Y.; Papandreou, G.; Schroff, F.; Adam, H. Encoder-Decoder with Atrous Separable Convolution for Semantic Image Segmentation. In Proceedings of the European Conference on Computer Vision (ECCV), Munich, Germany, 8–14 September 2018; pp. 801–818.
51. Strudel, R.; Garcia, R.; Laptev, I.; Schmid, C. Segmenter: Transformer for Semantic Segmentation. In Proceedings of the IEEE/CVF International Conference on Computer Vision, Nashville, TN, USA, 19–25 June 2021; pp. 7262–7272.
52. Zhao, H.; Shi, J.; Qi, X.; Wang, X.; Jia, J. Pyramid Scene Parsing Network. In Proceedings of the IEEE Conference on Computer Vision and Pattern Recognition, Honolulu, HI, USA, 21–26 July 2017; pp. 2881–2890.

53. Zhao, H.; Zhang, Y.; Liu, S.; Shi, J.; Loy, C.C.; Lin, D.; Jia, J. Psanet: Point-Wise Spatial Attention Network for Scene Parsing. In Proceedings of the European Conference on Computer Vision (ECCV), Munich, Germany, 8–14 September 2018; pp. 267–283.
54. Yu, W.; Luo, M.; Zhou, P.; Si, C.; Zhou, Y.; Wang, X.; Feng, J.; Yan, S. MetaFormer Is Actually What You Need for Vision. In Proceedings of the IEEE/CVF Conference on Computer Vision and Pattern Recognition, New Orleans, LA, USA, 18–24 June 2022.

remote sensing

Technical Note

Exploring Semantic Prompts in the Segment Anything Model for Domain Adaptation

Ziquan Wang, Yongsheng Zhang, Zhenchao Zhang *, Zhipeng Jiang, Ying Yu, Li Li and Lei Li

School of Geospatial Information, PLA Strategic Support Force Information Engineering University, Zhengzhou 450001, China; aresdrw@163.com (Z.W.); yszhang2001@vip.163.com (Y.Z.); jiangzp0803@163.com (Z.J.); yuying5559104@163.com (Y.Y.); lili315114@163.com (L.L.); 3110100798@zju.edu.cn (L.L.)

* Correspondence: zhzhc_1@163.com; Tel.: +86-150-9330-3012

Abstract: Robust segmentation in adverse weather conditions is crucial for autonomous driving. However, these scenes struggle with recognition and make annotations expensive, resulting in poor performance. As a result, the Segment Anything Model (SAM) was recently proposed to finely segment the spatial structure of scenes and to provide powerful prior spatial information, thus showing great promise in resolving these problems. However, SAM cannot be applied directly for different geographic scales and non-semantic outputs. To address these issues, we propose SAM-EDA, which integrates SAM into an unsupervised domain adaptation mean-teacher segmentation framework. In this method, we use a "teacher-assistant" model to provide semantic pseudo-labels, which will fill in the holes in the fine spatial structure given by SAM and generate pseudo-labels close to the ground truth, which then guide the student model for learning. Here, the "teacher-assistant" model helps to distill knowledge. During testing, only the student model is used, thus greatly improving efficiency. We tested SAM-EDA on mainstream segmentation benchmarks in adverse weather conditions and obtained a more-robust segmentation model.

Keywords: segment anything model (SAM); unsupervised domain adaptation; semantic road scene segmentation

Citation: Wang, Z.; Zhang, Y.; Zhang, Z.; Jiang, Z.; Yu, Y.; Li, L.; Li, L. Exploring Semantic Prompts in the Segment Anything Model for Domain Adaptation. *Remote Sens.* **2024**, *16*, 758. https://doi.org/10.3390/rs16050758

Academic Editors: Qian Du, Jiaojiao Li, Wei Li, Jocelyn Chanussot, Rui Song, Yunsong Li and Bobo Xi

Received: 20 November 2023
Revised: 1 February 2024
Accepted: 14 February 2024
Published: 21 February 2024

1. Introduction

The semantic segmentation [1–7] of road scenes is important for autonomous driving [5], particularly during scene data analyses and behavior decision-making [8]. This technology also has good applications in motion control planning [9,10] and multi-sensor fusion processing [11]. Furthermore, over the past decade, we have seen tremendous advancements in semantic segmentation technology [7,12–15]. Currently, intelligent semantic segmentation algorithms can even outperform humans in recognizing clear scenes [15]. However, these works mostly ignore the deterioration of image quality caused by adverse weather conditions such as fog, rain, and snow [16]. This leads to an obvious performance decline. Unfortunately, the reliable and safe operation of intelligent systems requires the underlying recognition processes to be highly robust under these adverse conditions. Thus, this issue is receiving increasing attention now.

Adverse weather conditions bring two main challenges to semantic segmentation. Firstly, important objects become blurred, which leads to higher uncertainty in the outputs of these intelligent algorithms. Although some studies have tried to restore these images [17] and have attempted to convert them into images with clear scenes, a domain gap still exists. Secondly, annotating these scenarios is more difficult than annotating clear ones, making it expensive to use supervised algorithms. Therefore, many studies have adopted unsupervised domain adaptation (UDA) strategies [18–20] in an attempt to transfer segmented knowledge from a clear annotated source domain to adverse weather scenes (the target domain). However, in the transfer process, a domain gap in the UDA methods inevitably leads to information loss, resulting in imprecise segmentation in the target domain scenario.

Recently, the Segment Anything Model (SAM) [1] has attracted much attention as it uses massive amounts of data to pre-train and conduct self-supervised learning, acquiring an extremely strong generalization ability. Such a generalization ability enables SAM to be directly applied to various vision-based tasks without task-oriented training, including camouflaged object detection [21] and image in-painting [22]. Concretely, SAM can finely segment all objects in an image, thus providing powerful prior spatial structure information. Even in adverse conditions, SAM remains robust [23]. Thanks to SAM's generalization ability, SAM-DA [24] can make predictions from nighttime images and has a large number of samples for training, which greatly improves the performance of the model. Thus, we can assume that applying SAM's spatial structure information to UDA methods, i.e., adding a powerful supervision signal to the UDA framework, will be beneficial.

However, currently, SAM cannot be integrated directly into the UDA framework for three main reasons: (1) As mentioned above, SAM is not a task-oriented model, and a well-designed access plugin is needed to adapt it to semantic segmentation tasks. (2) Limited by its computing power, SAM is difficult to mount on the platform of a vehicle. (3) The operational speed of SAM is very slow and is insufficient when applied to real scenarios. For problem (1), the SSA [25] method can be used to fuse the spatial structure information generated by SAM with the semantic information generated by a segmentation model. However, the SSA method exacerbates the problem of slow operation, taking 40–60 s to complete segmentation for just one image, and its original semantic branch has not been trained to adapt to adverse weather conditions, resulting in inaccurate information and, therefore, producing unsatisfactory results. For problems (2) and (3), some scholars put forward Fast-SAM [26] and Faster-SAM [26], which have greatly improved the operational efficiency of SAM and can be deployed from mobile terminals, thus further adding significance to the research in this paper, **as is shown in Figure 1**.

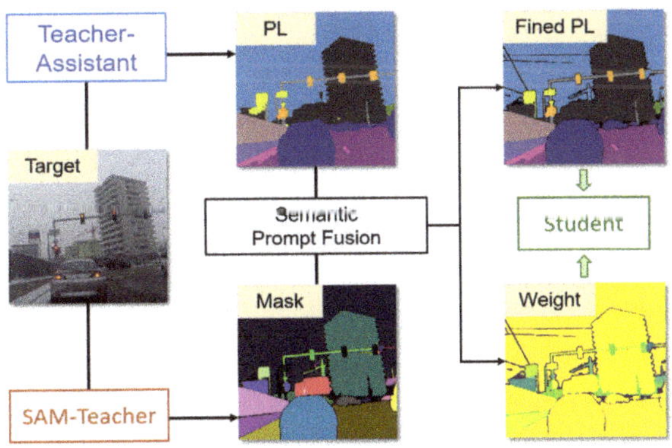

Figure 1. The main idea behind the proposed method. For images from the target domain, the teacher-assistant model and SAM-teacher generate semantic segmentation masks (called "semantic prompts") and spatial structure masks, respectively, and, then, use the algorithm mentioned in Section 2.2 for fusion. Due to SAM's strong generalization ability, this step can produce pseudo-labels that are more consistent with real scene distributions; so, the student model can completely explore the target domain knowledge, similar to the method of supervised learning.

To address the above issues, we propose a SAM-enhanced UDA method called **SAM-EDA as shown in Figure 2**, aiming to improve segmentation performance by utilizing the SAM knowledge while maintaining its original operational speed. Specifically, we plugged SAM (or its variants) into a mean-teacher's self-training domain adaptation architecture [19,27], dynamically carrying out SAM-enhanced learning on the target domain, as well as knowledge distillation.

The whole architecture and pipeline consist of three sub-modules: (1) the student model, (2) the teacher-assistant (TA) model, and (3) the SAM-teacher model. However, only the student segmentation model will be published for evaluation. In a single training iteration, the TA and SAM-teacher models generate semantic segmentation masks (called "semantic prompts") and spatial structure masks on the target domain, respectively, and, then, use the pseudo-label fusion algorithm mentioned in Section 2.2 for fusion. Due to SAM's strong generalization ability, this step can produce pseudo-labels that are more consistent with real scene distributions, so the student model can completely explore the target domain knowledge, similar to the method of supervised learning. After completing the training, neither the SAM-teacher nor TA models remain, thus maintaining the speed of the existing semantic segmentation network.

The contributions of this article can be summarized as follows:

(1) We propose a simple, but effective semantic filling and prompt method for SAM masks, which utilizes the output of existing semantic segmentation models to provide SAM with class information and explore methods to address the scale of the SAM segmentation results;

(2) To the best of our knowledge, we are the first to incorporate SAM into an unsupervised domain adaptation framework, which includes the SAM-teacher, teacher-assistant, and student models, achieving knowledge distillation in the case of completely inconsistent structures and output spaces between SAM and the main segmentation model, effectively improving its adaptability in adverse scenarios;

(3) Our method is applicable to different UDA frameworks and SAM variants, providing useful references for the application of large models in local professional fields.

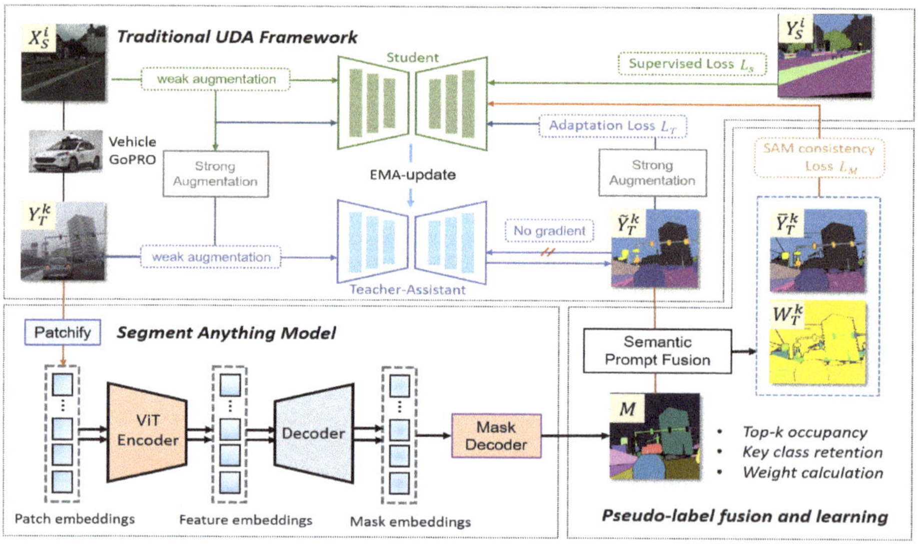

Figure 2. The pipeline of the proposed method. Both the source and target domain images used in this method were captured from a vehicle perspective camera. The target domain image Y_T was first fed into the teacher-assistant g_ϕ to generate coarse pseudo-labels $\widetilde{Y}_T$, which serve as semantic prompts. Then, Y_T was put into SAM to obtain a spatial structural segmentation map M, leveraging SAM's generalization. We merged $\widetilde{Y}_T$ and M to incorporate the semantic information. During the merging process, the top-k occupancy ratio method was mainly used to retain some key class pixels from $\widetilde{Y}_T$ while considering the holes in the SAM's missing segmentation. The weights were also calculated based on the proportion of semantic pixels to reduce the impact of uncertainty in SAM. The merged pseudo-label $\overline{Y}_T$ was close to the distribution of the real-world scene, thus enabling supervision of the student model.

2. Method

2.1. Unsupervised Domain Adaptation (UDA)

In order to perform an unsupervised domain adaptation for semantic segmentation, we utilized a student network f_θ and a teacher-assistant network g_ϕ based on the mean-teacher [19,27] pipeline. Given a set of labeled source domain data $\{(X_S^i, Y_S^i)\}_{i=1}^{N_S}$ (where Y_S^i is the pixel-wise semantic label of X_S^i), the student network directly learns from the source domain data using the cross-entropy loss function:

$$\mathcal{L}_i^{S,cls/seg} = \mathcal{H}(f_\theta(X_S^i), Y_S^i) \tag{1}$$

$$\mathcal{H}(\widetilde{y}, y) = -\sum_{h=1}^{H}\sum_{w=1}^{W}\sum_{c=1}^{C} y_{hwc} \log \widetilde{y}_{hwc} \tag{2}$$

However, models trained only on the source domain often lack generalization; thus, knowledge from the target domain needs to be extracted with N_T unlabeled images $\{(X_T^k)\}_{k=1}^{N_T}$. In our UDA pipeline, the teacher-assistant network g_ϕ needs to make predictions using the target domain images and needs to generate pseudo-labels $\{(\widetilde{Y}_T^k)\}_{k=1}^{N_T}$, so the learning loss function based on the pseudo-labels can be denoted as L_k^T, which is similar to the supervised $L_i^{S,cls/seg}$:

$$\mathcal{L}_k^T = \mathcal{H}(g_\phi(X_T^k), \widetilde{Y}_T^k) \tag{3}$$

The pseudo-labels generated by g_ϕ are often inaccurate (especially during the early stages of training), so it is necessary to set a dynamic weight λ to balance the impact of noise in the pseudo-labels. Generally, λ is set as the confidence pixel ratio exceeding a certain threshold τ:

$$\lambda_T^k = \frac{\sum_{p=1}^{H\times W}\left[\max_{c'} g_\phi(X_T^k)^{(p,c')} \geq \tau\right]}{H \times W} \tag{4}$$

Finally, the total loss function of our UDA architecture is the weighted sum of source domain loss and target domain loss:

$$\min_\theta \frac{1}{N_S}\sum_{i=1}^{N_S}\mathcal{L}_i^S + \frac{1}{N_T}\sum_{k=1}^{N_T}\lambda^T\mathcal{L}_k^T \tag{5}$$

2.2. Semantic Prompt Fusion and Learning

After standard UDA loss computation, we employed SAM [1] (or its variant) to make additional predictions on the target domain image X_T^k. Taking the standard SAM as an example, the input image was first patchified, automatically calculating the points of each patch as prompts. Then, SAM used a ViT-based [28] encoder and decoder head to obtain a feature embedding and a mask embedding. Finally, a mask decoder head identified several masks M without semantic information. Let h_φ denote SAM-series used in our pipeline. This process can be described as follows:

$$\{M_j\}_{j=1}^{N_k^m} = h_\varphi(X_T^k) \tag{6}$$

The distribution of M closely matches real-world scenarios, but it requires semantic information from pseudo-label $\widetilde{Y}_T$, referred to as a "semantic prompt". For a single mask M_j, the class ID can be obtained by calculating the most-frequent category ID within the corresponding region in $\widetilde{Y}_T$:

$$M_j^{cls} = \arg\max_c(count(M_j \odot \widetilde{Y}_T^k)_{p=c}) \tag{7}$$

However, existing dataset label systems are often restricted to the broadest instance-level labels (such as cars, buses, buildings, etc.), while SAM's segmentation has multi-scale outputs (e.g., window, car, etc.). This leads to some errors when preserving the SAM segmentation masks (see Section 4.2), meaning that certain parts of the SAM output are not fully representative of their objects, which makes it difficult to avoid using general rules. To mitigate this, we calculated the weights for each mask M_j based on the maximum occupied pixel's class ID proportion:

$$W_j^{cls} = \frac{\text{sum}(count(M_j \odot \widetilde{Y}_T^k)_{p=c})}{|M_j|} \tag{8}$$

Then, due to holes being present in small objects when using the SAM and the potential confusion between similar classes (such as roads and sidewalks), it is important to preserve some pixels to identify key classes. Taking the Cityscapes dataset [12] as an example, we selected a set of classes among $[0, 19]$, denoted as set K, through empirical judgment. The final obtained pseudo-label is a combination of the masks M_j, along with the inclusion of key class pixels from $\widetilde{Y}_T$:

$$\overline{Y}_T^k = \bigcup_{j=1}^{n_k^m} M_j^{cls} \cup \widetilde{Y}_T^k[c = c_d] \quad c_d \in K \tag{9}$$

For the mask filled using Equation (7), the weights are determined using Equation (8). For the remaining parts, the weights (which will participate in the loss function) were uniformly set to 1. The final weight matrix is as follows:

$$\overline{W}_T^k = \bigcup_{j=1}^{n_k^m} W_j^{cls} \cup \mathbf{1} \odot \widetilde{Y}_T^k[c = c_d] \quad c_d \in K \tag{10}$$

Thus, we can obtain the pseudo-labels enhanced by SAM, which can be used to construct a loss function similar to that in Equation (1), namely $\mathcal{L}_M = \mathcal{H}(f_\theta(X_T^k), \overline{Y}_T^k)$. Consequently, the final loss function is

$$\min_\theta \frac{1}{N_S} \sum_{i=1}^{N_S} \mathcal{L}_i^S + \frac{1}{N_T} \sum_{k=1}^{N_T} (\lambda^T \mathcal{L}_k^T + W_T^k \mathcal{L}_M) \tag{11}$$

3. Results

3.1. Implementation Details

3.1.1. Adverse Condition Semantic Segmentation Dataset

We used the Cityscapes dataset [12] as the source domain for training, which includes 2975 training images. The candidate target domain includes four different datasets—ACDC [16], Foggy Driving [29] and Foggy Driving Dense [30], Rainy Cityscapes [31], and Dark-Zurich (DZ) [32]—covering images with adverse conditions such as foggy, rainy, snowy, and nighttime scenes. Among them, ACDC contains 1600 images for training and 400 images for validation. Dark-Zurich contains 2416 training images and 151 test images. All datasets were labeled according to the Cityscapes standard, which includes 19 categories.

3.1.2. SAM-EDA Parameters

For the UDA architecture, we used DAFormer [19] as the baseline. Both the teacher-assistant and student models were SegFormer [7] with an MiT-B5 backbone. We followed DAFormer to set the EMA parameter $\alpha = 0.99$ and the confidence threshold $\tau = 0.968$. For the SAM mask generator, we used the largest SAM-ViT-H [1,28] and set the prediction IoU threshold δ_{iou} to 0.8. We also set the stability score threshold δ_{sta} to 0.8 and the minimum mask region $r_{\min}$ to 50 pixels. The settings of the SAM parameters directly affect the quality and quantity of segmentation masks and determine the geographical scale. All the experiments were conducted on a Tesla v100 graphic card with 32 GB of graphic memory, equipped with CUDA 10.2 and cudnn 7.6.5.

3.2. Performance Comparison

We compared our methods with prominent UDA methods for four kinds of adverse conditions, as well as with the segmentation method SSA [25] combined with SAM application. For foggy scenes, we compared CuDA-Net [20] and FIFO [33]; for nighttime scenes, we compared VBLC [34] and GCMA [35]. These methods are all specialized for specific scenes. As for universal domain adaptation methods, we compared DAFormer [19], CumFormer [36], and the SSA method combined with SAM. For the SSA method, we provide the results using different extractors (ViT-B, ViT-L, and ViT-H). All performance comparisons are shown in Table 1 and Figure 3. We not only provide comprehensive performance comparisons for each method, but also present their runtime and memory consumption. All evaluation metrics were calculated on the validation sets of each dataset. In Table 2, we show the improvement of our method to different UDA strategies. In Table 3 and Figure 4, we show the influence of different fusion methods between SAM-generated masks and original pseudo labels. In Table 4, we show the performance of replacing the original SAM to its variants.

Table 1. Performance comparison. Experiments were conducted on the ACDC, Foggy Driving, Foggy Driving Dense, Rainy Cityscapes, and Dark-Zurich validation sets and measured with the mean intersection over union (mIoU %) over all classes.

Model	Pub/Year	Backbone	Dataset								Speed/FPS	GPU/GB
			Fog			Rain		Snow	Night			
			ACDC-f	FD	FDD	ACDC-r	Rain-CS	ACDC-s	ACDC-n	DZ		
DAFormer [19]	CVPR 2022	SegFormer [7]	63.41	47.32	39.63	48.27	75.34	49.19	**46.13**	**43.80**		
CuDA-Net [20]	CVPR 2022	DeepLabv2 [13]	68.59	53.50	48.20	48.52	69.47	47.20	-	-		
FIFO [33]	CVPR 2022	Refinelw-101 [14]	70.36	50.70	48.90	-	-	-	-	-	6–10	Train: 16 GB Test: 8 GB
CumFormer [36]	TechRXiv 2023	SegFormer	74.92	56.25	**51.91**	57.14	79.34	62.42	44.75	43.20		
VBLC [34]	AAAI 2023	SegFormer	-	-	-	-	**79.80**	-	-	44.41		
GCMA [35]	ICCV 2019	DeepLabv2	-	-	-	-	-	-	-	42.01		
SegFormer (cs)	NeurIPS 2021	-	64.74	46.06	33.15	40.62	68.31	42.03	26.61	23.43	6–10	-
SSA + SAM + SegFormer	arXiv 2023 Github 2023	ViT-B [28]	60.57	39.02	25.33	43.17	67.51	42.93	24.97	22.36	<0.1	Train: 8–48 GB Test: 16–24 GB
		ViT-L [28]	66.78	48.02	31.33	52.94	68.69	51.47	27.69	26.73		
		ViT-H [28]	68.16	50.89	33.72	54.39	70.27	53.32	29.60	28.92		
OneFormer (cs) [15]	arXiv 2022	-	72.31	51.33	44.31	56.72	74.96	55.13	32.41	26.74	4–5	-
SSA + SAM + OneFormer	arXiv2023 GitHub2023	ViT-B	69.13	46.97	41.96	58.77	73.03	57.14	36.78	28.96	<0.1	Train: 8–48 GB Test: 16–24 GB
		ViT-L	75.94	53.14	46.78	64.25	75.62	64.21	40.14	34.25		
		ViT-H	77.87	55.61	48.41	69.25	76.31	66.22	41.22	37.43		
SAM-EDA(Ours)	-	ViT-B	68.10	50.74	43.66	54.20	71.01	55.47	33.62	27.63	6.7	Train: 8–48 GB Test: 8 GB
		ViT-L	75.30	55.49	46.98	64.68	73.41	58.12	41.30	35.45		
		ViT H	**78.25**	**56.37**	51.25	**69.38**	76.63	**68.17**	43.15	42.63		

Table 2. SAM-EDA for UDA methods. Experiments were conducted on the ACDC-Fog validation set and measured with the mean intersection over union (mIoU %) over all classes.

UDA Method	w/o SAM-EDA	w/ SAM-EDA	Diff.
DACS [37]	61.08	64.28	+3.20
ProDA [18]	65.17	68.74	+3.57
DAFormer [19]	67.93	71.61	+3.68
CuDA-Net [20]	68.56	72.37	+3.81
CumFormer [36]	74.92	77.89	+2.97

Table 3. Different semantic prompt fusion methods. Experiments were conducted on the ACDC and Dark-Zurich validation set and measured with the mean intersection over union (mIoU %) over all classes.

Method/Datasets	ACDC-F				Dark-Z	Mean	Gain (mIoU)
	Fog	Rain	Snow	Night			
DAFormer [19]	63.41	48.27	49.19	46.13	43.80	50.16	+0.00
IoU [38]	55.19	41.58	42.17	39.48	27.66	41.22	−8.94
SSA (SegFormer) [25]	68.16	54.39	53.32	29.60	28.92	46.88	−3.28
SAM-EDA w/o Weight	74.02	65.17	62.74	38.74	39.91	56.12	+5.96
SAM-EDA w/ Weight	78.25	69.38	68.17	43.15	42.63	60.32	+10.16

Table 4. SAM-EDA for SAM variants. Experiments were conducted on the ACDC-Rain validation set and measured with the mean intersection over union (mIoU %) over all classes.

	Performance	Time (s/iter)	Memory (GB)
SAM [1]	69.38	10	8–48
Fast-SAM [39]	68.22	0.5	16
Faster-SAM [26]	68.87	0.3	16

We found that, in bright scenes, such as foggy, rainy, and snowy ones, the SAM-enhanced algorithm outperformed the UDA algorithms. The SSA method performed better than DAFormer, and some methods even outperformed CumFormer, which was newly proposed by the authors, but our SAM-EDA was better than the SSA method. This is because SAM demonstrated strong generalization in bright scenes, providing sharper contour branches. Additionally, the teacher-assistant model can generate relatively accurate pseudo-labels, contributing to better fusion. For night scenes, however, the SAM itself has a significant bias (which will be shown in Section 4.2, thereby reducing the overall performance. However, our SAM-EDA still outperformed the two SSA algorithms for night scenes. Since we only kept the student model, the testing speed and memory consumption were the same as the fast SegFormer. In Figure 3, we show the qualitative comparison. Due to space limitations, we only show the results of ACDC. Based on our method, more-precise segmentation results were obtained in categories such as poles, traffic lights, and traffic signs with obvious shapes.

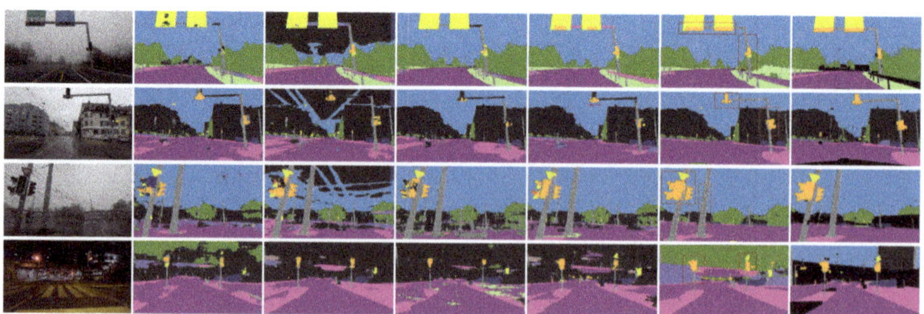

Figure 3. A qualitative comparison with other methods. From top to bottom, there are foggy, rainy, snowy, and nighttime scenes. Based on our method, more-precise segmentation results are obtained in the categories poles, traffic lights, and traffic signs with obvious shapes.

Figure 4. Different pseudo-label fusion methods. From left to right are the target domain image, the original pseudo-label (PL) generated by the teacher-assistant model, the original masks generated by the SAM, the pseudo-label fused using the IoU method, the SSA method, our SAM-EDA (S-E) method, SAM-EDA's weight, and the ground truth (GT).

4. Discussion

4.1. SAM-EDA for Different UDA Methods

We used the pseudo-labels generated by the teacher-assistant model as semantic prompts for filling in SAM's masks. In fact, SAM-EDA is suitable for any UDA segmentation method that utilizes pseudo-labels for self-training. We conducted ablation experiments on the ACDC-Fog validation set. Table 2 demonstrates the enhancement of different methods by SAM-EDA. We found that SAM-EDA can not only improve classic UDA methods (e.g., DACS [37], ProDA [18], and DAFormer [19]), but also improve methods specific to adverse scenes (CuDA-Net [20] and CumFormer [36]) by approximately 3%, indicating that SAM's information is generalizable. This shows that SAM-EDA is a good plugin, and through data-side processing, complex knowledge distillation or fine-tuning operations can be avoided, thus taking advantage of both SAM and domain-specific models.

4.2. Influence of Different Pseudo-Label Fusion Methods

Different semantic prompt fusion methods matter. We chose as many comprehensive fusion strategies as possible and present them in Figure 4. From left to right are the target domain image, the original pseudo-label generated by the teacher-assistant model, the original masks generated by SAM, the pseudo-label fused using the IoU method [38], the SSA method [25], our SAM-EDA method, and SAM-EDA's weight. From top to bottom are the four adverse-condition scenes. Among the three label fusion strategies, the simplest one is to directly assign the class that has the largest intersection over union (IoU) between the mask and category ID layer [38], which was successfully applied in weakly supervised semantic segmentation and saliency detection. However, this approach led to many holes (black areas in the fourth column of Figure 4). This is because the semantic segmentation task is at the "category level", while the SAM masks are at the instance level. When calculating the IoU, the instance-level mask takes the class-level label as the denominator, making the calculation ineffective. For example, if there are three cars in the image, in the "car" category layer, the pixels of the three cars will all be taken into account. Therefore, the proportion of pixels belonging to the "car" class in the mask of a car instance will decrease to 1/3 or 1/2 of the original proportion. If there are other classes present in the current area, it is likely that this area will be misclassified into another class. The SSA method relies entirely on the SAM mask and assigns instance-level pixel labels to all the masks output by SAM. This ensures that each mask has a definitive category label and does not generate large areas of holes. However, if the segmentation by SAM is inaccurate, it will directly result in large areas of errors.

When dealing with the SAM masks, we identified three shortcomings. Firstly, SAM struggled to differentiate classes with similar features, such as roads and sidewalks. In all scenarios, SAM uses the same mask for roads and sidewalks. This is unacceptable for autonomous driving. Secondly, SAM has difficulty distinguishing walls from railings or simply does not recognize them as objects, which could also be fatal for autonomous driving. Lastly, SAM performed poorly in nighttime conditions. For example, in the fourth row of Figure 4, SAM mistakenly assigned large areas of buildings to the sky, leading to errors in the label fusion region and undermining the performance brought by the original pseudo-labels.

To address these issues, we retained some critical categories from the original pseudo-labels (such as sidewalks, walls, and fences) to counter SAM's shortcomings. Then, we allowed the masks to be assigned to incorrect classes (which is difficult to avoid), and we calculated the weights for each mask and reduced them in the loss function, thus effectively optimizing the SSA method. In Table 3, we show the impact of different label fusion strategies. As seen, the SAM-EDA method, which incorporates weights, achieved real improvements and outperformed the SSA method and the case without weights.

4.3. SAM-EDA for SAM Variants

SAM-EDA is also applicable to SAM variants with different numbers of parameters, with the potential to accelerate training. In Table 4, we replaced SAM with the lighter Fast-SAM [39] and Faster-SAM [26], significantly reducing the duration and memory usage of each iteration. In the standard SAM-EDA, we do not need to include SAM in the final segmentation model, so different SAM variants have little impact. However, the emergence of Faster-SAM undoubtedly provided a better option for future methods to include SAM.

4.4. Influence of SAM's Hyper-Parameters

SAM's hyper-parameters are related to the quality, density, and porosity of the generated masks. We conducted tests on the effectiveness of two hyper-parameters: the prediction IoU threshold δ_{iou} and the stability score threshold δ_{sta} (Figure 5). The higher they were set, the more precise the mask contours, but the fewer the masks. We conducted separate experiments on the ACDC-Rain validation set and found that the best results were achieved when $\delta_{iou} = \delta_{sta} = 0.8$. This indicates that we need a stable quantity of masks to cover the entire image during label fusion rather than solely focusing on the quality of the masks.

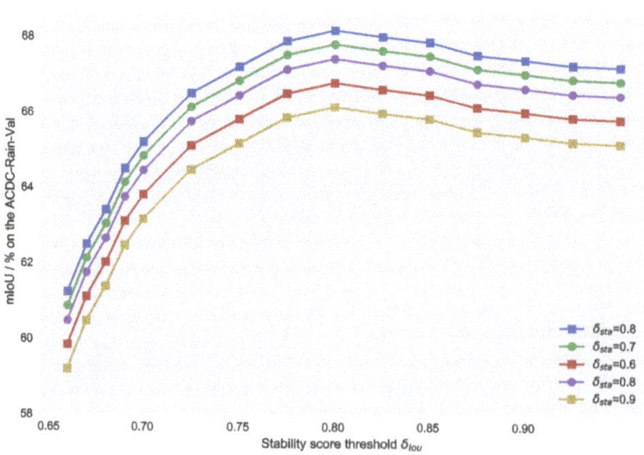

Figure 5. Influence of SAM's hyper-parameters. High δ_{iou} and δ_{sta} both result in performance degradation, and we found that the best results were achieved at $\delta_{iou} = \delta_{sta} = 0.8$.

5. Conclusions

We have presented SAM-EDA, a universal framework for using SAM in unsupervised semantic segmentation tasks. This method utilizes pseudo-labels generated by specific semantic segmentation models as prompts to fill in the spatial structure of SAM segmentation, thereby obtaining a more-accurate probability distribution of scene segmentation. The most-significant contribution of our method is the introduction of a more-accurate and fault-tolerant semantic prompt fusion approach. It can integrate the spatial structure provided by SAM with the semantic discernment generated by the original segmentation network. Our experiments showed that our method achieved better performance on semantic segmentation benchmarks under several adverse imaging conditions. Moreover, it can be implemented in a plug-and-play manner to enhance any unsupervised semantic segmentation algorithm based on pseudo-labels. After introducing a lightweight variant of SAM, our method obtained the ability to perform near real-time training and testing. We also explored the hyper-parameters of SAM.

The universality and generalizability of SAM are valuable resources. In future research, we plan to introduce SAM into tasks such as Test Time Adaptation, serving as a spatial structure anchor to combat the catastrophic forgetting that may occur during prolonged adaptation processes of the model.

Author Contributions: Conceptualization, Z.W. and Z.Z.; methodology, Z.W., Z.Z. and Z.J.; software, Z.W. and Z.J.; validation, Z.W., Y.Z. and Y.Y.; formal analysis, Y.Z. and L.L. (Li Li); investigation, Z.W., Z.Z., Y.Y. and L.L. (Li Li); data curation, Z.Z., Z.J., Y.Y., L.L. (Li Li) and L.L. (Lei Li); writing—original draft preparation, Z.W., Z.J. and L.L. (Lei Li); writing—review and editing, Z.W., Z.Z., Z.J., L.L. (Li Li) and L.L. (Lei Li); visualization, Y.Z. and Z.Z.; supervision, Y.Z. and Z.Z.; project administration, Y.Z.; funding acquisition, Y.Z. and Y.Y. All authors have read and agreed to the published version of the manuscript.

Funding: This work was supported by the National Natural Science Foundation of China under Grant 42071340 and a program of the Song Shan Laboratory (managed by the Major Science and Technology Department of Henan Province) under Grant 2211000211000-01 and 2211000211000-04.

Data Availability Statement: The original contributions presented in the study are included in the article, further inquiries can be directed to the corresponding author/s.

Conflicts of Interest: The authors declare no conflicts of interest.

References

1. Kirillov, A.; Mintun, E.; Ravi, N.; Mao, H.; Rolland, C.; Gustafson, L.; Xiao, T.; Whitehead, S.; Berg, A.C.; Lo, W.Y.; et al. Segment Anything. *arXiv* **2023**, arXiv:2304.02643.
2. Šarić, J.; Oršić, M.; Šegvić, S. Panoptic SwiftNet: Pyramidal Fusion for Real-Time Panoptic Segmentation. *Remote Sens.* **2023**, *15*, 1968. [CrossRef]
3. Lv, K.; Zhang, Y.; Yu, Y.; Zhang, Z.; Li, L. Visual Localization and Target Perception Based on Panoptic Segmentation. *Remote Sens.* **2022**, *14*, 3983. [CrossRef]
4. Dai, Y.; Li, C.; Su, X.; Liu, H.; Li, J. Multi-Scale Depthwise Separable Convolution for Semantic Segmentation in Street–Road Scenes. *Remote Sens.* **2023**, *15*, 2649. [CrossRef]
5. Liu, Q.; Dong, Y.; Jiang, Z.; Pei, Y.; Zheng, B.; Zheng, L.; Fu, Z. Multi-Pooling Context Network for Image Semantic Segmentation. *Remote Sens.* **2023**, *15*, 2800. [CrossRef]
6. Sun, Q.; Chao, J.; Lin, W.; Xu, Z.; Chen, W.; He, N. Learn to Few-Shot Segment Remote Sensing Images from Irrelevant Data. *Remote Sens.* **2023**, *15*, 4937. [CrossRef]
7. Xie, E.; Wang, W.; Yu, Z.; Anandkumar, A.; Alvarez, J.M.; Luo, P. SegFormer: Simple and efficient design for semantic segmentation with transformers. *Adv. Neural Inf. Process. Syst.* **2021**, *34*, 12077–12090.
8. Yang, W.; Wang, S.J.; Khanna, P.; Li, X. Pattern Recognition Techniques for Non Verbal Human Behavior (NVHB). *Pattern Recognit. Lett.* **2019**, *125*, 684–686. [CrossRef]
9. Chen, G.; Hua, M.; Liu, W.; Wang, J.; Song, S.; Liu, C.; Yang, L.; Liao, S.; Xia, X. Planning and tracking control of full drive-by-wire electric vehicles in unstructured scenario. *Proc. Inst. Mech. Eng. Part D J. Automob. Eng.* **2023**, 09544070231195233. [CrossRef]
10. Liu, W.; Hua, M.; Deng, Z.; Meng, Z.; Huang, Y.; Hu, C.; Song, S.; Gao, L.; Liu, C.; Shuai, B.; et al. A Systematic Survey of Control Techniques and Applications in Connected and Automated Vehicles. *IEEE Internet Things J.* **2023**, *10*, 21892–21916. [CrossRef]

11. Meng, Z.; Xia, X.; Xu, R.; Liu, W.; Ma, J. HYDRO-3D: Hybrid Object Detection and Tracking for Cooperative Perception Using 3D LiDAR. *IEEE Trans. Intell. Veh.* **2023**, *8*, 4069–4080. [CrossRef]
12. Cordts, M.; Omran, M.; Ramos, S.; Rehfeld, T.; Enzweiler, M.; Benenson, R.; Franke, U.; Roth, S.; Schiele, B. The cityscapes dataset for semantic urban scene understanding. In Proceedings of the IEEE Conference on Computer Vision and Pattern Recognition, Las Vegas, NV, USA, 27–30 June 2016; pp. 3213–3223.
13. Chen, L.C.; Papandreou, G.; Kokkinos, I.; Murphy, K.; Yuille, A.L. Deeplab: Semantic image segmentation with deep convolutional nets, atrous convolution, and fully connected crfs. *IEEE Trans. Pattern Anal. Mach. Intell.* **2017**, *40*, 834–848. [CrossRef]
14. Lin, G.; Milan, A.; Shen, C.; Reid, I. Refinenet: Multi-path refinement networks for high-resolution semantic segmentation. In Proceedings of the IEEE Conference on Computer Vision and Pattern Recognition, Honolulu, HI, USA, 21–26 July 2017; pp. 1925–1934.
15. Jain, J.; Li, J.; Chiu, M.; Hassani, A.; Orlov, N.; Shi, H. OneFormer: One Transformer to Rule Universal Image Segmentation. *arXiv* **2022**, arXiv:2211.06220.
16. Sakaridis, C.; Dai, D.; Van Gool, L. ACDC: The adverse conditions dataset with correspondences for semantic driving scene understanding. In Proceedings of the IEEE/CVF International Conference on Computer Vision, Montreal, QC, Canada, 10–17 October 2021; pp. 10765–10775.
17. Ren, W.; Ma, L.; Zhang, J.; Pan, J.; Cao, X.; Liu, W.; Yang, M.H. Gated fusion network for single image dehazing. In Proceedings of the IEEE Conference on Computer Vision and Pattern Recognition, Salt Lake City, UT, USA, 18–23 June 2018; pp. 3253–3261.
18. Zhang, P.; Zhang, B.; Zhang, T.; Chen, D.; Wang, Y.; Wen, F. Prototypical pseudo label denoising and target structure learning for domain adaptive semantic segmentation. In Proceedings of the IEEE/CVF Conference on Computer Vision and Pattern Recognition, Nashville, TN, USA, 20–25 June 2021; pp. 12414–12424.
19. Hoyer, L.; Dai, D.; Van Gool, L. Daformer: Improving network architectures and training strategies for domain-adaptive semantic segmentation. In Proceedings of the IEEE/CVF Conference on Computer Vision and Pattern Recognition, New Orleans, LA, USA, 18–24 June 2022; pp. 9924–9935.
20. Ma, X.; Wang, Z.; Zhan, Y.; Zheng, Y.; Wang, Z.; Dai, D.; Lin, C.W. Both style and fog matter: Cumulative domain adaptation for semantic foggy scene understanding. In Proceedings of the IEEE/CVF Conference on Computer Vision and Pattern Recognition, New Orleans, LA, USA, 18–24 June 2022; pp. 18922–18931.
21. Tang, L.; Xiao, H.; Li, B. Can SAM Segment Anything? When SAM Meets Camouflaged Object Detection. *arXiv* **2023**, arXiv:2304.04709.
22. Wang, X.; Wang, W.; Cao, Y.; Shen, C.; Huang, T. Images speak in images: A generalist painter for in-context visual learning. In Proceedings of the IEEE/CVF Conference on Computer Vision and Pattern Recognition, Vancouver, BC, Canada, 17–24 June 2023; pp. 6830–6839.
23. Shan, X.; Zhang, C. Robustness of Segment Anything Model (SAM) for Autonomous Driving in Adverse Weather Conditions. *arXiv* **2023**, arXiv:2306.13290.
24. Yao, L.; Zuo, H.; Zheng, G.; Fu, C.; Pan, J. SAM-DA: UAV Tracks Anything at Night with SAM-Powered Domain Adaptation. *arXiv* **2023**, arXiv:2307.01024.
25. Chen, J.; Yang, Z.; Zhang, L. Semantic Segment Anything. 2023. Available online: https://github.com/fudan-zvg/Semantic-Segment-Anything (accessed on 5 May 2023).
26. Zhang, C.; Han, D.; Qiao, Y.; Kim, J.U.; Bae, S.H.; Lee, S.; Hong, C.S. Faster Segment Anything: Towards Lightweight SAM for Mobile Applications. *arXiv* **2023**, arXiv:2306.14289.
27. Tarvainen, A.; Valpola, H. Mean teachers are better role models: Weight-averaged consistency targets improve semi-supervised deep learning results. In Proceedings of the 31st International Conference on Neural Information Processing Systems, Long Beach, CA, USA, 4–9 December 2017; Volume 30.
28. Dosovitskiy, A.; Beyer, L.; Kolesnikov, A.; Weissenborn, D.; Zhai, X.; Unterthiner, T.; Dehghani, M.; Minderer, M.; Heigold, G.; Gelly, S.; et al. An Image is Worth 16 × 16 Words: Transformers for Image Recognition at Scale. *arXiv* **2020**, arXiv:2010.11929.
29. Sakaridis, C.; Dai, D.; Van Gool, L. Semantic foggy scene understanding with synthetic data. *Int. J. Comput. Vis.* **2018**, *126*, 973–992. [CrossRef]
30. Sakaridis, C.; Dai, D.; Hecker, S.; Van Gool, L. Model adaptation with synthetic and real data for semantic dense foggy scene understanding. In Proceedings of the of the European Conference on Computer Vision (ECCV), Munich, Germany, 8–14 September 2018; pp. 687–704.
31. Lin, H.; Li, Y.; Fu, X.; Ding, X.; Huang, Y.; Paisley, J. Rain o'er me: Synthesizing real rain to derain with data distillation. *IEEE Trans. Image Process.* **2020**, *29*, 7668–7680. [CrossRef]
32. Dai, D.; Gool, L.V. Dark Model Adaptation: Semantic Image Segmentation from Daytime to Nighttime. *arXiv* **2018**, arXiv:1810.02575.
33. Lee, S.; Son, T.; Kwak, S. Fifo: Learning fog-invariant features for foggy scene segmentation. In Proceedings of the IEEE/CVF Conference on Computer Vision and Pattern Recognition, New Orleans, LA, USA, 18–24 June 2022; pp. 18911–18921.
34. Li, M.; Xie, B.; Li, S.; Liu, C.H.; Cheng, X. VBLC: Visibility Boosting and Logit-Constraint Learning for Domain Adaptive Semantic Segmentation under Adverse Conditions. *arXiv* **2022**, arXiv:2211.12256.
35. Sakaridis, C.; Dai, D.; Gool, L. Guided Curriculum Model Adaptation and Uncertainty-Aware Evaluation for Semantic Nighttime Image Segmentation. *arXiv* **2019**, arXiv:1901.05946.

36. Wang, Z.; Zhang, Y.; Ma, X.; Yu, Y.; Zhang, Z.; Jiang, Z.; Cheng, B. Semantic Segmentation of Foggy Scenes Based on Progressive Domain Gap Decoupling. *TechRxiv* **2023**. [CrossRef]
37. Tranheden, W.; Olsson, V.; Pinto, J.; Svensson, L. Dacs: Domain adaptation via cross-domain mixed sampling. In Proceedings of the IEEE/CVF Winter Conference on Applications of Computer Vision, Waikoloa, HI, USA, 3–8 January 2021; pp. 1379–1389.
38. Chen, T.; Mai, Z.; Li, R.; Chao, W.l. Segment anything model (sam) enhanced pseudo labels for weakly supervised semantic segmentation. *arXiv* **2023**, arXiv:2305.05803.
39. Zhao, X.; Ding, W.; An, Y.; Du, Y.; Yu, T.; Li, M.; Tang, M.; Wang, J. Fast Segment Anything. *arXiv* **2023**, arXiv:2306.12156.

remote sensing

MDPI

Technical Note

Optimizing Few-Shot Remote Sensing Scene Classification Based on an Improved Data Augmentation Approach

Zhong Dong [1], Baojun Lin [2,3,4] and Fang Xie [2,3,*]

1 Department of Automation, Tsinghua University, Beijing 100084, China; dongzhong1987@126.com
2 Innovation Academy for Microsatellites, Chinese Academy of Sciences, Shanghai 201210, China; linbaojun@aoe.ac.cn
3 Shanghai Engineering Center for Microsatellites, Shanghai 201304, China
4 School of Information Science and Technology, Shanghai Tech University, Shanghai 201210, China
* Correspondence: xief@microsate.com

Abstract: In the realm of few-shot classification learning, the judicious application of data augmentation methods has a significantly positive impact on classification performance. In the context of few-shot classification tasks for remote sensing images, the augmentation of features and the efficient utilization of limited features are of paramount importance. To address the performance degradation caused by challenges such as high interclass overlap and large intraclass variance in remote sensing image features, we present a data augmentation-based classification optimization method for few-shot remote sensing image scene classification. First, we construct a distortion magnitude space using different types of features, and we perform distortion adjustments on the support set samples while introducing an optimal search for the distortion magnitude (ODS) method. Then, the augmented support set offers a wide array of feature distortions in terms of types and degrees, significantly enhancing the generalization of intrasample features. Subsequently, we devise a dual-path classification (DC) decision strategy, effectively leveraging the discriminative information provided by the postdistortion features to further reduce the likelihood of classification errors. Finally, we evaluate the proposed method using a widely used remote sensing dataset. Our experimental results demonstrate that our approach outperforms benchmark methods, achieving improved classification accuracy.

Keywords: remote sensing scene classification; few-shot learning; data augmentation; feature distortion

Citation: Dong, Z.; Lin, B.; Xie, F. Optimizing Few-Shot Remote Sensing Scene Classification Based on an Improved Data Augmentation Approach. *Remote Sens.* **2024**, *16*, 525. https://doi.org/10.3390/rs16030525

Academic Editors: Jiaojiao Li, Qian Du, Jocelyn Chanussot, Wei Li, Bobo Xi, Rui Song and Yunsong Li

Received: 13 December 2023
Revised: 25 January 2024
Accepted: 29 January 2024
Published: 30 January 2024

1. Introduction

The field of remote sensing image classification holds a pivotal position in various application domains, including disaster detection [1], land use analysis [2], and environmental monitoring [3]. Early remote sensing scene classification methods predominantly relied on manually crafted features, encompassing texture features, structural features, and spectral features [4]. Correspondingly, a multitude of models based on these features emerged, such as the Bag of Words (BoWs) model [5] and sparse coding models [6]. Their fundamental strategies often revolved around enhancing or reducing certain aspects of the image, such as increasing the sparsity of features or reducing redundant image portions, aiming to improve classification performance. These methods are characterized by their simplicity and efficiency [4,7]. However, as the demand for improved performance has grown, these methods have shown limited feature representation capabilities and low utilization efficiency of data information, constraining their effectiveness in practical applications.

With the rapid evolution of parallel computing resources and advancements in artificial intelligence theory, deep learning algorithms have become the predominant trend in remote sensing image classification [8–10]. This approach involves using deep encoders, convolutional neural networks (CNNs), and similar architectures for end-to-end feature extraction, followed by analysis and processing using appropriate decoders. Several notable advancements have been made in model optimization. For instance, Chen et al. [11]

integrated local convolutional attention modules into the backbone network, resulting in significant target-highlighting effects in complex background remote sensing images. Ma et al. [12] introduced network evolution, training, and searching for better network structures using various remote sensing image datasets. Wang et al. [13] employed a target–background separation strategy, using background information beyond the effective target as decision support to enhance distinguishability between target similarity and background difference samples. They also combined texture and morphological features to guide feature learning, effectively reducing the impact of intraclass differences.

In practice, optimizing network architectures significantly enhances performance, assuming sufficient labeled data for structural training optimization. However, a major challenge in remote sensing scene classification is the scarcity of annotated data for model training, especially when faced with tasks involving unseen scene types. Few-shot learning, focusing on a limited number of samples, has gained prominence in addressing this challenge. The primary hurdle in few-shot learning is enabling deep models to quickly learn and infer from a small number of samples without extensive training on large-scale datasets [14–16].

There are two primary approaches to few-shot learning: meta-learning [17] and metric learning [18]. Meta-learning trains classifiers for quick adaptation to new tasks by sharing knowledge across multiple tasks, enhancing few-shot learning. In high-resolution satellite image scene classification, Zhai and colleagues introduced a lifelong few-shot learning approach [19], enabling easy adaptation to new datasets. Li et al. [20] improved intertask relevance by integrating more historical prior knowledge from partial intratask sequences. They also introduced a graph transformer to optimize the distribution of sample features in the embedding space. In contrast, similarity-based methods or metric learning methods are simpler and more effective. The core idea is to cluster similar samples and disperse dissimilar ones by measuring sample similarity. Deng et al. [21] proposed a deep metric learning-based feature embedding model using the nearest neighbor (NN) algorithm as a classifier, addressing classification tasks for high-spectral remote sensing images within and across scenes. Li et al. [22] introduced an adaptive matching network, concatenating support and query set discriminative features and assigning similarity scores to sample pairs. This method captures a more comprehensive range of image information and cues. The challenge for these methods lies in better representing sample features and measuring class similarity. They need to address the limitations of sample features and potential issues in handling similarity metrics, as shown in Figure 1.

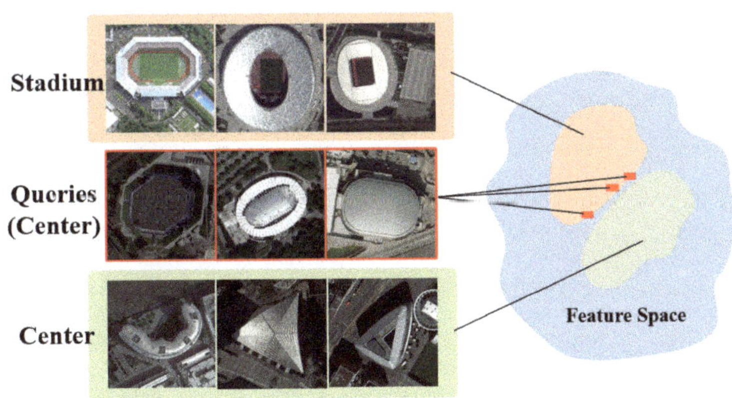

Figure 1. An illustration of a common issue in similarity computation for few-shot classification tasks using metric learning. The query image is classified as "stadium", but its true label is "center", leading to frequent similar misses. This significantly impacts the classification performance in few-shot learning for remote sensing images.

Previous studies confirm that effective data augmentation in few-shot learning significantly improves classification accuracy [23]. Data augmentation methods offer advantages by minimizing additional computational costs and being less constrained by training/testing framework designs. In addition to traditional techniques like rotation and color adjustments, researchers have innovated various augmentation methods, broadly categorized into two types. One type involves data generation. For instance, Antoniou et al. [24] introduced the data augmentation generative adversarial network (DAGAN) model. It extracts image data from a source domain, projects it into a lower-dimensional vector with an encoder, and concatenates the transformed random vector with a decoder to generate augmented images. Li et al. [25] proposed the adversarial feature hallucination network (AFHN) model, utilizing generative adversarial networks (cWGANs) for dataset expansion in few-shot learning. This model enhances discriminative capability and diversity by adding a classification regularizer and an anticollapse regularizer. Subedi et al. [26] presented a GAN-based data augmentation approach generating high-quality training data. Featuring an additional binary classifier in data and feature spaces, this approach controls the generator for optimized training data, improving classification performance. Chen et al. [27] simultaneously employed GAN and U-Net models to create medical images with additional information, elevating few-shot classification task performance. However, a challenge with such methods is the instability in the contribution of generated features to classification performance. The difficulty arises from evaluating whether the newly generated features possess adequate discriminability. Task-specific regularization may lead to the collapse of the synthesis process, resulting in a lack of diversity in generated samples [28].

The other type of method is based on feature enhancement, forming the foundation of this work. These methods assume that knowledge about relationships between samples within known visible categories can be acquired and transferred to unseen categories. Successfully establishing cross-associations between visible categories and learning these relationships allows the application of the knowledge to handle unseen categories with only a few labeled samples. Researchers believe that by increasing sample diversity, we can expand intraclass differences and better define classification boundaries between different categories [29]. Following this rationale, Chen et al. [30] proposed a semantic feature enhancement algorithm. This algorithm utilizes an encoder–decoder model to map samples to a semantic space, learning concepts of samples in the semantic space. By adding noise, extending samples in the semantic space, finding nearest neighbors, and mapping them back to the visual space, the algorithm achieves effective sample augmentation. Alfassy et al. [31] introduced a label-set operations (LaSOs) network for multilabel few-shot image classification tasks. LaSOs leverage relationships between label sets to extract potential semantic information, forming data augmentation at the feature space level. Such approaches introduce varying degrees of distortion to data, making it crucial to ensure that distorted samples maintain or increase discernibility; otherwise, achieving ideal classification performance becomes challenging. On the other hand, accurate delineation of classification decision boundaries depends on sufficient intraclass variance in labeled samples. Therefore, the process of feature enhancement can be understood as actively adding distortion to original features, with these distortions having limitations. Excessive distortion may lead to the loss of discernibility in numerous newly introduced features, increasing the risk of underfitting [32], as illustrated in Figure 2. The impact of image distortion levels on classification accuracy varies when different numbers of images are used for training (e.g., 10^3 and 10^4). The training data are randomly extracted in proportion, and the horizontal axis represents the distortion magnitude level, while the vertical axis shows the ratio of classifier accuracy when using additional distorted data compared to not using it.

It is noteworthy that frameworks or data optimization methods for specific tasks often lack generalizability [33,34]. Currently, a more universally applicable solution is the use of learned data augmentation policies [35]. The limited adoption of these methods is primarily due to the mostly discrete nature of the search space they construct. Each subpolicy within this space brings inconsistent gains to the model, with variations even far apart. Hence, the generation of these policy combinations is inherently challenging to generalize. Additionally, achieving optimal parameters involves independent and costly search and learning stages, resulting in unstable performance gains [36].

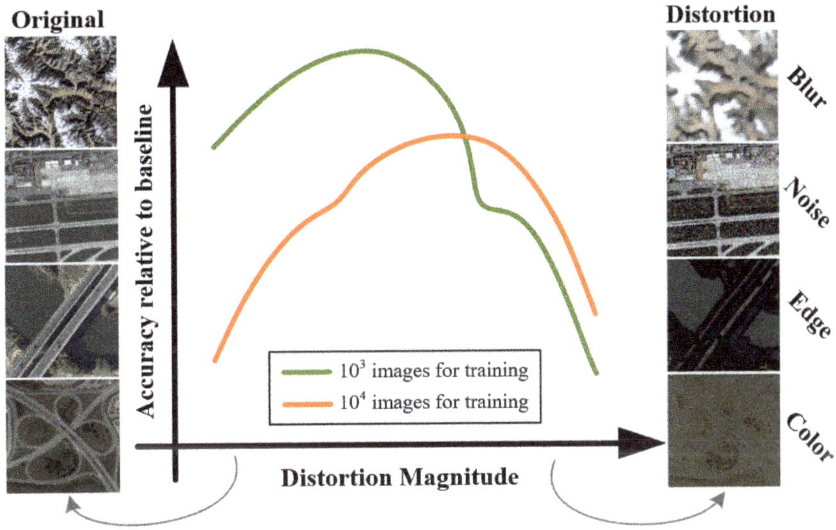

Figure 2. Illustration of how distortion magnitude influences classification accuracy.

For remote sensing images, few-shot classification tasks are significantly more challenging than those for ordinary object images. Remote sensing images have lower resolutions, less detail, and are more prone to confusion between images, making it relatively difficult to define interclass boundaries. In this work, to maximize the improvement of interclass boundaries in few-shot learning tasks, we first explored the impact of different types of feature distortions on learning and introduced a method to construct a continuous distortion space. Subsequently, we combined feature enhancement with metric learning, incorporating the distortion magnitude of features into the metric learning process. Through this amalgamation, we tried to construct a classification framework with a better generalization performance through the acquirement of more discriminable additional features and the support of an optimized learning network.

The specific contributions of this paper are as follows:

- For few-shot classification tasks in remote sensing images, we propose a data augmentation method based on distortion magnitude optimization. The core idea of this method is to introduce appropriate shifts in the feature space for limited samples across a distortion magnitude space, thereby probing and reconstructing interclass boundaries. This approach assimilates the strengths of feature enhancement and metric learning methods. By constructing a multidimensional feature distortion space and segmenting the search for distortion magnitude, it efficiently identifies the optimal distortion magnitude;
- We propose a dual-path classification strategy that optimizes the classification process by dynamically adjusting decision weights. This strategy is particularly suitable for few-shot classification tasks in remote sensing images, as it simultaneously considers the feature information provided by the overall sample distribution and individual samples, significantly reducing the probability of classification errors.

2. Proposed Method

The proposed methodology comprises three integral components: optimal distortion search, feature distortion space construction, and dual-path classification strategy. The optimal distortion search is employed to acquire samples with more discernible features, while the construction of the distortion magnitude space provides feature references for the optimization search. Ultimately, the dual-path classification strategy is employed to manage a more effective classification of augmented data.

2.1. Optimal Distortion Search

Assuming the current problem is a k-way, N-shot task, the overall framework of the optimal distortion search (ODS) is represented in Figure 3, where X_{ij} denotes a specific image's data in the support set, and X_{ij}^{D} represents the image after feature distortion. The core idea of this method is to introduce additional distortion to the sample features provided by the support set, allowing these features to generate appropriate shifts in the feature space.

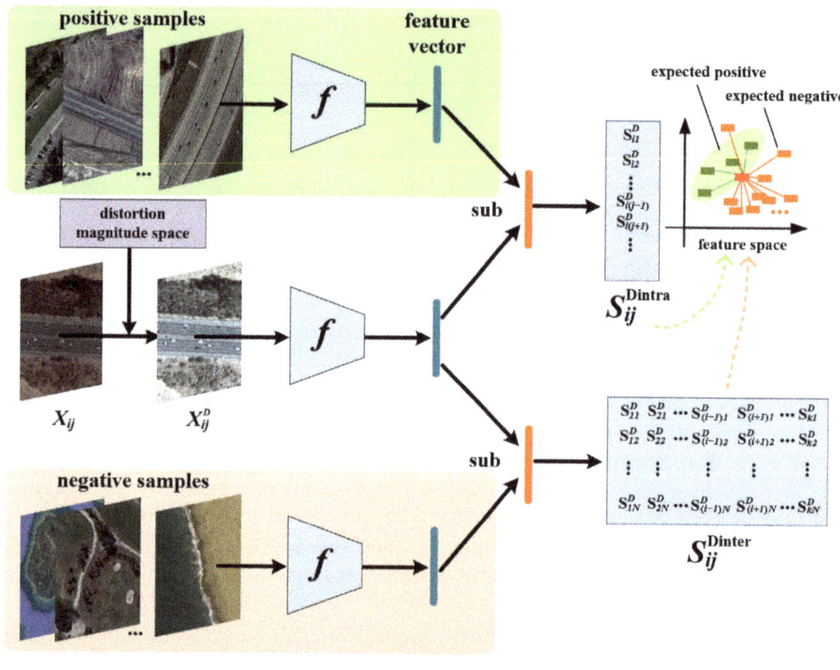

Figure 3. Schematic diagram of data augmentation method based on distortion magnitude optimization. Through the similarity comparison, the current distortion magnitude is gradually updated to a better level. During this workflow, optimizing the magnitude of feature shifts induced by distortion leads to better intraclass space and interclass boundaries.

In the initial stages of the method, we need to obtain a similarity matrix S_{ij} for each sample in the support set with all other samples. Here, i and j represent the indices of images in the support set, S represents the similarity between image pairs, and f denotes a convolutional neural network. From this matrix, we can find the maximum and minimum values of intraclass similarity S_{ij}^{intra} for each sample, as well as the maximum and minimum values of interclass similarity S_{ij}^{inter}. Similarly, S_{ij}^{intra} and S_{ij}^{inter}, respectively, represent the similarity matrices between the current distorted sample and samples of the same class and other classes, which can be represented as follows:

$$S_{ij}^{\mathrm{Dintra}} = \begin{bmatrix} S_{i1}^{D} \\ S_{i2}^{D} \\ \cdots \\ S_{i(j-1)}^{D} \\ S_{i(j+1)}^{D} \\ \cdots \\ S_{i(N-1)}^{D} \end{bmatrix}, S_{ij}^{\mathrm{Dinter}} = \begin{bmatrix} S_{11}^{D} & S_{21}^{D} & \cdots & S_{(i-1)1}^{D} & S_{(i+1)1}^{D} & \cdots & S_{(k-1)1}^{D} \\ S_{12}^{D} & S_{22}^{D} & \cdots & S_{(i-1)2}^{D} & S_{(i+1)2}^{D} & \cdots & S_{(k-1)2}^{D} \\ \cdots & \cdots & \ddots & \cdots & \cdots & \ddots & \cdots \\ S_{1N}^{D} & S_{1N}^{D} & \cdots & S_{(i-1)N}^{D} & S_{(i-1)N}^{D} & \cdots & S_{(k-1)N}^{D} \end{bmatrix} \quad (1)$$

Depending on the magnitude of the distortion, the feature vectors of the distorted samples may exhibit a significant shift from their original positions. Therefore, during the process of updating the distortion parameters, it is essential to impose reasonable constraints on this range. Therefore, we set $S_{\mathrm{min}}^{\mathrm{intra}}$ and $S_{\mathrm{max}}^{\mathrm{inter}}$ as the expected target values to achieve for S_{ij}. In previous few-shot learning methods, these two parameters were often set to 0 and 1, but in the distortion magnitude search process, such settings can lead to issues. For example, it may result in distorted sample features being too close to the original features, rendering the distortion itself meaningless. So, in our approach, we uniformly set the threshold for $S_{\mathrm{min}}^{\mathrm{intra}}$ to be 0.7 (set to 0.7 if it falls below this value) and the threshold for $S_{\mathrm{max}}^{\mathrm{inter}}$ to be 0.3 (set to 0.3 if it exceeds this value). This choice aims to ensure that the distortion of features increases the similarity between the current image and the sample with the minimum intraclass similarity while decreasing the similarity with the sample with the maximum interclass similarity. This way, the feature vector of the distorted image can approach the intraclass boundary as closely as possible while staying far away from the feature boundary of other sample classes in the support set. The loss function in the optimization process is defined as follows:

$$L_{\mathrm{d}} = \frac{1}{N}\sum_{i=1}^{N}\frac{1}{k}\sum_{j=1}^{k} Max\left(\left|S_{\mathrm{min}}^{\mathrm{intra}} - Min\left(S_{ij}^{\mathrm{Dintra}}\right)\right|, \left|S_{\mathrm{max}}^{\mathrm{inter}} - Max\left(S_{ij}^{\mathrm{Dinter}}\right)\right|\right) \quad (2)$$

This framework shares the same feature extractor f and fully connected layers with the earlier pretrained model. During the distortion magnitude search process, we need to freeze all the parameters of f and the fully connected layers until the entire distortion-based augmentation operation is complete. Once the search is finished, the final values of the distortion magnitudes will be directly used for data augmentation. These data, after undergoing feature distortion, will form a new support set along with the basic geometric transformation-based data augmentation (rotation, random clip, etc.). The augmented support set will provide data features that occupy more positions in the feature space compared to the original features, and the mean feature vector of individual classes will also exhibit varying degrees of shift.

2.2. Construction of Feature Distortion Space

In contrast to methods based on AutoAugment [36], the premise of the method proposed in this paper is to start with specified data augmentation strategies and then optimize their inner attributive parameters based on these strategies. Therefore, to search for the optimal distortion magnitudes, it is essential to construct an appropriate magnitude space. The preset distortion magnitude values are stored in registers, and as the iterative process proceeds, the current parameters are continuously updated based on the loss value. The last updated magnitude parameters represent the best magnitude for feature enhancement.

In the edge distortion section, we use simple operators (such as the Sobel operator) to first extract the edge information from the original image and then perform a dilation operation on the edges of the image. The size of the dilation matrix is $D \times D$, which determines the degree of thickening of the shape edges. Based on empirical evidence, the range of D is set within $D \in [1, 8]$. This parameter will be involved in the learning process.

For texture distortion, we refer to the six attributes of texture features proposed by Tamura et al. [37] (i.e., contrast, coarseness, directionality, regularity, linearity, and roughness, with the first three being more significant in feature representation). We use the addition of random pixel grains to control the roughness of the image. The parameter involved in the search process is the granularity level P, with P representing the number of times random pixels are added. Before adding, we can set the number of pixels to be added and the size of the pixel blocks artificially. The benefit of this approach is that it simplifies the model and makes the enhancement magnitude controllable.

To simplify the calculation in the color distortion section, we randomly (or at evenly spaced intervals) set C color combinations of RGB channels. In other words, we select values from the RGB channels to form C different combinations. In this section, there are C updatable values that determine the color.

This approach allows us to establish a discrete distortion magnitude space, denoted as $\Phi = (D, P, C)$. Assuming $P = 20$ and $C = 10$, this results in a potential pool of $8 \times 20 \times 10$ different distortion magnitude combinations. It is worth noting that the aforementioned augmentation strategies may not necessarily represent the optimal choices, as there can be multiple strategies to choose from. Furthermore, these diverse strategies entail different parameters for representing the distortion magnitude. It is important to clarify that our research focuses on exploring the distortion magnitude space and does not encompass learning within the strategy space.

During the exploration process, we employ a segmented search strategy by dividing the magnitude values for each feature into ρ subregions. The mean distortion magnitude from each subregion combination is introduced into the iterative process as the magnitude parameter. In each iteration, a set of distortion magnitudes for each feature is generated, which corresponds to a specific loss function, L_{d}. Subsequently, within the subregion associated with the smallest L_{d}, we conduct further searches. This subregion is then excluded, and the process is repeated iteratively. Figure 4 visually illustrates this process, where m denotes the index of the subspace. The best distortion magnitudes determined for X_{ij} during the current iteration are subsequently incorporated, replacing the original images in the support set for the next iteration. The value of ρ can be tailored to the size of the distortion magnitude space. Through this approach, we efficiently realize dynamic feature distortion selection while streamlining the search process, considerably reducing computational overhead in the iterations.

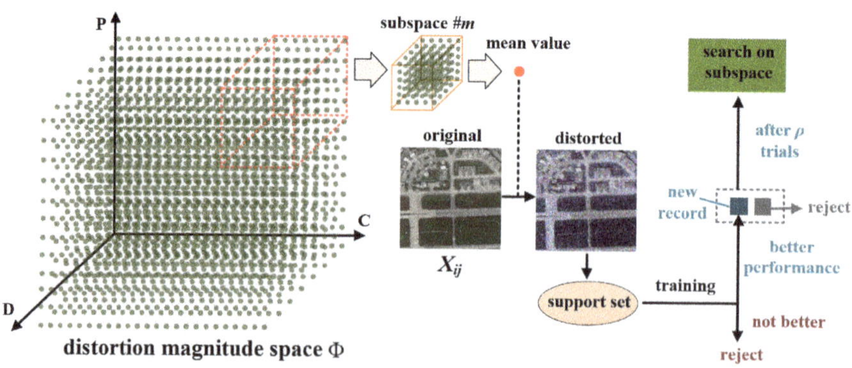

Figure 4. Illustration of search process of distortion magnitude space. The mean value of subspace nodes is utilized for distortion generation. The updated support set is employed for the subsequent round of model performance assessment, retaining support sets that yield higher accuracy. The ultimately retained subspace will be utilized for the final search.

2.3. Dual-Path Classification

Currently, the prevailing approach in few-shot classification involves comparing the query to the mean feature vectors of various classes within the support set. Cosine similarity has been demonstrated to be a highly effective method for measuring similarity [38], and it is typically computed using the formula expressed in Equation (3). Here, X represents the feature vector, s denotes the support set, and $\|\cdot\|_2$ signifies the calculation of the L2 norm.

$$\cos\theta = \frac{X_{query}^T X_s}{\|X_{query}\|_2 \cdot \|X_s\|_2} \quad (3)$$

In conventional classification, the mean feature vector represents the overall characteristics of the current class, essentially acting as its centroid. However, in few-shot remote sensing image classification tasks, relying solely on the overall features may lead to the loss of distinctive characteristics contributed by individual samples. Furthermore, the negative impact of this situation becomes more pronounced as the number of intraclass samples increases [39]. Through the effective feature expansion discussed in the previous two sections of this paper, each few-shot category now occupies a richer position in the feature space, providing clearer class boundaries. Based on this, we propose a dual-path classification (DC) strategy. The framework of this method is illustrated in Figure 5. In the figure, f denotes the feature extraction network, which remains entirely consistent with the one used in pretraining. A and B are the two-way classification output labels, λ_A and λ_B represent the weight parameters for the output probabilities, and μ indicates the number of samples added for each class after regular augmentation and feature distortion processing.

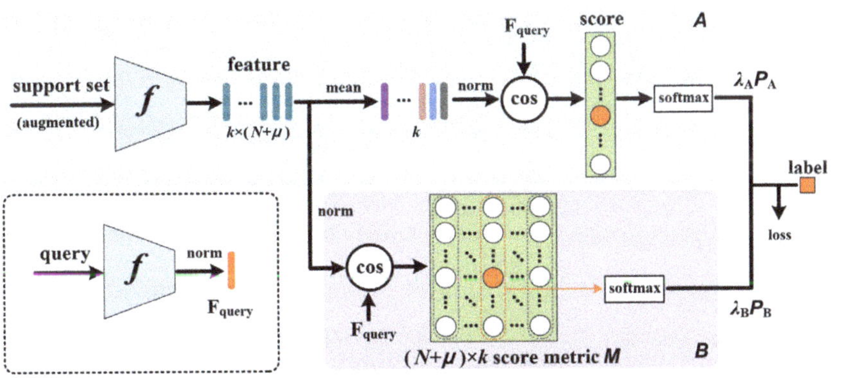

Figure 5. The work flow of the dual-path classification framework. Here, A represents the conventional classification pathway, focusing on overall category features. Branch B is an additional pathway added to emphasize richer individual features.

After extracting the features of the support set samples following data augmentation, we introduce an additional branch on top of the existing training branch. The feature vectors of the support set samples, after entering this additional branch, bypass the mean operation and are directly used to calculate their similarity with the query's feature vector, forming a similarity score matrix M. In this matrix, the scores computed between all samples of the same class and the query are randomly shuffled within each row. Each column represents a set of randomly composed samples with all class labels and their calculated similarity scores with the query. In this score matrix, the label corresponding to the highest similarity score is considered the classification result for that branch. Equations (4) and (5) illustrate the calculation process for the two-way outputs.

$$P_A(y = r|x) = \frac{\exp(\tau \cdot \langle f(x), w_r \rangle)}{\sum_{r'} \exp(\tau \cdot \langle f(x), w_{r'} \rangle)} \quad (4)$$

$$P_B(y=r|x) = \frac{\exp(\tau \cdot (Max\langle f(x), \widetilde{w}_r \rangle))}{\sum_{r'} \exp(\tau \cdot \langle f(x), \widetilde{w}_{r'} \rangle)} \quad (5)$$

Here, x represents the feature vector of the query image, w_r stands for the mean feature vector of the label r, $\langle \cdot \rangle$ denotes cosine similarity calculation, $\widetilde{w}_r$ is a vector within the set of feature vectors corresponding to the support set, labeled as r, and $\widetilde{w}_r$ forms a pair with a random sample from each of the other categories. τ represents the temperature hyperparameter, where a lower value of temperature leads to lower entropy, concentrating the distribution in a few high-confidence positions.

To compare the classification effectiveness between the two branches through learning, we add weights λ_A and λ_B at the output of both the original classification branch and the new classification branch. These two weights satisfy $\lambda_A + \lambda_B = 1$, and their initial values are both set to 0.5. Before making the comparison, the vector formed by selecting the maximum element column in the matrix is normalized, resulting in P_s as the normalized value of the maximum element in the matrix. During the process of sample learning, when the classification outcomes of the two branches are consistent with the labels, it indicates that both the mean features and the maximum similarity are effective. In this case, the weights remain unchanged. However, when one branch's classification outcome matches the label, and the other does not, the weight of the correct branch increases, while the incorrect one decreases. Furthermore, when the judgment based on the mean feature is not effective, it should be replaced by the predictions based on the similarity among individuals. So, when the classification outcomes of both branches do not match the labels, λ_A decreases while λ_B increases. Given a support set, assuming that for each category, a sample is randomly chosen from the support set to form the query set, the query set has a total of $(N + \mu)$ samples, and the support set contains $(k - 1) \times (N + \mu)$ samples for training. To encourage competition between branches A and B, we set the loss functions for the training of the A and B branches as follows:

$$L_A = -\sum_{i}^{N+\mu} \sum_{j}^{k} \log P_A\left(\left(y = y_j | x_i^q\right)\right) \quad (6)$$

$$L_B = -\sum_{i}^{N+\mu} \sum_{j}^{k} \log P_B\left(\left(y = y_j | x_i^q\right)\right) \quad (7)$$

In the equation, y_j represents the label of the sample, and the overall loss function of the network is defined as $L = \lambda_A\, L_A + \lambda_B\, L_B$. By optimizing L using gradient descent, end-to-end training of the network can be achieved. The weight parameters are updated based on the learning rate, with each update magnitude being $(\gamma \times lr)$, where γ is a learning rate coefficient. During training, the value of γ is set based on the number of samples in the support set, typically with smaller values for larger support sets. Additionally, to ensure that the parameters are initialized in an appropriate state, we initially conduct extra pretraining on the original A branch (without weight parameter λ_A) using the augmented support set [40]. Finally, the training is completed by combining both the A and B branches.

3. Experimental Results and Discussions

In this section, we first employed the three data augmentation methods mentioned in Section 2.2 to construct the distortion magnitude space. We evaluated the benefits of the distortion-based data augmentation method combined with the dual-path classification strategy for few-shot learning models. The datasets used for our experiments include the UCMerced Landuse dataset (UCM) [41], the Aerial Image dataset (AID) [42], and the NWPU-RESISC45 Remote Sensing Image Scene Classification dataset (NWPU) [43]. Finally, we conducted ablation experiments and analyzed the impact of feature distortion magnitude space dimensions on model improvements.

3.1. Dataset Description and Preprocessing

The UCM, AID, and NWPU datasets used in this study are publicly available general remote sensing image datasets. The UCM dataset originates from the National Map Urban Area Imagery series of the United States Geological Survey, offering labeled examples of diverse categories within typical urban remote sensing scenes. The AID, unveiled by the Huazhong University of Science and Technology and Wuhan University, constitutes an extensive aerial image dataset compiled from samples extracted from Google Maps imagery. The NWPU dataset, released by Northwestern Polytechnical University, stands as an openly accessible dataset showcasing notable variations across scene samples concerning translation, spatial resolution, and other factors. Table 1 provides the specific details for each of them.

Table 1. The comparison of experimental datasets.

Dataset	Image Size	Number of Scenes	Number of Samples	Samples per Class	Resolution (m)
UCM	256 × 256	21	2100	100	0.3
AID	600 × 600	30	10,000	220–420	0.5–0.8
NWPU	256 × 256	45	31,500	700	0.2–30

In our experiments, each remote sensing scene in the datasets was divided into three sets. Specifically, in the UCM dataset, we randomly selected 11 categories as the training set, 5 categories as the validation set, and 5 categories as the test set. Similarly, in the AID dataset, we randomly selected 16, 7, and 7 categories, and in the NWPU dataset, we randomly selected 23, 11, and 11 categories for training, validation, and testing, respectively. Our model was trained on two of these subsets and evaluated on the remaining one in a cross-validation fashion. For each testing task, we randomly sampled five scenes from the test set to simulate five new remote sensing scenes as encountered in the real world. Each scene was assigned only one or a few labeled samples for the scene classification task.

3.2. Experimental Settings

The experiments were conducted on a computer with an Intel(R) Core(TM) i5-13600KF CPU, with 64GB of RAM and an Nvidia GeForce RTX 3080Ti GPU. The distortion space parameters were set to $D = 8$, $P = 20$, and $C = 12$. For the purpose of comparison, we employed ResNet-12 as the backbone network and initialized the similarity function with pretraining on the UCM, AID, and NWPU datasets.

Before training, all image pixels were resized to 256 × 256. The hyperparameters τ and γ were set to 50 and 5, respectively. The initial learning rate was 0.001, the batch size was set to 32, and the number of epochs was set to 800. We utilized stochastic gradient descent (SGD) for optimization during both the pretraining and metric learning phases. During the first 200 epochs, only the original A branch was trained, and in the subsequent 600 epochs, the A and B branches were jointly trained. The learning rate was decayed by a factor of 0.5 every 100 epochs. Additionally, the traversal order within the distortion space was D, P, and C, with each parameter exploring its range from small to large. The reported classification accuracy results in all experiments are the averages of the accuracy results from 100 randomly sampled subsets from the test set, with a 95% confidence interval.

3.3. RS Scene Few-Shot Classification Results and Analysis

Figure 6 illustrates the change in accuracy during the distortion magnitude search process. A maximum accuracy value was recorded after every 200 attempts. If this accuracy was higher than the previously recorded maximum, it was updated as the current accuracy; otherwise, it remained unchanged. The solid line represents the mean accuracy obtained from 100 complete search processes, each utilizing different support sets randomly extracted from the test set. The shaded region shows the range in which the 100 search operations' curves appeared.

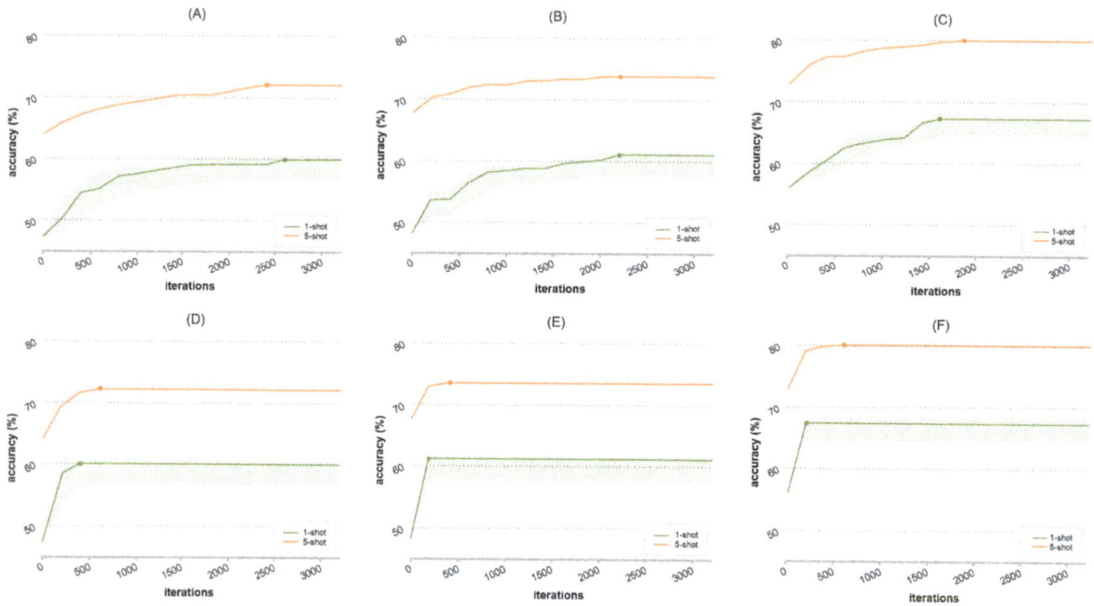

Figure 6. The mean accuracy update curves during the distortion magnitude search process (5-way). The vertical axis represents the accuracy of few-shot classification, and the horizontal axis represents the number of iterations. The iterations utilize combinations of distortion parameters (D, P, C), starting from the smallest values. (**A**,**D**) correspond to the UCM dataset, (**B**,**E**) correspond to the AID dataset, and (**C**,**F**) correspond to the NWPU dataset. The different colored curves in the graph represent scenarios with 1 and 5 samples per class, respectively. The nodes indicate the maximum average accuracy value, corresponding to the optimal distortion magnitude.

In the figures, we can observe that feature-based data augmentation significantly impacts the classifier's performance. Simultaneously, the segmented search approach substantially shortens the search process. Regardless of the different support sets used, the proposed method in this paper consistently identifies the optimal distortion magnitudes. Table 2 displays a comparison of the performance of the ODS method with advanced data augmentation models. It is evident that the proposed approach effectively leverages discriminative features, thereby enhancing the model's generalization performance.

Table 2. Performance comparison of the ODS method with advanced data augmentation methods in terms of classification accuracy on UCM, AID, and NWPU datasets (all using ResNet-12 as the backbone network).

Method	UCM (%)		AID (%)		NWPU (%)	
	1 Shot	5 Shot	1 Shot	5 Shot	1 Shot	5 Shot
Mixup [33]	49.51 ± 1.51	65.11 ± 1.26	49.99 ± 1.43	66.54 ± 1.30	58.03 ± 1.96	74.39 ± 1.77
CutMix [44]	52.69 ± 1.80	67.12 ± 1.45	53.52 ± 1.67	68.37 ± 1.59	61.64 ± 2.31	75.69 ± 2.01
DAGAN [24]	52.12 ± 1.16	66.59 ± 0.76	52.88 ± 1.15	66.97 ± 0.80	59.98 ± 1.60	75.22 ± 1.24
f-DAGAN [26]	53.25 ± 0.44	67.31 ± 0.35	55.89 ± 0.41	68.10 ± 0.33	63.86 ± 0.87	76.28 ± 0.64
AugGAN [45]	52.54 ± 0.53	66.76 ± 0.31	53.59 ± 0.64	67.80 ± 0.52	64.00 ± 0.89	76.05 ± 0.70
Style Aug. [46]	54.00 ± 1.33	68.33 ± 0.98	55.20 ± 1.23	69.05 ± 1.00	65.56 ± 1.52	77.23 ± 1.31
AutoAug. [36]	57.67 ± 0.65	68.89 ± 0.54	59.58 ± 0.57	70.99 ± 0.50	66.10 ± 0.82	77.80 ± 0.65
RandAug. [32]	58.76 ± 0.95	70.85 ± 0.62	60.09 ± 1.19	72.74 ± 0.90	67.94 ± 0.99	79.64 ± 0.67
MADAO [47]	59.40 ± 0.73	71.31 ± 0.56	61.06 ± 0.70	72.60 ± 0.47	66.87 ± 0.96	79.96 ± 0.58
ODS (Ours)	60.35 ± 1.02	72.67 ± 0.73	61.79 ± 1.26	74.31 ± 0.76	67.47 ± 1.17	80.59 ± 0.86

Figure 7 illustrates the variations in validation accuracy during the learning process of the dual-path classification network. It is evident that, in the training process of the five-way-five-shot scenario, the introduction of branch B significantly boosts the classification accuracy. Even in the five-way-one-shot training scenario, branch B provides performance gains. However, the five-shot curve demonstrates more pronounced gains compared to the one-shot scenario. Hence, in the context of few-shot classification, the greater the number of samples in the support set, the more pronounced the impact of ODS-DC on classification accuracy.

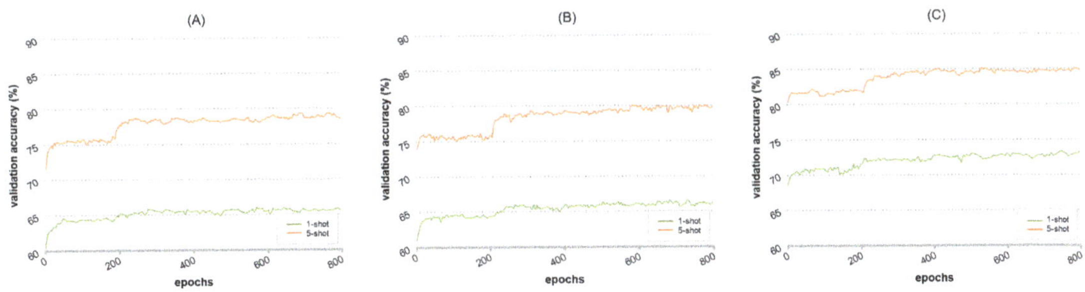

Figure 7. The variation curves of validation accuracy during the training process of the dual-path classification method (5-way). Specifically, (**A**) corresponds to the UCM dataset, (**B**) to the AID dataset, and (**C**) to the NWPU dataset. All curves are smoothed using a 0.2 ratio moving average for improved visualization.

Table 3 presents a performance comparison between the method proposed in this paper and currently advanced data augmentation-based few-shot learning methods. All the methods include data augmentation techniques such as random rotation, random cropping, and translation, and employ ResNet-12 as the backbone network for testing in a five-way scenario. It is evident that the method introduced in this paper outperforms other methods across the three widely used datasets.

Table 3. Performance comparison between the dual-path classification method in this paper and the current benchmark methods.

Method	**UCM (%)**		**AID (%)**		**NWPU (%)**	
	1 Shot	**5 Shot**	**1 Shot**	**5 Shot**	**1 Shot**	**5 Shot**
ProtoNet [48]	58.79 ± 0.81	72.82 ± 0.60	60.18 ± 0.78	74.00 ± 0.61	62.78 ± 0.85	80.19 ± 0.52
MAML [49]	54.97 ± 0.69	65.45 ± 0.70	56.50 ± 0.65	70.02 ± 0.50	56.01 ± 0.87	72.94 ± 0.63
RelationNet [50]	55.32 ± 0.87	72.59 ± 0.53	56.17 ± 0.80	73.94 ± 0.57	55.84 ± 0.88	75.78 ± 0.57
RS-MetaNet [51]	63.75 ± 0.51	76.94 ± 0.29	64.18 ± 0.49	76.68 ± 0.30	72.04 ± 0.43	82.69 ± 0.22
SGMNet [52]	64.17 ± 0.75	76.63 ± 0.59	64.32 ± 0.79	77.98 ± 0.42	73.01 ± 0.77	84.52 ± 0.50
ODS DC (ours)	65.93 ± 0.94	77.60 ± 0.72	66.28 ± 0.89	79.04 ± 0.69	73.93 ± 0.90	84.66 ± 0.76

In addition, Table 4 illustrates the variation in classification accuracy when different backbone networks are employed as feature extractors. As observed from the table, using deeper feature extractors leads to significantly better classification performance. However, it is worth noting that backbone networks with more layers tend to be more complex and demand greater computational resources. For example, transitioning from Conv-4 to ResNet-12 increases the number of layers by threefold, resulting in substantial accuracy improvement. On the other hand, substituting ResNet-12 with ResNet-50, which increases the number of layers by more than fourfold, yields only a minor accuracy gain. Hence, for the method proposed in this paper, the choice of the backbone network is not solely based on having more layers but rather involves a comprehensive consideration of factors such as gains in accuracy, computational resource utilization, and the efficiency of method reproduction.

Table 4. In the 5-way, 5-shot task, ODS-DC achieves average classification accuracy using different feature extraction networks.

Dataset	Conv-4 (%)	ResNet-12 (%)	ResNet-50 (%)
UCM	61.67 ± 0.83	77.60 ± 0.72	79.07 ± 0.33
AID	63.03 ± 0.90	79.04 ± 0.69	80.60 ± 0.36
NWPU	65.75 ± 0.91	84.66 ± 0.76	85.19 ± 0.42

3.4. Ablation Study

We assessed the performance of ODS-DC under different combinations of feature distortions using the AID dataset. Table 5 documents the average classification accuracy for each combination in the five-way few-shot classification. From the table, it is visually evident that the contribution of distortion optimization varies significantly for different types of features. Optimization of edge feature distortion yields the greatest performance gain, followed by texture features, with color features exhibiting the smallest gain. Moreover, as the dimension of feature distortion optimization increases, the model's performance shows varying degrees of gain change depending on the combination of different feature types.

Table 5. The average classification accuracy achieved by combining different distorted feature.

Feature Group	Average Accuracy (%)	
	1 Shot	5 Shot
Edge	61.02 ± 1.51	75.58 ± 1.32
Texture	60.85 ± 1.83	74.10 ± 1.50
Color	57.47 ± 1.94	73.70 ± 1.51
Edge and Color	62.91 ± 1.19	77.19 ± 1.04
Edge and Texture	65.11 ± 0.97	78.67 ± 0.80
Texture and Color	62.74 ± 1.72	76.15 ± 1.29
Edge, Texture, and Color	66.28 ± 0.89	79.04 ± 0.69

To better illustrate the benefits of the dual-path classification strategy, we compared the classification performance between the single path (without branch B and weighting coefficients) and the dual path by recording the classification accuracy for individual categories. Using the AID dataset as an example, we randomly selected five categories related to urban scenes and five unrelated to urban scenes for testing. Each category served as a query for few-shot classification, and after each round of classification for all test objects, a new round began by randomly selecting query images and a support set. This process continued for 200 rounds. Figure 8 presents the confusion matrix of the ODS-DC model for these scene categories under both single-path and dual-path scenarios (five-way, five-shot). As observed from the figure, dual-path classification exhibits a significant improvement over the single path, with substantial variations in the degree of enhancement for each category. The differences in gains are primarily related to the selection of scene types, where specific scenes are more influenced by feature types that exhibit better discriminability. For instance, in the comparison between (A) and (B), the dual-path classification model achieves greater performance improvement in the "Center" category, where edge features are more prominent. However, in the comparison between (C) and (D), the "Forest" category, where texture features are more pronounced, exhibits the greatest improvement in accuracy.

Furthermore, to better capture the variations in the operational performance of the model, we assessed the framework's average prediction time on the AID dataset under 1-shot, 5-shot, and 10-shot (5-way) scenarios. We randomly sampled 20 subsets of classes from the AID dataset for model training. Subsequently, using 100 randomly selected samples from each corresponding subset, we evaluated the model's predictions after each training iteration. Table 6 documents the average inference time for all sampled data under different configurations (95% confidence intervals included). Notably, dual-path classification exhibits a slight decrease in inference speed compared to single-path classification, and the model's inference time significantly elongates with an increase in the number of samples in the support set, as evident from the table.

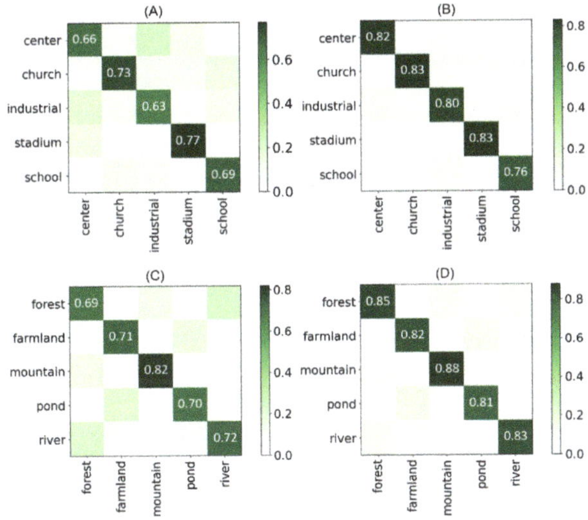

Figure 8. The confusion matrices for single-path (**A**) and dual-path (**B**) classification of 5 randomly selected urban scenes, as well as single-path (**C**) and dual-path (**D**) classification of 5 randomly selected nonurban scenes (5-way, 5-shot).

Table 6. Comparison of mean inference time (ms) of the proposed model in different settings.

Framework Type	1 Shot	5 Shot	10 Shot
Single path	72 ± 17	106 ± 19	134 ± 19
Dual path	98 ± 21	177 ± 28	685 ± 39

3.5. Discussion

Through the analysis of experimental results, the proposed ODS method demonstrates its effectiveness across three remote sensing scene datasets. In comparison to generative and policy-based approaches, ODS exhibits superior accuracy in five-way tasks with varying sample sizes. Notably, among the selected feature types, optimizing the distortion magnitude of edge features provides the model with the most significant gains. This suggests a substantial discrepancy in the contribution of distinguishable features generated by different types of feature distortions. Hence, exploring and optimizing combinations of distortion amplitudes for different features in the feature distortion space holds the potential for further accuracy improvement. However, constructing a higher dimensional feature distortion space will inevitably result in a significant increase in computational complexity, necessitating a specific task analysis and hardware ability assessment.

Simultaneously, the test results of ODS-DC on the three datasets indicate an enhancement in model robustness. In comparison to the single-path strategy, the dual-path strategy in five-way, five-shot tasks showed a potential improvement of approximately 7–18% in classification accuracy. Furthermore, the classification efficiency of ODS-DC is contingent on the number of support set samples. While an increase in the number of support set samples enhances model accuracy, it also leads to a substantial reduction in model inference efficiency. For instance, in a five-way scenario, the inference time for a single-path model only increased about twice from 1-shot to 10-shot tasks, while for a dual-path model, the inference time increased by over six times. Therefore, in tasks with fewer samples, the advantages of ODS-DC are more readily evident.

4. Conclusions

The impact of data augmentation on classification performance in few-shot learning is evident. Traditional feature enhancement methods have not explored the distinctiveness of features extensively, leading to unstable gains in classification performance. This issue is particularly common in the context of few-shot tasks with remote sensing images. Even with improvements in data augmentation techniques, it is challenging to provide effective support in the design of learning models. In this paper, we quantified feature distortion magnitudes and projected them onto a feature distortion magnitude space. Through the search of this distortion space, we optimized the distribution of sample features. Subsequently, to fully utilize this distribution, we proposed a classification model based on dual-path classification. The additional classification branch, through learning the comparison of intraclass and interclass similarities of all support samples, reinforced the classification process of the original branch while mitigating, to some extent, the shortcomings of the original branch in classifying challenging data. In the experimental section, we validated the effectiveness of the ODS-DC joint method using general remote sensing datasets. Furthermore, our comparative experiments revealed that the gains brought by the ODS-DC method surpass current State-of-the-Art data augmentation methods. In the ablation experiments, we explored the impact of changes in distortion magnitudes of different features on classification performance. Regrettably, due to hardware constraints and model efficiency, we were unable to conduct more in-depth investigations using higher-dimensional feature spaces comprising various feature types. However, in subsequent investigations, we will not only focus on expanding the dimensions of the feature space but may also introduce additional distortion parameters, thereby further exploring the potential of feature distortion in few-shot classification. Overall, this method's novel and valuable perspective on feature distortion and model optimization offers a more efficient way to utilize data for few-shot classification learning in remote sensing scenes. It also provides new insights into research on data augmentation in deep learning.

Author Contributions: Conceptualization, Z.D. and B.L.; methodology, Z.D.; software, Z.D. and F.X.; validation, Z.D. and F.X.; formal analysis, Z.D.; investigation, Z.D.; resources, Z.D.; data curation, Z.D. and F.X.; writing—original draft preparation, Z.D. and F.X.; writing—review and editing, Z.D. and F.X.; visualization, Z.D.; supervision, Z.D. and B.L.; project administration, Z.D.; funding acquisition, F.X. All authors have read and agreed to the published version of the manuscript.

Funding: This research received no external funding.

Data Availability Statement: The UCM dataset can be acquired from https://vision.ucmerced.edu/datasets/, accessed on 26 September 2023; the AID dataset can be acquired from https://www.kaggle.com/datasets/jiayuanchengala/aid-scene-classification-datasets, accessed on 8 October 2023; the NWPU dataset can be acquired from https://gcheng-nwpu.github.io/, accessed on 16 October 2023.

Conflicts of Interest: The authors declare no conflicts of interest.

References

1. Cheng, G.; Guo, L.; Zhao, T.; Han, J.; Li, H.; Fang, J. Automatic landslide detection from remote-sensing imagery using a scene classification method based on BOVW and PLSA. *Int. J. Remote Sens.* **2013**, *34*, 45–59. [CrossRef]
2. Mishra, N.B.; Crews, K.A. Mapping vegetation morphology types in a dry savanna ecosystem: Integrating hierarchical object-based image analysis with random forest. *Int. J. Remote Sens.* **2014**, *35*, 1175–1198. [CrossRef]
3. Huang, X.; Wen, D.; Li, J.; Qin, R. Multi-level monitoring of subtle urban changes for the megacities of China using high-resolution multi-view satellite imagery. *Remote Sens. Environ.* **2017**, *196*, 56–75. [CrossRef]
4. Cheng, G.; Han, J.; Zhou, P.; Guo, L. Multi-class geospatial object detection and geographic image classification based on collection of part detectors. *ISPRS J. Photogramm. Remote Sens.* **2014**, *98*, 119–132. [CrossRef]
5. Chen, S.; Tian, Y. Pyramid of spatial relations for scene-level land use classification. *IEEE Trans. Geosci. Remote Sens.* **2014**, *53*, 1947–1957. [CrossRef]
6. Zhu, Q.; Zhong, Y.; Zhang, L.; Li, D. Scene classification based on the fully sparse semantic topic model. *IEEE Trans. Geosci. Remote Sens.* **2017**, *55*, 5525–5538. [CrossRef]

7. Shao, W.; Yang, W.; Xia, G.S.; Liu, G. A hierarchical scheme of multiple feature fusion for high-resolution satellite scene categorization. In Proceedings of the International Conference on Computer Vision Systems, St. Petersburg, Russia, 16–18 July 2013; pp. 324–333.
8. Khan, S.D.; Basalamah, S. Multi-branch deep learning framework for land scene classification in satellite imagery. *Remote Sens.* **2023**, *15*, 3408. [CrossRef]
9. Xu, Q.; Shi, Y.; Yuan, X.; Zhu, X. Universal domain adaptation for remote sensing image scene classification. *IEEE Trans. Geosci. Remote Sens.* **2023**, *61*, 4700515. [CrossRef]
10. Thapa, A.; Horanont, T.; Neupane, B.; Aryal, J. Deep learning for remote sensing image scene classification: A review and meta-analysis. *Remote Sens.* **2023**, *15*, 4804. [CrossRef]
11. Chen, S.; Wei, Q.; Wang, W.; Tang, J.; Luo, B.; Wang, Z. Remote sensing scene classification via multi-branch local attention network. *IEEE Trans. Image Process.* **2021**, *31*, 99–109. [CrossRef]
12. Ma, A.; Wan, Y.; Zhong, Y.; Wang, J.; Zhang, L. SceneNet: Remote sensing scene classification deep learning network using multi-objective neural evolution architecture search. *ISPRS J. Photogramm. Remote Sens.* **2021**, *172*, 171–188. [CrossRef]
13. Wang, J.; Li, W.; Zhang, M.; Tao, R.; Chanussot, J. Remote Sensing Scene Classification via Multi-Stage Self-Guided Separation Network. *IEEE Trans. Geosci. Remote Sens.* **2023**, *61*, 5615312.
14. Song, Y.; Wang, T.; Cai, P.; Mondal, S.K.; Sahoo, J.P. A comprehensive survey of few-shot learning: Evolution, applications, challenges, and opportunities. *ACM Comput. Surv.* **2023**, *55*, 1–40. [CrossRef]
15. Sun, X.; Wang, B.; Wang, Z.; Li, H.; Li, H.C.; Fu, K. Research progress on few-shot learning for remote sensing image interpretation. *IEEE J. Sel. Top. Appl. Earth Obs. Remote Sens.* **2021**, *14*, 2387–2402. [CrossRef]
16. Cheng, G.; Yan, B.; Shi, P.; Li, K.; Yao, X.; Guo, L.; Han, J. Prototype-CNN for few-shot object detection in remote sensing images. *IEEE Trans. Geosci. Remote Sens.* **2021**, *60*, 5604610. [CrossRef]
17. Ren, M.; Triantafillou, E.; Ravi, S.; Snell, J.; Swersky, K.; Tenenbaum, J.B.; Larochelle, H.; Zemel, R.S. Meta-learning for semi-supervised few-shot classification. *arXiv* **2018**, arXiv:1803.00676.
18. Cakir, F.; He, K.; Xia, X.; Kulis, B.; Sclaroff, S. Deep metric learning to rank. In Proceedings of the IEEE/CVF Conference on Computer Vision and Pattern Recognition, Long Beach, CA, USA, 15–20 June 2019; pp. 1861–1870.
19. Zhai, M.; Liu, H.; Sun, F. Lifelong learning for scene recognition in remote sensing images. *IEEE Geosci. Remote Sens. Lett.* **2019**, *16*, 1472–1476. [CrossRef]
20. Li, F.; Li, S.; Fan, X.; Li, X.; Chang, H. Structural attention enhanced continual meta-learning for graph edge labeling based few-shot remote sensing scene classification. *Remote Sens.* **2022**, *14*, 485. [CrossRef]
21. Deng, B.; Jia, S.; Shi, D. Deep metric learning-based feature embedding for hyperspectral image classification. *IEEE Trans. Geosci. Remote Sens.* **2019**, *58*, 1422–1435. [CrossRef]
22. Li, L.; Han, J.; Yao, X.; Cheng, G.; Guo, L. DLA-MatchNet for few-shot remote sensing image scene classification. *IEEE Trans. Geosci. Remote Sens.* **2020**, *59*, 7844–7853. [CrossRef]
23. Liu, Y.; Zhang, H.; Zhang, W.; Lu, G.; Tian, Q.; Ling, N. Few-shot image classification: Current status and research trends. *Electronics* **2022**, *11*, 1752. [CrossRef]
24. Antoniou, A.; Storkey, A.; Edwards, H. Data augmentation generative adversarial networks. *arXiv* **2017**, arXiv:1711.04340.
25. Li, K.; Zhang, Y.; Li, K.; Fu, Y. Adversarial feature hallucination networks for few-shot learning. In Proceedings of the IEEE/CVF Conference on Computer Vision and Pattern Recognition, Seattle, WA, USA, 13–19 June 2020; pp. 13470–13479.
26. Subedi, B.; Sathishkumar, V.E.; Maheshwari, V.; Kumar, M.S.; Jayagopal, P.; Allayear, S.M. Feature learning-based generative adversarial network data augmentation for class-based few-shot learning. *Math. Probl. Eng.* **2022**, *2022*, 9710667. [CrossRef]
27. Chen, X.; Li, Y.; Yao, L.; Adeli, E.; Zhang, Y.; Wang, X. Generative adversarial U-Net for domain-free few-shot medical diagnosis. *Pattern Recognit. Lett.* **2022**, *157*, 112–118. [CrossRef]
28. Wang, Y.X.; Girshick, R.; Hebert, M.; Hariharan, B. Low-shot learning from imaginary data. In Proceedings of the IEEE Conference on Computer Vision and Pattern Recognition, Salt Lake City, UT, USA, 18–22 June 2018; pp. 7278–7286.
29. Zhang, R.; Che, T.; Ghahramani, Z.; Bengio, Y.; Song, Y. Metagan: An adversarial approach to few-shot learning. In *Advances in Neural Information Processing Systems*; NeurIPS: Montréal, QU, Canada, 2018; Volume 31.
30. Chen, Z.; Fu, Y.; Zhang, Y.; Jiang, Y.; Xue, X.; Sigal, L. Semantic feature augmentation in few-shot learning. *arXiv* **2018**, arXiv:1804.05298.
31. Alfassy, A.; Karlinsky, L.; Aides, A.; Shtok, J.; Harary, S.; Feris, R.; Giryes, R.; Bronstein, A.M. Laso: Label-set operations networks for multi-label few-shot learning. In Proceedings of the IEEE/CVF Conference on Computer Vision and Pattern Recognition, Long Beach, CA, USA, 15–20 June 2019; pp. 6548–6557.
32. Cubuk, E.D.; Zoph, B.; Shlens, J.; Le, Q.V. Randaugment: Practical automated data augmentation with a reduced search space. In Proceedings of the IEEE/CVF Conference on Computer Vision and Pattern Recognition Workshops, Seattle, WA, USA, 14–19 June 2020; pp. 702–703.
33. Zhang, H.; Cisse, M.; Dauphin, Y.N.; Lopez-Paz, D. Mixup: Beyond empirical risk minimization. *arXiv* **2017**, arXiv:1710.09412.
34. DeVries, T.; Taylor, G.W. Dataset augmentation in feature space. *arXiv* **2017**, arXiv:1702.05538.
35. Zoph, B.; Cubuk, E.D.; Ghiasi, G.; Lin, T.Y.; Shlens, J.; Le, Q.V. Learning data augmentation strategies for object detection. In *Proceedings of the Computer Vision–ECCV 2020: 16th European Conference, Glasgow, UK, 23–28 August 2020; Proceedings, Part XXVII*; Springer International Publishing: Berlin/Heidelberg, Germany, 2020; pp. 566–583.

36. Cubuk, E.D.; Zoph, B.; Mane, D.; Vasudevan, V.; Le, Q.V. Autoaugment: Learning augmentation policies from data. *arXiv* **2018**, arXiv:1805.09501.
37. Tamura, H.; Mori, S.; Yamawaki, T. Textural features corresponding to visual perception. *IEEE Trans. Syst. Man Cybern.* **1978**, *8*, 460–473. [CrossRef]
38. Luo, C.; Zhan, J.; Xue, X.; Wang, L.; Ren, R.; Yang, Q. Cosine normalization: Using cosine similarity instead of dot product in neural networks. In *Artificial Neural Networks and Machine Learning–ICANN 2018: 27th International Conference on Artificial Neural Networks, Rhodes, Greece, 4–7 October 2018, Proceedings, Part I*; Springer International Publishing: Berlin/Heidelberg, Germany, 2018; pp. 382–391.
39. Huang, W.; Yuan, Z.; Yang, A.; Tang, C.; Luo, X. TAE-net: Task-adaptive embedding network for few-shot remote sensing scene classification. *Remote Sens.* **2021**, *14*, 111. [CrossRef]
40. Dhillon, G.S.; Chaudhari, P.; Ravichandran, A.; Soatto, S. A baseline for few-shot image classification. *arXiv* **2019**, arXiv:1909.02729.
41. Yang, Y.; Newsam, S. Bag-of-visual-words and spatial extensions for land-use classification. In Proceedings of the 18th SIGSPATIAL International Conference on Advances in Geographic Information Systems, San Jose, CA, USA, 2–5 November 2010; pp. 270–279.
42. Xia, G.S.; Hu, J.; Hu, F.; Shi, B.; Bai, X.; Zhong, Y.; Zhang, L.; Lu, X. AID: A benchmark data set for performance evaluation of aerial scene classification. *IEEE Trans. Geosci. Remote Sens.* **2017**, *55*, 3965–3981. [CrossRef]
43. Cheng, G.; Han, J.; Lu, X. Remote sensing image scene classification: Benchmark and state of the art. *Proc. IEEE* **2017**, *105*, 1865–1883. [CrossRef]
44. Yun, S.; Han, D.; Oh, S.J.; Chun, S.; Choe, J.; Yoo, Y. Cutmix: Regularization strategy to train strong classifiers with localizable features. In Proceedings of the IEEE/CVF International Conference on Computer Vision, Seoul, Republic of Korea, 27–28 October 2019; pp. 6023–6032.
45. Huang, S.; Lin, C.; Chen, S.; Wu, Y.; Hsu, P.; Lai, S. Auggan: Cross domain adaptation with gan-based data augmentation. In Proceedings of the European Conference on Computer Vision, Munich, Germany, 8–14 September 2018; pp. 718–731.
46. Jackson, P.T.G.; Abarghouei, A.A.; Bonner, S.; Breckon, T.P.; Obara, B. Style augmentation: Data augmentation via style randomization. *CVPR Workshops* **2019**, *6*, 10–11.
47. Hataya, R.; Zdenek, J.; Yoshizoe, K.; Nakayama, H. Meta approach to data augmentation optimization. In Proceedings of the IEEE/CVF Winter Conference on Applications of Computer Vision, Waikoloa, HI, USA, 3–8 January 2022; pp. 2574–2583.
48. Snell, J.; Swersky, K.; Zemel, R. Prototypical networks for few-shot learning. In Proceedings of the Advances in Neural Information Processing Systems, NeurIPS, Long Beach, CA, USA, 4–9 December 2017; Volume 30.
49. Finn, C.; Abbeel, P.; Levine, S. Model-agnostic meta-learning for fast adaptation of deep networks. In Proceedings of the International Conference on Machine Learning, Sydney, Australia, 6–11 August 2017; pp. 1126–1135.
50. Sung, F.; Yang, Y.; Zhang, L.; Xiang, T.; Torr, P.H.S.; Hospedales, T.M. Learning to compare: Relation network for few-shot learning. In Proceedings of the IEEE Conference on Computer Vision and Pattern Recognition, Salt Lake City, UT, USA, 18–22 June 2018; pp. 1199–1208.
51. Li, H.; Cui, Z.; Zhu, Z.; Chen, L.; Zhu, J.; Huang, H.; Tao, C. RS-MetaNet: Deep meta metric learning for few-shot remote sensing scene classification. *arXiv* **2020**, arXiv:2009.13364. [CrossRef]
52. Zhang, B.; Feng, S.; Li, X.; Ye, Y.; Ye, R.; Luo, C.; Jiang, H. Sgmnet: Scene graph matching network for few-shot remote sensing scene classification. *IEEE Trans. Geosci. Remote Sens.* **2022**, *60*, 5628915. [CrossRef]

remote sensing

MDPI

Technical Note

Automated Detection and Analysis of Massive Mining Waste Deposits Using Sentinel-2 Satellite Imagery and Artificial Intelligence

Manuel Silva [1], Gabriel Hermosilla [1,*], Gabriel Villavicencio [2] and Pierre Breul [3]

[1] Escuela de Ingeniería Eléctrica, Pontificia Universidad Católica de Valparaíso, Avenida Brasil 2147, Valparaíso 2340000, Chile; manuel.silva@pucv.cl
[2] Escuela de Ingeniería de Construcción y Transporte, Pontificia Universidad Católica de Valparaíso, Avenida Brasil 2147, Valparaíso 2340000, Chile; gabriel.villavicencio@pucv.cl
[3] Département Génie Civil, Polytech Clermont, Institut Pascal UMR CNRS 6602, Université Clermont Auvergne, Av. Blaise Pascal SA 60206-63178 Aubière, CEDEX, 63000 Clermont Ferrand, France; pierre.breul@uca.fr
* Correspondence: gabriel.hermosilla@pucv.cl; Tel.: +56-322273688

Abstract: This article presents a method to detect and segment mine waste deposits, specifically waste rock dumps and leaching wasted dumps, in Sentinel-2 satellite imagery using artificial intelligence. This challenging task has important implications for mining companies and regulators like the National Geology and Mining Service in Chile. Challenges include limited knowledge of mine waste deposit numbers, as well as logistical and technical difficulties in conducting inspections and surveying physical stability parameters. The proposed method combines YOLOv7 object detection with a vision transformer classifier to locate mine waste deposits, as well as a deep generative model for data augmentation to enhance detection and segmentation accuracy. The ViT classifier achieved 98% accuracy in differentiating five satellite imagery scene types, while the YOLOv7 model achieved an average precision of 81% for detection and 79% for segmentation of mine waste deposits. Finally, the model was used to calculate mine waste deposit areas, with an absolute error of 6.6% compared to Google Earth API results.

Keywords: satellite imagery; scene segmentation; deep generative models; mine waste rock; leaching waste dumps; physical stability; closure planning

Citation: Silva, M.; Hermosilla, G.; Villavicencio, G.; Breul, P. Automated Detection and Analysis of Massive Mining Waste Deposits Using Sentinel-2 Satellite Imagery and Artificial Intelligence. *Remote Sens.* **2023**, *15*, 4949. https://doi.org/10.3390/rs15204949

Academic Editors: Qian Du, Jiaojiao Li, Wei Li, Jocelyn Chanussot, Rui Song, Yunsong Li and Bobo Xi

Received: 28 August 2023
Revised: 2 October 2023
Accepted: 9 October 2023
Published: 13 October 2023

1. Introduction

In the global mining sector, addressing the management and monitoring of massive mining waste deposits (MWDs) is critical, especially in countries like Chile, which leads in copper production worldwide [1–5]. The numerous phases of mining activities in Chile generate a significant amount of waste, which is stored in various forms such as tailing dams, waste rock dumps (WRDs, Figure 1a), and leaching waste dumps (LWDs, Figure 1b) [1–5]. This waste accumulation poses substantial challenges and requires intricate management and regulatory adherence, particularly during the closure and post-closure stages [6–10].

Addressing the challenges related to MWDs is pivotal due to the complexities involved in their management and the limited information available, which impacts entities like SERNAGEOMIN in their regulatory and monitoring roles [11,12]. The varied forms of MWDs, each with their unique characteristics and impacts, necessitate intricate management strategies and strict adherence to national legislation to ensure the safety and well-being of people and the environment [6–10].

The national legislation mandates adherence to the "Methodological Guide for the Evaluation of the Physical Stability of Remaining Mining Facilities" provided by SERNAGEOMIN [11]. This guide outlines the comprehensive methodologies and parameters for evaluating the potential failure mechanisms of MWDs, thereby optimizing the time, cost, and efficacy of physical stability (PS) studies and facilitating streamlined regulatory compliance and approval processes for closure [11].

Figure 1. Mine waste deposits (MWDs) located in the north region of Chile. (**a**) Waste rock dump (WRD), (**b**) Leaching waste dumps (LWDs).

The integration of satellite imagery, specifically from the Copernicus Sentinel series by ESA [13], and AI technologies offers advanced, innovative solutions in a variety of fields, including vegetation monitoring [14], urban planning [15], and land use classification [16].

This research aims to harness the capabilities of AI and Sentinel-2 satellite imagery to bridge the existing information gaps regarding the PS of MWDs during their closure and post-closure stages [11]. The objective is to create a comprehensive system for maintaining a national record of MWDs, enabling the extraction of crucial variables related to their PS through advanced DL algorithms such as image classification, deep generative models, and object detection. The innovative application of AI in analyzing satellite imagery for the detection and identification of MWDs is a significant advancement in the field, contributing to the establishment of a detailed, accurate national record of MWDs.

By providing a nuanced understanding of MWDs and their associated risks, this methodology supports the advancement of industry standards and regulatory frameworks. It aids entities like SERNAGEOMIN in their inspection and monitoring roles, enabling precise identification of MWD locations and condition assessments and facilitating risk-based prioritization and compliance processes, thereby enhancing operational and environmental safety protocols in the mining sector [11].

2. Related Works

In this section, a review of the literature most relevant to the research of this article is carried out. In particular, different works on satellite image classification, the use of deep generative models, and detection and segmentation algorithms for satellite images are detailed.

2.1. Image Classification

Image classification is a widely used technique for assigning predefined class labels to digital images based on their visual content. In the context of land classification, this technique can be used to automatically identify and map different land cover types, such as forests, croplands, urban areas, and water bodies, from satellite imagery. There are various popular image classification algorithms that are used in practice, including convolutional neural networks (CNNs) [17,18], support vector machines (SVMs) [19,20], and vision transformers (ViT) [21,22]. Each of these algorithms has their own unique strengths and weaknesses, and they have been shown to generalize well to unseen data. For our work, we considered as relevant the following studies that utilize the DL techniques mentioned previously. For instance, ref. [23] examines advancements in DL techniques for agricultural tasks such as plant disease detection, crop/weed discrimination, fruit counting, and land cover classification. Future directions for AI in agriculture are presented, emphasizing the potential of DL-based models to improve automation in the industry. In [24], a remote-sensing scene-classification method using vision transformers is proposed, resulting in high accuracy on various datasets. In [25], a lightweight ConvNet, MSDF-Net, is presented for aerial scene classification with competitive performance and reduced parameters. In [26], a new method, E-ReCNN, is presented for fine-scale change detection in satellite imagery, with improved results compared to other semi-supervised methods and with the potential for global application once trained.

2.2. Deep Generative Models

DGMs can improve DL models' performance and robustness when labeled data are scarce by generating synthetic data to augment the training dataset. These models can also generate new samples similar to real data, which is useful for data-intensive applications such as medical imaging and computer vision. Additionally, DGMs can be used to generate synthetic data when real-world data collection is difficult or costly.

The use of DGMs has been proposed for various medical and mining applications. In [27], researchers present a study on using GANs [28] for data augmentation in computed tomography segmentation tasks, showing that using CycleGAN [29] improves the performance of a U-Net model. In [30], the authors explore the use of the Stable Diffusion [31] model for generating synthetic medical images, finding that fine-tuning the U-Net [32] component can generate high-fidelity images. In [33], a method using AI algorithms and GANs was proposed to increase the number of samples for studying the PS of tailing dams in mining applications, resulting in an average F1-score of 97%. These studies highlight the potential of DGMs in expanding limited datasets and improving performance in various fields.

2.3. Image Detection and Segmentation

Image detection and segmentation for satellite imagery is a critical task in remote sensing, with applications such as land-use mapping and change detection. Recent techniques in the field are based on DL, specifically CNNs, which have shown superior performance compared to traditional methods. Popular algorithms for image detection and segmentation include RetinaNet [34], Mask R-CNN [35], and U-Net [32], each with their own advantages.

The authors of [36] investigate the use of CNNs for classifying and segmenting satellite orthoimagery and find that CNNs can achieve results comparable to state-of-the-art methods. Ref. [37] applies DL to detect and classify mines and tailing dams in Brazil using satellite imagery, demonstrating potential for low-cost, high-impact data science tools. Ref. [38] proposes a framework using YOLOv4 [39] and random forest [40] algorithms to extract tailings pond margins from high spatial resolution remote sensing images with high accuracy and efficiency. Ref. [41] presents a method for semantic segmentation of high-resolution satellite images using tree-based CNNs, which outperforms other techniques in terms of classification performance and execution time and which suggests that incorporating data augmentation techniques and deeper neural networks in future work could enhance the efficiency of the method.

This study proposes an innovative method for the precise localization, detection, and segmentation of MWDs in Chilean mining facilities. What sets this method apart from previous studies is the use of cutting-edge DL techniques such as YOLOv7, the ViT classifier, and generative models, which have not been applied before in this context. In addition, this study utilizes open access tools to obtain MWD information, making it cost-effective and accessible to other researchers and mining companies. The proposed method also addresses a current challenge in the Chilean mining context, which is the lack of accurate information in the area of MWDs. By leveraging the generated synthetic tiles of MWDs using deep generative models and the ViT classifier, this study is able to estimate the area of detected MWDs, which is a crucial factor in evaluating mining activities. Overall, this study provides a novel and practical approach to the characterization and assessment of MWDs, which can significantly improve safety and operational efficiency in the mining industry.

3. Methodology

The research methodology is outlined in two stages. Stage one involves the acquisition of satellite image datasets, while stage two encompasses tasks for detection, segmentation, and area estimation. Subsequently, a comprehensive explanation of the relevant metrics used to evaluate the models is provided.

3.1. Dataset Creation

The methodology employed for acquiring the dataset utilized in this research is described in Figure 2. The process consists of four distinct stages: (a) retrieval of satellite imagery from the European Space Agency's Copernicus Open Access Hub platform and subsequent processing utilizing TorchGeo v0.4.1 [42] to facilitate image analysis; (b) implementation of vision transformer (ViT) techniques for image classification; (c) utilization of deep generative models to generate synthetic maps, thereby augmenting the number of samples in the dataset; and (d) making the prepared dataset available for subsequent analytical stages.

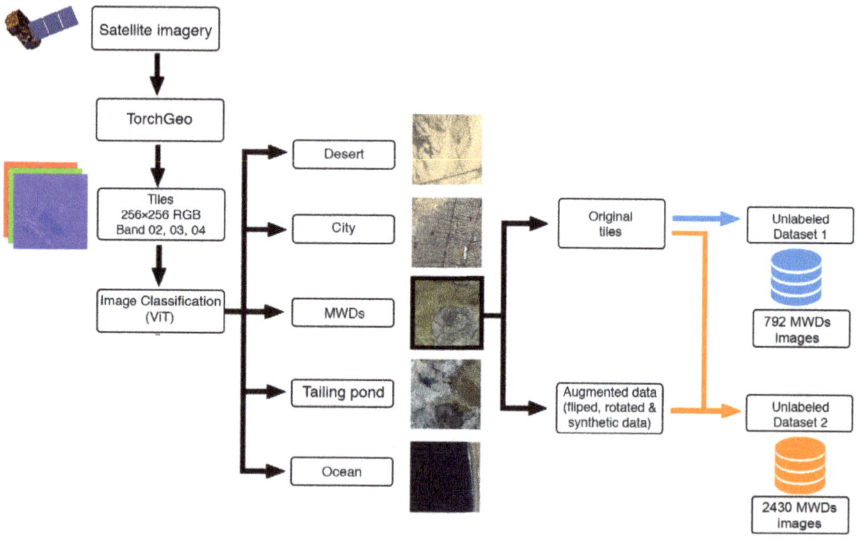

Figure 2. Methodology applied to obtain satellite imagery from the European Space Agency's (ESA) Copernicus site to create datasets for conducting experiments.

3.1.1. Satellite Imagery Acquisition

The areas of interest for the study were determined through the identification of the major mining facilities within the country, as sourced from the website of the National Mining Council of Chile [43] and depicted in Table 1. A total of 30 mining facilities were identified and used as a basis for the study.

The acquisition of satellite imagery was conducted through the Copernicus Open Access Hub, a platform that provides access to Sentinel data through an interactive graphical user interface. To ensure a high-quality dataset, only products (data items for satellite imagery [44]) with a cloud cover percentage of less than 9% were selected from the Sentinel-2A and Sentinel-2B platforms and S2MSI2A products with bottom-of-atmosphere reflectance. The products were downloaded within the time frame of 2019 to 2022 in SENTINEL-SAFE [45] format. The RGB bands (bands 02, 03, and 04; see Figure 3) were combined into a single image and then segmented into 256 × 256 pixel resolution tiles, with a 20% overlap on adjacent tiles, using the TorchGeo [42] software, for a 10 m spatial resolution. The metadata of the downloaded products were used to determine the vertices in decimal format coordinates in the resulting tiles, and this process was repeated for each of the determined zones.

Table 1. Major mining facilities in Chile. Figure adapted from [44].

Map of Chile	Zone	Region Name	Region Key	Mining Facilities
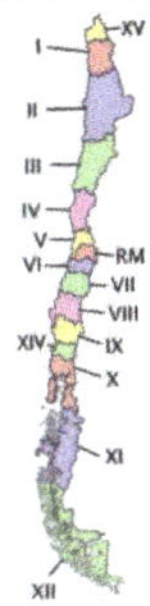	Northern	Tarapacá	I	Cerro Colorado, Quebrada Blanca, Collahuasi.
		Antofagasta	II	Antucoya, Chuquicamata, Ministro Hales, Spence, Sierra Gorda, Centinela, Gabriela Mistral, Lomas Bayas, Zaldivar, Escondida, Franke.
		Atacama	III	Cerro Negro Norte, Salvador, La Coipa, Lobo Marte, Maricunga, Ojos del Salado, Candelaria, Caserones, Los Colorados.
		Coquimbo	IV	El Romeral, Carmen de Andacollo, Los Pelambres.
	Central	Valparaíso	V	El Soldado, Andina.
		Metropolitana	RM	Los Bronces.
		Rancagua	VI	El Teniente.

Figure 3. Different satellite image bands of the Chuquicamata mining facility (Region II, Antofagasta, Chile). From left to right are the red, green, and blue bands, respectively.

3.1.2. Image Classification

Once the tiles from the mining facilities in Table 1 were obtained, and given the 1:159 relationship of tiles containing MWDs, a ViT image classifier was employed to select the MWD regions for analysis. The images were then categorized into five classes, namely city, desert, sea, tailings pond, and MWD, with each class consisting of 2200 images. This selection was made based on the most frequently occurring scenes from the analysis.

ViT utilizes self-attention mechanisms and the transformer architecture for learning spatial hierarchies of features without image-specific biases. This study adopted the approach of splitting images into positional embedding patches processed by the transformer encoder, which has proven to be highly effective. The results show the effectiveness of the ViT architecture in this context.

To improve performance and prevent overfitting, the selected images were augmented through horizontal and vertical flips and rotations of 90° and −90°, with each class containing 2200 images, including augmentations.

3.1.3. Data Augmentation

Data augmentation is a technique used to artificially increase the size of a dataset in DL and computer vision by applying various random transformations to the existing data, such as rotation, scaling, and flipping. This technique can improve the robustness and generalization of models by exposing them to different variations of the same data. Additionally, it can help to mitigate overfitting by providing the model with more diverse examples.

The procedure was carried out in two phases. The first phase was applied to the ViT classifier, while the second phase was applied to the original 769 MWD image dataset (Figure 4a). For both phases, two augmentations were employed: horizontal flip (Figure 4b) and −90° rotation (Figure 4c).

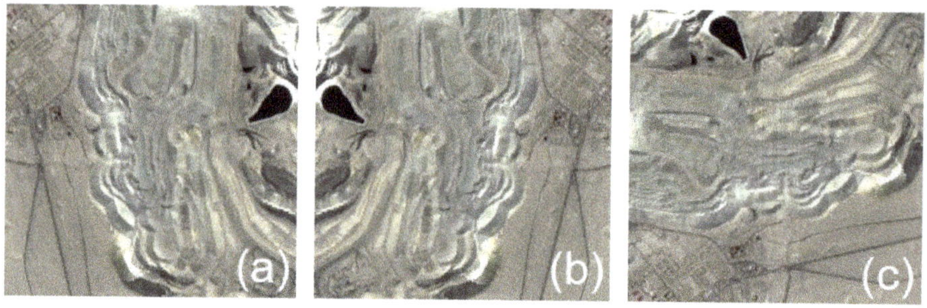

Figure 4. (**a**) Original MWD image used for data augmentation, (**b**) original image flipped horizontally, and (**c**) original image rotated −90°.

3.1.4. Deep Generative Model

For the generation of synthetic images of maps containing MWDs, we propose a pipeline based on denoising diffusion probabilistic models (DDPMs) [46]. The basic idea behind diffusion models is quite simple. They take the input image x_0 and gradually add Gaussian noise through a series of T steps (direct diffusion process). Subsequently, a neural network is trained to recover the original data by inverting the noise process. By modeling the inverse process, we can generate new data. This is called the inverse diffusion process or, in general, the sampling process of a generative model (see Figure 5).

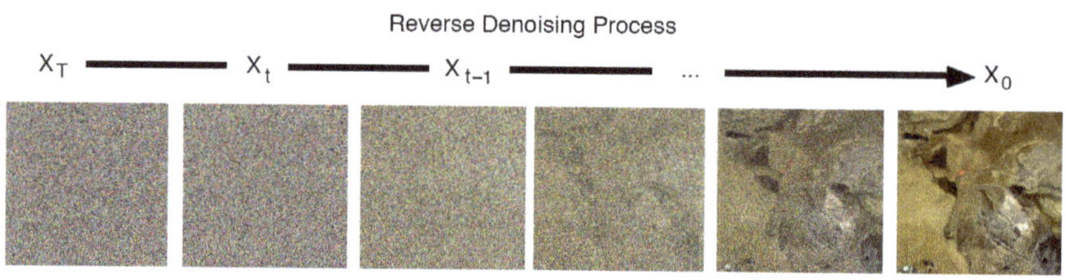

Figure 5. Reverse denoising process applied to generate a synthetic sample of MWDs.

The process is formulated using a Markov chain consisting of T steps, where each step depends solely on the previous one, a moderate assumption in diffusion models. Most diffusion models use architectures that are some variant of U-Net. The forward diffusion process executed at training is given by Equation (1):

$$q(\mathbf{x}_t|\mathbf{x}_{t-1}) = \mathcal{N}(\mathbf{x}_t; \sqrt{1-\beta_t}\mathbf{x}_{t-1}, \beta_t\mathbf{I}) \quad q(\mathbf{x}_{1:T}|\mathbf{x}_0) = \prod_{t=1}^{T} q(\mathbf{x}_t|\mathbf{x}_{t-1}). \tag{1}$$

This approach is useful as DDPMs are able to generate high-resolution images, preserving fine details and textures, making it suitable for data augmentation. This is essential for our application of generating synthetic maps containing MWD zones. As we saw in the previous stage, the ViT classifier is utilized to detect the patches corresponding to MWDs. The images classified as MWDs are then utilized as input to train the DDPM algorithm, using unconditional guidance. In our implementation, we train a DDPM model with a database of 792 MWD images. A U-Net architecture is used for the image denoising, configured with two ResNet layers for each U-Net block, with identical input and output channels corresponding to 3 channels for RGB and a resolution of 256 × 256 pixels. The noise scheduler process is configured to 1000 steps in order to add noise to the images.

3.1.5. Unlabeled Dataset

The images were arranged in preparation for the subsequent stage of locating, detecting, and segmenting MWDs. The images are assembled into two datasets as the final step of the procedure outlined in Figure 2 to perform the different experiments. These correspond to:

1. Unlabeled Dataset 1. Original dataset containing 792 MWD images.
2. Unlabeled Dataset 2. Original dataset plus increased data as described in data augmentation and synthetic MWDs tiles, totaling 2430 MWD images.

The composite image in Figure 6 displays the tiles (centered in the Centinela mining facility) that make up the dataset, showcasing the 20% overlap used. The image consists of unlabeled images.

Figure 6. Representation of the detected MWD-containing tiles superimposed on Google Earth in the Centinela mining facility, Region II, Antofagasta, Chile.

3.2. Detection and Segmentation of MWDs

Once the various image datasets have been obtained and arranged, the second part of the methodology enables detection, segmentation, and estimation of the areas of MWDs. This methodology is shown in Figure 7, where the following steps are observed: (a) the two obtained datasets are labeled with human-annotated labels for the MWD regions within each dataset; (b) the two datasets are then used to train three experiments with detection and segmentation algorithms; and (c) compute the surface area measurements based on the outputs of the detection and segmentation algorithms.

Figure 7. Methodology employed for the detection, segmentation, and calculation of the area of MWDs.

3.2.1. Human-Annotated Labels

In DL, human-annotated labels, which include object or image classes, bounding box coordinates, and other attributes, are used to train and evaluate machine learning models. Image classification recognizes objects and properties within an image, while object detection localizes objects through bounding boxes. Image segmentation allows for understanding of an image at the pixel level, with semantic segmentation assigning each pixel to a single class. In this research, the use of segmentation masks is employed to achieve the proposed objectives and application. The masks are labeled using Label Studio [47] software v1.5.0. The labels are confirmed by an expert who recognizes MWDs based on the geographical location of the mining facility and the characteristics of the MWDs, such as the distance from the mining pit, texture and color of the mine waste, the shape, and geometry. This process is repeated for all available datasets.

3.2.2. Detection and Segmentation of MWDs

Two segmentation algorithms were tested for extracting features from manually labeled tiles containing MWDs, resulting in 792 zones. The first algorithm is YOLOv7 [48], developed by Wong Kin-Yiu and Alexey Bochkovskiy, a state-of-the-art model for object detection and instance segmentation.

YOLOv7 is built upon the Efficient Layer Aggregation Network (ELAN) [49], optimizing several parameters and computational densities to design an efficient network, and is specifically extended to E-ELAN for more substantial learning ability. E-ELAN enhances the model's learning capability by using group convolution to expand the channels and cardinality of the computational block and by applying the same channel multiplier and group parameter to all the computational blocks in a computation layer. This model preserves the architecture of the transition layer while modifying the computational block in ELAN. The enhancements lead to the improvement of gradient flow paths and an increase in diverse feature learning, contributing to faster and more accurate inferences.

YOLOv7 employs advanced model scaling techniques, which are crucial for adjusting the model's depth, image resolution, and width to meet various application requirements. These adjustments are meticulously done to maintain the optimal structure and the initial properties of the architecture, even when concatenating with other layers.

In this study, the YOLOv7 pre-trained model was utilized with its default architecture. The chosen hyperparameters, including epochs, batch size, learning rate, and input image size, are described in Table 2. The image data were normalized and distributed into training (70%), validation (20%), and testing (10%) subsets to ensure a balanced evaluation.

The second algorithm is Mask R-CNN [35], another advanced object instance segmentation model that extends Faster R-CNN [50] by adding a fully convolutional network to predict object masks in parallel with bounding box and class predictions. For Mask R-CNN, two different backbones, namely ResNet50 and ResNet101 [51] from the Detectron2 [52] library, were employed, with fine-tuning performed on both configurations. The hyperparameters for these experiments are also detailed in Table 2, providing a consolidated view of the training configurations for both segmentation algorithms.

Table 2. Hyperparameters used in the detection and segmentation stages.

Hyperparameter	YOLOv7	Mask R-CNN
epochs	3000	3000
batch size	448	450
learning rate	0.002	0.0025
image input size	256 × 256	256 × 256

3.2.3. Area Estimation

Once the models were trained and their evaluation metrics obtained, they were applied to unseen images to make predictions regarding object detection and segmentation. This was done using tiles generated from RGB bands at 10 m of spectral resolution, with each pixel in the image representing a 10 m × 10 m square on the ground. Information on vertex coordinates in decimal format was also available from the tile generation process. By utilizing the fact that the image always has a resolution of 256 × 256 pixels, it was possible to establish a correlation between the identified mask and its representation in decimal format coordinates.

The calculation of the MWD area was performed using the Gaussian area formula. The area of the polygon P with vertices $(x_1, y_1), (x_2, y_2), \ldots, (x_n, y_n)$ is calculated using the Gaussian area formula, as shown in Equation (2).

$$A = \frac{1}{2}\left|\sum_{i=1}^{n}(x_i y_{i+1} - x_{i+1} y_i)\right| \tag{2}$$

Furthermore, the Google Earth (GE) API [53] can be employed to calculate the area of a polygon through its coordinates. This enables a comparison between the estimated area obtained through a mathematical approach and the measurements obtained through the API. The GE API is utilized to corroborate the results obtained, thereby obtaining verified values.

3.3. Evaluation Metrics

Evaluating the performance of the algorithms employed in this experiment requires an understanding of relevant metrics.

3.3.1. Metrics for Classification, Detection, and Segmentation

The most commonly used metrics for evaluating image classification, detection, and segmentation models are precision, recall, F1-score, accuracy, macro AVG, and weighted AVG.

$$Precision = \frac{TP}{TP + FP} \tag{3}$$

$$Recall = \frac{TP}{TP + FN} \tag{4}$$

$$F1 = \frac{2 \times Precision \times Recall}{Precision + Recall} = \frac{2 \times TP}{2 \times TP + FP + FN} \tag{5}$$

$$Accuracy = \frac{TP + TN}{TP + TN + FP + FN} \tag{6}$$

where TP—true positive, TN—true negative, FP—false positive, and FN—false negative. Precision (Equation (3)) measures the proportion of true positive detections among all positive detections made by the model. Recall (Equation (4)) measures the proportion of true positive detections among all actual positive instances in the dataset. F1-score (Equation (5)) is the harmonic mean of precision and recall. Accuracy (Equation (6)) is a measure of the model's overall performance, calculated as the proportion of correct predictions out of all predictions.

Additionally, for image detection and segmentation performance evaluation, the use of intersection over union (IoU) and mean average precision (mAP) are also considered relevant metrics in this context.

$$\text{IoU} = \frac{\text{Area of Overlap}}{\text{Area of Union}} \tag{7}$$

$$\text{mAP} = \frac{1}{n}\sum_{n}^{k=n} AP_k \tag{8}$$

IoU (Equation (7)) is a metric used to evaluate the accuracy of image segmentation models by comparing the overlap between the predicted and ground truth segments. mAP (Equation (8)) is a measure of the model's overall performance, used to measure the average precision of all classes by taking into account both true positive and false positive detections.

3.3.2. Metrics for Deep Generative Models

The Fréchet inception distance (FID) [54] is a method for evaluating the quality of images generated by generative models. It compares the statistics of the generated images to those of real images by measuring the Fréchet distance between the Inceptionv3 [55] features of the two distributions. The lower the FID score, the more similar the generated images are to the real images, indicating a higher quality of the generated images. It has been shown to correlate well with human judgment of image quality and is widely used in the literature to evaluate the performance of generative models. It is calculated by Equation (9):

$$FID = |\mu_r - \mu_g|^2 + Tr(\Sigma_r + \Sigma_g - 2(\Sigma_r \times \Sigma_g)^{\frac{1}{2}}), \tag{9}$$

where μ_r and μ_g represent the mean vectors of the feature activation in the Inception network for the real data and generated data, respectively. Σ_r and Σ_g represent the covariance matrices of the feature activation in the Inception network for the real data and generated data, respectively. Tr is the trace operator. $(\Sigma_r \times \Sigma_g)^{\frac{1}{2}}$ is the matrix square root of the product of the two covariance matrices.

4. Experiments and Results

In this section, the results obtained from the methodology presented in the previous section are presented. Initially, the results obtained in creating the two databases will be shown. For this, the results for each relevant stage of the process will be presented. In the second part, the results of the detection and area estimation system will be presented, with a comparison to the GE API.

4.1. Results for Dataset Creation

The creation of the datasets involved the utilization of two deep learning models: the ViT image classifier and the deep generative model, specifically, DDPM.

4.1.1. ViT Classifier

The ViT model utilizes the Transformer architecture to perform image classification. The input image is divided into smaller patches which are then flattened and fed as input sequences to the Transformer model. The Transformer performs self-attention operations on these sequences to capture global context and correlations between patches, ultimately producing a feature representation of the entire image. This representation is then fed through a classifier head to predict the class label of the image.

The ViT image classifier was configured with five balanced classes: desert, city, tailings pond, ocean, and MWD, and each class contained 2200 images. The hyperparameters from the model in [56], pre-trained on ImageNet-21k [57], were used. This model uses 12 attention heads, encoder layer dimensionality and hidden size set to 768, 12 hidden layers, and patches of 16 × 16 resolution. The data were split into training, validation, and testing subsets in ratios of 70%, 20%, and 10%, respectively. The model was trained for 300 epochs with a batch size of 240. The results of the ViT image classifier are presented in Table 3.

Table 3. ViT classifier results for the experiment proposed in Figure 2 to classify five different types of imagery.

Class	Precision	Recall	F1-Score
City	0.99	1	0.99
Desert	0.95	0.98	0.97
Ocean	1	1	1
Tailings Pond	0.98	0.96	0.97
MWDs	1	0.98	0.99
Accuracy			**0.98**
Macro average	0.98	0.98	0.98
Weighted average	0.98	0.98	0.98

The ViT classifier demonstrated high accuracy, with a result of 98.8%, in differentiating between five scenes of aerial imagery.

4.1.2. Deep Generative Model

The DDPM model was trained using images containing MWDs, with a configuration of 250 epochs and a batch size of 8, at a resolution of 256 × 256 pixels. The original denoising DDPM algorithm was employed to sample images for the model, as it generated the highest number of expert-verified samples of MWDs.

A total of 1125 maps were generated, which were then classified by the ViT classifier to identify those that visually resemble MWDs. Finally, the best **792** tiles were selected. The FID score was computed for 792 real and synthetic images using Equation (9), resulting in a score of 222.46. Potential avenues for improvement of the FID score include the generation of high-quality synthetic data for comparison with the original dataset and the implementation of a more diverse and extensive training set for the DDPM model. The implementation of these proposed solutions has the potential to result in a lower FID score.

The application of the ViT classifier and DDPM resulted in the successful creation of two datasets, composed of 792 and 2430 images of MWDs, respectively.

4.2. Results for Detection, Segmentation, and Area Estimation

This section contains the results of experiments using the application of Mask R-CNN and YOLOv7 detection and segmentation algorithms used for the purpose of estimating the areas of MWDs in satellite imagery. This procedure is repeated for each dataset created. A comprehensive analysis of the procedures and results of the area estimation was conducted using both the segmentation masks and the GE API.

4.2.1. Results for YOLOv7 and Mask R-CNN

The Mask R-CNN and YOLOv7 models were evaluated using a k-fold cross-validation approach with 5 folds. The experiments were performed on the dataset, and the reported results represent the average performance across the folds.

The Mask R-CNN models were configured with two backbone networks, namely ResNet50 and ResNet101, both of which were equipped with FPN for feature extraction, with 3× schedule [58]. Fine-tuning was performed on both configurations, utilizing the respective backbone weights. The experiment was executed with an epoch setting of 3000 and a batch size of 450.

The YOLOv7 model was fine-tuned using the official weights available in its repository, with an epoch set to 3000 and a batch size of 448. In both cases, the images were normalized and partitioned into three subsets, with 70% of the images designated for training, 20% for validation, and 10% for testing purposes, respectively.

The results of the experiments performed using the two generated datasets are presented in Table 4. The table displays the AP metrics for detection (AP box) and segmentation (AP Mask), with the best results emphasized in bold for all experiments performed. The AP calculation was conducted with an IoU of 1.

Table 4. Results of YOLOv7 and Mask R-CNN algorithms for MWD detection and segmentation.

Dataset	Algorithm	AP Box	AP Mask
Original dataset	Mask R-CNN 50 FPN 3×	0.30	0.15
	Mask R-CNN 101 FPN 3×	0.31	0.17
	YOLOv7	**0.72**	**0.71**
Original dataset + augmentation + synthetic	Mask R-CNN 50 FPN 3×	0.39	0.38
	Mask R-CNN 101 FPN 3×	0.42	0.40
	YOLOv7	**0.81**	**0.79**

As can be seen in Table 4, the AP score result for object detection, in all cases, is higher than for object segmentation, but by a small margin. It is observed that increasing the data with data augmentation techniques results in a significant improvement in the performance of MWD detection. Additionally, the YOLOv7 algorithm outperforms Mask R-CNN in all cases, based on its AP score. The disparity in results between YOLO and Mask R-CNN can be attributed to their utilization of distinct architectures. YOLO, a single-shot object detection model, predicts object bounding boxes through the utilization of a grid-based approach [59]. In contrast, Mask R-CNN operates in two stages, utilizing region proposal networks (RPNs) and a mask prediction stage to achieve object segmentation [35]. The differing architectures of the algorithms contribute significantly to their performance, with YOLO's simpler grid-based approach proving effective for objects with clear boundaries and well-defined shapes, while Mask R-CNN's more complex architecture excels in handling small or overlapping objects [60], which is not the case for our MWD detection and segmentation.

4.2.2. Results for Area Estimation

The objective of this experiment is to determine the accuracy of the estimated areas obtained from the applied area formula compared to the actual measurements obtained through the GE API.

The detected MWDs were obtained using YOLOv7, where the output of the segmentation model produces a polygonal mask. The Gaussian area formula was used to estimate the area of the identified MWDs. The same detected polygon was used with the GE API to obtain the area with this method.

The estimated areas of 10 random samples of detected MWDs, using both the Gaussian area formula and the GE API, are presented in Table 5.

The average absolute error for the samples in Table 5 was found to be **5.58%**, which is within the expected range. Further analysis of 100 detected MWDs revealed an average absolute error in area calculation using our proposed method of **6.6%**. While there was a slight increase in error, the results indicate that the performance of our system is highly satisfactory.

To validate the estimated areas using our solution, Figure 8 shows a detected MWD, Figure 8a shows the detected polygon area of size 623,886 m^2, and Figure 8b shows the projection of the area of size 647,681 m^2 in Google Earth.

Table 5. Areas of the 10 MWD random samples estimated using both the Gaussian area algorithm and the GE API. Subsequently, the absolute error, standard deviation, and variance of the two measurements were calculated.

MWD ID	Estimated Area (m^2)	Area GE API (m^2)	Absolute Error (%)
0	449,412	439,785	2.19
1	243,687	273,733	10.98
2	1,463,262	1,550,792	5.64
3	514,362	521,347	1.34
4	394,150	359,541	9.63
5	189,962	184,733	2.83
7	571,587	554,441	3.09
8	337,762	322,835	4.62
9	135,975	126,476	7.51
10	47,262	43,776	7.96
Average			5.58
Standard Deviation			3.14
Variance			11.01

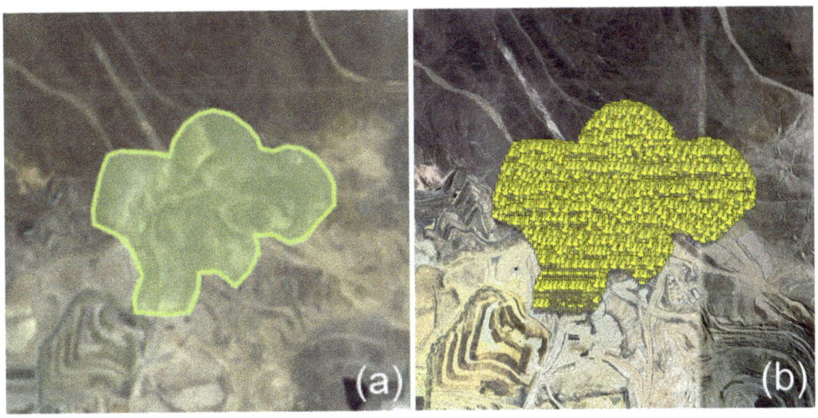

Figure 8. Detected MWDs using our proposed methodology. (**a**) shows the mask detected over a tile of Sentinel-2 imagery; (**b**) shows the same area exported to Google Earth.

The experiments demonstrated that the mathematical approach for estimating the area of MWDs is both accurate and valid, as the estimated area was in agreement with the measurements obtained through the GE API. This approach offers the advantage of providing accurate results while saving time compared to manual measurements. The utilization of the GE API also facilitated easy visualization and verification of the results, further confirming the validity of this approach as an alternative method for estimating the area of MWDs.

5. Conclusions and Future Work

In this study, we introduced a methodology aimed at the localization and estimation of the area of MWDs by employing advanced DL techniques. This methodology aspires to provide a foundational framework for analyzing PS, a crucial aspect during the closure and post-closure phases of mining operations.

Utilizing a ViT classifier, we achieved a classification accuracy of 98% across various aerial scenes. This demonstrates a promising avenue for processing satellite imagery in the context of mining waste management. Additionally, the employment of DGMs proved beneficial in augmenting the limited data available, showcasing a potential path for enhancing detection algorithms.

The application of YOLOv7 and Mask R-CNN algorithms on RGB imagery with a 10 m spectral resolution facilitated the accurate detection and segmentation of MWDs while preserving the location information derived from Sentinel-2 metadata. The experimental results indicate that the combination of YOLOv7 and diffusion models was effective in detecting and segmenting MWDs. Specifically, the YOLOv7 algorithm achieved an AP of 81% for detection and 79% for segmentation when integrating original, augmented, and synthetic data. This suggests that synthetic data can play a role in improving the accuracy of detection algorithms.

Furthermore, the methodology allowed for the estimation of areas of detected MWDs, offering a cost-effective alternative to the challenge of cataloging and accounting for the quantity of MWDs in a region.

The analysis of satellite images corresponding to the MWD areas could potentially provide variables associated with site sector, geomorphology, vegetation, populated sectors, and environmentally protected areas. This lays the groundwork for further exploration into machine learning algorithms for feature extraction, image processing, manual labeling, and deep learning in the context of mining waste management.

While this study presents an initial step towards addressing the challenges associated with the fiscalization process of MWDs, further research is warranted to refine the proposed methodology and explore other machine learning and deep learning algorithms for improved accuracy and efficiency.

Author Contributions: Conceptualization, M.S., G.H., G.V. and P.B.; methodology, M.S., G.H. and G.V.; software, M.S. and G.H.; validation, M.S., G.H. and G.V.; formal analysis, M.S. and G.H.; investigation, M.S., G.H., G.V. and P.B.; resources, P.B., G.V. and G.H.; data curation, M.S.; writing—original draft preparation, M.S.; writing—review and editing, P.B.; visualization, M.S., G.V. and G.H.; supervision, P.B.; project administration, G.V. and G.H.; funding acquisition, G.V., G.H. and P.B. All authors have read and agreed to the published version of the manuscript.

Funding: This research was funded and supported by Agencia Nacional de Investigación y Desarrollo (ANID): (i) grant number FONDEF IT20I0016: Plataforma Inteligente para la Evaluación Periódica de la Estabilidad Física en vista a un Cierre Progresivo y Seguro de Depósitos de Relaves de la Mediana Minería; (ii) grant number MEC 80190075; (iii) ANID Doctorado Nacional 2023-21232328; and DI-PUCV number 039.349/2023.

Data Availability Statement: Not applicable.

Acknowledgments: Our sincere thanks to Servicio Nacional de Geología y Minería (SERNAGEOMIN, Ministerio de Minería, Chile), Ministerio de Educación (Chile), Vicerrectoría de Investigación, Creación e Innovación de la Pontificia Universidad Católica de Valparaíso (Chile).

Conflicts of Interest: The authors declare no conflict of interest.

References

1. SERNAGEOMIN Site. Datos Públicos Depósitos de Relaves. Catastro de Depósitos de Relaves en Chile 2022. Available online: https://www.sernageomin.cl/datos-publicosdeposito-de-relaves/ (accessed on 30 January 2023).
2. Palma, C.; Linero, S.; Apablaza, R. Geotechnical Characterisation of Waste Material in Very High Dumps with Large Scale Triaxial Testing. In *Slope Stability 2007: Proceedings of the 2007 International Symposium on Rock Slope Stability in Open Pit Mining and Civil Engineering*; Potvin, Y., Ed.; Australian Centre for Geomechanics: Perth, Australia, 2007; pp. 59–75
3. Valenzuela, L.; Bard, E.; Campaña, J. Seismic considerations in the design of high waste rock dumps. In Proceedings of the 5th International Conference on Earthquake Geotechnical Engineering (5-ICEGE), Santiago, Chile, 10–13 January 2011 .
4. Bard, E.; Anabalón, M.E. Comportement des stériles Miniers ROM à Haute Pressions. Du Grain à l'ouvrage. 2008. Available online: https://www.cfms-sols.org/sites/default/files/manifestations/080312/2-Bard.pdf (accessed on 2 March 2023).
5. Fourie, A.; Villavicencio, G.; Palma, J.; Valenzuela, P.; Breul, P. Evaluation of the physical stability of leaching waste deposits for the closure stage. In Proceedings of the 20th International Conference on Soil Mechanics and Geotechnical Engineering, Sydney, Australia, 1–5 May 2022 .
6. Biblioteca del Congreso Nacional de Chile Site. Ley 19300. Ley Sobre Bases Generales del Medio Ambiente. Available online: https://www.bcn.cl/leychile/navegar?idNorma=30667&idParte=9705635&idVersion=2021-08-13 (accessed on 3 February 2023).
7. Biblioteca del Congreso Nacional de Chile Site. Decreto Supremo N° 132: Reglamento de Seguridad Minera. Available online: https://www.bcn.cl/leychile/navegar?idNorma=221064 (accessed on 3 February 2023).

8. Biblioteca del Congreso Nacional de Chile Site. Ley N° 20.551: Regula el Cierre de Faenas e Instalaciones Mineras. Available online: https://www.bcn.cl/leychile/navegar?idNorma=1032158 (accessed on 3 February 2023).
9. Biblioteca del Congreso Nacional de Chile site. Decreto N° 41: Aprueba el Reglamento de la Ley de Cierre de Faenas e Instalaciones Mineras. Available online: https://www.bcn.cl/leychile/navegar?idNorma=1045967&idParte=9314317&idVersion=2020-06-23 (accessed on 3 February 2023).
10. Biblioteca del Congreso Nacional de Chile Site. Ley 20.819: Modifica la Ley N° 20.551: Regula el Cierre de Faenas e Instalaciones Mineras. Available online: https://www.bcn.cl/leychile/navegar?i=1075399&f=2015-03-14 (accessed on 3 February 2023).
11. SERNAGEOMIN. Guía Metodológica para Evaluación de la Estabilidad Física de Instalaciones Mineras Remanentes. Available online: https://www.sernageomin.cl/wp-content/uploads/2019/06/GUIA-METODOLOGICA.pdf/ (accessed on 24 January 2023).
12. Hawley, P.M. *Guidelines for Mine Waste Dump and Stockpile Design*; CSIRO Publishing: Clayton, NC, USA , 2017; p. 370.
13. ESA: Sentinel-2 Mission. Available online: https://sentinels.copernicus.eu/web/sentinel/missions/sentinel-2/ (accessed on 25 January 2023).
14. McDowell, N.G.; Coops, N.C.; Beck, P.S.; Chambers, J.Q.; Gangodagamage, C.; Hicke, J.A.; Huang, C.; Kennedy, R.; Krofcheck, D.J.; Litvak, M.; et al. Global satellite monitoring of climate-induced vegetation disturbances. *Trends Plant Sci.* **2015**, *20*, 114–123. [CrossRef] [PubMed]
15. Van Etten, A.; Hogan, D.; Manso, J.M.; Shermeyer, J.; Weir, N.; Lewis, R. The Multi-Temporal Urban Development SpaceNet Dataset. In Proceedings of the IEEE/CVF Conference on Computer Vision and Pattern Recognition (CVPR), Nashville, TN, USA, 20–25 June 2021; pp. 6398–6407.
16. Talukdar, S.; Singha, P.; Mahato, S.; Shahfahad; Pal, S.; Liou, Y.A.; Rahman, A. Land-Use Land-Cover Classification by Machine Learning Classifiers for Satellite Observations—A Review. *Remote Sens.* **2020**, *12*, 1135. [CrossRef]
17. Chen, Y.; Ming, D.; Lv, X. Superpixel based land cover classification of VHR satellite image combining multi-scale CNN and scale parameter estimation. *Earth Sci. Inform.* **2019**, *12*, 341–363. [CrossRef]
18. Spectral Indexes Evaluation for Satellite Images Classification using CNN. *J. Inf. Organ. Sci.* **2021**, *45*, 435–449. [CrossRef]
19. Lantzanakis, G.; Mitraka, Z.; Chrysoulakis, N. X-SVM: An Extension of C-SVM Algorithm for Classification of High-Resolution Satellite Imagery. *IEEE Trans. Geosci. Remote Sens.* **2021**, *59*, 3805–3815. [CrossRef]
20. Abburu, S.; Golla, S. Satellite Image Classification Methods and Techniques: A Review. *Int. J. Comput. Appl.* **2015**, *119*, 20–25. [CrossRef]
21. Kaselimi, M.; Voulodimos, A.; Daskalopoulos, I.; Doulamis, N.; Doulamis, A. A Vision Transformer Model for Convolution-Free Multilabel Classification of Satellite Imagery in Deforestation Monitoring. *IEEE Trans. Neural Netw. Learn. Syst.* **2023**, *34*, 3299–3307. [CrossRef]
22. Horvath, J.; Baireddy, S.; Hao, H.; Montserrat, D.M.; Delp, E.J. Manipulation Detection in Satellite Images Using Vision Transformer. In Proceedings of the IEEE/CVF Conference on Computer Vision and Pattern Recognition (CVPR) Workshops, Nashville, TN, USA, 20–25 June 2021; pp. 1032–1041.
23. Saleem, M.H.; Potgieter, J.; Arif, K.M. Automation in Agriculture by Machine and Deep Learning Techniques: A Review of Recent Developments. *Precis. Agric.* **2021**, *22*, 2053–2091. [CrossRef]
24. Bazi, Y.; Bashmal, L.; Rahhal, M.M.A.; Dayil, R.A.; Ajlan, N.A. Vision Transformers for Remote Sensing Image Classification. *Remote Sens.* **2021**, *13*, 516. [CrossRef]
25. Yi, J.; Zhou, B. A Multi-Stage Duplex Fusion ConvNet for Aerial Scene Classification. *arXiv* **2022**, arXiv:2203.16325.
26. Camalan, S.; Cui, K.; Pauca, V.P.; Alqahtani, S.; Silman, M.; Chan, R.; Plemmons, R.J.; Dethier, E.N.; Fernandez, L.E.; Lutz, D.A. Change Detection of Amazonian Alluvial Gold Mining Using Deep Learning and Sentinel-2 Imagery. *Remote Sens.* **2022**, *14*, 1746. [CrossRef]
27. Sandfort, V.; Yan, K.; Pickhardt, P.J.; Summers, R.M. Data augmentation using generative adversarial networks (CycleGAN) to improve generalizability in CT segmentation tasks. *Sci. Rep.* **2019**, *9*, 16884. [CrossRef] [PubMed]
28. Goodfellow, I.J.; Pouget-Abadie, J.; Mirza, M.; Xu, B.; Warde-Farley, D.; Ozair, S.; Courville, A.; Bengio, Y. Generative Adversarial Networks. *arXiv* **2014**, arXiv:1406.2661.
29. Zhu, J.Y.; Park, T.; Isola, P.; Efros, A.A. Unpaired Image-to-Image Translation using Cycle-Consistent Adversarial Networks. *arXiv* **2017**, arXiv:1703.10593.
30. Chambon, P.; Bluethgen, C.; Langlotz, C.P.; Chaudhari, A. Adapting Pretrained Vision-Language Foundational Models to Medical Imaging Domains. *arXiv* **2022**, arXiv:2210.04133.
31. Rombach, R.; Blattmann, A.; Lorenz, D.; Esser, P.; Ommer, B. High-Resolution Image Synthesis with Latent Diffusion Models. In Proceedings of the 2022 IEEE/CVF Conference on Computer Vision and Pattern Recognition (CVPR), New Orleans, LA, USA, 18–24 June 2022; pp. 10674–10685. [CrossRef]
32. Ronneberger, O.; Fischer, P.; Brox, T. U-Net: Convolutional Networks for Biomedical Image Segmentation. *arXiv* **2015**, arXiv:1505.04597.
33. Pacheco, F.; Hermosilla, G.; Piña, O.; Villavicencio, G.; Allende-Cid, H.; Palma, J.; Valenzuela, P.; García, J.; Carpanetti, A.; Minatogawa, V.; et al. Generation of Synthetic Data for the Analysis of the Physical Stability of Tailing Dams through Artificial Intelligence. *Mathematics* **2022**, *10*, 4396. [CrossRef]
34. Lin, T.Y.; Goyal, P.; Girshick, R.; He, K.; Dollár, P. Focal Loss for Dense Object Detection. *arXiv* **2017**, arXiv:1708.02002.
35. He, K.; Gkioxari, G.; Dollár, P.; Girshick, R. Mask R-CNN. *arXiv* **2017**, arXiv:1703.06870.

36. Längkvist, M.; Kiselev, A.; Alirezaie, M.; Loutfi, A. Classification and Segmentation of Satellite Orthoimagery Using Convolutional Neural Networks. *Remote Sens.* **2016**, *8*, 329. [CrossRef]
37. Balaniuk, R.; Isupova, O.; Reece, S. Mining and Tailings Dam Detection in Satellite Imagery Using Deep Learning. *arXiv* **2020**, arXiv:2007.01076.
38. Lyu, J.; Hu, Y.; Ren, S.; Yao, Y.; Ding, D.; Guan, Q.; Tao, L. Extracting the Tailings Ponds from High Spatial Resolution Remote Sensing Images by Integrating a Deep Learning-Based Model. *Remote Sens.* **2021**, *13*, 743. [CrossRef]
39. Bochkovskiy, A.; Wang, C.Y.; Liao, H.Y.M. YOLOv4: Optimal Speed and Accuracy of Object Detection. *arXiv* **2020**, arXiv:2004.10934.
40. Breiman, L. Random Forests. *Mach. Learn.* **2001**, *45*, 5–32. [CrossRef]
41. Robinson, Y.H.; Vimal, S.; Khari, M.; Hernández, F.C.L.; Crespo, R.G. Tree-based convolutional neural networks for object classification in segmented satellite images. *Int. J. High Perform. Comput. Appl.* **2020**, 1094342020945026. [CrossRef]
42. Stewart, A.J.; Robinson, C.; Corley, I.A.; Ortiz, A.; Lavista Ferres, J.M.; Banerjee, A. TorchGeo: Deep Learning with Geospatial Data. In Proceedings of the 30th International Conference on Advances in Geographic Information Systems, Seattle, WA, USA, 1–4 November 2022; pp. 1–12. [CrossRef]
43. Consejo Minero. Mapa Minero. Available online: https://consejominero.cl/nosotros/mapa-minero/ (accessed on 12 January 2023).
44. ESA: Sentinel-2 Overview. Available online: https://sentinel.esa.int/web/sentinel/user-guides/sentinel-2-msi/overview (accessed on 17 January 2023).
45. ESA. Data Formats—User Guides—Sentinel-2 MSI—Sentinel Online—Sentinel Online. Available online: https://sentinels.copernicus.eu/web/sentinel/user-guides/sentinel-2-msi/data-formats (accessed on 20 January 2023).
46. Ho, J.; Jain, A.; Abbeel, P. Denoising Diffusion Probabilistic Models. *arXiv* **2020**, arXiv:2006.11239.
47. Tkachenko, M.; Malyuk, M.; Holmanyuk, A.; Liubimov, N. Label Studio: Data Labeling Software, 2020–2022. Available online: https://github.com/HumanSignal/label-studio (accessed on 11 October 2023).
48. Wang, C.Y.; Bochkovskiy, A.; Liao, H.Y.M. YOLOv7: Trainable Bag-of-Freebies Sets New State-of-the-Art for Real-Time Object Detectors. *arXiv* **2022**, arXiv:2207.02696.
49. Zhang, X.; Zeng, H.; Guo, S.; Zhang, L. Efficient Long-Range Attention Network for Image Super-resolution. In Proceedings of the European Conference on Computer Vision, Tel Aviv, Israel, 23–27 October 2022.
50. Ren, S.; He, K.; Girshick, R.; Sun, J. Faster R-CNN: Towards Real-Time Object Detection with Region Proposal Networks. *arXiv* **2015**, arXiv:1506.01497.
51. Szegedy, C.; Ioffe, S.; Vanhoucke, V.; Alemi, A. Inception-v4, Inception-ResNet and the Impact of Residual Connections on Learning. *arXiv* **2016**, arXiv:1602.07261.
52. Wu, Y.; Kirillov, A.; Massa, F.; Lo, W.Y.; Girshick, R. Detectron2. 2019. Available online: https://github.com/facebookresearch/detectron2 (accessed on 11 October 2023).
53. Gorelick, N.; Hancher, M.; Dixon, M.; Ilyushchenko, S.; Thau, D.; Moore, R. *Google Earth Engine: Planetary-Scale Geospatial Analysis for Everyone*; Elsevier: Amsterdam, The Netherlands, 2017. [CrossRef]
54. Heusel, M.; Ramsauer, H.; Unterthiner, T.; Nessler, B.; Hochreiter, S. GANs Trained by a Two Time-Scale Update Rule Converge to a Local Nash Equilibrium. *arXiv* **2017**, arXiv:1706.08500.
55. Szegedy, C.; Vanhoucke, V.; Ioffe, S.; Shlens, J.; Wojna, Z. Rethinking the Inception Architecture for Computer Vision. *arXiv* **2015**, arXiv:1512.00567.
56. Hugging Face. Google/Vit-Base-Patch16-224. Available online: https://huggingface.co/google/vit-base-patch16-224 (accessed on 30 January 2023).
57. Ridnik, T.; Ben-Baruch, E.; Noy, A.; Zelnik-Manor, L. ImageNet-21K Pretraining for the Masses. *arXiv* **2021**, arXiv:2104.10972.
58. He, K.; Girshick, R.; Dollár, P. Rethinking ImageNet Pre-training. *arXiv* **2018**, arXiv:1811.08883.
59. Redmon, J.; Divvala, S.; Girshick, R.; Farhadi, A. You Only Look Once: Unified, Real-Time Object Detection. *arXiv* **2015**, arXiv:1506.02640.
60. Girshick, R.B.; Donahue, J.; Darrell, T.; Malik, J. Rich feature hierarchies for accurate object detection and semantic segmentation. *arXiv* **2013**, arXiv:abs/1311.2524.

remote sensing

Technical Note

Semantic Segmentation of High-Resolution Remote Sensing Images Based on Sparse Self-Attention and Feature Alignment

Li Sun, Huanxin Zou *, Juan Wei, Xu Cao, Shitian He, Meilin Li and Shuo Liu

College of Electronic Science and Technology, National University of Defense Technology, Changsha 410073, China
* Correspondence: zouhuanxin@nudt.edu.cn; Tel.: +86-731-8700-3288

Abstract: Semantic segmentation of high-resolution remote sensing images (HRSI) is significant, yet challenging. Recently, several research works have utilized the self-attention operation to capture global dependencies. HRSI have complex scenes and rich details, and the implementation of self-attention on a whole image will introduce redundant information and interfere with semantic segmentation. The detail recovery of HRSI is another challenging aspect of semantic segmentation. Several networks use up-sampling, skip-connections, parallel structure, and enhanced edge features to obtain more precise results. However, the above methods ignore the misalignment of features with different resolutions, which affects the accuracy of the segmentation results. To resolve these problems, this paper proposes a semantic segmentation network based on sparse self-attention and feature alignment (SAANet). Specifically, the sparse position self-attention module (SPAM) divides, rearranges, and resorts the feature maps in the position dimension and performs position attention operations (PAM) in rearranged and restored sub-regions, respectively. Meanwhile, the proposed sparse channel self-attention module (SCAM) groups, rearranges, and resorts the feature maps in the channel dimension and performs channel attention operations (CAM) in the rearranged and restored sub-channels, respectively. SPAM and SCAM effectively model long-range context information and interdependencies between channels, while reducing the introduction of redundant information. Finally, the feature alignment module (FAM) utilizes convolutions to obtain a learnable offset map and aligns feature maps with different resolutions, helping to recover details and refine feature representations. Extensive experiments conducted on the ISPRS Vaihingen, Potsdam, and LoveDA datasets demonstrate that the proposed method precedes general semantic segmentation- and self-attention-based networks.

Keywords: semantic segmentation; high-resolution remote sensing; self-attention; context modeling; feature alignment

Citation: Sun, L. ; Zou, H.; Wei, J.; Cao, X.; He, T.; Li, M.; Liu, S. Semantic Segmentation of High-Resolution Remote Sensing Images Based on Sparse Self-Attention and Feature Alignment. *Remote Sens.* **2023**, *15*, 1598. https://doi.org/10.3390/rs15061598

Academic Editors: Qian Du, Jiaojiao Li, Wei Li, Jocelyn Chanussot, Rui Song, Yunsong Li and Bobo Xi

Received: 19 January 2023
Revised: 8 March 2023
Accepted: 13 March 2023
Published: 15 March 2023

1. Introduction

Semantic segmentation predicts the semantic labels for each pixel in an image. Semantic segmentation of high-resolution remote sensing images (HRSI) is the cornerstone of remote sensing interpretation. It is of great importance in many fields, such as mapping, navigation, land resource management, etc. [1–3]. Specifically, land cover maps depict local and overall landscape conditions, from which environmental change trends can be obtained. Semantic segmentation can be used to assess urban development and estimate the impact of natural disasters. Since remote sensing technology has advanced, HRSI with more complex pixel representation have become more readily available. Semantic segmentation is more crucial and challenging for HRSI. Traditional semantic segmentation methods [4–6] rely on expert experience and complex human-designs. Moreover, the segmentation performance relies on the accuracy and suitability of manually designed features. With robust feature modeling capabilities, deep learning technology has become an effective method used for semantic segmentation of HRSI, and researchers have applied deep learning technology to this operation. Specifically, a convolutional neural network (CNN) [7] has been widely used in semantic segmentation and achieved satisfactory results. To further enhance the accuracy of semantic segmentation, researchers focus on both contextual information fusion and the refinement of segmentation results.

To achieve contextual information fusion, several network variants are proposed to enhance contextual aggregation. PSPNet [8] developed spatial pyramid pooling to acquire a rich, multi-scale context. The Deeplab series [9–11] utilized the atrous spatial pyramid pooling (ASPP) to gather contextual clues, which consisted of parallel atrous convolutions with different dilated rates. GCN [12] removed the pooling in the network and developed a large decoupling convolution kernel to extract features. The large convolution kernel can obtain a large receptive field and is beneficial to the capture of long-range contextual information. However, the above methods fail to model the global contextual dependencies across an entire image. Recently, self-attention mechanisms commonly used in natural language processing (NLP) have been widely used for visual tasks with exciting results. Wang et al. [13] first proposed self-attention to capture global dependencies. Fu et al. [14] developed DANet to model non-local dependencies in position and channel dimensions. Instead of calculating self-attention at each point, EAMNet [15] utilized the expectation-maximization iteration manner to learn a more compact basis set, and then carried out self-attention. To model spatial long-range dependencies, CCNet [16] proposed recurrent a criss-cross attention module. Yuan et al. [17] developed OCNet with interlaced sparse self-attention. The above methods show that the self-attention operation is an effective way to capture global dependencies. Thus, several studies have used the self-attention mechanism for semantic segmentation of HRSI. Shi et al. [18] combined self-attention and atrous convolution with different atrous rates to capture spatially adaptive global context information. Li et al. [19] proposed kernel attention with linear complexity and combined it with the standard dot product attention. However, the above methods ignore a key problem: due to the complex background and rich details of HRSI, standard self-attention will introduce redundant information and interfere with semantic segmentation. To solve this problem, this paper proposes the sparse position self-attention module (SPAM) and sparse channel self-attention module (SCAM), which not only captures the global information, but also reduces the interference of redundant information.

For the refinement of segmentation results, the current semantic segmentation network uses several strategies. One is to obtain the high-level semantic information gradually via down-sampling and then integrate the features of various levels through the decoder to recover the details. For example, Long et al. [20] proposed fully convolutional networks (FCNs) that restored the original image size by incorporating the low-level features and high-level features. SegNet [21] retained the index of the maximum position when pooling, and the index was reused when upsampling. U-Net [22] adopted skip-connections to connect shallow layers and deep layers. RefineNet [23] utilized a Laplacian image pyramid to explicitly model the available information during downsampling and predictions from coarse to fine. Another potential strategy is to learn semantic information while maintaining high resolution feature maps. For example, HRNet [24] proposed a parallel structure backbone network, which maintained high resolution characteristics during the entire process. Additionally, several networks refine the segmentation edges to obtain more precise semantic segmentation results. Gated-SCNN [25] deconstructed the edge information from the regular features and used a shape branch to focus on semantic boundary information. SegFix [26] proposed a post-processing method to refine the boundaries of semantic segmentation results. ERN [27] developed the edge enhancement structure and the loss function used to supervise the edge to enhance the segmentation accuracy. Zheng et al. [28] developed a Dice-based edge-aware loss function to refine edge information directly from semantic segmentation prediction. Li et al. [29] highlighted the edge distribution of the feature map in a self-attention fashion. The above methods recover the details and improve the edge segmentation performance to some extent. However, the issue of feature maps with different resolutions being misaligned is ignored. To solve this problem, this paper proposes the feature alignment module (FAM), which generates a learnable offset map to align feature maps with different resolutions.

HRSI generally have complex background information and abundant details, which makes semantic segmentation more challenging. The standard self-attention and excessive fusion of long-range context information may introduce redundant information and cause interference to object segmentation. This paper proposes SPAM and SCAM to effectively model the position global context and channel-wise dependencies. Additionally, feature maps with different resolutions are not aligned. Features from shadow layers and deep layers are directly fused and, thus, fail to obtain higher-quality segmentation results. This paper proposes FAM, which combines low-level and high-level features with different resolutions. FAM is beneficial, as it refines segmentation results and improves the segmentation performance of an object edge. The contributions of this work are threefold:

1. The paper proposes SPAM and SCAM to efficiently model the position non-local information and channel-wise dependency, which reduces redundant information, contributing to the intraclass consistency of large objects and the segmentation accuracy of small objects.
2. The paper introduces FAM, which can align feature maps with different resolutions and further improve segmentation results.
3. Extensive experimental results demonstrate that SAANet achieves leading performance on ISPRS Vaihingen, Potsdam, and LoveDA datasets.

2. Materials and Methods

The particulars of the proposed semantic segmentation network based on sparse self-attention and feature alignment (SAANet) for semantic segmentation will be introduced. We first present the overall framework of our SAANet and then illustrate the details of the SPAM, SCAM, and FAM.

2.1. Overview

As shown in Figure 1, the proposed SAANet consists of a backbone, SPAM, SCAM, and FAM. Many studies have proved the good performance of a pretrained ResNet backbone in semantic segmentation tasks. First, the dilated ResNet-101 [30] is set as the backbone to extract features. The outputs of the dilated ResNet-101 in each stage are denoted as $\{S_1, S_2, S_3, S_4\}$. Due to the removal of down-sampling operations and adoption of dilated convolutions in the last two blocks, feature maps have strides of $\{4, 8, 8, 8\}$ pixels, with respect to the input image. Then, SPAM and SCAM take the feature map S_4 as input to model non-local dependencies in the position and channel dimensions. In addition, in order to achieve better feature representations, a feature pyramid network (FPN) [31] is used to combine low-level and high-level features and the outputs are denoted as $\{F_1, F_2, F_3, F_4\}$. Finally, feature maps F_2,F_3, and F_4 are up-sampled to the same size as feature map F_1 utilizing FAM. The four feature maps are concatenated to gain final pixel-level feature representations.

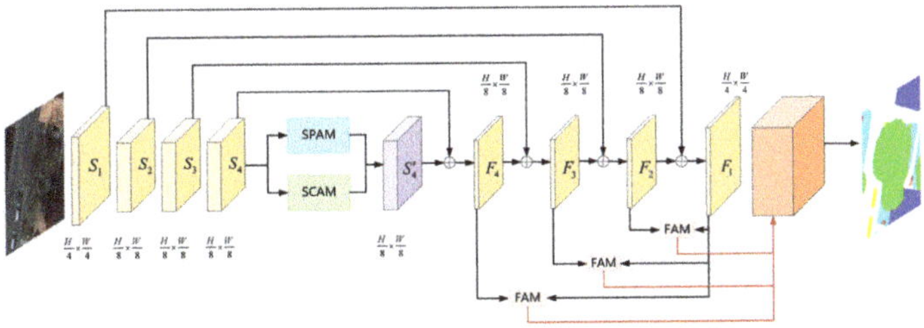

Figure 1. An overview of our proposed semantic segmentation network based on sparse self-attention and feature alignment (SAANet). H and W represent the height and width of the input image, respectively.

2.2. Sparse Position Self-Attention Module

Due to the complex scenes and rich details of HRSI, the implementation of a position self-attention module (PAM) on the whole image introduces redundant information and interferes with semantic segmentation. To capture long-range dependencies more efficiently and reduce redundant information, this paper proposes the SPAM, which is based on PAM.

2.2.1. Position Self-Attention Module

PAM is first introduced and shown in Figure 2. Given the feature map M, the features query (Q), key (K), and value (V) are first generated by convolutions, respectively, where $Q, K, V \in R^{C\times H\times W}$. C, H, and W denote the number of channels of the feature, image height, and image width, respectively. Then, they are reshaped to $Q_p, K_p, V_p \in R^{C\times N}$, where $N = H \times W$ is the number of pixels. Next, Q_p is multiplied by the transpose of K_p, and the softmax layer is applied to calculate the position attention map $A_p \in R^{N\times N}$:

$$A_p = \text{softmax}\left(K_p^T Q_p\right) \quad (1)$$

where A_p measures the influence between the two positions, and the more similar two pixel features are, the larger the value of A_p is. Then, V_p and the transpose of A_p are multiplied, and the resulting product is reshaped to $R^{C\times H\times W}$. Finally, to obtain the output $N_p \in R^{C\times H\times W}$, the feature map is multiplied by the scale coefficient α and sum with the feature map M.

$$N_p = \alpha V_p A_p^T + M \quad (2)$$

where α is a learnable parameter, which is initialized to 0.

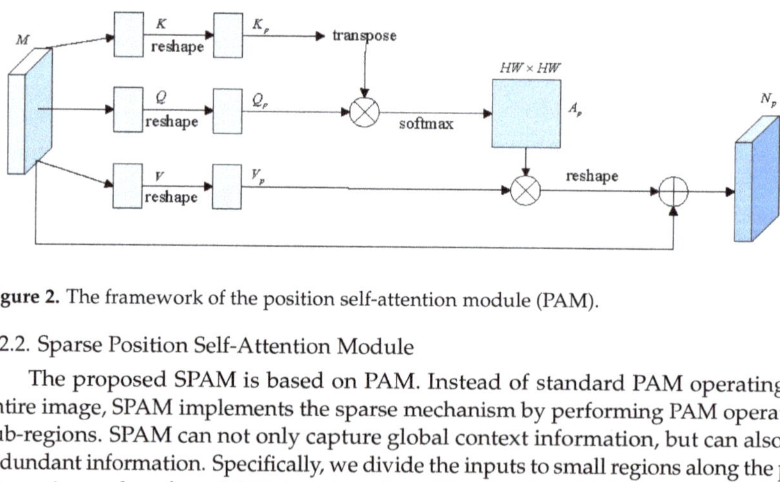

Figure 2. The framework of the position self-attention module (PAM).

2.2.2. Sparse Position Self-Attention Module

The proposed SPAM is based on PAM. Instead of standard PAM operating on the entire image, SPAM implements the sparse mechanism by performing PAM operations on sub-regions. SPAM can not only capture global context information, but can also reduce redundant information. Specifically, we divide the inputs to small regions along the position dimension and perform PAMs in sub-regions. The details of SPAM are shown in Figure 3. Given a feature map X with the spatial size of $H \times W$, the feature map X is grouped along the H and W dimensions and the spatial size of each group is $H\_n \times W\_n$. The feature map X is divided into $H/H\_n \times W/W\_n$ groups, named $\{X_1, X_2, X_3 \ldots\}$. Figure 3 illustrates the details of SPAM by taking $H, W = 8$ and $H\_n, W\_n = 2$ as an example. Then, the pixels at the same relative positions in each group are reorganized into new regions. The number of new regions is $H\_n \times W\_n$, and the pixels of each new region are $H/H\_n \times W/W\_n$. Meanwhile, PAMs are operated in new regions, and the feature map Y_1 is obtained, which is then set as the global operation. Finally, the pixel position of the feature map Y_1 is restored to the original combination, and PAMs are carried on $\{X_1, X_2, X_3 \ldots\}$. The feature Z_1 is obtained. The input of SPAM is S_4 in our SAANet. SPAM efficiently captures the long-range context information and models the pixel-wise relationship. The information propagation process of SPAM is shown in Figure 4. Take regions R_1, R_2 and pixels M, N, P,

Q as examples to illustrate details of the information propagation. Specifically, pixel M in the region R_1 and pixel N in the region R_2 first operate PAM, while pixel P in the region R_1 and pixel Q in the region R_2 operate PAM. Then, the region R_1 and the region R_2 continue PAM, respectively. Finally, pixels M, N, P, and Q aggregate the local and global contextual information. The above operations complete the information propagation between regions R_1 and R_2.

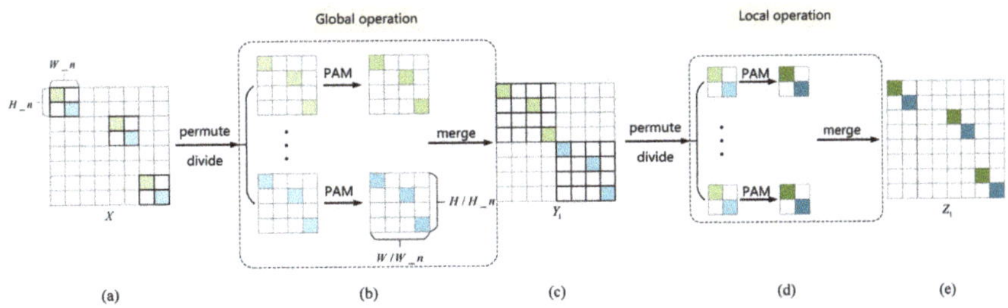

Figure 3. The structure of the proposed sparse position self-attention module (SPAM). (**a**) The input image is divided along position dimension. (**b**) PAMs are performed in rearranged small regions. (**c**) The pixel position of the feature map is restored to the original combination. (**d**) PAMs are performed in restored sub-regions. (**e**) The output of SPAM is obtained.

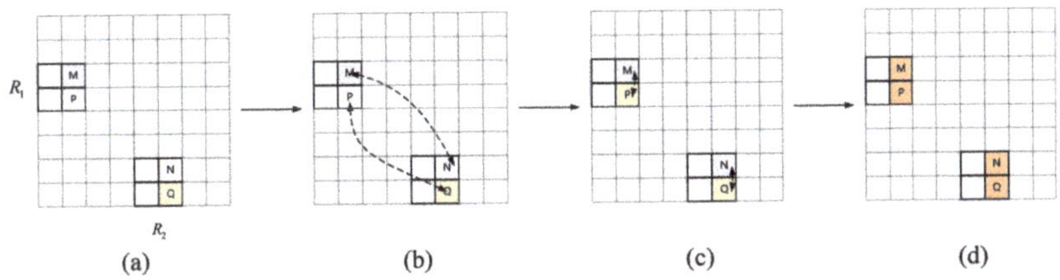

Figure 4. The information propagation process of the proposed SPAM. (**a**) M and P are the two pixels in region R_1; and N and Q are the two pixels in region R_2. (**b**) During the first PAM operation, the information is transmitted between M and N and between P and Q, respectively. (**c**) During the second PAM operation, the information is transmitted between M and P and between N and Q, respectively. (**d**) M, N, P, and Q contain global and local information.

2.3. Sparse Channel Attention Module

Due to the complexity of HRSI, there are large intra-class differences and small inter-class differences. Therefore, operating a standard self-attention module (CAM) on all channels introduces redundant information and causes category confusion. To model interdependent information between channels more efficiently and suppress redundant information, this paper proposes SCAM, which is based on CAM.

2.3.1. Channel Self-Attention Module

The architecture of CAM is shown in Figure 5. We first reshape the local feature map $M \in \mathbb{R}^{C\times H\times W}$ to $M_c \in \mathbb{R}^{C\times HW}$. Then, the matrix multiplication between M_c and the transpose of M_c is performed for the softmax layer, and the attention feature map $A_c \in \mathbb{R}^{C\times C}$ is obtained as follows:

$$A_c = \mathrm{softmax}\left(M_c {M_c}^T\right) \tag{3}$$

where A_c measures the influence of different channels. Then, M_c is multiplied by the transpose of A_c, and the multiplication is reshaped to $\mathbb{R}^{C\times H\times W}$. Finally, the product is multiplied by the scale coefficient β and added to the original feature M to obtain the final feature map $N_c \in \mathbb{R}^{C \times H \times W}$, as follows:

$$N_c = \beta A_c{}^T M_c + M \tag{4}$$

where β is a learnable parameter and is initialized to 0.

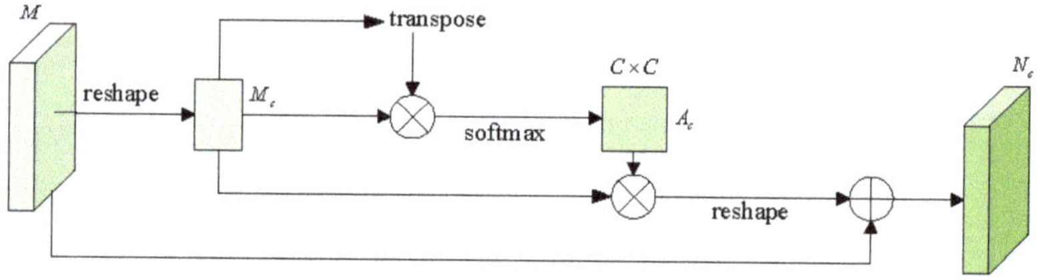

Figure 5. The architecture of the channel self-attention module (CAM).

2.3.2. Sparse Channel Self-Attention Module

The details of SCAM are shown in Figure 6. First, the feature map X is divided into C_n groups $\{C_1, C_2, C_3 \ldots\}$ in the channel dimension. The channel number of each group is C/C_n. Figure 6 illustrates the details of SCAM, taking $C\_n = 2$ as an example. Then, the groups $\{C_1, C_2, C_3 \ldots\}$ are further divided into C_n sub-groups, named $\{C_{11}, C_{12}, C_{13} \ldots\}, \{C_{21}, C_{22}, C_{23} \ldots\}, \{C_{31}, C_{32}, C_{33} \ldots\} \ldots$. The channel number of each sub-group is C/C^2_n. Next, for each channel group, sub-groups in the same relative position (for instance, C_{11}, C_{21}, C_{31} ...) are taken out to rearrange and generate new channel groups $\{C_{11}, C_{21}, C_{31} \ldots\}, \{C_{12}, C_{22}, C_{32} \ldots\}, \{C_{13}, C_{23}, C_{33} \ldots\} \ldots$. The feature map Y_2 is obtained by operating CAMs in C/C_n new groups. Finally, Y_2 is restored to the original channel arrangement, and Z_2 is acquired by performing CAMs in original groups $\{C_1, C_2, C_3 \ldots\}$.

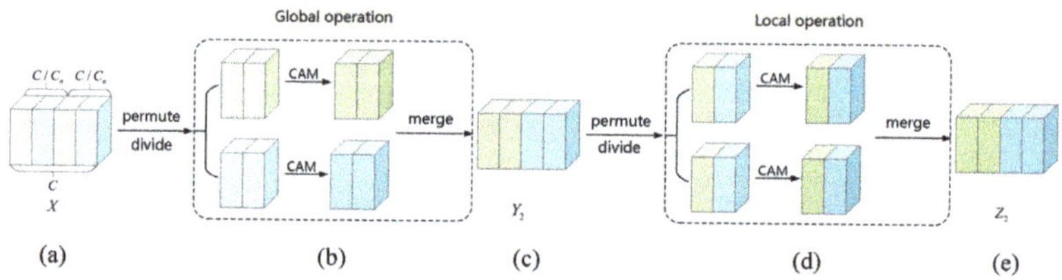

Figure 6. The structure of the proposed sparse channel self-attention module (SCAM). (**a**) The input image is divide along the channel dimension. (**b**) CAMs are performed in rearranged sub-channels. (**c**) The channels of the feature map are resorted to the original combination. (**d**) CAMs are performed in resorted sub-channels. (**e**) The output of SCAM is obtained.

2.4. Feature Alignment Module

Several methods are proposed to refine semantic segmentation results. However, the misalignment of features is ignored. To align features with different resolutions and refine semantic segmentation representations, this paper proposes an FAM. Specifically, the feature map S'_4 in Figure 1 from the last stage of ResNet fuses global context information and possesses enriched semantic information. However, the feature map S'_4 with coarse resolution lacks fine details. The proposed SAANet uses an FPN to fuse different resolution features from different stages. The FPN gradually fuses lower-level features with the details and higher-level features with abundant semantic information in a top-down pathway via 2× bilinear upsampling. However, the feature maps with different resolution are misaligned, which causes confusion in edges and small object segmentation. The misalignment has a great influence on the accuracy of semantic segmentation, especially on HRSI with complex scenes. After a series of operations in SAANet, such as downsampling, residual connection, self-attention, etc., the misalignment of the feature maps is more complicated. In the upsampling process, using bilinear interpolation alone fails to achieve better semantic segmentation results. The proposed SAANet develops a feature alignment module, which utilize convolutions to obtain a learnable offset map for feature alignment.

The details of the FAM are shown in Figure 7. The FAM is structured within the FPN framework. The inputs of the FAM are two feature maps with different spatial resolutions. It is assumed that F^l and F^{l-1} are the two input features of FAM, where $F^l \in R^{H_l \times W_l \times C}$ and $F^{l-1} \in R^{H_{l-1} \times W_{l-1} \times C}$. F^l is first upsampled via the standard regular grid sampling based bilinear interpolation. Then, the upsampled F^l and F^{l-1} are concatenated, and the feature map F' is obtained. The feature F' is passed through a 1×1 convolution, BN, and 3×3 convolution to predict an offset $\Delta_f \in R^{H_{l-1} \times W_{l-1} \times 2}$. Finally, the offset map is used to correct the upsampled F^l, which obtains the output feature map aligned with F^{l-1}. Mathematically, the above operations can be written as:

$$F' = concat(F^{l-1}, upsample(F^l)) \tag{5}$$

$$\Delta_f = conv_{3\times3}(BN(conv_{1\times1}(F'))) \tag{6}$$

where the *upsample* denotes the bilinear interpolation function, and Δ_f denotes offsets in horizontal and vertical directions. The FAM also involves less computation. SAANet uses three FAMs for the alignment of F_2, F_3, F_4, and F_1, respectively. FAM alleviates the feature misalignment and enhances the performance of semantic segmentation, especially for small objects and boundary regions.

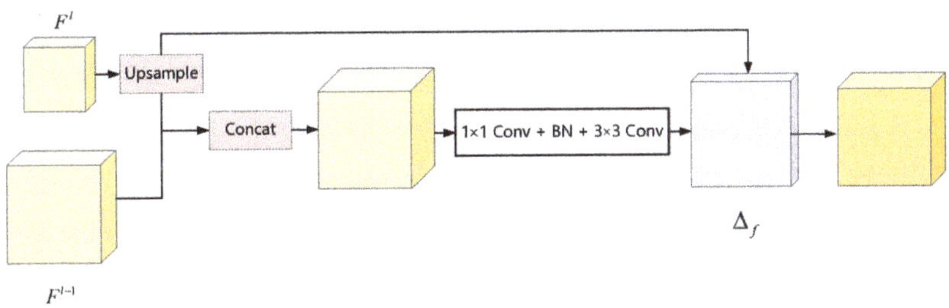

Figure 7. The framework of the proposed feature alignment module (FAM).

3. Experiments

We first introduce the datasets, evaluation metrics, and implementation details and then conduct ablation studies to validate the effectiveness of our framework. Finally, we compare the proposed network with several state-of-the-art methods on ISPRS Vaihingen, Potsdam [32], and LoveDA Urban [33] datasets.

3.1. Datasets and Evaluation Metrics

ISPRS Vaihingen dataset: ISPRS Vaihingen is a high-resolution remote sensing dataset used for semantic segmentation, which is composed of 33 images. The ground sampling distance (GSD) is 9 cm, and the average size of the images is 2496 × 2046 pixels. All images have corresponding semantic segmentation labels. The training and testing sets contain 17 and 16 images, respectively. There are six categories: impervious surface, building, low vegetation, tree, car, and clutter/background.

ISPRS Potsdam dataset: ISPRS Potsdam contains 38 images. The GSD is 5 cm, and the size of each image is 6000 × 6000 pixels. All images have corresponding semantic segmentation labels. The number of images in the training and testing sets is 21 and 17, respectively. As with the Vaihingen dataset, there are six categories.

LoveDA Urban dataset: The LoveDA dataset is constructed by Wang et al. [33]. The historical images were obtained from the Google Earth platform. LoveDA Urban dataset was obtained from urban areas in Wuhan, Changzhou, Nanjing, and other places in China. The size of each image is 1024 × 1024 pixels, and the GSD is 0.3m. The dataset was divided into three parts: a training set, a val set, and a test set, among which the training set and val set have semantic labels. In our experiment, 1156 training images were used as our training set, and 677 val set images were used as our test set. There are seven categories: background, building, road, water, barren, forest, and agricultural.

Evaluation Metrics: To evaluate the performance of semantic segmentation, this study sets the mean intersection over union (mIoU), F1-score (F1), and overall pixel accuracy (OA) [34] as its evaluation metrics. The aforementioned metrics are as follows.

$$mIOU = \frac{1}{N}\sum_{k=1}^{N}\frac{TP_k}{TP_k + FP_k + FN_k} \tag{7}$$

$$F_1 = 2 \times \frac{Precision \times Recall}{Precision + Recall}, Precision = \frac{TP}{TP + FP}, Recall = \frac{TP}{TP + FN} \tag{8}$$

$$OA = \frac{\sum_{k=1}^{N} TP_k}{\sum_{k=1}^{N} TP_k + FP_k + TN_k + FN_k} \tag{9}$$

where TP, FP, TN, FN, and k indicate the true positive, false positive, true negative, false negatives, and category, respectively.

3.2. Implementation Details

Due to limited computing resources, we cropped all images into 512 × 512 pixels. All experiments were implemented with PyTorch on a single NVIDIA GeForce RTX 2080Ti GPU, and the optimizer was set as standard the stochastic gradient descent (SGD). For different data sets, different learning rates were selected. The learning rates of the Vaihingen, Potsdam, and LoveDA Urban datasets were 0.001, 0.0008, and 0.0007, respectively. For all methods, cross-entropy loss is set as the loss function. For all datasets and networks, the maximum iteration period is 100 epochs.

3.3. Comparison to State-of-the-Art

To verify the superiority of our SAANet, we perform comparisons with several existing semantic segmentation methods, including self-attention-based and other general semantic segmentation networks. Aside from HRNet, whose backbone network is W48, other networks use the dilated ResNet-101 as the backbone. The experimental results on ISPRS Vaihingen, Potsdam, and LoveDA datasets are shown in Tables 1–3, respectively. The proposed SAANet achieves the best mIOU on ISPRS Vaihingen, Potsdam, and LoveDA Urban datasets.

Results on the Vaihingen dataset: Compared with the typical segmentation network FCN, our SSANet obtains 1.72%, 1.33%, and 0.76% improvement and achieves 68.50%, 80.22%, and 86.72% for mIOU, mF1, and OA, respectively. Moreover, the mIOU/F1/OA of our SAANet surpasses 0.92%/0.72%/0.37% by the network based on self-attention DANet. Thanks to SPAM, SCAM, and FAM, SSANet achieves more precise semantic segmentation results in all classes, especially on small objects. For example, SSANet outperforms the previous best one by 1.62% in the car category.

Results on the Potsdam dataset: Compared with a typical segmentation network based on self-attention CCNet, our SSANet obtains 1.20%, 1.02%, and 0.61% improvement and achieves 73.79%, 83.57%, and 88.22%, on mIOU, mF1, and OA, respectively. Moreover, the mIOU, mF1, and OA of our SAANet surpasses 1.07%/0.75%/0.76% by the typical network with multi-scale aggregation PSPNet. Meanwhile, SSANet achieves more precise semantic segmentation results in all classes, with the most significant improvement in the car category.

Table 1. Comparisons of different networks on ISPRS Vaihingen dataset. Note that we chose the IOU as the metric of each category. The best results are shown in boldface.

Method	Imp. Surf.	Building	Low Veg.	Tree	Car	Background	mIOU	mF1	OA
DeepLabv3+	77.15	85.40	61.22	74.54	57.46	26.44	63.70	75.90	85.30
HRNet	**79.12**	85.78	62.46	75.69	60.46	25.87	64.90	76.70	86.10
EMANet	77.83	85.73	62.61	75.37	60.95	32.18	65.78	77.87	85.88
PSPNet	77.94	85.77	62.90	75.65	60.40	34.59	66.21	78.34	85.97
CCNet	77.73	85.64	62.35	75.42	61.61	36.22	66.50	78.66	85.83
FCN	77.99	85.82	62.84	75.39	61.86	36.76	66.78	78.89	85.96
DANet	78.48	86.67	63.19	75.74	63.73	37.67	67.58	79.50	86.35
SAANet	79.00	**87.52**	**63.79**	**76.16**	**65.35**	**39.21**	**68.50**	**80.22**	**86.72**

Table 2. Comparisons of different networks on ISPRS Potsdam dataset. Note that we choose the IOU as the metric of each category. The best results are shown in boldface.

Method	Imp. Surf.	Building	Low Veg.	Tree	Car	Background	mIOU	mF1	OA
DeepLabv3+	81.87	90.32	71.75	73.58	83.12	34.19	72.47	82.50	87.45
CCNet	82.43	90.64	71.72	73.31	83.47	33.97	72.59	82.55	87.61
PSPNet	81.64	89.96	71.43	74.45	82.61	36.21	72.72	82.82	87.46
HRNet	82.65	89.99	72.17	74.16	83.58	35.06	72.94	82.87	87.76
DANet	82.80	**90.94**	72.23	74.42	83.70	33.87	72.99	82.80	87.96
FCN	82.30	90.66	71.62	74.37	83.55	36.03	73.09	83.03	87.73
EMANet	82.54	90.49	71.92	73.73	83.31	37.16	73.19	83.18	87.77
SAANet	**83.40**	90.78	**72.46**	**74.53**	**84.12**	**37.46**	**73.79**	**83.57**	**88.22**

Table 3. Comparisons of different networks on LoveDA Urban dataset. Note that we choose the IOU as the metric of each category. The best results are shown in boldface.

Method	Background	Building	Road	Water	Barren	Forest	Agricultural	mIOU	mF1	OA
DeepLabv3+	35.31	59.73	56.23	54.95	19.45	42.05	31.17	42.70	58.46	57.99
FCN	34.13	59.60	54.99	**68.42**	26.91	**47.90**	23.27	45.03	60.39	58.38
HRNet	37.96	60.03	**59.83**	68.33	25.07	44.63	30.59	46.63	62.11	60.87
PSPNet	38.72	58.80	53.00	60.30	23.18	44.36	48.13	46.64	62.64	63.41
DANet	38.67	**62.04**	58.93	66.52	23.26	43.92	34.37	46.82	62.33	61.54
CCNet	38.83	60.31	56.04	63.89	**39.74**	46.96	29.61	47.91	63.92	61.62
EMANet	40.46	60.02	58.18	64.55	30.36	47.74	46.22	49.65	65.58	65.19
SAANet	**42.09**	61.25	57.26	63.64	33.14	44.32	**48.38**	**50.01**	**66.03**	**65.45**

Results on the LoveDA Urban dataset: Compared with the ISPRS Vaihingen and Potsdam datasets, the LoveDA Urban dataset with lower GSD has more complex scenes, which makes semantic segmentation more challenging. Nevertheless, our SAANet still achieves the best mIOU, mF1, and OA. Particularly, for more challenging classes, the background class with greater intra-class variation, and the agricultural class with a small number of pixels, the proposed SAANet achieves the highest IOU. Specifically, compared with typical segmentation network FCN, our SSANet obtains 4.98%, 5.64%, and 7.07% improvement and achieves 50.01%, 66.03%, and 65.45% for mIOU, mF1, and OA, respectively. Moreover, the mIOU/F1/OA of our SAANet surpasses 2.10%/2.11%/3.83% for the network based on the self-attention CCNet.

Overall, our method achieves state-of-the-art semantic segmentation performance on the ISPRS Vaihingen, Potsdam, and LoveDA datasets. To qualitatively validate the effectiveness, several visualization results are shown in Figures 8 and 9. It is observed that the overall visual effect of our method outperforms other methods. Specifically, for large objects, our method contributes to the intra-class consistency. In the first group in Figure 8, for large buildings, other methods incorrectly predict that several pixels representing buildings are low vegetation or tree class. In the first group in Figure 9, other methods incorrectly predict pixels inside impervious surfaces. On the contrary, our SAANet can maintain category consistency. Additionally, for small objects, our method achieves more refined semantic segmentation results. For example, in the second group in Figures 8 and 9, several pixels representing cars are incorrectly predicted or have rough edges in the visual results of other methods. However, our SAANet obtains more accurate pixel classification and more precise edges. This suggests that our SAANet can obtain superior semantic segmentation performance and visual effects.

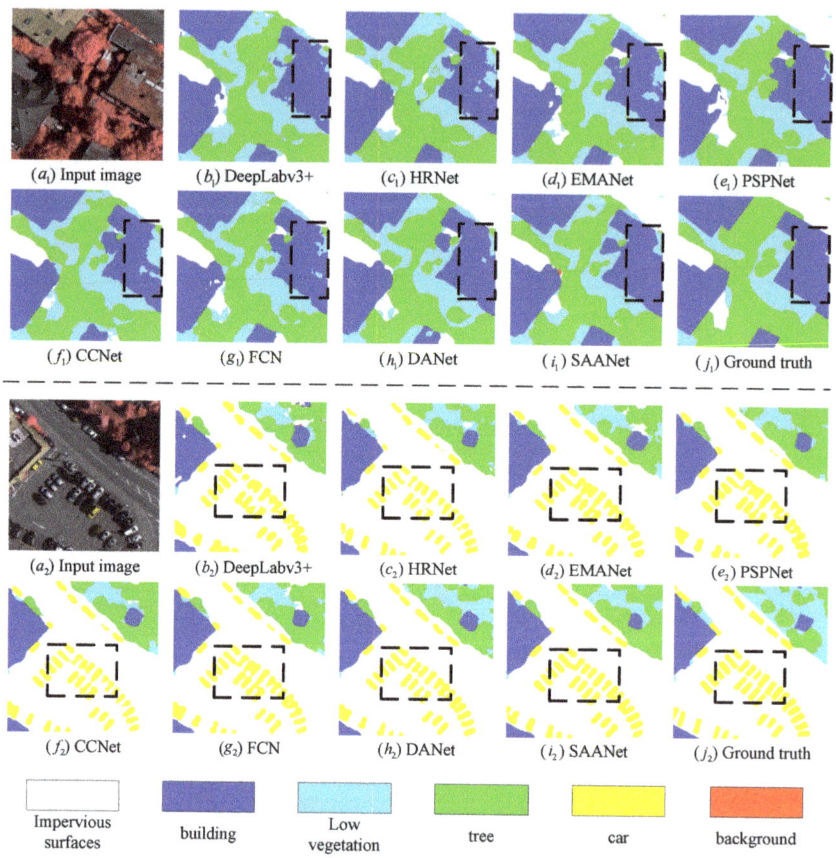

Figure 8. Visual results achieved by different networks on ISPRS Vaihingen dataset. For the first group, other methods incorrectly predict that several pixels representing buildings are low vegetation or tree class. For the second group, several pixels representing cars have rough edges. However, our SAANet can maintain category consistency and obtain more precise edges.

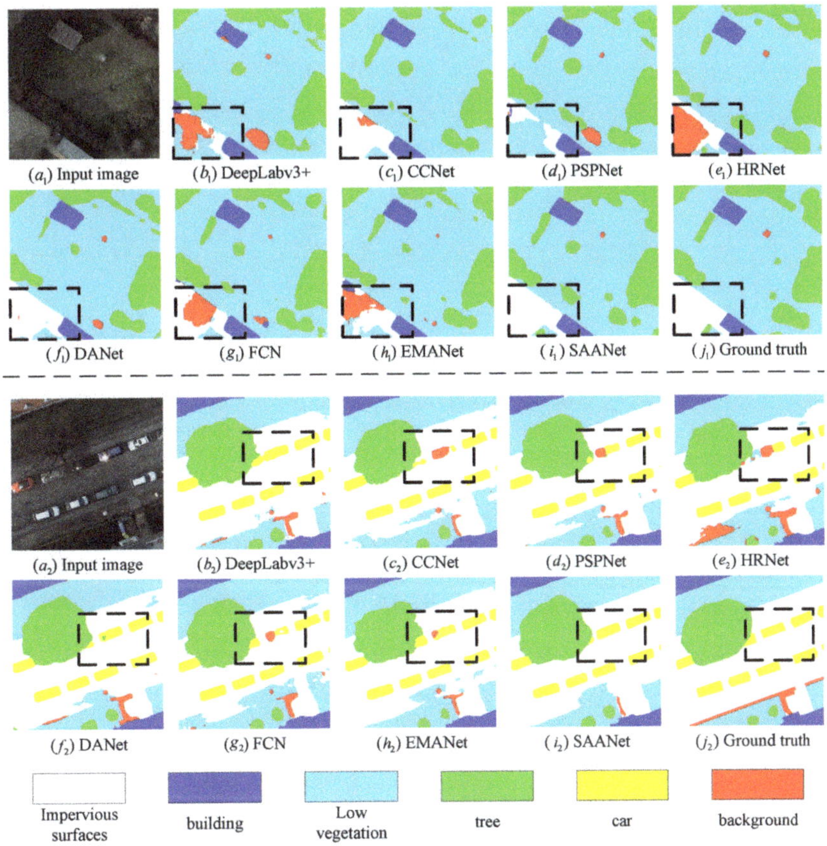

Figure 9. Visual results achieved by different networks on ISPRS Potsdam dataset. For the first group, other methods incorrectly predict pixels inside impervious surfaces. For the second group, several pixels representing cars are incorrectly predicted. However, our SAANet can maintain category consistency and obtain more accurate pixel classification.

3.4. Evaluation in Efficiency

We not only evaluate the segmentation accuracy and visualization results of different methods, but also measure the computational complexity and model parameters, in terms of giga floating-point operations per second (GFLOPs) (G) and the number of parameters with millions (Params) (M). All models are calculated with an input image size of $512 \times 512 \times 3$. The results are shown in Table 4. HRNet uses HRNetv2_W48 as the backbone network and has the lowest computational complexity. The backbone network of other methods is ResNet-101 with dilated convolution strategy. Compared with the self-attention-based networks DANet and CCNet, our method only increases the computational complexity by about 1.76% and the number of parameters by 0.6% to obtain better segmentation accuracy. Although our SAANet achieves better performance, it has a more complex structure and provides a small amount of computational complexity and parameters. We will focus on balancing the relationship between accuracy and complexity in future work.

Table 4. Comparison with other networks on GFLOPs and Params. The best results are shown in boldface.

Method	GFLOPs (G)	Params (M)
DeepLabv3+	254.56	60.21
HRNet	**93.73**	65.85
EMANet	246.63	**58.71**
PSPNet	256.63	65.60
CCNet	278.57	66.45
FCN	275.88	66.12
DANet	277.26	66.45
SAANet	283.46	66.85

4. Discussions

Previous work has focused on the fact that contextual information is important for semantic segmentation. PSPNet [8] uses the pooling operation of a pyramid structure to model the context information of different scales. Deeplabv3+ [11] combines the pyramid structure with the dilated convolution to capture the context information. In addition, several works [13–17] have proved that the self-attention mechanism is an effective way to model global context information. The self-attention mechanism captures context information through a sequence of matrix operations, which improves the accuracy of semantic segmentation. However, HRSI have complex scenes and rich details. The implementation of standard self-attention will introduce excessive redundant information and interfere with semantic segmentation. In this paper, SPAM and SCAM are proposed to model local and global context information, while avoiding the introduction of redundant information. In addition, FAM is proposed to improve the segmentation accuracy of edge regions and refine the semantic segmentation results. To better discuss and validate the effectiveness of each module of our SAANet, extensive ablation studies are conducted on the ISPRS Vaihingen and Potsdam datasets.

4.1. Sparse Position and Channel Attention Module

Both local and global context information is indispensable for the semantic segmentation task. In general, a larger receptive field can fuse a wider range of information, which is conducive to obtaining better feature representation. The standard self-attention operation is equivalent to fusing the information of each pixel of the image indistinguishably, which models long-range context information. The proposed SPAM and SCAM can capture local context information, as well as model long-range context information, and does so both sparsely and efficiently. To acquire a balance between local and global context, different H_n, W_n in SPAM and C_n in SCAM are set. Extensive experiments are conducted on the ISPRS Vaihingen and Potsdam datasets, and the results are shown in Table 5.

The pretrained ResNet-101 with the dilated strategy is adopted to initialize the backbone. The output of the last stage of ResNet-101 is used for semantic segmentation. Baseline based on ResNet101 obtains an mIOU of 65.19%, an mF1 of 77.32%, an OA of 85.60% on the ISPRS Vaihingen dataset. Baseline obtains an mIOU of 72.03%, an mF1 of 82.04%, an OA of 87.43% on the ISPRS Potsdam dataset. Compared with other H_n, W_n in SPAM and C_n in SCAM, SPAM with $H\_n, W\_n = 4$ and SCAM with $C\_n = 2$ achieves the best mIOU of 68.19% on the Vaihingen dataset and mIOU of 73.65% on the Potsdam dataset. The larger H_n and W_n are, the wider the region of capturing local information in the spatial dimension is, which will introduce more redundant information. Each channel map of high-level features is related to the category. By dividing channels into more groups (i.e., the larger C_n is), several channels with strong associations may be dispersed and rearranged, which is adverse to obtaining a better feature representation of each class. Therefore, H_n, W_n, and C_n are set as 4, 4, and 2, respectively, in follow-up experiments.

Table 5. Comparisons of different *H_n*, *W_n*, and *C_n* on ISPRS Vaihingen and Potsdam datasets. The best results are shown in boldface.

Dataset	H_n	W_n	C_n	mIOU	mF1	OA
Vaihingen	/	/	/	65.19	77.32	85.60
	4	4	2	**68.19**	**79.99**	**86.64**
	4	4	4	67.82	79.67	86.47
	8	8	2	67.73	79.71	86.24
	8	8	4	67.82	79.75	86.40
	16	16	2	67.67	79.52	86.42
	16	16	4	68.04	79.75	86.40
Potsdam	/	/	/	72.03	82.04	87.43
	4	4	2	**73.65**	**83.49**	88.09
	4	4	4	73.38	83.28	87.96
	8	8	2	73.44	83.22	**88.15**
	8	8	4	73.11	83.04	87.89
	16	16	2	73.54	83.45	87.98
	16	16	4	72.57	82.39	87.89

Sparse Position Attention Module: In order to efficiently model spatial long-range context information, SPAM is introduced to enhance the output of the backbone. The results are shown in Table 6. Compared with the baseline, SPAM provides an mIOU of 0.54% and 1.15% improvement and achieves an mIOU of 65.73% and 73.18%, an mF1 of 78.06% and 83.12%, and an OA of 85.61% and 87.88%, respectively, on the Vaigingen and Potsdam datasets. It is obvious that SPAM can effectively capture global context information and achieves a better segmentation performance.

Table 6. Comparisons of different versions of our network on ISPRS Vaihingen and Potsdam datasets. The best results are shown in boldface.

Dataset	SPAM	SCAM	FPN	FAM	mIOU	mF1	OA
Vaihingen					65.19	77.32	85.6
	✓				65.73	78.06	85.61
		✓			65.38	77.41	85.85
	✓	✓			68.19	79.99	86.64
	✓	✓	✓		66.82	78.88	86.11
	✓	✓	✓	✓	**68.50**	**80.22**	**86.72**
Potsdam					72.03	82.04	87.43
	✓				73.18	83.12	87.88
		✓			72.80	82.80	87.73
	✓	✓			73.65	83.49	88.09
	✓	✓	✓		73.63	83.45	88.07
	✓	✓	✓	✓	**73.79**	**83.57**	**88.22**

Sparse Channel Attention Module: In order to efficiently capture the interdependencies between channels, SCAM is introduced to enhance the output of the backbone. The results are shown in Table 6. Compared with the baseline, SCAM provides mIOU of 0.19% and 0.77% improvement and achieves an mIOU of 65.38% and 72.80%, an mF1 of 77.41% and 82.80%, and an OA of 85.85% and 87.73%, respectively, on the Vaigingen and Potsdam datasets. SCAM is of great significance when it comes to modeling the dependencies between channels.

We integrate SPAM and SCAM into the baseline to generate a network. Compared with the baseline, the SPAM models long-range context information, and SCAM efficiently captures the interdependencies between channels. SPAM and SCAM provide an mIOU of 3%, an mF1 of 2.67%, and an OA of 1.04% improvement and obtain an mIOU of 68.19%, an F1 of 79.99%, and an OA of 86.64% on the Vaihingen dataset. Additionally, SPAM and SCAM provide an mIOU of 1.62%, an mF1 of 1.45%, and an OA of 0.66% improvement and obtain an mIOU of 73.65%, an F1 of 83.49%, and an OA of 88.09% on the Potsdam dataset. Extensive experiments demonstrate that SPAM and SCAM enhance the semantic segmentation performance of HRSI.

4.2. Feature Alignment Module

We compare the segmentation results of the method with SPAM and SCAM with labels. Several results are shown in Figure 10, demonstrating that most of the regions with inaccurate segmentation are boundary regions. In this paper, the FPN structure is used to integrate the high-level features with semantic information and the low-level features with detail information to obtain a finer semantic segmentation result. However, feature maps with different resolutions are misaligned. Utilizing the FPN structure to fuse features from shadow layers and deep layers fails to obtain better results. Therefore, it is vital that the FAM aligns and fuses features with different resolutions. The results are shown in Table 6. The network with the FPN obtains an mIoU of 66.82%, an mF1 of 78.88%, and an OA of 86.11% on the Vaigingen dataset. The network with FAM achieves the best mIoU of 68.5%, mF1 of 80.22%, and OA of 86.72%. Meanwhile, the network with FAM achieves the best mIoU of 73.79%, an mF1 of 83.57%, and an OA of 88.22% on the Potsdam dataset. The results prove that feature alignment is essential and the proposed FAM is effective. Additionally, FAM is beneficial, as it refines the boundaries. To prove the effectiveness of FAM for boundary regions, the mIOU, mF1, and OA are calculated on the edge region. Since there is no standard edge region, a neighborhood in which different classes are connected is selected as the edge region in this paper. Specifically, we first extract the boundary of different objects in the label and then perform the dilation operation in morphology operations to obtain the edge region. Note that the pixels closer to the object boundary are more likely to be confused, and the pixels closer to the object interior are more likely to be classified. As the dilation kernel increases, the edge region expands and the pixels grow closer to the interior of the object. The mIOU, mF1, and OA in a larger area cannot fully highlight the improvement in the edge region. Therefore, in this paper, a kernel of 3×3 is selected for the dilation operation to obtain the edge region. The results are shown in Table 7. The network without FAM obtains an mIoU of 35.68%, an mF1 of 51.88%, and an OA of 55.19% on the Vaigingen dataset. FAM provides an mIOU of 1.04%, an mF1 of 0.84%, and an OA of 0.23% improvement. Meanwhile, the method without FAM obtains an mIoU of 38.32%, an mF1 of 54.34%, and an OA of 56.48% on the Potsdam dataset. The method with FAM obtains an mIoU of 38.81%, an mF1 of 54.82%, and an OA of 56.90%. The results demonstrate that FAM refines the edge regions of semantic segmentation results and further explains the necessity and effectiveness of feature alignment.

In general, the experiments and visual results illustrate that SPAM, SCAM, and FAM achieve better semantic segmentation results. As shown in Tables 1–3, the proposed method achieves optimal OA, mF1, and mIOU on the ISPRS Vaihingen, Potsdam, and LoveDA Urban datasets. Specifically, the accuracy of small object cars is significantly improved. Additionally, as shown in Figures 8 and 9, other networks incorrectly the predicted pixels inside large objects, such as impervious surfaces and buildings. For small objects, such as cars, incorrect pixel classifications occur, as well as inaccurate edges. In contrast, our SAANet can maintain intra-class consistency for large objects and accuracy for small objects. Meanwhile, the experimental results show that global context information enhancement on HRSI with complex backgrounds introduces redundant information. The researchers further explore more adaptive global context information fusion methods to suppress redundant information as much as possible.

Table 7. Quantitative results achieved by different variants of our network on boundaries. The best results are shown in boldface.

Dataset	Method	mIOU	F1	OA
Vaihingen	baseline + SPAM + SCAM	35.68	51.88	55.19
	baseline + SPAM + SCAM + FAM	**36.52**	**52.92**	**55.42**
Potsdam	baseline + SPAM + SCAM	38.32	54.34	56.48
	baseline + SPAM + SCAM + FAM	**38.81**	**54.82**	**56.90**

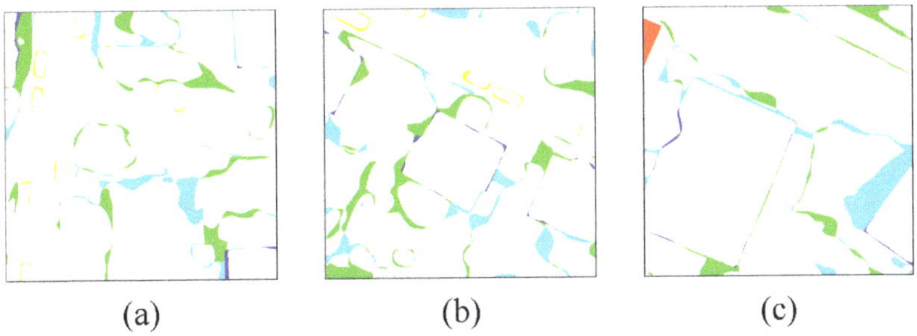

Figure 10. Visualization results of the difference between predictions and labels. (**a**–**c**) from the test set of the Vaihingen dataset. Note that most of the regions with inaccurate segmentation are boundary regions.

5. Conclusions

In this paper, we present a network based on sparse self-attention and feature alignment for semantic segmentation of HRSI. Specifically, SPAM is developed to capture long-range context information. SCAM is adopted to model interdependencies between channels more efficiently, while FAM is introduced to align features with different resolutions and refine semantic segmentation results. Moreover, extensive ablation experiments demonstrate the effectiveness of our method on the ISPRS Vaihingen and Potsdam datasets. Comparative experiments on the ISPRS Vaihingen, Potsdam, and LoveDA Urban datasets demonstrate that our SAANet obtains finer semantic segmentation results and achieves outstanding performance. Although our SAANet enhances the context information and details, there are still problems in the field of semantic segmentation of HRSI, such as large intra-class differences and small inter-class differences. For example, trees and low vegetation in the Vaihingen and Potsdam datasets are easily confused. In subsequent research, we will use comparative learning in the semantic segmentation of HRSI to obtain better feature embedding space and more easily distinguished feature representation.

Author Contributions: H.Z. determined the research direction and revised the expression of the article; L.S. came up with innovative ideas, developed the SAANet, conducted experiments, and completed the manuscript. J.W. helped to modify the conception and provided suggestions for expression. X.C., S.H., M.L. and S.L. checked out the article's writing. All authors have read and agreed to the published version of the manuscript.

Funding: This work was supported by the Natural Science Foundation of China under grant 62071474.

Data Availability Statement: Not applicable.

Conflicts of Interest: The authors declare no conflict of interest.

References

1. Tokarczyk, P.; Wegner, J.D.; Walk, S.; Schindler, K. Features, Color Spaces, and Boosting: New Insights on Semantic Classification of Remote Sensing Images. *IEEE Trans. Geosci. Remote Sens.* **2014**, *53*, 280–295. [CrossRef]
2. Tang, Y.; Zhang, L. Urban change analysis with multi-sensor multispectral imagery. *Remote Sens.* **2017**, *9*, 252. [CrossRef]

3. Wu, L.; Lu, M.; Fang, L. Deep Covariance Alignment for Domain Adaptive Remote Sensing Image Segmentation. *IEEE Trans. Geosci. Remote Sens.* **2022**, *60*, 1–11. [CrossRef]
4. Liu, M.Y.; Tuzel, O.; Ramalingam, S.; Chellappa, R. Entropy rate superpixel segmentation. In Proceedings of the The 24th IEEE Conference on Computer Vision and Pattern Recognition, CVPR 2011, Colorado Springs, CO, USA, 20–25 June 2011.
5. Radman, A.; Zainal, N.; Suandi, S.A. Automated segmentation of iris images acquired in an unconstrained environment using HOG-SVM and GrowCut. *Digit. Signal Process.* **2017**, *64*, 60–70. [CrossRef]
6. Thanh Noi, P.; Kappas, M. Comparison of random forest, k-nearest neighbor, and support vector machine classifiers for land cover classification using Sentinel-2 imagery. *Sensors* **2017**, *18*, 18. [CrossRef] [PubMed]
7. Krizhevsky, A.; Sutskever, I.; Hinton, G.E. Imagenet classification with deep convolutional neural networks. *Commun. ACM* **2017**, *60*, 84–90. [CrossRef]
8. Zhao, H.; Shi, J.; Qi, X.; Wang, X.; Jia, J. Pyramid Scene Parsing Network. In Proceedings of the IEEE Computer Society, Las Vegas, NV, USA, 26 June–1 July 2016.
9. Chen, L.-C.; Papandreou, G.; Kokkinos, I.; Murphy, K.; Yuille, A.L. Deeplab: Semantic image segmentation with deep convolutional nets, atrous convolution, and fully connected crfs. *IEEE Trans. Pattern Anal. Mach. Intell.* **2017**, *40*, 834–848. [CrossRef] [PubMed]
10. Chen, L.C.; Papandreou, G.; Schroff, F.; Adam, H. Rethinking atrous convolution for semantic image segmentation. *arXiv* **2017**, arXiv:1706.05587.
11. Chen, L.C.; Zhu, Y.; Papandreou, G.; Schroff, F.; Adam, H. Encoder-decoder with atrous separable convolution for semantic image segmentation. In Proceedings of the European Conference on Computer Vision (ECCV), Munich, Germany, 8–14 September 2018; pp. 801–818.
12. Peng, C.; Zhang, X.; Yu, G.; Luo, G.; Sun, J. Large kernel matters–improve semantic segmentation by global convolutional network. In Proceedings of the IEEE Conference on Computer Vision and Pattern Recognition, Honolulu, HI, USA, 21–26 July 2017; pp. 4353–4361.
13. Wang, X.; Girshick, R.; Gupta, A.; He, K. Non-local neural networks. In Proceedings of the Proceedings of the IEEE conference on computer vision and pattern recognition, 2018, pp. 7794–7803.
14. Fu, J.; Liu, J.; Tian, H.; Li, Y.; Bao, Y.; Fang, Z.; Lu, H. Dual Attention Network for Scene Segmentation. In Proceedings of the 2019 IEEE/CVF Conference on Computer Vision and Pattern Recognition (CVPR), Long Beach, CA, USA, 15–20 June 2019.
15. Li, X.; Zhong, Z.; Wu, J.; Yang, Y.; Lin, Z.; Liu, H. Expectation-maximization attention networks for semantic segmentation. In Proceedings of the IEEE/CVF International Conference on Computer Vision, Seoul, Republic of Korea, 27–28 October 2019; pp. 9167–9176.
16. Huang, Z.; Wang, X.; Huang, L.; Huang, C.; Wei, Y.; Liu, W. Ccnet: Criss-cross attention for semantic segmentation. In Proceedings of the IEEE/CVF international conference on computer vision, 2019, pp. 603–612.
17. Yuan, Y.; Huang, L.; Guo, J.; Zhang, C.; Chen, X.; Wang, J. OCNet: Object context for semantic segmentation. *Int. J. Comput. Vis.* **2021**, *129*, 2375–2398. [CrossRef]
18. Shi, H.; Fan, J.; Wang, Y.; Chen, L. Dual attention feature fusion and adaptive context for accurate segmentation of very high-resolution remote sensing images. *Remote Sens.* **2021**, *13*, 3715. [CrossRef]
19. Li, R.; Zheng, S.; Zhang, C.; Duan, C.; Su, J.; Wang, L.; Atkinson, P.M. Multiattention network for semantic segmentation of fine-resolution remote sensing images. *IEEE Trans. Geosci. Remote Sens.* **2021**, *60*, 1–13. [CrossRef]
20. Long, J.; Shelhamer, E.; Darrell, T. Fully convolutional networks for semantic segmentation. In Proceedings of the IEEE Conference on Computer Vision and Pattern Recognition, Boston, MA, USA, 7–12 June 2015; pp. 3431–3440.
21. Badrinarayanan, V.; Kendall, A.; Cipolla, R. Segnet: A deep convolutional encoder-decoder architecture for image segmentation. *IEEE Trans. Pattern Anal. Mach. Intell.* **2017**, *39*, 2481–2495. [CrossRef] [PubMed]
22. Ronneberger, O.; Fischer, P.; Brox, T. *U-Net: Convolutional Networks for Biomedical Image Segmentation*; Springer: Munich, Germany, 2015.
23. Lin, G.; Milan, A.; Shen, C.; Reid, I. Refinenet: Multi-path refinement networks for high-resolution semantic segmentation. In Proceedings of the IEEE Conference on Computer Vision and Pattern Recognition, Honolulu, HI, USA, 21–26 July 2017; pp. 1925–1934.
24. Sun, K.; Zhao, Y.; Jiang, B.; Cheng, T.; Wang, J. High-Resolution Representations for Labeling Pixels and Regions. *arXiv* **2019**, arXiv:1904.04514.
25. Takikawa, T.; Acuna, D.; Jampani, V.; Fidler, S. Gated-scnn: Gated shape cnns for semantic segmentation. In Proceedings of the IEEE/CVF International Conference on Computer Vision, Seoul, Republic of Korea, 27–28 October 2019; pp. 5229–5238.
26. Yuan, Y.; Xie, J.; Chen, X.; Wang, J. Segfix: Model-agnostic boundary refinement for segmentation. In Proceedings of the European Conference on Computer Vision, Glasgow, UK, 23–28 August 2020; pp. 489–506.
27. Liu, S.; Ding, W.; Liu, C.; Liu, Y.; Wang, Y.; Li, H. ERN: Edge loss reinforced semantic segmentation network for remote sensing images. *Remote Sens.* **2018**, *10*, 1339. [CrossRef]
28. Zheng, X.; Huan, L.; Xia, G.S.; Gong, J. Parsing very high resolution urban scene images by learning deep ConvNets with edge-aware loss. *ISPRS J. Photogramm. Remote Sens.* **2020**, *170*, 15–28. [CrossRef]
29. Li, X.; Li, T.; Chen, Z.; Zhang, K.; Xia, R. Attentively learning edge distributions for semantic segmentation of remote sensing imagery. *Remote Sens.* **2021**, *14*, 102. [CrossRef]

30. He, K.; Zhang, X.; Ren, S.; Sun, J. Deep residual learning for image recognition. In Proceedings of the IEEE Conference on Computer Vision and Pattern Recognition, Las Vegas, NV, USA, 26 June–1 July 2016; pp. 770–778.
31. Lin, T.Y.; Dollár, P.; Girshick, R.; He, K.; Hariharan, B.; Belongie, S. Feature pyramid networks for object detection. In Proceedings of the IEEE Conference on Computer Vision and Pattern Recognition, Honolulu, HI, USA, 21–26 July 2017; pp. 2117–2125.
32. Available online: https://www2.isprs.org/commissions/comm2/wg4/benchmark/2d-sem-label-vaihingen/ (accessed on 1 March 2021).
33. Wang, J.; Zheng, Z.; Ma, A.; Lu, X.; Zhong, Y. LoveDA: A remote sensing land-cover dataset for domain adaptive semantic segmentation. *arXiv* **2021**, arXiv:2110.08733.
34. Zhao, Q.; Liu, J.; Li, Y.; Zhang, H. Semantic Segmentation with Attention Mechanism for Remote Sensing Images. *IEEE Trans. Geosci. Remote Sens.* **2021**, *60*, 1–13. [CrossRef]

MDPI AG
Grosspeteranlage 5
4052 Basel
Switzerland
Tel.: +41 61 683 77 34

Remote Sensing Editorial Office
E-mail: remotesensing@mdpi.com
www.mdpi.com/journal/remotesensing

www.ingramcontent.com/pod-product-compliance
Lightning Source LLC
Chambersburg PA
CBHW061458220926
51850CB00004B/120
9783725813650